Genetics and Molecular Biology

Genetics and Molecular Biology

SECOND EDITION

Robert Schleif

Department of Biology
The Johns Hopkins University
Baltimore, Maryland

The Johns Hopkins University Press   Baltimore and London

The Johns Hopkins University Press
2715 North Charles Street
Baltimore, Maryland 21218-4319
The Johns Hopkins Press Ltd., London

Library of Congress Cataloging-in-Publication Data
Schleif, Robert F.
  Genetics and molecular biology / by Robert Schleif.—2nd ed.
  p.   cm.
  Includes bibliographical references and index.
  ISBN 0-8018-4673-0 (acid-free paper).—ISBN 0-8018-4674-9 (pbk : acid-free paper)
  1. Molecular genetics.   I. Title
QH442.S34 1993
574.87'328—dc20   93-15652

A catalog record for this book is available from the British Library.

Published from electronic files prepared by the author in Ventura Publisher 3.0 and
Adobe Illustrator 3.0 software

Preface

This book evolved from a course in molecular biology which I have been teaching primarily to graduate students for the past twenty years. Because the subject is now mature, it is possible to present the material by covering the principles and encouraging students to learn how to apply them. Such an approach is particularly efficient as the subject of molecular genetics now is far too advanced, large, and complex for much value to come from attempting to cover the material in an encyclopedia-like fashion or teaching the definitions of the relevant words in a dictionary-like approach. Only the core of molecular genetics can be covered by the present approach. Most of the remainder of the vast subject, however, is a logical extension of the ideas and principles presented here. One consequence of the principles and analysis approach taken here is that the material is not easy. Thinking and learning to reason from the fundamentals require serious effort but, ultimately, are more efficient and more rewarding than mere memorization.

An auxiliary objective of this presentation is to help students develop an appreciation for elegant and beautiful experiments. A substantial number of such experiments are explained in the text, and the cited papers contain many more.

The book contains three types of information. The main part of each chapter is the text. Following each chapter are references and problems. References are arranged by topic, and one topic is "Suggested Readings." The additional references cited permit a student or researcher to find many of the fundamental papers on a topic. Some of these are on topics not directly covered in the text. Because solving problems helps focus one's attention and stimulates understanding, many thought-provoking problems or paradoxes are provided. Some of these require use of material in addition to the text. Solutions are provided to about half of the problems.

Although the ideal preparation for taking the course and using the book would be the completion of preliminary courses in biochemistry, molecular biology, cell biology, and physical chemistry, few students have such a background. Most commonly, only one or two of the above-mentioned courses have been taken, with some students coming from a more physical or chemical background, and other students coming from a more biological background.

My course consists of two lectures and one discussion session per week, with most chapters being covered in one lecture. The lectures often summarize material of a chapter and then discuss in depth a recent paper that extends the material of the chapter. Additional readings of original research papers are an important part of the course for graduate students, and typically such two papers are assigned per lecture. Normally, two problems from the ends of the chapters are assigned per lecture.

Many of the ideas presented in the book have been sharpened by my frequent discussions with Pieter Wensink, and I thank him for this. I thank my editors, James Funston for guidance on the first edition and Yale Altman and Richard O'Grady for ensuring the viability of the second edition. I also thank members of my laboratory and the following who read and commented on portions of the manuscript: Karen Beemon, Howard Berg, Don Brown, Victor Corces, Jeff Corden, David Draper, Mike Edidin, Bert Ely, Richard Gourse, Ed Hedgecock, Roger Hendrix, Jay Hirsh, Andy Hoyt, Amar Klar, Ed Lattman, Roger McMacken, Howard Nash, and Peter Privalov.

Contents

10 Advanced Genetic Engineering   297

11 Repression and the *lac* Operon   331

An Overview of Cell Structure and Function

1

In this book we will be concerned with the basics of the macromolecular interactions that affect cellular processes. The basic tools for such studies are genetics, chemistry, and physics. For the most part, we will be concerned with understanding processes that occur within cells, such as DNA synthesis, protein synthesis, and regulation of gene activity. The initial studies of these processes utilize whole cells. These normally are followed by deeper biochemical and biophysical studies of individual components. Before beginning the main topics we should take time for an overview of cell structure and function. At the same time we should develop our intuitions about the time and distance scales relevant to the molecules and cells we will study.

Many of the experiments discussed in this book were done with the bacterium *Escherichia coli*, the yeast *Saccharomyces cerevisiae*, and the fruit fly *Drosophila melanogaster*. Each of these organisms possesses unique characteristics making it particularly suitable for study. In fact, most of the research in molecular biology has been confined to these three organisms. The earliest and most extensive work has been done with *Escherichia coli*. The growth of this oranism is rapid and inexpensive, and many of the most fundamental problems in biology are displayed by systems utilized by this bacterium. These problems are therefore most efficiently studied there. The eukaryotic organisms are necessary for study of phenomena not observed in bacteria, but parallel studies on other bacteria and higher cells have revealed that the basic principles of cell operation are the same for all cell types.

Cell's Need for Immense Amounts of Information

Cells face enormous problems in growing. We can develop some idea of the situation by considering a totally self-sufficient toolmaking shop. If we provide the shop with coal for energy and crude ores, analogous to a cell's nutrient medium, then a very large collection of machines and tools is necessary merely to manufacture each of the parts present in the shop. Still greater complexity would be added if we required that the shop be totally self-regulating and that each machine be self-assembling. Cells face and solve these types of problems. In addition, each of the chemical reactions necessary for growth of cells is carried out in an aqueous environment at near neutral pH. These are conditions that would cripple ordinary chemists.

By the tool shop analogy, we expect cells to utilize large numbers of "parts," and, also by analogy to factories, we expect each of these parts to be generated by a specialized machine devoted to production of just one type of part. Indeed, biochemists' studies of metabolic pathways have revealed that an *E. coli* cell contains about 1,000 types of parts, or small molecules, and that each is generated by a specialized machine, an enzyme. The information required to specify the structure of even one machine is immense, a fact made apparent by trying to describe an object without pictures and drawings. Thus, it is reasonable, and indeed it has been found that cells function with truly immense amounts of information.

DNA is the cell's library in which information is stored in its sequence of nucleotides. Evolution has built into this library the information necessary for cells' growth and division. Because of the great value of the DNA library, it is natural that it be carefully protected and preserved. Except for some of the simplest viruses, cells keep duplicates of the information by using a pair of self-complementary DNA strands. Each strand contains a complete copy of the information, and chemical or physical damage to one strand is recognized by special enzymes and is repaired by making use of information contained on the opposite strand. More complex cells further preserve their information by possessing duplicate DNA duplexes.

Much of the recent activity in molecular biology can be understood in terms of the cell's library. This library contains the information necessary to construct the different cellular machines. Clearly, such a library contains far too much information for the cell to use at any one time. Therefore mechanisms have developed to recognize the need for particular portions, "books," of the information and read this out of the library in the form of usable copies. In cellular terms, this is the regulation of gene activity.

Rudiments of Prokaryotic Cell Structure

A typical prokaryote, *E. coli*, is a rod capped with hemispheres (Fig. 1.1). It is 1–3 μ (10^{-4} cm = 1 μ = 10^4 Å) long and 0.75 μ in diameter. Such a

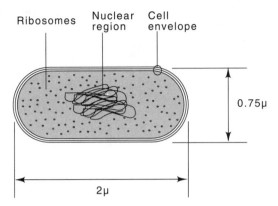

Figure 1.1 The dimensions of a typical *E. coli* cell.

cell contains about 2×10^{-13} g of protein, 2×10^{-14} g of RNA that is mostly ribosomal RNA, and 6×10^{-15} g of DNA.

The cell envelope consists of three parts, an inner and outer membrane and an intervening peptidoglycan layer (Fig. 1.2). The outer surface of the outer membrane is largely lipopolysaccharides. These are attached to lipids in the outer half of the outer membrane. The polysaccharides protect the outer membrane from detergent-like molecules found in our digestive tract.outer membrane The outer membrane also consists of matrix proteins that form pores small enough to exclude the detergent-like bile salts, but large enough to permit passage of small molecules and phospholipids.

Figure 1.2 Schematic drawing of the structure of the envelope of an *E. coli* cell.

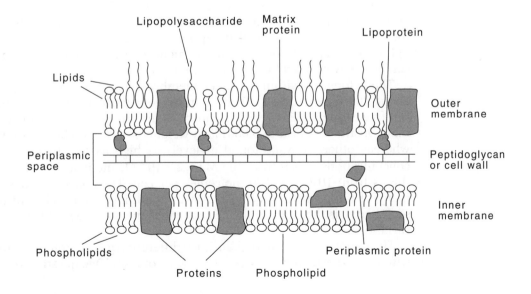

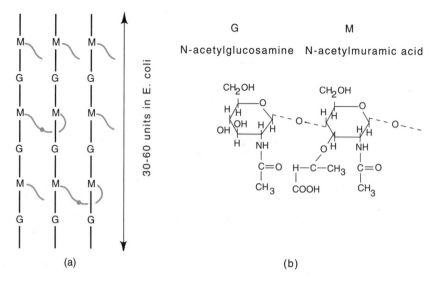

(a)   (b)

Figure 1.3 Structure of the cell wall showing the alternating N-acetylglu-cosamine N-acetylmuramic acid units. Each N-acetylmuramic acid possesses a peptide, but only a few are crosslinked in *E. coli*.

The major shape-determining factor of cells is the peptidoglycan layer or cell wall (Fig. 1.3). It lies beneath the outer membrane and is a single molecule containing many polysaccharide chains crosslinked by short peptides (Fig. 1.4). The outer membrane is attached to the peptidoglycan layer by about 10^6 lipoprotein molecules. The protein end of each of these is covalently attached to the diaminopimelic acid in the peptidoglycan. The lipid end is buried in the outer membrane.

The innermost of the three cell envelope layers is the inner or cytoplasmic membrane. It consists of many proteins embedded in a phospholipid bilayer. The space between the inner membrane and the outer membrane that contains the peptidoglycan layer is known as the periplasmic space. The cell wall and membranes contain about 20% of the cellular protein. After cell disruption by sonicating or grinding, most of this protein is still contained in fragments of wall and membrane and can be easily pelleted by low-speed centrifugation.

The cytoplasm within the inner membrane is a protein solution at about 200 mg/ml, about 20 times more concentrated than the usual cell-free extracts used in the laboratory. Some proteins in the cytoplasm may constitute as little as 0.0001% by weight of the total cellular protein whereas others may be found at levels as high as 5%. In terms of concentrations, this is from 10^{-8} M to 2×10^{-4} M, and in a bacterial cell this is from 10 to 200,000 molecules per cell. The concentrations of many of the proteins vary with growth conditions, and a current research area is the study of the cellular mechanisms responsible for the variations.

The majority of the more than 2,000 different types of proteins found within a bacterial cell are located in the cytoplasm. One question yet to

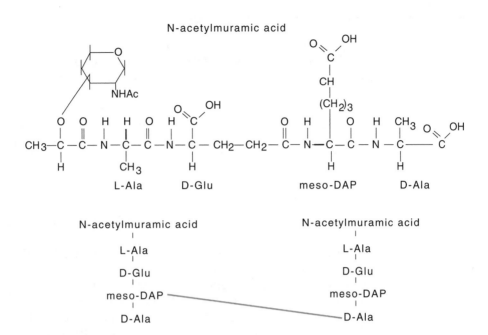

Figure 1.4 Structure of the peptide crosslinking N-acetylmuramic acid units. DAP is diaminopimelic acid.

be answered about these proteins is how they manage to exist in the cell without adhering to each other and forming aggregates since polypeptides can easily bind to each other. Frequently when a bacterium is engineered for the over-synthesis of a foreign protein, amorphous precipitates called inclusion bodies form in the cytoplasm. Sometimes these result from delayed folding of the new protein, and occasionally they are the result of chance coprecipitation of a bacterial protein and the newly introduced protein. Similarly, one might also expect an occasional mutation to inactivate simultaneously two apparently unrelated proteins by the coprecipitation of the mutated protein and some other protein into an inactive aggregate, and occasionally this does occur.

The cell's DNA and about 10,000 ribosomes also reside in the cytoplasm. The ribosomes consist of about one-third protein and two-thirds RNA and are roughly spherical with a diameter of about 200 Å. The DNA in the cytoplasm is not surrounded by a nuclear membrane as it is in the cells of higher organisms, but nonetheless it is usually confined to a portion of the cellular interior. In electron micrographs of cells, the highly compacted DNA can be seen as a stringy mass occupying about one tenth of the interior volume, and the ribosomes appear as granules uniformly scattered through the cytoplasm.

Rudiments of Eukaryotic Cell Structure

A typical eukaryotic cell is 10 μ in diameter, making its volume about 1,000 times that of a bacterial cell. Like bacteria, eukaryotic cells contain cell membranes, cytoplasmic proteins, DNA, and ribosomes, albeit of somewhat different structure from the corresponding prokaryotic elements (Fig. 1.5). Eukaryotic cells, however, possess many structural features that even more clearly distinguish them from prokaryotic cells. Within the eukaryotic cytoplasm are a number of structural proteins that form networks. Microtubules, actin, intermediate filaments, and thin filaments form four main categories of fibers found within eukaryotic cells. Fibers within the cell provide a rigid structural skeleton, participate in vesicle and chromosome movement, and participate in changing the cell shape so that it can move. They also bind the majority of the ribosomes.

The DNA of eukaryotic cells does not freely mix with the cytoplasm, but is confined within a nuclear membrane. Normally only small proteins of molecular weight less than 20 to 40,000 can freely enter the nucleus through the nuclear membrane. Larger proteins and nuclear RNAs enter the nucleus through special nuclear pores. These are large structures that actively transport proteins or RNAs into or out of the nucleus. In each cell cycle, the nuclear membrane dissociates, and then later reaggregates. The DNA itself is tightly complexed with a class of proteins called histones, whose main function appears to be to help DNA retain a condensed state. When the cell divides, a special apparatus called the spindle, and consisting in part of microtubules, is necessary to pull the chromosomes into the daughter cells.

Eukaryotic cells also contain specialized organelles such as mitochondria, which perform oxidative phosphorylation to generate the cell's needed chemical energy. In many respects mitochondria resemble bacteria and, in fact, appear to have evolved from bacteria. They contain DNA, usually in the form of a circular chromosome like that of *E. coli*

Figure 1.5 Schematic drawing of a eukaryotic cell.

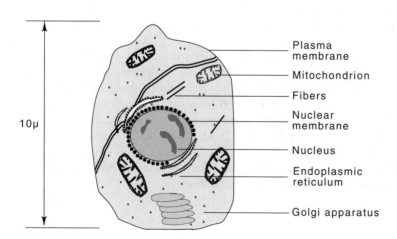

10μ

Plasma membrane

Mitochondrion

Fibers

Nuclear membrane

Nucleus

Endoplasmic reticulum

Golgi apparatus

and ribosomes that often more closely resemble those found in bacteria than the ribosomes located in the cytoplasm of the eukaryotic cell. Chloroplasts carry out photosynthesis in plant cells, and are another type of specialized organelle found within some eukaryotic cells. Like mitochondria, chloroplasts also contain DNA and ribosomes different from the analogous structures located elsewhere in the cell.

Most eukaryotic cells also contain internal membranes. The nucleus is surrounded by two membranes. The endoplasmic reticulum is another membrane found in eukaryotic cells. It is contiguous with the outer nuclear membrane but extends throughout the cytoplasm in many types of cells and is involved with the synthesis and transport of membrane proteins. The Golgi apparatus is another structure containing membranes. It is involved with modifying proteins for their transport to other cellular organelles or for export out of the cell.

Packing DNA into Cells

The DNA of the *E. coli* chromosome has a molecular weight of about 2×10^9 and thus is about 3×10^6 base pairs long. Since the distance between base pairs in DNA is about 3.4 Å, the length of the chromosome is 10^7 Å or 0.1 cm. This is very long compared to the 10^4 Å length of a bacterial cell, and the DNA must therefore wind back and forth many times within the cell. Observation by light microscopy of living bacterial cells and by electron microscopy of fixed and sectioned cells shows that often the DNA is confined to a portion of the interior of the cell with dimensions less than 0.25 μ.

To gain some idea of the relevant dimensions, let us estimate the number of times that the DNA of a bacterium winds back and forth within a volume we shall approximate as a cube 0.25 μ on a side. This will provide an idea of the average distance separating the DNA duplexes and will also give some idea of the proportion of the DNA that lies on

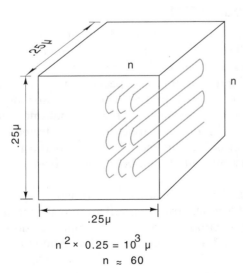

Figure 1.6 Calculation of the number of times the *E. coli* chromosome winds back and forth if it is confined within a cube of edge 0.25 μ. Each of the n layers of DNA possesses n segments of length 0.25 μ.

$$n^2 \times 0.25 = 10^3 \ \mu$$
$$n \approx 60$$

the surface of the chromosomal mass. The number of times, N, that the DNA must wind back and forth will then be related to the length of the DNA and the volume in which it is contained. If we approximate the path of the DNA as consisting of n layers, each layer consisting of n segments of length 0.25 μ (Fig. 1.6), the total number of segments is n^2. Therefore, $2,500n^2$ Å $= 10^7$ Å and $n = 60$. The spacing between adjacent segments of the DNA is 2,500 Å/60 = 40 Å.

The close spacing between DNA duplexes raises the interesting problem of accessibility of the DNA. RNA polymerase has a diameter of about 100 Å and it may not fit between the duplexes. Therefore, quite possibly only DNA on the surface of the nuclear mass is accessible for transcription. On the other hand, transcription of the lactose and arabinose operons can be induced within as short a time as two seconds after adding inducers. Consequently either the nuclear mass is in such rapid motion that any portion of the DNA finds its way to the surface at least once every several seconds, or the RNA polymerase molecules do penetrate to the interior of the nuclear mass and are able to begin transcription of any gene at any time. Possibly, start points of the arabinose and lactose operons always reside on the surface of the DNA.

Compaction of the DNA generates even greater problems in eukaryotic cells. Not only do they contain up to 1,000 times the amount of the DNA found in bacteria, but the presence of the histones on the DNA appears to hinder access of RNA polymerase and other enzymes to the DNA. In part, this problem is solved by regulatory proteins binding to regulatory regions before nucleosomes can form in these positions. Apparently, upon activation of a gene additional regulatory proteins bind, displacing more histones, and transcription begins. The DNA of many eukaryotic cells is specially contracted before cell division, and at this time it actually does become inaccessible to RNA polymerase. At all times, however, accessibility of the DNA to RNA polymerase must be hindered.

Moving Molecules into or out of Cells

Small-molecule metabolic intermediates must not leak out of cells into the medium. Therefore, an impermeable membrane surrounds the cytoplasm. To solve the problem of moving essential small molecules like sugars and ions into the cell, special transporter protein molecules are inserted into the membranes. These and auxiliary proteins in the cytoplasm must possess selectivity for the small-molecules being transported. If the small-molecules are being concentrated in the cell and not just passively crossing the membrane, then the proteins must also couple the consumption of metabolic energy from the cell to the active transport.

The amount of work consumed in transporting a molecule into a volume against a concentration gradient may be obtained by considering the simple reaction where A_o is the concentration of the molecule outside the cell and A_i is the concentration inside the cell:

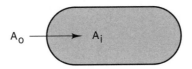

$$A_o \rightleftharpoons A_i$$

This reaction can be described by an equilibrium constant

$$K_{eq} = \frac{A_i}{A_o}$$

The equilibrium constant K_{eq}, is related to the free energy of the reaction by the relation

$$\Delta G = RT \ln K_{eq}$$

where R is about 2 cal/deg·mole and T is 300° K (about 25° C), the temperature of many biological reactions. Suppose the energy of hydrolysis of ATP to ADP is coupled to this reaction with a 50% efficiency. Then about 3,500 of the total of 7,000 calories available per mole of ATP hydrolyzed under physiological conditions will be available to the transport system. Consequently, the equilibrium constant will be

$$K_{eq} = e^{-\frac{\Delta G}{RT}}$$

$$= e^{\frac{3,500}{600}}$$

$$= 340.$$

One interesting result of this consideration is that the work required to transport a molecule is independent of the absolute concentrations; it depends only on the ratio of the inside and outside concentrations. The transport systems of cells must recognize the type of molecule to be transported, since not all types are transported, and convey the molecule either to the inside or to the outside of the cell. Further, if the molecule is being concentrated within the cell, the system must tap an energy source for the process. Owing to the complexities of this process, it is not surprising that the details of active transport systems are far from being fully understood.

Four basic types of small-molecule transport systems have been discovered. The first of these is facilitated diffusion. Here the molecule

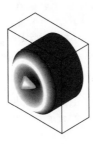

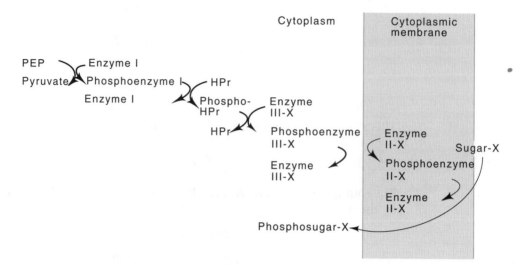

Figure 1.7 The cascade of reactions associated with the phosphotransferase sugar uptake system of *E. coli*.

must get into or out of the cell on its own, but special doors are opened for it. That is, specific carriers exist that bind to the molecule and shuttle it through the membrane. Glycerol enters most types of bacteria by this mechanism. Once within the cell the glycerol is phosphorylated and cannot diffuse back out through the membrane, nor can it exit by using the glycerol carrier protein that carried the glycerol into the cell.

A second method of concentrating molecules within cells is similar to the facilitated diffusion and phosphorylation of glycerol. The phosphotransferase system actively rather than passively carries a number of types of sugars across the cell membrane and, in the process, phosphorylates them (Fig. 1.7). The actual energy for the transport comes from phosphoenolpyruvate. The phosphate group and part of the chemical energy contained in the phosphoenolpyruvate is transferred down a series of proteins, two of which are used by all the sugars transported by this system and two of which are specific for the particular sugar being transported. The final protein is located in the membrane and is directly responsible for the transport and phosphorylation of the transported sugar.

Protons are expelled from *E. coli* during the flow of reducing power from NADH to oxygen. The resulting concentration difference in H^+ ions between the interior and exterior of the cell generates a proton motive force or membrane potential that can then be coupled to ATP synthesis or to the transport of molecules across the membrane. Active transport systems using this energy source are called chemiosmotic systems. In the process of permitting a proton to flow back into the cell, another small molecule can be carried into the cell, which is called symport, or carried out of the cell, which is called antiport (Fig. 1.8).

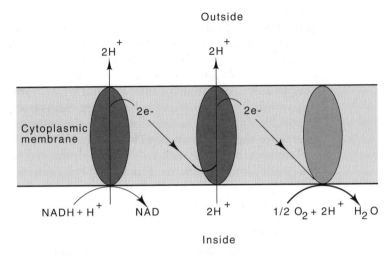

In many eukaryotic cells, a membrane potential is generated by the sodium-potassium pump. From the energy of hydrolysis of one ATP molecule, 3 Na^+ ions are transported outside the cell and 2 K^+ ions are transported inside. The resulting gradient in sodium ions can then be coupled to the transport of other molecules or used to transmit signals along a membrane.

Study of all transport systems has been difficult because of the necessity of working with membranes, but the chemiosmotic system has been particularly hard due to the difficulty of manipulating membrane potentials. Fortunately the existence of bacterial mutants blocked at

Figure 1.8 Coupling the excess of H^+ ions outside a cell to the transport of a specific molecule into the cell, symport, or out of the cell, antiport, by specific proteins that couple the transport of a proton into the cell with the transport of another molecule. The ATPase generates ATP from ADP with the energy derived from permitting protons to flow back into the cell.

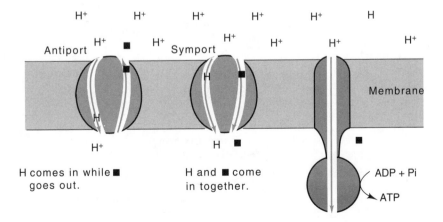

various steps of the transport process has permitted partial dissection of the system. We are, however, very far from completely understanding the actual mechanisms involved in chemiosmotic systems.

The binding protein systems represent another type of transport through membranes. These systems utilize proteins located in the periplasmic space that specifically bind sugars, amino acids, and ions. Apparently, these periplasmic binding proteins transfer their substrates to specific carrier molecules located in the cell membrane. The energy source for these systems is ATP or a closely related metabolite.

Transporting large molecules through the cell wall and membranes poses additional problems. Eukaryotic cells can move larger molecules through the membrane by exocytosis and endocytosis processes in which the membrane encompasses the molecule or molecules. In the case of endocytosis, the molecule can enter the cell, but it is still separated from the cytoplasm by the membrane. This membrane must be removed in order for the membrane-enclosed packet of material to be released into the cytoplasm. By an analogous process, exocytosis releases membrane-enclosed packets to the cell exterior.

Releasing phage from bacteria also poses difficult problems. Some types of filamentous phage slip through the membrane like a snake. They are encapsidated as they exit the membrane by phage proteins located in the membrane. Other types of phage must digest the cell wall to make holes large enough to exit. These phage lyse their hosts in the process of being released.

An illuminating example of endocytosis is the uptake of low density lipoprotein, a 200 Å diameter protein complex that carries about 1,500 molecules of cholesterol into cells. Pits coated with a receptor of the low density lipoprotein form in the membrane. The shape of these pits is guided by triskelions, an interesting structural protein consisting of three molecules of clathrin. After receptors have been in a pit for about

Figure 1.9 Endocytosis of receptor-coated pits to form coated vesicles and the recycling of receptor that inserts at random into the plasma membrane and then clusters in pits.

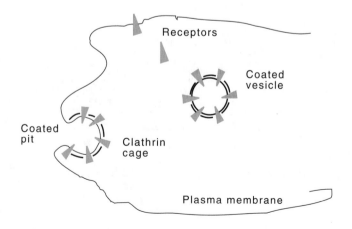

ten minutes, the pit pinches off and diffuses through the cytoplasm (Fig. 1.9). Upon reaching the lysosome, the clathrin cage of triskelions is disassembled, cholesterol is released, and the receptors recycle.

Diffusion within the Small Volume of a Cell

Within several minutes of adding a specific inducer to bacteria or eukaryotic cells, newly synthesized active enzymes can be detected. These are the result of the synthesis of the appropriate messenger RNA, its translation into protein, and the folding of the protein to an active conformation. Quite obviously, processes are happening very rapidly within a cell for this entire sequence to be completed in several minutes. We will see that our image of synthetic processes in the cellular interior should be that of an assembly line running hundreds of times faster than normal, and our image for the random motion of molecules from one point to another can be that of a washing machine similarly running very rapidly.

The random motion of molecules within cells can be estimated from basic physical chemical principles. We will develop such an analysis since similar reasoning often arises in the design or analysis of experiments in molecular biology. The mean squared distance $\overline{R}^2$ that a molecule with diffusion constant D will diffuse in time t is $\overline{R}^2 = 6Dt$ (Fig. 1.10). The diffusion constants of many molecules have been measured and are available in tables. For our purposes, we can estimate a value for a diffusion constant. The diffusion constant is $D = {}^{KT}\!/_f$, where K is the Boltzmann constant, 1.38×10^{-16} ergs/degree, T is temperature in degrees Kelvin, and f is the frictional force. For spherical bodies, $f = 6\pi\eta r$, where r is the radius in centimeters and η is the viscosity of the medium in units of poise.

The viscosity of water is 10^{-2} poise. Although the macroviscosity of the cell's interior could be much greater, as suggested by the extremely high viscosity of gently lysed cells, the viscosity of the cell's interior with

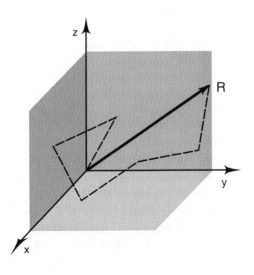

Figure 1.10 Random motion of a particle in three dimensions beginning at the origin and the definition of the mean squared distance $\overline{R}^2$.

respect to motion of molecules the size of proteins or smaller is more likely to be similar to that of water. This is reasonable, for small molecules can go around obstacles such as long strands of DNA, but large molecules would have to displace a huge tangle of DNA strands. A demonstration of this effect is the finding that small molecules such as amino acids readily diffuse through the agar used for growing bacterial colonies, but objects as large as viruses are immobile in the agar, yet diffuse normally in solution.

Since $D = KT/6\pi\eta r$, then $D = 4.4 \times 10^{-7}$ for a large spherical protein of radius 50 Å diffusing in water, and the diffusion constant for such a protein within a cell is not greatly different. Therefore $\overline{R}^2 = 6 \times 4.4 \times 10^{-7}t$, and the average time required for such protein molecules to diffuse the length of a 1 μ bacterial cell is 1/250 second and to diffuse the length of a 20 μ eukaryotic cell is about 2 seconds. Analogous reasoning with respect to rotation shows that a protein rotates about 1/8 radian (about 7°) in the time it diffuses a distance equal to its radius.

Exponentially Growing Populations

Reproducibility from one day to the next and between different laboratories is necessary before meaningful measurements can be made on growing cells. Populations of cells that are not overcrowded or limited by oxygen, nutrients, or ions grow freely and can be easily reproduced. Such freely growing populations are almost universally used in molecular biology, and several of their properties are important. The rate of increase in the number of cells in a freely growing population is proportional to the number of cells present, that is,

$$\frac{dN}{dt} = \mu N, \quad \text{or} \quad N(t) = N(0)e^{\mu t}.$$

In these expressions μ is termed the exponential growth rate of the cells.

The following properties of the exponential function are frequently useful when manipulating data or expressions involving growth of cells.

$$e^{a\ln x} = x^a$$

$$\frac{d}{dx}e^{ax} = ae^{ax}$$

$$e^x = \sum_{n=0}^{\infty} \frac{x^n}{n!}$$

Quantities growing with the population increase as $e^{\mu t}$. Throughout this book we will use μ as the exponential growth rate. The time required for cells to double in number, T_d, is easier to measure experimentally as well as to think about than the exponential growth rate. Therefore we often need to interconvert the two rates T_d and μ. Note that the number of cells or some quantity related to the number of cells in freely growing populations can be written as $Q(t) = 2^{t/T_d}$, and since $2 = e^{\ln 2}$, $Q(t)$ can also be written as $Q(t) = e^{(\ln 2/T_d)t}$, thereby showing that the relation between T_d and μ is $\mu = \ln 2/T_d$.

Composition Change in Growing Cells

In many experiments it is necessary to consider the time course of the induction of an enzyme or other cellular component in a population of growing cells. To visualize this, suppose that synthesis of an enzyme is initiated at some time in all the cells in the population and thereafter the synthesis rate per cell remains constant. What will the enzyme level per cell be at later times?

The Relationship between Cell Doublings, Enzyme Doublings, and Induction Kinetics

Time	$t=0$	$t=T_d$	$t=2T_d$	$t=3T_d$
Cell Mass	1	2	4	8
Enzyme present if synthesis began long ago	A	2A	4A	8A
Enzyme synthesized during one doubling time		A	2A	4A
Enzyme present if synthesis begins at t=0	0	A	3A	7A

One way to handle this problem is to consider a closely related problem we can readily solve. Suppose that synthesis of the enzyme had begun many generations earlier and thereafter the synthesis rate per cell had remained constant. Since the synthesis of the enzyme had been initiated many cell doublings earlier, by the time of our consideration, the cells are in a steady state and the relative enzyme level per cell remains constant. As the cell mass doubles from 1 to 2 to 4, and so on, the amount of the enzyme, A, also doubles, from A to 2A to 4A, and so on. The differences in the amount of the enzyme at the different times give the amounts that were synthesized in each doubling time. Now consider the situation if the same number of cells begins with no enzyme but instead begins synthesis at the same rate per cell as the population that had been induced at a much earlier time (see the last row in the table). At the beginning, no enzyme is present, but during the first doubling time, an amount A of the enzyme can be synthesized by the cells. In the next doubling time, the table shows that the cells can synthesize an amount 2A of the enzyme, so that after two doubling times the total amount of enzyme present is 3A. After another doubling time the amount of enzyme present is 7A. Thus at successive doublings after induction the enzyme level is $\frac{1}{2}$, $\frac{3}{4}$, $\frac{7}{8}$,... of the final asymptotic value.

Age Distribution in Populations of Growing Cells

The cells in a population of freely growing cells are not all alike. A newly divided cell grows, doubles in volume, and divides into two daughter

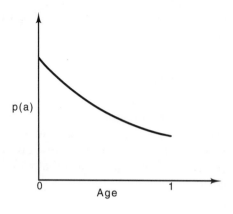

Figure 1.11 Age distribution in an exponentially growing population in which all cells divide when they reach age 1. Note that the population contains twice as many zero-age cells as unit-age cells.

cells. Consequently, freely growing populations contain twice as many cells that have just divided as cells about to divide. The distribution of cell ages present in growing populations is an important consideration in a number of molecular biology experiments, one of which is mentioned in Chapter 3. Therefore we will derive the distribution of ages present in such populations.

Consider an idealized case where cells grow until they reach the age of 1, at which time they divide. In reality most cells do not divide at exactly this age, but the ages at which cell division occurs cluster around a peak. To derive the age distribution, let $N(a,t)da$ be the number of cells with age between a and $a + da$ at time t. For convenience, we omit writing the da. Since the number of cells of age a at time t must be the same as the number of zero-age cells at time $t-a$, $N(a,t) = N(0,t-a)$. Since the numbers of cells at any age are growing exponentially, $N(0,t) = N(0,0)e^{\mu t}$, and $N(a,t) = N(0,t-a) = N(0,t)e^{-\mu a}$. Therefore the probability that a cell is of age a, $p(a)$, is $p(0)e^{-\mu a} = p(0)2^{-a/Td}$ (Fig. 1.11).

Problems

1.1. Propose an explanation for the following facts known about *E. coli*: appreciable volume exists between the inner membrane and the peptidoglycan layer; the inner membrane is too weak to withstand the osmotic pressure of the cytoplasm and must be supported by a strong, rigid structure; and no spacers have been discovered that could hold the inner membrane away from the peptidoglycan layer.

1.2. If the *E. coli* interior were water at pH 7, how many H^+ ions would exist within the cell at any instant?

1.3. If a population of cells growing exponentially with a doubling time T_d were contaminated at one part in 10^7 with cells whose doubling time is $0.95\ T_d$, how many doublings will be required until 50% of the cells are contaminants?

1.4. If an enzyme is induced and its synthesis per cell is constant, show that there is a final upper-bound less than 100% of cellular protein that this enzyme can constitute. When the enzyme has reached this level, what is the relation between the rate of synthesis of the enzyme and the rate of dilution of the enzyme caused by increase of cellular volume due to growth?

1.5. In a culture of cells in balanced exponential growth, an enzyme was induced at time $t = 0$. Before induction the enzyme was not present, and at times very long after induction it constituted 1% of cell protein. What is the fraction of cellular protein constituted by this protein at any time $t > 0$ in terms of the cell doubling time? Ignore the 1 min or so lag following induction until the enzyme begins to appear.

1.6. In a culture of cells in balanced exponential growth, an enzyme was fully induced at some very early time, and the level of enzyme ultimately reached 1% of total protein. At time $t = 0$ the synthesis of enzyme was repressed. What fraction of cellular protein is constituted by the enzyme for $t > 0$ (a) if the repressed rate of synthesis is 0 and (b) if the repressed rate of synthesis is 0.01 of the fully induced rate?

1.7. If the concentration of a typical amino acid in a bacterium is 10^{-3} M, estimate how long this quantity, without replenishment, could support protein synthesis at the rate that yields 1×10^{-13} g of newly synthesized protein with a cell doubling time of 30 min.

1.8. If a typical protein can diffuse from one end to the other of a cell in 1/250 sec when it encounters viscosity the same as that of water, how long is required if the viscosity is 100 times greater?

1.9. A protein of molecular weight 30,000 daltons is in solution of 200 mg/ml. What is the average distance separating the centers of the molecules? If protein has a density of 1.3, what fraction of the volume of such a solution actually is water?

1.10. How can the existence of the Na^+-K^+ pump in eukaryotic cells be demonstrated?

1.11. How can valinomycin be used to create a temporary membrane potential in cells or membrane vesicles?

1.12. Suppose the synthesis of some cellular component requires synthesis of a series of precursors P_1, P_2, P_n proceeding through a series of pools S_i.

$$P_1 \rightarrow P_2 \rightarrow P_3 \rightarrow \;\rightarrow P_n$$

$$S_1 \rightarrow S_2 \rightarrow S_3 \rightarrow \;\rightarrow S_n$$

Suppose the withdrawal of a precursor molecule P_i from pool S_i, and its maturation to S_{i+1} is random. Suppose that at $t = 0$ all subsequently synthesized precursors P_1 are radioactively labeled at constant specific activity. Show that at the beginning, the radioactive label increases proportional to t^n in pool S_n.

1.13. Consider cells growing in minimal medium. Suppose a radio-active amino acid is added and the kinetics of radioactivity incorporation into protein are measured for the first minute. Assume that upon addition of the amino acid, the cell completely stops its own synthesis of the amino acid and that there is no leakage of the amino acid out of the cell. For about the first 15 sec, the incorporation of radioactive amino acid into protein increases as t^2 and thereafter as t. Show how this delayed entry of radioactive amino acids into protein results from the pool of free nonradioactive amino acid in the cells at the time the radioactive amino acid was added. Continue with the analysis and show how to calculate the concentration of this internal pool. Use data of Fig. 2 in J. Mol. Bio. *27*, 41 (1967) to calculate the molarity of free proline in *E. coli* B/r.

1.14. Consider a more realistic case for cell division than was considered in the text. Suppose that cells do not divide precisely when they reach age 1 but that they have a probability given by the function *f(a)* of dividing when they are of age *a*. What is the probability that a cell is of age *a* in this case?

References

Recommended Readings

Role of an Electrical Potential in the Coupling of Metabolic Energy to Active Transport by Membrane Vesicles of *Escherichia coli*, H. Hirata, K. Altendorf, F. Harold, Proc. Nat. Acad. Sci. USA *70*, 1804-1808 (1973).

Coated Pits, Coated Vesicles, and Receptor-Mediated Endocytosis, J. Goldstein, R. Anderson, M. Brown, Nature *279*, 679-685 (1979).

Osmotic Regulation and the Biosynthesis of Membrane-Derived Oligosaccharides in *Escherichia coli*, E. Kennedy, Proc. Nat. Acad. Sci. USA *79*, 1092-1095 (1982).

Cell Structure

Sugar Transport, I. Isolation of A Phosphotransferase System from *E. coli*, W. Kundig, S. Roseman, J. Biol. Chem. *246*, 1393-1406 (1971).

Localization of Transcribing Genes in the Bacterial Cell by Means of High Resolution Autoradiography, A. Ryter, A. Chang, J. Mol. Biol. *98*, 797-810 (1975).

The Relationship between the Electrochemical Proton Gradient and Active Transport in *E. coli* Membrane Vesicles, S. Ramos, H. Kaback. Biochem. *16*, 854-859 (1977).

Escherichia coli Intracellular pH, Membrane Potential, and Cell Growth, D. Zilberstein, V. Agmon, S. Schuldiner, E. Padan, J. Bact. *158*, 246-252 (1984).

Ion Selectivity of Gram-negative Bacterial Porins, R. Benz, A. Schmid, R. Hancock, J. Bact. *162*, 722-727 (1985).

Measurement of Proton Motive Force in *Rhizobium meliloti* with *Escherichia coli lacY* Gene Product, J. Gober, E. Kashket, J. Bact. *164*, 929-931 (1985).

Internalization-defective LDL Receptors Produced by Genes with Nonsense and Frameshift Mutations That Truncate the Cytoplasmic Domain, M. Lerman, J. Goldstein, M. Brown, D. Russell, W. Schneider, Cell *41*, 735-743 (1985).

Escherichia coli and *Salmonella typhimurium*, Cellular and Molecular Biology, eds. F. Neidhardt, J. Ingraham, K. Low, B. Magasanik, M. Schaechter, H. Umbarger, Am. Society for Microbiology (1987).

Introduction of Proteins into Living Bacterial Cells: Distribution of Labeled HU Protein in *Escherichia coli*, V. Shellman, D. Pettijohn, J. Bact. *173*, 3047-3059 (1991).

Characterization of the Cytoplasm of *Escherichia coli* as a Function of External Osmolarity, S. Cayley, B. Lewis, H. Guttman, M. Record, Jr., J. Mol. Biol. *222*, 281-300 (1991).

Estimation of Macromolecule Concentrations and Excluded Volume Effects for the Cytoplasm of *Escherichia coli*, S. Zimmerman, S. Trach, J. Mol. Biol. *222*, 599-620 (1991).

Inside a Living Cell, D. Goodsell, Trends in Biological Sciences, *16*, 203-206 (1991).

Nucleic Acid and
Chromosome Structure

2

Thus far we have considered the structure of cells and a few facts about their functioning. In the next few chapters we will be concerned with the structure, properties, and biological synthesis of the molecules that have been particularly important in molecular biology—DNA, RNA, and protein. In this chapter we consider DNA and RNA. The structures of these two molecules make them well suited for their major biological roles of storing and transmitting information. This information is fundamental to the growth and survival of cells and organisms because it specifies the structure of the molecules that make up a cell.

Information can be stored by any object that can possess more than one distinguishable state. For example, we could let a stick six inches long represent one message and a stick seven inches long represent another message. Then we could send a message specifying one of the two alternatives merely by sending a stick of the appropriate length. If we could measure the length of the stick to one part in ten thousand, we could send a message specifying one of ten thousand different alternatives with just one stick. Information merely limits the alternatives.

We will see that the structure of DNA is particularly well suited for the storage of information. Information is stored in the linear DNA molecule by the particular sequence of four different elements along its length. Furthermore, the structure of the molecule or molecules—two are usually used—is sufficiently regular that enzymes can copy, repair, and read out the stored information independent of its content. The duplicated information storage scheme also permits repair of damaged information and a unified mechanism of replication.

One of the cellular uses of RNA, discussed in later chapters, is as a temporary information carrier. Consequently, RNA must also carry information, but ordinarily it does not participate in replication or repair activities. In addition to handling information, some types of RNA molecules have been found to have structural or catalytic activities. The ability of RNA to perform all these roles has led to the belief that in the evolution of life, RNA appeared before DNA or protein.

The Regular Backbone of DNA

The chemical structure of DNA is a regular backbone of 2'-deoxyriboses, joined by 3'-5' phosphodiester bonds (Fig. 2.1). The information carried by the molecule is specified by bases attached to the 1' position of the deoxyriboses. Four bases are used: the purines adenine and guanine, and the pyrimidines cytosine and thymine. The units of base plus ribose or deoxyribose are called nucleosides, and if phosphates are attached to the sugars, the units are called nucleotides.

The chemical structure of RNA is similar to that of DNA. The backbone of RNA uses riboses rather than 2'-deoxyriboses, and the methyl group on the thymine is absent, leaving the pyrimidine uracil.

Clearly the phosphate-sugar-phosphate-sugar along the backbones of DNA and RNA are regular. Can anything be done to make the information storage portion of the molecule regular as well? At first glance this seems impossible because the purines and the pyrimidines are different sizes and shapes. As Watson and Crick noticed however, pairs of these molecules, adenine-thymine and guanine-cytosine, do possess regular shapes (Fig. 2.2). The deoxyribose residues on both A-T and G-C pairs are separated by the same distance and can be at the same relative orientations with respect to the helix axis. Not only are these pairs regular, but they are stabilized by strong hydrogen bonds. The A-T pair generally can form two hydrogen bonds and the G-C base pair can form

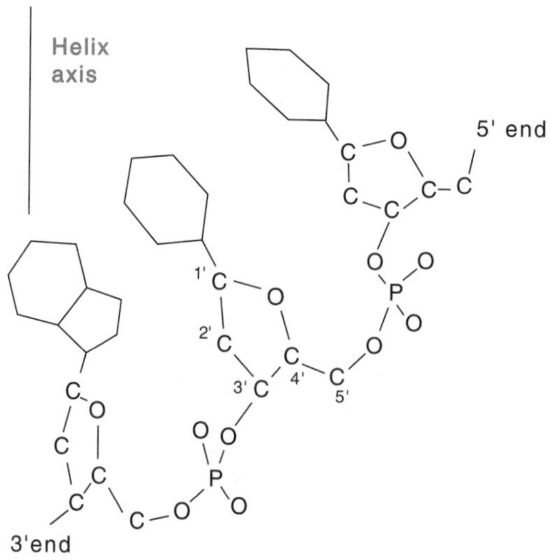

Figure 2.1 The helical backbone of DNA showing the phosphodiester bonds, the deoxyribose rings which are nearly parallel to the helix axis and perpendicular to the planes of the bases, and the bases.

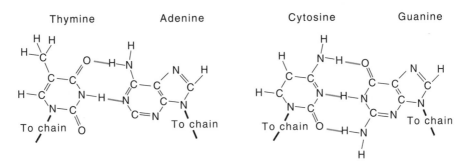

Figure 2.2 Hydrogen bonding between adenine-thymine and guanine-cytosine pairs.

three hydrogen bonds between the respective bases. Finally, the base pairs A-T and G-C can stack via hydrophobic interactions.

Hydrogen bonds can form when a hydrogen atom can be shared by a donor such as an amino group and an acceptor such as a carbonyl group. The hydrogen bonds between the bases of DNA are strong because in all cases the three atoms participating in hydrogen bond formation lie in nearly straight lines. In addition to the familiar Watson-Crick pairings of the bases, other interactions between the bases have been observed and are also biologically important. These alternative structures frequently occur in tRNA and also are likely to exist in the terminal structures of chromosomes, called telomeres.

Grooves in DNA and Helical Forms of DNA

Watson and Crick deduced the basic structure of DNA by using three pieces of information: X-ray diffraction data, the structures of the bases, and Chargaff's findings that, in most DNA samples, the mole fraction of guanine equals that of cytosine, as well as the mole fraction of adenine equals that of thymine. The Watson-Crick structure is a pair of oppositely oriented, antiparallel, DNA strands that wind around one another in a right-handed helix. That is, the strands wrap clockwise moving down the axis away from an observer. Base pairs A-T and G-C lie on the interior of the helix and the phosphate groups on the outside.

In semicrystalline fibers of native DNA at one moisture content, as well as in some crystals of chemically synthesized DNA, the helix repeat is 10 base pairs per turn. X-ray fiber diffraction studies of DNA in different salts and at different humidities yield forms in which the repeats vary from 9 1/3 base pairs per turn to 11 base pairs per turn. Crystallographers have named the different forms A, B, and C (Table 2.1).

More recent diffraction studies of crystals of short oligonucleotides of specific sequence have revealed substantial base to base variation in

Table 2.1 Parameters of Some DNA Helices

Helix	Base Pairs/turn	Rotation per Base Pair	Rise, Å/Base Pair	Diameter
A	11	32.7	2.56	23
B	10	36	3.38	19
C	9 ⅓	38.6	3.32	19
Z	12	30	3.63	18

the twist from one base to the next. Therefore, it is not at all clear whether it is meaningful to speak of the various forms of the DNA. Nonetheless, on average, natural DNA, that is DNA with all four bases represented in random sequence over short distances, has a conformation most closely represented by B-form DNA.

The A, B, and C forms of DNA are all helical. That is, as the units of phosphate-deoxyribose-base:base-deoxyribose-phosphate along the DNA are stacked, each succeeding base pair unit is rotated with respect to the preceding unit. The path of the phosphates along the periphery

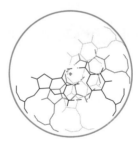

of the resulting structure is helical and defines the surface of a circular cylinder just enclosing the DNA. The base pair unit is not circular, however, and from two directions it does not extend all the way to the enclosing cylinder. Thus a base pair possesses two indentations. Because the next base pair along the DNA helix is rotated with respect to the preceding base, its indentations are also rotated. Thus, moving along the DNA from base to base, the indentations wind around the cylinder and form grooves.

Fig. 2.3 shows a helix generated from a rectangle approximating the base pair unit of B DNA. Note that the rectangle is offset from the helix axis. As a result of this offset, the two grooves generated in the helix are of different depths and slightly different widths. The actual widths of the two grooves can be seen more clearly from the side view of the three dimensional helix in which the viewpoint is placed so that you are looking directly along the upper pair of grooves.

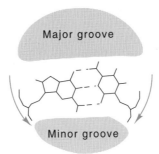

In actual DNA, the deoxyribose-phosphate units are not aligned parallel to the base pairs as they were represented in the rectangle approximation above. The units are both oriented toward one of the grooves. This narrows one of the grooves on the helical DNA molecule and widens the other. The two grooves are therefore called the minor and major grooves of the DNA. Thus, the displacement of the base pair from the helix axis primarily affects the relative depth of the two grooves, and the twisted position of the phosphates relative to the bases primarily affects the widths of the grooves.

The A-form of DNA is particularly interesting because the base pairs are displaced so far from the helix axis that the major groove becomes very deep and narrow and the minor groove is barely an indentation. Helical RNA most often assumes conformations close to the A-form. Additional factors such as twist and tilt of the base pairs which are

Figure 2.3 Generation of a helix by the stacking of rectangles. Each rectangle is rotated 34° clockwise with respect to the one below. Left, the viewpoint and the generating rectangle; right, a view along the major and minor grooves showing their depths and shapes.

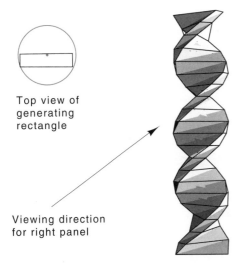

Top view of generating rectangle

Viewing direction for right panel

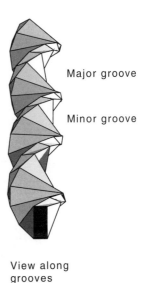

Major groove

Minor groove

View along grooves

present in the A- and C-forms of DNA, but largely absent in the B-form, have lesser effects on the width and depth of the grooves. Unexpectedly, even a left-helical form can form under special conditions, but thus far this Z-form DNA has not been found to play any significant biological role.

Helical structures can also form from a single strand of RNA or DNA if it folds back upon itself. The two most common structures are hairpins and pseudoknots. In a pseudoknot, bases in the loop form additional base pairs with nucleotides beyond the hairpin region.

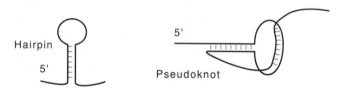

Dissociation and Reassociation of Base-paired Strands

Heating DNA in solution breaks the hydrogen bonds between the A-T and G-C base pairs, unstacks the bases, and destroys the double-helical structure of the DNA. Such a process is called melting. Generally, not all the bonds break at one temperature, and DNA exhibits a transition zone between fully double-stranded DNA and fully melted DNA that often is 15° wide. The midpoint of this melting zone is defined as the melting temperature, which occurs at about 95° in 0.1 M NaCl. The actual value of the melting temperature, however, depends on the base composition of the DNA, for the three hydrogen bonds in G-C base pairs provide more stability than the two found in A-T base pairs. The ionic composition of the solution also affects the melting temperature. The higher the concentration of an ion such as sodium, the greater the shielding between the negatively charged phosphates and the higher the melting temperature. A divalent ion such as magnesium is still more effective in raising the melting temperature of DNA.

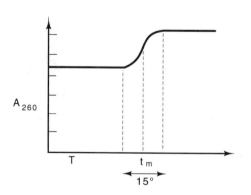

Figure 2.4 Increase in absorbance at 260 nm, A_{260}, of a DNA sample as the temperature is increased through the melting temperature.

Melting can be observed in a variety of ways. One of the simplest is based on the fact that unstacked bases have a higher absorbance in the ultraviolet region of the spectrum than stacked, paired bases (Fig. 2.4). Therefore the absorbance of DNA in the ultraviolet region increases as the DNA melts, and by following the optical density of a DNA solution as a function of temperature, a melting curve can be obtained.

A most remarkable property of denatured DNA is its ability to renature *in vitro* to re-form double-stranded DNA. This re-formation of double-stranded DNA usually is very precise and exactly in register. Two strands may renature to the native double-helical form if their sequences are complementary, that is, if their sequences permit extensive formation of hydrogen-bonded base pairs.

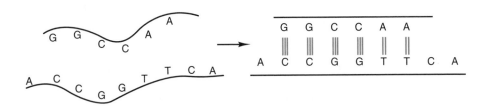

The ability of self-complementary sequences to hybridize together and form a double helix is not unique to DNA; RNA can also do this, and it is possible *in vitro* to form RNA-DNA hybrids or RNA-RNA duplexes. The ability of nucleic acids to renature has been extremely important in the development of molecular biology as it has provided ways of detecting the presence of small quantities of specific sequences of RNA or DNA and in some cases to determine their intracellular locations through their ability to form sequence-specific duplexes.

Reading Sequence without Dissociating Strands

Can the sequence of the DNA be recognized without destroying its double helical structure? Since thousands of regulatory proteins must bind to their cognate regulatory sequences near the genes they regulate, it is crucial that these proteins be able to recognize their binding sequences without requiring that the DNA strands be separated.

Sequence-dependent effects can be seen in the slight structural differences found in crystallized oligonucleotides. It is possible that proteins could utilize these structural differences and ignore the chemical differences between the bases. For example, a protein might recognize its correct binding site strictly by the locations of phosphates in space. The regulator of the *trp* operon in *Escherichia coli* appears to recognize its binding site utilizing such principles because it appears to make almost no base-specific hydrogen bonds.

The second possibility for recognition of sequences is to read the chemical structures of the bases. Hydrogen bonds can be made to donors and acceptors in both the major and minor grooves as shown in

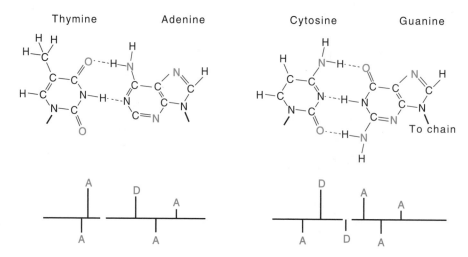

Figure 2.5 Above, T-A and C-G base pairs. Below, schematic representation of the bases in which the locations of the hydrogen bond donors (D), and acceptors (A), in the major groove are shown above the line and those in the minor groove are shown below the line.

the schematic (Fig. 2.5). Considering the typical flexibility in proteins, it becomes apparent that distinguishing the four base pairs solely by the presence or absence of hydrogen bonding capabilities of the four bases requires a minimum of two hydrogen bonds per base pair in the major groove. A-T and T-A base pairs cannot be distinguished in the minor groove. In nature we can expect some proteins to recognize sequence by structure determination, some proteins to recognize sequence by hydrogen bonding to the portions of the bases exposed in the major groove, some to utilize additional interactions to the methyl group of thymine, and many to utilize a combination of all these methods.

Electrophoretic Fragment Separation

The phosphate backbones of DNA and RNA molecules give them a uniform charge per unit length. Therefore, upon electrophoresis through polyacrylamide or agarose gels, molecules will migrate at rates largely independent of their sequences. The frictional or retarding forces the gels exert on the migrating molecules increase sharply with the length of the DNA or RNA so that the larger the molecule, the slower it migrates through a gel. This is the basis of the exceptionally valuable technique of electrophoresis. In general, two molecules whose sizes differ by 1% can be separated. Polyacrylamide gels are typically used for molecules from five to perhaps 5,000 base pairs, and agarose gels are used for molecules 1,000 base pairs and larger.

Following electrophoresis, the locations of specific DNA fragments can be located by staining or by autoradiography. Ethidium bromide is a most useful stain for this purpose. The molecule is nonpolar and readily intercalates between bases of DNA. In the nonpolar environment between the bases, its fluorescence is increased about 50 times. Therefore a gel can be soaked in a dilute solution of ethidium bromide and illumination with an ultraviolet lamp reveals the location of DNA as bands glowing cherry-red. As little as 5 ng of DNA in a band can be detected by this method. For detection of smaller quantities of DNA, the DNA can be radioactively labeled before electrophoresis. A simple enzymatic method of doing this is to use the enzyme polynucleotide kinase to transfer a phosphate group from ATP to the 5'-OH of a DNA molecule. After electrophoresis, a radioactive DNA band is located by exposing a photographic film to the gel and developing. The radioactive decay of the ^{32}P sensitizes the silver halide crystals in the film so that upon development, black particles of silver remain to reveal positions of radioactive DNA or RNA in the gel.

Above about 50,000 base pairs long, all DNA migrates in gels at about the same rate. This results from the DNA assuming a conformation in which its charge to frictional force ratio is independent of its length as it snakes through the gel in a reptilian fashion. It was empirically found, however, that brief periodic changes in the direction of the electric field or polarity reverses often will separate still larger DNA molecules. This techniques is called pulsed field electrophoresis. Overall, the major motion of the DNA is in one direction, but it is punctuated by reversals or changes in direction from once per second to once per minute. The change in migration direction destroys the structure of the species whose migration rates are independent of size, and for a short while, the long DNA molecules migrate at rates related to their sizes. Additional size separation is achieved in these electrophoretic techniques because the larger the molecule, the longer it takes to achieve the steady-state snaking state. By these means, molecules as large as 1,000,000 base pair chromosomes can be separated by size.

Bent DNA Sequences

If a series of small bends in DNA are added coherently, together they generate a significant bend. Most surprisingly, a short DNA fragment of

Random adding   Coherent adding

several hundred base pairs containing such a bend migrates anomalously slowly in electrophoresis. Such abnormally migrating fragments were recognized soon after DNA sequencing became possible, but some time passed before it was realized that the major contributor to bending is a run of three or four A's. Significant bends can be generated by

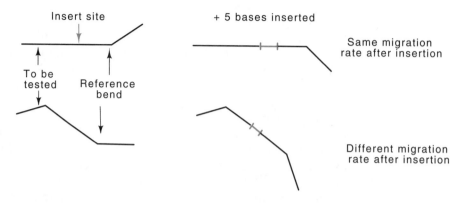

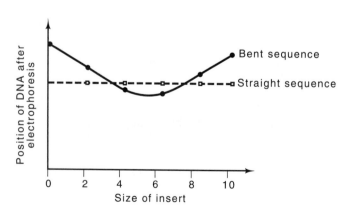

Figure 2.6 Determining whether a DNA segment to be tested is straight or bent. Top, segment is straight and a five base insertion generates no change in migration rate. Bottom, insertion changes the migration rate if the segment is bent.

spacing several AAAATTT elements an average of 10 to 11 base pairs apart.

Bending of DNA is biologically important. Some DNA-binding proteins bend DNA, and the natural bend of other sequences like some origins of DNA replication appears to be essential. Many of the binding sites of proteins are naturally bent, and the binding of the protein further bends the DNA. Thus, the presence of a bend in the DNA helps the binding of a protein which further bends the DNA.

The bending generated by a specific DNA sequence or by the binding of a protein to a specific site can be tested even if the bend is small. Suppose the DNA in question is connected to another segment of "reference" DNA that is known to contain a bend. These bends may add as shown in Fig. 2.6, or they may cancel. Now consider the consequences of adding two, four, six, eight, and ten base pairs between the test and reference regions. As the additional spacing DNA is added, one segment is rotated with respect to the other. At some point in the process the two bends are in the same direction and add and the migration rate of the

Figure 2.7 Relative migration positions in a gel of straight and bent DNA fragments connected to a bent segment of DNA by various lengths of a linker region.

segment will be low. At another position of relative rotation, the two bends cancel one another and the migration rate of the DNA is faster. Plotting the migration rate as a function of the relative rotation or spacer DNA added (Fig. 2.7) can prove the existence of a bend, as well as determine its direction and magnitude.

Measurement of Helical Pitch

It is not straightforward to determine the helical repeat of DNA under *in vivo* conditions. Such measurements have been made, and will be described later. Here we shall consider measuring the helical pitch *in vitro* of linear DNA not bound to any proteins.

Klug and co-workers found that DNA can bind tightly to the flat surface of mica or calcium phosphate crystals. While bound to such surfaces, only a portion of the cylindrical DNA is susceptible to cleavage by DNAse I, an enzyme that hydrolyzes the phosphodiester backbone of DNA (Fig. 2.8). Consider the consequences of: 1. utilizing an homogenous population of DNA molecules, 2. radioactively labeling each molecule on one end with $^{32}PO_4$, 3. rotationally orienting all the DNA molecules similarly, i.e. the 5' end of the labeled strand begins in contact with the solid support, 4. performing a partial digestion with DNAse I, so that on average, each DNA molecule is cleaved only once.

In the population, the labeled strand will be cleaved more frequently at those positions where it is on the part of the helix up away from the support, and thus cleavages will be concentrated at positions ½, 1½, 2½ etc. helical turns from the labeled end. A similar population of labeled DNA digested while in solution will possess some molecules

Figure 2.8 Determination of the helical pitch of DNA while bound to a solid support. While bound to the support, and while free in solution, the DNA is lightly digested with DNAse, the denatured fragments are separated according to size by electrophoresis and an autoradiograph is made of the gel. When the DNA is on the solid support, DNAse has only limited access.

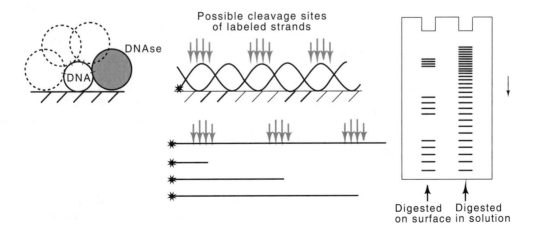

cleaved at every position. Electrophoretic separation of the populations yields the pattern shown (Fig. 2.8). In the sample cleaved while on the flat support most cleavages, and hence the darkest bands, are separated by 10 or 11 base pairs. The DNA cleaved in solution shows all sizes of DNA fragments.

Topological Considerations in DNA Structure

Topology introduces a structural feature in addition to the base-paired and helical aspects of DNA. The origin of this structure can best be understood by considering a mathematical property of two closed rings. The number of times that one ring encircles or links the other must be an integral number. It cannot be changed without physically opening

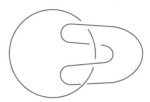

one of the rings. That is, their linking number is a topological invariant. Many types of DNA molecules found in cells are covalently closed circles because each strand is circular. Hence the concept of a linking number applies to DNA molecules obtained from many sources. The concept also applies to linear DNA if the ends are prevented from freely rotating, either because of the extreme length of the DNA or because the DNA is attached to something else.

The forces tending to hold double-stranded DNA in a right-handed helix with about 10.5 base pairs per turn add a dimension to the analysis of the structures of covalently closed circles. These forces are sufficiently great that the linking number, Lk, generally resolves itself into two easily distinguished components: the twist, Tw, which in DNA's usual right-helical form has a value of 1 per each 10.5 base pairs, and the writhe, Wr. Twist is the local wrapping of one of the two strands about the other. If Lk does not equal Tw, then the discrepancy must be made up from a global writhing of the molecule as such global effects can alter the actual number of times one strand encircles the other. These global effects are called supercoiling or superhelical turns. Their computation is most difficult because the entire path of the DNA duplex must be considered. To repeat, for any covalently closed double-stranded DNA molecule, no matter how it is distorted, unless its phosphodiester backbone is broken, $Lk = Tw + Wr$. This equation sometimes is written $\tau = \alpha + \beta$ where $\tau = Lk$, $\alpha = Tw$, and $\beta = Wr$. It is curious that the topological invariant Lk equals the sum of two terms, each of which is not invariant.

It is convenient to normalize the deviation of the linking number from the normal, unconstrained value, Lk_0. Since the normal linking is 1 per

10.5 base pairs of DNA, linking values other than this must change either the twist, the writhe, or both. Rather than just give the linking number deviation for a DNA molecule, it is more informative to give the linking number deviation per unit length of the DNA, $Lk - Lk_0$ divided by Lk_0. This value is denoted by σ. Typical values of σ in DNA extracted from bacterial cells are -0.02 to -0.06. Colloquially σ is called supercoiling density, but it must be remembered that part of the linking number deviation goes into changing the twist of the DNA molecules and part goes into generating writhe.

Generating DNA with Superhelical Turns

To understand how we may experimentally vary Lk, and consequently the degree of supercoiling, let us consider the lambda phage DNA. The molecules of this DNA are about 50,000 base pairs long and possess what are called sticky ends; that is, the ends of the DNA duplex are not flush. As shown in Fig. 2.9, the 5' ends protrude in a single-stranded region of 12 bases. The sequence of the left end is complementary to the sequence of the right end. These sticky ends can be reassociated together to form a circle, which sometimes is called a Hershey circle after its discoverer. The phosphodiester bonds are not contiguous around the Hershey circle; hence its other name, a nicked circle. Circles having a break in only one of their backbones also are called nicked.

Nicks can be covalently sealed with DNA ligase. This enzyme seals the phosphodiester backbone of DNA between nicks that have a 5'-phosphate and a 3'-hydroxyl. Following ligation which forms circles, Lk cannot be altered without breaking the backbone of one of the two strands. Hence, the sum of Tw, the right-helical turns, and Wr, the number of superhelical turns, is fixed. If under fixed buffer and temperature conditions, we were to anneal the ends of the lambda DNA together and then seal with ligase, the number of superhelical turns would be zero and Lk would be about 5,000, about one turn per ten base

Figure 2.9 Association of the self-complementary single-stranded ends of lambda phage DNA to form a nicked circle

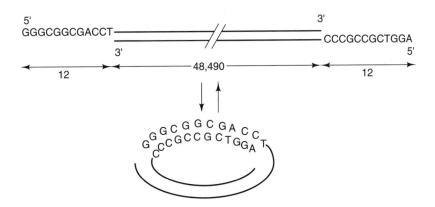

Figure 2.10 Intercalation of a molecule of ethidium bromide between two bases in the DNA duplex reduces their relative twist.

pairs. For convenience let us say that the number is exactly 5,000. Furthermore, if we were to introduce distortion or even to wrap this DNA around a protein, the sum of Tw and Wr must still remain 5,000.

Suppose instead, the annealing and sealing had been done in the presence of ethidium bromide. Its intercalation between bases pushes the bases apart and partly untwists the DNA in this region because the phosphodiester backbone of the DNA cannot lengthen (Fig. 2.10). Hence the amount by which one strand wraps around another is decreased by the intercalation of the ethidium bromide. In the common B form of DNA, the bases are twisted about 34° per base, but the intercalation of an ethidium bromide molecule removes 24° of this twist. The number of helical turns in a lambda DNA molecule sealed in the presence of a particular concentration of ethidium bromide might be about 100 less than the number contained in a lambda DNA molecule sealed in the absence of ethidium bromide. Treating with DNA ligase under these conditions would produce a molecule with no writhe, $Wr = 0$, and with $Lk = Tw = 4,900$. If the ethidium bromide were then removed by extraction with an organic solvent, Tw would return to near its standard value of 5,000; but because of the requirement that $Tw + Wr = 4,900$ be a constant, Wr would become -100 and the circular DNA would writhe. It would have 100 negative superhelical turns, or σ or -100/5000 = -0.02.

Measuring Superhelical Turns

Superhelical turns in DNA may introduce distortions or torsion in the molecules that assist or hinder processes we would like to study such as recombination or the initiation of transcription. Supercoiling must be easily measurable in order to be productively studied. One way to measure superhelical turns might be to observe the DNA in an electron microscope and see it twisted upon itself. Quantitation of the superheli-

Wr = 0   Wr ≅ 1/2   Wr > 8

cal turns in DNA with more than a few turns is difficult however. More convenient measurement methods exist. Consider the DNA molecule

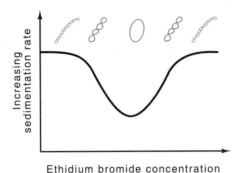

Figure 2.11 Sedimentation rate of a covalently closed circular DNA molecule as a function of the ethidium bromide concentration in the centrifugation solution.

just described with 100 negative superhelical turns. Because it is so twisted upon itself, the molecule is rather compact and sediments in the ultracentrifuge at a high rate. If the sedimentation is performed in the presence of a low concentration of ethidium bromide, a few molecules will intercalate into the DNA. This will reduce the number of negative superhelical turns, thereby opening up the DNA, which will sediment more slowly than it would in the absence of ethidium bromide.

Consider a series of sedimentation measurements made in the presence of increasing concentrations of ethidium bromide. At higher and higher concentrations of ethidium bromide, more and more will intercalate into the DNA and unwind the DNA more and more. Consequently the DNA will become less and less compact and sediment more and more slowly (Fig. 2.11). Finally a concentration of ethidium bromide will be reached where the molecule is completely free of superhelical turns. At this concentration, the DNA will sediment most slowly. If the centrifugation is done in the presence of still higher concentrations of ethidium bromide, the molecule will be found to sediment more rapidly as the DNA acquires positive superhelical turns and becomes more compact again. The concentration of ethidium bromide required to generate the slowest sedimentation rate can then be related to the number of superhelical turns originally in the DNA via the affinity of ethidium bromide for DNA and the untwisting produced per intercalated ethidium bromide molecule.

Even more convenient than centrifugation for quantitation of superhelical turns has been electrophoresis of DNA through agarose. Under some conditions DNA molecules of the same length but with different linking numbers can be made to separate from one another upon electrophoresis. The separation results from the fact that two molecules with different linking numbers will, on the average during the electrophoresis, possess different degrees of supercoiling and consequently different compactness. Those molecules that are more greatly supercoiled during the electrophoresis will migrate more rapidly. Not only can agarose gels be used for quantitating species with different numbers of superhelical turns, but any particular species can be extracted out of the gel and used in subsequent experiments.

The agarose gels show an interesting result. If DNA is ligated to form covalently closed circles and then subjected to electrophoresis under conditions that separate superhelical forms, it is found that not all of the DNA molecules possess the same linking number. There is a distribution centered about the linking number corresponding to zero superhelical turns, Lk_0. This is to be expected because the DNA molecules in solution are constantly in motion, and a molecule can be ligated into a covalently closed circle at an instant when it possesses a linking number unequal to Lk_0. These molecules are frozen in a slightly higher average energy state than those with no superhelical turns. Their exact energy depends on the twisting spring constant of DNA. The stiffer the DNA, the smaller the fraction of molecules that will possess any superhelical turns at the time of sealing. Quantitation of the DNA molecules in the bands possessing different numbers of superhelical turns permits evaluation via statistical mechanics of the twisting spring constant of DNA.

The ability to measure accurately the number of superhelical turns in DNA allows a determination of the amount of winding or unwinding produced by the binding of molecules. For example, unwinding measurements first indicated that RNA polymerase melts about 8 bases of DNA when it binds tightly to lambda DNA. Later, more precise measurements have shown that the unwinding is closer to 15 base pairs. This unwinding was shown directly by binding RNA polymerase to nicked circular DNA and then sealing with ligase to form covalently closed circles, removing the RNA polymerase, and measuring the number of superhelical turns in the DNA. The first measurements were done by accurately comparing the sedimentation velocity of the DNA sealed in the presence and in the absence of RNA polymerase. Later experiments have used a better DNA substrate and have used gel electrophoresis.

Another way to measure the winding produced by binding of a molecule to DNA is to measure the affinity of a molecule for DNA samples containing different numbers of superhelical turns. This method is based on the fact that a protein which introduces negative superhelical turns as it binds to DNA will bind much more tightly to a DNA molecule already containing negative superhelical turns. From the thermodynamics of the situation, this type of approach is very sensitive.

Determining *Lk*, *Tw*, and *Wr* in Hypothetical Structures

A number of the fundamental processes in molecular biology involve the binding and interactions of proteins with DNA, and the covalent cutting and rejoining of DNA. Much insight into these processes is provided by learning whether a portion of the DNA duplex is melted, and how the cutting and rejoining is performed. Often these processes affect the linking number or twist of the DNA, and both can be measured before and after the reaction. Then the effects of possible models on these numbers can be compared to experimental results. The determination of linking number, twist, and superhelical turns in a structure can sometimes be tricky, and in the general situation is a relatively difficult mathematical problem. At our level of analysis, we can proceed

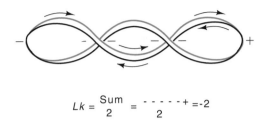

$$Lk = \frac{Sum}{2} = \frac{-\ -\ -\ -\ -\ +}{2} = -2$$

Figure 2.12 Example of a structure with zero twist but a linking number of 2 and hence containing two superhelical turns.

with straightforward approximations based on the fact that the linking number is the number of times that one DNA strand encircles the other. It is useful for this number to have a sign that is dependent on the orientation of the strands. Therefore, to determine the linking number of a structure, draw arrows on the two strands pointing in opposite directions, for example, pointing in the 5' to 3' direction on each. At each point where the two different strands cross, assign a + or - value dependent on the orientation. If the upper strand at a crossover can be brought into correspondence with the lower strand with a clockwise

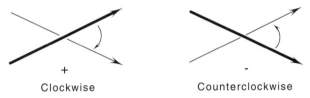

rotation, assign a +; if a counterclockwise rotation is required, assign a -. The linking number equals the sum of these values divided by 2, as shown in Fig. 2.12.

Viewing the DNA duplex as a ribbon provides one way to picture twist. Consider a straight line drawn perpendicular to the axis of the ribbon and through both edges. The rotation of this line as it is moved down the axis gives the twist. For a given structure, the writhing or superhelical turns is most easily given by whatever value Wr must have to make $Lk = Tw + Wr$.

Altering Linking Number

Cells have enzymes that alter the linking numbers of covalently closed DNA molecules. Wang, who found the first such protein, called it omega, but now it is often called DNA topoisomerase I. Astoundingly, this enzyme removes negative superhelical turns, one at a time, without hydrolyzing ATP or any other energy-rich small molecule. It can act on positively supercoiled DNA, but only if a special trick is used to generate a single-stranded stretch for the enzyme to bind to. No nicks are left in

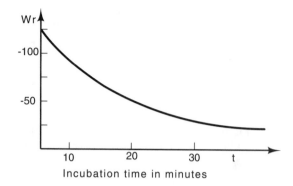

Figure 2.13 Reduction in the number of negative superhelical turns contained in lambda phage DNA as a function of the time of incubation with DNA topoisomerase I.

the DNA after omega has acted. Purified, covalently closed lambda phage DNA extracted from lambda-infected cells is found to possess about 120 negative superhelical turns. Incubation of omega with this DNA increases the linking number until only about 20 negative superhelical turns remain (Fig. 2.13). The enzyme appears to remove twists, and hence superhelical turns, in a controlled way. It binds to the DNA and then interrupts the phosphodiester backbone of one strand by a phosphotransfer reaction that makes a high-energy phosphate bond to the enzyme. In this state the enzyme removes one twist and then re-forms the phosphodiester bond in the DNA.

Additional enzymes capable of altering the linking number of DNA have also been discovered. The topoisomerase from eukaryotic cells can apparently bind double-stranded DNA and will remove positive superhelical turns as well as negative. Cells possess another remarkable enzyme. This enzyme, DNA gyrase or DNA topoisomerase II, adds negative superhelical turns; as expected, energy from the hydrolysis of ATP is required for this reaction.

Studies of the activity of topoisomerases I and II, omega and gyrase, have shown that these enzymes can best be thought of as functioning by strand passage mechanisms. Where two DNA duplexes cross, gyrase, with the consumption of an ATP molecule, can permit the duplex lying below to be cut, to pass the first, and to be rejoined. Topoisomerase I

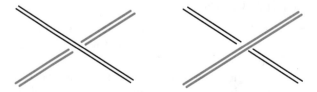

can proceed by the same sort of pathway, but it passes one strand of a duplex through the other strand, having first denatured a region where

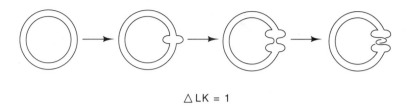

$$\triangle \text{LK} = 1$$

Figure 2.14 The strand inversion pathway as applied by DNA topoisomerase I in which a region of DNA is melted and one strand is broken, passed by the other, and rejoined.

strand crossing is to occur (Fig. 2.14). These are known as sign-inversion pathways because they change the sign of the linking number that is contributed by a point of crossing of the two DNA strands.

The sign-inversion pathway has a particularly useful property: it permits DNA duplexes to interpenetrate one another. Gyrase is an enzyme that can untangle knots in DNA! Undoubtedly this property is of great value to the cell as the DNA is compressed into such a small volume that tangles seem inevitable. In general, enzymes that cut and rejoin DNA are topoisomerases.

Biological Significance of Superhelical Turns

As explained above, supercoiling results from a linking number deficit in covalently closed, double-stranded circles. If such DNA did not form supercoils, it would possess a twist of less than one per 10.5 base pairs. It might be one twist per 11 base pairs, for example. Since DNA can wrap upon itself globally as it tries to attain a local twisting of once per 10.5 base pairs, it forms supercoils. Of course, the DNA resists the introduction of too many superhelical turns. Therefore, not all the linking number deficit is taken up by supercoiling. The deficit is partitioned between supercoiling and reducing the local twist of the DNA. The greater the linking number deficit, the greater the supercoiling and the greater the untwisting of the DNA. Untwisting DNA helps separate its strands. Therefore, negative supercoiling assists the formation of melted sections of DNA. This is the situation *in vitro* with pure DNA possessing a linking number deficit. What about *in vivo*?

Does the same DNA *in vivo* feel such a torsion, or are there unmelted regions of the DNA here and there, perhaps formed by bound proteins, that generate the overall linking number deficit? When these proteins are removed, we would find the linking number deficit. Despite its topological linking number deficit, such DNA *in vivo* would not feel the torsion described above.

The considerations discussed above can also apply to linear DNA molecules if its ends are prevented from free rotation. DNA may be constrained from free rotation because it is attached to a cellular structure or because free rotation is hindered due to its great length or the bulkiness of proteins that may be bound to it.

Several experiments suggest that DNA in bacteria not only possesses superhelical turns but also is under a superhelical torsion. The *in vitro* integration reaction of lambda phage, in which a special set of enzymes catalyzes the insertion of covalently closed lambda DNA circles into the chromosome, proceeds only when the lambda DNA possesses negative superhelical turns. In fact, tracking down what permitted the *in vitro* reaction to work led to the discovery of DNA gyrase. Presumably the enzymology of the *in vitro* and *in vivo* integration reactions is the same, and the supercoiling requirement means that *in vivo* the chromosome possesses superhelical turns.

A second experiment also suggests that DNA in normally growing *E. coli* contains superhelical torsion. Adding inhibitors of the DNA gyrase, such as nalidixic acid or oxolinic acid, which block activity of the A subunit of the enzyme, or novobiocin or coumermycin, which inhibit the B subunit, alters the rates of expression of different genes. The activities of some genes increase while the activities of others decrease. This shows that the drug effects are not a general physiological response and that the DNA must be supercoiled *in vivo*. Yet another indication of the importance of supercoiling to cells is shown by the behavior of DNA topoisomerase I mutants. Such mutants grow slowly, and faster-growing mutants frequently arise. These are found to possess mutations that compensate for the absence of the topoisomerase I by a second mutation that reduces the activity of gyrase, topoisomerase II. A third line of experimentation also suggests that the DNA in bacteria, but not eukaryotic cells, experiences an unwinding torsion from the linking number deficit. This is the rate at which an intercalating drug called psoralen intercalates and reacts with DNA when irradiated with UV light. The reaction rate is torsion-dependent. Altogether, it seems reasonable to conclude that the DNA in bacterial cells not only is supercoiled but also is under a supercoiling torsion.

The Linking Number Paradox of Nucleosomes

A paradox is raised by the structure of nucleosomes. As mentioned in Chapter 1, the DNA in eukaryotic cells is wrapped around nucleosomes. DNA in the B conformation wraps about 1.8 times around a core consisting of pairs of the four histones, H2A, H2B, H3 and H4. In the

presence of the histone H1, the wrapping is extended to just about two complete turns. This histone both serves to complete the wrapping as well as to connect from one nucleosome to the next. Superficially, the wrapped nucleosome structure appears to possess two superhelical turns, and yet when the protein is removed, the DNA is found to possess only about one superhelical turn for each nucleosome it had contained.

One explanation for the paradox could be that the path of the DNA between nucleosomes negates part of the writhe generated by the wrapping. Because electron microscopy suggests that the connection from one nucleosome to the next is regular, tricky topology connecting nucleosomes seems not to be the explanation.

Another explanation for the paradox is that while the DNA is wrapped on the nucleosome, it is overwound. Upon removal of the nucleosome, the winding of the DNA returns to normal, reducing the twist of the DNA, so that writhe or negative supercoiling is reduced in magnitude from an average of two per nucleosome to an average of one negative superhelical turn per nucleosome. The linking number, twist, and writhe might be the following while the DNA is wrapped on one nucleosome, $Lk = 20$, $Tw = 22$, $Wr = -2$, and after removing from the nucleosome, the same DNA might have the following values, $Lk = 20$, $Tw = 21$, $Wr = -1$. The evidence, in fact, suggests this is part of the explanation.

Analysis of the sequences of DNA found on nucleosomes indicates that the bends introduced by runs of A's as described earlier tend to lie with the minor grooves of such runs in contact with the nucleosomes. Thus nucleosomes appear to bind to DNA to regions that already are partially bent. Analysis of the locations of these runs of A's shows that they are spaced an average of 10.17 base pairs apart, not 10.5 base pairs apart. This then partially, but not fully, explains the linking number paradox. When this overwound DNA returns to its natural twist of 10.5 base pairs per turn, part of the supercoiling is eliminated. Although this reduces the supercoiling discrepancy, a new question is raised about the cause of the overwinding of the DNA.

General Chromosome Structure

The DNA in eukaryotic cells is largely contained in nucleosomes. Sets of these nucleosomes form solenoids, and the solenoids wrap together to form yet a larger structure. Within all this compaction, the DNA must remain accessible to regulatory proteins, to RNA polymerase for transcription, to DNA repair enzymes, and to any other proteins that have a need for access to the DNA. If a nucleosome is bound to the promoter for a given gene, RNA polymerase would be denied access, and necessary transcription would not occur. Are there any special mechanisms that either remove nucleosomes from important regions of DNA or which prevent their binding there in the first place? The next section shows how nucleosome positions on DNA can be determined.

Southern Transfers to Locate Nucleosomes on Genes

Southern transfers are a versatile technique of molecular genetics that combines electrophoresis and DNA-DNA or RNA-DNA hybridization

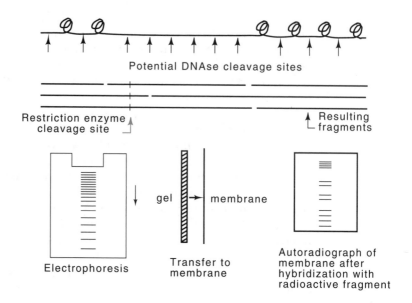

Potential DNAse cleavage sites

Restriction enzyme cleavage site

Resulting fragments

Electrophoresis

gel → membrane

Transfer to membrane

Autoradiograph of membrane after hybridization with radioactive fragment

Figure 2.15 Use of Southern transfer technology to determine the locations of nuclease hypersensitive and hyposensitive sites in the vicinity of a gene. Gently extracted DNA is lightly digested with DNAse, cleaved with a restriction enzyme at specific sites, denatured, separated by electrophoresis, transferred to a membrane, and then hybridized to a short radioactive oligonucleotide that hybridizes near the gene in question. Finally, an autoradiograph is made of the membrane.

(Fig. 2.15). The application of Southern transfers to various questions will be mentioned a number of this throughout the book. The power of the transfer and hybridization technology as applied to examining nucleosome positioning is that it permits determination of the suscep- tibility to DNAse cleavage in the region of whatever gene we are interested in. This can be done in the presence of DNA from thousands of other genes. Here we shall consider the application of this technology to the problems of ascertaining whether nucleosomes near a specific gene occupy fixed positions and whether nucleosomes cover regulatory sequences just ahead of genes. The approach has shown that in front of many genes are areas apparently not occupied by nucleosomes. This leaves these regions hypersensitive to hydrolysis by nucleases added to gently lysed nuclei. The nuclease sensitivity within the genes is much less due to the presence of nucleosomes. Often such nucleosomes tend to occupy specific positions.

The DNA for nucleosome position measurements is gently extracted from nuclei and lightly treated with a nuclease like DNAse I to generate about one nick per one thousand base pairs. Different molecules will be nicked in different places, but very few molecules will be nicked in areas covered by nucleosomes. After the digestion, protein is removed by extraction with phenol and all the DNA molecules are digested by an enzyme that cleaves DNA at specific sequences. These enzymes are known as restriction enzymes and will be discussed more fully later.

Suppose that such a cleavage site lies several hundred base pairs in front of the gene we are considering. After the cleavage steps, the DNA fragments are denatured and the single-stranded fragments are separated according to size by electrophoresis. After the electrophoresis, the fragments are transferred to a sheet of nylon membrane. The transfer to the membrane preserves the pattern of size-separated fragments. The membrane can then be incubated in a solution containing radioactive oligonucleotide possessing a sequence complementary to sequence from the gene of interest near to the cleavage site. The oligonucleotide will hybridize to just those DNA fragments possessing this complementary sequence. Hence the membrane will be radioactive in the areas containing the fragments. In any area of the DNA that was protected from DNAse I nicking by the presence of a nucleosome, no cleavages will occur. Hence, there will be no fragments of the size extending from the position of the restriction enzyme cleavage site to the area occupied by the nucleosome. Conversely, in areas readily cleaved by the nuclease, many different molecules will be cleaved, and therefore many DNA fragments will exist of a length equal to the distance from the restriction cleavage site to the nuclease-sensitive, nucleosome free, region.

Nucleosome protection experiments show that several hundred nucleotides in regions ahead of genes in which regulatory proteins are expected to bind frequently are devoid of nucleosomes. Two factors are responsible. First, regulatory proteins can bind to these regions and prevent nucleosomes from binding there. Another reason is natural bending of the DNA. As discussed earlier, DNA is not straight, and most DNA possesses minor bends. Such bends greatly facilitate the wrapping of DNA around the histones in the formation of a nucleosome. Thus, bends in the DNA can position a nucleosome, and this in turn partially positions its neighbors, generating a region of phased nucleosomes. Such phasing can leave gaps where they are necessary for the binding of regulatory proteins.

ARS Elements, Centromeres, and Telomeres

Survival of a chromosome requires three basic properties–replication, proper segregation upon DNA replication and cell division, and replication and protection of the ends of the chromosome. Multiple origins of replication exist in the chromosomes of cells. These origins are called autonomously replicating sequences, ARS, because they can be cloned into DNA that will replicate on its own in other cells. Such DNA, however, does not properly partition itself into daughter cells because it lacks the necessary signals for segregation. Frequently the daughters fail to receive a copy of the DNA replicating under ARS control.

Classical cell biology has identified the portion of the chromosome that is responsible for segregation of the chromosomes into daughter cells. This is the centromere. As cells divide, the centromeres are pulled into the two daughter cells by microtubules. It has been possible to identify a centromere by seeking a DNA segment from a chromosome that confers the property of more correct segregation on a DNA element containing an ARS element.

A third necessary part to a normal chromosome is the telomere. Telomeres also have been identified by classical biology as being special. First, most chromosomes of eukaryotic cells are linear. This poses a problem in DNA replication as the normal DNA polymerase cannot elongate to the ends of both strands because it replicates only in a 5' to 3' direction. The end of one strand can't be reached. Something else must extend the portion of the strand that cannot be completely replicated. Secondly, since chromosome breakage occasionally occurs, and has dire consequences to cells, they have evolved a way to try to rescue broken chromosomes by fancy recombination processes. The normal ends of chromosomes are inert in these rescue processes by virtue of special markers called telomeres. These telomeres have been identified by their properties of permitting the existence of linear artificial chromosomes that contain ARS elements and centromeres. Interestingly, telomeres are repeated sequences of five to ten bases, largely of C's and G's. A special enzyme adds these sequences onto single-stranded DNA possessing the same telomeric sequence. These unusual enzymes must first recognize the sequence to which they will make additions, and then they add nucleotides, one at a time, to generate the correct telomeric structure. They do this making use of an internal RNA molecule that provides the sequence information needed for the additions.

Problems

2.1. In textbooks it is possible to find the phosphodiester backbone of DNA or RNA drawn in either of two ways (Fig. 2.16.). Which is correct?

2.2. DNA is stable at pH 11, whereas RNA is degraded by alkali to nucleosides. Look up the reason in chemistry or biochemistry textbooks.

2.3. Estimate in centimeters the total length of the DNA in a human cell.

2.4. How much base pair discrimination can be performed from the minor groove of B-form DNA, that is, can A-T be distinguished from T-A and distinguished from G-C and C-G, etc?

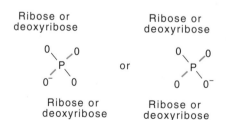

Figure 2.16 Two textbook drawings of the structure of the phosphate in phosphodiester backbones of DNA and RNA.

Leading   Trailing
edge   edge
sample   sample

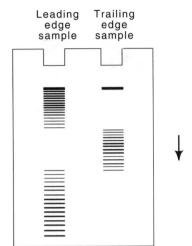

Figure 2.17 Data for problem 2.7.

2.5. During transcription RNA polymerase winds around the DNA. If the trailing RNA cannot wind around the DNA, show that transcription tends to generate positive supercoils ahead of the polymerase and negative supercoils behind the polymerase.

2.6. Consider DNA 50,000 base pairs long containing sticky ends, such as phage lambda. If the sticky ends are annealed together and the circle is covalently closed by ligase, several species of superhelically twisted DNA are formed. These differ from one another by single superhelical twists, and they can be separated from one another by electrophoresis through agarose in the presence of moderate concentrations of ethidium bromide. How does the resultant ladder pattern change if lambda DNA containing an internal deletion of five base pairs is used in the experiment? How can this method be extended to determine the helical repeat distance of DNA in solution?

2.7. A double-stranded DNA fragment 300 base pairs long was labeled at the 5' end of one of the strands. This sample was then subjected to a treatment that removed one nucleotide from a random location from each duplex. Upon electrophoresis the sample generated a significantly broadened band. The front quarter of this band and the trailing quarter of this band were then run on high resolution denaturing gels with the results shown in Fig. 2.17. Explain.

Figure 2.18 Bases for problem 2.11.

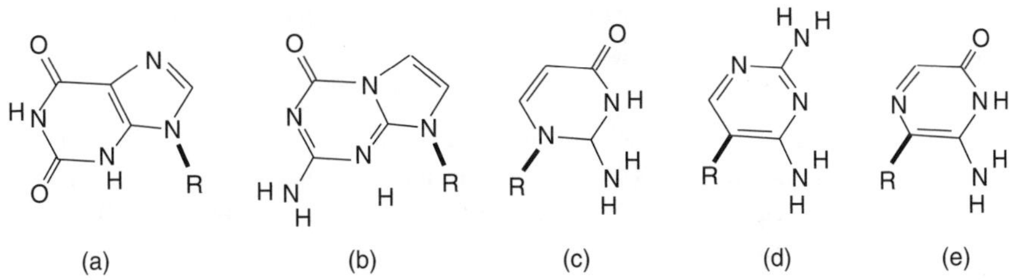

(a)   (b)   (c)   (d)   (e)

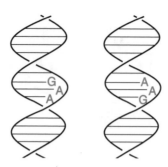

Figure 2.19 Drawing for problem 2.18.

2.8. Is the activity of DNA topoisomerase I on negative superhelical DNA and not on positive superhelical DNA proper for this enzyme to act as the protein to unwind the twisting ahead of the replication fork that is introduced by DNA replication?

2.9. Show that negative superhelicity but not positive superhelicity increases the number of base pairs per helical turn and assists the melting out of a few base pairs.

2.10. In a lambda phage DNA molecule, a sequence of nine base pairs was followed by three base pairs, and then the complement of the sequence was repeated in reverse order. Show the results of intrastrand base pairing of these sequences, known as a cruciform. In a covalently closed duplex, how much would the formation of the cruciform, as defined here, change the number of superhelical turns?

2.11. Which pair of bases in Figure 2.18 could be used to extend the genetic alphabet from A, G, C, and T?

2.12. Show that a sign-inversion pathway such as that proposed for gyrase changes the number of superhelical twists in a covalently closed circular DNA molecule by two.

2.13. One method for determining the approximate number of super-helical turns in a proposed structure is to build it without twists from a ribbon, to pull the ends taut, and to count the twists. Why does this work?

2.14. Ethidium bromide fluoresces 50 to 100 times more strongly when it has intercalated into DNA. How could this property be used to determine whether DNA from a phage was a covalently closed circle?

2.15. How many bonds per base pair are there about which rotation is possible, neglecting those of the deoxyribose ring?

2.16. Assume that the B form of DNA possesses 10.4 base pairs per turn and the Z form possesses 12 base pairs. Suppose that a large plasmid contains a 156-base pair insert of alternating dG, dC. Show that the conversion of the insert DNA from the B to the Z form will add 28 superhelical turns to the plasmid.

2.17. Suppose electrophoresis on agarose is performed such that the twist of DNA is the same as it is in solution. A sample of circular DNA molecules of the same length but possessing -1, 0, and +1 superhelical turns is to be electrophoresed. How many bands will appear? How can the +1 and -1 species be separated?

2.18. DNA sequences are normally written 5' to 3', e.g., 5'-AAG-3'. In Figure 2.19 this sequence has been imposed on the lower strand in the major groove. By reference to a detailed structural drawing of DNA, determine whether it should be represented as on the left or the right. This is equivalent to determining the 5' and 3' ends of the two DNA strands.

2.19. Devise an assay for catenation and an assay for decatenation of DNA.

2.20. Sketch the top and side view of a DNA molecule that has a writhe of about 0.5. Don't worry about the sign. It could be + 0.5 or - 0.5.

References

Recommended Readings

DNA Flexibility Studied by Covalent Closure of Short Fragments into Circles, D. Shore, J. Langoushi, R. Baldwin, Proc. Nat. Acad. Sci. USA *78*, 4833-4837 (1981).

Electrophoretic Separations of Large DNA Molecules by Periodic Inversion of the Electric Field, C. Carle, M. Frank, M. Olson, Science *232*, 65-68 (1986).

In vivo Alteration of Telomere Sequences and Senescence Caused by Mutated *Tetrahymena* Telomerase RNAs, G. Yu, J. Bradley, L. Attardi, E. Blackburn, Nature *344*, 126-132 (1990).

DNA and RNA Structure

Molecular Structure of Deoxypentose Nucleic Acids, J. Watson, F. Crick, Nature *171*, 737-738 (1953).

Improved Estimation of Secondary Structure in Ribonucleic Acids, I. Tinoco Jr., P. Borer, B. Dengler, M. Levine, O. Uhlenbeck, Nature New Biol. *246*, 40-41 (1973).

Sequence-specific Recognition of Double Helical Nucleic Acids by Proteins, H. Seeman, J. Rosenberg, A. Rich, Proc. Nat. Acad. Sci. USA *73*, 804-808 (1976).

Molecular Structure of a Left-handed Double Helical DNA Fragment at Atomic Resolution, A. Want, G. Quigley, F. Kolpak, J. Crawford, J. van Boom, G. van der Marel, A. Rich, Nature *282*, 680-686 (1980).

Crystal Structure Analysis of a Complete Turn of B-DNA, R. Wing, H. Drew, T. Takano, C. Broka, S. Tanaka, K. Itakura, R. Dickerson, Nature *287*, 755-758 (1980).

Polymorphism of DNA Double Helices, A. Leslie, S. Arnott, R. Chandrasekaran, R. Ratliff, J. Mol. Biol. *143*, 49-72 (1980).

Structure of a B-DNA Dodecamer: Conformation and Dynamics, H. Drew, R. Wing, T. Takana, C. Broka, S. Tanaka, K. Itakura, R. Dickerson, Proc. Nat. Acad. Sci. USA *78*, 2179-2183 (1981).

Sequence Dependence of the Helical Repeat of DNA in Solution, L. Peck, J. Wang, Nature *292*, 375-378 (1981).

Sequence-dependent Helical Periodicity of DNA, D. Rhodes, A. Klug, Nature *292*, 378-380 (1981).

Bent Helical Structure in Kinetoplast DNA, J. Marini, S. Levene, D. Crothers, P. Englund, Proc. Nat. Acad. Sci. USA *79*, 7664-7668 (1982).

Dependence of Z-DNA Antibody Binding to Polytene Chromosomes on Acid Fixation and DNA Torsional Strain, R. Hill, B. Stollar, Nature *305*, 338-340 (1983).

Evidence for the Existence of Stable Curvature of DNA in Solution, P. Hagerman, Proc. Nat. Acad. Sci. USA *81*, 4632-4636 (1984).

Sequence-specific Cleavage of Single-stranded DNA: Oligodeoxynucleotide-EDTA·Fe(II), G. Dreyer, P. Dervan, Proc. Nat. Acad. Sci. USA *82*, 968-972 (1985).

DNA Bending at Adenine·Thymine Tracts, H. Koo, H. Wu, D. Crothers, Nature *320*, 501-506 (1986).

Curved DNA: Design, Synthesis, and Circularization, L. Ulanovsky, B. Bodner, E. Trifonov, M. Chodner, Proc. Natl. Acad. Sci. USA *83*, 862-866 (1986).

Sequence-directed Curvature of DNA, P. Hagerman, Nature *321*, 449-450 (1986).

The Structure of an Oligo (dA).oligo(dT) Tract and Its Biological Implications, H. Nelson, J. Finch, B. Luisi, A. Klug, Nature *330*, 221-226 (1987).

Calibration of DNA Curvature and a Unified Description of Sequence-directed Bending, H. Koo, D. Crothers, Proc. Natl. Acad. Sci. USA *85*, 1763-1767 (1988).

Telomeric DNA Oligonucleotides Form Novel Intramolecular Structures Containing Guanine-guanine Base Pairs, E. Henderson, C. Hardin, S. Walk, I. Tinoco Jr., E. Blackburn, Cell *51*, 899-908 (1988).

Structural Details of an Adenine Tract that Does Not Cause DNA to Bend, A. Burkhoff, T. Tullius, Nature *331*, 455-457 (1988).

Monovalent Cation-induced Structure of Telomeric DNA: The G Quartet Model, J. Williamson, M. Raghuraman, T. Cech, Cell *59*, 871-880 (1989).

Enzymatic Incorporation of a New Base Pair into DNA and RNA Extends the Genetic Alphabet, J. Piccirilli, T. Krauch, S. Moroney, S. Benner, Nature *343*, 33-37 (1989).

Dependence of the Torsional Rigidity of DNA on Base Composition, R. Fujimoto, J. Schurr, Nature *344*, 175-178 (1990).

A New Model for the Bending of DNAs Containing the Oligo(dA) Tracts Based on NMR Observations, M. Katahira, H. Sugeta, Y. Kyogoku, Nuc. Acids Res. *18*, 613-618 (1990).

Nucleosome and Chromosome Structure and Formation

Repeated Sequences in DNA, R. Britten, D. Kohne, Science, *161*, 529-540 (1968).

On the Structure of the Folded Chromosome of *E. coli*, A. Worcel, E. Burgi, J. Mol. Biol. *71*, 127-147 (1972).

A Low Resolution Structure for the Histone Core of the Nucleosome, A. Klug, D. Rhodes, J. Smith, J. Finch, J. Thomas, Nature *287*, 509-515 (1980).

The Chromatin Structure of Specific Genes, C. Wu, P. M. Bingham, K. J. Livak, R. Holmgren, S. Elgin, Cell *16*, 797-806 (1979).

Isolation of a Yeast Centromere and Construction of Functional Small Circular Chromosomes, L. Clarke, J. Carbon, Nature *287*, 504-509 (1980).

Chromosomes in Living *E. coli* Cells are Segregated into Domains of Supercoiling, R. Sinden, D. Pettijohn, Proc. Nat. Acad. Sci. *78*, 224-228 (1981).

X-ray Diffraction Study of a New Crystal Form of the Nucleosome Core Showing Higher Resolution, J. Finch, R. Brown, D. Rhodes, T. Richmond, B. Rushton, L. Lutter, A. Klug, J. Mol. Biol. *145*, 757-769 (1981).

Cloning Yeast Telomeres on Linear Plasmid Vectors, J. Szostak, Cell *29*, 245-255 (1982).

Construction of Artifical Chromosomes in Yeast, A. Murray, J. Szostak, Nature *305*, 189-193 (1983).

Characteristic Folding Pattern of Polytene Chromosomes in *Drosophila* Salivary Gland Nuclei, D. Mathog, M. Hochstrasser, Y. Gruenbaum, H. Saumweber, J. Sedat, Nature *308*, 414-421 (1984).

Identification of a Telomere-binding Activity from Yeast, J. Berman, C. Tachibana, B. Tye, Proc. Nat. Acad. Sci. USA *83*, 3713-3717 (1986).

Sequence Periodicities in Chicken Nucleosome Core DNA, S. Satchwell, H. Drew, A. Travers, J. Mol. Biol. *191*, 659-675 (1986).

The Telomere Terminal Transferase of Tetrahymena is a Ribonucleoprotein Enzyme with Two Kinds of Primer Specificity, C. Greider, E. Blackburn, Cell *51*, 887-898 (1987).

Telomeric DNA-protein Interactions of *Oxytricha* Macronuclear DNA, C. Price, T. Cech, Genes and Development *1*, 783-793 (1987).

A Telomeric Sequence in the RNA of *Tetrahymena* Telomerase Required for Telomere Repeat Synthesis, C. Greider, E. Blackburn, Nature *337*, 331-337 (1989).

In vivo Alteration of Telomere Sequences and Senescence Caused by Mutated Tetrahymena Telomerase RNAs, G. Yu, J. Bradley, L. Attardi, E. Blackburn, Nature *344*, 126-132 (1990).

Yeast Centromere Binding Protein CBF1, of the Helix-loop-helix Protein Family, Is Required for Chromosome Stability and Methionine Prototrophy, M. Cai, R. Davis, Cell *61*, 437-446 (1990).

Histone-like Protein H1 (H-NS), DNA Supercoiling, and Gene Expression in Bacteria, C. Hulton, A. Seirafi, J. Hinton, J. Sidebotham, L. Waddel, G. Pavitt, T. Owens-Hughes, A. Spassky, H. Buc, C. Higgins, Cell *63*, 631-642 (1990).

Genetic Evidence for an Interaction between SIR3 and Histone H4 in the Repression of the Silent Mating Loci in *Saccharomyces cerevisiae*, L. Johnson, P. Kayne, E. Kahn, M. Grunstein, Proc. Natl. Acad. Sci. USA *87*, 6286-6290 (1990).

Position Effect at *Saccharomyces cerevisiae* Telomeres: Reversible Repression of PolII Transcription, D. Gottschling, O. Aparicso, B. Billington, V. Zakian, Cell *63*, 751-762 (1990).

Effects of DNA Sequence and Histone-histone Interactions on Nucleosome Placement, T. Shrader, D. Crothers, J. Mol. Biol. *216*, 69-84 (1990).

Physical Chemistry of Nucleic Acids

Kinetics of Renaturation of DNA, J. Wetmur, N. Davidson, J. Mol. Biol. *31*, 349-370 (1968).

Effects of the Conformation of Single-stranded DNA on Renaturation and Aggregation, F. Studier, J. Mol. Biol. *41*, 199-209 (1969).

Effect of DNA Length on the Free Energy of Binding of an Unwinding Ligand to a Supercoiled DNA, N. Davidson, J. Mol. Biol. *66*, 307-309 (1972).

Effects of Microscopic and Macroscopic Viscosity on the Rate of Renaturation of DNA, C. Chang, T. Hain, J. Hutton, J. Wetmur, Biopolymers *13*, 1847-1858 (1974).

Energetics of DNA Twisting, I. Relation between Twist and Cyclization Probability, D. Shore, R. Baldwin, J. Mol. Biol. *170*, 957-981 (1983).

Energetics of DNA Twisting, II. Topoisomer Analysis, D. Shore, R. Baldwin, J. Mol. Biol. *170*, 983-1007 (1983).

Cruciform Formation in a Negatively Supercoiled DNA May be Kinetically Forbidden Under Physiological Conditions, A. Courey, J. Wang, Cell *33*, 817-829 (1983).

Ring Closure Probabilities for DNA Fragments by Monte Carlo Simulation, S. Levene, D. Crothers, J. Mol. Biol. *189*, 61-72 (1986).

Influence of Abasic and Anucleosidic Sites on the Stability, Conformation, and Melting Behavior of a DNA Duplex: Correlations of Thermodynamics and Structural Data, G. Vesnaver, C. Chang, M. Eisenberg, A. Grollman, K. Breslauer, Proc. Natl. Acad. Sci. USA *86*, 3614-3618 (1989).

Topoisomerases

Interactions between Twisted DNAs and Enzymes. The Effect of Superhelical Turns, J. Wang, J. Mol. Biol. *87*, 797-816 (1974).

DNA Gyrase: An Enzyme that Introduces Superhelical Turns Into DNA, M. Gellert, K. Mizuuchi, M. O'Dea, H. Nash, Proc. Nat. Acad. Sci. USA *73*, 3872-3876 (1976).

Novobiocin and Coumermycin Inhibit DNA Supercoiling Catalyzed by DNA Gyrase, M. Gellert, M. O'Dea, T. Itoh, J. Tomizawa, Proc. Nat. Acad. Sci. USA *73*, 4474-4478 (1976).

Type II DNA Topoisomerases: Enzymes That Can Unknot a Topologically Knotted DNA Molecule via a Reversible Double-Strand Break, L. Liu, C. Liu, B. Alberts, Cell *19*, 697-707 (1980).

DNA Gyrase Action Involves the Introduction of Transient Double-strand Breaks into DNA, K. Mizuuchi, L. Fisher, M. O'Dea, M. Gellert, Proc. Nat. Acad. Sci. USA *77*, 1847-1851 (1980).

Catenation and Knotting of Duplex DNA by Type 1 Topoisomerases: A Mechanism Parallel with Type 2 Topoisomerases, P. Brown, N. Cozzarelli, Proc. Nat. Acad. Sci. USA *78*, 843-847 (1981).

Escherichia coli DNA Topoisomerase I Mutants Have Compensatory Mutations in DNA Gyrase Genes, S. DiNardo, K. Voelkel, R. Sternglanz, A. Reynolds, A. Wright, Cell *31*, 43-51 (1982).

Bacterial DNA Topoisomerase I Can Relax Positively Supercoiled DNA Containing a Single-stranded Loop, K. Kirkegaard, J. Wang, J. Mol. Biol. *185*, 625-637 (1985).

Superhelical Structures

Conformational Fluctuations of DNA Helix, R. Depew J. Wang, Proc. Nat. Acad. Sci. USA *72*, 4275-4279 (1975).

Action of Nicking-closing Enzyme on Supercoiled and Non-supercoiled Closed Circular DNA: Formation of a Boltzmann Distribution of Topological Isomers, D. Pulleyblank, M. Shure, D. Tang, J. Vinograd, H. Vosberg, Proc. Nat. Acad. Sci. USA 72, 4280-4284 (1975).

Helical Repeat of DNA in Solution, J. Wang, Proc. Nat. Acad. Sci. USA 76, 200-203 (1979).

Torsional Tension in the DNA Double Helix Measured with Trimethylpsoralen in Living E. coli Cells: Analogous Measurements in Insect and Human Cells, R. Sinden, J. Carlson, D. Pettijohn, Cell 21, 773-783 (1980).

Torsional Rigidity of DNA and Length Dependence of the Free Energy of DNA Supercoiling, D. Horowitz, J. Wang, J. Mol. Biol. 173, 75-91 (1984).

Supercoiling of the DNA Template During Transcription, L. Liu, J. Wang, Proc. Natl. Acad. Sci. USA 84, 7024-7027 (1987).

Description of the Topological Entanglement of DNA Catenanes and Knots by a Powerful Method Involving Strand Passage and Recombination, J. White, K. Millett, N. Cozzarelli, J. Mol. Biol. 197, 585-603 (1987).

Use of Site-specific Recombination as a Probe of DNA Structure and Metabolism in vivo, J. Bliska, N. Cozzarelli, J. Mol. Biol. 194, 205-218 (1988).

Helical Repeat and Linking Number of Surface-wrapped DNA, J. White, N. Cozzarelli, W. Bauer, Science 241, 323-327 (1988).

The Helical Repeat of Double-stranded DNA Varies as a Function of Catenation and Supercoiling, S. Wasserman, J. White, N. Cozzarelli, Nature 334, 448-450 (1988).

Transcription Generates Positively and Negatively Supercoiled Domains in the Template, H. Wu, S. Shyy, T. Want, L. Liu, Cell 53, 433-440 (1988).

Curved Helix Segments Can Uniquely Orient the Topology of Supertwisted DNA, C. Landon, J. Griffith, Cell 52, 545-549 (1988).

Molecular Mechanics Model of Supercoiled DNA, R. Tan, S. Harvey, J. Mol. Biol. 205, 573-591 (1989).

Torsionally Tuned Cruciform and Z-DNA Probes for Measuring Unrestrained Supercoiling at Specific Sites in DNA of Living Cells, G. Zheng, T. Kochel, R. Hoepfner, S. Timmons, R. Sinden, J. Mol. Biol. 221, 107-129 (1991).

Conformational and Thermodynamic Properties of Supercoiled DNA, A. Vologodskii, S. Levene, K. Klenin, M. Frank-Kamenetskii, N. Cozzarelli, J. Mol. Biol. 227, 1224-1243 (1992).

Experimental Techniques

Electron Microscopic Visualization of the Folded Chromosome of E. coli, H. Delius, A. Worcel, J. Mol. Biol. 82, 107-109 (1974).

Determination of the Number of Superhelical Turns in Simian Virus 40 DNA by Gel Electrophoresis, W. Keller, Proc. Nat. Acad. Sci. USA 72, 4876-4880 (1975).

Separation of Yeast Chromosome-sized DNA's by Pulsed Field Gradient Gel Electrophoresis, D. Schwartz, C. Cantor, Cell 37, 67-75 (1984).

Base-base Mismatches, Thermodynamics of Double Helix Formation for dCA$_3$XA$_3$G+dCT$_3$YT$_3$G, F. Aboul-ela, D. Koh, I. Tinoco, Jr., Nuc. Acids Res. *13*, 4811-4824 (1985).

Iron(II) EDTA Used to Measure the Helical Twist Along Any DNA Molecule, T. Tullius, B. Dombroski, Science *230*, 679-681 (1985).

Temperature and Salt Dependence of the Gel Migration Anomoly of Curved DNA Fragments, S. Diekmann, Nuc. Acids Res. *15*, 247-265 (1987).

Empirical Estimation of Protein-induced DNA Bending Angles: Applications to Lambda Site-specific Recombination Complexes, J. Thompson, A. Landy, Nucleic Acids Res. *16*, 9687-9705 (1988).

Conformational Dynamics of Individual DNA Molecules During Gel Electrophoresis, D. Schwartz, M. Koval, Nature *338*, 520-522 (1989).

DNA Synthesis

3

One fundamental approach in the study of complex systems is to determine the minimal set of purified components that will carry out the process under investigation. In the case of DNA synthesis, the relatively loose association of the proteins involved created problems. How can one of the components be assayed so that its purification can be monitored if all the components must be present for DNA synthesis to occur? We will see in this chapter that the problem was solved, but the purification of the many proteins required for DNA synthesis was a monumental task that occupied biochemists and geneticists for many years. By contrast, the machinery of protein synthesis was much easier to study because most of it is bound together in a ribosome.

A basic problem facing an organism is maintaining the integrity of its DNA. Unlike protein synthesis, in which one mistake results in one altered protein molecule, or RNA synthesis, in which one mistake ultimately shows up just in the translation products of a single messenger RNA, an uncorrected mistake in the replication of DNA can last forever. It affects every descendant every time the altered gene is expressed. Thus it makes sense for the mechanism of DNA synthesis to have evolved to be highly precise. There is only one real way to be precise, and that is to check for and correct any errors a number of times. In the replication of DNA, error checking of an incorporated nucleotide could occur before the next nucleotide is incorporated, or checking for errors could occur later. Apparently, checking and correcting occurs at both times. In the case of bacteria, and at least in some eukaryotes, the replication machinery itself checks for errors in the process of nucleotide incorporation, and an entirely separate machinery detects and corrects errors in DNA that has already been replicated. Retroviruses like HIV are an interesting exception. These have small genomes and they need a high spontaneous mutation rate in order to evade their host's immunological surveillance system.

Generally, DNA must also maintain its structure against environmental assaults. Damage to the bases of either DNA strand could lead to incorrect base-pairing upon the next round of DNA replication. A number of enzymes exist for recognizing, removing, and replacing damaged bases.

Since many cell types can grow at a variety of rates, sophisticated mechanisms have developed to govern the initiation of DNA replication. In both bacteria and eukaryotic cells, it is the initiation of replication that is regulated, not the elongation. Although such a regulation system seems difficult to coordinate with cell division, the alternative mechanisms for regulating the rate of DNA synthesis are more complicated. In principle, the DNA elongation rate could be adjusted by changing the concentrations of many different substrates within the cell. This, however, would be most difficult because of the interconnected pathways of nucleotide biosynthesis. Alternatively, the elongation rate of the DNA polymerase itself could be variable. This too, would be most difficult to manage and still maintain high fidelity of replication. Another problem closely tied to DNA replication is the segregation of completed chromosomes into daughter cells. Not surprisingly, this process requires a complicated and specialized machinery.

In this chapter we begin with the process of DNA synthesis itself. After examining the basic problems generated by the structure of DNA, we discuss the enzymology of DNA synthesis. Then we mention the methods cells use to maximize the stability of information stored in DNA. The second half of the chapter concerns physiological aspects of DNA synthesis. Measurement of the number of functioning replication areas per chromosome, the speed of DNA replication, and the coupling of cell division to DNA replication are covered.

A. Enzymology

Proofreading, Okazaki Fragments, and DNA Ligase

As already discussed in Chapter 2, a single strand of DNA possesses a polarity resulting from the asymmetric deoxyribose-3'-phosphate-5'-deoxyribose bonds along the backbone. Most DNA found in cells is double-stranded; a second strand is aligned antiparallel to the first strand and possesses a sequence complementary to the first. This self-complementary structure solves problems in replication because

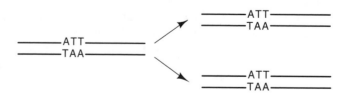

the product of replication is two daughter molecules, each identical to one of its parents. As a result of the structural similarity between the parent and daughter duplexes, the mechanisms necessary for readout of the genetic information or replication of the DNA need not accommodate multiple structures. Further, the redundancy of the stored information permits DNA with damage in one strand to be repaired by reference to the sequence preserved in the undamaged, complementary strand.

As is often the case in biology, numerous illustrative exceptions to the generalizations exist. Single-stranded DNA phage exist. These use a double-stranded form for intracellular replication but encapsidate only one of the strands. Apparently, what they lose in repair abilities they gain in nucleotides saved.

Generally, DNA is double-stranded, and both strands are replicated almost simultaneously by movement of a replication fork down the

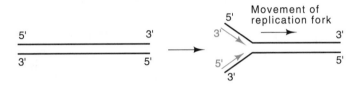

DNA. In this case, the opposite polarities of the strands require that as a replication fork moves down the DNA, overall, one daughter strand grows in a 3'-to-5' direction and the other daughter strand grows in a

Figure 3.1 DNA could theoretically be elongated in both the 5'-to-3' and the 3'-to-5' directions with the use of only nucleoside 5' triphosphates.

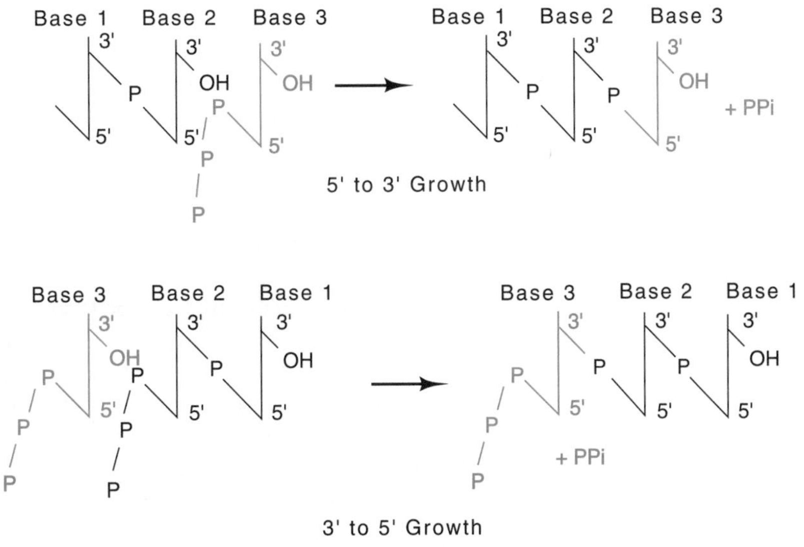

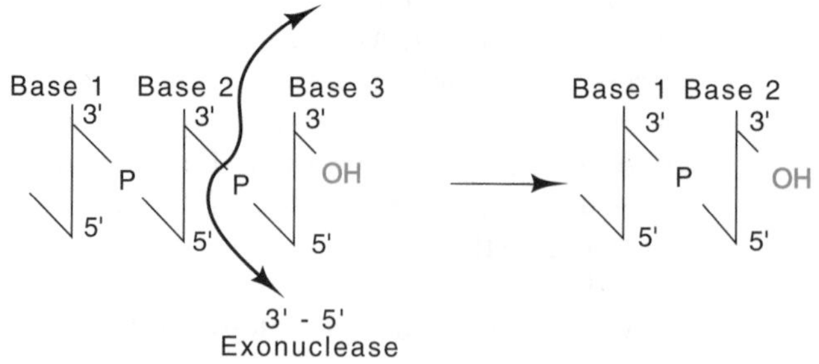

Figure 3.2 Exonucleolytic removal of the final nucleotide from a chain elongating in the 5'-to-3' direction, a 3'-5' exonucleolytic activity, regenerates the 3'-OH.

5'-to-3' direction. We shall see that the requirements of high speed and high accuracy in the replication process complicates matters.

Although cells have been found to possess 5'-nucleoside triphosphates and not 3'-nucleoside triphosphates, the two daughter DNA strands could be elongated nucleotide by nucleotide in both the 5'-to-3', and the 3'-to-5' directions using just 5'-nucleoside molecules (Fig. 3.1). This latter possibility is largely excluded, however, by the need for proofreading to check the accuracy of the latest incorporated nucleotide.

If the nucleotide most recently incorporated into the elongating strand does not correctly pair with the base on the complementary strand, the misincorporated nucleotide ought to be excised. Ultimately such editing to remove a misincorporated base must generate a DNA end precisely like the end that existed before addition of the incorrectly paired nucleotide. Removal of the final nucleotide from the strand growing in the 5'-to-3' direction immediately regenerates the 3'-OH that is normally found at the end (Fig. 3.2). A strand growing in the 3'-to-5' direction utilizing 5' triphosphates will possess a triphosphate on its 5' end. Simple excision of the final nucleotide from such a strand does not regenerate the 5' triphosphate end. Creation of the end normally seen by the polymerase elongating such a strand would then require another enzymatic activity. This, in turn, would require dissociation of the polymerase from the DNA and the entry of the other enzyme, a process that would drastically slow the process of DNA elongation.

It is worthwhile examining why DNA polymerase must remain bound to the complex of template strand and elongating strand through thousands of elongation cycles. Such a processive behavior is essential since elongation rates per growing chain must be hundreds of nucleotides per second. If the polymerase dissociated with the addition of each nucleotide, it would have to bind again for the next nucleotide, but even with moderately high concentrations of polymerase, the binding rate of a protein to a site on DNA in cells is about one per second to one per 0.1 second, which is far below the necessary elongation rate. Consequently, accurate DNA synthesis prohibits elongation in the 3'-to-5' direction.

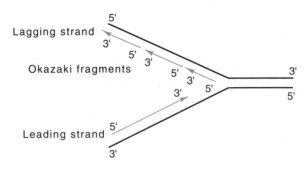

Figure 3.3 Nomenclature of the leading and lagging strands.

A variety of experiments show that both strands are synthesized at replication forks. This means that overall, one strand grows in the 5'-to-3' direction and the other in the 3'-to-5' direction. One strand can be synthesized continuously in the 5'-to-3' direction. The other strand cannot. It must grow in the 3'-to-5' direction by the synthesis of short segments which themselves are elongated in the 5'-to-3' direction. These are ligated together so that the net growth of this strand is 3'-to-5' (Fig. 3.3). These fragments are called Okazaki fragments after their discoverer. The strand synthesized continuously is called the leading strand, and the discontinuously synthesized strand is the lagging strand.

The above-mentioned considerations suggest that DNA polymerases should have the following properties. They should use 5'-nucleoside triphosphates to elongate DNA strands in a 5'-to-3' direction, and they should possess a 3'-to-5' exonuclease activity to permit proofreading. Additionally, cells should possess an enzyme to join the fragments of DNA that are synthesized on the lagging strand. This enzyme is called DNA ligase.

Detection and Basic Properties of DNA Polymerases

Use of a purified enzyme is one requirement for careful study of DNA synthesis. In the early days of molecular biology, important experiments by Kornberg demonstrated the existence in cell extracts of an enzyme that could incorporate nucleoside triphosphates into a DNA chain. This enzymatic activity could be purified from bacterial extracts, and the resultant enzyme was available for biochemical studies. Naturally, the first question to be asked with such an enzyme was whether it utilized a complementary DNA strand to direct the incorporation of the nucleotides into the elongating strand via Watson-Crick base-pairing rules. Fortunately the answer was yes. As this enzyme, DNA pol I, was more carefully studied, however, some of its properties appeared to make it unlikely that the enzyme synthesized the majority of the cellular DNA.

Cairns tried to demonstrate that pol I was not the key replication enzyme by isolating a bacterial mutant lacking the enzyme. Of course, if the mutant could not survive, his efforts would have yielded nothing. Remarkably, however, he found a mutant with much less than normal

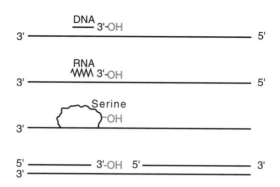

Figure 3.4 Four ways of generating 3'-OH groups necessary for DNA elongation, a deoxyribonucleotide oligomer, a ribonucleotide oligomer, a hydroxyl provided by a protein, and a nick.

activity. Such a result appeared to prove that cells must possess other DNA-synthesizing enzymes, but until a mutant could be found completely lacking DNA pol I, the demonstration was not complete.

Another way to show the existence of the DNA polymerases other than DNA pol I is to detect and purify them biochemically. Previously such attempts had proved futile because DNA pol I masked the presence of other polymerases. Once Cairns's mutant was available, however, it was a matter of straightforward biochemistry to examine bacterial extracts for the presence of additional DNA-polymerizing enzymes. Two more such enzymes were found: DNA pol II and DNA pol III.

None of the three polymerases is capable of initiating DNA synthesis on a template strand. They are able to elongate growing polynucleotide chains, but they cannot initiate synthesis of a chain. This inability is not surprising, however, because initiation must be carefully regulated and could be expected to involve a number of other proteins that would not be necessary for elongation. For initiation, all three polymerases require the presence of a hydroxyl group in the correct position. The hydroxyl group can derive from a short stretch of DNA or RNA annealed to one strand, from cleavage of a DNA duplex, or even from a protein, in which the hydroxyl utilized is on a serine or threonine residue (Fig. 3.4).

The Okazaki fragments from which the lagging strand is built have been found to be initiated by a short stretch of RNA. This can be demonstrated by slowing elongation by growing cells at 14°. To maximize the fraction of radioactive label in newly synthesized Okazaki fragments, the DNA is labeled with radioactive thymidine for only fifteen seconds. Then it is extracted, denatured, and separated according to density by equilibrium centrifugation in CsCl. RNA is more dense than DNA in such gradients, and the ten to fifteen ribonucleotides at the end of the Okazaki fragments somewhat increase their density. Digestion with RNAse or alkali removes the RNA and the density of the fragments shifts to that of normal DNA. Therefore, one function of the

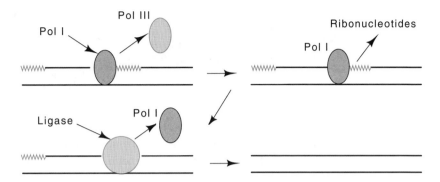

Figure 3.5 Conversion of elongating Okazaki fragments to a completed chain by the dissociation of DNA pol III, nick translation through the RNA by DNA pol I and sealing by DNA ligase.

replication machinery is to lay down the RNA primers of the Okazaki fragments. This activity is called primase.

In bacteria, DNA pol III is the main DNA replication enzyme. DNA pol I fills in the gaps in lagging strand synthesis and assists during repair of damaged DNA. No function is known for DNA pol II. Both DNA pol I and DNA pol III possess the 3'-to-5' exonuclease activity that is necessary for proofreading removal of misincorporated nucleotides.

DNA pol I also possesses a 5'-to-3' exonuclease activity not found in the other polymerases. This activity permits pol I to bind to a nick in the DNA, remove the nucleotide on the 3' side, incorporate a nucleotide on the 5' side and perform this process over and over again without dissociating from the DNA after each nucleotide. Thus pol I can processively translate a nick down the DNA in the 5'-to-3' direction. This process is used to remove the RNA that primes synthesis of the Okazaki fragments (Fig. 3.5). When pol III reaches such a primer it dissociates from the DNA. Then DNA pol I can bind and nick translate through the stretch of RNA. DNA pol I is less processive than DNA pol III, so that when pol I dissociates at some point after nick translating through the RNA primer, the resulting DNA-DNA nick can be sealed by DNA ligase. This ends the elongation process for this Okazaki fragment.

The two exonuclease activities and the polymerizing activity of DNA pol I reside on the single polypeptide chain of the enzyme. DNA pol III is not so simple. Two slightly different forms of the enzyme appear to function as a dimer, one which synthesizes the leading strand, and one which synthesizes the lagging strand. Most likely the complete complex, the holoenzyme, actually consists of at least nine different polypeptide chains. The α subunit possesses the polymerizing activity, and the ε subunit contains the 3'-5' exonuclease activity. The unit of α, ε, and θ forms an active core that is capable of synthesizing short stretches of DNA. Remaining associated with the template strand and being highly processive is conferred by the complex of γ, δ, δ', χ, and ψ which facilitate the addition of a dimer of β. The dimer of β encircles the DNA and forces the complex to remain associated with the same DNA molecule throughout the replication cycle.

Not surprisingly, eukaryotic cells also possess multiple DNA polymerases which contain multiple subunits. These polymerases, not the subunits, often are named α, β, γ, δ, and ϵ. The α polymerase was long thought to be the general polymerizing enzyme, with the small β enzyme considered to be a repair enzyme. The γ enzyme is mitochondrial. We now know that α is likely to initiate synthesis on both the leading and lagging strands since only it possesses priming activity. Further elongation on the leading strand is performed by δ, and on the lagging strand by ϵ or α. The β enzyme could complete the lagging strands by filling the same role as is played by DNA pol I in bacteria.

In vitro DNA Replication

The mere incorporation of radiolabeled nucleotides into polymers of DNA is far from the complete biological process of DNA replication. The initial experiments seeking polymerization activities from cell extracts used nicked and gapped DNA as a template. This yielded polymerases capable of elongating DNA, but did not provide an assay for any of the DNA initiating components. To seek the cellular machinery necessary for initiating replication, DNA templates were required that contained sequences specifying origins of replication. The most convenient source of such origins was small DNA phages since each molecule must contain an origin. The results of experiments with several different phage templates revealed the astounding fact that the proteins required for initiating replication varied from one DNA origin to another. At first it was not possible to discern the biochemical principles underlying initiation of replication. Therefore, when DNA cloning became possible, attention turned to a replication origin of greater generality and importance, the origin of replication of the *E. coli* chromosome. Later, when it became possible to work with animal viruses and to isolate and study replication origins from eukaryotic cells, these also were studied.

Kornberg and his collaborators were able to find conditions in which a cell extract prepared from *E. coli* could replicate DNA from the *E. coli* origin, *oriC*. Such an extract undoubtedly possessed many different proteins acting in concert to replicate the DNA. Once this step was working, it was then possible to seek to identify specific proteins involved in the reaction. Geneticists assisted this difficult step through their isolation of temperature-sensitive mutations that blocked DNA synthesis in growing cells. For example, extracts prepared from cells with a temperature-sensitive *dnaA* mutation were inactive. This, of course, is a biochemist's dream, for it provides a specific assay for the DnaA protein. Extracts prepared from wild-type cells can supplement extracts prepared from temperature-sensitive *dnaA* mutant cells. This supplementation results from the wild-type DnaA protein in the wild-type extract. Next, the wild-type extract can be fractionated and the *in vitro* complementation assay detects which fraction contains the DnaA protein. With such an assay the DnaA protein was purified.

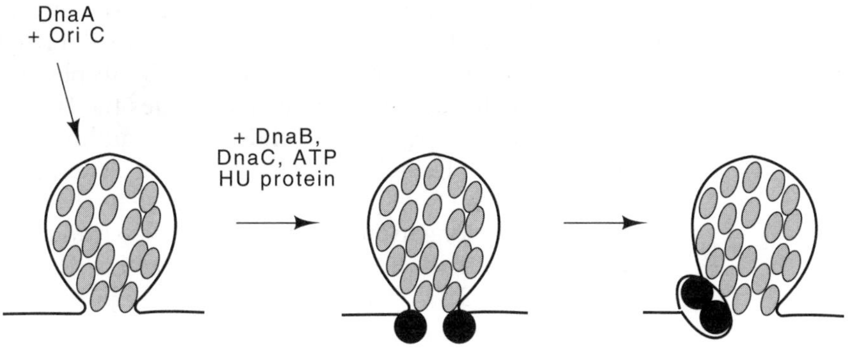

Figure 3.6 The initiation of DNA replication from OriC by the combined activities of DnaA, DnaB, DnaC, and Dna pol III.

The strategy used with the DnaA protein is straightforward, and it can be used to purify proteins by making use of any replication mutant whose cell extracts are inactive. The work required for this approach is immense, and therefore it helps to try to guess proteins required for replication and to add these as purified components. If the assay doesn't replicate DNA, cell extracts are added and the resulting activity can be used to guide purification of the remaining components. Ultimately, the following components were identified as necessary for *in vitro* replication from *oriC*: DnaA protein, DnaB protein, DnaC protein, DnaG protein, DNA polymerase III holoenzyme, DNA gyrase, single-stranded binding protein, and ATP. Analogous experiments using replication origins from animal viruses have also permitted the detection, purification and study of the complete set of proteins necessary for their activity.

In vitro initiation and synthesis reactions have permitted the replication process at the *E. coli oriC* to be dissected into several steps. Initiation begins with 20 to 40 molecules of DnaA protein binding to four sites in the 260 base pair *oriC* region (Fig. 3.6). The complex of DnaA and *oriC* contains the DNA wrapped around the outside and the protein in the middle. The large number of protein molecules required

Figure 3.7 The activities in the vicinity of a DNA replication fork. Also shown are the rotations generated by movement of a replication fork.

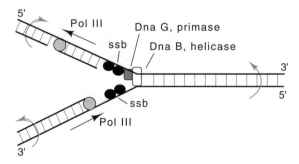

for the first step of initiation makes the step critically dependent on the concentration of the protein. Such a step is ideal for tight regulation of the initiation process. In the second step, DnaB plus DnaC bind. DnaB possesses a helicase activity which separates the DNA strands with the consumption of energy. DnaA binds to three additional sites in *oriC* and several regions of single-stranded DNA are generated. Finally, DnaG protein lays down RNA primers that are used by the pol III holoenzyme.

During the elongation process the single-stranded binding protein SSB binds to single-stranded regions opened by the helicase activities of DnaB (Fig. 3.7). DNA gyrase also is required for elongation. It untwists the rotations that are generated ahead of the moving replication fork. *In vivo* the twists that are generated behind the moving replication fork are removed by topisomerase I.

Several other proteins are also involved with replication. The histone-like protein, HU, seems to assist the process, but its mechanism is unknown. RNAse H digests the RNA from RNA-DNA hybrids left over from transcription and prevents initiation from points other than the origin.

Error and Damage Correction

The bacterial DNA polymerases I and III, but not the eukaryotic DNA polymerases, as they are commonly purified, possess the ability to correct mistakes immediately upon misincorporation of a nucleoside triphosphate. The subunit of the eukaryotic DNA polymerases required for this 3'-to-5' exonuclease activity must be similar to the prokaryotic error correcting unit, but is less tightly bound to the polymerizing subunit. We know this because adding the bacterial protein to the eukaryotic polymerase confers an error correcting 3'-to-5' exonuclease activity to the eukaryotic enzyme.

Cells also possess the ability to correct replication mistakes that have eluded the editing activity of the polymerases. Enzymes recognize mispaired bases and excise them. The gap thus produced is filled in by DNA polymerase I, or in eukaryotic cells by DNA polymerase β, and sealed with DNA ligase. At first glance, such a repair mechanism would not seem to have much value. Half the time it would correct the nucleotide from the incorrect strand. How can the cell restrict its repair to the newly synthesized strand? The answer is that DNA repair enzymes

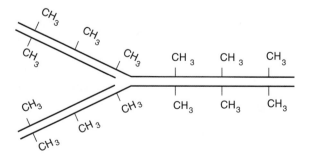

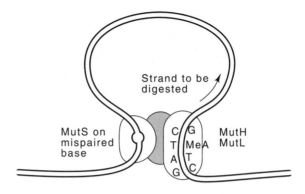

Figure 3.8 The interaction of MutS at a mispaired base with MutH and MutL activates resynthesis of the newly synthesized daughter strand from the GATC sequence past the mispaired base.

distinguish old from new DNA strands with methyl groups. Only after a newly synthesized strand has been around for some time does it become methylated and thereby identified as "old." By this means the repair enzymes can tell which of the strands should be repaired. Since the methyl groups involved are not densely spaced along the DNA, the repair enzymes find the mispaired bases by binding at appropriate sites and moving along the DNA to the mispaired base, all the while keeping track of which strand is parental and which is the newly synthesized daughter that should be corrected. In *Escherichia coli*, the ages of strands are identified by the attachment of methyl groups to the adenines of the sequence GATC.

At least one class of mismatched bases that may exist in *Escherichia coli* following DNA replication is detected by the MutS protein. This protein binds directly to the mismatched base (Fig. 3.8). Two other proteins of the Mut system, MutH and MutL, bind to the hemimethylated GATC sequence. Apparently, MutH, MutL, and MutS then interact, generating a nick at the GATC site on the unmethylated strand. The unmethylated daughter strand is digested from this point and resynthesized by nick translation past the mispaired base, thereby correctly repairing the original mismatch.

The information stored in the DNA can also be compromised by damage that occurs subsequent to synthesis. For example, DNA can be alkylated by chemicals in the environment. As a number of positions in DNA may be alkylated, a number of proteins exist for removing these groups. This class of repair system was discovered by the observation that prior exposure of cells to low concentrations of the mutagen nitrosoguanidine, which alkylates DNA, greatly reduced the lethality and mutagenicity caused by subsequent exposure to higher concentrations of the agent. This shows that resistance was induced by the first exposure. Furthermore, the resistance could not be induced in the presence of protein synthesis inhibitors, also showing that synthesis of a protein had to be induced to provide resistance.

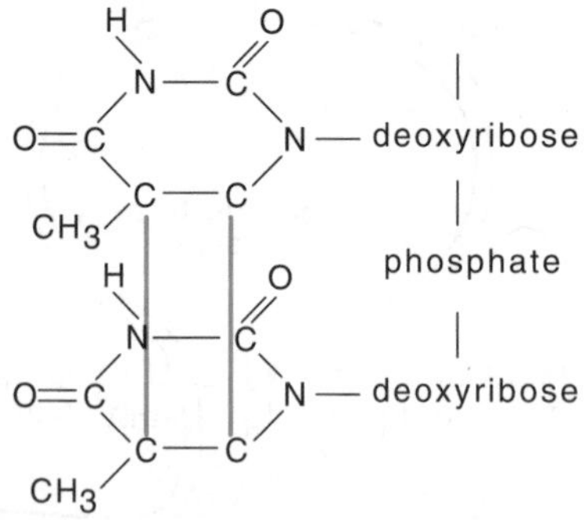

Figure 3.9 Chemical structure of a thymine dimer.

Additional investigation of the alkylation repair system has shown that *E. coli* normally contains about 20 molecules of a protein that can remove methyl groups from O^6-methylguanine. The protein transfers the methyl group from DNA to itself, thereby committing suicide, and at the same time generating a conformational change so that the protein induces more synthesis of itself. The methylated protein acts as a transcription activator of the gene coding for itself. Another domain of this same protein can transfer methyl groups from methyl phosphotriester groups on the DNA. Additional proteins exist for removal of other adducts.

Ultraviolet light can also damage DNA. One of the major chemical products of UV irradiation is cyclobutane pyrimidine dimers formed between adjacent pyrimidines, for example between thymine and thymine as shown in Fig. 3.9. These structures inhibit transcription and replication and must be removed. The UvrA, UvrB, and UvrC gene products perform this task in *E. coli*. In humans, defects in analogs of these enzymes lead to a condition called Xeroderma pigmentosum, which produces unusual sensitivity to sunlight.

The repair process for UV damaged DNA requires making nicks in the damaged DNA strand flanking the lesion, removing the damaged section, and repair synthesis to fill in the gap. Exposure to UV has long been known to generate mutations. Although the repair process itself could be inaccurate and generate the mutations, the actual situation is a bit more complicated. It appears that the existence of the damaged DNA turns off at least one of the normal error correction pathways. Thus mutations accumulate, even in portions of chromosomes not damaged by irradiation, if somewhere in the cell there exists damaged DNA. Ordinarily, pol III utilizes its 3'-5' exonuclease activity to correct mispaired bases, but RecA protein can turn off this repair activity. Apparently, RecA protein binds to a damaged site in the DNA, and as replication passes the point, the protein binds to the polymerase. The bound RecA then inhibits the normal 3'-5' exonuclease activity of the enzyme. Perhaps the relaxation of the fidelity of pol III helps the enzyme replicate past sites of UV damage. On the other hand, it is possible that the cell merely utilizes the opportunity of excessive DNA damage to

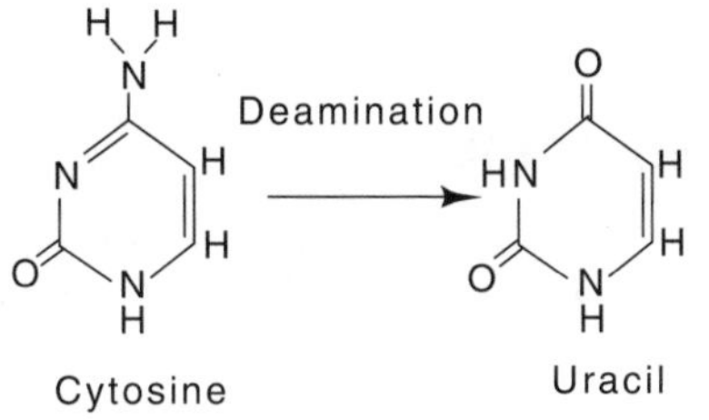

Figure 3.10 Deamination of cytosine produces uracil.

increase its spontaneous mutation rate. Such a strategy of increasing mutation rates during times of stress would be of great evolutionary value. Recent experiments suggest that similar increased mutation frequencies occur during times of nutrient starvation.

Simple chemical degradation can also compromise the information stored in DNA. The amino group on cytosine is not absolutely stable. Consequently, it can spontaneously deaminate to leave uracil (Fig. 3.10). DNA replication past such a point would then convert the former G-C base pair to an A-T base pair in one of the daughters because uracil has

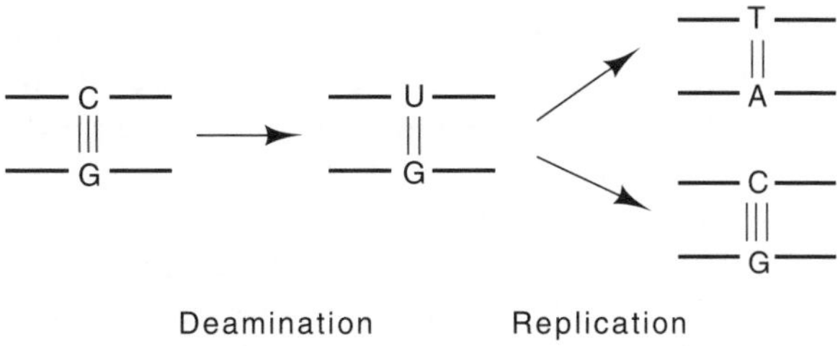

the base-pairing properties of thymine. Enzymes exist to repair deaminated cytosine. The first enzyme of the pathway is a uracil-DNA glycosidase that removes the uracil base, leaving the deoxyribose and phosphodiester backbone. Then the deoxyribose is removed and the resulting gap in the phosphodiester backbone is filled in. The fact that deaminated cytosine can be recognized as alien may be the reason thymine rather than uracil is used in DNA. If uracil were a natural component, cytosine deaminations could not be detected and corrected.

An interesting example demonstrates the efficiency of the deaminated cytosine repair system. As mentioned above, and as we examine more carefully in Chapter 10, some cells methylate bases found in particular sequences. In *E. coli* one such sequence is CCAGG, which is methylated on the second C. Should this cytosine deaminate, its extra methyl group blocks the action of the uracil-DNA glycosidase. As a result, spontaneous deamination at this position cannot be repaired. Indeed, this nucleotide has been found to be at least ten times as susceptible to spontaneous mutation as adjacent cytosine residues.

B. Physiological Aspects

DNA Replication Areas in Chromosomes

After considering the enzymology of the DNA replication and repair processes, we turn to more biological questions. As a first step, it is useful to learn the number of DNA synthesis regions per bacterial or eukaryotic chromosome. To see why this is important, consider the two extremes. On one hand, a chromosome could be duplicated by a single replication fork traversing the entire stretch of a DNA molecule. On the other hand, many replication points per chromosome could function simultaneously. The requisite elongation rates and regulation mechanisms would be vastly different in the two extremes. Furthermore, if many replication points functioned simultaneously, they could be either scattered over the chromosome or concentrated into localized replication regions.

The most straightforward method for determining the number of replication regions on a chromosome is electron microscopy. This is possible for smaller bacteriophage or viruses, but the total amount of DNA contained in a bacterial chromosome is far too great to allow detection of the few replication regions that might exist. The situation is even worse for chromosomes from eukaryotes because they contain as much as one hundred times the amount of DNA per chromosome as bacteria. The solution to this problem is not to look at all the DNA but to look at just the DNA that has been replicated in the previous minute. This can easily be done by autoradiography. Cells are administered highly radioactive thymidine, and a minute later the DNA is extracted and gently spread on photographic film to expose a trail which, upon development, displays the stretches of DNA that were synthesized in the presence of the radioactivity.

The results of such autoradiographic experiments show that cultured mammalian cells contain DNA synthesis origins about every 40,000 to 200,000 base pairs along the DNA. In bacteria, the result was different;

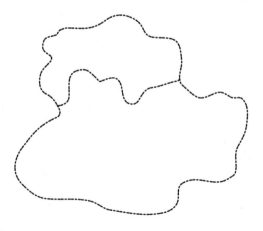

Figure 3.11 Sketch of an electron micrograph of an autoradiograph of a partially replicated bacterial chromosome that has been labeled with radioactive precursor for 40 minutes before extraction of DNA and preparation.

administration of a short pulse of radioactive thymidine was unnecessary. When thymidine was provided for more than one doubling time, the entire chromosome of the bacterium could be visualized via the exposed photographic grains. Two startling facts were seen: the chromosome was circular, and it possessed only one or two replication regions (Fig. 3.11).

The existence of a circular DNA molecule containing an additional segment of the circle connecting two points, the theta form, was interpreted as showing that the chromosome was replicated from an origin by one replication region that proceeded around the circular chromosome. It could also have been interpreted as demonstrating the existence of two replication regions that proceeded outward in both directions from a replication origin. Some of the original autoradiographs published by Cairns contain suggestions that the DNA is replicated in both directions from an origin. This clue that replication is bidirectional was overlooked until the genetic data of Masters and Broda provided solid and convincing data for two replication regions in the *E. coli* chromosome.

Bidirectional Replication from *E. coli* Origins

The Masters and Broda experiment was designed to locate the origin of replication on the genetic map. It is primarily a genetic experiment, but it utilizes the fact discussed in the first chapter that an exponentially growing population contains more young individuals than old individuals. Chromosomes can be considered in the same way. Correspondingly, a population of growing and dividing chromosomes contains more members just beginning replication than members just finishing replication. The Cairns autoradiograph experiments show that the chromo-

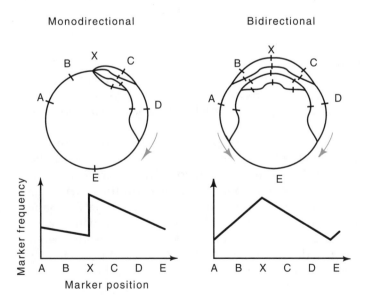

Figure 3.12 Left: Monodirectional replication from point X generates a step plus gradient of marker frequencies. Right: Bidirectional replication originating from X generates two gradients of marker frequencies. In the drawings, successive initiations have taken place on the single chromosome.

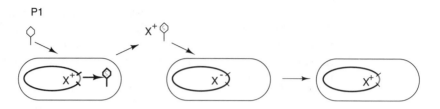

Figure 3.13 Transduction of genetic markers by phage P1 infecting an X$^+$ cell and carrying the X$^+$ marker into an X$^-$ cell.

some is replicated sequentially. A population of cells or chromosomes that is growing exponentially will contain more copies of genes located near the origin of replication than genes located at the terminus of replication. This same idea was used to locate the replication origin and demonstrate bidirectional replication of the SV40 animal virus.

Determining whether the bacterial chromosome is replicated in one direction from a unique origin or in both directions from a unique origin becomes a question of counting gene copies. If we could count the number of copies of genes A, B, C, D, and E, we could determine whether the cell uses monodirectional or bidirectional DNA replication initiating from point X (Fig. 3.12).

Several methods exist for counting the relative numbers of copies of different genes or chromosome regions. Here we shall consider a biological method for performing such counting that utilizes the phage P1. This method depends upon the fact that after P1 infection, a cell synthesizes about 100 new P1 particles. Most of these package their own DNA. A few phage particles package *E. coli* DNA instead. If a P1 lysate that was prepared on one type of cells is then used to infect a second culture of cells, most of the infected cells will proceed to make new phage P1. Those few cells that are infected with a P1 coat containing *E. coli* DNA from the first cells may be able to recombine that particular stretch of *E. coli* DNA into their chromosomes. By this means they can replace stretches of chromosomal DNA with chromosomal DNA brought into them by the phage particles (Fig. 3.13). This process is termed transduction.

The numbers of these defective phage particles carrying different genes from the infected cells is related to the numbers of copies of these genes present at the time of phage infection. Transduced cells can be made to reveal themselves as colonies, thereby permitting their simple and accurate quantitation. Consequently the use of phage P1 enabled counting of the relative numbers of copies within growing cells of various genes located around the chromosome. The results together with the known genetic map indicated that *E. coli* replicates its chromosome bidirectionally and determined the genetic location of the replication origin.

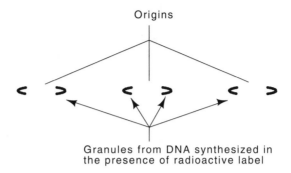

Origins

Granules from DNA synthesized in
the presence of radioactive label

Figure 3.14 Schematic of an electron micrograph of an autoradiograph prepared from mouse DNA extracted after 30 seconds of labeling with radioactive
thymidine. DNA is replicated bidirectionally from the multiple origins.

On the opposite side of the chromosome from the origin lies a
terminus region. A protein called Tus binds to the terminus. It blocks
elongation by inactivating the oncoming helicase of the replication fork.

Autoradiographic experiments have also suggested that mammalian
DNA also replicates bidirectionally from origins (Fig. 3.14). Multiple
replication forks or "eyes" could be shown to derive from a single DNA
duplex. The replication trails indicated that two forks originated from
a single origin. Furthermore, the rate of elongation of labeled strands
showed that the synthesis rate of mammalian DNA is about 200 nucleotides per second.

The DNA Elongation Rate

It seems reasonable that half the single bacterial chromosome should
be replicated by a single replication region traversing the chromosome
in about one doubling time. If such a replication region does not contain
multiple points of DNA elongation, then the speed of DNA chain elongation must be on the order of 1,000 nucleotides per second to replicate
the chromosome's 3×10^6 bases in a typical doubling time for rapidly
growing bacteria of 30 minutes. Measuring such a rate is very difficult,
but fortunately it is possible to reduce its value by about a factor of 5 by
growing the cells at 20° instead of 37°, the temperature of most rapid
growth.

If radioactive DNA precursors are added to the cell growth medium
and growth is stopped soon thereafter, the total amount of radioactivity,
T, incorporated into DNA equals the product of four factors: a constant
related to the specific activity of the label, the number of growing chains,
the elongation rate of a chain, and the time of radioactive labeling
$T = c \times N \times R \times t$. Similarly, the total amount of radioactivity incorporated into the ends of elongating chains, E, equals the product of two

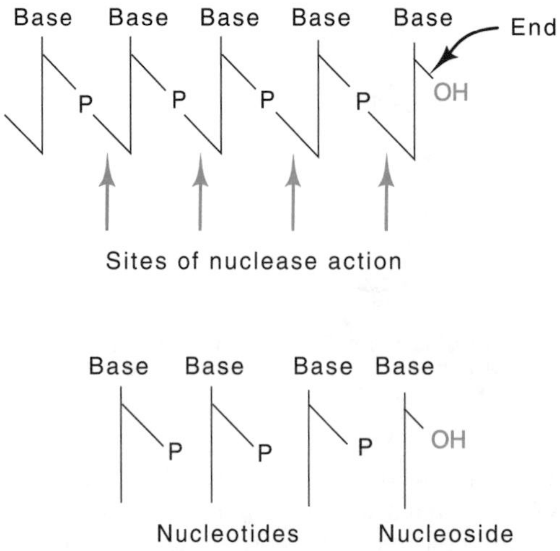

Figure 3.15 Cleavage of DNA between the phosphate and the 5' position leaves a nucleoside (no phosphate) from the terminal nucleotide of the elongating chain and nucleotides from all other positions.

factors: the same constant related to the specific activity and the number of elongating chains $E = c \times N$. The ratio of T to E yields

$$\frac{T}{E} = R \times t, \ or \ R = \frac{T}{t \times E}.$$

Determination of the T radioactivity is straightforward, but determination of the end radioactivity is not so obvious. Furthermore, careful experimentation would be required so that losses from either T or E samples do not introduce errors. These problems can be solved by digesting the DNA extracted from the labeled cells with a nuclease that cleaves so as to leave a phosphate on the 3' position of the deoxyribose. After the digestion, the terminal deoxyribose from the elongating chain lacks a phosphate, whereas the internal deoxyriboses all possess phosphate groups (Fig. 3.15). Hence separation and quantitation of the radioactive nucleosides and nucleotides in a single sample prepared from cells following a short administration of the four radioactive DNA precursors yield the desired T and E values.

If the elongation rate is several hundred bases per second, then one second of synthesis will label several hundred bases, and a separation of nucleosides and nucleotides will then need to be better than one part in several hundred. Also, it is difficult to add label suddenly and then to stop the cells' DNA synthesis quickly. Finally, the specific activity of intracellular nucleoside triphosphate pools does not immediately jump to the same specific activity as the label added to the medium. Fortunately the effect of a changing specific activity can be simply accounted for by taking a series of samples for analysis at different times after the addition of radioactive label.

The sample from the first point that could be taken from the cells after the addition of radioactive label possessed little radioactivity. It had counts of 17-20 cpm in ends and 2-20 $\times$ 10^5 cpm in total DNA. Samples taken later possessed greater amounts of radioactivity. This experiment yielded elongation rates of 140-250 bases/sec in cells with a doubling time of 150 minutes. This corresponds to a rate of 400 to 800 bases/sec in cells growing at 37° with a doubling time of 45 minutes. Therefore, only about two elongation points exist per replication region. The effects on this type of measurement of discontinuous DNA replication on the lagging strand are left as a problem for the reader.

The primary conclusion that should be drawn from the direct measurement of the DNA elongation rate is that a small number of enzyme molecules are involved. The cell does not utilize an extensive factory with many active DNA polymerase molecules in a growing region. What happens when cells grow at 37° but at different growth rates due to the presence of different growth media?

Constancy of the *E. coli* DNA Elongation Rate

Determining the strategy that cells use to adjust DNA replication to different growth rates once again utilizes measurement of the numbers of gene copies. Consider two cell types, one with a doubling time of 1 hour and one with a doubling time of 2 hours. If each cell type requires the full doubling time to replicate its chromosome, then the distributions of structures of replicating chromosomes extracted from random populations of cells growing at the two different rates would be identical (Fig. 3.16.). However, if both cell types replicate their chromosomes in 1 hour, then the cells with the 1-hour doubling time will possess a different distribution of chromosome structures than the cells with the 2-hour doubling time (Fig. 3.17). The problem, once again, is that of counting copies of genes. Measuring the number of copies of genes located near the origin and of genes located near the terminus of replication permits these two possible DNA replication schemes to be distinguished.

Instead of measuring numbers of copies of genes by transduction frequencies, the more precise method of DNA hybridization could be used because the locations of the origin and terminus were known. DNA

Figure 3.16 States of chromosomes of cells with 1-hour and 2-hour doubling times assuming 1 hour or 2 hours are required to replicate the DNA. The ratio of copies of genes near the replication origin and replication terminus are the same for cells growing with the two rates.

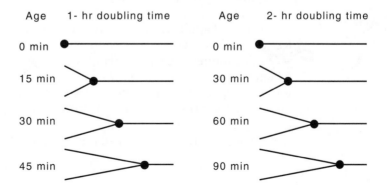

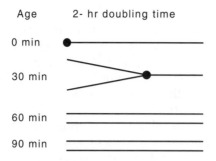

Figure 3.17 States of chromo-somes in cells with a 2-hour doubling time, assuming only 1 hour is required for replication of the chromosome. The ratio of the number of copies of genes near the replication origin and terminus is different than if 2 hours were re-quired for replication of the DNA.

was extracted from the cells, denatured, and immobilized on filter paper. Then an excess of a mixture of ^{3}H-labeled DNA fragment from the origin and ^{14}C-labeled DNA fragment from the terminus region was added and allowed to hybridize to the single-stranded DNA on the filters. When all hybridization to the immobilized DNA was completed, the filters were washed free of unannealed DNA and the ratio of bound ^{3}H to ^{14}C radioactivity was determined by liquid scintillation counting. This ratio reflects the ratio of the number of origins and termini in the culture of cells from which the DNA was extracted. The ratio could be measured using DNA extracted from cells growing at various growth rates. It showed that for cell doubling times in the range of 20 minutes to 3 hours the chromosome doubling time remained constant at about 40 minutes.

The constancy of the chromosome doubling time raises new prob-lems, however. How do the cells manage to keep DNA replication and cell division precisely coordinated, and how can DNA replication, which requires 40 minutes, manage to keep up with cell division if cells are dividing in less than 40 minutes?

Regulating Initiations

Helmstetter and Cooper have provided an explanation for the problem of maintaining a strain of *Escherichia coli* cells in balanced growth despite a difference between the cellular division time and the chromo-some replication time. Even though other strains and organisms may differ in the details of their regulation mechanisms, the model is of great value because it summarizes a large body of data and provides a clear understanding of ways in which cell division and DNA replication can be kept in step.

The model applies most closely to cells growing with a doubling time less than 1 hour. One statement of the model is that a cell will divide I + C + D minutes after the start of synthesis of an initiator substance, which is DnaA protein itself. I can be thought of as the time required for the I protein to accumulate to a level such that replication can initiate on all origins present in the cell. In our discussions, we will call this critical level 1. That is, when a full unit of I has accumulated, all chromosomes present in the cell initiate replication, a round of replica-

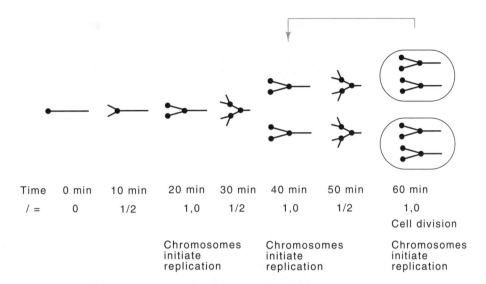

Time	0 min	10 min	20 min	30 min	40 min	50 min	60 min
I =	0	1/2	1,0	1/2	1,0	1/2	1,0

Cell division

Chromosomes initiate replication   Chromosomes initiate replication   Chromosomes initiate replication

Figure 3.18 Application of the Helmstetter-Cooper model to cells with a doubling time of 20 minutes. At 60 minutes, cell division is possible and the resulting state of the cells is the same as cells shown at 40 minutes in the diagram.

tions begins, and the *I* substance present at this time is consumed so that its accumulation begins again from a value of zero.

C is the time required for complete synthesis of the chromosome. It is independent of the growth rate of the cells as long as they grow at 37°, and it has a value of 40 minutes. *D* is a constant equal to 20 minutes. This is the time following completion of a round of DNA synthesis until the cell divides. It should be emphasized that in this model a cell must divide *D* minutes after completion of a round of DNA replication. *D* can be considered to be the time required for segregation of the daughter chromosomes into opposite ends of the cell and for growth of the septum that separates the cells.

The only parameter in the model that is responsive to the growth rate of the cells is the rate at which *I* accumulates. If the growth rate doubles as a result of a richer nutrient medium, the rate of accumulation of *I* doubles. The time required for accumulation of a full unit of *I* is the doubling time of the cells.

To illustrate the model, we will consider cells growing with a doubling time of 20 minutes (Fig. 3.18). For convenience, consider the bidirectional replication of the cell's circular chromosome to be abstracted to a forked line. We use a dot to represent the replication fork, and we encircle the chromosome to represent cell division. Since the model is stable, analysis of a cell division cycle can begin from a point not in a normal division cycle, and continued application of the model's rules should ultimately yield the states of cells growing with the appropriate doubling times.

We begin in Fig. 3.18 from a point at which replication of a chromosome has just begun and the value of *I* is zero. After 10 minutes, the chromosome is one-quarter replicated, and $I=\frac{1}{2}$. At 20, 40, and 60 minutes after the start of replication, full units of 1 have accumulated, new rounds of replication begin from all the origins present, and the chromosomes become multiforked. The first cell division can occur at 60 minutes after the start of replication, and divisions occur at 20-minute intervals thereafter. The DNA configuration and quantity of I present in cells just after division at 60 minutes is the same as in cells at 40 minutes. Therefore the cell cycle in this medium begins at 40 minutes and ends at 60 minutes. This means that zero-age cells in this medium possess two half-replicated chromosomes.

How can the multiple origins which exist within rapidly growing cells all initiate at precisely the same instant, or do they initiate at the same instant? A related question is what keeps an origin of replication from being reused immediately after it has initiated a round of synthesis? The answer is known in general terms, and could well apply to the analogous problems in eukaryotic cells. We know that a sizeable number of DnaA protein molecules are required to initiate replication from an origin. This large number makes the reaction critically dependent upon the concentration of DnaA. When it is a little too low, there is a very low probability of initiation, but the moment the critical concentration is reached, initiation can occur. Then, to keep the origin from being used again, it is promptly buried in the membrane. The signal for burying is that it be half methylated. Before initiation, both strands of the origin are methylated on the multiple GATC sequences contained in bacterial origins. After initiation, the new daughter strand is not methylated, and the hemimethylated DNA is bound to the membrane and therefore inaccessible to the initiation machinery. On a time scale of ten minutes or so, the newly synthesized strand becomes methylated and the origin is released from the membrane. In the meantime, either the level of DnaA or some other critical component has been reduced, and the origin is in no danger of firing again until the proper time.

Gel Electrophoresis Assay of Eukaryotic Replication Origins

It has been possible to construct artificial yeast chromosomes by combining isolated centromeres, telomeres, and autonomously replicating sequences known as ARS sequences. Although ARS sequences function in the artificial chromosomes, it is interesting to know whether they also normally function as origins. A gel electrophoresis technique based on the Southern transfer technology as described in the previous chapter permits examination of this question.

Experiments have shown that ordinarily gel electrophoresis separates DNA according to molecular weight. If, however, the voltage gradient is increased five-fold above normal to about 5 V/cm, and higher than normal concentrations of agarose are used, then the separation becomes largely based on shape of the DNA molecules rather than their total molecular weight. Normal electrophoresis and this shape-sensitive electrophoresis can be combined in a two dimensional electrophoretic separation technique that is particularly useful in the analysis of repli-

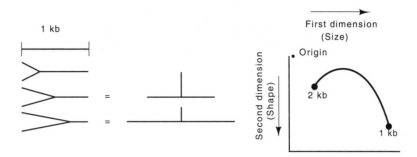

Figure 3.19 The various structures resulting from a replication fork entering the 1 kb region from the left. The arc shown indicates the positions of the various structures after 2D electrophoresis.

cation origins. The DNA fragments necessary for the analysis are generated by cleaving DNA extracted from cells with restriction enzymes. These cut at specific sequences, and will be discussed more extensively in a later chapter. A DNA sample obtained by cutting chromosomal DNA with such a restriction enzyme is first separated according to size by electrophoresis in one direction. Then it is separated according to shape by electrophoresis in a direction perpendicular to the first. After the electrophoresis, the locations of the fragments containing the sequence in question are determined by Southern transfer.

Let us first consider the two-dimensional electrophoretic pattern of DNA fragments that would be generated from DNA extracted from a large number of growing cells if replication origins were to enter a 1000 base pair region from the left (Fig. 3.19). In the majority of the cells no replication origin will be on the 1000 base pair stretch of DNA. There-

Figure 3.20 The various structures resulting from a replication fork originating from the center of the 1 kb ARS region. The arc of the positions reached by the various structures does not reach the 2 kb point because none of the structures approaches a linear 2 kb molecule.

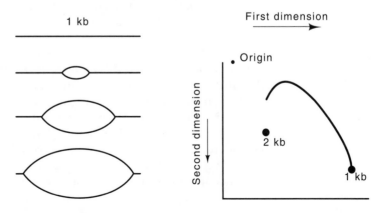

fore, after cutting with the restriction enzyme these stretches will be simple 1000 base pair pieces of DNA. After the Southern transfer, these molecules will show up as a spot at 1000 base pairs. Some of the cells would contain replication origins in the 1000 base pair region. After cutting with the restriction enzyme, these molecules not only would possess a mass larger than the 1000 base pair molecules, but also they would be more asymmetric. The extreme of asymmetry would occur in those molecules in which the replication origin was 500 base pairs from the end. These would generate the peak shown in Fig. 3.19. Those molecules for which the replication origin was nearly at the right end would, again, be like simple DNA molecules, but molecules 2000 base pairs long. Thus, the collection of molecular species would generate the arc shown upon two dimensional electrophoresis.

If the region of DNA we are considering contains a replication origin, quite a different pattern is generated. Suppose the origin is exactly in the middle of the region. Then the various replication forms are as shown in Fig. 3.20 and the pattern that will be found after Southern transfer is as shown. It is left as a problem to deduce the pattern expected if the origin is located to one side of the center of the DNA segment.

How Fast Could DNA Be Replicated?

In the preceding sections we have seen that the bacterial chromosome is replicated by two synthesis forks moving away from a replication origin at a chain elongation speed of about 500 nucleotides per second for cells growing at 37°. How does this rate compare to the maximum rate at which nucleotides could diffuse to the DNA polymerase? This question is one specific example of a general concern about intracellular conditions. Often it is important to know an approximate time required for a particular molecule to diffuse to a site.

Consider a polymerase molecule to be sitting in a sea of infinite dimensions containing the substrate. Even though the polymerase moves along the DNA as it synthesizes, we will consider it to be at rest since the processes of diffusion of the nucleotides which we are considering here are much faster. We will consider that the elongation rate of the enzyme is limited by the diffusion of nucleotides to its active site. Under these conditions, the concentration of substrate is zero on the surface of a sphere of radius r_0 constituting the active site of the enzyme (Fig. 3.21). Any substrate molecules crossing the surface into this region disappear. At great distances from the enzyme, the concentration of substrate remains unaltered. These represent the boundary conditions of the situation, which requires a mathematical formulation to determine the concentrations at intermediate positions.

The basic diffusion equation relates time and position changes in the concentration C of a diffusible quantity. As diffusion to an enzyme can be considered to be spherically symmetric, the diffusion equation can be written and solved in spherical coordinates involving only the radius r, the concentration C, the diffusion coefficient D, and time t:

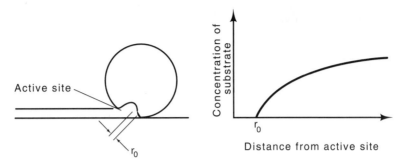

Figure 3.21 A DNA polymerase possessing an active site of radius r_0 into which nucleotides disappear as fast as they can reach the enzyme and the expected nucleotide concentration as a function of distance from the enzyme.

$$\frac{dC}{dt} = D\frac{d^2C}{dr^2} + \frac{2D}{r}\frac{dC}{dr}.$$

The solution to this equation, which satisfies the conditions of being constant at large r and zero at $r = r_0$, is

$$C(r) = C_0(1 - \frac{r_0}{r}).$$

The flow rate J of substrate to the enzyme can now be calculated from the equation, giving the flux J:

$$J = -D\frac{dC}{dr}.$$

The total flow through a sphere centered on the enzyme yields

$$\text{Flow} = 4\pi r^2 J$$

$$= 4\pi D r_0 C_0.$$

The final result shows that the flow is independent of the size of the sphere chosen for the calculation. This is as it should be. The only place where material is being destroyed is at the active site of the enzyme. Everywhere else, matter must be conserved. In steady state there is no change of the concentration of substrate at any position, and hence the net amount flowing through the surface of all spheres must be equal.

To calculate the flow rate of nucleoside triphosphates to the DNA polymerase, we must insert numerical values in the final result. The concentration of deoxynucleoside triphosphates in cells is between 1 mM and 0.1 mM. We will use 0.1 mM, which is 10^{-4} moles per liter or 10^{-7} moles per cm^3. Taking the diffusion constant to be 10^{-7} cm^2/sec and r_0 to be 10 Å, we find that the flow is 10^{-20} moles/sec or about 6,000 molecules/sec. The rate of DNA synthesis per enzyme molecule of about 500 nucleoside triphosphates per second is less than 10% of the upper limit on the elongation rate set by the laws of diffusion. Considering that the actual active site for certain capture of the triphosphate might be much smaller than 10 Å, the elongation rate of DNA seems remarkably high.

Problems

3.1. Why is it sensible for regulatory purposes to have a large rather than a small number of DnaA protein molecules involved in initiating DNA replication?

3.2. By looking up the appropriate papers, write a brief summary of how the DNA polymerase I mutant in *E. coli* was found.

3.3. Design a substrate that would be useful in assaying the editing activity of DNA polymerase.

3.4. What would you say about a claim that the only effect of a mutation was an increase in the 3'-5' exonuclease activity of DNA pol I, both for a correctly and an incorrectly paired nucleotide and that this reduced the spontaneous mutation frequency?

3.5. How best could an autoradiography experiment with *E. coli* be performed to demonstrate bidirectional replication of the chromosome?

3.6. Masters and Broda, Nature New Biol. *232*, 137 (1970), report that the normalization of gene transduction frequency by P1 is unnecessary; in fact, the raw data support their conclusions better than the normalized data. What is a possible explanation for this result and what experiment could test your proposal?

3.7. Cells that have been growing for many generations with a doubling time of 20 minutes are shifted to medium supporting a doubling time of 40 minutes. Assume that there are no nutritional starvations as a result of this shift. Sketch the DNA configurations and indicate cell divisions at 10-minute intervals of cells whose age at the time of the shift was 0 minutes and 10 minutes. Continue the analysis until cells have fully adapted to growth at the slower growth rate.

3.8. If the *dam* methylase methylates DNA for use in error correction of newly synthesized DNA strands, why would mutations increasing or decreasing *dam* methylase synthesis be mutagenic?

3.9. A DNA elongation rate experiment yielded the data in the following table. What is the DNA chain elongation rate in these cells?

t(sec)	cpm(nucleosides)	cpm(nucleotides)
5	43	24,000
10	110	91,000
15	140	200,000
20	205	410,000
25	260	650,000

3.10. Why should or shouldn't the existence of Okazaki fragments interfere with a Manor-Deutscher measurement as DNA labeling times become short?

3.11. In the chromosome replication time measurements of Chandler, Bird, and Caro, a constant replication time of 40 minutes was found. Does their result depend on when in the cell cycle this 40 minutes of replication occurs?

3.12. An exponentially growing population was used to start a synchrony experiment. After loosely bound cells have been washed off the filter and only newly divided cells are being eluted in a synchrony experiment, the concentration of cells in the eluate is found to oscillate for several cycles with a period about equal to the cell doubling time. What is the explanation for this oscillation? You should assume that the probability that a cell initially binds to the filter is independent of its age.

3.13. What is the shape expected in a two-dimensional size-shape separation experiment as described in the text for an ARS element containing a bidirectional replication element asymmetrically located in a 1,000 base pair fragment?

3.14. Show by differentiation that the solution $C = C_0(1 - r_0/r)$ satisfies the diffusion equation in spherical coordinates.

3.15. *Escherichia coli* deleted of their origin, OriC, are inviable unless they possess a second mutation in RNAseH. How does being RNAseH defective permit viability?

3.16. What kind of a substrate would you use to assay the 3'-to-5' exonuclease activity of a DNA polymerase?

References

Recommended Readings

Chromosome Replication and the Division Cycle of *Escherichia coli* B/r, S. Cooper, C. Helmstetter, J. Mol. Biol. *31*, 519-540 (1968).

Requirement for d(GATC) Sequences in *Escherichia coli mutHLS* Mismatch Correction, R. Lahue, S. Su, P. Modrich, Proc. Natl. Acad. Sci. USA *84*, 1482-1486 (1987).

The Localization of Replication Origins on ARS Plasmids in S. *cerevisiae*, B. Brewer, W. Fangman, Cell *51*, 463-471 (1987).

E. *coli oriC* and the *dnaA* Gene Promoter are Sequestered from *dam* Methyltransferase Following the Passage of the Chromosomal Replication Fork, J. Campbell, N. Kleckner, Cell *62*, 967-979 (1990).

DNA Polymerase Activities

Isolation of an E. *coli* Strain with a Mutation Affecting DNA Polymerase, P. De Lucia, J. Cairns, Nature *224*, 1164-1166 (1964).

Analysis of DNA Polymerases II and III in Mutants of *Escherichia coli* Thermosensitive for DNA Synthesis, M. Gefter, Y. Hirota, T. Kornberg,

J. Wechsler, C. Barnoux, Proc. Nat. Acad. Sci. USA *68*, 3150-3153 (1971).

Rates of DNA Chain Growth in *E. coli*, H. Manor, M. Deutscher, U. Littauer, J. Mol. Biol. *61*, 503-524 (1971).

RNA-Linked Nascent DNA Fragments in *Escherichia coli*, A. Sugino, S. Hirose, R. Okazaki, Proc. Natl. Acad. Sci. USA *69*, 1863-1867 (1972).

Mechanism of DNA Chain Growth XV. RNA-Linked Nascent DNA Pieces in *E. coli* Strains Assayed with Spleen Exonuclease, Y. Kurosawa, T. Ogawa, S. Hirose, T. Okazaki, R. Okazaki, J. Mol. Biol. *96*, 653-664 (1975).

Contribution of 3'-5' Exonuclease Activity of DNA Polymerase III Holoenzyme from *Escherichia coli* to Specificity, A. Fersht, J. Knill-Jones, J. Mol. Biol. *165*, 669-682 (1983).

A Separate Editing Exonuclease for DNA Replication: The Epsilon Subunit of *Escherichia coli* DNA Polymerase III Holoenzyme, R. Scheuermann, H. Echols, Proc. Nat. Acad. Sci. USA *81*, 7747-7751 (1984).

Evidence that Discontinuous DNA Replication in *Escherichia coli* is Primed by Approximately 10 to 12 Residues of RNA Starting with a Purine, T. Kitani, K. Yoda, T. Ogawa, T. Okazaki, J. Mol. Biol. *184*, 45-52 (1985).

A Cryptic Proofreading 3'-5' Exonuclease Associated with the Polymerase Subunit of the DNA Polymerase-primase from *Drosophila melanogaster*, S. Cotterill, M. Teyland, L. Loeb, I. Lehman, Proc. Natl. Acad. Sci. USA *84*, 5635-5639 (1987).

DNA Polymerase III Holoenzyme of Escherichia coli, H. Maki, S. Maki, A. Kornberg, J. Biol. Chem. 263, 6570-6578 (1988).

DNA Polymerase III, A Second Essential DNA Polymerase, Is Encoded by the *S. cerevisiae* CDC2 Gene, K. Sitney, M. Budd, J. Campbell, Cell *56*, 599-605 (1989).

Proofreading by the Epsilon Subunit of *Escherichia coli* DNA Polymerase III Increases the Fidelity by Calf Thymus DNA Polymerase Alpha, F. Perrino, L. Loeb, Proc. Natl. Acad. Sci. USA *86*, 3085-3088 (1989).

A Third Essential DNA Polymerase in *S. cerevisiae*, A. Morrison, H. Araki, A. Clark, R. Hamatake, A. Sugino, Cell *62*, 1143-1151 (1990).

DNA polymerase II is Encoded by the Damage-inducible *dinA* Gene of *Escherichia coli*, C. Bonner, S. Hays, K. McEntee, M. Goodman, Proc. Natl. Acad. Sci. USA *87*, 7663-7667 (1990).

Three-dimensional Structure of the β Subunit of *E. coli* DNA Polymerase III Holoenzyme: A Sliding DNA Clamp, X. Kong, R. Onrust, M. O'Donnell, J. Kuriyan, Cell, *69*, 425-437 (1993).

Initiating from Origins

Transient Accumulation of Okazaki Fragments as a Result of Uracil Incorporation into Nascent DNA, B. Tye, P. Nyman, I. Lehman, S. Hochhauser, B. Weiss, Proc. Nat. Acad. Sci. USA *74*, 154-157 (1977).

Mini-chromosomes: Plasmids Which Carry the *E. coli* Replication Origin, W. Messer, H. Bergmans, M. Meijer, J. Womack, F. Hansen, K. von Meyenburg, Mol. Gen. Genetics *162*, 269-275 (1978).

Nucleotide Sequence of the Origin of Replication of the *E. coli* K-12 Chromosome, M. Meijer, E. Beck, F. Hansen, H. Bergmans, W. Messer, K. von Meyenburg, H. Schaller, Proc. Nat. Acad. Sci. USA *76*, 580-584 (1979).

Nucleotide Sequence of the *Salmonella typhimurium* Origin of DNA Replication, J. Zyskind, D. Smith, Proc. Nat. Acad. Sci. USA 77, 2460-2464 (1980).

Unique Primed Start of Phage φX174 DNA Replication and Mobility of the Primosome in a Direction Opposite Chain Synthesis, K. Arai, A. Kornberg, Proc. Nat. Acad. Sci. USA 78, 69-73 (1981).

Inhibition of ColE1 RNA Primer Formation by a Plasmid-specified Small RNA, J. Tomizawa, T. Itoh, G. Selzer, T. Som, Proc. Nat. Acad. Sci. USA 78, 1421-1425 (1981).

Enzymatic Replication of the Origin of the *Escherichia coli* Chromosome, R. Fuller, J. Kaguni, A. Kornberg, Proc. Nat. Acad. Sci USA 78, 7370-7374 (1981).

Replication of φX174 DNA with Purified Enzymes II. Multiplication of the Duplex Form by Coupling of Continuous and Discontinuous Synthetic Pathways, N. Arai, L. Polder, K. Arai, A. Kornberg, J. Biol. Chem. 256, 5239-5246 (1981).

Specific Cleavage of the p15A Primer Precursor by Ribonuclease H at the Origin of DNA Replication, G. Selzer, J. Tomizawa, Proc. Natl. Acad. Sci. 79, 7082-7086 (1982).

Structure of Replication Origin of the *Escherichia coli* K-12 Chromosome: The Presence of Spacer Sequences in the *ori* Region Carrying Information for Autonomous Replication, K. Asada, K. Sugimoto, A. Oka, M. Takanami, Y. Hirota, Nucleic Acids Res. 10, 3745-3754 (1982).

ARS Replication During the Yeast S Phase, W. Fandman, R. Hice, E. Chlebowicz-Sledziewska, Cell 32, 831-838 (1983).

Replication Initiated at the Origin (oriC) of the *E. coli* Chromosome Reconstituted with Purified Enzymes, J. Kaguni, A. Kornberg, Cell 38, 183-190 (1984).

The *rnh* Gene is Essential for Growth of *Escherichia coli*, S. Kanaya, R. Crouch, Proc. Nat. Acad. Sci. USA 81, 3447-3451 (1984).

Specialized Nucleoprotein Structures at the Origin of Replication of Bacteriophage Lambda: Localized Unwinding of Duplex DNA by a Six Protein Reaction, M. Dodson, H. Echols, S. Wickner, C. Alfano, K. Mensa-Wilmont, B. Gomes, J. LeBowitz, J. Roberts, R. McMacken, Proc. Nat. Acad. Sci. USA 83, 7638-7642 (1986).

Enhancer-Origin Interaction in Plasmid R6K Involves a DNA Loop Mediated by Initiator Protein, S. Mukherjee, H. Erickson, D. Bastia, Cell 52, 375-383 (1988).

The Replication Origin of the *E. coli* Chromosome Binds to Cell Membranes Only when Hemimethylated, G. Ogden, M. Pratt, M. Schaechter, Cell 54, 127-135 (1988).

Role of the *Escherichia coli* DnaK and DnaJ Heat Shock Proteins in the Initiation of Bacteriophage Lambda DNA Replication, K. Liberek, C. Georgopoulos, M. Zylicz, Proc. Natl. Acad. Sci. USA 85, 6632-6636 (1988).

Transcriptional Activation of Initiation of Replication from the *E. coli* Chromosomal Origin: An RNA-DNA Hybrid Near *oriC*, T. Baker, A. Kornberg, Cell 55, 113-123 (1988).

Transcriptional Activatory Nuclear Factor I Stimulates the Replication of SV40 Minichromosomes *in vivo* and *in vitro*, L. Chong, T. Kelly, Cell 59, 541-551 (1989).

The DnaA Protein Determines the Initiation Mass of *Escherichia coli* K-12, A. Lobner-Olesen, K. Skarstad, F. Hansen, K. von Meyenburg, E. Boye, Cell *57*, 881-889 (1989).

In vivo Studies of DnaA Binding to the Origin of Replication of *Escherichia coli*, C. Samitt, F. Hansen, J. Miller, M. Schaechter, EMBO Journal *8*, 989-993 (1989).

Control of ColE1 Plasmid Replication, Intermediates in the Binding of RNAI and RNAII, J. Tomizawa, J. Mol. Biol. *212*, 683-694 (1990).

Complex Formed by Complementary RNA Stem-Loops and Its Stabilization by a Protein: Function of ColE1 Rom Protein, Y. Equchi, J. Tomizawa, Cell *60*, 199-209 (1990).

Chromosome Replication and its Control

The Replication of DNA in *Escherichia coli*, M. Meselson, F. Stahl, Proc. Nat. Acad. Sci. USA *44*, 671-682 (1958).

The Bacterial Chromosome and its Manner of Replication as Seen by Autoradiography, J. Cairns, J. Mol. Biol. *6*, 208-213 (1963).

On the Regulation of DNA Replication in Bacteria, F. Jacob, S. Brenner, F. Cuzin, Cold Spring Harbor Symp. Quant. Biol. *28*, 329-348 (1963).

DNA Synthesis during the Division Cycle of Rapidly Growing *Escherichia coli* B/r, C. Helmstetter, S. Cooper, J. Mol. Biol. *31*, 507-518 (1968).

Evidence for the Bidirectional Replication of the *E. coli* Chromosome, M. Masters, P. Broda, Nature New Biology *232*, 137-140 (1971).

Bidirectional Replication of the Chromosome in *Escherichia coli*, D. Prescott, P. Kuempel, Proc. Nat. Acad. Sci. USA *69*, 2842-2845 (1972).

Properties of a Membrane-attached Form of the Folded Chromosome of *E. coli*, A. Worcel, E. Burgi, J. Mol. Biol. *82*, 91-105 (1974).

Electron Microscopic Visualization of the Folded Chromosome of *E. coli*, H. Delius, A. Worcel, J. Mol. Biol. *82*, 107-109 (1974).

The Replication Time of the *E. coli* K-12 Chromosome as a Function of Cell Doubling Time, M. Chandler, R. Bird, L. Caro, J. Mol. Biol. *94*, 127-132 (1975).

Map Position of the Replication Origin on the *E. coli* Chromosome, O. Fayet, J. Louarn, Mol. Gen. Genetics *162*, 109-111 (1978).

An *Escherichia coli* Mutant Defective in Single-strand Binding Protein is Defective in DNA Replication, R. Meyer, J. Glassberg, A. Kornberg, Proc. Nat. Acad. Sci. USA *76*, 1702-1705 (1979).

E. coli Mutants Thermosensitive for DNA Gyrase Subunit A: Effects on DNA Replication, Transcription and Bacteriophage Growth, K. Kreuzer, N. Cozzarelli, J. Bact. *140*, 424-435 (1979).

Partition Mechanism of F Plasmid: Two Plasmid Gene-Encoded Products and a Cis-acting Region are Involved in Partition, T. Ogura, S. Hiraga, Cell *32*, 351-360 (1983).

Effects of Point Mutations on Formation and Structure of the RNA Primer for ColE1 Replication, H. Masukata, J. Tomizawa, Cell *36*, 513-522 (1984).

Bacterial Chromosome Segregation: Evidence for DNA Gyrase Involvement in Decatenation, T. Steck, K. Drlica, Cell *36*, 1081-1088 (1984).

Escherichia coli DNA Distributions Measured by Flow Cytometry and Compared with Theoretical Computer Simulations, K. Skarstad, H. Steen, E. Boye, J. Bact. *163*, 661-668 (1985).

Timing of Initiation of Chromosome Replication in Individual *Escherichia coli* Cells, K. Skarstad, E. Boye, and H. Steen, Eur. J. Mol. Biol. *5*, 1711-1717 (1986).

Control of ColE1 Plasmid Replication: Binding of RNAI to RNAII and Inhibition of Primer Formation, J. Tomizawa, Cell *47*, 89-97 (1986).

Requirement for Two Polymerases in the Replication of Simian Virus 40 DNA *in vitro*, D. Weinberg, T. Kelly, Proc. Natl. Acad. Sci. USA *86*, 9742-9746 (1989).

Viable Deletions of a Telomere from a *Drosophila* Chromosome, R. Levis, Cell *58*, 791-801 (1989).

Antiparallel Plasmid-Plasmid Pairing May Control P1 Plasmid Replication, A. Abeles, S. Austin, Proc. Natl. Acad. Sci. USA *88*, 9011-9015 (1991).

ATP-Dependent Recognition of Eukaryotic Origins of DNA Replication by a Multiprotein Complex, S. Bell, B. Stillman, Nature *357*, 128-134 (1992).

Error Detection and Correction

Escherichia coli Mutator Mutants Deficient in Methylation-instructed DNA Mismatch Correction, B. Glickman, M. Radman, Proc. Natl. Acad. Sci. USA *77*, 1063-1067 (1980).

Methyl-directed Repair of DNA Base-pair Mismatches *in vitro*, A. Lu, S. Clark, P. Modrich, Proc. Natl. Acad. Sci. USA *80*, 4639-4643 (1983).

Changes in DNA Base Sequence Induced by Targeted Mutagenesis of Lambda Phage by Ultraviolet Light, R. Wood, T. Skopek, F. Hutchinson, J. Mol. Biol. *173*, 273-291 (1984).

Removal of UV Light-induced Pyrimidine-pyrimidone (6-4) Products from *Escherichia coli* DNA Requires *uvrA*, *uvrB*, and *uvrC* Gene Products, W. Franklin, W. Haseltine, Proc. Nat. Acad. Sci. USA *81*, 3821-3824 (1985).

One Role for DNA Methylation in Vertebrate Cells is Strand Discrimination in Mismatch Repair, J. Hare, J. Taylor, Proc. Nat. Acad. Sci. USA *82*, 7350-7354 (1985).

Induction and Autoregulation of *ada*, a Positively Acting Element Regulating the Response of *Escherichia coli* to Methylating Agents, P. Lemotte, G. Walker, J. Bact. *161*, 888-895 (1985).

The Intracellular Signal for Induction of Resistance to Alkylating Agents in *E. coli*, I. Teo, B. Sedgwick, M. Kilpatrick, T. McCarthy, T. Lindahl, Cell *45*, 315-324 (1986).

Escherichia coli mutS-encoded Protein Binds to Mismatched DNA Base Pairs, S. Su, P. Modrich, Proc. Natl. Acad. Sci. USA *83*, 5057-5061 (1986).

Characterization of Mutational Specificity within the *lacI* Gene for a *mutD5* Mutation Strain of *Escherichia coli* Defective in 3'-5' Exonuclease (Proofreading) Activity, R. Fowler, R. Schaaper, B. Glickman, J. Bact. *167*, 130-137 (1986).

Effect of Photoreactivation on Mutagenesis of Lambda Phage by Ultraviolet Light, F. Hutchinson, K. Yamamoto, J. Stein, J. Mol. Biol. *202*, 593-601 (1988).

RecA-mediated Cleavage Activates UmuD for Mutagenesis: Mechanistic Relationship between Transcriptional Derepression and Posttranslational Activation, T. Nohmi, J. Battista, L. Dodson, G. Walker, Proc. Natl. Acad. Sci. USA *85*, 1816-1820 (1988).

Escherichia coli mutY Gene Product is Required for Specific A-G to C-G Mismatch Correction, K. Au, M. Cabrera, J. Miller, P. Modrich, Proc. Natl. Acad. Sci. USA *85*, 9163-9166 (1988).

Strand-Specific Mismatch Correction in Nuclear Extracts of Human and *Drosophila melanogaster* Cell Lines, J. Holmes Jr., S. Clark, P. Modrich, Proc. Natl. Acad. Sci. USA *87*, 5837-5841 (1990).

MutT Protein Specifically Hydrolyzes a Potent Mutagenic Substrate for DNA Synthesis, H. Maki, M. Sekiguchi, Nature *355*, 273-275 (1992).

Replication Termination

Map Positions of the Replication Terminus on the *E. coli* Chromosome, J. Louarn, J. Patte, J. Louarn, Mol. Gen. Genetics *172*, 7-11 (1979).

Specialized Nucleoprotein Structures at the Origin of the Terminus Region of the *Escherichia coli* Chromosome Contains Two Separate Loci that Exhibit Polar Inhibition of Replication, T. Hill, J. Henson, P. Kuempel, Proc. Natl. Acad. Sci. USA *84*, 1754-1758 (1987).

Identification of the DNA Sequence from the *E. coli* Terminus Region that Halts Replication Forks, T. Hill, A. Pelletier, M. Tecklenburg, P. Kuempel, Cell *55*, 459-466 (1988).

The Replication Terminator Protein of *E. coli* Is a DNA Sequence-specific Contra-Helicase, G. Khatri, T. MacAllister, P. Sista, D. Bastia, Cell *59*, 667-674 (1989).

Tus, The Trans-acting Gene Required for Termination of DNA Replication in *Escherichia coli*, Encodes a DNA-binding Protein, T. Hill, M. Tecklenberg, A. Pelletier, P. Kuempel, Proc. Natl. Acad. Sci. USA *86*, 1593-1597 (1989).

Escherichia coli Tus Protein Acts to Arrest the Progression of DNA Replication Forks *in vitro*, T. Hill, K. Marians, Proc. Natl. Acad. Sci. USA *87*, 2481-2485 (1990).

RNA Polymerase and RNA Initiation

4

The previous two chapters discussed DNA and RNA structures and the synthesis of DNA. In this chapter we consider RNA polymerase and the initiation of transcription. The next chapter considers elongation, termination, and the processing of RNA.

Cells must synthesize several types of RNA in addition to the thousands of different messenger RNAs that carry information to the ribosomes for translation into protein. The protein synthesis machinery requires tRNA, the two large ribosomal RNAs, and the small ribosomal RNA. Additionally, eukaryotic cells contain at least eight different small RNAs found in the nucleus. Because these contain protein also, they are called small ribonucleoprotein particles, snRNPs. Eukaryotic cells use three types of RNA polymerase to synthesize the different classes of RNA, whereas *E. coli* uses only one. All these polymerases, however, are closely related.

Experiments first done with bacteria and then with eukaryotic cells have shown that the basic transcription cycle consists of the following: binding of an RNA polymerase molecule at a special site called a promoter, initiation of transcription, further elongation, and finally termination and release of RNA polymerase. Although the definition of promoter has varied somewhat over the years, we will use the term to mean the nucleotides to which the RNA polymerase binds as well as any others that are necessary for the initiation of transcription. It does not include disconnected regulatory sequences which are discussed below, that may lie hundreds or thousands of nucleotides away.

Promoters from different prokaryotic genes differ from one another in nucleotide sequence, in the details of their functioning, and in their overall activities. The same is true of eukaryotic promoters. On some prokaryotic promoters, RNA polymerase by itself is able to bind and initiate transcription. In eukaryotic cells, the analog of such an unregulated system is a promoter that requires no auxiliary proteins bound to sites separated from the promoter. On many promoters, RNA polymerase requires the assistance of one or more auxiliary proteins for binding, to displace histones, or for the initiation of transcription. That other proteins should be involved is logical as the activities of some promoters have to be regulated in response to changing conditions in the cell or in the growth medium. These auxiliary proteins sense these conditions and appropriately modulate the activity of RNA polymerase in initiating from some promoters. Occasionally, the regulated step is after initiation. Part of this chapter discusses measurements of the differences between promoters because this information ultimately should assist learning the functions of these auxiliary proteins.

Measuring the Activity of RNA Polymerase

Studies on protein synthesis in *E. coli* in the 1960s revealed that a transient RNA copy of DNA is sent to ribosomes to direct protein synthesis. Therefore cells had to contain an RNA polymerase that was capable of synthesizing RNA from a DNA template. This property was sufficient to permit enzymologists to devise assays for detection of the enzyme in crude extracts of cells.

The original assay of RNA polymerase was merely a measurement of the amount of RNA synthesized *in vitro*. The RNA synthesized was easily determined by measuring the incorporation of a radioactive RNA precursor, usually ATP, into a polymer (Fig. 4.1). After synthesis, the radioactive polymer was separated from the radioactive precursor nucleotides by precipitation of the polymer with acid, and the radioactivity in the polymer was determined with a Geiger counter.

The precipitation procedure measures total incorporated radioactivity. It is adequate for the assay of RNA polymerase activity and can be used to guide steps in the purification of the enzyme, but it is indiscrimi-

Figure 4.1 Assay of RNA polymerase by incorporation of radioactive nucleotides into RNA in a reaction containing buffer, NaCl, $MgSO_4$, and triphosphates. After transcription, radioactive RNA and radioactive nucleotides remain. These are separated by addition of trichloroacetic acid and filtration. The filter paper then contains the RNA whose radioactivity is quantitated in a Geiger counter or scintillation counter.

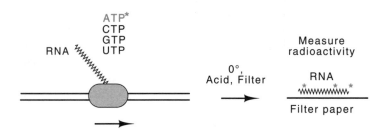

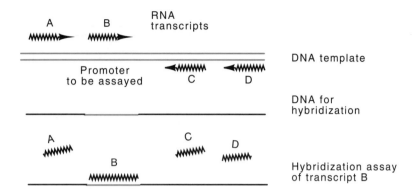

Figure 4.2 Assaying for specific transcription by using a template and DNA for hybridization whose only sequences in common are the region whose transcript is to be assayed.

nate. Only the total amount of RNA synthesized in the reaction tube is quantitated. Since many of the convenient DNA templates contain more than one promoter and since *in vitro* transcription frequently initiates from random locations on the DNA in addition to initiating from the promoters, a higher-resolution assay of transcription is required to study specific promoters and the proteins that control their activities.

Several basic methods are used to quantitate the activities of specific promoters. One is to use DNA that contains several promoters and specifically fish out and quantitate just the RNA of interest by RNA-DNA hybridization (Fig. 4.2). Run-off transcription is another method that is often used for examining the activity of promoters. This permits the simultaneous assay of several promoters as well as nonspecific transcription. Small pieces of DNA 200 to 2,000 base pairs long and contain-

Figure 4.3 Assaying specific transcription by electrophoresis on polyacrylamide gels. The radioactive RNAs of different sizes are synthesized using radioactive nucleoside triphosphates. Then the DNAs are separated according to size by electrophoresis and their positions in the gel found by autoradiography.

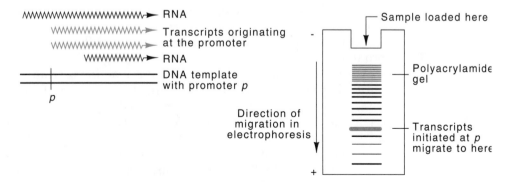

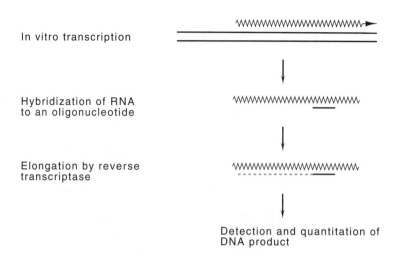

In vitro transcription

Hybridization of RNA
to an oligonucleotide

Elongation by reverse
transcriptase

Detection and quantitation of
DNA product

Figure 4.4 Assay for the presence of an RNA sequence by elongation with reverse transcriptase.

ing the promoter in question can be isolated. Transcription initiating from the promoter on these templates begins at the promoter and usually extends to the end of the DNA. This produces a small transcript of a unique size. Transcription initiating from other sites on the DNA generates other sizes of RNA transcripts. The resultant RNA molecules, whose sizes vary from 10 to 1,000 nucleotides, may easily be separated from one another by electrophoresis in polyacrylamide gels in the presence of high concentrations of urea (Fig. 4.3). These denaturing agents reduce the formation of transient hairpins in the RNA resulting from partial complementarity between portions of the molecules. Therefore the RNA polymers migrate at velocities dependent on their length and independent of their sequences. If α-$^{32}PO_4$ triphosphates have been used during the transcription, the locations of RNA molecules in the gel can subsequently be determined by autoradiography.

The third basic approach for examining the activity of a particular promoter became possible through improvements in recombinant DNA technology. Small DNA fragments several hundred nucleotides long can be isolated and used as templates in transcription assays. Quantitation of the total RNA synthesized in these reactions permits assay of a single promoter since the small DNA templates usually contain only one promoter.

Assaying transcription from an *in vitro* reaction is relatively easy if purified proteins can be used. Obtaining the purified proteins is not so easy, however, since often their very purification requires assaying crude extracts for the transcriptional activity. The presence of nucleotides in the extracts precludes radioactive labeling of the RNA synthesized *in vitro*. Therefore an assay is required that can utilize nonradioactive RNA. Primer extension meets the requirements for such an assay (Fig. 4.4). RNA that was initiated *in vitro* and which transcribed off the end of a DNA template is purified away from proteins and some of the DNA. This RNA is hybridized to a short DNA fragment. Then reverse transcriptase, an enzyme utilized by RNA viruses to make DNA copies of themselves, and radioactive deoxynucleotide triphosphates are added. The reverse transcriptase synthesizes a radioactive DNA copy of the RNA. This can be separated from extraneous and nonspecific DNA fragments by electrophoresis and be quantitated by autoradiography.

Concentration of Free RNA Polymerase in Cells

It is necessary to know the concentration of free intracellular RNA polymerase to design meaningful *in vitro* transcription experiments. One method of determining the concentration utilizes the fact that the β and β' subunits of the *E. coli* RNA polymerase are larger than most other polypeptides in the cell. This permits them to be easily separated from other cellular proteins by SDS polyacrylamide gel electrophoresis. Consequently, after such electrophoresis, the amount of protein in the β and β' bands is compared to the total amount of protein on the gel. The results are that a bacterial cell contains about 3,000 molecules of RNA polymerase. A calculation using the cell doubling time and amounts of messenger RNA, tRNA, and ribosomal RNA in a cell leads to the conclusion that about 1,500 RNA molecules are being synthesized at any instant. Hence half the cell's RNA polymerase molecules are synthesizing RNA. Of the other 1,500 RNA polymerase molecules, fewer than 300 are free of DNA and are able to diffuse through the cytoplasm. The remainder are temporarily bound to DNA at nonpromoter sites.

How do we know these numbers? On first consideration, a direct physical measurement showing that 300 RNA polymerase molecules are free in the cytoplasm seems impossible. The existence, however, of a special cell division mutant of *E. coli* makes this measurement straightforward (Fig. 4.5). About once per normal division, these mutant cells divide near the end of the cell and produce a minicell that lacks DNA. This cell contains a sample of the cytoplasm present in normal cells. Hence, to determine the concentration of RNA polymerase free of DNA in cells, it is necessary only to determine the concentration of ββ' in the DNA-less minicells. Such measurements show that the ratio of ββ' to total protein has a value in minicells of one-sixth the value found in

Figure 4.5 The generation of a minicell lacking a chromosome.

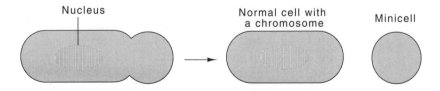

Nucleus

Normal cell with a chromosome

Minicell

whole cells. Thus, at any instant, less than 20% of the RNA polymerase in a cell is free within the cytoplasm.

The RNA Polymerase in *Escherichia coli*

Once biochemists could assay and purify an RNA polymerase from *E. coli*, it was important to know the biological role of the enzyme. For example, the bacterial cell might possess three different kinds of RNA polymerase: one for the synthesis of messenger RNA, one for the synthesis of tRNA, and one for the synthesis of ribosomal RNA. If that were the case, much effort could have been wasted in studying *in vitro* transcription from a gene if the wrong RNA polymerase had been used. Enzymologists' failures to find more than one type of RNA polymerase in *E. coli* were no proof that more did not exist, for, as we have seen in Chapter 3, detection of an enzyme in cells can be difficult. In other words, what can be done to determine the biological role of the enzyme that can be detected and purified?

Fortunately, a way of determining the role of the *E. coli* RNA polymerase finally appeared. It was in the form of a very useful antibiotic, rifamycin, which blocks bacterial cell growth by inhibiting transcription initiation by RNA polymerase. If many cells are spread on agar medium containing rifamycin, most do not grow. A few do, and these rifamycin-resistant mutants grow into colonies. Such mutants exist in populations of sensitive cells at a frequency of about 10^{-7}. Examination of the resistant mutants shows them to be of two classes. Mutants of the first class are resistant because their cell membrane is less permeable to rifamycin than the membrane in wild-type cells. These are of no interest to us here. Mutants of the second class are resistant by virtue of an alteration in the RNA polymerase. This can be demonstrated by the fact that the RNA polymerase purified from such rifamycin-resistant cells has become resistant to rifamycin.

Since rifamycin-resistant cells now contain a rifamycin-resistant polymerase, it would seem that this polymerase must be the only type present in cells. Such need not be the case, however. Consider first the hypothetical possibility that cells contain two types of RNA polymerase, one that is naturally sensitive to rifamycin, and one that is naturally resistant. We might be purifying and studying the first enzyme when we should be studying the naturally resistant polymerase. This possibility of this situation can be excluded by showing that rifamycin addition to cells stops all RNA synthesis. Therefore cells cannot contain a polymerase that is naturally resistant to rifamycin. A second possibility is that cells contain two types of polymerase and both are sensitive to rifamycin. Because mutants resistant to rifamycin can be isolated, both types of polymerase would then have to be mutated to rifamycin resistance. Such an event is exceedingly unlikely, however. The probability of mutating both polymerases is the product of the probability of mutating either one. From other studies we know that the mutation frequency for such an alteration in an enzyme is on the other of 10^{-7}. Therefore, the probability of mutating two polymerases to rifamycin resistance would

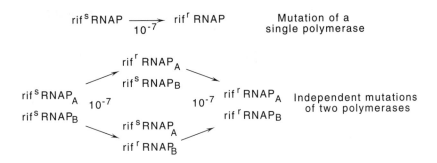

Figure 4.6 Mutation of a single RNA polymerase to rifamycin resistance occurs at a frequency of 10^{-7}, whereas if two independent mutational events are required to mutate two different polymerases to rifamycin resistance, the frequency is $10^{-7} \times 10^{-7} = 10^{-14}$.

be about $(10^{-7})^2$ (Fig. 4.6), which is far below the frequency of 10^{-7} that is actually observed.

Thus far, then, we know these facts: The target of rifamycin is a single type of RNA polymerase in bacteria. This RNA polymerase synthesizes at least one essential class of RNA, and this polymerase is the one that biochemists purify. How do we know that this RNA polymerase synthesizes all the RNAs? Careful physiological experiments show that rifamycin addition stops synthesis of all classes of RNA, mRNA, tRNA, and rRNA. Therefore the same RNA polymerase molecule must be used for the synthesis of these three kinds of RNA, and this RNA polymerase must be the one that the biochemists purify.

Unfortunately there is an imperfection in the reasoning leading to the conclusion that *E. coli* cells contain only a single type of RNA polymerase molecule. That imperfection came to light with the discovery that the prokaryotic RNA polymerase is not a single polypeptide but in fact contains four different polypeptide chains. Therefore, the rifamycin experiment proves that the same polypeptide is used by whatever polymerases synthesize the different classes of RNA. Much more arduous biochemical reconstruction experiments have been required to exclude the possibility that bacteria contain more than one single basic core RNA polymerase.

Three RNA Polymerases in Eukaryotic Cells

Investigation of eukaryotic cells shows that they do contain more than one type of RNA polymerase. The typical protein fractionation schemes used by biochemists to purify proteins yield three distinct types of RNA polymerase from a variety of higher cells. These three species are called RNA polymerases I, II, and III for the order in which they elute from an ion exchange column during their purification.

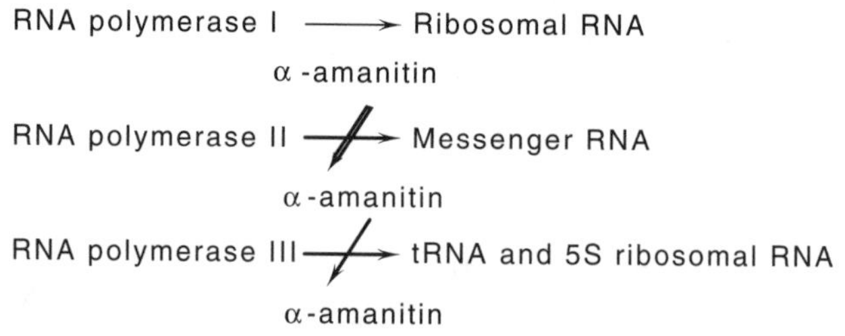

RNA polymerase I   ⟶   Ribosomal RNA

α-amanitin

RNA polymerase II —⫽→ Messenger RNA

α-amanitin

RNA polymerase III—⫽→ tRNA and 5S ribosomal RNA

α-amanitin

Figure 4.7 Three eukaryotic RNA polymerases, their sensitivities to α-amanitin, and the products they synthesize.

RNA polymerase I synthesizes ribosomal RNA. It is found in the nucleolus, an organelle in which ribosomal RNA is synthesized. Further confirmation of this conclusion is the finding that only purified RNA polymerase I is capable of correctly initiating transcription of ribosomal RNA *in vitro*. A simple experiment demonstrating this is to use DNA containing ribosomal RNA genes as a template and then to assay the strand-specificity of the resulting product. RNA polymerase I synthesizes RNA predominantly from the correct strand of DNA, whereas RNA polymerases II and III do not.

RNA polymerase II is the polymerase responsible for most synthesis of messenger RNA. *In vitro* experiments show that, of the three, this polymerase is the most sensitive to a toxin from mushrooms, α-amanitin (Fig. 4.7). The addition of low concentrations of α-amanitin to cells or to isolated nuclei blocks additional synthesis of just messenger RNA. Further, α-amanitin-resistant cells possess a toxin-resistant RNA polymerase II.

RNA polymerase III is less sensitive to α-amanitin than is RNA polymerase II, but it is sufficiently sensitive that its *in vivo* function can be probed. The sensitivity profile for synthesis of tRNA and 5S RNA parallels the sensitivity of RNA polymerase III. *In vitro* transcription experiments with purified polymerase III also show that this enzyme synthesizes tRNA and 5S ribosomal RNA as well as some of the RNAs found in spliceosomes as discussed in the next chapter.

Multiple but Related Subunits in Polymerases

How can one be sure that an enzyme contains multiple subunits? One of the best methods for detection of multiple species of polypeptides in a sample is electrophoresis through a polyacrylamide gel. If the protein has been denatured by boiling in the presence of the detergent sodium dodecyl sulfate, SDS, and the electrophoresis is performed in the presence of SDS, polypeptides separate according to size. This results from the fact that the charged SDS anions that bind to the polypeptides completely dominate the charge as well as force all polypeptides to adopt a rodlike shape whose length is proportional to the molecular

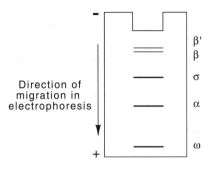

Figure 4.8 The polypeptide band pattern found by SDS polyacrylamide gel electrophoresis of purified *E. coli* RNA polymerase.

weight of the protein. Therefore, in most cases two polypeptides of the same size will migrate at the same rate and two of different molecular weight will migrate at different rates. Following electrophoresis, the positions of proteins in the gel can be visualized by staining. Each band on a gel derives from a different-sized polypeptide species.

When purified *E. coli* RNA polymerase is subjected to SDS polyacrylamide gel electrophoresis, five distinct bands are seen (Fig. 4.8). The mere presence of multiple polypeptides in purified enzyme doesn't prove that all the peptides are necessary for activity. Do all the bands on the gel represent subunits of RNA polymerase or are some of the bands extraneous proteins that adventitiously copurify with RNA polymerase? A reconstitution experiment provides the most straightforward demonstration that the four largest polypeptides found in RNA polymerase are all essential subunits of the enzyme. The four bands from an SDS polyacrylamide gel are cut out, the proteins eluted, and SDS removed. RNA polymerase activity can be regained only if all four of the proteins are included in the reconstitution mixture.

RNA polymerase from *E. coli* consists of subunits β' and β of molecular weights 155,000 and 151,000, two subunits of α whose molecular weight is 36,000, a low molecular weight subunit ω whose presence is not necessary for activity, and one somewhat less tightly-bound subunit, σ of 70,000 molecular weight. Measurement of the amounts of each of the five proteins on SDS polyacrylamide gels shows that the enzyme contains two copies of the α subunit for every single copy of the others, that is, the subunit structure of RNA polymerase is $\sigma\alpha_2\beta\beta'\omega$.

The reconstitution experiments permit pinpointing the actual target of rifamycin. RNA polymerase from rifamycin-sensitive and rifamycin-resistant cells is subjected to SDS polyacrylamide gel electrophoresis. Then reconstitution experiments with the two sets of proteins can be performed in all possible combinations to determine which of the four subunits from the rifamycin-resistant polymerase confers resistance to the reconstituted enzyme. The β subunit was found to be the target of rifamycin.

If we view the RNA polymerase as a biochemical engine, then it is reasonable to expect each subunit to have a different function. As discussed below, rifamycin inhibits initiation by RNA polymerase, but

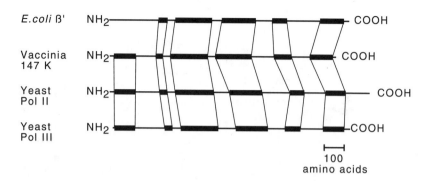

Figure 4.9 The dark bands indicate the similarities among the β' subunits of RNA polymerase from *E. coli*, vaccinia virus, and yeast polymerases II, and III.

it has no effect on the steps of elongation of the polynucleotide chain. A different antibiotic, streptolydigin, has also been found to inhibit RNA polymerase. This blocks elongation steps, and therefore we might have expected to find a subunit other than β to be the target of this drug. Alas, however, the β subunit is also the target of streptolydigin. Some specialization exists. The β subunit binds ribonucleotides and possesses the catalytic site while the β' subunit binds DNA. Most likely the larger two subunits are comprised of a number of domains, each playing a different role in the initiation and elongation of RNA. Evolution seems to have conserved the structures and functions of some of these different domains. The larger subunits from prokaryotic and the three types of eukaryotic RNA polymerase all share significant homology. Regions of homology are also found amongst the other subunits as well.

The combined molecular weights of the subunits of RNA polymerase total nearly one half million, but from a mechanistic viewpoint it is not at all clear why the polymerase should be so large. Phage T7, which grows in *E. coli*, encodes its own RNA polymerase, and this enzyme has a molecular weight of only about 100,000. Apparently the actual RNA initiation and elongation steps do not require an enzyme as large as the *E. coli* polymerase. Perhaps the large size of the cellular polymerases permits them to initiate from a wider variety of promoters and to interact with a variety of auxiliary regulatory proteins.

The eukaryotic RNA polymerases are also large and possess multiple subunits. RNA polymerase II from many different organisms has been shown to contain 12 different polypeptides. The largest three are homologous to β', β, and α of the *E. coli* RNA polymerase. Fig. 4.9 shows the shared homology of the β' subunit among the *E. coli*, vaccinia virus, and *Saccharomyces cerevisiae* polymerases II and III. The RNA polymerases I and III possess five subunits in common with RNA polymerase II.

The eukaryotic polymerases contain more subunits than the *E. coli* RNA polymerases. Part of the differences may merely be in the tightness with which subunits cling together. For example, a protein from *E. coli* that is involved with RNA chain termination, the *nusA* gene product, could be considered a part of RNA polymerase because it binds to the polymerase after initiation has occurred and after the σ subunit has been released from the core complex of β, β', and α. Since however, it does not copurify with the RNA chain elongating activity that is contained in the core, it usually is not classified as part of RNA polymerase. Some of the peptides in the eukaryotic polymerase might just stick together more tightly.

One notable difference between the prokaryotic and eukaryotic polymerases is that the largest subunit of RNA polymerase II possesses at its C-terminal end a heptad of (Tyr-Ser-Pro-Thr-Ser-Pro-Ser) repeated 25 to 50 times. This C-terminal domain, or CTD, can be multiply phosphorylated by any of several proteins known to activate transcription on various promoters. It appears that the unphosphorylated form of the CTD helps RNA polymerase bind and interact with the auxiliary

proteins necessary for transcription initiation, but that the CTD must be phosphorylated before it can release from the initiation proteins and allow the polymerase to elongate freely. Cells constructed so as to lack the CTD, or cells in which the CTD is too short are sick or inviable.

Multiple Sigma Subunits

In vitro transcription experiments with the *E. coli* RNA polymerase have shown that the σ subunit is required for initiation at promoters, but that σ is not required for elongation activity. In fact, the σ subunit comes off the polymerase when the transcript is between 2 and 10 nucleotides long. Polymerase lacking the σ subunit, core polymerase, binds randomly to DNA and initiates nonspecifically or from nicks, but it rarely initiates from promoters. These results raise an interesting question: If the σ subunit is required for promoter recognition, then could different σ subunits be used to specify transcription from different classes of genes? The answer is yes.

Although many years of searching were required, sigma subunits specific for more than five different specific classes of genes have been found in *E. coli*. Most transcription is initiated by the σ^{70} subunit. Not only in *E. coli*, but in all types of cells, heat shock induces the synthesis of about 40 proteins that aid in surviving the stressful conditions. One of the proteins induced in *E. coli* by heat shock is a σ factor that recognizes promoters located in front of other heat shock responsive genes. Other σ factors are used for transcription of nitrogen regulated genes, the flagellar and chemotaxis genes, and genes induced under oxidative stress.

Developmentally regulated genes are another area in which the specificity provided by σ factors proves useful. *Bacillus subtilis* forms spores under some conditions. Synthesizing a spore requires conversion of one cell into two drastically different cells. One ultimately becomes the spore, and the other completely surrounds the maturing spore and synthesizes the protective cell wall of the spore. Cascades of different sigma factors turn on the appropriate genes in these two cell-types.

Eukaryotic polymerases appear to contain an analog of the σ subunit. This protein is required for correct initiation, reduces nonspecific initiation, and competes for binding to the polymerase with the bacterial σ subunit.

The Structure of Promoters

Some proteins in the cell are needed in high quantities and their messenger RNAs are synthesized at a high rate from active promoters. Other proteins are required in low levels and the promoters for their RNAs have low activity. Additionally, the synthesis of many proteins must respond to changing and unpredictable conditions inside or outside the cell. For this last class, auxiliary proteins sense the various conditions and appropriately modulate the activities of the promoter. Such promoters must not possess significant activity without the assistance of an auxiliary protein. Not surprisingly, then, the sequences of promoters show wide variations. Yet behind it all, there could be elements of a basic structure or structures contained in all promoters.

As the first few bacterial promoters were sequenced, considerable variation was observed between their sequences, and their similarities could not be distinguished. When the number of sequenced promoters reached about six, however, Pribnow noticed that all contained at least part of the sequence TATAAT about six bases before the start of messengers, that is, 5'**XXXTATAATXXXXXAXXXX**-3', with the messenger often beginning with *A*. This sequence is often called the Pribnow box. Most bacterial promoters also possess elements of a second region of conserved sequence, TTGACA, which lies about 35 base pairs before the start of transcription. Examination of the collection of *E. coli* σ^{70} promoters reveals not only the bases that tend to be conserved at the -35 and -10 regions, but also three less well conserved bases near the transcription start point (Fig. 4.10). The figure also shows that the spacing between the -35 and -10 elements is not rigidly retained, although the predominant spacing is 17 base pairs. To a first approximation, the more closely a promoter matches the -10 and -35 sequences and possesses the correct spacing between these elements, the more active the promoter. Those promoters that utilize regulatory proteins to activate transcription deviate from the consensus sequences in these regions. These activating proteins must help the polymerase through steps of the initiation process that it can do by itself on "good" promoters. The promoters that are activated by other σ factors possess other

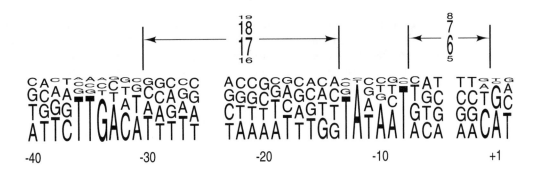

-40   -30   -20   -10   +1

Figure 4.10 Consensus of all *E. coli* promoters in which the height of a base at each position is proportional to the frequency of occurrence of that base. Similarly, the heights of the numbers represent the frequency of the indicated spacings between the elements.

sequences in both the -35 and -10 regions. This implies that the σ subunit contacts both regions of the DNA. Direct experiments have shown this is the case. Perhaps the most elegant was the generation of a hybrid promoter with a -35 region specific for one type of σ and a -10 region specific for a different σ. The hybrid promoter could be activated only by a hybrid σ subunit containing the appropriate regions from the two normal σ subunits.

Protection of promoters from DNAse digestion, chemical modification, and electron microscopy have shown that polymerase indeed is bound to the -35 and -10 regions of DNA before it begins transcription and that it contacts the DNA in these areas. Experiments of the type described in Chapters 9 and 10 have permitted direct determination of

Figure 4.11 The consensus sequences found in *E. coli* promoters and the helical structure of the corresponding DNA. The majority of the contacts between RNA polymerase and DNA are on one side of the DNA.

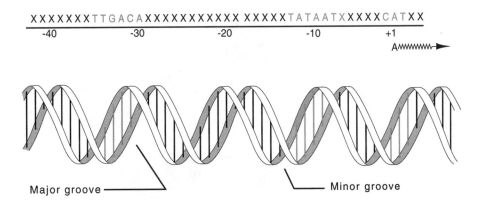

the bases and phosphates contacted in the promoter region that *E. coli* RNA polymerase contacts. These areas are clustered in the regions of the conserved bases in the -35, -10, and +1 areas (Fig. 4.11). These regions all may be contacted by RNA polymerase binding to one face of the DNA. More complicated experiments have attempted to answer the reverse question of which polymerase subunits contact these bases.

Study of eukaryotic promoters reveals relatively few conserved or consensus sequences. Most eukaryotic promoters possess a TATA sequence located about 30 base pairs before the transcription start site. In yeast, however, this sequence is found up to 120 base pairs ahead of

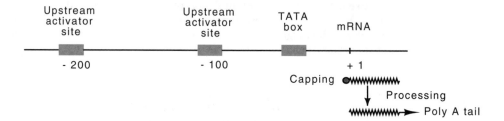

the transcription start point, but its presence does not affect strongly the overall activity of the promoter.

The eukaryotic minimal RNA polymerase II promoter consists of the TATA box and additional nucleotides around the transcription start

Figure 4.12 The binding order and approximate location of the transcription factors TFIIA, B, D, and E, and RNA pol II.

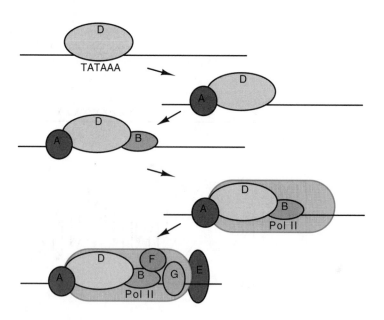

point. The apparatus required for initiation is the polymerase plus at least six additional proteins or protein complexes (Fig. 4.12). The first to bind is TFIID. This is a complex of about ten different proteins. The one that binds the TATA sequence is called TATA-binding protein, and the others are called TATA associated factors or TATA associated proteins. Although TFIID was first recognized as necessary for initiating transcription by RNA polymerase II, at least the TATA-binding protein of TFIID is now known to be required for initiation by all three types of RNA polymerase, I, II, and III. At the promoters served by RNA polymerase II, after TFIID binds, then TFIIA and B bind followed by RNA polymerase II. After this TFIIE, F, G and still others bind. The order of protein binding *in vitro* and the approximate location of binding of the proteins was assayed using DNAse footprinting and the migration retardation assay. In this assay, a piece of DNA about 200 base pairs long is incubated with various proteins and then subjected to electrophoresis under conditions that increase the tightness of proteins binding to the DNA. The DNA migrates at one rate, and the protein-DNA complex migrates more slowly through the gel, thereby permitting detection and quantitation of protein binding to DNA.

Enhancers

The proteins mentioned in the previous section are termed basal factors. They are required on all promoters and constitute a core of the transcription activity. Additionally, eukaryotic promoters often also possess one or more 8-to-30 base-pair elements located 100 to 10,000 base pairs upstream or downstream from the transcription start point. Similar elements are found with prokaryotic promoters, but less frequently. These elements enhance the promoter activity by five to a thousand-fold. They were first found in animal viruses, but since then have been found to be associated with nearly all eukaryotic promoters. The term, enhancer, is shifting slightly in meaning. Originally it meant a sequence possessing such enhancing properties. As these enhancers were dissected, frequently they were found to possess binding sites for not just one, but sometimes up to five different proteins. Each of the proteins can possess enhancing activity, and now the term enhancer can mean just the binding site for a single enhancer protein.

Remarkably, enhancer elements still function when their distances to the promoter are altered, and frequently they retain activity when the enhancers are turned around, or even when they are placed downstream from the promoter. Proteins bound to the enhancer sequences must communicate with the RNA polymerase or other proteins at the promoter. There are two ways this can be done, either by sending signals along the DNA between the two sites as was first suggested, or by looping the DNA to permit the two proteins to interact directly (Fig. 4.13). Most of the existing data favor looping as the method of communication between most enhancers and their associated promoter.

A second remarkable property of enhancers is their interchangeability. Enhancers frequently confer the appropriate regulation properties

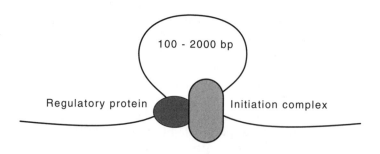

Figure 4.13 How DNA looping permits a protein bound to DNA to contact an initiation complex bound hundreds of nucleotides away.

on multiple promoters. That is, an enhancer-binding protein senses the cellular conditions and then stimulates appropriately whatever promoter is nearby. Enhancers confer specific responses that are sensitive to tissue-type, developmental stage, and environmental conditions. For example, when an enhancer that provides for steroid-specific response of a gene is placed in front of another gene and its promoter, the second gene acquires a steroid-specific response. That is, enhancers are general modulators of promoter activity, and in most cases a specific enhancer need not be connected to a specific promoter. Most enhancers are able to function in association with almost any promoter. Not surprisingly, a promoter that must function in many tissues, for example, the promoter of a virus that grows in many tissues, has many different enhancers associated with it.

Enhancer-Binding Proteins

Some enhancer-binding proteins like the glucocorticoid receptor protein have been purified and studied *in vitro*. Other enhancer-binding proteins, like the GAL4 and GCN4 proteins from yeast, can be engineered and studied *in vivo* without ever purifying the protein. Both types of studies indicate that enhancer proteins possess multiple independent domains. The DNA-binding domain of the GAL4 enhancer protein can be replaced with the DNA-binding domain of a bacterial repressor protein, LexA (Fig. 4.14). When the LexA-binding sequence is placed in front of some other yeast gene, the gene acquires the ability to be induced by the GAL4-LexA hybrid protein when galactose is present. The glucocorticoid receptor protein possesses DNA-binding, steroid-binding, and activation domains. These too can be separated and interchanged.

The regions of enhancer proteins necessary for activation can be explored. By progressively deleting protein from an activating domain, both Ptashne and Struhl have found that stretches of negatively charged amino acids on some activator proteins are required for activation. In

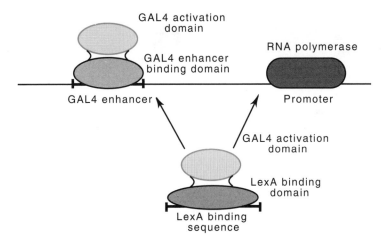

Figure 4.14 Hybrid enhancer proteins can activate transcription if they are held in the right areas of DNA.

GCN4, two such regions are required for full activation abilities. It appears necessary that these negatively charged amino acids all lie on one face of an alpha helix. The other side of the helix can be hydrophobic. Activating helices can be designed *de novo* if they follow these principles, but if the charged amino acids are shuffled, they do not activate. Other structures in addition to negatively charged surfaces of α-helices also function to activate RNA polymerase. Some enhancer-binding proteins lack significant negatively charged regions and instead possess large quantities of proline or glutamine.

Many enhancer proteins may be relatively simple. They can possess nearly independent domains for DNA-binding, for binding a small molecule like a hormone, and for activating RNA polymerase or the basal machinery. The binding of a hormone may unmask the DNA-binding domain or the activation domain of the protein. In some cases the activating domain may be little more than a high concentration of negative charge whose interaction with TFIID activates transcription.

Activation in eukaryotes has to contend with the presence of histones tightly-bound to the DNA. Undoubtedly, their presence interferes with transcription. Some activator proteins therefore overcome the repressive effects of bound histones. Other activator proteins can be expected to go beyond overcoming repression, and will stimulate transcription.

The number of different enhancer-binding proteins appears to be remarkably small in nature. Repeatedly, researchers are finding that some of the enhancer proteins from one gene are highly similar to one that controls another gene, either in that same organism or in a different organism. Not only are the sequences of such proteins similar, but they are functionally interchangeable. Heterologous *in vitro* transcription systems can be constructed in which enhancer proteins from yeast activate transcription from a human system. The yeast GCN4 protein is similar to the AP-1, c-myc, c-jun, and c-fos proteins. AP-1 is a mammal-

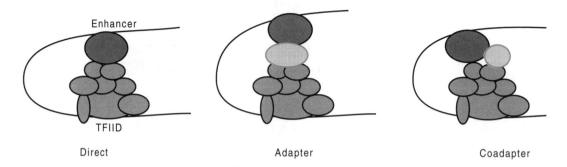

Figure 4.15 Enhancer proteins can interact with the TFIID complex directly or through other proteins, shown in red.

ian protein that binds upstream from promoters, and c-myc, c-jun, and c-fos are proteins that sometimes mutate and induce oncogenic growth of cells.

Many of the enhancer-binding proteins interact with the TFIID complex. Because gene regulation is so important to the cell and because many different genes in a cell must be regulated, we can expect a wide diversity of interaction modes (Fig. 4.15). The enhancer-binding proteins can interact directly, via adapters, or in cooperation with coadapters. Additionally, proteins that are part of the TFIID complex or the other proteins may be required for some interactions and interfere with others. That is, a protein may play an activating role in the expression of some genes and a repressing role in the expression of others.

DNA Looping in Regulating Promoter Activities

DNA looping is a reasonable way enhancers can interact with the transcription apparatus. The available data say that this, indeed, is one of the ways they work. For example, an enhancer can be placed on one DNA circle, and the promoter it stimulates on another DNA circle. When the DNA rings are linked, the enhancer functions. This shows that the enhancer must be close to the promoter in three-dimensional space. The linking experiment also shows that a protein or signal doesn't move down the DNA from the enhancer to the promoter.

DNA looping solves two physical problems in gene regulation. The first concerns space. Regulatory proteins must do two things. They sense intracellular conditions, for example, the presence of a growth hormone. They then must turn on or turn off the expression of only those genes appropriate to the conditions. These responses require that a signal be transmitted from a sensor part of the regulatory protein to the cellular apparatus responsible for transcribing or initiating transcription from the correct gene. The phrase "correct gene" is the key here. How can the regulatory protein confine its activity to the correct gene?

RNA polymerase

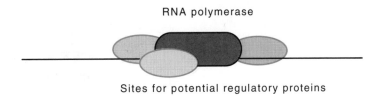

Sites for potential regulatory proteins

Figure 4.16 The limited number of sites immediately adjacent to an RNA polymerase molecule.

The easiest, and virtually the only general way for a regulatory protein to recognize the correct gene is for the protein to recognize and bind to a DNA sequence near or within the correct gene.

If a regulatory protein is bound adjacent to an RNA polymerase molecule or adjacent to an auxiliary protein required for initiating transcription, we can imagine direct protein-protein contacts for communication of the necessary signals. The space problem arises since only a limited number of proteins may bind immediately adjacent to the transcription initiation complex (Fig. 4.16). The limit seems to be two to four proteins. Since the regulatory pattern of many genes is complex and likely to require the combined influence of more than two or three regulatory proteins, we have a problem.

How can more than a couple of proteins directly influence the RNA polymerase? DNA looping is one answer. A regulatory protein can be bound within several hundred or several thousand base pairs of the initiation complex and directly touch the complex by looping the DNA. With DNA looping a sizeable number of proteins can simultaneously affect transcription initiation via multiple loops. Additional possibilities exist. For example, proteins could regulate by helping or hindering loop formation or alternative looping could exist in the regulation scheme of a gene.

A second reason for DNA looping is the cooperativity generated by a system that loops. Consider a system in which a protein can bind to two DNA sites separated by several hundred base pairs and then the proteins can bind each other, thus forming a DNA loop. A different reaction pathway can also be followed. A molecule of the protein could bind to one of the sites and a second molecule of the protein could bind to the first. By virtue of the potential for looping, the concentration of the second protein in the vicinity of the second DNA site has been increased.

Such a concentration change increases the occupancy at the second site above the value it would have in the absence of looping. Hence, the presence of one site and looping increases occupancy of the second site. Such a cooperativity can substantially facilitate binding at low concentrations of regulatory proteins. It also eliminates any time lags upon gene induction associated with diffusion of a protein to its DNA-binding site.

Increasing the local concentration of a regulatory protein near its binding site solves a serious problem for cells. Thousands of regulatory

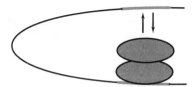

proteins must be present in a bacterial cell and tens of thousands of regulatory proteins may have to be present in the nucleus of some eukaryotic cells. Since the total protein concentration possible in the cell or nucleus is limited to about 200 mg/ml, and since the same space must be shared with the chromosome and housekeeping proteins as well, the concentration possible for any one type of regulatory protein is strictly limited.

How then can the requisite binding of the regulatory protein to its DNA target sequence be achieved? Basically, the effective concentration of the protein must be high relative to the dissociation constant from the site. If the affinity is too high, however, the tightness of binding may make the dissociation rate from the site so slow that the protein's presence interferes with normal cellular activities like DNA replication, recombination, and repair. A nice solution to these contradictory requirements is to build a system in which the affinity of the protein for the site is not too high, the overall concentration of the protein in the cell is not too high, but the local concentration of the protein just in the vicinity of the binding sites is high. DNA looping provides a simple mechanism for increasing the local concentrations of regulatory proteins.

Steps of the Initiation Process

The beginning state for initiation by RNA polymerase is a polymerase molecule and a promoter free in solution, and the end state is a polymerase molecule bound to DNA elongating an RNA chain. In this state the DNA is partially melted so that base-pairing of ribonucleotides to the template strand of the DNA can be used to determine the nucleotides to be incorporated into the RNA. The initiation process that separates the two states must be continuous, but it may well be approximated as consisting of discrete steps. Can any of these be detected and quantitated and, if so, can measurement of the rates of proceeding from one such state to the next during the initiation process provide useful information? Ultimately we would hope that studies like these could explain the differences in activities between promoters of different sequence as well as provide the information necessary to design promoters with specific desired activities or properties.

The first biochemical characterizations of the binding and initiation rates of RNA polymerase on promoters were performed on highly active promoters of bacterial origin. Such a choice for a promoter is natural because it maximizes the signal-to-noise ratio in the data. The data

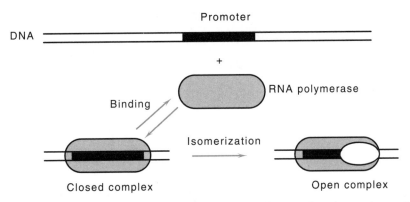

Figure 4.17 Binding of an RNA polymerase to form a closed complex and the near irreversible formation of an open complex.

obtained by Chamberlin with these promoters indicated that the initiation process could be dissected into two steps: a rapid step during which polymerase binds to DNA and a slower "isomerization" step in which RNA polymerase shifts to an active form capable of immediately initiating transcription (Fig. 4.17). Later studies on a wide variety of promoters indicates that this approximation is generally useful. Some promoters possess additional discernible steps in the initiation process.

It is natural to expect that the first step would be the point for regulation of transcription. Since cells need to have promoters of widely differing activity as well as promoters whose activities can be controlled, it seems sensible that the binding step would vary among promoters, and if auxiliary proteins are required for initiation at a promoter, they would alter the binding rate. If the regulation of initiation by auxiliary proteins occurred after the binding step, for example in the isomerization step, then many polymerase molecules in cells would be nonproductively bound at promoters. This would be a waste of substantial numbers of polymerase molecules. Apparently we lack important information, for nature actually has it both ways. Some weak promoters have good binding of polymerase, but slow isomerization, whereas the reverse is true with others. Similarly, on some regulated promoters regulation is at the binding step, and on others the regulation occurs at the isomerization step or at a step subsequent to isomerization.

Measurement of Binding and Initiation Rates

What experiments can be done to resolve unambiguously questions about binding and initiation rates? One of the easiest measurements to perform with RNA polymerase is quantitation of the total RNA synthesized. It is tempting to try to adapt such measurements to the determination of binding and activation rates of RNA polymerase. The execution of the experiments and the interpretation of the data are difficult, however, and many misleading experiments have been done.

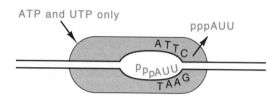

Figure 4.18 Generation of abortive initiation products by provision of only the first two nucleoside triphosphates required by a promoter.

Somewhat more direct measurements on the initiation of transcription are possible. Such measurements were first performed by McClure, who discovered that as RNA polymerase initiates *in vitro* it often goes through several cycles of abortive initiation in which a short two or three nucleotide, or sometimes longer polynucleotide is synthesized. The sequence of these abortive initiation products is the same as the 5' end of the normal RNA transcript. RNA, like DNA, is elongated in the 5'-to-3' direction. After one short polynucleotide is made, RNA polymerase does not come off the DNA, but it remains bound at the promoter, and another attempt is made at initiation. Additionally, two conditions lead to the exclusive synthesis of the short polynucleotides: the presence of rifamycin or the absence of one or more of the ribonucleoside triphosphates (Fig. 4.18). If nucleotides are omitted, those present must permit synthesis of a 5' portion of the normal transcript. Under these conditions, a polymerase molecule bound to a promoter and in the initiation state will continuously synthesize the short, abortive initiation polynucleotides.

Assay of the short polynucleotides synthesized under conditions where longer polynucleotides cannot be synthesized provides a good method of assaying initiation by RNA polymerase on many promoters. The measurement is free of complexities generated by multiple initiations at a single promoter or of the difficulties in interpretation generated by the addition of inhibitors. All that is necessary is the omission of the appropriate nucleoside triphosphates. Once RNA polymerase initiates on a promoter, it begins production of short polynucleotides, but nothing else happens at this copy of the promoter.

Let us consider the possibility that the time required for RNA polymerase molecules to locate and bind to the promoters is a significant fraction of the total time required for binding and initiation. An experiment can be performed by mixing RNA polymerase, DNA containing a promoter, and two or three of the ribonucleoside triphosphates and measuring the kinetics of appearance of the short polynucleotides that signal initiation. Suppose that the same experiment is repeated but with the RNA polymerase at twice the original concentration (Fig. 4.19). In this case the time required for RNA polymerase to find and bind to the promoter should be half as long as in the first case, but the time required for the isomerization part of the initiation process will remain the same.

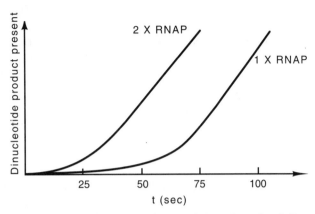

Figure 4.19 The kinetics of synthesis of dinucleotides following the mixing RNA polymerase and DNA at two polymerase concentrations differing by a factor of two.

Overall, the synthesis of the short polynucleotides should begin sooner when polymerase is provided at a higher concentration.

By varying the concentration of polymerase in a set of experiments and appropriately plotting the results, we can extrapolate to determine the results that would have been seen if polymerase had been added at infinite concentration. In such a situation, the delay of polymerase binding to promoter should be zero. The residual delay in the appearance of polynucleotides is the time required for isomerization of the polymerase and DNA to the active state. Knowing this, we can then work back and determine the rate of binding of RNA polymerase to the promoter.

Consider now the possibility that the initial binding of RNA polymerase to promoters is fast and that the conversion of the bound polymerase to the active state is slow. If only a fraction of the promoters are occupied by RNA polymerase. Increasing the concentration of RNA polymerase in this situation will also increase the rate of the initial appearance of the polynucleotides. The reason is that the initial concentration of polymerase in the bound state increases with increasing concentrations of polymerase in the reaction. The polynucleotide assay permits quantitation of the apparent binding constant and of the isomerization rate for this situation as well. Most importantly, the assay may also be used to determine how an auxiliary protein assists initiation on some promoters.

Relating Abortive Initiations to Binding and Initiating

The binding and initiation reactions explained in the previous section are described by the following equation:

$$R + P \underset{k_{-1}}{\overset{k_1}{\rightleftharpoons}} RP_c \underset{k_{-2}}{\overset{k_2}{\rightleftharpoons}} RP_o$$

where R is RNA polymerase free in solution, P is uncomplexed promoter, RP_c is promoter with RNA polymerase bound in an inactive state defined as "closed," and RP_0 is promoter with RNA polymerase bound in an active state called "open" because it can immediately begin transcription if provided nucleotides.

If RNA polymerase and DNA containing a promoter are mixed together, then the concentration of RP_0 at all times thereafter can be calculated in terms of the initial concentrations R and P and the four rate constants; however, the resulting solution is too complex to be of much use. A reasonably close mathematical description of the actual situation can be found by making three approximations. The first is that R be much greater than P_o. This is easily accomplished because the concentrations of R and P added are under the experimentalist's control. The second is known to enzymologists as the steady-state assumption. Frequently the rate constants describing reactions of the type written above are such that, during times of interest, the rate of change in the amount of RP_c is small, and the amount of RP_c can be considered to be in equilibrium with R, P, and RP_c. That is,

$$\frac{dRP_c}{dt} = k_1 R \times P - k_{-1} RP_c - k_2 RP_c + k_{-2} RP_o = 0.$$

The third assumption is that k_{-2} is much smaller than k_2. Experiments show this to be a very good approximation. RNA polymerase frequently takes hours or days to dissociate from a promoter. Straightforward solution of the equations then yields RP_o as a function of time in a useful form:

$$RP_o = (P_{\text{initial}})(1 - e^{-k_{obs}t})$$

$$k_{obs} = \frac{k_1 k_2 R}{k_1 R + k_{-1} + k_2}, \text{ or}$$

$$\frac{1}{k_{obs}} = \frac{1}{k_2} + \frac{k_{-1} + k_2}{k_1 k_2 R}.$$

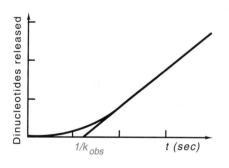

Figure 4.20 Determination of the parameter k_{obs} from the kinetics of incorporation of radioactivity into dinucleotides.

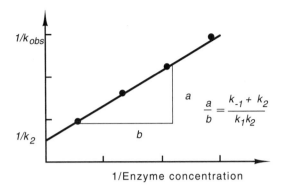

Figure 4.21 Determination of the kinetic parameters describing initiation by DNA polymerase from a series of k_{obs} values obtained at different RNA polymerase concentrations.

After starting the binding initiation assay, the total amount of oligonucleotides synthesized by any time can be measured by removing a sample from the synthesis mixture and chromatographically separating nucleoside triphosphates from the short oligonucleotides. Since the rate of oligonucleotide production is proportional to RP_0, the total amount of oligonucleotides synthesized as a function of time is given by

$$\int_0^t RP_o(t')dt' = (P_{\text{initial}})(t - 1/k_{\text{obs}} + 1/k_{\text{obs}}e^{-k_{\text{obs}}t}).$$

At very large t this increases linearly as $P_{initial}$ ($t - 1/k_{\text{obs}}$). Hence extrapolating the linear portion of the curve to the point of zero oligonucleotides gives $1/k_{obs}$ (Fig. 4.20). As seen above, $1/k_{obs}$ in the limit of high R yields $1/k_2$. At other concentrations, $(k_{-1} + k_2/k_1k_2)$ is a linear function of $1/R$. Performing the abortive initiation reaction at a variety of concentrations of R and measuring the kinetics of synthesis of oligonucleotides permits straightforward evaluation of k_2 (Fig. 4.21). Often, k_{-1} is much greater than k_2, in which case the ratio k_1/k_{-1}, called K_B, or equivalently, k_{-1}/k_1 which is called K_d, is obtained as well.

A variety of promoters have been examined by these techniques. Highly active promoters must bind polymerase well and must perform the initiation "isomerization" quickly. Less active promoters are poor in binding RNA polymerase or slow in isomerization.

Roles of Auxiliary Transcription Factors

A strong promoter must have both a high affinity for polymerase and also a high isomerization rate. The reverse is true of the weakest promoters. They have low affinity and slow isomerization rates. Medium-activity promoters are weak binders or slow at isomerization (Fig. 4.22). An auxiliary factor that stimulates RNA polymerase can change either the binding constant or the isomerization rate. The *lac* operon

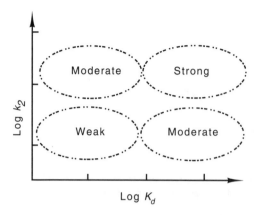

Figure 4.22 Requirements on K_d and k_2 for strong, moderate, and weak promoter activity.

promoter is stimulated by the cyclic AMP receptor protein called CAP or CRP. Characterization of the *lac* promoter with and without CAP shows that the protein primarily accelerates the isomerization step.

Melted DNA Under RNA Polymerase

One of the first steps of transcription is the binding of RNA polymerase to the proper sequence on the DNA. The best evidence at present supports the notion that RNA polymerase reads the sequence of the DNA and identifies the promoter in a double-stranded unmelted structure. It is theoretically possible that the bases of the growing RNA could be specified by double-stranded unmelted DNA, but it seems vastly easier for these bases to be specified by Watson-Crick base pairing to a partially melted DNA duplex.

Direct experimental evidence shows that during the initiation process, RNA polymerase melts at least 11 base pairs of DNA. For example, positions on adenine rings normally occupied in base pairs become available for chemical reaction if the pairs are disrupted, and their exact positions along a DNA molecule can then be determined by methods analogous to those used in DNA sequencing. Results obtained from this type of measurement reveal that 11 base pairs of DNA from about the middle of the Pribnow box to the start site of transcription are melted when the RNA polymerase binds to a promoter.

A different method has also been used to measure the amount of DNA that is melted by the binding of RNA polymerase. This method consists of binding RNA polymerase to a nicked circular DNA molecule, sealing the nick with polymerase still bound, and determining the change in the supercoiling generated by the presence of the polymerase. If we assume that the melted DNA strands are held parallel to the helix axis, this method yields 17 base pairs melted. The problem is that no method has been developed to determine whether the melted region contains any twist. If it does, then the size of the region that is melted cannot be precisely determined (Fig. 4.23).

The melting of 10 to 15 base pairs of DNA under physiological conditions requires appreciable energy because this length of oligomers

Apparent
base pairs
melted

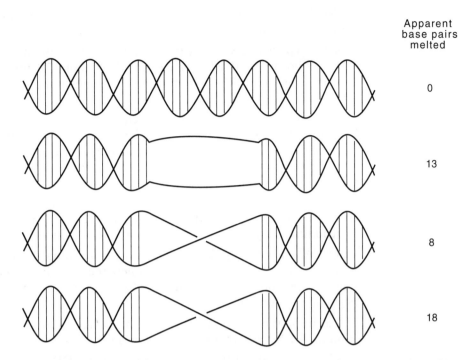

0

13

8

18

Figure 4.23 Topological measurements cannot give the exact number of base pairs opened by binding of a protein. In each of the three cases, 13 base pairs are broken, but only if the melted region is not twisted does the DNA contain one less twist.

hybridize together with a very high binding constant. The tight hybridization derives from the same source that pushes DNA to the double helical structure, base stacking interactions and hydrogen bonds. Since a large amount of energy is required to melt the DNA and a limited amount of energy is available from the binding of RNA polymerase, thermal motion in the solution provides the activation energy for the melting. Once this occurs, polymerase binds tightly to parts of the separated strands and maintains the bubble. At low temperatures, and therefore lower thermal motion, an RNA polymerase-DNA duplex is much less likely to possess the requisite activation energy, and the melting rate at lower temperatures is much reduced. At 0° virtually no RNA polymerase bound to phage T7 DNA is able to initiate in reasonable periods of time, while at 30° almost 100% has isomerized and is able to initiate within a few minutes. Similarly, the melting rate is affected by the salt concentration.

Problems

4.1. If RNA polymerase subunits β and β' constitute 0.005 by weight of the total protein in an *E. coli* cell, how many RNA polymerase molecules are there per cell, assuming each β and β' within the cell is found in a complete RNA polymerase molecule?

4.2. A typical *E. coli* cell with a doubling time of 50 minutes contains 10,000 ribosomes. If RNA is elongated at about 70 nucleotides per second, how many RNA polymerase molecules must be synthesizing ribosomal RNA at any instant?

4.3. Proteins can protect the DNA sequence to which they have bound from modification by some chemicals. What is a logical conclusion of the fact that some bases in the -35 region of a promoter are not protected by the presence of RNA polymerase, and yet their modification prior to addition of RNA polymerase prevents open complex formation?

4.4. In Summers, Nature *223*, 1111 (1969), what are the experimental data that suggest the model proposed in the paper is incorrect and instead that phage T7 synthesizes its own RNA polymerase?

4.5. Why should DNA in front of a transcribing RNA polymerase molecule be less negatively supercoiled than in the absence of the polymerase, and why should the DNA behind the polymerase be more negatively supercoiled?

4.6. In light of the size and necessary DNA contacts made by the σ subunit of RNA polymerase, comment on its probable shape.

4.7. From the definition of optical density, OD, and the fact that passage of a charged particle through a crystal of silver halide renders it capable of being developed into a silver particle, estimate the number of P^{32} decays that are necessary in an autoradiograph experiment to blacken an area equal to the size of a penny to an OD of 1.0. To solve this, it will be necessary to look up some fundamental information on photographic emulsions.

4.8. Why does urea or methyl mercury denature RNA?

4.9. When one copy of an enhancer sequence was placed in any position upstream from a reporter gene, the expression level of the reporter was one. The enhancer sequence was fully occupied by the enhancer protein. When two copies of the enhancer were placed upstream of the reporter, the expression level was not two as expected, but was five to ten. What would the value of two have meant, and what does the expression value of five to ten mean about the target of the enhancer-binding proteins?

4.10. In one orientation an enhancer could work when placed any distance from a promoter, but when inverted, it could work only if it was located at greater than a critical minimum distance from the promoter. Why?

4.11. Suppose *in vitro* transcription from a promoter located on a 400-base-pair piece of DNA requires an auxiliary protein A in addition to RNA polymerase. In experiments to determine A's mechanism of action, order of addition experiments were performed (Fig. 4.24). Propose an explanation consistent with these data and an experiment to confirm your hypothesis.

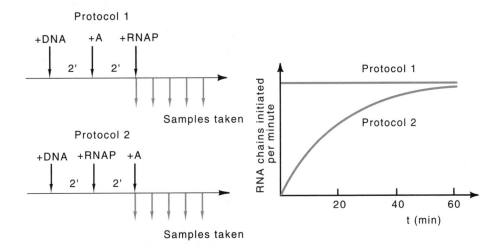

Figure 4.24 Data for Problem 4.11. Additions are made to the reaction mixtures as indicated on the horizontal time lines, and samples are taken at various times according to the two protocols. The resultant kinetics of RNA chains initiated per minute are indicated.

4.12. How might you attack the question of whether transcription of DNA in eukaryotes displaces nucleosomes from the transcribed DNA?

4.13. Suppose we have a repressor that can prevent the binding of RNA polymerase if it is bound to a site overlapping the RNA polymerase-binding site, and which can completely block the progress of an RNA polymerase if an elongating polymerase runs into it. Why would you expect the effectiveness of the repressor to be higher when its binding site overlaps the polymerase binding site than when its binding site is located at, say, +50? Assume all the relevant kinetic parameters are identical in the two cases.

References

Recommended Readings

The Molecular Topography of RNA Polymerase Promoter Interaction, R. Simpson, Cell *18*, 277-285 (1979).

Topography of Transcription: Path of the Leading End of Nascent RNA Through the *Escherichia coli* Transcription Complex, M. Hanna, C. Meares, Proc. Nat. Acad. Sci. USA *80*, 4238-4242 (1983).

Five Intermediate Complexes in Transcription Initiation by RNA Polymerase II, S. Buratowski, S. Hahn, L. Guarente, and P. Sharp, Cell *56*, 549-561 (1989).

A Transcriptional Enhancer Whose Function Imposes a Requirement that Proteins Track Along DNA, D. Hesendeen, G. Kassavetis, E. Geiduschek, Science *256*, 1298-1303 (1992).

RNA Polymerase and Core Transcription Factors

Amber Mutations of *E. coli* RNA Polymerase, S. Austin, I. Tittawella, R. S. Hayward, J. Scaife, Nature New Biology *232*, 133-136 (1971).

In vivo Distribution of RNA Polymerase Between Cytoplasm and Nucleoid in *E. coli*, W. Runzi, H. Matzura, J. Bact. *125*, 1237-1239 (1976).

Systematic Nomenclature for the RNA Polymerase Genes of Prokaryotes, R. Hayward, J. Scaife, Nature *260*, 646-647 (1976).

Extensive Homology Among the Largest Subunits of Eukaryotic and Prokaryotic RNA Polymerases, L. Allison, M. Moyle, M. Shales, C. Ingles, Cell *42*, 599-610 (1985).

A Unique Structure at the Carboxyl Terminus of the Largest Subunit of Eukaryotic RNA Polymerase II, J. Corden, D. Cadena, J. Ahearn, M. Dahmus, Proc. Natl. Acad. Sci. USA *82*, 7934-7938 (1985).

Homology between RNA Polymerases of Poxviruses, Prokaryotics, and Eukaryotes: Nucleotide Sequence and Transcriptional Analysis of Vaccinia Virus Genes Encoding 147-kDa and 22-kDa Subunits, S. Broyles, B. Moss, Proc. Natl. Acad. Sci. USA *83*, 3141-3145 (1986).

Phosphorylation of RNA Polymerase by the Murine Homologue of the Cell-cycle Control Protein CDC2, J. Cisek, J. Corden, Nature *339*, 679-684 (1989).

A Protein Kinase that Phosphorylates the C-terminal Repeat Domain of the Largest Subunit of RNA Polymerase II, J. Lee, A. Greenleaf, Proc. Natl. Acad. Sci. USA *86*, 3624-3628 (1989).

rpoZ, Encoding the Omega Subunit of *Escherichia coli* RNA Polymerase, is in the Same Operon as spoT, D. Gentry, R. Burgess, J. Bact. *171*, 1271-1277 (1989).

Altered Promoter Recognition by Mutant Forms of the σ^{70} Subunit of *Escherichia coli* RNA Polymerase, D. Siegele, J. Hu, W. Walter, C. Gross, J. Mol. Biol. *206*, 591-603 (1989).

Cloning and Structure of a Yeast Gene Encoding a General Transcription Initiation Factor TFIID that Binds to the TATA Box, M. Horikoshi, C. Wang, H. Fujii, T. Cromlish, P. Weil, R. Roeder, Nature *341*, 299-303 (1989).

RNA Polymerase II C-terminal Repeat Influences Response to Transcriptional Enhancer Signals, C. Scafe, D. Chao, J. Lopes, J. Hirsch, S. Henry, R. Young, Nature *347*, 491-494 (1990).

Structural Study of the Yeast RNA Polymerase A, P. Schultz, H. Célia, M. Riva, S. Darst, P. Colin, R. Kornberg, A. Sentenac, P. Oudet, J. Mol. Biol. *216*, 353-362 (1990).

Spatial Arrangement of σ-Factor and Core Enzyme of *Escherichia coli* RNA Polymerase, A Neutron Solution Scattering Study, H. Lederer, K. Mortensen, R. May, G. Baer, H. Crespi, D. Dersch, H. Heumann, J. Mol. Biol. *219*, 747-755 (1991).

Three-Dimensional Structure of Yeast RNA Polymerase II at 16 Å Resolution, S. Darst, A. Edwards, E. Kabalek, R. Kornberg, Cell *66*, 121-128 (1991).

DNA Repair Helicase: A Component of BTF2(TFIIH) Basic Transcription Factor, L. Schaeffer, R. Roy, S. Humbert, V. Moncollin, W. Vermenlen, J. Hoeijmakers, P. Chambon, J. Egly, Science *250*, 58-63 (1993)

Identification of a 3'-5' Exonuclease Activity Associated with Human RNA Polymerase II, D. Wang, D. Hawley, Proc. Natl. Acad. Sci. USA *90*, 843-847 (1993).

Transcript Cleavage Factors from *E. coli*, S. Borukhov, V. Sagitov, A. Goldfarb, Cell *72*, 459-466 (1993).

Promoters

Nucleotide Sequence of an RNA Polymerase Binding Site at an Early T7 Promoter, D. Pribnow, Proc. Nat. Acad. Sci. USA *72*, 784-788 (1975).

Bacteriophage T7 Early Promoters: Nucleotide Sequences of Two RNA Polymerase Binding Sites, D. Pribnow, J. Mol. Biol. *99*, 419-443 (1975).

Distinctive Nucleotide Sequences of Promoters Recognized by RNA Polymerase Containing a Phage-Coded "Sigma-like" Protein, C. Talkington, J. Pero, Proc. Nat. Acad. Sci. USA *76*, 5465-5469 (1979).

Sequence Determinants of Promoter Activity, P. Youderian, S. Bouvier, M. Susskind, Cell *30*, 843-853 (1982).

A *lac* Promoter with a Changed Distance between -10 and -35 Regions, W. Mandecki, W. Reznikoff, Nuc. Acids Res. *10*, 903-911 (1982).

Compilation and Analysis of *Escherichia coli* Promoter DNA Sequences, D. Hawley, W. McClure, Nucleic Acids Res. *11*, 2237-2255 (1983).

Each of the Three "TATA Elements" Specifies a Subset of the Transcription Initiation Sites of the CYC-1 Promoter of *Saccharomyces cerevisiae*, S. Hahn, E. Hoar, L. Guarente, Proc. Nat. Acad. Sci. USA *82*, 8562-8566 (1985).

Several Distinct "CCAAT" Box Binding Proteins Coexist in Eukaryotic Cells, M. Raymondjean, S. Cereghini, M. Yaniv, Proc. Nat. Acad. Sci. USA *85*, 757-761 (1988).

Periodic Interactions of Yeast Heat Shock Transcriptional Elements, R. Cohen, M. Messelson, Nature *332*, 856-857 (1988).

Human CCAAT-Binding Proteins Have Heterologous Subunits, L. Chodosh, A. Baldwin, R. Carthew, P. Sharp, Cell *53*, 11-24 (1988).

Defining the Consensus Sequences of *E. coli* Promoter Elements by Random Selection, Nuc. Acids Res. *16*, 7673-7683 (1988).

Synthetic Curved DNA Sequences Can Act as Transcriptional Activators in *Escherichia coli*, L. Bracco, D. Kotlarz, A. Kolb, S. Diekmann, H. Buc, EMBO Journal *8*, 4289-4296 (1989).

Weight Matrix Descriptions of Four Eukaryotic RNA Polymerase II Promoter Elements Derived from 502 Unrelated Promoter Sequences, P. Bucher, J. Mol. Biol. *212*, 563-578 (1990).

Enhancers, Regulatory Proteins, and Function

Expression of a β-Globin Gene is Enhanced by Remote SV40 DNA Sequences, J. Banerji, S. Rusconi, W. Schaffner, Cell *27*, 299-308 (1981).

A Small Segment of Polyoma Virus DNA Enhances the Expression of a Cloned β-Globin Gene Over a Distance of 1400 Base Pairs, J. de Villiers, L. Olson, J. Banerji, W. Schaffner, Nuc. Acids Res. *9*, 6251-6264 (1981).

A 12-base-pair DNA Motif that is Repeated Several Times in Metallothionein Gene Promoters Confers Metal Regulation to a Heterologous Gene, G. Stuart, P. Searle, H. Chen, R. Brinster, R. Palmiter, Proc. Nat. Acad. Sci. USA *81*, 7318-7322 (1984).

Distinctly Regulated Tandem Upstream Activation Sites Mediate Catabolite Repression in the *CYC1* Gene of *S. cerevisiae*, L. Guarente, B. Salonde, P. Gifford, E. Alani, Cell *36*, 503-511 (1984).

GCN4 Protein, Synthesized *In vitro*, Binds HIS3 Regulatory Sequences: Implications for General Control of Amino Acid Biosynthetic Genes in Yeast, I. Hope, K. Struhl, Cell *43*, 177-188 (1985).

Products of Nitrogen Regulatory Genes *ntrA* and *ntrC* of Enteric Bacteria Activate *glnA* Transcription *in vitro*: Evidence that the *ntrA* Product is a σ Factor, J. Hirschman, P. Wong, K. Keener, S. Kustu, Proc. Nat. Acad. Sci. USA *82*, 7525-7529 (1985).

Transcription of the Human β-Globin Gene Is Stimulated by an SV40 Enhancer to Which it is Physically Linked but Topologically Uncoupled, S. Plon, J. Wang, Cell *45*, 575-580 (1986).

Heat Shock Regulatory Elements Function as an Inducible Enhancer in the *Xenopus* hsp 70 Gene and When Linked to a Heterologous Promoter, M. Bienz, H. Pelham, Cell *45*, 753-760 (1986).

Transcription of *glnA* in *E. coli* is Stimulated by Activator Bound to Sites Far from the Promoter, L. Reitzer, B. Magasanik, Cell *45*, 785-792 (1986).

Functional Dissection of a Eukaryotic Transcriptional Activator Protein, GCN4 of Yeast, I. Hope, K. Struhl, Cell *46*, 885-894 (1986).

Multiple Nuclear Factors Interact with the Immunoglobulin Enhancer Sequences, R. Sen, D. Baltimore, Cell *46*, 705-716 (1986).

A Cellular DNA-binding Protein that Activates Eukaryotic Transcription and DNA Replication, K. Jones, J. Kadonaga, P. Rosenfeld, T. Kelly, R. Tjian, Cell *48*, 79-89 (1986).

Diversity of Alpha-fetoprotein Gene Expression in Mice is Generated by a Combination of Separate Enhancer Elements, R. Hammer, R. Krumlanf, S. Campter, R. Brinster, S. Tilghman, Science *235*, 53-58 (1987).

Transcription in Yeast Activated by a Putative Amphipathic alpha Helix Linked to a DNA Binding Unit, E. Giniger, M. Ptashne, Nature *330*, 670-672 (1987).

The JUN Oncoprotein, a Vertebrate Transcription Factor, Activates Transcription in Yeast, K. Struhl, Nature *332*, 649-650 (1987).

GAL4 Activates Gene Expression in Mammalian Cells, H. Kakidoni, M. Ptashne, Cell *52*, 161-167 (1988).

The Yeast UAS$_G$ is a Transcriptional Enhancer in Human HeLaCells in the Presence of the GAL4 *Trans*-activator, N. Webster, J. Jin, S. Green, M. Hollis, P. Chambon, Cell *52*, 169-178 (1988).

v-jun Encodes a Nuclear Protein with Enhancer Binding Properties of AP-1, T. Bos, D. Bohmann, H. Tsuchie, R. Tjian, P. Vogt, Cell *52*, 705-712 (1988).

The JUN Oncoprotein, a Vertebrate Transcription Factor, Activates Transcription in Yeast, K. Struhl, Nature *332*, 649-650 (1988).

A Yeast and a Human CCAAT-binding Protein Have Heterologous Subunits That are Functionally Interchangeable, L. Chodosh, J. Olesen, S. Hahn, A. Baldwin, L. Guarente, P. Sharp, Cell *53*, 25-35 (1988).

Yeast Activators Stimulate Plant Gene Expression, J. Ma, E. Przibilla, J. Hu, L. Bogorad, M. Ptashne, Nature *334*, 631-633 (1988).

Chromosomal Rearrangment Generating a Composite Gene for Developmental Transcription Factor, P. Stragier, B. Kuenkel, L. Kroos, R. Losick, Science *243*, 507-512 (1989).

Yeast GCN4 Transcriptional Activator Protein Interacts with RNA Polymerase II *in vitro*, C. Brandl, K. Struhl, Proc. Natl. Acad. Sci. USA *86*, 2652-2656 (1989).

An Enhancer Stimulates Transcription in *trans* When Attached to the Promoter via a Protein Bridge, H. Muller, J. Sogo, W. Schaffner, Cell *56*, 767-777 (1989).

Direct and Selective Binding of an Acidic Transcriptional Activation Domain to the TATA-box Factor TFIID, K. Stringer, J. Ingles, J. Greenblatt, Nature *345*, 783-786 (1990).

Evidence for Interaction of Different Eukaryotic Transcriptional Activators with Distinct Cellular Targets, K. Martin, J. Lille, M. Green, Nature *346*, 199-202 (1990).

Stringent Spacing Requirements for Transcription Activation by CRP, K. Gaston, A. Bell, A. Kolb, H. Buc, S. Busby, Cell *62*, 733-743 (1990).

A Mediator Required for Activation of RNA Polymerase II Transcription *in vitro*, P. Flanagan, R. Kelleher III, M. Sayre, H. Tschochner, R. Kornberg, Nature *350*, 436-438 (1991).

A New Mechanism for Coactivation of Transcription Initiation: Repositioning of an Activator Triggered by the Binding of a Second Activator, E. Richet, D. Vidal-Ingigliardi, O. Raibaud, Cell *66*, 1185-1195 (1991).

Suppressor Mutations in *rpoA* Suggest that OmpR Controls Transcription by Direct Interaction with the α Subunit of RNA Polymerase, J. Slauch, F. Russo, T. Silhavy, J. Bact. *173*, 7501-7510 (1991).

Sequence-specific Antirepression of Histone H1-mediated Inhibition of Basal RNA Polymerase II Transcription, G. Croston, L. Kerrigan, L. Lira, D. Marshak, J. Kadonaga, Science *251*, 643-649 (1991).

TFIID Binds in the Minor Groove of the TATA Box, B. Star, D. Hawley, Cell *67*, 1231-1240 (1991).

The TATA-binding Protein is Required for Transcription by All Three Nuclear RNA Polymerases in Yeast Cells, B. Cormack, K. Struhl, Cell *69*, 685-696 (1992).

Variants of the TATA-binding Protein Can Distinguish Subsets of RNA Polymerase I, II, and III Promoters, M. Schultz, R. Reeder, S. Hahn, Cell *69*, 697-702 (1992).

A Carboxyl-terminal-domain Kinase Associated with RNA Polymerase II Transcription Factor δ from Rat Liver, H. Serizawa, R. Conaway, J. Conaway, Proc. Natl. Acad. Sci. USA *89*, 7476-7480 (1992).

Human General Transcription Factor IIH Phosphorylates the C-terminal Domain of RNA Polymerase II, H. Lu, L. Zawel, L. Fisher, J. Egly, D. Reinberg, Nature *358*, 641-645 (1992).

Stimulation of Phage λP$_L$ Promoter by Integration Host Factor Requires the Carboxyl Terminus of the α-subunit of RNA Polymerase, H. Giladi, K. Igarashi, A. Ishihama, A. Oppenheim, J. Mol. Biol. *227*, 985-990 (1992).

Genetic Evidence that an Activation Domain of GAL4 Does not Require Acidity and May Form a β Sheet, K. Leuther, J. Salmeron, S. Johnston, Cell *72*, 575-585 (1993).

The Initiation Process

Studies of Ribonucleic Acid Chain Initiation by *E. coli* Ribonucleic Acid Polymerase Bound to T7 Deoxyribonucleic Acid I. An Assay for the Rate and Extent of RNA Chain Initiation, W. F. Mangel, M. J. Chamberlin, J. Biol. Chem. *249p*, 2995-3001 (1974).

Physiochemical Studies on Interactions Between DNA and RNA Polymerase: Unwinding of the DNA Helix by *E. coli* RNA Polymerase, J. Wang, J. Jacobsen, J. Saucier, Nuc. Acids Res. *4*, 1225-1241 (1978).

Physiochemical Studies on Interactions between DNA and RNA Polymerase: Ultraviolet Absorption Measurements, T. Hsieh, J. Wang, Nuc. Acids Res. *5*, 3337-3345 (1978).

E. coli RNA Polymerase Interacts Homologously with Two Different Promoters, U. Siebenlist, R. Simpson, W. Gilbert, Cell *20*, 269-281 (1980).

Rate-limiting Steps in RNA Chain Initiation, W. McClure, Proc. Nat. Acad. Sci. USA *77*, 5634-5638 (1980).

Mechanism of Activation of Transcription Initiation from the Lambda P_{RM} Promoter, D. Hawley, W. McClure, J. Mol. Biol. *157*, 493-525 (1982).

A Topological Model for Transcription Based on Unwinding Angle Analysis of *E. coli* RNA Polymerase Binary Initiation and Ternary Complexes, H. Gamper, J. Hearst, Cell *29*, 81-90 (1982).

Separation of DNA Binding from the Transcription-activating Function of a Eukaryotic Regulatory Protein, L. Keegan, G. Gill, M. Ptashne, Science *231*, 699-704 (1986).

Dynamic and Structural Characterization of Multiple Steps During Complex Formation between *E. coli* RNA Polymerase and the *tetR* Promoter from pSC101, G. Duval-Valentin, R. Ehrlich, Nuc. Acids Res. *15*, 575-594 (1987).

Transcriptional Slippage Occurs During Elongation at Runs of Adenine or Thymine in *Escherichia coli*, L. Wagner, R. Weiss, R. Driscoll, D. Dunn, R. Gesteland, Nuc. Acids Res. *18*, 3529-3535 (1990).

Development of RNA Polymerase-promoter Contacts During Open Complex Formation, J. Mecsas, D. Cowing, C. Gross, J. Mol. Biol. *220*, 585-597 (1991).

The Phosphorylated Form of the Enhancer-binding Protein NTRC Has an ATPase Activity that is Essential for Activation of Transcription, D. Weiss, J. Batat, K. Klose, J. Keener, S. Kustu, Cell *67*, 155-167 (1991).

Techniques

The Reliability of Molecular Weight Determinations by Dodecyl Sulfate-Polyacrylamide Gel Electrophoresis, K. Weber, M. Osborn, J. Biol. Chem. *244*, 4406-4412 (1969).

Reconstitution of Bacterial DNA Dependent RNA Polymerase from Isolated Subunit as a Tool for Elucidation of the Role of the Subunits in Transcription, A. Heil, W. Zillig, FEBS Letters *11*, 165-168 (1970)

Binding of Dodecyl Sulfate to Proteins at High Binding Ratios, Possible Implications for the State of Proteins in Biological Membranes, J. Reynolds, C. Tanford, Proc. Nat. Acad. Sci. USA *66*, 1002-1007 (1970).

A Procedure for the Rapid, Large-Scale Purification of *E. coli* DNA-Dependent RNA Polymerase Involving Polymin P Precipitation and DNA-cellulose Chromotography, R. Burgess, J. Jendrisak, Biochem. *14*, 4634-4638 (1975).

A Steady State Assay for the RNA Polymerase Initiation Reaction, W. McClure, C. Cech, D. Johnston, J. Biol. Chem. *253*, 8941-8948 (1978).

Laser Crosslinking of *E. coli* RNA Polymerase and T7 DNA, C. Harrison, D. Turner, D. Hinkle, Nuc. Acids Res. *10*, 2399-2414 (1982).

Transcription, Termination, and RNA Processing

5

In the previous chapter we considered the structure of RNA polymerases, the transcription initiation process, the structure of promoters and enhancers, and their functions. In this chapter we shall continue with the transcription process. We shall briefly consider the elongation process and then discuss the termination of transcription. Finally, we shall discuss the processing of RNA that occurs after transcription. This includes both the simple modification of RNA by cleavage or the addition of groups and bases, and the more complex cutting, splicing, and editing that occurs more often in eukaryotic cells than in prokaryotic cells.

Polymerase Elongation Rate

Even more than in DNA synthesis, it is sensible for cells to regulate RNA synthesis at the initiation steps so that the elaborate machinery involved in independently regulating thousands of genes need not be built into the basic RNA synthesis module. Once RNA synthesis has been initiated, it proceeds at the same average rate on most, independent of growth conditions. Can this be demonstrated? Another need for knowing the RNA elongation rate is in the interpretation of physiological experiments. How soon after the addition of an inducer can a newly synthesized mRNA molecule appear?

RNA elongation rate measurements are not too hard to perform *in vitro*, but they are appreciably more difficult to perform on growing

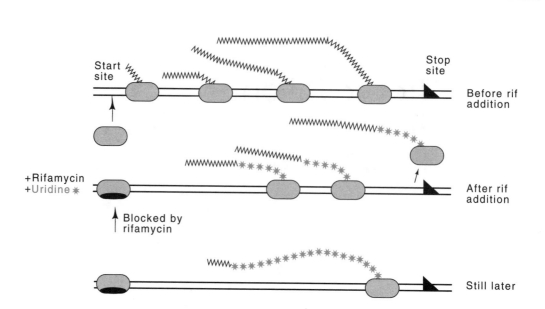

Start site

Stop site

Before rif addition

+Rifamycin
+Uridine ✻

Blocked by rifamycin

After rif addition

Still later

Figure 5.1 Effects of rifamycin addition on transcription of a large operon. Upon the addition of rifamycin, no more RNA polymerase molecules may initiate transcription. Those polymerase molecules that were transcribing continue to the end of the operon. Finally, the polymerase molecule that had initiated transcription just before the addition of rifamycin completes transcription of the operon.

cells. Here we shall explain one method that has been used to determine the *in vivo* RNA elongation rate in *Escherichia coli*.

The measurement used rifamycin, an antibiotic that inhibits RNA polymerase only at the initiation step. It has no effect on RNA polymerase molecules engaged in elongation. Rifamycin and radioactive uridine were simultaneously added to bacteria; thus only those RNA chains that were in the process of elongation at the time of the additions were radioactively labeled, and no new ones could be initiated (Fig. 5.1). At various times after the rifamycin and uridine addition, samples were taken from the culture and their RNA was separated according to size

Figure 5.2 Structure of the ribosomal RNA operon used to determine the RNA elongation rate in *E. coli*.

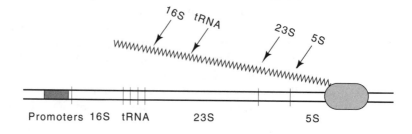

16S   tRNA   23S   5S

Promoters 16S tRNA   23S   5S

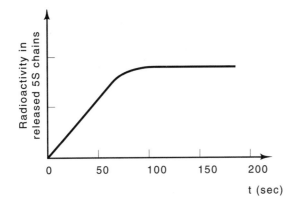

Figure 5.3 Radioactivity incorporation kinetics into 5S RNA following the simultaneous addition of radioactive uridine and rifamycin.

by electrophoresis on polyacrylamide gels. Suppose that a specific species of RNA molecule is well separated from all other species by the electrophoresis. Then, the radioactivity in this size class will increase with time for as long as RNA polymerase molecules transcribe the corresponding gene, but once the last polymerase molecule to initiate has crossed the region, there can be no additional increase in radioactivity. The interval between the addition of rifamycin and the end of the period over which radioactivity increases is the time required for an RNA polymerase molecule to transcribe from the promoter to the end of the transcribed region.

The ribosomal RNA gene complexes were a convenient system for these measurements. Each of these seven nearly identical gene complexes consists of two closely spaced promoters, a gene for the 16S ribosomal RNA, a spacer region, a tRNA gene, the gene for the 23S ribosomal RNA, and the gene for the 5S ribosomal RNA (Fig. 5.2). The total length of this transcriptional unit is about 5,000 nucleotides. The 16S RNA, spacer tRNA, 23S RNA, and 5S RNA are all generated by cleavage from the growing polynucleotide chain.

The interval between the time of rifamycin addition and the time at which the last RNA polymerase molecule transcribes across the end of the 5S gene is the time required for RNA polymerase to transcribe the 5,000 bases from the promoter to the end of the ribosomal gene complex. This time is found from the radioactive uridine incorporation measurements. Transcription across the 5S gene ends when the radioactivity in 5S RNA stops increasing. This happens about 90 seconds after rifamycin and uridine addition (Fig. 5.3). This yields an elongation rate of about 60 nucleotides per second. This type of elongation rate measurement has been performed on cells growing at many different growth rates, and as expected, the results show that the RNA chain growth rate is independent of the growth rate of cells at a given temperature.

Transcription Termination at Specific Sites

If transcription of different genes is to be regulated differently, then the genes must be transcriptionally separated. Such transcriptional isolation could be achieved without explicit transcriptional barriers between

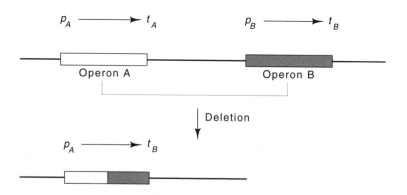

Figure 5.4 Fusion between two operons to place some of the genes of operon B under control of the promoter of operon A. The fusion removes the transcription termination signal t_A and the promoter p_B.

genes if genes were widely spaced and if RNA polymerase occasionally randomly terminated transcription. A more efficient method to separate transcriptional units is merely to have transcription termination signals at their ends.

Transcription termination signals can be shown to exist by several types of experiments. One that was first done in bacteria is genetic. As mentioned earlier, transcriptional units are called operons. Even though genes in two different operons may be located close to one another on the chromosome, only by deleting the transcription termination signal at the end of one operon can genes of the second operon be expressed under control of the first promoter (Fig. 5.4).

A second type of demonstration uses *in vitro* transcription. Radioactive RNA is synthesized *in vitro* from a well-characterized DNA template and separated according to size on polyacrylamide gels. Some templates yield a discrete class of RNA transcripts produced by initiation at a promoter and termination at a site before the end of the DNA molecule. Thus these templates must contain a transcription termination site. Of course, cleavage of an RNA molecule could be mistaken for termination.

Termination

Simple experiments modifying the beginning and middle parts of genes show that most often it is just the sequence at the end of the transcribed region that specifies transcription termination. One simple mechanism for transcription termination in bacteria utilizes just the RNA polymerase and requires only a special sequence near the 3' end of the RNA. The termination signal consists of a region rich in GC bases that can form a hairpin loop closely followed by a string of U's. This class of terminators functions without the need for auxiliary protein factors from the cell. Frequently termination in eukaryotic systems occurs in a

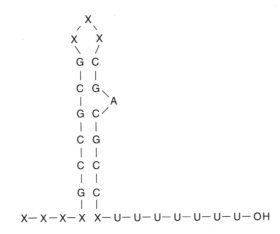

Figure 5.5 Typical sequence and the likely hairpin structure at the 3'-OH end of the RNA whose transcription terminates without the assistance of rho protein.

run of U's as in the prokaryotic case, but no hairpin upstream is apparent.

Most likely as the RNA is elongated past the region rich in GC, it base pairs with itself to form a hairpin (Fig. 5.5). This hairpin may fit so poorly in the transcript groove or canyon in the polymerase (Fig. 5.6), that it weakens the binding of polymerase to the transcription bubble and also causes the RNA polymerase to pause in this region. Release then occurs from the run of U's. Direct physical measurements have shown that oligo (rU:dA) hybridizes with exceptional weakness compared to other oligonucleotides. The combination of these factors changes the RNA elongation complex from being extremely stable to being so unstable that transcription termination usually occurs.

A second class of prokaryotic terminators is much different. This class requires the presence of the rho protein for termination. The termination activity is further stimulated by a second protein, the *nusA* gene product. During transcription, RNA polymerase pauses near a termination sequence, most likely aided in pausing by the NusA protein, and then rho terminates the transcription process and releases the RNA and RNA polymerase. Analysis of the 3' ends of rho-dependent transcripts reveals them to contain no discernible sequence patterns or significant secondary structures. They have even less secondary structure than expected for completely random sequences.

Most likely, when the nascent RNA extending from the RNA polymerase is free of ribosomes and lacks significant secondary structure, the

Figure 5.6 The melted DNA and RNA likely fit within canyons on RNA polymerase. The walls, however, may close in after polymerase binds to DNA to help hold onto the DNA.

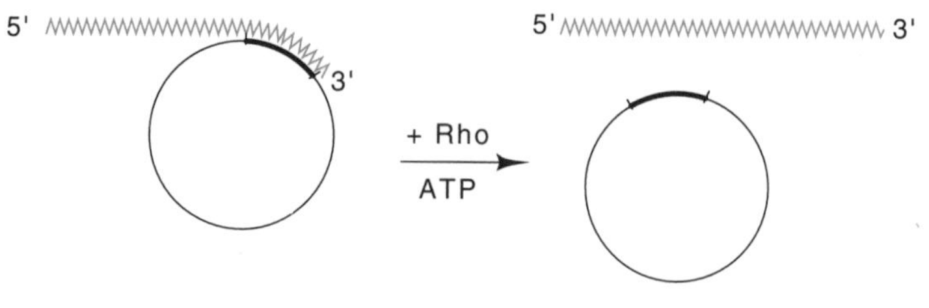

Figure 5.7 RNA hybridized to a circular DNA molecule is a substrate for rho protein plus ATP which separates the two nucleic acids.

rho protein can bind and move with the consumption of energy along the RNA up to the polymerase. When it reaches the polymerase, it separates the growing transcript from the template and terminates transcription. The separation of the two strands is accomplished by an RNA-DNA helicase (Fig. 5.7).

The discovery of rho factor was accidental. Transcription of lambda phage DNA in an *in vitro* system produced a large amount of incorrect transcript. This inaccuracy was revealed by hybridizing the RNA to the two separated strands of lambda phage. Correct transcripts would have hybridized predominantly to only one strand. Apparently the conditions being used for transcription did not faithfully reproduce those existing within the cell and the rho factor somehow reduced the amount of incorrect transcription. This is a biochemist's dream for it means that something must exist and is waiting to be found. Therefore Roberts looked for and found a protein in cell extracts that would enhance the fidelity of *in vitro* transcription. Upon completing the purification and in studying the properties of his "fidelity" factor, he discovered that it terminated transcription. Rho factor shares suggestive properties with a DNA helicase required in DNA replication, DnaB. Both bind to nucleic acid and move along the nucleic acid with the consumption of ATP. In the process of this movement, a complementary strand can be displaced. Further, both helicases are hexameric, and both hydrolyze significant amounts of ATP when in the presence of a single-stranded oligonucleotide.

Although transcription termination and its regulation is usually determined only by events occurring near the 3' end of the transcript, sometimes the 5' end is also involved. The transcription in *E. coli* of ribosomal RNA and of some of the genes of phage lambda depends upon modifying the transcription complex shortly after the polymerase crosses a special sequence near the promoter. At this point several proteins bind to the RNA copy of the sequence and bind to the RNA polymerase as well. After such a modification, elongation will proceed to the end of the transcription unit and ignore some opportunities for termination that would be utilized by an unmodified RNA polymerase. It would be amazing if eukaryotic cells did not also utilize this as a mechanism for gene regulation.

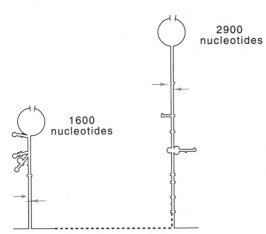

Figure 5.8 Schematic of the base paired stems surrounding the 16S and 23S ribosomal RNAs. The arrows mark the points of RNAse III cleavage.

Processing Prokaryotic RNAs After Synthesis

It is not altogether surprising that intact transcripts produced by transcription of some operons are not always suited for all the biological roles of RNA. In *E. coli* the ribosomal and tRNAs are processed following transcription. For example, RNAse III first cleaves in the portions of the rRNA that are folded back on themselves to form hairpins (Fig. 5.8). Then other nucleases make additional cuts. RNAse III also cleaves the early transcripts of phage T7. A number of methyl groups are also added to the rRNA after its synthesis.

How do we know RNAse III cleaves the ribosomal and phage T7 RNAs? The effects of mutations are one way to attack this general problem of demonstrating the *in vivo* role of an enzyme. Investigators first isolated a mutant defective in RNAse III. They did this by screening extracts prepared from isolated colonies of mutagenized cells for the activity of the enzyme RNAse III (Fig. 5.9). One out of 1,000 colonies yielded cells with greatly depressed levels of the enzyme. Subsequently,

Figure 5.9 An isolation scheme for RNAse III mutants. Mutagenized cells are diluted and spread on plates so as to yield colonies derived from individual cells. Cultures are grown from these colonies and then tested for the presence of RNAse III.

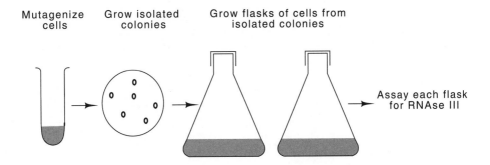

these cells were found to possess appreciable quantities of uncleaved ribosomal RNA. Furthermore, this RNA could be cleaved *in vitro* with purified RNAse III to yield the smaller ribosomal RNAs.

Although it occurs very rarely, messenger RNA in prokaryotic cells can also be cut and rejoined. As the basic principles of such a splicing reaction apply equally well to eukaryotic mRNA, the phenomenon is described more fully in the sections which follow.

S1 Mapping to Locate 5' and 3' Ends of Transcripts

Early in the investigation of a gene and its biological action, it is helpful to learn its transcription start and stop points. One simple method for doing this is S1 mapping, which was developed by Berk and Sharp. S1 is the name of a nuclease that digests single-stranded RNA and DNA. S1 mapping shows on the DNA the endpoints of homology of RNA molecules. That is, the method can map the 5' and 3' ends of the RNA. Consider locating the 5' end of a species of RNA present in cells. Messenger RNA is isolated from the organism, freed of contaminating protein and DNA, then hybridized to end-labeled, single-stranded DNA that covers the region of the 5' end. After hybridization, the remaining single-stranded RNA and DNA tails are removed by digestion with S1 nuclease. The exact size of the DNA that has been protected from nuclease digestion can then be determined by electrophoresis on a DNA sequencing gel (Fig. 5.10). This size gives the distance from the labeled end of the DNA molecule to the point corresponding to the 5' end of the RNA molecule.

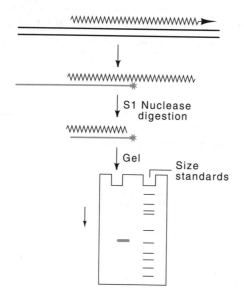

Figure 5.10 S1 mapping to locate the 5' transcription start site. RNA extracted from cells is hybridized to a radioactive fragment covering the transcription start region, then digested with S1 nuclease. The denatured complex is subjected to electrophoresis alongside size standards.

S1 Nuclease digestion

Gel

Size standards

Caps, Splices, Edits, and Poly-A Tails on Eukaryotic RNAs

Eukaryotic RNAs are processed at their beginning, middle, and terminal sections. Although the structures resulting from these posttranscriptional modifications are known, why the modifications occur is not clear.

The 5' ends of nearly all cellular messenger RNAs contain a "cap," a guanine in a reversed orientation plus several other modifications (Fig. 5.11). More precisely, the cap is a guanine methylated on its 7 position joined through 5'-5' pyrophosphate linkage to a base derived from transcription. Both the first and second bases after the capping nucleotide usually are methylated. These modifications were discovered by tracking down why and where viral RNA synthesized *in vitro* could be methylated. The trail led to the 5' end whose structure was then determined by Shatkin. The cap helps stabilize the RNA, is sometimes involved in export to the cytoplasm, may be involved with splicing, and also assists translation. Ribosomes translate cap-containing messengers more efficiently than messengers lacking a cap.

A sizable fraction of the RNA that is synthesized in the nucleus of eukaryotic cells is not transported to the cytoplasm. This RNA is of a variety of sizes and sequences and is called heterogeneous nuclear RNA. Within this class of RNA are messenger RNA sequences that ultimately are translated into protein, but, as synthesized, the nuclear RNAs are not directly translatable (Fig. 5.12). They contain extraneous sequences that must be removed by cutting and splicing to form the continuous

Figure 5.11 The structure of the cap found on eukaryotic messenger RNAs. The first base is 7-methylguanylate connected by a 5'-5' triphosphate linkage to the next base. The 2' positions on bases 1 and 2 may or may not be methylated.

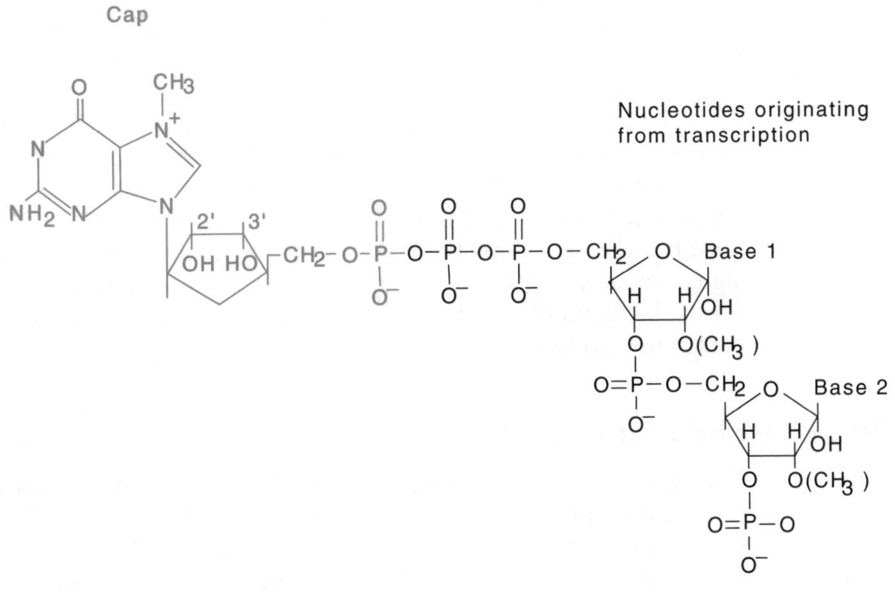

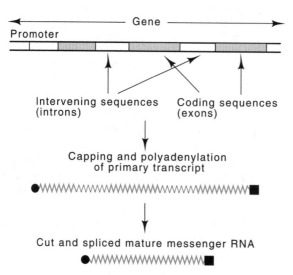

Figure 5.12 Maturation of a primary transcript to yield a translatable message.

coding sequence that is found in the mature messenger in the cytoplasm. The intervening sequences, which are removed from mature RNA, are called introns, and the other parts of the messenger sequence are often referred to as exons.

Another type of modification in RNA after its synthesis is called editing. Although most RNAs are spliced, editing occurs very rarely. One prominent example of editing is the apolipoprotein. Humans contain a single copy of this gene. In the liver its gene product is 512,000 daltons molecular weight, but in intestinal cells its gene product is only 242,000 daltons molecular weight. Examination of the mRNA shows that a specific cytosine of the RNA is converted to a uracil or uracil-like nucleotide in intestinal cells, but not in hepatic cells. This conversion creates a translation stop codon, hence the shorter gene product in intestinal cells. The conversion process involved is called RNA editing. It is also found in some animal viruses. In a few extreme cases, hundreds of nucleotides in an RNA are edited. In these cases, special short guide RNA molecules direct the choice of inserted or deleted nucleotides.

The final type of RNA modification found in eukaryotic cells is the posttranscriptional addition to the 3' end of 30 to 500 nucleotides of polyadenylic acid. This begins about 15 nucleotides beyond the poly-A signal sequence AAUAAA. Transcription itself appears to terminate somewhat beyond the poly-A signal and processing quickly removes the extra nucleotide before the poly-A addition.

The Discovery and Assay of RNA Splicing

The discovery by Berget and Sharp and by Roberts and co-workers that segments within newly synthesized messenger RNA could be eliminated before the RNA was exported from the nucleus to the cytoplasm was a great surprise. Not only was the enzymology of such a reaction unfamiliar, but the biological need for such a reaction was not apparent. Although splicing is a possible point for cells to regulate expression, there seemed no particular need for utilizing this possibility. One possible reason for splicing could be in the evolution of proteins. Intervening sequences place genetic spacers between coding regions in a gene. As a result, recombination is more likely to occur between coding regions than within them. This permits a coding region to be inherited as an independent module. Not surprisingly then, the portions of proteins that such modules encode often are localized structural domains within proteins. As a result, during the evolution of a protein, domains may be shuffled. It is hard to imagine, however, that this need was great enough to drive the evolution of splicing.

The most plausible explanation for the existence of intervening sequences is that they are the remnants of a parasitic sequence that spread through the genome of some early cell-type. In order that the sequence not inactivate a coding region into which it inserted itself, the coding region arranged that it splice itself out of mRNA. Thus, although the parasitic sequence might have inserted into the middle of an essential gene, the gene was not inactivated. After transcription, and before translation, the RNA copy of the gene containing the sequence was cut and spliced to recover the intact uninterrupted gene. Now that the intervening sequences are there, the cell is beginning to make use of them. One example is regulating the use of alternative splice sites to generate one or another gene product from a single gene.

Both the discovery of messenger splicing and a clear demonstration of splicing used adenovirus RNA. The virus provided a convenient source of DNA for hybridization reactions and RNA extracted from adeno-infected cultured cells was a rich source of viral RNA. Electron microscopy of hybrids formed between adenovirus mRNA and a fragment of the adenovirus genome coding for the coat hexon protein was performed to locate the transcriptional unit. Curiously, the hybrids which formed lacked the expected structure. Many nucleotides at one

Figure 5.13 RNA-DNA hybrid between a single-stranded fragment of adenovirus DNA and hexon mRNA extracted from cells. The RNA and DNA are not complementary at the 5' end of the mRNA.

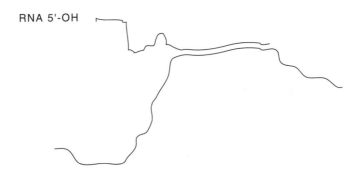

RNA 5'-OH

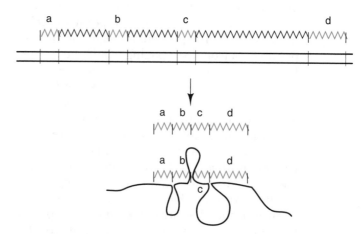

Figure 5.14 Hybridization of adenovirus hexon mRNA to adenovirus DNA from a region well upstream from the hexon gene. The regions a, b, c and d of viral RNA hybridize, but the regions between these segments are missing from the RNA, and the DNA accordingly loops out.

end of the RNA failed to hybridize to the hexon DNA (Fig. 5.13). One of the ways splicing was discovered was tracking down the source of this extraneous RNA. Further experiments with longer DNA fragments located DNA complementary to this RNA, but it was far away from the virus hexon gene (Fig. 5.14). The RNA as extracted from the cells had to have been synthesized, and then cut and spliced several times.

Gel electrophoresis provides a convenient assay to detect splicing and to locate the spliced regions. Hybridization of DNA to messenger RNA

Figure 5.15 Endonuclease S1 is used to digest single-stranded regions of RNA and DNA, and exonuclease Exo VII is used to digest single-stranded DNA exonucleolytically. The DNA is labeled in these experiments. After digestions, electrophoresis on neutral gels analyzes duplexes, and electrophoresis on alkaline gels denatures and analyzes the sizes of fragments.

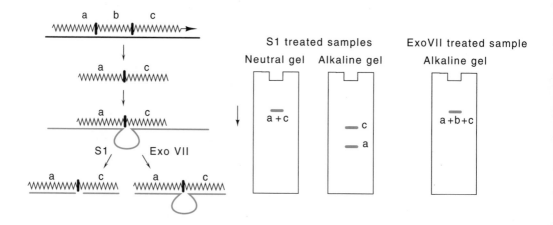

that has been spliced to remove introns yields duplexes and loops as shown in Fig. 5.15. The locations and sizes can be determined by digestion with S1 and ExoVII nucleases followed by electrophoresis on denaturing gels and on gels that preserve the integrity of RNA-DNA duplexes. Since S1 nuclease digests single-stranded RNA and DNA and can act endonucleolytically, it removes all single-stranded or unpaired regions from RNA and DNA. ExoVII, however, digests only single-stranded DNA, and only from the ends. This enzyme therefore provides the additional information necessary to determine intron and exon sizes and locations. After digestion with these enzymes, the oligonucleotides can be separated according to size by electrophoresis on polyacrylamide gels. Gels run with normal buffer near neutral pH provide the sizes of the double-stranded duplex molecules, and gels run at alkaline pH provide the molecular weights of the denatured, single-stranded oligonucleotides.

Involvement of the U1 snRNP Particle in Splicing

Initially, progress in the study of the biochemical mechanisms of splicing was slow. One of the first clues about the mechanism of splicing came from the sequencing of many splice sites. It was noticed that the sequence flanking the 5' splice site was closely complementary to the sequence of RNA found at the 5' end of an RNA found in the U1 class of small nuclear ribonucleoprotein particles called snRNPs. In addition to the U1 particles, U2, U4, U6, and others exist, each containing 90 to 150 nucleotides and about 10 different proteins. The complementarity between U1 RNA and the splice site of pre-mRNA suggested that base pairing occurred between the two during splicing. Stronger evidence for the proposal was the discovery by Steitz and Flint that splicing in nuclei could be blocked by antibodies against U1 particles.

The experiment showing inactivation of splicing by anti-U1 antibodies is not definitive since their specificity may not be high and because other antibodies could also be present. One ingenious method for specifically inactivating U1 particles was to remove nucleotides from their 5' ends with RNAse H. This enzyme digests RNA from RNA-DNA

Figure 5.16 How compensating mutations in two interacting structures can restore activity.

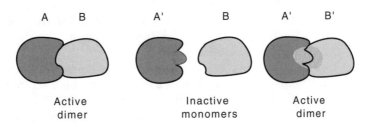

A	B	A'	B	A'	B'

Active dimer Inactive monomers Active dimer

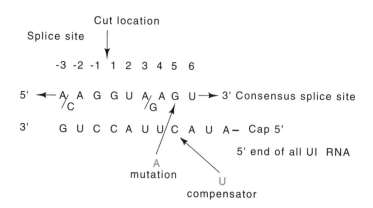

Figure 5.17 The consensus of the sequence at the 5' end of an exon-intron junction. The sequence is highly complementary to the 5' end of U1 RNA. Alterations in the splice sequence reduce splicing, and compensating alterations as shown in the U1 sequence restore splicing.

duplexes. DNA oligonucleotides complementary to the 5' end of U1 RNA were hybridized under gentle conditions to U1 particles in cell extracts and then RNAseH was added. These treatments left the extracts unable to catalyze intron removal whereas extracts which had received oligonucleotides of different sequence were not inhibited.

One of the nicest demonstrations of biologically significant interactions between two macromolecules are those in which compensating mutations can be utilized. First, a mutation is isolated that interferes with a particular interaction between A and B (Fig. 5.16). That is, A' no longer interacts with B *in vivo*. Then a compensating mutation is isolated in B that potentiates the *in vivo* interaction between A' and B'. The isolation of the mutation and the compensating mutations in components of the splicing apparatus could not be done using traditional genetics tools since the splicing reactions occur in animals for which only rudimentary genetics exists. Genetic engineering methods had to be utilized.

Two steps are necessary to perform the experiment. The first is inducing the cells to synthesize messenger with an altered splice site as well as synthesize U1 RNA with an altered sequence, and the second is assaying for splicing of the special messenger in the presence of the normal cellular levels of messenger and pre-messenger RNAs. Weiner introduced into the cells a segment of adenovirus sequence coding for the E1a protein with wild-type or variant 5' splice sites as well as the gene for an altered U1. Only when DNA was introduced that encoded a variant U1 gene that compensated for the splice site mutation and restored Watson-Crick base pairing across the region was the variant splice site utilized (Fig. 5.17).

The use of an adenovirus sequence for the experiment permitted altering the splicing of a messenger which was not important to the cells. Similarly, introducing a new gene for U1 RNA avoided disruption of the ongoing cellular splicing processes. Finally, examining splicing in the

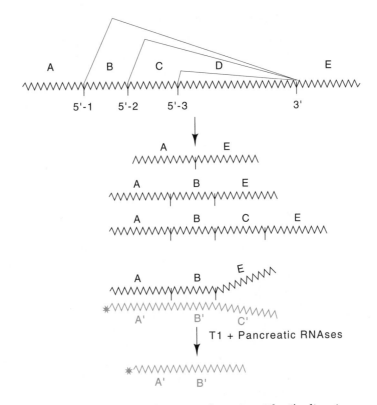

Figure 5.18 The pattern of splicing in adenovirus. The 5' splice sites separating regions A, B, and C are all joined to the 3' splice site separating regions D and E. Thus the three RNAs shown can be produced. The amount of the species ABE can be determined by hybridizing to radioactive RNA complementary to ABC and digesting with T1 and pancreatic RNAses. The amount of product of length AB is determined by electrophoresis.

E1a gene from the virus increased the sensitivity since this gene contains three different 5' splice sites, each utilizing the same 3' splice site. Damaging the activity of one 5' splice site diverted splicing to the other splice sites, whereas had there been only one 5' splice site, altering it might only have slowed kinetics of the splicing process without changing the actual amounts of RNA eventually spliced.

An RNAse protection assay was used to monitor the splicing. RNA was isolated from the cells and hybridized with radioactive RNA complementary to the adenovirus mRNA. The regions of the RNA which are not base paired are sensitive to T1 and pancreatic RNAses and are digested away (Fig. 5.18). This leaves a radioactive RNA molecule whose size indicates the splice site utilized. A variant sequence at the middle E1a splice site eliminated splicing from this location, but the introduction of the variant U1 gene to the cells containing the compensating mutation restored use of this splice site.

Splicing Reactions and Complexes

In addition to requiring U1 snRNP particles, pre-mRNA splicing requires at least three other snRNPs, U2, U5, and U4/U6, as well as a number of soluble proteins. Together these form a large complex that can be observed in the electron microscope, and which can be biochemically purified. The complex forms in the nucleus even while the RNA is being elongated, and exons near the 5' end of the RNA can be removed even before synthesis of the RNA is complete. Formation of the complex requires that the regions to which both U1 and U2 bind be present. Scanning by the splicing apparatus from 5' to the 3' end may help explain the paradoxically high degree of specificity to splicing. The donor and acceptor splice sites contain only two essential nucleotides, too few to ensure specificity in RNA of random sequence. It is likely that the spacing between introns and exons also helps the splicing apparatus choose sites appropriately. One purification method of spliceosomes is to synthesize substrate RNA *in vitro*. This RNA is synthesized with ordinary nucleotides plus biotin-substituted uridine. After the RNA has been added to a splicing extract, the biotin can be used to fish out this RNA selectively with streptavidin bound to a chromatography column. Along with the RNA are found the U1, U2, U4/U6, and U5 snRNAs.

The reaction of the mammalian splicing components is at least partly ordered. U1 binds by base pairing to the 5' splice site and U2 binds by base pairing to a sequence within the intervening sequence containing a nucleotide called the branch point that participates in the splicing reaction. The RNAs of U4 and U6 are extensively base paired whereas a shorter region of base pairing is formed between the U6 and U2 RNAs (Fig. 5.19). In the course of the splicing reaction the U4 particle is released first. Splicing in yeast is similar to that found in mammals, but

Figure 5.19 Base pairing among U1, U2, U4, U6, and the pre-mRNA showing the branch site and the 5' cut site.

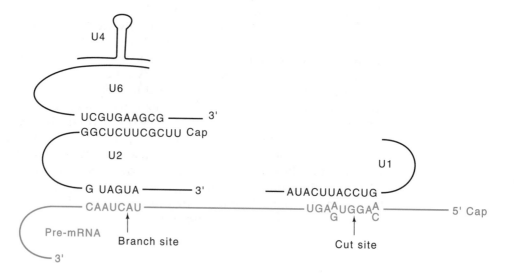

differs in many small details. The same snRNPs are involved, but most of the U RNAs are considerably larger than their mammalian counterparts. Only U4 and U6 are closely homologous in both organisms.

Sometimes extensive splicing is required to regenerate a single intact mammalian gene. For example, there are genes containing one million bases and 60 splice sites. Only a few genes in yeast are spliced, and they contain only a few introns. Great specificity is required in order that all of the splicing reactions proceed with sufficient fidelity that most of the pre-messenger RNA molecules ultimately yield correctly spliced messenger RNA. In part, we still do not know the reasons for the high fidelity. Although a consensus sequence at the 5' and 3' splice sites can be derived by aligning many splice sites, there are only two invariant and essential nucleotides present in each of the sites. This hardly seems like enough information to specify correct splicing.

Once *in vitro* splicing reactions could be performed with unique substrates, it was straightforward to examine the products from the reactions. Amazingly, the sizes of the products as determined by electrophoretic separations did not add up to the size of the substrate pre-mRNA. The structures were then determined by chemical means and by electron microscopy of the resultant RNA molecules. The excised RNA was found to be in a lariat form.

This results from the reaction of the nucleotide at the branch point within the intron attacking the phosphodiester at the 5' splice site. Subsequent attack of the 3'-OH at the 5' splice site on the phosphodiester bond at the 3' splice site releases the intron in a lariat form and completes the splicing process. The freed introns in lariat form are rapidly degraded within the nucleus.

The Discovery of Self-Splicing RNAs

Cech found that the nuclear ribosomal RNA from *Tetrahymena* contains an intervening sequence. In efforts to build an *in vitro* system in which splicing would occur, he placed unspliced rRNA in reactions containing and lacking extracts prepared from cells. Amazingly, the control reactions for the splicing reactions, those lacking the added extract, also spliced out the intervening sequence. Naturally, contamination was suspected, and strenuous efforts were made to remove any possible *Tetrahymena* proteins from the substrate RNA, but splicing in the absence of *Tetrahymena* extract persisted. Finally, Cech placed the gene for the rRNA on a plasmid that could replicate in *E. coli*, prepared the DNA from *E. coli* cells, a situation that had to be devoid of any hypothetical *Tetrahymena* protein, and he still found splicing. This proved, to even the most skeptical, that the *Tetrahymena* rRNA was

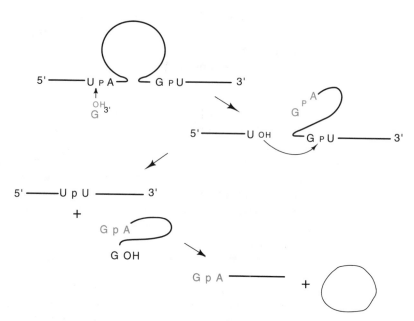

Figure 5.20 The series of self-splicing reactions followed by *Tetrahymena* rRNA.

performing splicing on itself without the action of any *Tetrahymena* proteins.

As shown in Fig. 5.20, guanosine is essential to this self-splicing reaction, but it does not contribute chemical energy to the splicing products. Further studies on the *Tetrahymena* self-splicing reaction show that not only is a 480 nucleotide section of RNA removed from the middle of the ribosomal RNA, but the removed portion then goes on to close on itself forming a circle and releasing a short linear fragment. At first it seems surprising that neither ATP nor any other energy source is required for the cutting and splicing reactions. The reason is that external energy is not required. Chemically, all the reactions are transesterifications and the numbers of phosphodiester bonds are conserved. One might ask, then, why the reaction proceeds at all. One answer is that in some cases the products of the reactions are three polynucleotides where initially only one plus guanosine existed before. Together these possess higher entropy than the starting molecules, and therefore their creation drives the reaction forward.

The transesterification reactions involved in the self-splicing proceed at rates many orders of magnitude faster than they could with ordinary transesterifications. Only two reasons can explain the stimulated rate of these reactions. First, the secondary structure of the molecules can hold the reactive groups immediately adjacent to one another. This increases their effective collision frequencies far above their normal solution values. The second reason is that the probability of a reaction occurring with a collision can be greatly improved if the bonds involved are strained. Studies with very small self–cutting RNAs and also molecular dynamics calculations indicate that such a strain is crucial to the reactions. Undoubtedly, self–splicing utilizes both principles.

Self–splicing has been found in two of the messenger RNAs of the bacteriophage T4, and in the splicing of mRNA in the mitochondrion of yeast. The mitochondrial self–splicing introns comprise a second group of self–splicing introns. Their secondary structure differs from those of the Group I self–splicing introns of which the *Tetrahymena* rRNA intron is an example. The Group II introns do not use a free guanosine to initiate the splicing process. They use an internal nucleotide, and in this respect use a reaction mechanism more like that used in processing pre–mRNA. The existence of splicing in bacteria and eukaryotes suggests that splicing in general has existed in the common precursor to both organisms. The scarcity of splicing events in prokaryotes might be a result of the greater number of generations they have had in which to select for the loss of introns. Eukaryotes may still be struggling with the "infection."

A Common Mechanism for Splicing Reactions

One early difficulty in studying splicing of mRNA was obtaining the RNA itself. Cells contain large amounts of rRNA, but most of the mRNA has been processed by splicing. Further, only a small fraction of the unspliced pre–mRNA present at any moment is from any one gene. One convenient source of pre-mRNA for use in splicing reactions came from genetic engineering. The DNA for a segment of a gene containing an intervening sequence could be placed on a small circular plasmid DNA molecule that could be grown in the bacterium *Escherichia coli* and easily purified. These circles could be cut at a unique location and then they could be transcribed *in vitro* from special phage promoters placed just upstream of the eukaryotic DNA (Fig. 5.21). By this route, large quantities of unspliced substrate RNA could be obtained.

Figure 5.21 The use of SP6 or T7 phage promoters on small DNA molecules to generate sizeable amounts *in vitro* of RNAs suitable for study of splicing reactions.

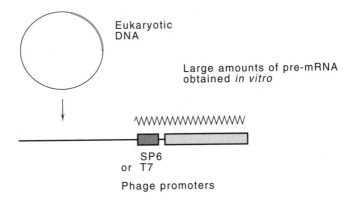

Eukaryotic
DNA

Large amounts of pre-mRNA
obtained *in vitro*

SP6
or T7

Phage promoters

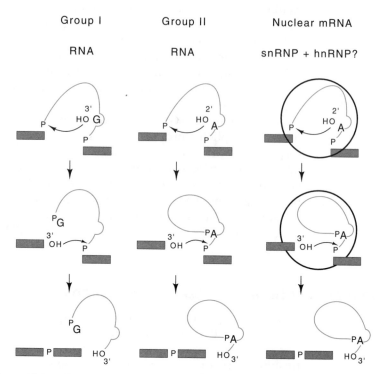

Figure 5.22 The two classes of self–splicing RNA and the pathway of nuclear mRNA splicing, all drawn to emphasize their similarities.

Both of the two self-splicing reactions and the snRNP catalyzed splicing reactions can be drawn similarly (Fig. 5.22). In the self-splicing cases a hydroxyl from a guanosine nucleotide or an adenine in the chain attacks the phosphodiester and a transesterification ensues in which the 5′ end of the RNA is released. For the Group I self-splicing reaction, a tail is formed, and for the Group II and mRNA splicing reactions, a ring with a tail is formed. Then a hydroxyl from the end of the 5′ end of the molecule attacks at the end of the intervening sequence and another transesterification reaction joins the head and tail exons and releases the intron.

In the case of pre–mRNA splicing, the same reaction occurs, but it must be assisted by the snRNP particles. In some intervening sequences of yeast, internal regions are involved in excision and bear some resemblance to portions of the U1 sequences.

The similarities among the splicing reactions suggest that RNA was the original molecule of life since it can carry out the necessary functions on its own, and only later did DNA and protein evolve. The splicing reactions that now require snRNPs must once have proceeded on their own.

Other RNA Processing Reactions

The precursor tRNAs in yeast contain an intervening sequence. This is removed in a more traditional set of enzymatic cleavage and ligation reactions (Fig. 5.23).

Plant viruses frequently have RNA genomes. These viruses can themselves have viruses. These are known as virusoids, and they can grow in cells only in the presence of a parental virus. Virusoids do not encode any proteins, but they are replicated. Part of their replication cycle requires the specific cleavage of their RNA molecules. This they do in a self–cutting reaction. Symons and Uhlenbeck have investigated the

```
                              3'
                              U
                              G  5'
                              C -G
                              A -U  C   UGAGCG UG
                          A          C  |||||| ||     A
                      A       A        C  ACUCGC C      U
             A  GGGCCG                        C    C C  A
          U    |||||| G                    U       C C
          GG CCCGGU                     G
               C      A        G    A
                      G   A
                          A
```

minimal nucleotide requirements for self–cleavage of these molecules. Remarkably, it is a scant 25 nucleotides that can form into a functional

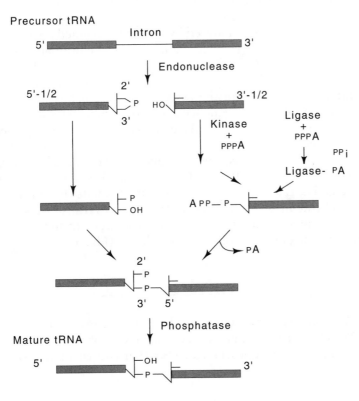

Precursor tRNA

Figure 5.23 The pathway for cutting and splicing tRNA followed by the yeast *Saccharomyces*.

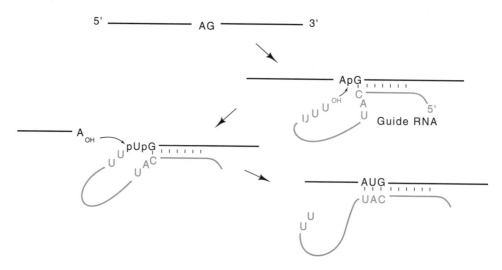

Figure 5.24 Editing of pre-mRNA by a guide RNA using transesterifications.

hammer-headed structure. Another general structure for self-cutting RNA molecules is a simple hairpin.

Earlier in the chapter editing reactions were mentioned. Simple editing of a nucleotide or two has been observed in a mammalian RNA, but in the mitochondrion of some protozoa far more extensive editing has been observed. This raises the question of just where the information for the editing is stored. Changing a single base can conceivably be a result of a set of reactions catalyzed by enzymes designed for just the sequence at which the change occurs, but in the more dramatic examples of editing, in which more than 50 U's are inserted to produce the final edited sequences, far too many different enzymes would be required. Initial examinations of the DNA of the organisms, both with computer searches of known sequence and hybridization studies, failed to reveal any sequences that could have encoded the edited sequence. Eventually it was found that the information for the edited sequences was carried in short RNAs complementary to segments of the final edited sequence. These are called guide sequences. Although cutting and re-ligation could be the pathway for editing, intermediates in editing are found that indicate instead, that the guide sequences transfer U's from their 3' ends to the necessary positions in the mRNA by transesterification reactions analogous to those used in splicing (Fig. 5.24).

Problems

5.1. RNA is extracted from cells, hybridized with an excess of genomic sequence DNA from a particular gene, digested with DNAse S1, and run on a denaturing gel. What does it mean if Southern transfer and probing with radioactive fragments from part of the gene yield two radioactive bands?

5.2. What does possible secondary structure in mRNA (hairpins and pseudoknots) have to do with the fact that eukaryotes use the bind and slide mechanism for finding the first AUG of mRNA and this is used to begin translation, whereas translation of mRNA in prokaryotes begins at AUG codons which often are not the first occurrence of the codon in the mRNA?

5.3. Suppose radioactive uridine and rifamycin are simultaneously added to growing cells. Assume that both the uridine and the rifamycin instantaneously enter the cells and that the radioactive uridine immediately begins entering RNA chains that are in the process of being elongated. Messenger RNA will be completed and will later decay, whereas ribosomal RNA will be synthesized and will be stable. Sketch the kinetics of radioactivity incorporation into RNA in such an experiment and indicate how it may be used to determine the fraction of RNA synthesis devoted to mRNA and to rRNA.

5.4. How would you go about determining whether rho protein acts by reading DNA sequences, by reading RNA sequences coming out of the RNA polymerase, or by interacting directly with RNA polymerase?

5.5. An important question about capped eukaryotic messenger RNA is whether the cap location is a processing site produced by cleavage from a longer RNA or whether the cap site is the actual start site of transcription. How could β-labeled ATP be used *in vitro* to answer this question for an RNA that begins GpppApXp?

5.6. How could you determine the average physical half-life of mRNA in cells from data obtained by simultaneously adding radioactive uridine and an inhibitor of the initiation of RNA synthesis?

5.7. Without looking at individual mRNAs, what is a general way to examine a cell extract for the existence of any Group I self–splicing introns?

5.8. Why is self-cutting and splicing important to the viability of a virusoid?

5.9. Why should one nucleotide in a lariat possess three phosphodiester bonds, and how would you look for such a nucleotide?

5.10. Why might one wish to include G(5')ppp(5')G nucleotides in *in vitro* transcription mixtures whose RNA products will be used in splicing reactions?

5.11. The hammerhead self-cutting RNA molecules that are found in virusoids not only cut RNA. They can ligate once after a cutting step. Predict where the energy for the ligation comes from subject to the constraints that the hammerhead molecules do not utilize a free nucleotide like the *Tetrahymena* intron, they don't use any external molecules as a source of energy for the ligation, and they do not generate branched molecules as are found in pre-mRNA splicing reactions.

References

Recommended Readings

A Small Nuclear Ribonucleoprotein is Required for Splicing of Adenoviral Early RNA Sequences, V. Yang, M. Lerner, J. Steitz, J. Flint, Proc. Nat. Acad. Sci. USA *78*, 1371-1375 (1981).

A Compensatory Base Change in U1 snRNA Suppresses a 5' Splice Site Mutation, Y. Zhuang, A. Weiner, Cell *46*, 827-835 (1986).

The Control of Alternative Splicing at Genes Regulating Sexual Differentiation in *D. Melanogaster*, R. Nagoshi, M. McKeown, K. Burtis, J. Belote, B. Baker, Cell *53*, 229-236 (1988).

Transcriptional Antitermination in the *bgl* Operon of E. coli is Modulated by a Specific RNA Binding Protein, F. Houman, M. Diaz-Torres, A. Wright, Cell *62*, 1153-1163 (1990).

Elongation, Cutting, Ligating, and Other Reactions on RNA

T7 Early RNAs and *Escherichia coli* Ribosomal RNAs are Cut from Large Precursor RNAs *in vivo* by Ribonuclease III, J. Dunn, W. Studier, Proc. Nat. Acad. Sci. USA *70*, 3296-3300 (1973).

Reovirus Messenger RNA Contains a Methylated, Blocked 5'-Terminal Structure: $m^7G(5')ppp(5')G^mpCp$-, Y. Fukruichi, M. Morgan, S. Muthukrishnan, A. Shatkin, Proc. Natl. Acad. Sci. USA *72*, 362-366 (1975).

Ribosomal RNA Chain Elongation Rates in *Escherichia coli*, S. Molin, in Control of Ribosome Synthesis, ed. N. Kjeldgaard, N. Maaloe, Alfred Benzon Symposium IX, Munksgaard, 331-339 (1976).

Capping of Eucaryotic mRNAs, A. Shatkin, Cell *9*, 645-653 (1976).

Complementary Sequences 1700 Nucleotides Apart Form a Ribonuclease III Cleavage Site in *Escherichia coli* Ribosomal Precursor RNA, R. Young, J. Argetsinger-Steitz, Proc. Nat. Acad. Sci. USA *75*, 3593-3597 (1978).

Processing of the 5' End of *Escherichia coli* 16S Ribosomal RNA, A. Dahlberg, J. Dahlberg, E. Lund, H. Tokimatsu, A. Ribson, P. Calvert, F. Reynolds, M. Zahalak, Proc. Nat. Acad. Sci. USA *75*, 3598-3602 (1978).

The Ribonuclease III Site Flanking 23S Sequences in the 30S Ribosomal Precursor RNA of *E. coli*, R. Bram, Y. Young, J. Steitz, Cell *19*, 393-401 (1980).

Mechanism of Action of a Yeast RNA Ligase in tRNA Splicing, C. Greer, C. Peebles, P. Gegenheimer, J. Abelson, Cell *32*, 537-546 (1983).

The RNA Moiety of Ribonuclease P is the Catalytic Subunit of the Enzyme, C. Guerrier-Tokada, K. Gardiner, T. Marsh, N. Pace, S. Altman, Cell *35*, 849-857 (1983).

An Unwinding Activity that Covalently Modifies its Double-stranded RNA Substrate, B. Bass, H. Weintraub, Cell *55*, 1089-1093 (1988).

The Secondary Structure of Ribonuclease P RNA, the Catalytic Element of a Ribonucleoprotein Enzyme, B. James, G. Olsen, J. Liu, N. Pace, Cell *52*, 19-26 (1988).

Ribosomal RNA Antitermination, Function of Leader and Spacer Region BoxB-BoxA Sequences, K. Berg, C. Squires, C. Squires, J. Mol. Biol. *209*, 345-358 (1989).

A Double-Stranded RNA Unwinding Activity Introduces Structural Alterations by Means of Adenosine to Inosine Conversions in Mammalian Cells and *Xenopus* Eggs, R. Wagner, J. Smith, B. Cooperman, K. Nishikura, Proc. Natl. Acad. Sci. USA *86*, 2647-2651 (1989).

Structural Analysis of Ternary Complexes of *Escherichia coli* RNA Polymerase, B. Krummel, M. Chamberlin, J. Mol. Biol. *225*, 221-237 (1992).

GreA Protein: A Transcription Elongation Factor from *Escherichia coli*, S. Borwkhov, A. polyakov, V. Nikiforov, A. Goldfarb, Proc. Natl. Acad. Sci. USA *89*, 8899-8902 (1992).

Recognition of *boxA* Antiterminator by the *E. coli* Antitermination Factors NusB and Ribosomal Protein S10, J. Nodwell, J. Greenblatt, Cell *72*, 261-268 (1993).

Splicing

Spliced Segments at the 5' Terminus of Adenovirus 2 Late mRNA, S. Berget, C. Moore, P. Sharp, Proc. Natl. Acad. Sci. USA *74*, 3171-3175 (1977).

An Amazing Sequence Arrangement at the 5' Ends of Adenovirus 2 Messenger RNA, L. Chow, R. Gelinas, T. Broker, R. Roberts, Cell *12*, 1-8 (1977).

Secondary Structure of the Tetrahymena Ribosomal RNA Intervening Sequence: Structural Homology with Fungal Mitochondrial Intervening Sequences. T. Cech, H. Tanner, I. Tinoco Jr., B. Weir, M. Zuker, P. Perlman, Proc. Nat. Acad. Sci. USA *80*, 5230-5234 (1983).

Splicing of Adenovirus RNA in a Cell-free Transcription System, R. Padgett, S. Hardy, P. Sharp, Proc. Nat. Acad. of Sci. USA *80*, 5230-5234 (1983).

The U1 Small Nuclear RNA-protein Complex Selectively Binds a 5' Splice Site *in vitro*, S. Mount, I. Pattersson, M. Hinterberger, A. Karmas, J. Steitz, Cell *33*, 509-518 (1983).

Yeast Contains Small Nuclear RNAs Encoded by Single Copy Genes, J. Wise, D. Tollervey, D. Maloney, H. Swerdlow, E. Dunn, C. Guthrie, Cell *35*, 743-751 (1983).

The 5' Terminus of the RNA Moiety of U1 Small Nuclear Ribonucleoprotein Particles is Required for the Splicing of Messenger RNA Precursors, A. Krämer, W. Keller, B. Appel, R. Lührmann, Cell *38*, 299-307 (1984).

Intervening Sequence in the Thymidylate Synthase Gene of Bacteriophage T4, F. Chu, G. Maley, F. Maley, M. Belfort, Proc. Nat. Acad. Sci. USA *81*, 3049-3053 (1984).

Point Mutations Identify the Conserved, Intron-contained TACTAAC Box as an Essential Splicing Signal Sequence in Yeast, C. Langford, F. Klinz, C. Donath, D. Gallwitz, Cell *36*, 645-653 (1984).

Excision of an Intact Intron as a Novel Lariat Structure During Pre-mRNA Splicing *in vitro*, B. Ruskin, A. Krainer, T. Maniatis, M. Green, Cell *38*, 317-331 (1984).

U2 as Well as U1 Small Nuclear Ribonucleoproteins are Involved in Premessenger RNA Splicing, D. Black, B. Chabot, J. Steitz, Cell *42*, 737-750 (1985).

Multiple Factors Including the Small Nuclear Ribonucleoproteins U1 and U2 are Necessary for Pre-mRNA Splicing *in vitro*, A. Krainer, T. Maniatis, Cell *42*, 725-736 (1985).

Characterization of the Branch Site in Lariat RNAs Produced by Splicing of mRNA Precursors, M. Konarska, P. Grabowski, R. Padgett, P. Sharp, Nature *313*, 552-557 (1985).

Processing of the Intron-containing Thymidylate Synthase (td) Gene of Phage T4 is at the RNA Level, M. Belfort, J. Pedersen-Lane, D. West, K. Ehrenman, G. Maley, F. Chu, F. Maley, Cell *41*, 375-382 (1985).

A Multicomponent Complex is Involved in the Splicing of Messenger RNA Precursors, P. Grabowski, S. Seiler, P. Sharp, Cell *42*, 345-353 (1985).

U1, U2, and U4/U6 Small Nuclear Ribonucleoproteins are Required for *in vitro* Splicing but Not Polyadenylation, S. Berget, B. Robberson, Cell *46*, 691-696 (1986).

Affinity Chromatography of Splicing Complexes: U2, U5, and U4+U6 Small Nuclear Ribonucleoprotein Particles in the Spliceosome, P. Grabowski, P. Sharp, Science *233*, 1294-1299 (1986).

Formation of the 3' End of U1 snRNA Requires Compatible snRNA Promoter Elements, N. Hernandez, A. Weiner, Cell *47*, 249-258 (1987).

An Essential snRNA from *S. cerevisiae* Has Properties Predicted for U4, Including Interaction with a U6-Like snRNA, P. Siliciano, D. Brow, H. Roiha, C. Guthrie, Cell *50*, 585-592 (1987).

Splicesomal RNA U6 is Remarkably Conserved from Yeast to Mammals, D. Brow, C. Guthrie, Nature *334*, 213-218 (1988).

The Mw 70,000 Protein of the U1 Small Nuclear Ribonucleoprotein Particle Binds to the 5'Stem-Loop of U1 RNA and Interacts with the Sm Domain Proteins, J. Patton, T. Pederson, Proc. Natl. Acad. Sci. USA *85*, 747-751 (1988).

Trans-spliced Leader RNA Exists as Small Nuclear Ribonucleoprotein Particles in *Caenorhabditis elegans*, K. VanDoren, D. Hirsh, Nature *335*, 556-591 (1988).

Protein Encoded by the Third Intron of Cytochrome b Gene of *Saccharomyces cerevisiae* is an mRNA Maturase, J. Lazowska, M. Claisse, A. Gargouri, Z. Kotylak, A. Spyridakis, P. Slonimsky, J. Mol. Biol. *205*, 275-289 (1989).

Drosophila doublesex Gene Controls Somatic Sexual Differentiation by Producing Alternatively Spliced mRNAs Encoding Related Sex-Specific Polypeptides, K. Burtis, B. Baker, Cell *56*, 997-1010 (1989).

Scanning from an Independently Specified Branch Point Defines the 3' Splice Site of Mammalian Introns, C. Smith, E. Porro, J. Patton, B. Nadal-Ginard, Nature *342*, 243-247 (1989).

Sex-Specific Alternative Splicing of RNA from the Transformer Gene Results from Sequence-dependent Splice Site Blockage, B. Sosnowski, J. Belote, M. McKeown, Cell *58*, 449-459 (1989).

UACUAAC is the Preferred Branch Site for Mammalian mRNA Splicing, Y. Zhaung, A. Goldstein, A. Weiner, Proc. Natl. Acad. Sci. USA *86*, 2752-2756 (1989).

Identification of Nuclear Proteins that Specifically Bind to RNAs Containing 5'Splice Sites, D. Stolow, S. Berget, Proc. Natl. Acad. Sci. USA *88*, 320-324 (1991).

Genetic Evidence for Base Pairing between U2 and U6 snRNA in Mammalian mRNA Splicing, B. Datta, A. Weiner, Nature *352*, 821-824 (1991).

Requirement of the RNA Helicase–like Protein PR22 for Release of Messenger RNA from Spliceosomes, M. Company, J. Arenas, J. Abelson, Nature *349*, 487-493 (1991).

Base Pairing between U2 and U6 snRNAs is Necessary for Splicing of Mammalian pre–mRNA, J. Wu, J. Manley, Nature *352*, 818-821 (1991).

Genetic Evidence for Base Pairing between U2 and U6 snRNA in Mammalian mRNA Splicing, B. Datta, A. Weiner, Nature *352*, 821-824 (1991).

Self-Splicing

Self-splicing RNA: Autoexcision and Autocyclization of the Ribosomal RNA Intervening Sequence of *Tetrahymena*, K. Kruger, P. Grabowski, A. Zaug, J. Sands, D. Gottschling, T. Cech, Cell *31*, 147-157 (1982).

Self-splicing of Yeast Mitochondrial Ribosomal and Messenger RNA Precursors, G. van der Horst, H. Tabak, Cell *40*, 759-766 (1985).

A Self-splicing RNA Excises an Intron Lariat, C. Peebles, P. Perlman, K. Mecklenburg, M. Petrillo, J. Tabor, K. Jarrell, H. Cheng, Cell *44*, 213-223 (1986).

The Tetrahymena Ribozyme Acts Like an RNA Restriction Endonuclease, A. Zaug, M. Been, T. Cech, Nature *324*, 429-433 (1986).

Self-cleavage of Virusoid RNA is Performed by the Proposed 55-Nucleotide Active Site, A. Forster, R. Symons, Cell *50*, 9-16 (1987).

Simple RNA Enzymes with New and Highly Specific Endoribonuclease Activities, J. Haseloff, W. Gerlach, Nature *334*, 585-591 (1988).

Reversible Cleavage and Ligation of Hepatitis Delta Virus RNA, H. Wu, M. Lai, Science *243*, 652-654 (1989).

A Catalytic 13-mer Ribozyme, A. Jeffries, R. Symons, Nucleic Acids Res. *17*, 1371-1377 (1989).

Defining the Inside and Outside of a Catalytic RNA Molecule, J. Latham, T. Cech, Science *245*, 276-282 (1989).

A Computational Approach to the Mechanism of Self-Cleavage of Hammerhead RNA, H. Mei, T. Koaret, T. Bruice, Proc. Natl. Acad. Sci. USA *86*, 9727-9731 (1989).

A Mobile Group I Intron in the Nuclear rDNA of *Physarum polycephalum*, D. Muscarella, V. Vogt, Cell *56*, 443-454 (1989).

PolyA Addition

Polyadenylic Acid Sequences in the Heterogenous Nuclear RNA and Rapidly-labeled Polyribosomal RNA of HeLa Cells: Possible Evidence for a Precursor Relationship, M. Edmonds, M. Vaughn, H. Nakazato, Proc. Natl. Acad. Sci. USA *68*, 1336-1340 (1971).

3' Non-coding Region Sequences in Eukaryotic Messenger RNA, N. Proudfoot, G. Brownlee, Nature *263*, 211-214 (1976).

A Small Nuclear Ribonucleoprotein Associates with the AAUAAA Polyadenylation Signal *in vitro*, C. Hashimoto, J. Steitz, Cell *45*, 601-610 (1986).

Compensatory Mutations Suggest that Base-pairing with a Small Nuclear RNA is Required to Form the 3' End of H3 Messenger RNA, F. Schaufele, G. Gilmartin, W. Bannwarth, M. Birnstiel, Nature *323*, 777-781 (1986).

Identification of the Human U7 snRNP as One of Several Factors Involved in the 3' End Maturation of Histone Premessenger RNA's, K. Mowry, J. Steitz, Science *238*, 1683-1687 (1987).

A 64 Kd Nuclear Protein Binds to RNA Segments That Include the AAUAAA Polyadenylation Motif, J. Wilusz, T. Shenk, Cell *52*, 221-228 (1988).

Termination

Termination Factor for RNA Synthesis, J. Roberts, Nature 224, 1168-1174 (1969).

Termination of Transcription and its Regulation in the Tryptophan Operon of *E. coli*, T. Platt, Cell *24*, 10-23 (1981).

The *nusA* Gene Protein of *Escherichia coli*. G. Greenblatt, J. Li, J. Mol. Biol. *147*, 11-23 (1981).

Termination of Transcription by *nusA* Gene Protein of *E. coli*, J. Greenblatt, M. McLimont, S. Hanly, Nature *292*, 215-220 (1981).

Termination Cycle of Transcription, J. Greenblatt, J. Li, Cell *24*, 421-428 (1981).

Transcription-terminations at Lambda t_{R1} in Three Clusters, L. Lau, J. Roberts, R. Wu, Proc. Nat. Acad. Sci. USA *79*, 6171-6175 (1982).

Termination of Transcription in *E. coli*, W. Holmes, T. Platt, M. Rosenberg, Cell *32*, 1029-1032 (1983).

lac Repressor Blocks Transcribing RNA Polymerase and Terminates Transcription, U. Deuschle, R. Gentz, H. Bujard, Proc. Nat. Acad. Sci. USA *83*, 4134-4137 (1986).

Transcription Termination Factor Rho Is an RNA-DNA Helicase, C. Brennan, A. Dombroski, T. Platt, Cell *48*, 945-952 (1987).

Purified RNA Polymerase II Recognizes Specific Termination Sites During Transcription *in vitro*, R. Dedrick, C. Kane, M. Chamberlin, J. Biol. Chem. *262*, 9098-9108 (1987).

NusA Protein is Necessary and Sufficient *in vitro* for Phage Lambda N Gene Product to Suppress a Rho-independent Terminatory Placed Downstream of *nutL*, W. Whalen, B. Ghosh, A. Das, Proc. Natl. Acad. Sci. USA *85*, 2494-2498 (1988).

Structure of Rho Factor: An RNA-binding Domain and a Separate Region with Strong Similarity to Proven ATP-binding Domains, A. Dombroski, T. Platt, Proc. Natl. Acad. Sci. USA *85*, 2538-2542 (1988).

Direct Interaction between Two *Escherichia coli* Transcription Antitermination Factors, NusB and Ribosomal Protein S10, S. Mason, J. Li, J. Greenblatt, J. Mol. Biol. *223*, 55-66 (1992).

Ribosomal RNA Antitermination *in vitro*: Requirement for Nus Factors and One or More Unidentified Cellular Components, C. Squires, J. Greenblatt, J. Li, C. Condon, C. Squires, Proc. Natl. Acad. Sci. USA *90*, 970-974 (1993).

Editing

Major Transcript of the Frameshifted cixII Gene from Trypanosome Mitochondira Contains Four Nucleotides that are Not Encoded in the DNA, R. Benne, J. VanDenBurg, J. Brakenhoff, P. Sloof, J. van Boom, M. Tromp, Cell *46*, 819-826 (1986).

Editing of Kinetoplastid Mitochondrial mRNA by Uridine Addition and Deletion Generates Conserved Amino Acid Sequences and AUG Initiation Codons, J. Shaw, J. Feagin, K. Stuart, L. Simpson, Cell *53*, 401-411 (1988).

An *in vitro* System for the Editing of Apolipoprotein B mRNA, D. Driscoll, J. Wynne, S. Wallis, J. Scott, Cell *58*, 519-525 (1989).

A Model for RNA Editing in Kinetoplastic Mitochondria: "Guide" RNA Molecules Transcribed from Maxicircle DNA Provide the Edited Information, B. Blum, N. Bakalara, L. Simpson, Cell *60*, 189-198 (1990).

RNA Editing Involves Indiscriminate U Changes Throughout Precisely Defined Editing Domains, C. Decker, B. Sollner-Webb, Cell *61*, 1001-1011 (1990).

Chimeric gRNA-mRNA Molecules with Oligo(U) Tails Covalently Linked at Sites of RNA Editing Suggest That U Addition Occurs by Transesterification, B. Blum, N. Sturm, A. Simpson, L. Simpson, Cell *65*, 543-550 (1991).

Protein Structure

6

Proteins carry out most, but not all, of the interesting cellular processes. Enzymes, structural components of cells, and even secreted cellular adhesives are almost always proteins. One important property shared by most proteins is their ability to bind molecules selectively. How do proteins assume their structures and how do these structures give the proteins such a high degree of selectivity? Many of the principles are known and are discussed in this chapter.

Ultimately we want to understand proteins so well that we can design them. That is, our goal is to be able to specify an amino acid sequence such that, when synthesized, it will assume a desired three-dimensional structure, bind any desired substrate, and then carry out any reasonable enzymatic reaction. Furthermore, if our designed protein is to be synthesized in cells, we must know what necessary auxiliary DNA sequences to provide so that the protein will be synthesized in the proper quantities and at appropriate times.

The most notable advances of molecular biology in the 1980s involved nucleic acids, not proteins. Nonetheless, since DNA specifies the amino acid sequence of proteins, our ability to synthesize DNA of arbitrary sequence and put it back into cells means that the amino acid sequences of proteins can also be specifically altered. Consequently, the pace of research investigating protein structure dramatically increased around 1990. Systematic studies of the structure and activity of proteins resulting from specific amino acid substitutions are now increasing greatly our understanding of protein structure and function.

In this chapter we examine the fundamentals of protein structure. Much of this information is discussed more completely in biochemistry or physical biochemistry texts. We review the material here to develop our intuitions on the structures and properties of proteins so as to have a clearer feel for how cells function. First we discuss the components of proteins, the amino acids. Then we consider the consequences of the linking of amino acids via peptide bonds.

A variety of forces are possible between amino acids. Their origins are discussed and explained. These include electrostatic forces, dispersion forces, hydrogen bonds, and hydrophobic forces. These forces plus steric constraints lead the amino acids along many portions of the polypeptide backbone to adopt, to a first approximation, relatively simple, specific orientations known as alpha-helices, beta-sheets and beta bends. Motifs within proteins are recognizable structural elements. The structures of a number of DNA-binding motifs will be discussed. Independent folding units of proteins are called domains, and these will also be covered. Finally, physical methods that can be used to determine the identity and strength of specific amino acid residue-base interactions of DNA-binding proteins will be covered.

The Amino Acids

Proteins consist of α-L-amino acids linked by peptide bonds to form polypeptide chains (Fig. 6.1). At neutral pH, the carboxyl group of a free amino acid is negatively charged and the amino group is positively charged. In a protein, however, these charges are largely, but not completely absent from the interior amino acids owing to the formation of the peptide bonds between the amino groups and carboxyl groups. Of course, the N-terminal amino group of a protein is positively charged and the C-terminal carboxyl group is negatively charged.

Twenty different types of α-L-amino acids are commonly found in proteins (Fig. 6.2). Except for proline, which technically is an imino acid, these differ from one another only in the structure of the side group attached to the alpha carbon. A few other types of amino acids are occasionally found in proteins, with most resulting from modification of one of the twenty after the protein has been synthesized. Frequently these modified amino acids are directly involved with chemical reactions catalyzed by the protein. Each of the basic twenty must possess unique and invaluable properties since most proteins contain all twenty different amino acids (Table 6.1).

Even though we must understand the individual properties of each of the amino acids, it is convenient to classify the twenty into a smaller

Figure 6.1 An α-L-amino acid with negative charge on the carboxyl and positive charge on the amino group and three amino acids linked by peptide bonds.

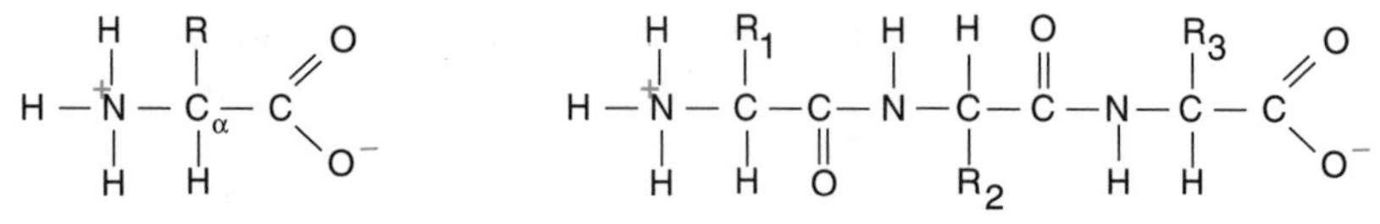

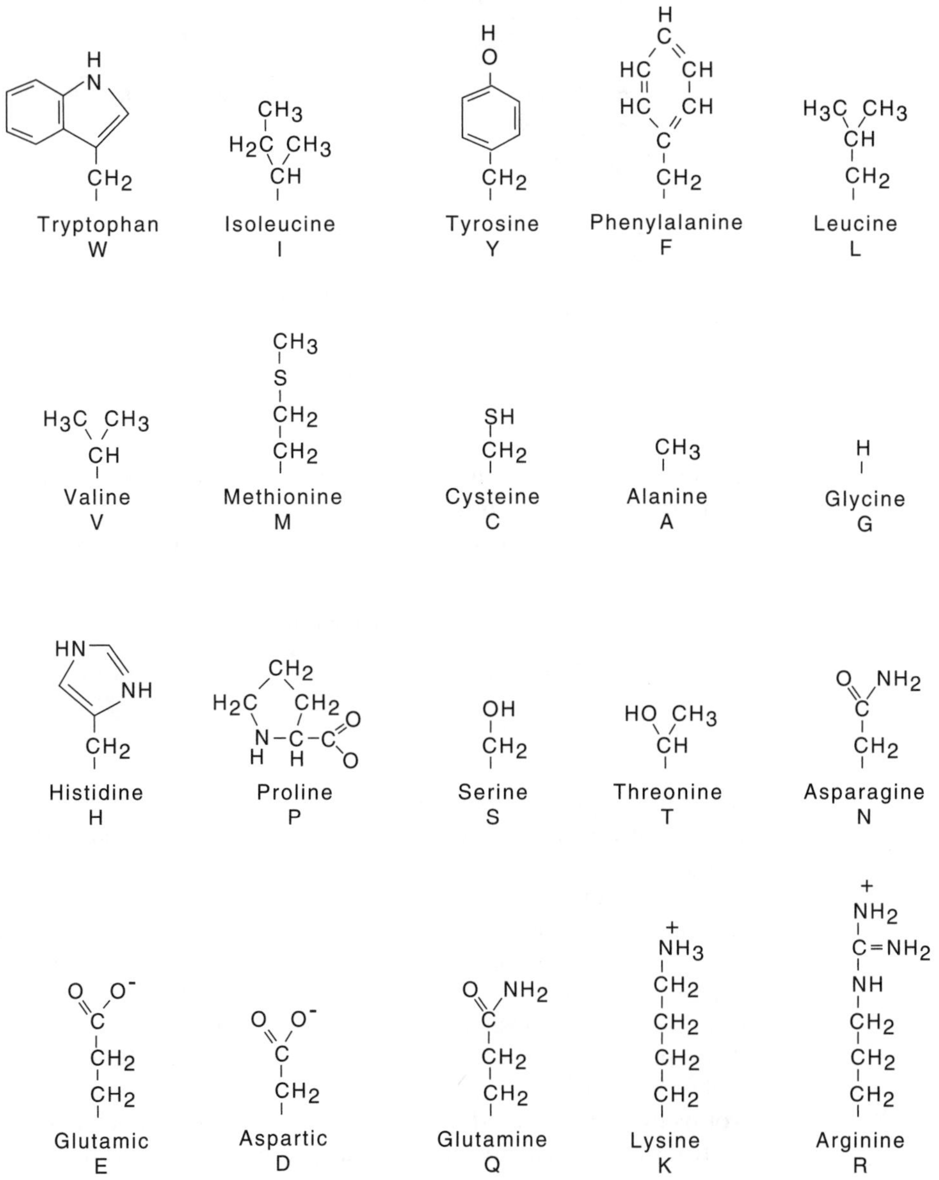

Figure 6.2 The side chains of amino acids and their single letter abbreviations. The complete structure of proline is shown. The most hydrophobic amino acids are at the top and the most hydrophilic are at the bottom.

number of groups and to understand common properties of the groups. One of the most important such groups is the hydrophobics. The side groups of the aliphatic amino acids are hydrophobic and prefer to exist in a nonaqueous, nonpolar environment like that found in the contact

Table 6.1 Properties of Amino Acids

Property	Amino acids
Hydrophobic	Ala, Ile, Leu, Val, Phe, Met, Pro
Positive charge	Arg, Lys
Negative charge	Asp, Glu
Polar	Ser, Thr, Tyr, Cys, Asn, Gln, His
Small	Gly, Ala
pK near neutrality	His
Aromatic	Phe, Tyr, Trp
Hydroxyl side chain	Ser, Thr, Tyr
Helix bend or break	Pro

region between two subunits, in the portion of a protein bound to a membrane, or in the interior of a globular protein. A contiguous area of such amino acids on a portion of the surface can make a protein bind

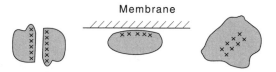

to a similar hydrophobic patch on the surface of another protein, as in the oligomerization of protein subunits, or it can make the protein prefer to bind to or even enter a membrane. Hydrophobic amino acids on the interior of a protein prefer the company of one another to the exclusion of water. This is one of the major forces that maintains the structure of a folded protein.

The basic amino acid side groups of amino acids like lysine and arginine possess a positive charge at neutral pH. If located on the surface of the protein, such positive charges can assist the binding of a negatively charged ligand, for example DNA. The acidic amino acid side groups of glutamic acid and aspartic acid possess a negative charge at neutral pH. Neutral amino acid side groups possess no net charge, and

Figure 6.3 Two reduced cysteine residues and their oxidized state, which forms a disulfide bond.

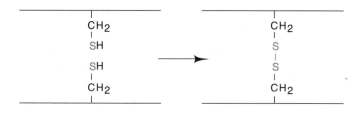

polar amino acid side groups possess separated charges like those found on glutamine. Separated charges lead to dipole interactions with other amino acids or with ligands binding to the protein.

Cysteine is a notable unique amino acid since in an oxidizing extracellular environment, but not in the intracellular environment, two cysteine residues in a protein can spontaneously oxidize to form a rather stable disulfide bond (Fig. 6.3). In the isolated protein, this bond can be reduced by the presence of an excess of a reducing reagent to regenerate cysteines.

The Peptide Bond

A peptide bond links successive amino acids in a polypeptide chain. The mere linking of amino acids to form a polypeptide chain, however, is insufficient to ensure that the joined amino acids will adopt a particular three-dimensional structure. The peptide bond possesses two extraordinarily important properties that facilitate folding of a polypeptide into a particular structure.

First, as a consequence of the partial double-bond character of the peptide bond between the carbonyl carbon and nitrogen, the unit

$$C_\alpha - \overset{\overset{\displaystyle O}{\|} (\delta-)}{C} - \underset{\underset{\displaystyle H}{|} (\delta+)}{N} - C_\alpha \rightleftharpoons C_\alpha - \overset{\overset{\displaystyle O}{|} (\delta-)}{C} = \underset{\underset{\displaystyle H}{|} (\delta+)}{N} - C_\alpha$$

bounded by the alpha carbon atoms of two successive amino acids is constrained to lie in a plane. Therefore, energy need not be consumed from other interactions to generate the "proper" orientation about the C-N bond in each amino acid. Rotation is possible about each of the two peptide backbone bonds from the C_α atom of each amino acid (Fig. 6.4). Angles of rotation about these two bonds are called ϕ and ψ, and their specification for each of the amino acids in a polypeptide completely describes the path of the polypeptide backbone. Of course, the side chains of the amino acids are free to rotate and may adopt a number of conformations so that the ϕ and ψ angles do not completely specify the structure of a protein.

The second consequence of the peptide bond is that the amide hydrogen from one amino acid may be shared with the carbonyl oxygen from another amino acid in a hydrogen bond. Since each amino acid in

$$\overset{\displaystyle \delta- \quad \delta+ \quad \delta-}{{}_{\diagdown}N-H \quad O=C_{\diagup}^{\diagdown}} \rightleftharpoons {}_{\diagdown}N-H_{\text{\tiny{IIII}}}O=C_{\diagup}^{\diagdown}$$

a polypeptide chain possesses both a hydrogen bond donor and an acceptor, many hydrogen bonds may be formed, and in fact are formed

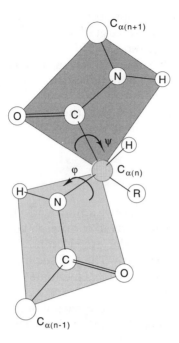

Figure 6.4 Two amino acid units in a polypeptide chain illustrating the planar structure of the peptide bond, the two degrees of rotational freedom for each amino acid unit in a polypeptide, and the angles φ and ψ.

in a polypeptide. Due to their positions on the amino acids these bonds have to be between different amino acids in the protein. Therefore they provide many stabilizing and structure-forming interactions. Although the individual hydrogen bonds are weak, the large number that can form in a protein contributes substantially to maintaining the three-dimensional structure of a protein.

Electrostatic Forces that Determine Protein Structure

Proteins have a difficult time surviving. If we heat them a little above the temperatures normally found in the cells from which they are isolated, most are denatured. Why should this be? On first consideration it would make sense for proteins to be particularly stable and to be able to withstand certain environmental insults like mild heating. One explanation for the instability is that proteins just cannot be made more stable. A second possibility is that the instability is an inherent part of proteins' activities. This latter possibility seems more likely since enzymes extracted from bacteria that thrive at temperatures near the boiling point of water frequently are inactive at temperatures below 40°. It could be that to act as catalysts in chemical reactions or to participate in other cellular activities, proteins must be flexible, and such flexibility means that proteins must exist on the verge of denaturing. A final possibility is that rapid fluctuations in structure of the folding intermediates are necessary for a protein to find the correct folded conformation. The existence of such meta-stable states may preclude the existence of a highly stable folded state. Future research should illuminate this

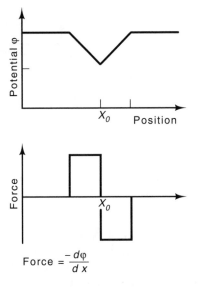

Figure 6.5 A potential φ as a function of distance x and the force it generates, which is proportional to the derivative of the potential function.

question. For the present, we will examine the origins of the weak forces that barely manage to give proteins specific shapes.

Often it is helpful to think of the interactions between amino acids in terms of forces. In some cases, physicists and physical chemists have also found it convenient to consider the interactions between objects in terms of potentials (Fig. 6.5). Some of our discussion will be more streamlined if we, also, use potentials. Forces and potentials are easily interconvertible since forces are simply related to potentials. The steepness of an object's potential at a point is proportional to the force on the object while at that point,

$$-\frac{d\varphi(r)}{dr} = F.$$

Alternatively, the potential difference between two points is proportional to the work required to move an object between the two points.

Figure 6.6 The electrostatic force between two charges of value Q_1 and Q_2 separated by a distance r and the potential φ produced by a single charge Q_1.

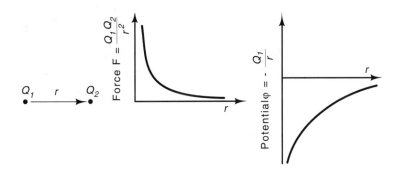

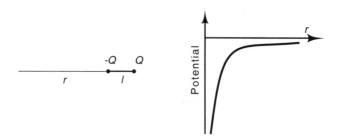

Figure 6.7 An electrical dipole and the potential generated at a distance r from the dipole. Angular dependencies have been ignored.

Electrostatics are the basis of several of the forces that determine protein structure. Charges of unlike sign attract each other with a force proportional to each of the charges and inversely proportional to the square of the charge separation (Fig. 6.6). The force is also inversely proportional to the dielectric constant of the medium, but for purposes of our discussion, this factor will not be considered. Thus, the potential generated at a point by a charge is proportional to the magnitude of the charge and inversely proportional to the distance of the point from the charge

$$\varphi(r) = -\frac{Q}{r}.$$

The electric field at a point is defined as the force a unit charge would feel at that point. Therefore, the magnitude of the electric field at a point r generated by a charge Q at the origin is

$$E = -\frac{d}{dr}\frac{Q}{r}$$

$$= \frac{Q}{r^2}.$$

Since both positively and negatively charged amino acids exist, direct electrostatic attractions are possible in proteins. These generate what are called salt bridges. A number of proteins contain such salt bridges.

When two equal and opposite charges or partial charges are located near one another, as is found in the polar amino acids and in each peptide bond, it is convenient to view their combined effect on other atoms and molecules as a whole rather than considering each charge individually. A dipole with charges $+Q$ and $-Q$ separated by a distance l (Fig. 6.7) generates a potential and electric field proportional to

$$\varphi(r) = \frac{Q}{r} - \frac{Q}{r+l}$$

$$\approx \frac{Q}{r}\left(1 - \left(1 - \frac{l}{r}\right)\right)$$

$$\approx \frac{Ql}{r^2},$$

$$E \approx \frac{Ql}{r^3}.$$

Interactions between pairs of dipoles are also common in proteins. By the same reasoning as used above, the potential between the two dipoles ql and QL can be shown to be proportional to

$$\frac{qlQL}{r^3}.$$

Even more important to protein structure and function than the interactions between permanent dipoles in proteins are the momentary interactions between temporary dipoles that have been created by a brief fluctuation in the positions of charges. The forces generated by interactions between such dipoles are called London dispersion forces. They are weak, short-ranged, up to an Angstrom or two, attractive forces that exist between all molecules. These form the basis of most of the selectivity in the binding of other molecules to proteins. If the shapes of a protein and another molecule are complementary, then many of these attractive forces can act and hold the two molecules tightly together. If

the shapes are not exactly complementary, then because the forces are short-ranged, only the small areas in contact are subject to dispersion attractive forces and the two molecules do not strongly bind to one another.

Dispersion forces are particularly short-ranged since their attractive potential falls off with the sixth power of the distance separating the molecules. Although these forces are best understood in the framework of quantum mechanics, we can understand the origin of the sixth-power dependence. Consider an electrically neutral nonpolar molecule. Thermal fluctuations can generate a momentary separation of its plus and

Dipole, D_1

minus charges. That is, a dipole of strength $D_1 = ql$ is briefly generated. The electric field produced by this dipole can induce a dipole in an adjacent susceptible molecule. The strength of the induced dipole is

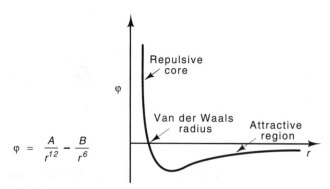

$$\varphi = \frac{A}{r^{12}} - \frac{B}{r^6}$$

Figure 6.8 A Van der Waals potential.

directly proportional to the strength of the local electric field. That is, the induced dipole has a strength, D_2, proportional to

$$\frac{D_1}{r^3}.$$

Since, as discussed above, the potential between two dipoles is proportional to the product of their strength and inversely proportional to the cube of their separation,

$$\varphi = \frac{D_1 D_2}{r^3}.$$

Substituting the value for D_2 in the potential yields the result that the potential between the two dipoles is inversely proportional to the sixth power of their separation,

$$\varphi = c\,\frac{D_1^2}{r^6}.$$

Due to the inverse sixth-power dependence, the dispersion forces become much stronger as the distance separating two molecules becomes smaller. The force cannot become too great, for once the electronic cloud of one molecule begins to interpenetrate the cloud of the other molecule, a very strong repulsive interaction sets in. It is computationally convenient to approximate this repulsive potential as an inverse twelfth power of the separation of the centers of the atoms. The combination of the two potentials is known as a Van der Waals potential. The radius at which the strong repulsion begins to be significant is the Van der Waals radius (Fig. 6.8).

Hydrogen Bonds and the Chelate Effect

A hydrogen atom shared by two other atoms generates a hydrogen bond. This sharing is energetically most important when the three atoms are in a straight line and the atom to which the hydrogen is covalently bonded, the hydrogen bond donor, possesses a partial negative charge

$$\overset{\delta^-}{\underset{}{\diagdown}}\underset{}{\overset{\delta^+}{N-H}} \text{ⅢⅢ} \overset{\delta^-}{O} \diagup \overset{H}{\diagdown_H} \qquad \text{Stronger than} \qquad \diagdown N - H \equiv O \diagdown_{H\ \ H}$$

and the partner atom, a hydrogen bond acceptor, also possesses a partial negative charge. Then the atoms may approach each other quite closely and the electrostatic attractive forces and the dispersion forces will be appreciable. Since the amide of the peptide bond can be a hydrogen donor, and the carboxyl can be a hydrogen acceptor, proteins have a potential for forming a great many hydrogen bonds. In addition, more than half the side groups of the amino acids usually participate in hydrogen bonding.

A paradox is generated by the existence of hydrogen bonds in proteins. Studies with model compounds show that a hydrogen bond to water should be stronger than a hydrogen bond between amino acids. Why then don't proteins denature and make all their hydrogen bonds to water? A part of the answer is the chelate effect. That is, two objects appear to bind to one another far more strongly if something else holds them in the correct binding positions than if their own attractive forces must correctly position the objects. In a protein with a structure that holds amino acids in position, any single bond between amino acids within the protein is entropically more favorable than altering the structure of the protein and making the bond to water. Another way of looking at this is that the formation of one hydrogen bond holds other amino acids in position so that they may more easily form hydrogen bonds themselves.

The chelate effect is important in understanding many phenomena of molecular biology. A different example, explained more fully later, concerns proteins. Much of the work required for two macromolecules to bind to one another is correctly positioning and orienting them. Consider the binding of a protein to DNA. If the protein and DNA have been correctly positioned and oriented, then all of their interaction energy can go into holding the two together. In the binding of a dimeric protein to DNA, once the first subunit has bound, the second subunit is automatically positioned and oriented correctly. Therefore the second subunit appears to have the larger effect in binding the protein to DNA than the first one. Equivalently, the dimer appears to bind more tightly than would be predicted by simply doubling the ΔG of the binding reaction of the monomer.

Hydrophobic Forces

The structures of the many proteins that have been determined by X-ray diffraction and nuclear magnetic resonance reveal that, in general, the polar and charged amino acids tend to be found on the surface and the aliphatic amino acids tend to be found in the interior. Hydrophobic

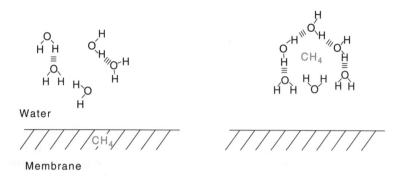

Water

Membrane

Figure 6.9 The creation of a water cage around a hydrocarbon in water, when it moves from membrane into water.

forces make aliphatic amino acids try to escape from a water environment and to cluster together in the center of a protein away from water.

The precise definition of hydrophobic force and methods of its measurement are currently under rapid development. One way of considering the phenomenon begins by considering the energy and entropy change in moving a neutral, nonpolar amino acid out of the interior of a protein and into the surrounding water (Fig. 6.9). The entry of a hydrocarbon into water facilitates the formation of structured cages of water molecules around the hydrocarbon molecule. These surround the hydrocarbon but do not significantly interact with it. The energy of formation of these structures actually favors their generation, but the translational and rotational entropy loss required to form the structured water cages inhibits their production. From considerations at this level, we cannot deduce the magnitude of the effects. Those are determined by measuring the relative solubility of different hydrocarbons in water and organic solvents at various temperatures. The results show that the state of the system in which these cages are absent, that is, with the nonpolar amino acids in the interior of the protein, is more probable than the state in which they are present on the protein's surface.

Hydrophobic forces can be expected to be strongest at some intermediate temperature between freezing and boiling. Near freezing temperatures, the water throughout the solution becomes more structured, and thus there is little difference between the status of a water molecule in solution or a water molecule in a cage around a hydrophobic group. Alternatively, at high temperatures, little of the water around a hydrophobic group can be be structured. It is melted out of structure. The difference between water around a hydrophobic group and water elsewhere in the solution is maximized at some intermediate temperature. As this difference is important to protein structure, some proteins possess maximum stability at intermediate temperatures. A few are actually denatured upon cooling. A more common manifestation of the hydrophobic forces is the fact that some polymeric structures are destabilized by cooling and depolymerize because the hydrophobic forces holding them together are weaker at lower temperatures.

Thermodynamic Considerations of Protein Structure

Thermodynamics provides a useful framework for calculation of equilibrium constants of reactions. This also applies to the "reaction" of protein denaturation. Consider a protein denaturing from a specific native conformation, N, to any of a great many nonspecific, random conformations characteristic of denatured proteins, D. The reaction can be described by an equilibrium constant that relates the amount of the protein found in each of the two states if the system has reached equilibrium,

$$N \leftrightarrows D$$

$$K_{eq} = \frac{D}{N}.$$

Thermodynamics provides a way of calculating K_{eq} as

$$K = e^{-\Delta G/RT}$$

$$= e^{-(\Delta H - T\Delta S)/RT}$$

where ΔG is the change in Gibbs free energy; R is the universal gas constant; T is the absolute temperature in degrees Kelvin; ΔH is the enthalpy change of the reaction, which in biological systems is equivalent to binding energy when volume changes can be neglected; and ΔS is the entropy change of the reaction. Entropy is related to the number of equivalent states of a system. The state of a protein molecule confined to one conformation without any degrees of freedom possesses much lower entropy than a denatured protein that can adopt any of a great number of conformations all at the same energy. For clarity, we will neglect the contributions of the surrounding water in further considerations, but in physically meaningful calculations these too must be included.

Let us examine why proteins denature when the temperature is raised. If the protein is in the folded state at the lower temperature, K_{eq} is less than 1, that is,

$$\frac{-\Delta G}{RT} < 0$$

$$-\Delta H + T\Delta S < 0,$$

$$\Delta H > T\Delta S.$$

As the temperature increases, neglecting the small temperature-dependent changes that occur in the interaction energies and entropy change, the term $T\Delta S$ increases, and eventually exceeds ΔH. Then the equilibrium shifts to favor the denatured state.

The temperature dependence of the denaturing of proteins provides the information necessary for determination of ΔH of denaturing. It is very large! This means that ΔS for denaturing is also very large, just as we inferred above, and at temperatures near the denaturing point, the difference of these two large numbers barely favors retention of the structure of the protein. Hence the binding energies of the many interactions that determine protein structure, hydrogen bonds, salt bridges, dipole-dipole interactions, dispersion forces, and hydrophobic forces just barely overcome the disruptive forces. Thus we see the value of the peptide bond. If rotations about the C-N bond were not restricted, the increased degrees of freedom available to the protein would be enormous. Then the energy and entropy balance would be tipped in the direction of denatured proteins.

Structures within Proteins

It is useful to focus attention on particular aspects of protein structures. The primary structure of a protein is its linear sequence of amino acids. The local spatial structure of small numbers of amino acids, independent of the orientations of their side groups, generates a secondary structure. The alpha helix, beta sheet, and beta turn are all secondary structures that have been found in proteins. Both the arrangement of the secondary structure elements and the spatial arrangement of all the atoms of the molecule are referred to as the tertiary structure. Quaternary structure refers to the arrangement of subunits in proteins consisting of more than one polypeptide chain.

A domain of a protein is a structure unit intermediate in size between secondary and tertiary structures. It is a local group of amino acids that have many fewer interactions with other portions of the protein than they have among themselves (Fig. 6.10). Consequently, domains are independent folding units. Interestingly, not only are the amino acids of a domain near one another in the tertiary structure of a protein, but they usually comprise amino acids that lie near one another in the primary structure as well. Often, therefore, study of a protein's structure can be done on a domain-by-domain basis. The existence of semi-independent domains should greatly facilitate the study of the folding of polypeptide chains and the prediction of folding pathways and structures.

Particularly useful to the ultimate goal of prediction of protein structure has been the finding that many alterations in the structure of proteins produced by changing amino acids tend to be local. This has

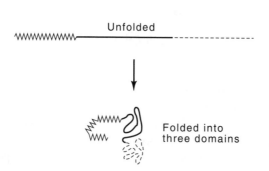

Figure 6.10 Folding of adjacent amino acids to form domains in a protein.

Unfolded

Folded into three domains

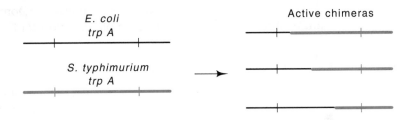

Figure 6.11 Substitution of portions of the *E. coli* tryptophan synthetase α subunit with corresponding regions from the *Salmonella typhimurium* synthetase subunit.

been found in exhaustive genetic studies of the *lac* and lambda phage repressors, in the thermodynamic properties of mutant proteins, and in the actual X-ray or NMR determined structures of a number of proteins. In the *lac* and lambda repressors, the majority of the amino acid changes that alter the ability to bind to DNA lie in the portion of the protein that makes contact with the DNA. Similar results can be inferred from alterations in the amino acid sequence of the tryptophan synthetase protein generated by fusing two related but nonidentical genes. Despite appreciable amino acid sequence differences in the two parental types, the fusions that contain various amounts of the N-terminal sequence from one of the proteins and the remainder of the sequence from the other protein retain enzymatic activity (Fig. 6.11). This means that the amino acid alterations generated by formation of these chimeric proteins do not need to be compensated by special amino acid changes at distant points in the protein.

The results obtained with repressors and tryptophan synthetase mean that a change of an amino acid often produces a change in the tertiary structure that is primarily confined to the immediate vicinity of the alteration. This, plus the finding that protein structures can be broken down into domains, means that many of the potential long-range interactions between amino acids can be neglected and interactions over relatively short distances of up to 10 Å play the major role in determining protein structure.

The proteins that bind to enhancer sequences in eukaryotic cells are a particularly dramatic example of domain structures in proteins. These proteins bind to the enhancer DNA sequence, often bind to small molecule growth regulators, and activate transcription. In the glucocorticoid receptor protein, any of these three domains may be independently inactivated without affecting the other two. Further, domains may be interchanged between enhancer proteins so that the DNA-binding specificity of one such protein can be altered by replacement with the DNA-binding domain from another protein.

As we saw earlier in discussing mRNA splicing, DNA regions encoding different domains of a protein can be appreciably separated on the chromosome. This permits different domains of proteins to be shuffled so as to accelerate the rate of evolution by building new proteins from new assortments of preexisting protein domains. Domains rather than amino acids then become building blocks in protein evolution.

The Alpha Helix, Beta Sheet, and Beta Turn

The existence of the alpha helix was predicted by Pauling and Cory from careful structural studies of amino acids and peptide bonds. This prediction came before identification of the alpha helix in X-ray diffraction patterns of proteins. Even though the data were all there, it was overlooked. The alpha helix is found in most proteins and is a fundamental structural element. In the alpha helix, hydrogen bonds are formed between the carbonyl oxygen of one peptide bond and the amide hydrogen of the amino acid located three and a third amino acids away.

3.3 Residue separation

$$H-N-\overset{\overset{H}{|}}{\underset{\underset{H}{|}}{C}}-\overset{\overset{R}{|}}{\underset{\underset{H}{|}}{C}}-\overset{\overset{O}{||}}{C}-N-\overset{\overset{R}{|}}{\underset{\underset{H}{|}}{C}}-\overset{\overset{O}{||}}{C}-N-\overset{\overset{R}{|}}{\underset{\underset{H}{|}}{C}}-\overset{\overset{O}{||}}{C}-N-\overset{\overset{R}{|}}{\underset{\underset{H}{|}}{C}}-\overset{\overset{O}{||}}{C}-N-\overset{\overset{R}{|}}{\underset{\underset{H}{|}}{C}}$$

The side chains of the amino acids extend outward from the helix, and the hydrogen bonds are nearly parallel to the helix axis (Fig. 6.12). If they were precisely parallel to the axis, the helix pitch would be 3.33 amino acids per turn, but due to steric constraints, the hydrogen bonds are somewhat skewed, and the average pitch is found to be 3.6 to 3.7 amino acids per turn.

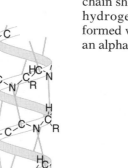

Figure 6.12 A polypeptide chain showing the backbone hydrogen bonds that are formed when the chain is in an alpha helix.

3.3 Amino acids per turn

3.6 Amino acids per turn

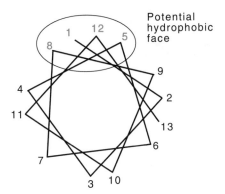

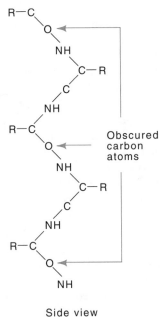

Figure 6.13 An alpha helical wheel showing the positions of successive amino acids as viewed from the end of a helix.

If we look down the axis of an alpha helix, we see the amino acids winding around in a circle. Every third and then every fourth amino acid lies on one side of the helix (Fig. 6.13). This pattern follows from the fact that the alpha helix is nearly 3.5 amino acids per turn. If every third and then every fourth amino acid were hydrophobic, two such helices could bind together through their parallel strips of hydrophobic amino acids. This occurs in structures called coiled coils. These are found in structural proteins like myosin as well as in a class of transcriptional regulators that dimerize by these interactions. These activators are called leucine-zipper proteins. They possess leucine residues seven amino acids apart. Strips of hydrophobic amino acids along one face of alpha helices are frequently found in bundles containing two, three, or four alpha helices.

The beta-strand is a second important structural element of proteins. In it the polypeptide chains are quite extended (Fig. 6.14). From a top

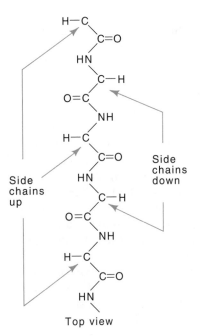

Figure 6.14 A portion of a beta strand viewed from the top and from the side.

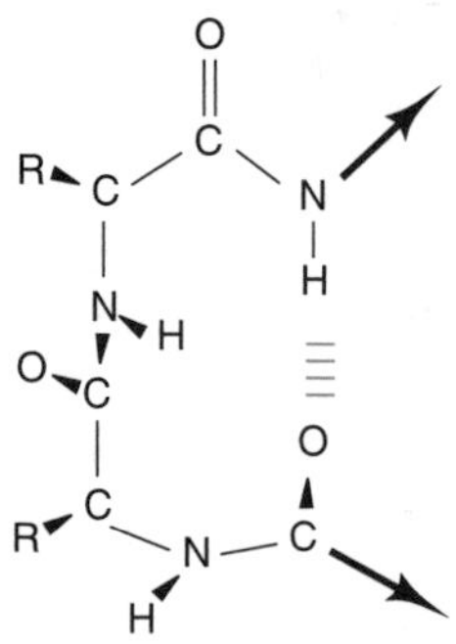

Figure 6.15 Three amino acids forming a beta bend.

view the peptide backbone is relatively straight, but in a side view the peptide backbone is pleated. The side chains of the amino acids are relatively unconstrained since alternate groups are directed straight up and straight down. The amide hydrogens and the carboxyl groups are directed to either side and are available for hydrogen bonding to another beta-strand lying alongside to form a beta sheet. This second strand can be oriented either parallel or antiparallel to the first.

The third readily identified secondary structural element is the reverse or beta bend (Fig. 6.15). A polypeptide chain must reverse direction many times in a typical globular protein. The beta bend is an energy-effective method of accomplishing this goal. Three amino acids often are involved in a reverse bend.

Calculation of Protein Tertiary Structure

The sequence of amino acids in a protein and the environment of the protein usually determines the structure of the protein. That is, most proteins are capable of folding to their correct conformations without the assistance of any folding enzymes. This is known from the fact that many proteins can be denatured by heat or the addition of 6 M urea and will renature if slowly returned to nondenaturing conditions. Since the sequence is sufficient to determine structure, can we predict the structure? The correct folding of some proteins, however, appears to require the assistance of auxiliary proteins called chaperonins.

We can imagine several basic approaches to the prediction of protein structure. The first is simply to consider the free energy of every possible conformation of the protein. We might expect that the desired structure of the protein would be the conformation with the lowest potential energy. The approach of calculating energies of all possible conformations possesses a serious flaw. Computationally, it is completely infeasible since a typical protein of 200 amino acids has 400 bonds along the peptide backbone about which rotations are possible. If we consider each 36° of rotation about each such bond in the protein to be a new state, there are 10 states per bond, or 10^{400} different conformational states of the protein. With about 10^{80} particles in the universe, a calculational speed of 10^{10} floating point operations per second (flops,

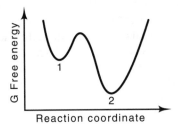

Figure 6.16 A system in state 1 is metastable and, when equilibrium is reached, should be found in state 2. Some proteins as isolated in their active form could be in meta-stable states. Similarly, in calculation of conformation, a calculation seeking an energy minimum might become trapped in state 1, when, in fact, state 2 is the correct conformation.

a unit of speed measure of computers), an age of the universe of roughly 10^{18} seconds, with one superfast computer for each particle in the universe and beginning to calculate at the origin of the universe, we would have had time to list, let alone calculate the energy of, only an infinitesimal fraction of the possible states of one protein.

The preceding example is known as the Levinthal paradox. It illustrates two facts. First, that we cannot expect to predict the folded structure of a protein by examining each possible conformation. Second, it seems highly unlikely that proteins sample each possible conformational state either. More likely they follow a folding pathway in which at any moment the number of accessible conformations is highly limited. We can try to fold a protein by an analogous method. This can be done by varying, individually, the structure variables like angles. As long as changing an angle or distance in one direction continues to lower the total energy of the system, movement in this direction is permitted to continue. When minima have been found for all the variables, the protein ought to be in a state of lowest energy. Unfortunately, the potential energy surface of proteins does not contain just one local minimum. Many exist. Thus, when the protein has "fallen" into a potential energy well, it is very unlikely to be in the deepest well (Fig. 6.16). This energy minimization approach has no convenient way to escape from a well and sample other conformation states so as to find the deepest well. One approach to avoiding this problem might be to try to fold the protein by starting at its N-terminus by analogy to the way natural proteins are synthesized. Unfortunately, this does not help much in avoiding local mimima or achieving the correct structures.

Yet a third way for us to calculate structure might be to mimic what a protein does. Suppose we calculate the motion of each atom in a protein simply by making use of Newton's law of motion

$$F = ma.$$

From chemistry we know the various forces pushing and pulling on an atom in a molecule. These are the result of stretching, bending, and twisting ordinary chemical bonds, plus the dispersion forces or Van der Waals forces we discussed earlier, electrical forces, and finally hydrogen bonds to other atoms. Of course, we cannot solve the resulting equations analytically as we do in some physics courses for particularly simple idealized problems. Solving has to be done numerically. At one instant positions and velocities are assumed for each atom in the structure. From the velocities we can calculate where each atom will be 10^{-14} second later. From the potentials we can calculate the average forces acting on each atom during this interval. These alter the velocities according to Newton's law, and at the new positions of each atom, we adjust the velocities accordingly and proceed through another round of calculations. This is done repeatedly so that the structure of the protein develops in segments of 10^{-14} second. The presence of local minima in the potential energy function is not too serious for protein dynamics calculations since the energies of the vibrations are sufficient to jump out of the local minima.

The potential function required to describe a protein, while large, can

$$E = \sum_{Bonds} \frac{K_b}{2}(b - b_0)^2 + \sum_{Angles} \frac{K_\theta}{2}(\theta - \theta_0)^2 + \sum_{Torsions} \frac{K_\varphi}{2}(1 + \cos(n\varphi - \delta))$$

$$\sum_{Pairs} \left(\frac{A}{r^{12}} - \frac{C}{r^6} + \frac{Dq_1q_2}{r}\right) + \sum_{H\ bonds} \left(\frac{A'}{r^{12}} - \frac{C'}{r^6}\right).$$

be handled by large computers. These calculations take many hours on the largest computers and can simulate the motions of a protein only for times up to 10 to 100 picoseconds. This interval is insufficient to model the folding of a protein or even to examine many of the interesting questions of protein structure.

Another useful approach with molecular dynamics is to begin with the coordinates of a protein derived from X-ray crystallography. Each of the atoms is then given a random velocity appropriate to the temperature being simulated. Soon after the start of the calculations, the protein settles down and vibrates roughly as expected from general physics principles. During the course of such simulations the total energy in the system ought to remain constant, and the calculations are done with sufficient accuracy that this constraint is satisfied. The vibrations seen in these simulations can be as large as several angstroms. Frequently sizeable portions of the protein engage in cooperative vibrations.

Secondary Structure Predictions

Less ambitious than calculating the tertiary structure of a protein is predicting its secondary structure. There is some hope that this is a much simpler problem than prediction of tertiary structure because most of the interactions determining secondary structure at an amino acid residue derive from amino acids close by in the primary sequence. The problem is how many amino acids need to be considered and how likely is a prediction to be correct? We can estimate both by utilizing information from proteins whose structures are known. The question is how long a stretch of amino acids is required to specify a secondary structure. For example, if a stretch of five amino acids were sufficient, then the same sequence of five amino acids ought to adopt the same structure, regardless of the protein in which they occur.

The tertiary structures available from X-ray diffraction studies can be used as input data. Many examples now exist where a sequence of five amino acids appears in more than one protein. In about 60% of these cases, the same sequence of five amino acids is found in the same secondary structure. Of course, not all possible five amino acid sequences are represented in the sample set, but the set is sufficiently large that it is clear that we should not expect to have better than about 60% accuracy in any secondary structure prediction scheme if we consider only five amino acids at a time.

Several approaches have been used to determine secondary structure prediction rules. At one end is a scheme based on the known conformations assumed by homopolymers and extended by analysis of a small number of known protein structures. The Chou and Fasman approach is in this category. A more general approach is to use information theory to generate a defined algorithm for predicting secondary structure. This overcomes many of the ambiguities of the Chou-Fasman prediction scheme.

Recently, neural networks have been applied to predicting secondary structure. While usually implemented on ordinary computers, these simulate on a crude scale some of the known properties of neural connections in parts of the brain. Depending on the sum of the positive and negative inputs, a neuron either does not fire, or fires and sends activating and inhibiting signals on to the neurons its output is connected to.

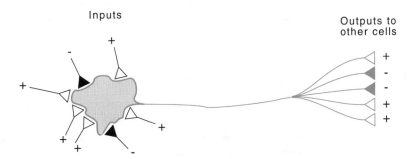

In predicting secondary structure by a neural network, the input is the identity of each amino acid in a stretch of ten to fifteen amino acids (Fig. 6.17). Since each of these can be any of the twenty amino acids, about 200 input lines or "neurons" are on this layer. Each of these activates or inhibits each neuron on a second layer by a strength that is adjusted by training. After summing the positive and negative signals reaching it, a neuron on the second layer either tends to "fire" and sends a strong activating or inhibiting signal on to the third layer, or it tends not to fire. In the case of protein structure prediction, there would be three neurons in the third layer. One corresponds to predicting α-helix, one to β-sheet, and one to random coil. For a given input sequence, the network's secondary structure prediction for the central amino acid of

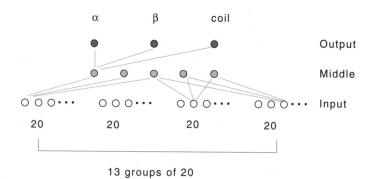

Figure 6.17 A three layer neural network of 20×13 inputs, a middle layer, and an output layer. Each of the input "neurons" is connected to each of the middle layer neurons and each of the middle layer neurons is connected to the three output neurons. The strengths of the interactions are not all equal.

the sequence is considered to correspond to the neuron of the third layer with the highest output value. "Training" such a network is done by presenting various stretches of amino acids whose secondary structure is known and adjusting the strengths of the interactions between neurons so that the network predicts the structures correctly.

No matter what scheme is used, the accuracy of the resulting structure prediction rules never exceeds about 65%. Note that a scheme with no predictive powers whatsoever would be correct for about 33% of the amino acids in a protein. The failure of these approaches to do better than 65% means that in some cases, longer-range interactions between amino acids in a protein have a significant effect in determining secondary structure (See problem 6.18).

Structures of DNA-Binding Proteins

A protein that regulates the expression of a gene most often recognizes and binds to a specific DNA sequence in the vicinity of that gene. Bacteria contain at least several thousand different genes, and most are likely to be regulated. Eukaryotic cells contain at least 10,000 and perhaps as many as 50,000 different regulated genes. Although combinatorial tricks could be used to reduce the number of regulatory proteins well below the total number of genes, it seems likely that cells contain at least several thousand different proteins that bind to specific sequences. What must the structure of a protein be in order that it bind with high specificity to one or a few particular sequences of DNA? Does nature use more than one basic protein structure for binding to DNA sequences?

Chapter 2 discussed the structure of DNA and pointed out that the sequence of DNA can be read by hydrogen bonding to groups within the major groove without melting the DNA's double-stranded structure. Each of the four base pair combinations, A-T, T-A, G-C, and C-G,

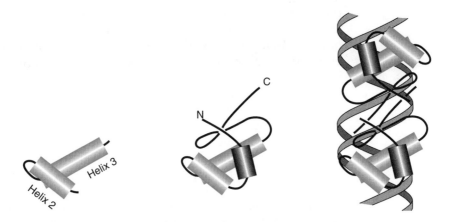

Figure 6.18 Left, a helix-turn-helix. Center, the helix-turn-helix of *cro* repressor in the context of the remainder of the polypeptide chain. Right, a dimer of *cro* repressor with the recognition helices fitting into the major groove of the DNA.

generates a unique pattern of hydrogen bond donors and acceptors. Therefore, the identity of a base pair can be read by a single, properly positioned, amino acid. Of course, an amino acid need not be constrained to bond only to a single base pair or just to the hydrogen bond donors and acceptors in the major groove. Bonding from an amino acid residue is also possible to multiple base pairs, to the deoxyribose rings, or the phosphate groups.

Since the width of the major groove of DNA nicely accepts an alpha-helix, it seems likely that such helices play a major role in DNA-binding proteins, and this has been found. We might also expect proteins to maximize their sequence selectivity by constructing their recognition surfaces to be as rigid as possible. If the surface is held in the correct shape while the protein is free in solution, then none of the protein-DNA binding energy needs to be consumed in freezing the surface in the correct shape. All the interaction energy between the protein and DNA can go into holding the protein on the DNA and none needs to be spent holding the protein in the correct conformation. Further, the DNA-contacting surface must protrude from the protein in order to reach into the major groove of the DNA. These considerations lead to the idea that proteins may utilize special mechanisms to stiffen their DNA-binding domains, and indeed, this expectation is also met.

The high interest and importance in gene regulatory proteins means that a number have been purified and carefully studied. Sequence analysis and structure determination have revealed four main families of DNA-binding domains. These are the helix-turn-helix, zinc domain, leucine zipper, and the beta-ribbon proteins.

The helix-turn-helix domains possess a short loop of four amino acids between the two helical regions (Fig. 6.18). Because the connection between the helices is short and the helices partially lie across one another, they form a rigid structure stabilized by hydrophobic interactions between the helices. For historical reasons the first of these two helices is called Helix 2, and the following helix is called Helix 3. In these proteins Helix 3 lies within the major groove of the DNA, but its orientation within the groove varies from one DNA-binding protein to another. In some, the helix lies rather parallel to the major groove, but in others, the helix sticks more end-on into the groove. The actual positioning of the helix in the groove is determined by contacts between the protein and phosphate groups, sugar groups, and more distal parts of the protein. Contacts between the protein and DNA need not be direct. *Trp* repressor makes a substantial number of contacts indirectly via water molecules. Most prokaryotic helix-turn-helix proteins are homodimeric, and they therefore bind a repeated sequence. Since the subunits face one another, the repeats are inverted, and form a symmetrical sequence, for example AAAGGG-CCCTTT. Developmental genes in eukaryotes often are regulated by proteins that are a close relative of the helix-turn-helix protein. These homeodomain proteins possess structures highly similar to the helix-turn-helix structures, but either of the helices may be longer than their prokaryotic analogs. The homeodomain proteins usually are monomeric and therefore bind to asymmetric sequences.

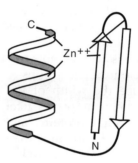

The zinc finger proteins are the most prominent members of the zinc-containing proteins that bind to DNA. The zinc finger is a domain of 25 to 30 amino acids consisting of a 12 residue helix and two beta-ribbons packed against one another. In these proteins, Zn ligands to four amino acids and forms a stiffening cross-bridge. Often the Zn is held by two cysteines at the ends of the ribbons beside the turn, and by two histidines in the alpha-helix. This unit fits into the major groove of the DNA and contacts three base pairs. Usually these Zn-finger proteins

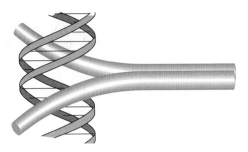

Figure 6.19 A leucine zipper protein binding to DNA.

contain multiple fingers, and the next finger can contact the following three base pairs. Other Zn-containing proteins that bind to DNA contain multiple Zn ions in more complex structures. These proteins also use residues in the alpha-helix and at its end to contact the DNA. Zn fingers can also be used to make sequence-specific contacts to RNA.

The leucine zipper proteins are of a particularly simple design. They contain two alpha-helical polypeptides dimerized by hydrophobic faces on each. Characteristically, each polypeptide contains leucine residues seven amino acids apart that form the dimerization faces. Beyond the dimerization domains the helices diverge as in a "Y", with each arm passing through a major groove (Fig. 6.19). Few leucine zipper proteins have been identified in prokaryotes, but they are common in eukaryotes. They are notable in forming heterodimers. For example, the *fos-jun* transcription factor is a leucine zipper protein. It is easy to see how specific dimerization could be determined by patterns of positive and negative charges near the contact regions of the two dimerization regions.

The beta-sheet domains contact either the major or minor groove of DNA via two antiparallel beta-strands. A few prokaryotic and eukaryotic examples of these proteins are known, the most notable being the X-ray determined structure of the MetJ repressor-operator complex. At least part of the DNA contacts made by transcription factor TFIID are in the minor groove and are made by such beta-strands.

Salt Effects on Protein-DNA Interactions

Many of the macromolecular interactions of interest in molecular biology involve the binding of proteins to DNA. Experiments show that often the affinity of a protein for DNA is sharply reduced as the concentration of NaCl or KCl in the buffer is increased. It might seem that the source of such behavior would be attractive forces between positive charges in the protein and the negatively charged phosphates of the DNA. The presence of high salt concentrations would then shield the attractive forces and weaken the binding by affecting both the association rate and the dissociation rate. Experimentally, however, the concentration of salts in the buffer primarily affects only the dissociation rates!

Consider the binding of a protein to DNA. Before binding of the protein, positively charged ions must reside near the negatively charged phosphate groups. Some of these will be displaced as the protein binds, and this displacement, even though it is not the breaking of a covalent bond, must be considered in the overall binding reaction. Let us consider these to be sodium ions. The reaction can be written

$$P + nNa^+ \cdot D \leftrightarrows nNa^+ + P \cdot D$$

$$K_{eq} = \frac{[P] \times [nNa^+ \cdot D]}{[Na^+]^n \times [P \cdot D]},$$

where P is protein concentration, D is DNA concentration, and an effective number n of Na^+ ions are displaced in the binding reaction. The fact that n is often greater than five makes the binding affinity sharply dependent upon salt concentration. The value of n can be easily extracted from experimental data by plotting the log of K_d against the log of the salt concentration. Many regulatory proteins appear to displace a net of four to ten ions as they bind.

The physical basis for the ion strength dependence is straightforward. The dissociation of a protein from DNA would require the nearly simultaneous and precise binding of the n sodium ions to their former positions on the DNA. The higher the concentration of sodium ions in the solution, the more easily this may be accomplished.

The above explanation can also be couched in thermodynamic terms. The entropic contribution of the sodium ions to the binding-dissociation reaction is large. Before the binding of the protein, the ions are localized near the backbone of the DNA. Upon the binding of the protein, these ions are freed, with a substantial entropy increase. The greater the sodium concentration in the buffer, the smaller the entropy increase upon protein binding and therefore the weaker the protein binding. Loosely bound, and therefore poorly localized, ions do not make a substantial contribution to the changes in binding as the buffer composition is changed because their entropies do not change substantially during the course of the reaction.

Locating Specific Residue-Base Interactions

In Chapter 2 we discussed how a histone bound to DNA can protect the nucleotides it contacts from DNAse cleavage of phosphodiester bonds. Biochemical methods similar to those DNAse footprinting experiments can even identify interactions between specific amino acid residues and specific bases. In some cases the biochemical approaches provide as much resolution as X-ray crystallography.

First, consider how to test a guess of a specific residue-base interaction. In this approach a correct guess can be confirmed, but a poor guess yields little information. The idea is that if the amino acid residue in question is substituted by a smaller residue, say a glycine or alanine,

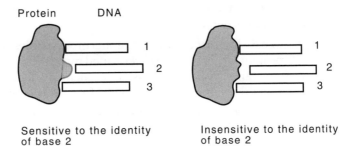

Figure 6.20 How replacing a DNA contacting residue of a protein with a smaller residue like alanine or glycine makes the protein indifferent to the identity of the base that formerly was contacted by the residue.

then the former interaction between the residue and the base will be eliminated. As a result, the strength of the protein's binding becomes independent of the identity of the base formerly contacted, whereas the wild-type unsubstituted protein would vary in its affinity for the DNA as the contacted base is varied (Fig. 6.20).

The work required for this missing contact experiment is high. Genetic engineering techniques must be used to alter the gene encoding the protein. Similarly, both the wild-type sequence and a sequence for each base being tested for contacting must also be synthesized. Finally, the protein must be synthesized *in vivo*, purified, and its affinity for the various DNAs tested.

The testing for specific contacts can be streamlined. Instead of having to guess correctly both the base and the residue, the technique finds whatever base is contacted by a residue. As before, a variant protein is made containing a glycine or alanine substitution in the residue suspected of contacting the DNA. The DNA sample is probed biochemically to find all the residues contacted by the wild-type protein and by the mutant protein. The difference between the set of bases contacted by the wild-type protein and the mutant protein is the base or two that are normally contacted by the altered residue. The procedure for locating the bases which contact the protein is more fully described in Chapter 10.

Problems

6.1. Which amino acid side chains can serve as hydrogen bond donors and/or acceptors?

6.2. Use the thermodynamic relations

$$K = e^{-\Delta G/RT}$$

$$= e^{-(\Delta H - T\Delta S)/RT}$$

to explain more precisely what was claimed in the text concerning the strength of hydrophobic forces being maximal at some temperature intermediate 0° and 100°. Use this result to explain why some oligomeric proteins denature on cooling.

6.3. The discussion of the thermodynamics of proteins assumed that protein structures are determined by equilibrium constants and not rate constants. Some proteins spontaneously denature or become inactive. What does this mean about the former assumption?

6.4. Why might secondary structure predictions be expected to be more successful with small domains than with large domains?

6.5. Why is proline disruptive to an alpha helix?

6.6. If a novice X-ray crystallographer had the temerity to suggest that his or her protein was complementary to left helical DNA, how by simple examination of the proposed high resolution structure might you be certain that the crystallographer was wrong?

6.7. Consider an α-helix consisting of L-amino acids. Suppose one of the residues was changed to the D form. What would such a change do to the basic pattern of hydrogen bonds in the α-helix?

6.8. Draw the wiring or connection diagram you would use in building a neural network to predict whether or not a particular nucleotide was a donor in a splicing reaction.

6.9. Suppose a protein contains about 300 amino acids that are mostly in alpha helix conformation. Estimate a likely value for the number of times the polypeptide chain would make sharp bends so that the protein could be roughly spherical.

6.10. What amino acid side chains might make specific hydrogen bonds within the major groove of DNA to permit detection of specific DNA sequences by proteins?

6.11. Why would a homodimeric coiled-coil protein of the sequence $(VxxLxxx)_8$ be expected to possess parallel rather than antiparallel α-helices?

6.12. In the next chapter we will see that proteins are synthesized from their N-terminal amino acid to their C-terminus. What is the likely significance of the finding that secondary structure predictions are better for the N-terminal half of many proteins?

6.13. Why can't subunits of an oligomeric protein be related to one another by reflections across a plane or inversion through a point as atoms can in crystals?

6.14. Repressor proteins must respond to their environment and must dissociate from their operators under some conditions. Furthermore, the DNA-binding domains of repressors constructed like lambda repressor protein could be completely isolated from the domains involved in the binding of small-molecule inducers. Indeed, suppose that the binding of a small-molecule inducer was shown not to alter the structure of the DNA-binding domains at all. What then could cause such a repressor to dissociate from the DNA?

6.15. Roughly, proteins possess a density of 1.3 g/cm^3. Suppose a protein is a dimer with each monomer containing about 300 amino acids. Approximately what overall shape would you expect the protein to possess, if when it binds to DNA it makes contact in four adjacent major grooves of the DNA all located on one side of the helix? That is, the protein does not reach around the cylinder to make contacts on the back. Can the protein be spherical or must it be highly elongated?

6.16. Why is the substitution of glycine at a specific position in a protein more likely to destabilize the protein than the substitution of alanine?

6.17. Refer to J. Mol. Biol. *202*, 865-885 (1988). What amino acid homopolymers are most likely to form alpha-helices? What amino acid homopolymers are least likely to form alpha-helices? Write a sequence of 13 amino acids that maximizes the probability the central amino acid will be found in an alpha helical conformation.

6.18. How do you reconcile the seemingly contradictory conclusions mentioned in the text that amino acid changes in proteins generally produce local structural changes, with our inability to predict secondary structure with greater than 65% accuracy? The first implies that long-range interactions are unimportant, and the second that they are highly important. Hint, see Proc. Natl. Acad. Sci. USA *90*, 439-441 (1993).

References

Recommended Readings

The Operator-binding Domain of Lambda Repressor: Structure and DNA Recognition, C. Pabo, M. Lewis, Nature *298*, 443-447 (1982).

Predicting the Secondary Structure of Globular Proteins Using Neural Network Models, N. Qian, T. Sejnowski, J. Mol. Biol. *202*, 865-884 (1988).

X-ray Structure of the GCN4 Leucine Zipper, a Two-stranded, Parallel Coiled Coil, E. O'Shea, J. Klemm, P. Kim, T. Alber, Science *254*, 539-544 (1991).

Secondary Structure

The Structure of Proteins: Two Hydrogen-bonded Helical Configurations of the Polypeptide Chain, L. Pauling, R. Corey, H. Branson, Proc. Natl. Acad. Sci. USA *37*, 205-211 (1951).

Handedness of Crossover Connections in β Sheets, J. Richardson, Proc. Natl. Acad. Sci. USA *73*, 2619-2633 (1976).

β-Turns in Proteins, P. Chou, G. Fasman, J. Mol. Biol. *115*, 135-175 (1977).

Prediction of β-Turns, P. Chou, G. Fasman, Biophys. J. *26*, 367-384 (1979).

Analysis of Sequence-similar Pentapeptides in Unrelated Protein Tertiary Structures, P. Argos, J. Mol. Biol. *197*, 331-348 (1987).

Use of Techniques Derived from Graph Theory to Compare Secondary Structure Motifs in Proteins, E. Mitchell, P. Artymiuk, D. Rice, P. Willett, J. Mol. Biol. *212*, 151-166 (1990).

Tertiary Structure

The Theory of Interallelic Complementation, F. Crick, L. Orgel, J. Mol. Biol. *8*, 161-165 (1964).

On the Conformation of Proteins: The Handedness of the β-Strand-α-Helix-β-Strand Unit, M. Sternberg, J. Thornton, J. Mol. Biol. *105*, 367-382 (1976).

The Taxonomy of Protein Structure, M. Rossmann, P. Argos, J. Mol. Biol. *109*, 99-129 (1977).

Procedure for Production of Hybrid Genes and Proteins and its Use in Assessing Significance of Amino Acid Differences in Homologous Tryptophan Synthetase α Polypeptides, W. Schneider, B. Nichols, C. Yanofsky, Proc. Natl. Acad. Sci. USA *78*, 2169-2173 (1981).

Thermodynamic and Kinetic Examination of Protein Stabilization by Glycerol, K. Gekko, S. Timasheff, Biochemistry *20*, 4677-4686 (1981).

Modular Structural Units, Exons, and Function in Chicken Lysozyme, M. Go, Proc. Natl. Acad. Sci. USA *80*, 1964-1968 (1983).

Solvation Energy in Protein Folding and Binding, D. Eisenberg, A. McLachlan, Nature *319*, 199-203 (1986).

Combinatorial Cassette Mutagenesis as a Probe of Informational Content of Protein Sequences, J. Reidhaar-Olson, R. Sauer, Science *241*, 53-57 (1988).

The Role of the Leucine Zipper in the fos-jun Interaction, T. Kouzarides, E. Ziff, Nature *336*, 646-651 (1988).

Investigating Protein-Protein Interaction Surfaces Using a Reduced Stereochemical and Electrostatic Model, J. Warwicker, J. Mol. Biol. *206*, 381-395 (1989).

Trimerization of a Yeast Transcriptional Activator via a Coiled-Coil Motif, P. Sorger, H. Nelson, Cell *59*, 807-813 (1989).

Transient Folding Intermediates Characterized by Protein Engineering, A. Matouschek, J. Kellis Jr., L. Serrano, M. Bycroft, A. Fersht, Nature *346*, 488-490 (1990).

Suggestions for "Safe" Residue Substitutions in Site-directed Mutagenesis, D. Bordo, P. Argos, J. Mol. Biol. *217*, 721-729 (1991).

Hydrophobic Interactions

A Simple Method for Displaying the Hydropathic Character of a Protein, J. Kyte, R. Doolittle, J. Mol. Biol. *157*, 105-132 (1982).

Hydrophobic Stabilization in T4 Lysozyme Determined by Multiple Substitutions of Ile3, M. Matsumura, W. Becktel, B. Matthews, Nature *334*, 406-410 (1988).

GroE Heat-shock Proteins Promote Assembly of Foreign Prokaryotic Ribulose Bisphosphate Carboxylase Oligomers in *Escherichia coli*, P. Goloubinoff, A. Gatenby, G. Larimer, Nature *337*, 44-47 (1989).

Changing Fos Oncoprotein to a Jun-independent DNA-binding Protein with GCN4 Dimerization Specificity by Swapping "Leucine Zippers," J. Sellers, K. Struhl, Nature *341*, 74-76 (1989).

Protein Folding in Mitochondria Requires Complex Formation with hsp60 and ATP Hydrolysis, J. Ostermann, A. Howrich, W. Neupert, F. Hartl, Nature *341*, 125-130 (1989).

Reconsitution of Active Dimeric Ribulose Bisphosphate Carboxylase from an Unfolded State Depends on Two Chaperonin Proteins and Mg-ATP,

P. Goloubinoff, J. Christiler, A. Gatenby, G. Lorimer, Nature *342*, 884-888 (1989).

Reverse Hydrophobic Effects Relieved by Amino-acid Substitutions at a Protein Surface, A. Pakula, R. Sauer, Nature *344*, 363-364 (1990).

The *E. coli dnaK* Gene Product, the hsp70 Homolog, Can Reactivate Heat-inactivated RNA Polymerase in an ATP Hydrolysis-Dependent Manner, D. Skowyra, C. Georgopoulos, M. Zylicz, Cell *62*, 939-944 (1990).

Structure Determination, Prediction, and Dynamics

Dynamics of Folded Proteins, J. McCammon, B. Gelin, M. Karplus, Nature *267*, 585-590 (1977).

Side-chain Torsional Potentials: Effect of Dipeptide, Protein, and Solvent Environment, B. Gelin, M. Karplus, Biochemistry *18*, 1256-1268 (1979).

Protein Dynamics in Solution and in a Crystalline Environment: A Molecular Dynamics Study, W. Gunsteren, M. Karplus, Biochem. *21*, 2259-2273 (1982).

Ion-pairs in Proteins, D. Barlow, J. Thornton, J. Mol. Biol. *168*, 867-885 (1983).

The Use of Double Mutants to Detect Structural Changes in the Active Site of the Tyrosyl-tRNA Synthetase (*Bacillus Stearothermophilus*), P. Carter, G. Winter, A. Wilkinson, A. Fersht, Cell *38*, 835-840 (1984).

An Evaluation of the Combined Use of Nuclear Magnetic Resonance and Distance Geometry for the Determination of Protein Conformations in Solution, T. Havel, K. Wüthrich, J. Mol. Biol. *182*, 281-294 (1985).

Complex of *lac* Repressor Headpiece with a 14 Base-pair Operator Fragment Studied by Two-dimensional Nuclear Magnetic Resonance, R. Boelens, R. Scheck, J. van Boom, R. Kaptein, J. Mol. Biol. *193*, 213-216 (1987).

Further Developments of Protein Secondary Structure Prediction Using Information Theory, J. Bibrat, J. Farnier, B. Robson, J. Mol. Biol. *198*, 425-443 (1987).

Accurate Prediction of the Stability and Activity Effects of Site-Directed Mutagenesis on a Protein Core, C. Lee, M. Levitt, Nature *352*, 448-451 (1991).

Computational Method for the Design of Enzymes with Altered Substrate Specificity, C. Wilson, J. Marc, D. Agard, J. Mol. Biol. *220*, 495-506 (1991).

Effect of Alanine Versus Glycine in α-Helices on Protein Stability, L. Serrano, J. Neira, J. Sancho, A. Fersht, Nature *356*, 453-455 (1992).

Kinetics of Protein-Protein Association Explained by Brownian Dynamics Computer Simulation, S. Northrup, H. Erickson, Proc. Natl. Acad. Sci. USA *89*, 3338-3342 (1992).

Protein Design, Mutagenesis, Structures of Specific Proteins

Design, Synthesis and Characterization of a 34-Residue Polypeptide that Interacts with Nucleic Acids, B. Gutte, M. Daumigen, E. Wittschieber, Nature *281*, 650-655 (1979).

Structure of the *cro* Repressor from Bacteriophage Lambda and its Interaction with DNA, W. Anderson, D. Ohlendorf, Y. Takeda, B. Matthews, Nature *290*, 754-758 (1981).

Structural Similarity in the DNA Binding Domains of Catabolite Gene Activator and Cro Repressor Proteins, T. Steitz, D. Ohlendorf, D. McKay, W. Anderson, B. Matthews, Proc. Natl. Acad. Sci. USA *79*, 3097-3100 (1982).

The N-Terminal Arms of Lambda Repressor Wrap Around the Operator DNA, C. Pabo, W. Krovatin, A. Jeffrey, R. Sauer, Nature *298*, 441-443 (1982).

Homology Among DNA-binding Proteins Suggests Use of A Conserved Super-Secondary Structure, R. Sauer, R. Yocum, R. Doolittle, M. Lewis, C. Pabo, Nature *298*, 447-451 (1982).

The Molecular Basis of DNA-protein Recognition Inferred from the Structure of *cro* Repressor, D. Ohlendorf, W. Anderson, R. Fisher, Y. Takeda, B. Matthews, Nature *298*, 718-723 (1982).

Structure of Catabolite Gene Activator Protein at 2.9 Å Resolution: Incorporation of Amino Acid Sequence and Interactions with Cyclic-AMP, D. McKay, I. Weber, T. Steitz, J. Biol. Chem. *257*, 9518-9524 (1982).

The 3 Å Resolution Structure of a D-galactose-binding Protein for Transport and Chemotaxis in *Escherichia coli*, N. Vyas, F. Quicho, Proc. Natl. Acad. Sci. USA *80*, 1792-1796 (1983).

Cocrystals of the DNA-binding Domain of Phage 434 Repressor and a Synthetic Phage 434 Operator, J. Anderson, M. Ptashne, S. Harrison, Proc. Nat. Acad. Sci. USA *81*, 1307-1311 (1984).

3Å-Resolution Structure of a Protein with Histone-like Properties in Prokaryotes, I. Tanaka, K. Appelt, J. Dijk, S. White, R. Wilson, Nature *310*, 376-381 (1984)

Structure of Large Fragment of *Escherichia coli* DNA Polymerase I Complexed with dTMP, D. Ollis, P. Brick, R. Hamlin, N. Xuong, T. Steitz, Nature *313*, 762-766 (1985).

Importance of the Loop at Residues 230-245 in the Allosteric Interactions of *Escherichia coli* Aspartate Carbamoyltransferase, S. Middleton, E. Kantrowitz, Proc. Nat. Acad. Sci. USA *83*, 5866-5870 (1986).

Crystal Structure of *trp* Repressor/Operator Complex at Atomic Resolution, Z. Otwinowski, R. Schevitz, R. Zhang, C. Lawson, A. Joachimiak, R. Marmorstein, B. Luisi, P. Sigler, Nature *335*, 321-329 (1987).

Cocrystal Structure of an Editing Complex of Klenow Fragment with DNA, P. Freemont, J. Friedman, L. Beese, M. Sanderson, T. Steitz, Proc. Natl. Acad. Sci. USA *85*, 8924-8928 (1988).

Crystal Structure of an Engrailed Homeodomain-DNA Complex at 2.8Å Resolution: A Framework for Understanding Homeodomain-DNA Interactions, C. Kissinger, B. Liu, E. Martin-Blanco, T. Kornberg, C. Pabo, Cell *63*, 579-590 (1990).

Sequence Requirements for Coiled-coils: Analysis with Lambda Repressor-GCN4 Leucine Zipper Fusions, J. Hu, E. O'Shea, P. Kim, R. Sauer, Science *250*, 1400-1402 (1990).

Systematic Mutation of Bacteriophage T4 Lysozyme, D. Rennell, S. Bouvier, L. Hardy, A. Poteete, J. Mol. Biol. *222*, 67-87 (1991).

Crystal Structure of a MATα2 Homeodomain-operator Complex Suggests a General Model for Homeodomain-DNA Interactions, C. Wolberger, A. Vershon, B. Liu, A. Johnson, C. Pabo, Cell *67*, 517-528 (1991).

Mechanism of Specificity in the Fos-jun Oncoprotein Heterodimer, E. O'Shea, R. Rutkowski, P. Kim, Cell *68*, 699-708 (1992).

DNA-Protein Interactions

Substituting an α-Helix Switches the Sequence-Specific DNA Interactions of a Repressor, R. Wharton, E. Brown, M. Ptashne, Cell *38*, 361-369 (1984).

Changing the Binding Specificity of a Repressor by Redesigning an α-helix, R. Wharton, M. Ptashne, Nature *316*, 601-605 (1985).

Lambda Repressor Mutations that Increase Affinity and Specificity of Operator Binding, H. Nelson, R. Sauer, Cell *42*, 549-558 (1985).

How Lambda Repressor and Lambda Cro Distinguish between O_R1 and O_R3, A. Hochschild, J. Douhan III, M. Ptashne, Cell *47*, 807-816 (1986).

Evidence for a Contact between Glutamic-18 of *lac* Repressor and Base Pair 7 of *lac* Operator, R. Ebright, Proc. Nat. Acad. Sci. USA *83*, 303-307 (1986).

Missing Contact Probing of DNA-protein Interactions. A. Brunelle, R. Schleif, Proc. Natl. Acad. Sci. USA *84*, 6673-6676 (1987).

Structure of the Repressor-Operator Complex of Bacteriophage 434, J. Anderson, M. Ptashne, S. Harrison, Nature *326*, 846-852 (1987).

Effect of Non-contacted Bases on the Affinity of 434 Operator for 434 Repressor and Cro, G. Koudelka, S. Harrison, M. Ptashne, Nature *326*, 886-888 (1987).

Zinc-finger Motifs Expressed in *E. coli* and Folded *in vitro* Direct Specific Binding to DNA, K. Nagai, Y. Nakaseko, K. Nasmyth, D. Rhodes, Nature *332*, 284-286 (1988).

Structure of the Lambda Complex at 2.5 Angstrom Resolution: Details of the Repressor-operator Interactions, S. Jordan, C. Pabo, Science *242*, 893-899 (1988).

Zinc Finger-DNA Recognition: Crystal Structure of a Zif268-DNA Complex at 2.1 Å, N. Pauletich, C. Pabo, Science *252*, 809-817 (1991).

Crystal Structure of the *met* Repressor-operator Complex at 2.8 Å Resolution Reveals DNA Recognition by β-Strands, W. Somers, S. Phillips, Nature *359*, 387-393 (1992).

Protein Synthesis

7

Having studied the synthesis of DNA and RNA and the structure of proteins, we are now prepared to examine the process of protein synthesis. We will first be concerned with the actual steps of protein synthesis. Then, to develop further our understanding of cellular processes, we will discuss the rate of peptide elongation, how cells direct specific proteins to be located in membranes, and how the machinery that translates messenger RNA into protein in cells is regulated in order to use most efficiently the limited cellular resources. The major part of the translation machinery is the ribosomes. A ribosome consists of a larger and smaller subunit, each containing a major RNA molecule and more than twenty different proteins. The synthesis and structure of ribosomes will be considered in a later chapter.

In outline, the process of protein synthesis is as follows. Amino acids are activated for protein synthesis by amino acid synthetases which attach the amino acids to their cognate tRNA molecules. The smaller ribosomal subunit and then the larger ribosomal subunit attach to messenger RNA at the 5' end or near the initiating codon. Translation then begins at an initiation codon with the assistance of initiation factors. During the process of protein synthesis, the activated amino acids to be incorporated into the peptide chain are specified by three-base codon-anticodon pairings between the messenger and aminoacyl-tRNA. Elongation of the peptide chain terminates on recognition of one of the three termination codons, the ribosomes and messenger dissociate, and the newly synthesized peptide is released. Some proteins appear to fold spontaneously as they are synthesized, but others appear to utilize auxiliary proteins to help in the folding process.

The actual rate of peptide elongation in bacteria is just sufficient to keep up with transcription; a ribosome can initiate translation immediately behind an RNA polymerase molecule and keep up with the transcription. In eukaryotic cells, however, the messenger is modified and transported from the nucleus to the cytoplasm before it can be translated.

Although the cytoplasm in bacteria contains most of the cell's protein, the inner membrane, the periplasmic space, and even the outer membrane contain appreciable amounts of protein. Eukaryotic cells also must direct some proteins to organelles and membranes. How do cells do this? One mechanism is with a signal peptide. These are the N-terminal 20 amino acids or so of some proteins that appear largely responsible for directing the protein away from the cytoplasm and into or through the membrane.

Finally, it is necessary to discuss the regulation of the level of ribosomes. Since ribosomes constitute a large fraction of a cell's total protein and RNA, only as many ribosomes as are needed should be synthesized. Consequently, complicated mechanisms have developed coupling protein synthesis to ribosome synthesis.

A. Chemical Aspects

Activation of Amino Acids During Protein Synthesis

In Chapter 6 we learned that proteins possess a definite sequence of amino acids that are linked by peptide bonds. Aminoacyl-tRNA molecules participate in the formation of these bonds in two ways. First, they bring activated amino acids to the reaction. They also serve as adapters between the various three-base codons in the messenger RNA and the actual amino acids to be incorporated into the growing polypeptide chain. Because different tRNA molecules must be distinguishable at the

Figure 7.1 The cloverleaf structure of a charged tRNA. Filled circles represent variable bases.

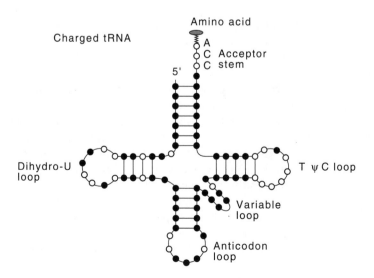

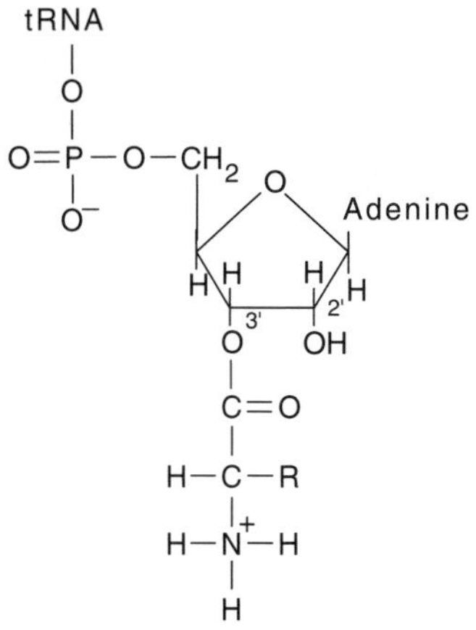

Figure 7.2 The structure of the aminoacylated adenine at the end of a charged tRNA molecule.

step of attaching amino acids, they differ in structure (Fig. 7.1). On the other hand, the tRNA molecules must also possess important structural features in common so that the tRNA molecules may be used in peptide bond formation at the ribosome. Formation of a peptide bond is energetically unfavorable and is assisted at the step of bond formation by the energy in the amino acid-tRNA bond. This is an ester to the 3' hydroxyl at the invariant -C-C-A end of the tRNA. Aminoacyl-tRNA synthetases, one for each amino acid, form these bonds. The formation of the ester bonds, activation, occurs in two steps. First the enzyme links

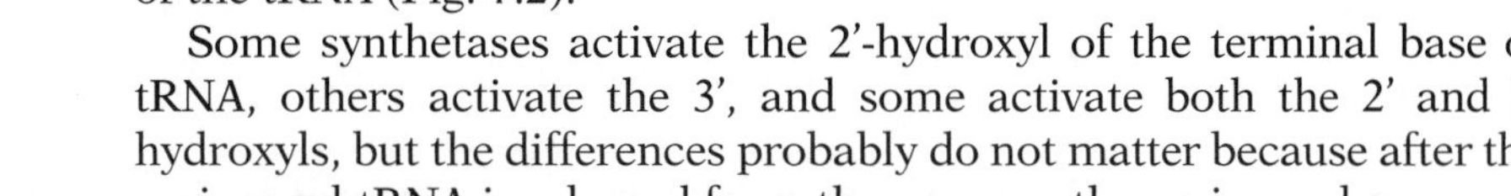

the amino acid to AMP, and then transfers it to the 3' terminal adenosine of the tRNA (Fig. 7.2).

Some synthetases activate the 2'-hydroxyl of the terminal base of tRNA, others activate the 3', and some activate both the 2' and 3' hydroxyls, but the differences probably do not matter because after the aminoacyl-tRNA is released from the enzyme, the aminoacyl group on the tRNA migrates back and forth.

Fidelity of Aminoacylation

The aminoacyl-tRNA synthetases are remarkable enzymes since they recognize amino acids and their cognate tRNA molecules and join them together. Inaccuracies in either recognition process could be highly deleterious because choosing the wrong amino acid or the wrong tRNA would ultimately yield a protein with an incorrect sequence. We know,

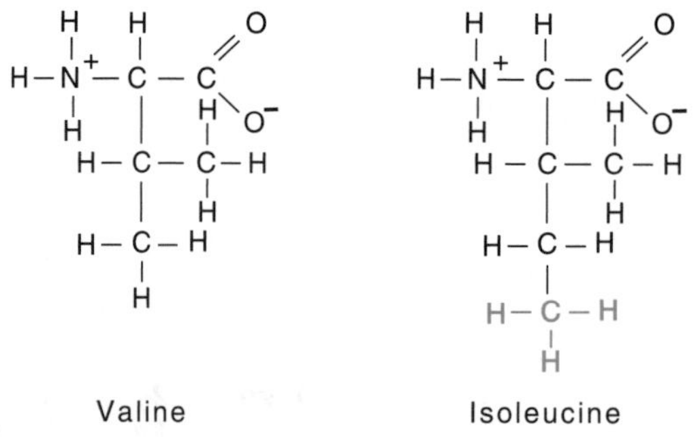

Valine

Isoleucine

Figure 7.3 The structures of valine and isoleucine.

however, from measurements on peptides highly purified from proteins of known sequence, that the overall frequency of misincorporation, at least of charged amino acids, is only about 1/1000.

Let us first consider the process of choosing the correct amino acid. The greatest difficulty in accurate translation appears to be in discriminating between two highly similar amino acids. Valine and isoleucine are an example since replacing a hydrogen on valine with a methyl group yields isoleucine (Fig. 7.3). The valyl-tRNA synthetase should not have trouble in discriminating against isoleucine because isoleucine is larger than valine and probably does not fit into the active site on the enzyme. The reverse situation is more of a problem. Valine will form all of the contacts to the enzyme that isoleucine can form except for those to the missing methyl group. How much specificity could the absence of these contacts provide? Estimates of the differences in binding energy predict about a 200-fold discrimination, but since the actual error rate is found to be much lower, something in addition to a simple discrimination based on one binding reaction must contribute to specificity. An additional step in the overall reaction in the form of editing by the synthetase increases the accuracy.

Although isoleucyl-tRNA synthetase can form a valyl adenylate complex, upon the addition of tRNA^{Ile} the tRNA is activated and then the

$$\text{tRNA}^{Ile}\text{-Val} \longrightarrow \text{tRNA}^{Ile} + \text{Val}$$

complex is immediately hydrolyzed. One way to think of this process is that activation is a two-step sieving process (Fig. 7.4). It permits the correct amino acid and similar but smaller amino acids to be activated. Then all amino acids smaller than the correct amino acid have a hydrolytic pathway available for removal of the misacylated amino acid. DNA synthesis and DNA cutting by restriction enzymes also use two-step error checking to achieve high accuracy. In the case of protein synthesis, fidelity is increased by identifying the amino acid several times, and for the DNA cutting enzymes, the nucleotide sequence is read more than once.

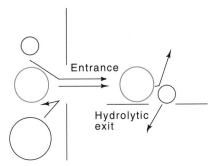

Figure 7.4 A schematic of the double-sieving action of some aminoacyl-tRNA synthetases. Only the correct amino acid is activated and then avoids the hydrolytic exit.

How Synthetases Identify the Correct tRNA Molecule

A second problem in specificity of protein synthesis is the selection of the tRNA molecule by the synthetase. In principle this selection could be done by reading the anticodon of the tRNA. A wide variety of experiments have revealed, however, that only for tRNA^{Met} is the anticodon the sole determinant of the charging specificity. For about half of the tRNAs, the anticodon is involved in the recognition process, but it is not the sole determinant. For the remaining half of the tRNAs, the anticodon is not involved at all.

Two extreme possibilities exist for the other charging specificity determinants. They could be the identity of one or more nucleotides somewhere in the tRNA. On the other hand, the charging specificity could be determined by part or all of the overall structure of the tRNA molecule. Of course, this structure is determined by the nucleotide sequence, but the structure as dictated by the overall sequence may be more important than the chemical identity of just a couple of amino or carboxyl groups. In view of the diversity of nature, it is reasonable to expect different aminoacyl-tRNA synthetases will utilize different structural details in their identification of their cognate tRNA molecules.

Just as in the study of RNA splicing, the development of genetic engineering has greatly accelerated the rate of progress in understanding the specificity determinants on tRNA molecules. This resulted from facilitating the synthesis *in vitro* of tRNA molecules of any desired sequence. Such a synthesis utilizes the phage T7 RNA polymerase to initiate transcription from a T7 promoter that can be placed near the end of a DNA molecule (Fig. 7.5). Essentially any DNA sequence downstream from the promoter can be used so that any tRNA sequence can be synthesized. The RNA molecules resulting from such reactions can

Figure 7.5 *In vitro* synthesis of an artificial tRNA molecule using T7 RNA polymerase.

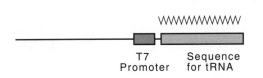

T7
Promoter

Sequence
for tRNA

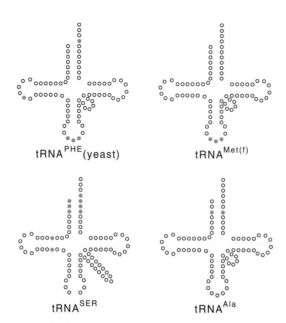

Figure 7.6 The positions of nucleotides that determine the charging identity of several tRNAs.

be aminoacylated and utilized in translation despite the fact that they lack the specialized chemical modifications that are found on tRNA molecules synthesized *in vivo*. Apparently these modifications are not essential to the process of protein synthesis and they exist more for fine-tuning.

Genetic engineering also enables us to alter the gene encoding a tRNA molecule, reinsert the gene in a cell, and examine the *in vivo* charging and translation properties of the altered molecule. The ability to be charged by alanine synthetase is specified by the identity of just two nucleotides (Fig. 7.6). This was determined by identifying the smallest common subset of nucleotide changes that permitted the molecule to be charged with alanine. These proved to be two nucleotides in the acceptor stem, a G and a U that form a non Watson-Crick base pair. Providing these two nucleotides in any tRNA molecule enables the molecule to be charged with alanine. The specificity determinants of other tRNA molecules have been found to be three or more nucleotides scattered around the molecule.

The structure of the crystallized glutamyl-tRNA synthetase-tRNA complex permitted direct examination of the contacts between the enzyme and the tRNA. These showed, as expected, that this enzyme read the anticodon of the tRNA plus several nucleotides located elsewhere on the tRNA.

Decoding the Message

What decodes the messages? Clearly, base pairing between a codon of message and an anticodon of the aminoacyl-tRNA decodes. This is not the complete story however. Since the ribosome pays no attention to the

Table 7.1 The Genetic Code

First Position, 5' End	Second Position				Third Position
	U	C	A	G	
U	Phe	Ser	Tyr	Cys	U
	Phe	Ser	Tyr	Cys	C
	Leu	Ser	Ochre	End	A
	Leu	Ser	Amber	Trp	G
C	Leu	Pro	His	Arg	U
	Leu	Pro	His	Arg	C
	Leu	Pro	Gln	Arg	A
	Leu	Pro	Gln	Arg	G
A	Ile	Thr	Asn	Ser	U
	Ile	Thr	Asn	Ser	C
	Ile	Thr	Lys	Arg	A
	Met	Thr	Lys	Arg	G
G	Val	Ala	Asp	Gly	U
	Val	Ala	Asp	Gly	C
	Val	Ala	Glu	Gly	A
	Val	Ala	Glu	Gly	G

correctness of the tRNA charging, the aminoacyl-tRNA synthetases are just as important in decoding, for it is also essential that the tRNA molecules be charged with the correct amino acid in the first place.

Once the amino acids have been linked to their cognate tRNAs, the process of protein synthesis shifts to the ribosome. The "code" is the correspondence between the triplets of bases in the codons and the amino acids they specify (Table 7.1). In a few exciting years molecular biologists progressed from knowing that there must be a code, to learning that each amino acid is encoded by three bases on the messenger, to actually determining the code. The history and experiments of the time are fascinating and can be found in *The Eighth Day of Creation* by Horace Freeland Judson.

In the later stages of solving the code it became apparent that the code possessed certain degeneracies. Of the 64 possible three-base codons, in most cells, 61 are used to specify the 20 amino acids. One to six codons may specify a particular amino acid. As shown in Table 7.1, synonyms generally differ in the third base of the codons. In the third position, U is equivalent to C and, except for methionine and tryptophan, G is equivalent to A.

Under special circumstances likely to reflect early evolutionary history, one of the three codons that code for polypeptide chain termination also codes for the insertion of selenocysteine. The same codon at the end of other genes specifies chain termination. Thus, the context surrounding this codon also determines how it is read. Generally such

Table 7.2 Wobble Base Pairs

First Anticodon Base	Third Codon Base
C	G
A	U
U	A or G
G	U or C
I	U, C, G

context effects change only the rates of insertion of amino acids, but not their identity.

With the study of purified tRNAs, several facts became apparent. First, tRNA contains a number of unusual bases, one of which is inosine,

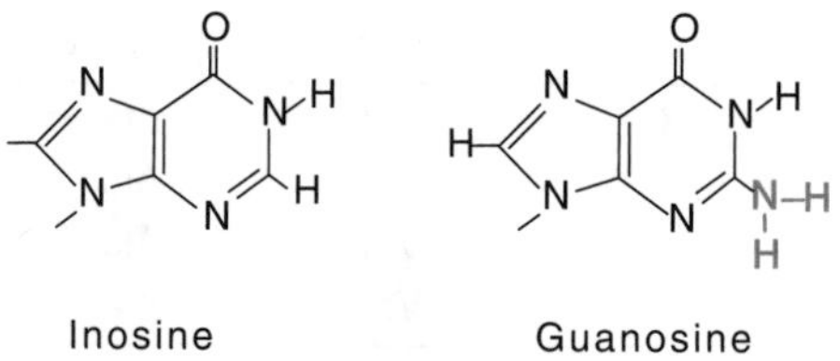

Inosine   Guanosine

which is occasionally found in the first position of the anticodon. Second, more than one species of tRNA exists for most of the amino acids. Remarkably, however, the different species of tRNA for any amino acid all appear to be charged by the same synthetase. Third, strict Watson-Crick base pairing is not always followed in the third position of the codon (Table 7.2). Apparently the third base pair of the codon-anticodon complex permits a variety of base pairings (Fig. 7.7). This phenomenon is called wobble. The G-U base pair of alanine tRNA mentioned above is an example of wobble.

Figure 7.7 Structure of the wobble base pairs.

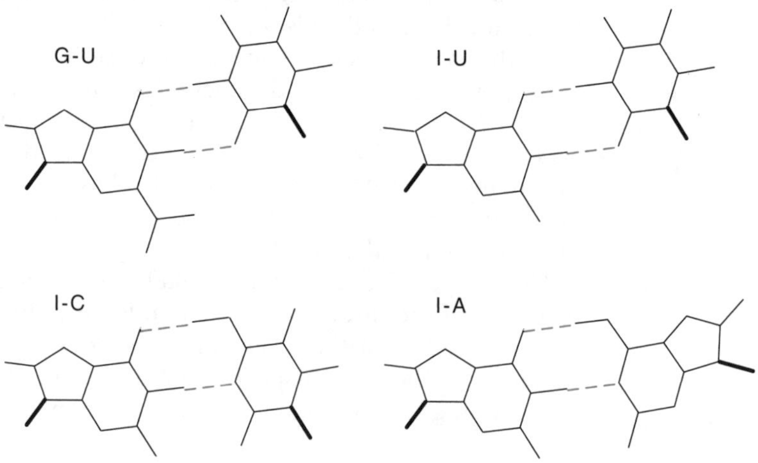

G-U   I-U

I-C   I-A

In mitochondria the genetic code might be slightly different from that described above. There the translation machinery appears capable of translating all the codons used with only 22 different species of tRNA. One obstacle to understanding translation in mitochondria is the RNA editing that can occur in these organelles. Because the sequence of mRNA can be changed after its synthesis, we cannot be sure that the sequence of genes as deduced from the DNA is the sequence that actually is translated at the ribosome. Hence deductions about the use or lack of use of particular codons as read from the DNA sequence cannot be made with reliability.

Base Pairing between Ribosomal RNA and Messenger

Ribosomes must recognize the start codons AUG or GUG on the messenger RNA to initiate protein synthesis. Characterization of the proteins that are synthesized when the *lac* operon or other operons are induced show that normally only one AUG or GUG of a gene is utilized to initiate protein synthesis. Many of the internal AUG or GUG codons are not used to initiate protein synthesis. This means that something in addition to the initiating codon itself must signal the point at which translation begins.

Studies of bacterial translation show that the first step in initiation is the binding of messenger to the smaller of the two ribosomal subunits, the 30S subunit. The absence of a strictly conserved sequence preceding start codons suggested that whatever first bound the messenger to the 30S subunit might be an RNA-RNA interaction between mRNA and ribosomal RNA. The originators of this idea, Shine and Dalgarno, were so confident of their proposal that they proceeded to sequence the 3' end of the 16S rRNA which is found in the smaller ribosomal subunit. They found that the rRNA sequence provided strong support for their idea. The sequence on mRNA which binds to the 16S ribosomal RNA is called the Shine-Dalgarno sequence or the ribosome binding site.

Bacterial messengers contain a ribosome-binding sequence slightly ahead of an initiating AUG. This sequence base pairs well with a region

```
                        3'-OH                          5'

16S rRNA          A U U C C U C C A C U A G . . . .
                          ‖ ‖‖ ‖‖ ‖
LacZ mRNA         . . . A G G A A A C A G C U A U G . . .

                        5'                             3'
```

near the 3' end of the 16S ribosomal RNA. This upstream region has been examined in more than hundreds of messenger start sequences. Typically it is three or four bases and is centered about ten nucleotides ahead of the start codon. Despite the data to be described below, the ribosome-binding sequence lying ahead of the AUG initiation codon is not the whole story. Undoubtedly, secondary structure in the mRNA also can alter translation efficiency. Occasionally, a sequence upstream or downstream from the Shine-Dalgarno sequence pairs with it and blocks translation initiation. Examination of the sequence in the vicinity of the AUG and ribosome-binding sequence has shown the existence of preferences for some nucleotides. Were no other factors involved, the identity of all these other nucleotides would be random.

Experimental Support for the Shine-Dalgarno Hypothesis

Since the time of the original proposal by Shine and Dalgarno, four lines of evidence have provided firm support for the idea that the 3' end of the 16S rRNA base pairs with a three- to seven-base stretch of the mRNA lying ahead of the translation initiation site. The first line of evidence is inhibition of an *in vitro* protein synthesis system by an oligonucleotide. A polynucleotide of sequence very similar to that which is found pre-

Figure 7.8 Base pairing between the 3' end of 16S rRNA and radioactive messenger RNA. After incubation and before electrophoresis, protein was denatured by addition of sodium dodecyl sulfate. The mRNA barely entered the gel if it participated in forming an initiation complex. Cutting the mRNA-16S rRNA complex with colicin E3 released mRNA complexed to a fragment of 16S rRNA. Heating dissociated the mRNA fragment and 16S rRNA fragments.

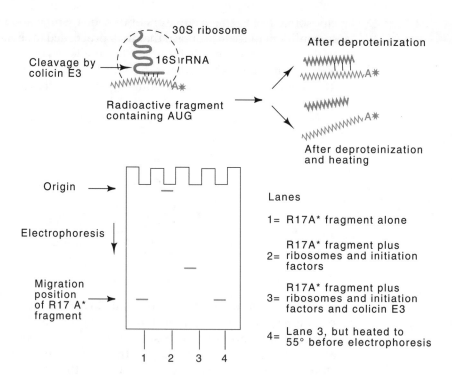

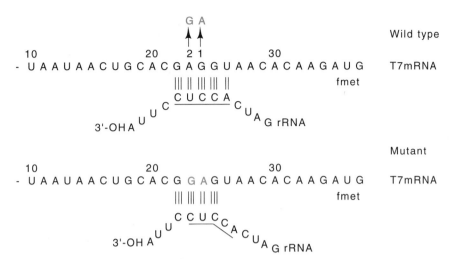

Figure 7.9 The base pairings possible between normal phage T7 messenger and 16S rRNA. Mutation 1 reduces the binding. Mutation 2, a revertant, restores the ability to base pair with the rRNA, but it pairs in a different position.

ceding the AUG of many messengers inhibits mRNA from binding to the ribosome. Presumably, this inhibition results from the binding of the polynucleotides near the end of the 16S rRNA, thereby blocking the ability of the ribosomes to bind properly to messenger.

The second line of evidence for the Shine-Dalgarno proposal is direct physical evidence for base pairing between the messenger and the 3' end of 16S ribosomal RNA (Fig. 7.8). This experiment used a bacteriocidal agent called colicin E3 which is released from some strains of bacteria. E3 kills by inactivating the ribosomes in sensitive cells by cleaving their 16S rRNA molecules 40 bases from the 3' end. To test the ribosome-binding site idea Jakes and Steitz first bound a fragment of phage R17 messenger *in vitro* to ribosomes in an initiation complex and then cut the ribosomal RNA by the addition of colicin E3. They demonstrated the presence of base pairing between the messenger and the 3'-terminal 40 bases of ribosomal RNA by co-electrophoresis of the fragment of R17 RNA with the fragment of 16S rRNA. Repeating the experiment but preventing formation of the mRNA-ribosome initiation complex prevented formation of the hybrid between the two RNAs.

Genetics provides two additional arguments for the utilization of the ribosome-binding site. A base change in phage T7 messenger in the ribosome-binding site reduced the translation efficiency of the messenger. The proof came with the isolation of a revertant that restored the high translation efficiency. The revertant did not recreate the original sequence; it created another ribosome-binding sequence in the mRNA, two nucleotides further upstream (Fig. 7.9).

An elegant demonstration of pairing between mRNA and rRNA became possible with genetic engineering. One gene in *E. coli* was altered to possess a totally different ribosome-binding site. An additional gene coding for the ribosomal 16S RNA was added to the same cells. Its recognition region near the 3'-OH end was changed to be complementary to the altered ribosome-binding site on the one gene. Only when both the altered gene and altered rRNA gene were present in cells was the protein from the altered gene synthesized. This proved beyond any doubt that the ribosome-binding sequence actually does base pair with a portion of the 16S rRNA.

Eukaryotic Translation and the First AUG

The sequences preceding the initiation codons in eukaryotic messengers do not contain significant regions of complementarity to the 18S RNA from the smaller ribosomal subunit. Additionally, translation almost always begins on these RNAs at the first AUG codon. In bacteria, a number of AUG triplets can precede the actual initiation codon. Translation of most eukaryotic messengers begins with the binding of a cap-recognizing protein to the 5' end of the mRNA. The translation efficiency of most, but not all, messengers in eukaryotic systems is much higher when the messenger contains the cap structure discussed in Chapter 5. Additional proteins bind followed by the 40S ribosomal subunit. With the consumption of ATP, the complex moves down the RNA to the first AUG, at which point another protein binds to the complex and finally the 60S ribosomal subunit binds (Fig. 7.10). Translation begins from this point. The preinitiation complex can open and slide right through moderately stable regions of base paired RNA to reach the first AUG of the messenger. This is unlike the prokaryotic translation machinery, which has difficulty reaching an initiation codon that is buried in extraneous secondary structure of the mRNA.

In special situations when only a low level of protein is required, translation begins not at the first AUG, but at a later one. Similarly, an

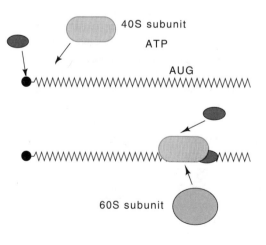

40S subunit

ATP

AUG

60S subunit

Figure 7.10 The initiation process in eukaryotes in which the smaller subunit slides from the 5' end of the mRNA to the first AUG, associates with the larger subunit and translation begins.

AUG codon immediately followed by a termination signal and then another AUG codon, also produces a low translation efficiency.

The need for a mechanism that can deposit the translation machinery at the initiation codon despite the existence of secondary structure in the mRNA arises from the eukaryotic translation pathway. In eukaryotes, the RNA is synthesized, spliced, transported to the cytoplasm and then translated. Undoubtedly, regions of secondary structure obscuring a potential ribosome-binding site would exist on many species of messengers. To circumvent this problem, the translation apparatus recognizes the capped 5' end, which cannot be involved in base pairing. After binding, the apparatus slides down the mRNA until it reaches a starting AUG. In contrast, prokaryotes lack a need for binding and sliding since ribosomes attach to a ribosome recognition sequence on mRNA as soon as it protrudes from the RNA polymerase. Thus, prokaryotic mRNA has little opportunity to fold and hide the start region of a protein.

Tricking the Translation Machinery into Initiating

Once messenger has bound to the smaller ribosomal subunit and then a larger subunit has been added, translation can begin. *Escherichia coli* proteins initiate primarily at AUG codons with a peptide chain analog in which part of the initiating amino acid looks like a peptide bond (Fig. 7.11). This permits the system to form the first real peptide bond with essentially the same machinery as it uses to form subsequent peptide bonds.

The initiation analog is N-formyl methionine. Rather than utilize an entirely separate charging pathway with a separate tRNA synthetase, the cells formylate a methionine after it has been put on a tRNA^{Met}. This is, however, a special tRNA called tRNA$_f^{Met}$ and it is used only for initiation. Methionine bound to the second type of tRNA^{Met} cannot be formylated, and this is used only for elongation at internal AUG codons. This scheme prevents the difficulties that would arise from attempting to initiate a protein with an unformylated methionine or attempting to put a formylated methionine on the interior of a protein. A new question

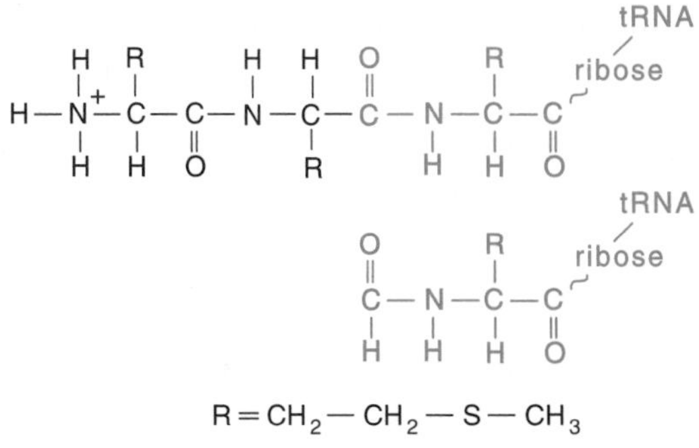

Figure 7.11 The structure of a polypeptide and the similar structure generated by formylation of an amino acid. The initiation amino acid is methionine, whose side group R is shown.

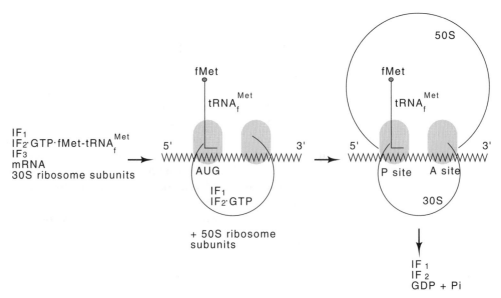

IF$_1$
IF$_2$·GTP·fMet-tRNA$_f^{Met}$
IF$_3$
mRNA
30S ribosome subunits

Figure 7.12 Formation of a translation initiation complex.

is raised by the two tRNAs. How are the two methionine tRNAs distinguished during translation? AUG codons occur both at the beginning of protein coding sequences and within coding sequences.

The two Met-tRNAs are distinguished by a set of proteins used in the initiation and elongation processes. These proteins carry the charged tRNA molecules into the ribosome (Fig. 7.12). Interactions between the proteins and the ribosome-messenger complex help hold the tRNA-protein complex in the ribosome. The most important of these interactions is that between the three-base anticodon of the tRNA and the three-base codon in the messenger. It is these codon-anticodon interactions that specify the f-Met initiating amino acid and successive amino acids along the polypeptide chain. The protein IF$_2$, initiation factor 2, carries the f–met–tRNA$_f^{Met}$ into the site normally occupied by the growing peptide chain, the P site, whereas all other charged tRNAs, including met-tRNA, are carried into the other site, the acceptor, A, site, by the elongation factors. Thus N-formyl methionine can be incorporated only at the beginning of a polypeptide.

In addition to the initiation factor IF$_2$, two other proteins, IF$_1$ and IF$_3$, are also used during the initiation steps. Factor IF$_1$ accelerates the initiation steps but is not absolutely required, and it can be assayed *in vitro* by its acceleration of ^{3}H-Met-tRNA^{Met} binding to ribosomes. IF$_3$ binds to the 30S subunit to assist the initiation process.

The initiation of translation in eukaryotes shows some similarity to the process used in bacteria. Two methionine tRNAs are used, one for initiation and one for elongation, but the methionine on the initiating tRNA is not formylated. Methionine on this tRNA, however, can be formylated by the bacterial formylating enzyme. It appears as if the eukaryotic initiation system has evolved from the bacterial system to the point that the formyl group is no longer necessary for protein synthesis, but the tRNA$_f^{Met}$ is still similar to its progenitor bacterial tRNA$_f^{Met}$.

Protein Elongation

Although it would appear that the charged tRNAs could diffuse into the ribosome and bind to the codons of mRNA, they are in fact carried into the binding sites on a protein. The protein that serves this function during elongation was originally called Tu (unstable) but is now sometimes called EF$_1$ (elongation factor 1). A rather complex cycle is used for carrying the charged tRNAs to the ribosome A site (Fig. 7.13). First, GTP binds to EF$_1$, then an aminoacyl tRNA binds, and this complex enters the ribosome A site containing a complementary codon. There GTP is hydrolyzed to GDP, and EF$_1$-GDP is ejected from the ribosome. The completion of the cycle takes place in solution. GDP is displaced from EF$_1$-GDP by EF$_2$, which in turn is displaced by GTP. The binding of GDP to EF$_1$ is tight; K$_d$ equals approximately 3×10^{-9} M. This plus the fact that EF$_1$ binds to filters permits a simple filter-binding assay to be used to quantitate the protein. Originally the very tight binding for GDP generated confusion because the commercial preparation of GTP contained minor amounts of contaminating GDP.

Both the initiation factor IF$_2$ and the elongation factor EF$_1$ carry charged tRNA molecules into the ribosome. Not surprisingly, they possess considerable amino acid sequence homology. Additionally, cells which incorporate selenocysteine into the one or two proteins containing this amino acid use yet a third factor. This carries the charged tRNA into the ribosome, and, as expected, also possesses significant homology with the other two factors. All three of the proteins are members of the broad and important class of proteins known as G proteins as they bind GTP. In eukaryotic cells the G proteins most often are part of signal transduction pathways from receptor proteins bound in the membrane to gene regulation or other intracellular points of regulation.

Figure 7.13 The cycle by which EF$_1$ carries GTP and aa-tRNA to the ribosome during protein synthesis.

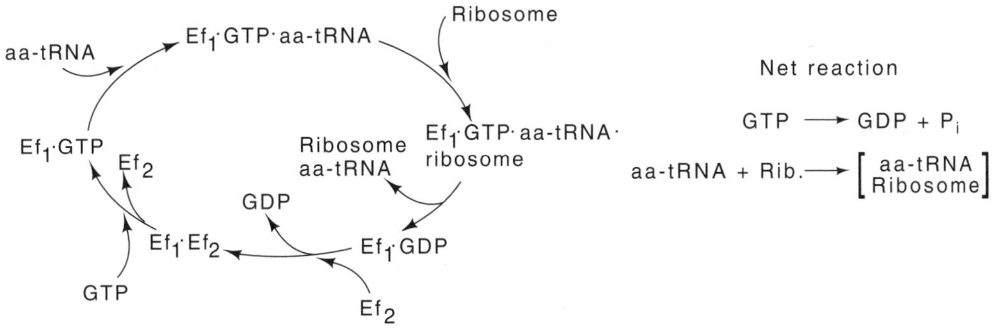

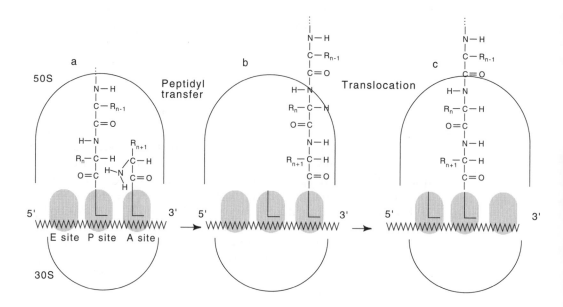

Figure 7.14 The process of protein synthesis. (a) The polypeptide occupies the P site, and the incoming aa-tRNA occupies the A site. (b) After formation of the next peptide bond, the polypeptide occupies the A site and the tRNA in the P site is not acylated. (c) The ribosome following translocation. The elongated polypeptide now occupies the P site, the A site is empty, awaiting arrival of EF_1 with another aa-tRNA, and the discharged tRNA occupies the E site.

Peptide Bond Formation

Elongation of the polypeptide chain occurs on the ribosome. The growing chain attached to a tRNA occupies the peptidyl, P, site of the ribosome (Fig. 7.14). Alongside is another charged tRNA, also with its anticodon base paired to a codon on the mRNA. This second site is termed the acceptor, or A site since the amino acid here will act as an acceptor as the peptide chain is transferred to it. This transfer is catalyzed by the peptidyl transferase activity of the large subunit. Although this activity is stimulated by ribosomal proteins, *in vitro* it can be catalyzed by the ribosomal RNA itself. This step does not require an external source of energy like GTP.

Translocation

Following formation of a peptide bond, the P site of the ribosome contains an uncharged tRNA, and the A site contains a tRNA linked to the growing peptide chain. Translocation is the process of recocking the elongation mechanism. The uncharged tRNA in the P site is moved to the exit, or E site, messenger translocates three bases toward the P site, thereby moving the tRNA with the peptide chain into the P site. The translocation process itself requires hydrolysis of a GTP molecule that has been carried to the ribosome by the EF-G or G factor. Since the elongation factors are used once for each amino acid added, a large number of molecules of each must be present in the cell to support protein synthesis. It is also logical that their level should parallel the level of ribosomes, and, indeed, as growth rate varies, their levels do keep pace with the levels of ribosomes. With the entry of a charged tRNA into the A site of the ribosome, the uncharged tRNA in the E site is released.

At some time during the growth of the peptide chains, the N-terminal amino acid is modified. Approximately 40% of the proteins isolated from *E. coli* are found to begin with methionine, but since all initiate with N-formyl methionine, the remaining 60% must lose at least the N-terminal methionine. Similarly, the 40% of the proteins that do begin with methionine all lack the formyl group. Thus the formyl group must be removed after protein synthesis has initiated. Examination of nascent polypeptide chains on ribosomes shows that the formyl group is missing if they are larger than about 30 amino acids, and therefore the

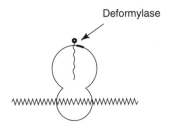

deformylase could well be a part of the ribosome. It could act when the growing peptide chain is long enough to reach the enzyme.

The deformylase is a very labile enzyme that is exceedingly sensitive to sulfhydryl reagents. Since many other enzymes isolated from the same cells require the same sulfhydryl reagents for stability, it may be that the deformylase is normally bound to some structure that contributes to its stability, and when it is isolated from extracts and partially purified, it is particularly labile in its unnatural environment.

Termination, Nonsense, and Suppression

How is the elongation process ended and the completed polypeptide released from the final tRNA and from the ribosome? The signal for ending the elongation process is any one of the three codons UGA, UAA, and UAG. Of the 64 possible three-base codons, 61 code for amino acids, and are "sense," and 3 code for termination, and are "nonsense." In termination, as in the other steps of protein synthesis, specialized proteins come to the ribosome to assist the process. Apparently upon chain termination, the ribosome is unlocked but not immediately released from the messenger. For a short while before it can fully dissociate from messenger, it can drift phaselessly forward and backward

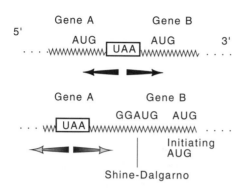

Figure 7.15 Translation termination and reinitiation. Top: A ribosome terminating at the UAA codon could directly reinitiate at either of the nearby AUG codons if either possesses a ribosome binding site. Bottom: A ribosome terminating at the UAA codon would not be able to drift to the more distant AUG codon on the right. A new initiation complex would have to form using the GGAG sequence to bind 16S rRNA.

short distances and can reinitiate translation if it encounters a ribosome-binding site and an initiation codon before it dissociates (Fig. 7.15).

Three proteins are involved in termination: a protein factor R_1, which is necessary for termination at UAA and UAG codons; a protein factor R_2, which is necessary for termination at UAA and UGA codons; and a

Termination codons

UAG
UAA — R_1 Termination factors
UGA — R_2

protein R_3, which accelerates the termination process. Cells contain approximately one molecule of R per five ribosomes, a number consistent with the usage of these molecules.

The existence of chain termination codons is responsible for an interesting phase in the growth of molecular biology. A mutation within a gene can change one of the 61 sense codons into one of the polypeptide chain terminating, or nonsense codons. This result shows that only the three bases are necessary to code for chain termination, no others are required, and no special secondary structure of the mRNA is required. As a result of a nonsense codon within a gene, the protein encoded by the mutated gene will be prematurely terminated during translation. Usually the shortened polypeptide possesses no enzymatic activity, and it is frequently degraded by proteases within the cell. An additional effect of a nonsense mutation is depressed translation of a following gene in an operon. This polar effect results from termination of transcription due to the sizeable portion of the barren mRNA following the nonsense codon and preceding the next ribosome initiation site. Quite surprisingly, some bacterial strains were found that could suppress the effects of a nonsense mutation. Although the suppressors rarely restored the levels of the "suppressed" protein to prior levels, the cell often possessed sufficient amounts of the protein to survive. In the case of a nonsense mutation in a phage gene, the suppressor strains permitted the phage to grow and form plaques.

Capecchi and Gussin showed that a suppressing strain inserted a particular amino acid at the site of the nonsense mutation. It did so by "mistranslating" the nonsense codon as a codon for an amino acid. They also showed that the mistranslation resulted from a change in one of the tRNAs for the inserted amino acid. Subsequently, sequencing of suppressor tRNAs has shown that, except in a special case, their anticodons have been altered so as to become complementary to one of the termination codons. Apparently one of two different events can then occur when a ribosome reaches a nonsense codon in a suppressing strain. Termination can occur via the normal mechanism, or an amino acid can be inserted into the growing polypeptide chain and translation can proceed.

Using genetic selections, suppressors have been found that insert tyrosine, tryptophan, leucine, glutamine, and serine. Except in unusual cases, such suppressors must be derived from the original tRNAs by single nucleotide changes. By chemical means and the utilization of genetic engineering, an additional half dozen or so suppressors have been synthesized.

The termination codon UAG has come to be called amber and UAA called ochre. No generally used name for the UGA codon exists, although it is sometimes called opal. Amber-suppressing tRNAs read only the UAG codon, and ochre-suppressing tRNAs read both UAA and UAG codons as a result of the "wobble" in translation. Since the R factors are protein and cannot be constructed like tRNA, it is no surprise that R_2 does not "wobble" and does not recognize the UGG (trp) codon.

How do normal proteins terminate in suppressing cells? If a suppressor were always to insert an amino acid in response to a termination codon instead of terminating, then many of the cellular proteins in suppressor-containing cells would be fused to other proteins or at least be appreciably longer than usual. The problem of terminating normal proteins could be solved in part by the presence of several different termination signals at the end of every gene. Then only the introduction of several different suppressors to a cell could create problems. Few genes, however, have been found to be ended by tandem translation terminators.

The more likely explanation for the viability of nonsense-suppressing strains is that the efficiency of suppression never approaches 100%. Typically it is 10% to 40%. Hence suppression of normal translation termination codons can fuse or lengthen some proteins in a cell, but most terminate as usual. On the other hand, the gene possessing the nonsense mutation would occasionally yield a suppressed instead of terminated protein. A suppression efficiency of 20% could reduce the amount of some cellular proteins from 100% to 80%, relatively speaking, not a substantial reduction. On the other hand, the existence of this suppressor would raise the amount of the suppressed protein from 0%

to 20% of normal. Relative to the nonsuppressed level, this is an enormous increase.

Chaperones and Catalyzed Protein Folding

In the 1960s Anfinson showed that pancreatic ribonuclease could be denatured and, when placed in buffers that resemble intracellular solvent conditions, would renature. This finding led to the belief that all proteins fold *in vivo* without the assistance of other proteins. Consequently it has come as a second surprise to find that virtually all types of cells, from bacteria to higher eukaryotes, possess proteins that appear to assist the folding of nascent proteins. Although the majority of the cell's proteins do fold on their own, an important number utilize auxiliary folding proteins.

In the *E. coli* cytoplasm, some newly synthesized and therefore unfolded proteins first interact with DnaK and then DnaJ. Binding to these two proteins prevents premature misfolding or aggregation. Then, with the assistance of GrpE, and the hydrolysis of ATP, the oligomeric protein GroEL/ES binds. This complex recognizes secondary structure of polypeptides and appears to stabilize conformational intermediates as the newly synthesized proteins settle from what is called the molten globule state into their final compact folded state.

Eukaryotic cells possess analogs of DnaK and GroEL. These are known as the heat shock proteins Hsp70 and Hsp60. The synthesis of these 70,000 and 60,000 dalton proteins is dramatically increased by exposure of the cells to heat or other agents that denature proteins. Members of these families help maintain polypeptides in the extended state for import into mitochondria and then help fold the imported polypeptide. The proteins are called chaperones for their roles in assisting the transport process.

Resolution of a Paradox

Hybridization is possible between nucleic acids of complementary sequence. Of course, there is a lower limit to the length of the participating polymers. A single adenine in solution is not normally seen to base pair with a single thymine. The lower length limit for specific hybridization between two nucleic acid molecules in typical buffers containing 0.001 M to 0.5 M salt at temperatures between 10°C and 70°C is about ten bases. How then can protein synthesis have any degree of accuracy since only three base pairs form between the codon and anticodon?

One part of the answer to how triplet base pairing can provide accuracy in protein synthesis is simply that additional binding energy is provided by contacts other than base pairing. A second part of the answer is that by holding the codon and anticodon rigid and in complementary shapes, binding energy is not consumed in bringing all three bases of the codon and anticodon into correct positions (Fig. 7.16). Once pairing occurs at the first base, the second and third bases are already

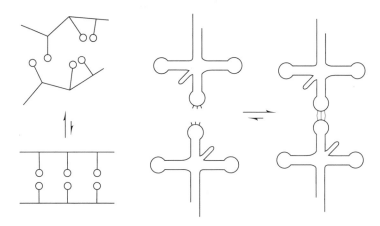

Figure 7.16 Under the same conditions, complementary anticodons in tRNA molecules will hybridize whereas the same isolated trinucleotides will not.

in position to pair. When two complementary trinucleotides bind to each other, this is not the case. Even after one base pair has formed, the second and third bases of each polymer must be brought into the correct positions before they can base pair. Additional binding energy is consumed in properly orienting them. This is another example of the chelate effect.

A simple hybridization experiment with tRNA provides a direct demonstration of the consequences of correctly positioning and orienting bases. Two tRNA molecules with self-complementary anticodons will hybridize! Since the structure of the entire tRNA molecule serves to hold the anticodon rigid, the binding energy derived from forming the second and third base pairs need not go into holding these in the proper position. They already are in the correct position. As a result, the two tRNA molecules will hybridize via their anticodons whereas self-complementary trinucleotides will not hybridize.

B. Physiological Aspects

Messenger Instability

The information for the sequence of amino acids in a protein is carried from the DNA to ribosomes in the messenger RNA. Once the necessary proteins have been synthesized from a messenger, it is necessary that the translation of the messenger cease. In principle, exponentially growing bacterial cells can eliminate unneeded mRNA merely by dilution due to growth of the cells. Cells that have ceased overall growth, such as eukaryotic cells in a fully grown organism, cannot use this approach.

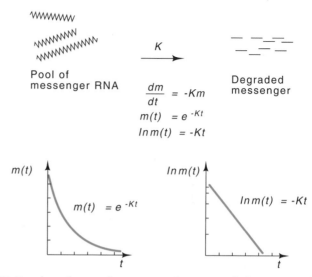

Figure 7.17 Random decay of messenger from a pool characterized by a rate constant K gives rise to an exponential decrease in the amount of messenger in the pool if it is not replenished by new synthesis. Left: Kinetics if plotted on rectilinear coordinates. Right: Kinetics if plotted on semilogarithmic coordinates.

Complicated mechanisms can be imagined for destruction of messenger once it has been used a fixed number of times. Cells appear not to use strict bookkeeping on translation, however. Once a messenger RNA has been synthesized, it has a fixed probability per unit time of being degraded by nucleases (Fig. 7.17). This is a random decay process. The population of such molecules then can be characterized as having a half-life and will show an exponential decay in levels if synthesis stops. Some molecules in the population will survive for long times while others will be degraded soon after their synthesis.

Cells that must adapt to a changing environment must vary their enzyme synthesis and consequently possess many messengers with relatively short lifetimes. Most bacterial messengers have half-lives of about two minutes, although some messengers have half-lives of over ten minutes. Some messengers in eukaryotic cells have half-lives of several hours and other eukaryotic messengers have half-lives of several weeks or even longer in stored forms.

Protein Elongation Rates

In bacteria, the protein elongation rate is about 16 amino acids per second. This means the ribosomes are moving about 48 nucleotides per second along the messenger RNA. This value is very close to the corresponding transcription rate of 50 to 60 bases per second. Therefore once a ribosome begins translation it can keep up with the transcribing RNA polymerase. In several of the better-studied operons, the rate of ribosome attachment is sufficiently fast that the ribosomes are rather

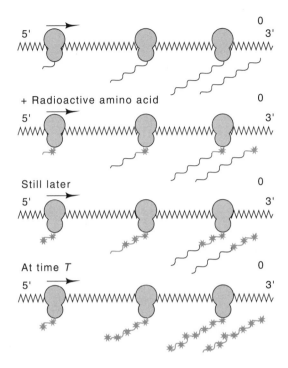

Figure 7.18 Synthesis of a protein by ribosomes traversing a messenger. Ribosomes initiate at the 5' end of the messenger, and the protein is completed and released at the 3' end. Addition of radioactive amino acids produces partially labeled polypeptides, indicated by the stars, and a completed protein with radioactive amino acids near its carboxy terminus. Longer labeling intervals lead to longer radioactive regions until radioactive amino acids have been present for the length of time, T, required to synthesize the complete protein. After this time all released polypeptides are labeled over their entire lengths.

closely spaced on the messenger, with about 100 Å of free space between ribosomes. In other operons, however, the spacing is greater.

How can the *in vivo* rate of protein elongation be measured? One approach would be to use methods analogous to those used to measure RNA elongation rates as described in Chapter 4. Unfortunately, no adequate inhibitors of protein initiation analogous to rifamycin are known, and slightly more complicated experiments must be done. The most general method for rate measurement uses an idea originally developed for measurement of RNA elongation rates before rifamycin was available.

Consider an experiment in which a radioactive amino acid is added to a culture of growing cells (Fig. 7.18). Let us focus our attention on a class of protein of one particular size and consider intervals that are short compared with the cell doubling time. The number of polypeptide chains in the size class completed after the addition of radioactive label is proportional to the time of labeling, $n \propto t$.

Also, for labeling intervals shorter than the time required to synthesize this size of polypeptide chain from one end to the other, the average amount of label incorporated into each chain is proportional to the time that label has been present. Hence during this early period, the total amount of radioactivity in the particular size class increases in proportion to the number of chains released multiplied by their average radioactivity. Both of these are proportional to the time of labeling, t.

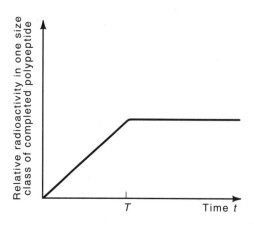

Figure 7.19 The fraction of radioactivity in one size class of polypeptide. The time required to synthesize this peptide is T.

Thus the radioactivity in the particular size class of protein increases in proportion to t^2.

After label has been present for the length of time, T, necessary to synthesize completely the polypeptide, the radioactivity per completed peptide chain can no longer increase. After this time, the radioactivity in the particular polypeptide size class can only increase in proportion to the number of chains completed. That is, the radioactivity now increases in proportion to t.

For experimental quantitation, it is convenient to compare the radioactivity in a particular size class of polypeptide to the total radioactivity in all sizes of polypeptide. The total amount of radioactivity in all size classes of polypeptide must increase in proportion to the time of labeling, t. Therefore the fraction of radioactivity in a particular size class of protein increases in proportion to t until the radioactive label has been present long enough for the entire length of the protein to become radioactively labeled. Thereafter, the fraction of label in the size class remains constant (Fig. 7.19). Hence, determining the transition time from linear increase to being constant provides the synthesis time for the particular size class of polypeptide.

The experimental protocol for performing the elongation rate measurement is simply to add a radioactive amino acid to a growing culture of cells. At intervals thereafter, samples are withdrawn and the protein

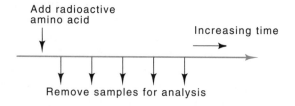

from the entire sample is denatured with sodium dodecyl sulfate and electrophoresed on a polyacrylamide gel. This provides the requisite size separation of the polypeptides as described in Chapter 4. The results found in *E. coli* and other bacteria are that the protein elongation rate is about 16 amino acids per second when cells are growing at 37°. This value is independent of the length of the polypeptide chain. A similar value is found by measuring the time until the appearance of N-terminal label in free completed β-galactosidase or by measuring the induction kinetics of enzymatically active β-galactosidase.

Directing Proteins to Specific Cellular Sites

Cells must direct proteins to several different locations. In addition to the cytoplasm in which most proteins in bacteria are found, proteins are also found in membranes, the cell wall, in the periplasmic space, and secreted altogether. Proteins in eukaryotic cells are found in the cytoplasm and in various cell organelles such as the mitochondria, chloroplasts, lysosomes. Proteins may be excreted as well. Proteins are synthesized in the cytoplasm, but directing them to enter or cross the cell membranes occurs during or shortly after synthesis of the protein. How is this done?

Crucial first observations on protein localization were made on immunoglobulin secretion. There it was observed that ribosomes synthesizing immunoglobulin were bound to the endoplasmic reticulum. Furthermore, when messenger from the membrane-bound ribosomes was extracted and translated *in vitro*, immunoglobulin was synthesized; however, it was slightly larger than the normal protein. This protein possessed about 20 extra amino acids at its N-terminus. Finally, when translation that had initiated *in vivo* was completed *in vitro*, immunoglobulin of the normal length was synthesized. These observations led Blobel to propose the signal peptide model for excretion of proteins.

The signal peptide model utilizes observations on immunoglobulin excretion and has the following parts (Fig. 7.20): first, the N-terminal

Figure 7.20 Using an N-terminal peptide of a protein as a signal for protein export. As ribosomes begin translation, the hydrophobic signal peptide signals attachment to a site on the membrane. The protein is exported through the membrane as it is synthesized, and at some point the signal peptide is cleaved.

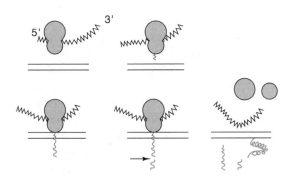

sequence of a protein to be excreted contains a signal that specifies its transport; second, during translation of messenger coding for an excreted protein, the N-terminal amino acids are required for binding the ribosomes to the membrane; third, during the synthesis of the remainder of the protein, the growing polypeptide chain is directly excreted through the membrane; fourth, often during synthesis the signal peptide is cleaved from the remainder of the protein. This leads to the fifth element of the model: there should exist in or on the membrane a protease to cleave the signal peptide from excreted proteins.

Verifying the Signal Peptide Model

Often it is easier to make observations on eukaryotic cells and then to prove the resulting ideas with experiments done on bacteria. One direct demonstration of concomitant translation and excretion through a membrane was made in bacteria. Chains of a periplasmic protein in the process of being synthesized were labeled by a chemical that was excluded from the cytoplasm by the inner membrane. The experiment was done by adding the labeling chemical to spheroplasts. These are cells lacking their outer membrane and peptidyl-glycan layer. Shortly after adding the reactive labeling compound, membrane-bound ribosomes were isolated and their nascent polypeptide chains were examined and were found to be labeled. Proteins that are normally found in the cytoplasm were not similarly labeled.

Does the N-terminal sequence on a protein signal that the remainder of the protein is to be transported into or through a membrane? In principle, this could be tested by tricking a cell into synthesizing a new protein in which the N-terminal sequence from an excreted protein has been fused to a protein normally found in the cytoplasm. If the hybrid protein is excreted, then the new N-terminal sequence must be signaling export.

Fortunately, *E. coli* has been sufficiently well studied that a number of candidates exist whose N-terminal sequences might be used in such a project. The *malF* gene product is a protein involved in the uptake of maltose into cells. It is located in the periplasmic space. This should be an excellent source of an "excretion-coding" N-terminal sequence. The ideal situation to test the excretion hypothesis would be to fuse the N-terminal sequence of *malF* to an easily assayed cytoplasmic protein. One very good candidate for this fusion is β-galactosidase, a protein for which the genetics have also been fully developed.

Remarkably, the fusion of the N-terminal portion of the *malF* gene to β-galactosidase was performed *in vivo* without using recombinant DNA techniques (Fig. 7.21). Through clever genetic manipulations, Silhavy and Beckwith moved the β-galactosidase gene, *lacZ*, near *malF* and then generated a deletion that fused the N-terminal portion of the *malF* gene to β-galactosidase. A postulate of the signal peptide model could therefore be tested. Indeed, a sizable fraction of the β-galactosidase from some of the fusion strains was not located in the cytoplasm

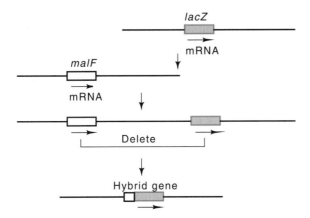

Figure 7.21 Construction of a *malF-lacZ* fusion. The β-galactosidase gene is first brought near the *malF* gene, and then a deletion between the two fuses the N-terminus of *malF* to *lacZ*.

but instead was bound to or located within the inner membranes of the cells.

How do you show that a protein is located in the inner membrane of a cell? One method is to disrupt cells, to collect membrane fragments, and then to separate the inner and outer membrane fragments and assay each for the protein. Inner membrane is less dense than outer membrane, and therefore the two can be separated according to density by isopycnic centrifugation in a tube containing a gradient of sucrose concentration (Fig. 7.22). The two membrane fractions sediment down the tube into higher and higher sucrose concentrations until they reach positions where their densities equal the density of the sucrose. There they come to rest. By assay of the fractions collected from such a sucrose gradient, the β-galactosidase from a *malF-lacZ* fusion could be demon-

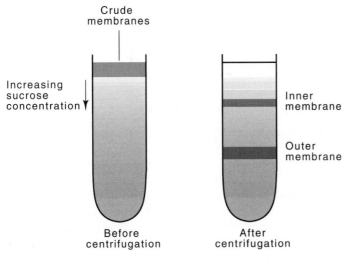

Figure 7.22 Isopycnic separation of membrane fractions by centrifugation on a preformed sucrose gradient.

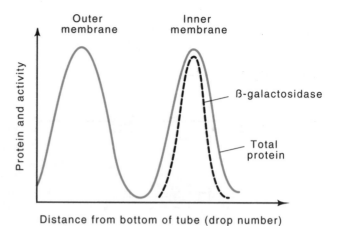

Figure 7.23 Results of isopycnic centrifugation in which outer membrane came to rest near the bottom of the centrifuge tube and inner membrane came to rest closer to the top. The β-galactosidase was associated with the inner membrane fraction.

strated to be located largely in the inner membrane of the cells (Fig. 7.23).

Is the N-terminal sequence from *malF* actually necessary for export of the β-galactosidase into or through the membrane? Some of the fusions of *malF* to *lacZ* that were generated in the study yielded proteins that remained in the cytoplasm. Genetic studies showed that the process of producing the fusions in these strains had removed most of the signal peptide of the *malF* gene. Other studies have shown that mutations which abolish export of fusion proteins alter the signal peptide. These experiments show that the signal peptide plays an important role in export. It is clear, however, that the signal peptide is not the whole story. The *malF-lacZ* fusion protein was stuck to the inner membrane. It was not located in the periplasmic space, the final destination of the native *malF* protein. Therefore something in the structure of the β-galactosidase protein prevents the hybrid from reaching the periplasmic space. Hence, reaching the periplasmic space requires the leader sequence as well as a compatible structure in the remainder of the protein.

The Signal Recognition Particle and Translocation

In some types of eukaryotic cells, most of the proteins translocated into or across membranes utilize the signal recognition particles. These are small ribonucleoprotein particles containing a 300 nucleotide RNA molecule and five proteins, ranging in size from 14 KDa to 72 KDa. As the elongating signal peptide protrudes from the ribosome, the signal recognition particles bind and arrest translation. After binding of the signal recognition particle to the endoplasmic reticulum, translation resumes and the protein is translocated across the membrane during its synthesis. The generality of this process is not known, and some eukaryotic cells seem not to arrest translation. Controversy has waxed hot on the matter of whether bacteria also utilize the same sort of pathway. They contain a small ribonucleoprotein particle containing an RNA that sediments at 4.5S. This possesses significant homology to the eukaryotic signal recognition particle, but its role in protein secretion is not yet clear.

Can the roles of the signal recognition particle components be identified? Genetics frequently can be utilized in simpler organisms to obtain mutants unable to perform certain reactions. The complexity of eukaryotic systems blocks the use of mutants for functional dissection of the signal recognition particle. Instead, a biochemical approach had to be used.

There are two requirements for a biochemical dissection. First, it must be possible to assay for each of the steps in the process. This is possible. Binding of signal recognition particles can easily be measured because after binding they cosediment with the translating ribosomes. By using homogenous messenger, translation arrest can be seen by the failure of the proteins to be completed and by the accumulation of short incomplete proteins. Translocation into membranes can be quantitated by adding membrane vesicles to a translation mixture. Translocation of the protein into the vesicles leaves them resistant to protease digestion.

The second requirement is the ability to take the signal recognition particle apart and then to reassemble it. When the particle had been dissociated and the various protein components were separated, they were individually inactivated by reaction with N-ethylmaleimide. Particles were reassembled containing one reacted component and the remainder unreacted. By this means, the proteins responsible for signal recognition, translation arrest, and translocation were identified. In

Function	Components Required
Signal Recognition	54 Kd
Elongation Arrest	9 Kd, 14 Kd, RNA
Translocation	68 Kd, 72 Kd

contrast to many systems, each of these functions was determined by a separate component of the particle. Many times, functions are shared.

Expectations for Ribosome Regulation

The final topic we consider in this chapter is the regulation of the level of the protein synthesis machinery itself. Not surprisingly, we will find that the synthesis of this machinery is regulated by the intensity of its use.

A ribosome is a large piece of cellular machinery. It consists of about 55 proteins, two large pieces of RNA, and one or two smaller RNAs.

Hence it is natural to expect a cell to regulate levels of ribosomes so that they are always used at the highest possible efficiency. Since bacterial cells and some eukaryotic cells may grow with a wide variety of rates, a sophisticated regulation mechanism is necessary to ensure that ribosomes are fully utilized and synthesizing polypeptides at their maximal rate under most growth conditions.

In an earlier section we discussed the finding that proteins in bacteria are elongated at about 16 amino acids per second. In the next sections we will find that the rate of cellular protein synthesis averaged over all the ribosomes is also about 16 amino acids per second. This means that virtually no ribosomes sit idle. All are engaged in protein synthesis.

Proportionality of Ribosome Levels and Growth Rates

The dramatic ability of bacteria to grow with a wide variety of rates prompts the question of how they manage to maintain balanced synthesis of their macromolecules. In a study of this question, Schaechter, Maaløe, and Kjeldgaard made the discovery that ribosomes are used at constant efficiency, independent of the cell growth rate. To appreciate their contribution fully, it will be helpful first to examine a related question: what is the average rate of protein synthesis per ribosome? As a first step we will estimate this value using typical cellular parameters, then we will calculate this value more carefully and include the effects of increase in the number of ribosomes during a cell doubling time.

An average bacterial cell with a doubling time of 50 minutes contains about 1×10^{-13} g protein and about 10,000 ribosomes. Approximating the molecular weight of amino acids to be 100,

1×10^{-13} g protein is $1 \times 10^{-13}/10^2 = 10^{-15}$ moles amino acid;

10^{-15} moles amino acid is $6 \times 10^{23} \times 10^{-15} = 6 \times 10^8$ molecules;

10^4 ribosomes polymerize these 6×10^8 amino acids in 50 min or 3×10^3 sec.

Thus the average rate of protein synthesis per ribosome is $6 \times 10^8/3 \times 10^3 = 20$ amino acids per second per ribosome. Compared to the typical turnover number of enzymes, greater than 1,000 per second, this is a low number. We have seen already, however, the process of addition of a single amino acid to the growing polypeptide chain is complex and involves many steps.

To calculate accurately the rate of protein synthesis per ribosome during steady-state growth, we must include the growth of the cells in the calculation. This can be done in the following way. Define α_r as the relative rate of synthesis of ribosomal protein, that is,

$$\alpha_r = \frac{\dfrac{dP_r}{dt}}{\dfrac{dP_t}{dt}}$$

where P_r, is ribosomal protein and P_t is total protein. Let $R(t)$ be the number of ribosomes in a culture at time t. Since ribosome number will increase like cell number, $R(t) = R(0)\,e^{\mu t}$, and hence, $\dfrac{dR}{dt} = \mu R(t)$. Also, the rate of ribosome synthesis, dR/dt, equals the rate of ribosomal protein synthesis in amino acids per unit time divided by the number of amino acids in the protein of one ribosome, C:

$$\frac{dR}{dt} = \alpha_r \times \frac{dP_t}{dt} \times \frac{1}{C}.$$

Then dP_t/dt equals the number of ribosomes times the average elongation rate per ribosome, K. That is,

$$\frac{dP_t}{dt} = KR.$$

We have the following two expressions for dR/dt

$$\frac{dR}{dt} = \mu R(t), \text{ and } \frac{dR}{dt} = \alpha_r \frac{KR(t)}{C},$$

which yields $K = \mu C/\alpha_r$, our desired relation. Note that if α_r is roughly proportional to the growth rate, as has been found for bacteria except at the slowest growth rates (Fig. 7.24), then the term μ/α_r is a constant and hence K, average activity of a ribosome, is independent of the growth rate.

For *E. coli* B/r growing at 37°, with a doubling time of 48 minutes,

$$\mu = \frac{ln2}{T_d} = \frac{0.693}{2.9 \times 10^3} \text{ sec}^{-1} = 2.4 \times 10^{-4} \text{ sec}^{-1},$$

$$C = \frac{\text{Molecular weight of ribosomal protein}}{\text{Average molecular weight of amino acids}}$$

$$= \frac{9.5 \times 10^5}{1.1 \times 10^2} = 8.3 \times 10^3 , \text{ and}$$

$$\alpha_r = 0.12.$$

Therefore

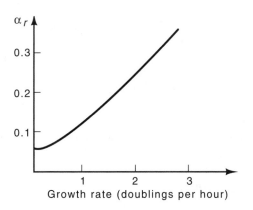

Figure 7.24 The value of α, as a function of growth rate. Except at slow growth rates, α_r is proportional to growth rate.

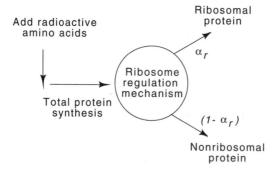

Figure 7.25 Schematic of the ribosomal protein regulation system. The mechanism determines the fraction, α, of total protein synthesis to be devoted to ribosomal protein synthesis.

$$K = \frac{2.4 \times 10^{-4}\,\text{sec}^{-1} \times 8.3 \times 10^3}{0.12} \quad \frac{\text{amino acids}}{\text{ribosome second}}$$

$$= 17 \text{ amino acids per ribosome per second.}$$

This value is close to the elongation rate of polypeptides, showing that most ribosomes in the bacterial cell are engaged in protein synthesis and are not sitting idle.

A measurement of α_r at any time during cell growth can be accomplished by adding a radioactive amino acid to growing cells for a short interval (Fig. 7.25). Then an excess of the nonradioactive form of the amino acid is added and cells are allowed to grow until all the radioactive ribosomal proteins have been incorporated into mature ribosomes. The value of α_r, is the fraction of radioactivity in ribosomal protein compared to total radioactivity in all the cellular protein. This fraction can be determined by separating ribosomal protein from all other cellular protein by electrophoresis or sedimentation and measuring the radioactivity in each sample.

Regulation of Ribosome Synthesis

We already know that α_r, during balanced exponential growth, is proportional to the growth rate. Let us consider the response of α_r during a transition between growth rates. Originally this type of data was sought in an effort to place constraints on the possible feedback loops in the ribosome regulation system. By analogy to electronic circuits, the response of the ribosome regulation system could be highly informative. The most straightforward growth rate shift is generated by addition of nutrients supporting faster growth. The resulting response in α_r is a quick increase to the α_r characteristic of the new growth conditions (Fig. 7.26). Superimposed on this shift are small oscillations. The existence of these oscillations shows that a number of cellular components are involved in the regulatory system. The ten-minute period of the oscillations suggests that at least one of the regulating components is not a

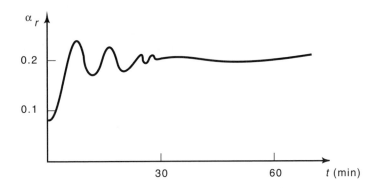

Figure 7.26 The response of α_r, to a growth rate increase.

molecule like ATP that is completely turned over on a time scale appreciably less than ten minutes.

In addition to regulating ribosome synthesis in response to differing growth rates, bacterial cells, and likely others as well, display a second type of ribosome regulation. This is the stringent response, in which the synthesis of ribosomal RNA and tRNA completely stops if protein synthesis stops. One obvious question about ribosome regulation is whether the stringent response itself is joined to the growth rate response system. Studies have shown that mutants in the stringent system, which are called relaxed, regulate their ribosome levels the same as wild-type cells as growth rate is varied. Consequently, these two systems are separate, as is verified by more recent experiments by Gourse.

The halt in the accumulation of rRNA in amino acid-starved, stringent cells appears to result from decreased initiations by RNA polymerase rather than from degradation of rRNA or blockage of elongation by RNA polymerase. This is not surprising in light of what we have already discussed about regulation of DNA synthesis. One demonstration of this fact is that no radioactive rRNA is synthesized if amino acids,

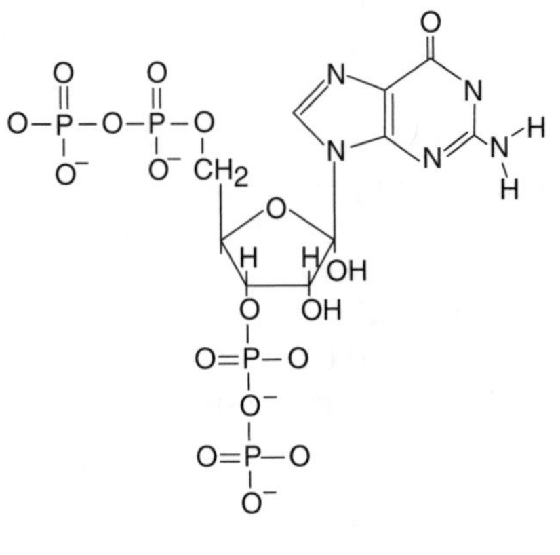

Figure 7.27 The structure of ppGpp.

Table 7.3 Conditions for Synthesis of ppGpp and pppGpp

Reaction Mixture	Synthesis of ppGpp and pppGpp
Complete	+
- 30S	-
- 50S	-
- mRNA	-
- tRNA	-
- Stringent factor	-

rifamycin, and labeled uridine are simultaneously added to amino acid-starved, stringent cells. The same conclusion is also reached by observing the kinetics of synthesis of the 16S and 23S rRNAs immediately after restoring the missing amino acid to an amino acid-requiring strain.

During the stringent response, guanosine tetra- and pentaphosphate, ppGpp (Fig. 7.27) and pppGpp, accumulate in amino acid-starved, stringent cells but not in amino acid-starved, relaxed cells. These compounds may directly interfere with transcription of the ribosomal RNA genes. *In vitro* experiments show that synthesis of ppGpp requires a particular protein, ribosomes, messenger, and an uncharged tRNA corresponding to the codon of messenger in the A site of a ribosome (Table 7.3). The required protein is normally rather tightly associated with the ribosomes and is the product of a gene called *relA*.

Balancing Synthesis of Ribosomal Components

Although the synthesis of the individual components of ribosomes may be rather well regulated, a slight imbalance in the synthesis of one component could eventually lead to elevated and potentially toxic levels of that component. Synthesis of the ribosomal RNAs in bacteria are kept in balance by a simple mechanism. As we have already seen in Chapter 5, these RNAs are synthesized in one piece by an RNA polymerase that initiates at a promoter and transcribes across the genes for the three RNAs. Different mechanisms are used to maintain balanced synthesis of some of the ribosomal proteins. In one case, one of the proteins encoded in a ribosomal protein operon reduces translation of all the proteins in that operon. This effect is called translational repression.

The finding that a ribosomal protein represses translation of proteins only from the same operon provides an efficient means for the cell to maintain balanced synthesis of all the ribosomal proteins. Suppose that some ribosomal proteins began to accumulate because their synthesis is a little faster than the other proteins and the rRNA. Then, as the level of these proteins begins to rise in the cytoplasm, they begin to repress

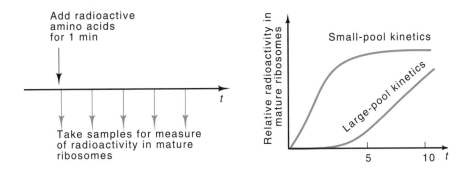

Figure 7.28 Determination of the pool size of ribosomal protein by the kinetics of a pulse of radioactive amino acids into ribosomal protein.

their own synthesis, and the system rapidly comes back to a balanced state.

How do we know about translational repression? The main clue came from careful measurements on cells with an increased number of genes coding for some of the ribosomal proteins. The increased copy number might have been expected to increase the synthesis of the corresponding proteins, but it did not. The synthesis of the mRNA for these proteins did increase as expected, and therefore it appeared that extra ribosomal proteins in the cell inhibited translation of their own mRNA. Proof of the idea of translational repression came from *in vitro* studies in which levels of individual free ribosomal proteins could be adjusted at will. The addition of DNA containing genes for some of the ribosomal proteins and properly prepared cell extract permits transcription and translation to yield ribosomal proteins synthesized *in vitro*. Nomura found that addition of the appropriate free ribosomal proteins to such a system repressed synthesis of the proteins encoded by the same operon as the added protein.

Not surprisingly, a ribosomal protein that regulates synthesis of a group of proteins binds to the mRNA to effect the repression. The structure of the binding region on the mRNA for some of these proteins is the same as the structure the protein binds to in the rRNA in the ribosome.

Global limits can be placed on the accuracy with which the synthesis of ribosomal components is balanced. A short pulse of radioactive amino acid is provided to the cells, and the total pool of all ribosomal proteins can be determined by measuring the kinetics of incorporation of label into mature ribosomes (Fig. 7.28). The results show that the pool contains less than a five-minute supply of ribosomal proteins. Similarly, the pool size of each individual ribosomal protein can be measured. The results of these experiments show that most of the ribosomal proteins also have very small intracellular pools.

We have seen that the mechanisms regulating ribosome synthesis have good reason to be sophisticated, and, indeed those aspects that have been investigated have turned out to be complicated. Much of the biochemistry and perhaps even much of the physiology of ribosome regulation remain to be worked out. In bacteria and other single-celled organisms such as yeast, it is likely that most of the regulation mechanism can be dissected by a combination of physiology, genetics, and biochemistry. It will be interesting to see if analogous problems in higher organisms can also be solved without the availability of genetics.

Problems

7.1. Look up the necessary data and calculate the turnover rate of charged tRNA in growing *E. coli*. That is, how long would it take the charged tRNA to become completely uncharged at the normal rates of protein synthesis if charging were suddenly to stop?

7.2. On the basis of finding ribosomes with the enzymatic activity of a nascent but incomplete peptide chain of an induced enzyme, it is sometimes claimed that proteins fold up as they are being synthesized. Why is this reasoning inadequate and what experiments could definitively determine the time between completion of synthesis of a polypeptide and its attainment of enzymatic activity? [Hint: Lactose is cleaved to glucose and galactose by the inducible enzyme β-galactosidase. Galactose itself is metabolized by enzymes of another inducible operon.]

7.3. In the experiment in which translation of Qβ RNA was blocked by the prior binding of a oligonucleotide homologous to the Shine-Dalgarno sequence, it was found that f-met tRNA could not bind to the ribosomes. However, the trinucleotide, AUG, was able to bind. What is the meaning of this finding and does it contradict the Shine-Dalgarno hypothesis?

7.4. How did N-formyl methionine come to be discovered as the initiating amino acid? Look up the references and describe the experiments.

7.5. Predict several tRNA synthetases that will activate incorrect amino acids and hydrolytically edit to reduce misacylation. Identify the amino acids likely to be misactivated.

7.6. How do we know of the existence and properties of the A, P, and E sites on ribosomes? Look up the necessary papers and describe the experiments.

7.7. Why have no amber suppressors inserting tryptophan been found?

7.8. A frameshift mutation is created by the insertion or deletion of one or more bases that alter the triplet reading frame. The only mutagens that generate mutations suppressing the effects of some frameshift mutations are frameshift mutagens. Some of the resulting suppressors do not lie in the original gene. Propose an explanation and an experiment to test your idea.

7.9. Suppressors for UAA, UAG, and UGA termination codons can be created by single-base changes in the anticodon regions of tRNA genes. What amino acids could be inserted by suppressors to each of these termination signals? Despite energetic searches, not all of these possible suppressors have been found. What are the suppressors that have been found, and what is an explanation for the failure to find the rest?

7.10. With wobble permitted, what is the minimum number of tRNA types that would be required to read all 61 sense codons?

7.11. In view of the fact that ribosomes can translate messenger as fast as it appears, perhaps they could translate still faster if they didn't run up against the RNA polymerase. What kinds of experiments could show this doesn't happen?

7.12. If transport of lactose into cells is unnecessary in a strain with β-galactosidase relocated to the outer membrane by fusion to a maltose transport protein, it appears strange that nature even bothers to have a transport system for the *lac* system. Why wouldn't it have been simpler for nature to use a signal peptide to place β-galactosidase on the outer membrane?

7.13. If the level of ribosomes is proportional to the growth rate, show that the rate of synthesis of ribosomes is proportional to the square of the growth rate.

7.14. At the ribosome translation efficiency normally seen in *E. coli*, what is the absolute maximum growth rate possible?

7.15. Why is it reasonable that exposure of cells to low concentrations of methanol would induce the synthesis of many of the same proteins that are induced by exposure to excessively high temperatures?

7.16. Sketch a graph of the synthesis of additional rRNA in amino acid-starved stringent cells on the restoration of amino acids, assuming (a) that the block of synthesis was at the RNA polymerase initiation step, and (b) that the block of synthesis was at an RNA chain elongation step, that RNA polymerase molecules are randomly distributed through the rRNA operons, and that they remain on the DNA during the starvation.

7.17. Overproduction of ribosomal proteins for structural studies was found to make cells so sick that they could hardly grow. What is the most likely explanation?

7.18. Why might it be reasonable for a UGA (termination) codon to exist within the gene encoding the R_2 release factor?

References

Recommended Readings

Codon-Anticodon Pairing: The Wobble Hypothesis, F. H. C. Crick, J. Mol. Biol. *19*, 548-555 (1966).

How Ribosomes Select Initiator Regions in mRNA: Base Pair Formation Between the 3'-Terminus of 16S rRNA and the mRNA During Initiation of Protein Synthesis in *E. coli.*, J. Steitz, K. Jakes, Proc. Nat. Acad. Sci. USA *72*, 4734-4738 (1975).

Each of the Activities of Signal Recognition Particle (SRP) Is Contained within a Distinct Domain: Analysis of Biochemical Mutants of SRP, V. Siegel, P. Walter, Cell *52*, 39-49 (1988).

Absolute *in vivo* Translation Rates of Individual Codons in *Escherichia coli*, M. Sorensen, S. Pedersen, J. Mol. Biol. *222*, 265-280 (1991).

Unusual Resistance of Peptidyl Transferase to Protein Extraction Procedures, H. Noller, V. Hoffarth, L. Zimniak, Science *256*, 1416-1419 (1992).

tRNA and Charging tRNA

Evidence for the Double-sieve Editing Mechanism for Selection of Amino Acids in Protein Synthesis: Steric Exclusion of Isoleucine by valyl-tRNA Synthetases, A. Fersht, C. Dingwall, Biochemistry *18*, 2627-2631 (1979).

A Covalent Adduct between the Uracil Ring and the Active Site on an Aminoacyl tRNA Synthetase, R. Starzyk, S. Koontz, P. Schimmel, Nature *298*, 136-140 (1982).

Changing the Identity of a Transfer RNA, J. Normanly, R. Ogden, S. Horvath, J. Abelson, Nature *321*, 213-219 (1986).

Gene for a Novel tRNA Species that Accepts L-serine and Cotranslationally Inserts Selenocysteine, W. Leinfelder, E. Zehelein, M. Mandrand-Berthelot, A. Böck, Nature *331*, 723-725 (1988).

A Simple Structural Feature is a Major Determinant of the Identity of a Transfer RNA, Y. Hou, P. Schimmel, Nature *333*, 140-145 (1988).

Aminoacylation of RNA Minihelices with Alanine, C. Francklyn, P. Schimmel, Nature *337*, 478-481 (1989).

Nucleotides in Yeast tRNA^{Phe} Required for the Specific Recognition by its Cognate Synthetase, J. Sampson, A. DiRenzo, L. Behlen, O. Uhlenbeck, Science *243*, 1363-1366 (1989).

Structure of *E. coli* Glutaminyl-tRNA Synthetase Complexed with tRNA^{Glu} and ATP at 2.8 Å Resolution, M. Rould, J. Perona, D. Söll, Science *246*, 1153-1154 (1989).

The Anticodon Contains a Major Element of the Identity of Arginine Transfer RNAs, L. Schulman, H. Pelka, Science *246*, 1595-1597 (1989).

Structural Basis of Anticodon Loop Recognition by Glutaminyl-tRNA Synthetase, M. Rould, J. Perona, T. Steitz, Nature *352*, 213-218 (1991).

Initiating Protein Synthesis

N-formylmethionyl-sRNA as the Initiatior of Protein Synthesis, J. Adams, M. Capecchi, Proc. Nat. Acad. Sci. USA *55*, 147-155 (1966).

A Mutant which Reinitiates the Polypeptide Chain after Chain Termination, A. Sarabhai, S. Brenner, J. Mol. Biol. *27*, 145-162 (1967).

Removal of Formyl-methionine Residue from Nascent Bacteriophage f2 Protein, D. Housman, D. Gillespie, H. Lodish, J. Mol. Biol. *65*, 163-166 (1972).

The 3'-Terminal Sequence of *Escherichia coli* 16S Ribosomal RNA: Complementary to Nonsense Triplets and Ribosome Binding Sites, J. Shine, L. Dalgarno, Proc. Nat. Acad. Sci. USA *71*, 1342-1346 (1974).

How Do Eucaryotic Ribosomes Select Initiation Regions in Messenger RNA?, M. Kozak, Cell *15*, 1109-1123 (1978).

Mutations of Bacteriophage T7 that Affect Initiation of Synthesis of the Gene 0.3 Protein, J. Dunn, E. Buzash-Pollert, F. Studier, Proc. Nat. Acad. Sci. USA *75*, 2741-2745 (1978).

Inhibition of Qβ RNA 70S Ribosome Initiation Complex Formation by an Oligonucleotide Complementary to the 3' Terminal Region of *E. coli* 16S Ribosomal RNA, T. Taniguchi, C. Weissmann, Nature *275*, 770-772 (1978).

Efficient Cap-dependent Translation of Polycistronic Prokaryotic mRNAs is Restricted to the First Gene in the Operon, M. Rosenberg, B. Paterson, Nature *279*, 696-701 (1979).

Compilation and Analysis of Sequences Upstream from the Translational Start Site in Eukaryotic mRNAs, M. Kozak, Nucleic Acids Res. *12*, 857-872 (1984).

mRNA Cap Binding Proteins: Essential Factors for Initiating Translation, A. Shatkin, Cell *40*, 223-224 (1985).

Initiation of Translation is Impaired in *E. coli* Cells Deficient in 4.5S RNA, D. Bourgaize, M. Fournier, Nature *325*, 281-284 (1987).

Specialized Ribosome System: Preferential Translation of a Single mRNA Species by a Subpopulation of Mutated Ribosomes in *Escherichia coli*, A. Hui, H. de Boer, Proc. Nat. Acad. Sci. USA *84*, 4762-4766 (1987).

Biochemical Aspects of Protein Synthesis, and Elongation, Termination, and Processing

Some Relationships of Structure to Function in Ribonuclease, F. White, Jr., C. Anfinsen, Ann. N. Y. Acad. Sci. *81*, 515-523 (1959).

Assembly of the Peptide Chains of Hemoglobin, H. M. Dintzis, Proc. Nat. Acad. Sci. USA *47*, 247-261 (1961).

Release Factors Mediating Termination of Complete Proteins, M. Capecchi, H. Klein, Nature, *226* 1029-1033 (1970).

The Accumulation as Peptidyl-transfer RNA of Isoaccepting Transfer RNA Families in *Escherichia coli* with Temperature-sensitive Peptidyl-transfer RNA Hydrolase, J. Menninger, J. Biol. Chem. *253*, 6808-6813 (1978).

Measurement of Suppressor Transfer RNA Activity, J. Young, M. Capecchi, L. Laski, U. RajBhandary, P. Sharp, P. Palese, Science *221*, 873-875 (1983).

Hemoglobin Long Island is Caused by a Single Mutation (Adenine to Cytosine) Resulting in a Failure to Cleave Amino-terminal Methionine, J. Prchal, D. Cashman, Y. Kan, Proc. Nat. Acad. Sci. USA *83*, 24-27 (1985).

Point Mutations Define a Sequence Flanking the AUG Initiator Codon that Modulates Translation by Eukaryotic Ribosomes, M. Kozak, Cell *44*, 283-292 (1986).

Stereochemical Analysis of Ribosomal Transpeptidation, Conformation of Nascent Peptide, V. Lim, A. Spirin, J. Mol. Biol. *188*, 565-577 (1986).

Mechanism of Ribosomal Translocation, J. Robertson, C. Urbanke, G. Chinali, W. Wintermeyer, A. Parmeggiani, J. Mol. Biol. *189*, 653-662 (1986).

Unique Pathway of Expression of an Opal Suppressor Phosphoserine tRNA, B. Lee, P. de la Peña, J. Tobian, M. Zasloff, D. Hatfield, Proc. Nat. Acad. Sci. USA *84*, 6384-6388 (1987).

Reading Frame Selection and Transfer RNA Anticodon Loop Stacking, J. Curran, M. Yarus, Science *238*, 1545-1550 (1987).

Interaction of Elongation Factors Ef-G and Ef-Tu with a Conserved Loop in 23S RNA, D. Moazed, J. Robertson, H. Noller, Nature *334*, 362-364 (1988).

Interaction of tRNA with 23S rRNA in the Ribosomal A, P, and E Sites, D. Moazed, H. Noller, Cell *57*, 585-597 (1989).

Extent of N-terminal Methionine Excision from *Escherichia coli* Proteins is Governed by the Side-chain Length of the Penultimate Amino Acid. P. Hirel, J. Schmitter, P. Dessen, G. Fayat, S. Blanquet, Proc. Natl. Acad. Sci. USA *86*, 8247-8251 (1989).

Identification of a Novel Translation Factor Necessary for the Incorporation of Selenocysteine into Protein, K. Forchammer, W. Leinfelder, A. Böck, Nature *342*, 453-456 (1989).

Time of Action of 4.5 S RNA in *Escherichia coli* Translation, S. Brown, J. Mol. Biol. *209*, 79-90 (1989).

Construction of *Escherichia coli* Amber Suppressor tRNA Genes II, L. Kleina, J. Masson, J. Normanly, J. Abelson, J. Miller, J. Mol. Biol. *213*, 705-713 (1990).

Relaxation of a Transfer RNA Specificity by Removal of Modified Nucleotides, V. Perret, A. Garcia, H. Grosjean, J. Ebel, C. Florentz, R. Giege', Nature *344*, 787-789 (1990).

The Allosteric Three-site Model for the Ribosomal Elongation-cycle: Features and Future, K. Nierhaus, Biochemistry *29*, 4997-5008 (1990).

Catalytic Properties of an *Escherichia coli* Formate Dehydrogenase Mutant in Which Sulfur Replaces Selenium, M. Axley, A. Böck, T. Stadtman, Proc. Natl. Acad. Sci. USA *88*, 8450-8454 (1991).

Escherichia coli Alkaline Phosphatase Fails to Acquire Disulfide Bonds When Retained in the Cytoplasm, A. Derman, J. Beckwith, J. Bact. *173*, 7719-7722 (1991).

Identification of a Protein Required for Disulfide Bond Formation *in vivo*, J. Bardwell, K. McGovern, J. Beckwith, Cell *67*, 581-589 (1991).

Structure and Function of Messenger RNA

Coding Properties of an Ochre-suppressing Derivative of *E. coli*, tRNA[Tyr], S. Feinstein, S. Altman, J. Mol. Biol. *112*, 453-470 (1977).

Novel Features in the Genetic Code and Codon Reading Patterns in *Neurospora crassa* Mitochondria based on Sequences of Six Mitochondrial tRNAs, J. Heckman, J. Sarnoff, B. Alzner-DeWeerd, S. Yin, U. RajBhandary, Proc. Nat. Acad. Sci. USA 77, 3159-3163 (1980).

Different Pattern of Codon Recognition by Mammalian Mitochondrial tRNAs, B. Barrell, S. Anderson, A. Bankier, M. DeBruijn, E. Chen, A. Coulson, J. Drouin, I. Eperon, D. Nierlich, B. Roe, F. Sanger, P. Schreier, A. Smith, R. Staden, I. Young, Proc. Nat. Acad. Sci. USA 77, 3164-3166 (1980).

Codon Recognition Rules in Yeast Mitochondria, S. Bonitz, R. Berlani, G. Coruzzi, M. Li, G. Macino, F. Nobrega, M. Nobrega, B. Thalenfeld, A. Tzagoloff, Proc. Nat. Acad. Sci. USA 77, 3167-3170 (1980).

Effects of Surrounding Sequence on the Suppression of Nonsense Codons, J. Miller, A. Albertinia, J. Mol. Biol. *164*, 59-71 (1983).

Context Effects: Translation of UAG Codon by Suppressor tRNA is Affected by the Sequence Following UAG in the Message, L. Bossi, J. Mol. Biol. *164*, 73-87 (1983).

Deviation from the Universal Genetic Code Shown by the Gene for Surface Protein 51A in *Paramecium*, J. Preer, L. Preer, B. Rudman, A. Barnett, Nature *314*, 188-190 (1985).

Messenger RNA Conformation and Ribosome Selection of Translational Reinitiation Sites in the *lac* Repressor RNA, K. Cone, D. Steege, J. Mol. Biol. *186*, 725-732 (1986).

Influences of mRNA Secondary Structure on Initiation by Eukaryotic Ribosomes, M. Kozak, Proc. Nat. Acad. Sci. USA *83*, 2850-2854 (1986).

Identification of a Positive Retroregulator that Stabilizes mRNAs in Bacteria, H. Wong, S. Chong, Proc. Nat. Acad. Sci. USA *83*, 3233-3237 (1986).

Protein Targeting and Secretion

Transfer of Proteins Across Membranes, G. Blobel, B. Dobberstein, J. of Cell Biol. *67*, 835-851 (1975).

Conversion of β-Galactosidase to a Membrane-bound State By Gene Fusion, T. Silhavy, M. Casadaban, H. Shuman, J. Beckwith, Proc. Nat. Acad. Sci. USA *73*, 3423-3427 (1976).

Extracellular Labeling of Nascent Polypeptides Traversing the Membrane of *Escherichia coli*, W. Smith, P-C. Tai, R. Thompson, B. Davis, Proc. Nat. Acad. Sci. USA *74*, 2830-2834 (1977).

Detection of Prokaryotic Signal Peptidase in an *Escherichia coli* Membrane Fraction: Endoproteolytic Cleavage of Nascent f1 Pre-coat Protein, C. Chang, G. Blobel, P. Model, Proc. Nat. Acad. Sci. USA *75*, 361-365 (1978).

Escherichia coli Mutants Accumulating the Precursor of a Secreted Protein in the Cytoplasm, P. Bassford, J. Beckwith, Nature *277*, 538-541 (1979).

Chicken Ovalbumin Contains an Internal Signal Sequence, V. Lingappa, J. Lingappa, G. Blobel, Nature *281*, 117-121 (1979).

Secretion of Beta-lactamase Requires the Carboxy End of the Protein, D. Koshland, D. Botstein, Cell *20*, 749-760 (1980).

expA: A Conditional Mutation Affecting the Expression of a Group of Exported Proteins in *E. coli* K-12, E. Dassa, P. Boquet, Mol. and Gen. Genetics *181*, 192-200 (1981).

Signal Recognition Particle Contains a 7S RNA Essential for Protein Translocation Across the Endoplasmic Reticulum, P. Walter, G. Blobel, Nature *299*, 691-698 (1982).

Diverse Effects of Mutations in the Signal Sequence on the Secretion of β-Lactamase in *Salmonella typhimurium*, D. Koshland, R. Sauer, D. Botstein, Cell *30*, 903-914 (1982).

Demonstration by a Novel Genetic Technique that Leader Peptidase is an Essential Enzyme of *Escherichia coli*, T. Data, J. Bact. *154*, 76-83 (1983).

Disassembly and Reconstitution of Signal Recognition Particles, P. Walter, G. Blobel, Cell *34*, 525-533 (1983).

A Genetic Approach to Analyzing Membrane Protein Topology, C. Manoil, J. Beckwith, Science *233*, 1043-1048 (1986).

Binding Sites of the 19-Kda and 68/72-Kda Signal Recognition Particle (SRP) Proteins on SRP RNA as Determined by Protein-RNA Footprinting, V. Siegel, P. Walter, Proc. Nat. Acad. Sci. USA *85*, 1801-1805 (1988).

The Antifolding Activity of SecB Promotes the Export of the *E. coli* Maltose-binding Protein, D. Collier, V. Bankaitis, J. Weiss, P. Bassford Jr., Cell *53*, 273-283 (1988).

Transient Association of Newly Synthesized Unfolded Proteins with the Heat-shock GroEL Protein, E. Bochkareva, N. Lissin, A. Girshovich, Nature *336*, 179-181 (1988).

HumanSRP RNA and *E. coli* 4.5S RNA Contain a Highly Homologous Structural Domain, M. Poritz, K. Strub, P. Walter, Cell *55*, 4-6 (1988).

Mitochondrial Heat-shock Protein hsp60 is Essential for Assembly of Proteins Imported into Yeast Mitochondria, M. Cheng, F. Hartl, J. Martin. R. Pollock, F. Falousek, W. Neuport, E. Hallberg, R. Hallberg, A. Horwich, Nature *337*, 620-625 (1989).

Identification of a Mitochondrial Receptor Complex Required for Recognition and Membrane Insertion of Precursor Proteins, M. Kiebler, R. Pfaller, T. Söllner, G. Griffiths, H. Horstmann, N. Pfanner, W. Neupert, Nature *348*, 610-616 (1990).

Membrane Protein Structure Prediction, Hydrophobicity Analysis and the Positive-inside Rule, G. van Heijne, J. Mol. Biol. *225*, 487-494 (1992).

Sigman-sequence Recognition by an *Escherichia coli* Ribonucleoprotein Complex, J. Luirink, S. High, H. Wood, A. Giner, D. Tollervey, B. Dobberstein, Nature *359*, 741-743 (1992).

Accuracy and Frameshifting

The Frequency of Errors in Protein Biosynthesis, R. Loftfield, D. Vanderjagt, Biochem. J. *128*, 1353-1356 (1972).

Frameshift Suppression: A Nucleotide Addition in the Anticodon of a Glycine Transfer RNA, D. Riddle, J. Carbon, Nature New Biology *242*, 230-234 (1973).Mistranslation in *E. coli*, P. Edelmann, J. Gallant, Cell *10*, 131-137 (1977).

The Accuracy of Protein Synthesis is Limited by its Speed: High Fidelity Selection by Ribosomes of Aminoacyl-tRNA Ternary Complexes Containing GTP(γS), R. Thompson, A. Karin, Proc. Nat. Acad. Sci. USA *79*, 4922-4926 (1982).

Lysis Gene Expression of RNA Phage MS2 Depends on a Frameshift During Translation of the Overlapping Coat Protein Gene, R. Kastelein, E. Remaut, W. Fiers, J. van Duin, Nature *295*, 35-41 (1982).

An Estimate of the Global Error Frequency in Translation, N. Ellis, J. Gallant, Mol. Gen. Genet. *188*, 169-172 (1982).

Molecular Model of Ribosome Frameshifting, R. Weiss, Proc. Nat. Acad. Sci. USA *81*, 5797-5801 (1984).

Expression of the Rous Sarcoma Virus *pol* Gene by Ribosomal Frameshifting, T. Jacks, H. Varmus, Science *230*, 1237-1242 (1985).

Expression of Peptide Chain Release Factor 2 Requires High-efficiency Frameshift, W. Craigen, T. Caskey, Nature *322*, 273-275 (1986).

A Persistent Untranslated Sequence Within Bacteriophage T4 DNA Topoisomerase Gene 60, W. Huang, S. Ao, S. Casjens, R. Orlandi, R. Zeikus, R. Weiss, D. Winge, M. Fang, Science *239*, 1005-1012 (1988).

Efficient Translational Frameshifting Occurs within a Conserved Sequence of the Overlap Between Two Genes of a Yeast Ty1 Transposon, J. Clare, M. Belcourt, P. Farabaugh, Proc. Natl. Acad. Sci. USA *85*, 6816-6820 (1988).

Signals for Ribosomal Frameshifting in the Rous Sarcoma Virus gag-pol Region, T. Jacks, H. Madhani, F. Masiarz, H. Varmous, Cell *55*, 447-458 (1988).

Characterization of an Efficient Coronavirus Ribosomal Frameshifting Signal Requirement for an RNA Pseudoknot, I. Brierley, P. Digard, S. Inglis, Cell *57*, 537-547 (1989).

Ribosomal Frameshifting in the Yeast Retrotransposon Ty: tRNAs Induce Slippage on a 7 Nucleotide Minimal Site, M. Belcourt, P. Farabaugh, Cell *62*, 339-352 (1990).

Evidence that a Downstream Pseudoknot is Required for Translational Read-through of the Moloney Murine Leukemia Virus *gag* Stop Codon, N. Sills, R. Gestland, J. Atkins, Proc. Natl. Acad. Sci. USA *88*, 8450-8454 (1991).

An RNA Pseudoknot and an Optimal Heptameric Shift Site are Required for Highly Efficient Ribosomal Frameshifting on a Retroviral Messenger RNA, M. Chamorro, N. Parkin, H. Varmus, Proc. Natl. Acad. Sci. USA *89*, 713-717 (1992).

Frameshifting in the Expression of the *E. coli trpR* Gene Occurs by the Bypassing of a Segment of its Coding Sequence, I. Benhar, H. Engelberg-Kulka, Cell *72*, 121-130 (1993).

Cell Physiology and Protein Synthesis

Dependency on Medium and Temperature of Cell Size and Chemical Composition During Balanced Growth of *Salmonella typhimurium*, M. Schaechter, O. Maaloe, N. Kjeldgaard, J. Gen. Microbiol. *19*, 592-606 (1958).

Control of Production of Ribosomal Protein, R. Schleif, J. Mol. Biol. *27*, 41-55 (1967).

Synthesis of 5S Ribosomal RNA in *Escherichia coli* after Rifampicin Treatment, W. Doolittle, N. Pace, Nature *228*, 125-129 (1970).

Accumulation and Turnover of Guanosine Tetraphosphate in *Escherichia coli*, N. Fiil, K. von Meyenburg, J. Friesen, J. Mol. Biol. *71*, 769-783 (1972).

Synthesis of Guanosine Tetra- and Pentaphosphate Requires the Presence of a Codon-specific, Uncharged tRNA in the Acceptor Site of Ribosomes, W. Haseltine, R. Block, Proc. Nat. Acad. Sci. USA *70*, 1564-1568 (1973).

Codon Specific, tRNA Dependent *in vitro* Synthesis of ppGpp and pppGpp, F. Pedersen, E. Lund, N. Kjeldgaard, Nature New Biol. *243*, 13-15 (1973).

Chain Growth Rate of β-galactosidase During Exponential Growth and Amino Acid Starvation, F. Engbaek, N. Kjeldgaard, O. Maaloe, J. Mol. Biol. *75*, 109-118 (1973).

Patterns of Protein Synthesis in *Escherichia coli*: A Catalog of the Amount of 140 Individual Proteins at Different Growth Rates, S. Pedersen, P. Bloch, S. Reeh, F. Neidhardt, Cell *14*, 179-190 (1978).

Tandem Promoters Direct *E. coli* Ribosomal RNA Synthesis, R. Young, J. Steitz, Cell *17*, 225-235 (1979).

In vitro Expression of *E. coli* Ribosomal Protein Genes: Autogenous Inhibition of Translation, J. Yates, A. Arfstein, M. Nomura, Proc. Nat. Acad. Sci. USA *77*, 1837-1841 (1980).

The Distal End of the Ribosomal RNA Operon *rrnD* of *E. coli* Contains a tRNA^{Thr} Gene, Two 5S rRNA Genes and a Transcription Terminator, G. Duester, W. Holmes, Nucleic Acid Res. *8*, 3793-3807 (1980).

Regulation of the Synthesis of *E. coli* Elongation Factor Tu, F. Young, A. Furano, Cell *24*, 695-706 (1981).

Identification of Ribosomal Protein S7 as a Repressor of Translation within the *str* Operon of *E. coli*, D. Dean, J. Yates, M. Nomura, Cell *24*, 413-419 (1981).

Injected Anti-sense RNAs Specifically Block Messenger RNA Translation *in vivo*, D. Melton, Proc. Nat. Acad. Sci. USA *82*, 144-148 (1985).

Autoregulated Instability of β-tubulin mRNAs by Recognition of the Nascent Amino Terminus of β-tubulin, T. Yen, P. Machlin, D. Cleveland, Nature *334*, 580-585 (1988).

Codon Usage Determines Translation Rate in *Escherichia coli*, M. Sorensen, C. Kurland, S. Pedersen, J. Mol. Biol. *207*, 365-377 (1989).

Rates of Aminoacyl-tRNA Selection at 29 Sense Codons *in vivo*, J. Curran, M. Yarus, J. Mol. Biol. *209*, 359-378 (1989).

The *E. coli dnaK* Gene Product, the hsp70 Homolog, Can Reactivate Heat-Inactivated RNA Polymerase in an ATP Hydrolysis-Dependent Manner, D. Skowyra, C. Georgopoulos, M. Zylicz, Cell *62*, 939-944 (1990).

Identification of a Protein Required for Disulfide Bond Formation *in vivo*, J. Bardwell, K. McGovern, J. Beckwith, Cell *67*, 581-589 (1991).

Successive Action of DnaK, DnaJ and GroEL Along the Pathway of Chaperone-mediated Protein Folding, T. Langer, C. Lu, H. Echols, J. Flanagan, M. Hayer, F. Hartl, Nature *356*, 683-689 (1992).

Identification and Characterization of an *Escherichia coli* Gene Required for the Formation of Correctly Folded Alkaline Phosphatase, a Periplasmic Enzyme, S. Komitani, Y. Akiyama, K. Ito, EMBO Journal *11*, 57-62 (1992).

A Pathway for Disulfide Bond Formation in vivo, J. Bardwell, J. Lee, G. Jander, N. Martin, D. Belin, J. Beckwith, Proc. Natl. Acad. Sci. USA *90*, 1038-1042, (1993).

Genetics

8

Thus far we have covered the structure of cells and the structure, properties, and synthesis of the components of major interest to molecular biologists–DNA, RNA, and proteins. We will now concern ourselves with genetics. Historically, the study and formulation of many genetic principles preceded an understanding of their chemical basis. By inverting the order, however, major portions of genetics become easier to understand and can be covered in a single chapter.

Genetics was central to the development of the ideas presented in this book for three reasons. First, the exchange of genetic information between cells or organisms and the ability to recombine this DNA by cutting and splicing is widespread in nature. This means that these phenomena must confer high survival value and therefore are of great biological importance. Second, for many years genetics was at the center of research in molecular biology, first serving as an object of study and then as an aid to the study of the biochemistry of biological processes. Now genetics in the form of genetic engineering has become indispensable in the study of biological systems and in physical studies of systems.

Mutations

Historically, one reason for the study of genetics was to discover the chemical basis of heredity. Naturally, the existence of mutations was necessary to the execution of the classical experiments in genetics, and an understanding of mutations will facilitate our study of these experiments. We have already covered the chemical basis of heredity and the basics of gene expression. Perhaps here we should explicitly state that a gene refers to a set of nucleotides that specifies the sequence of an RNA or protein. We will define mutation and in the next section mention the three basic types of mutations. In the following section we will review the classical genetic experiments before turning to recombination.

A mutation is merely an inheritable alteration from the normal. It is an alteration in the nucleotide sequence of the DNA or, in the case of RNA viruses, an alteration in the nucleotide sequence of its genomic RNA. We already know that changes in coding portions of DNA may alter the amino acid sequences of proteins and that changes in noncoding regions of DNA have the potential for changing the expression of genes, for example by altering the strength of a promoter. Of course, any cellular process that makes use of a sequence of DNA can be affected by a mutation. The existence of mutations implies that the sequence of DNA in living things, including viruses, is sufficiently stable that most individuals possess the same sequence but sufficiently unstable that alterations do exist and can be found.

The terms wild-type, mutant, mutation, and allele are closely related but must be distinguished. Wild-type is a reference, usually found naturally. It can mean an organism, a set of genes, a gene, a gene product like a protein, or a nucleotide sequence. A mutation is an inheritable change from that reference. A mutant is the organism that carries the mutation. Two mutations are said to be allelic if they lie in the same gene. However, now that genes can be analyzed at the nucleotide level, in some situations alleles refers to nucleotides rather than to genes.

Until it became possible to sequence DNA easily, mutations could readily be identified only by their gross effects on the appearance of the cell or the shape, color, or behavior of an organism. Some of the most easily studied biological effects of mutations in bacteria and viruses were changes in the colony or plaque morphology. Other easily studied effects of mutations were the inability of cells to grow at low or high temperatures or the inability to grow without the addition of specific chemicals to the growth medium. Such readily observed properties of cells constitute their phenotype. The status of the genome giving rise to the phenotype is called the genotype. For example, the Lac⁻ phenotype is the inability to grow on lactose. It can result from mutations in lactose transport, β-galactosidase enzyme, *lac* gene regulation, or the cells' overall regulation of classes of genes that are not well induced if cells are grown in the presence of glucose. Such cells would have a mutation in any of the following genes: *lacY*, *lacZ*, *lacI*, *crp*, or *cya*.

Point Mutations, Deletions, Insertions, and Damage

The structure of DNA permits only three basic types of alteration or mutation at a site: the substitution of one nucleotide for another, the deletion of one or more nucleotides, and the insertion of one or more nucleotides. A nucleotide substitution at a point is called a transition if one purine is substituted for the other or one pyrimidine is substituted for the other and is called a transversion if a purine is substituted for a pyrimidine or vice versa.

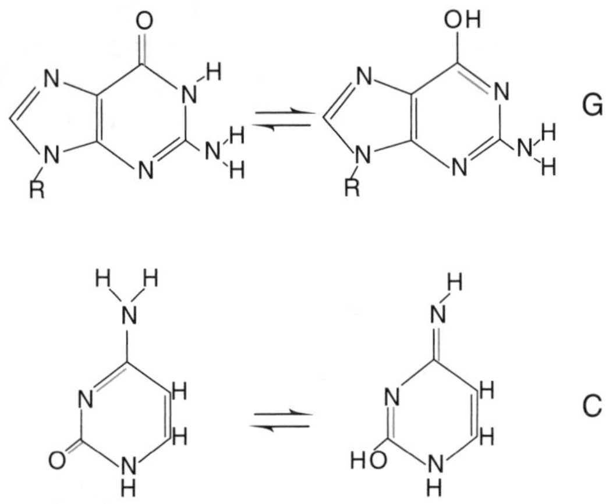

Figure 8.1 Tautomeric forms of guanine and cytosine base pair differently due to alternations of hydrogen bond donating and accepting groups.

In addition to substitutions of one nucleotide for another in single-stranded DNA or one base pair for another in double-stranded DNA, nucleotides are susceptible to many types of chemical modification. These can include tautomerizations and deamination or more extensive damage such as the complete loss of a base from the ribose phosphate backbone (Fig. 8.1). The cellular repair mechanisms, however, remove many such modified bases so that the gap can be refilled with normal nucleotides. Those modified bases that escape repair cannot themselves be passed on to the next generation because, on DNA replication, one of the usual four nucleotides is incorporated into the daughter strand opposite the altered base. Frequently the base so incorporated will be incorrect, and consequently, a mutation is introduced at such a position.

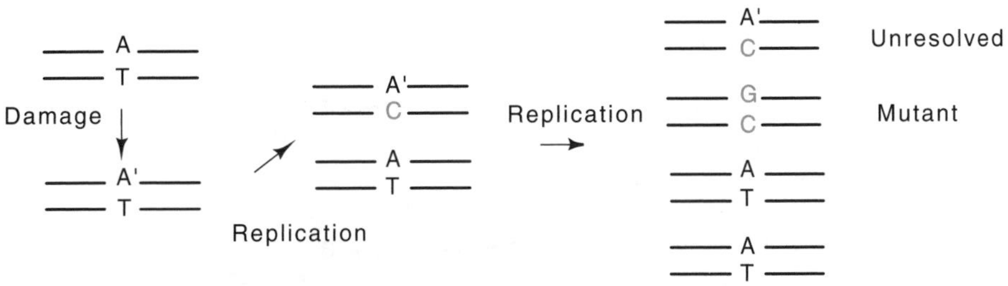

Mutations arise from a variety of sources. As discussed in Chapter 3, point mutations can occur spontaneously during replication of the DNA through the misincorporation of a nucleotide and the failure of the editing mechanisms to correct the mistake or through the chemical instability of the nucleotides. For example, cytosine can deaminate to form uracil, which is then recognized as thymine during DNA replication.

The frequency of spontaneous appearance of point mutations often is too low for convenient experimentation, and mutagens are therefore used to increase the frequency of mutants in cultures 10 to 1,000 times above the spontaneous frequency. A variety of mutagens have been discovered, some by rational considerations and some by chance. Many are nucleotide analogs that are incorporated into the DNA instead of the normal nucleotides. These increase the frequency of mispairing in subsequent rounds of DNA replication. Other mutagens are chemically reactive molecules that damage or modify bases in DNA. Ultraviolet light is also a mutagen as discussed earlier. The damage it creates ultimately leads to mutations either through a reduced fidelity of synthesis or an elevated probability of mistaken repair of the original damage. In one way or another, mutagens increase the frequency of generating mispaired bases or increase the frequency that mispaired bases escape repair. Ultimately, either leads to an increase in the probability that the original DNA sequence will be changed.

The mechanisms generating deletions and insertions are not as well understood. Errors in DNA replication provide plausible mechanisms for the generation of one or two base insertions or deletions. Most likely slippage, perhaps stimulated by an appropriate sequence, will permit a

Figure 8.2 Two mechanisms for generating deletions between repeated sequences. The first is looping with recombination between points within a single chromosome, and the second is unequal crossing over between two chromosomes.

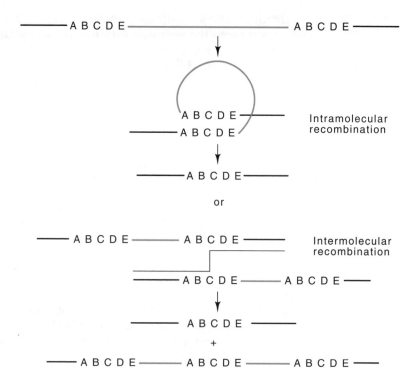

daughter strand to possess a different number of bases than the parent strand.

Insertions and deletions larger than a few bases arise by a different mechanism. The end points of a number of deletions in bacteria are located at short, repeated or almost repeated sequences. The deletion removes one of the repeats and the intervening sequence. We can picture two plausible events that could create such deletions (Fig. 8.2). The first is a looping of a single chromosome followed by elimination of the material between the two repeats. The second is similar to the first, but it occurs between two chromosomes and transfers the material from one chromosome to the other. One chromosome suffers a deletion and the other an insertion.

The creation of deletions is also stimulated by the presence of some genetic elements called insertion sequences or transposons. These elements transpose themselves or copies of themselves into other sites on the chromosome. In the process they often generate deletions in their vicinity.

Classical Genetics of Chromosomes

We should not proceed to a detailed discussion of molecular genetics without a brief review of Mendelian genetics. Chromosomes in eukaryotes consist mainly of DNA and histones. During some stages of the cell's division cycle in plants and animals, chromosomes can be observed with light microscopes, and they display beautiful and fascinating patterns. Careful microscopic study of such chromosomes sets the stage for subsequent molecular experiments that have revealed the exact chemical nature of heredity. We are now approaching a similar level of understanding of genetic recombination.

The basis of many of the classical studies is that most types of eukaryotic cells are diploid. This means that each cell contains pairs of identical or almost identical homologous chromosomes, one chromosome of each pair deriving from each of the parents. There are exceptions. Some types of plants are tetraploid or even octaploid, and some variants of other species possess alternate numbers of one or more of the chromosomes.

During normal cell growth and division, the pairs of chromosomes in each dividing cell are duplicated and distributed to the two daughter cells in a process called mitosis. As a result, each daughter cell receives the same genetic information as the parent cell contained. The situation, however, must be altered for sexual reproduction. During this process, special cells derived from each of the parents fuse and give rise to the new progeny. To maintain a constant amount of DNA per cell from one generation to the next, the special cells, which are often called gametes, are generated. These contain only one copy of each chromosome instead of the two copies contained by most cells. The normal chromosome number of two is called diploid, and a chromosome number of half this is called haploid. The cell divisions giving rise to the haploid gametes in animals and haploid spores in plants is called meiosis.

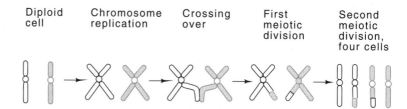

Figure 8.3 The classical view of meiosis.

During the process of meiosis, a pair of chromosomes doubles, genetic recombination may occur between homologous chromosomes, and the cell then divides (Fig. 8.3). Each of the daughter cells then divides again without duplication of the chromosomes. The net result is four cells, each containing only one copy of each chromosome. Subsequent fusion of a sperm and egg cell from different individuals yields a diploid called a zygote that grows and divides to yield an organism containing one member of each chromosome pair from each parent.

The chromosomes from each of the parents may contain mutations that produce recognizable traits or phenotypes in the offspring. Let us consider just one chromosome pair of a hypothetical organism. Let gene *A* produce trait A, and, if it is mutant, let it be denoted as gene *a* and its trait be a. In genetic terminology, *A* and *a* are alleles. We can describe the genetic state of an individual by giving its genetic composition or genotype. For example, both copies of the chromosome in question could contain the *A* allele. For convenience, denote this as (*A/A*). Such a cell is called homozygous for gene *A*. A mating between organisms containing diploid cells of type (*A/A*) and (*a/a*) must produce offspring of the type (*a/A*), which is, of course, identical to (*A/a*). That is, the chromosomes in the offspring are copies of each of the parental chromosomes. These offspring are said to be heterozygous for gene *A*.

Figure 8.4 The haploids produced from diploid parent cells, the combinations of haploids possible upon their fusion, and the apparent phenotypes if *A* is totally or partially dominant to *a*.

		A/a		Phenotypes of offspring
Haploids		A	a	If A (red) is dominant to a (white)
	A	A/A	A/a	3/4 red, A/A or a/A or A/a 1/4 white, a/a
A/a				If A (red) is partially dominant to a (white)
	a	a/A	a/a	1/4 red, A/A 2/4 pink, A/a or a/A 1/4 white, a/a

The interesting results come when two heterozygous individuals mate and produce offspring. A gamete can inherit one or the other of each of the homologous chromosomes from each chromosome pair. This generates a variety of gamete types. When large numbers of offspring are considered, many representatives are found of every possible combination of assortment of the chromosomes, and the results become predictable. It is easiest to systematize the possibilities in a square matrix (Fig. 8.4). For evaluation of experimental results, however, the appearance of heterozygotes must be known or deduced. The appearance of a heterozygote (*a*/*A*) is that of trait A if *A* is dominant, which means automatically that *a* is recessive. Strict dominance need not be seen, and a heterozygote may combine the traits displayed by the two alleles. For example, if the trait of gene *A* were the production of red pigment in flowers and the trait of gene *a* were the absence of production of the pigment, the heterozygote (*a*/*A*) might produce half the normal amount of red pigment and yield pink flowers.

Complementation, *Cis, Trans*, Dominant, and Recessive

Complementation, *cis*, *trans*, dominant, and recessive are commonly used genetic terms. Their meaning can be clarified by considering a simple set of two genes that are transcribed from a single promoter, an operon. We shall consider the *lac* operon of *Escherichia coli*. Denote the promoter by *p* and the three genes that code for proteins that diffuse through the cell's cytoplasm or membrane by *lacZ*, *lacY*, and *lacA* (Fig. 8.5). The *lacZ* gene codes for β-galactosidase. This is an enzyme that hydrolyzes lactose to produce glucose and galactose. The *lacY* gene codes for a membrane protein that transports lactose into the cell, and *lacA* codes for an enzyme that transfers acetyl groups to some galactosides, thereby reducing their toxicity.

Diploids heterozygous for genes may easily be constructed by genetic crosses in diploid eukaryotes and even in prokaryotes by special tricks. If the genes are on phage genomes, cells may be simultaneously infected with both phage types. Consider the *lac* operon and possibilities for mutations in genes Z and Y. If we introduce the operon $p^+Z^-Y^+$ into cells that are $p^+Z^+Y^-$, the diploid $p^+Z^-Y^+/p^+Z^+Y^-$ will possess good β-galactosidase and transport protein, and phenotypically will be able to grow on lactose. The Z^+ gene complements the Z^- gene and the Y^+ gene complements the Y^- gene. Both Z^+ and Y^+ act in *trans* and are *trans* dominant, or simply dominant. Analogously, Z^- and Y^- are recessive.

Figure 8.5 Genetic structure of the *lac* operon.

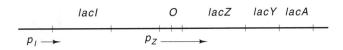

A mutation in the promoter generates more complexity. A p^- mutation also appears to be Z^- and Y^- even though the Z and Y genes retain their normal sequences. This results from the fact that no β-galactosidase or lactose transport protein can be synthesized if the *lac* promoter is defective. In genetics terminology, such a strain is *phenotypically* Z^- and Y^- since, for growth purposes, it behaves as though it lacks the Z and Y activities. However, it is *genotypically* p^-, Z^+, and Y^+ since the Z and Y genes actually remain intact, as may be revealed in other types of experiments. In our example, the p element is *cis* dominant to the Z^+ and Y^+ genes, that is, its effect is only over genes on the same piece of DNA. Also, p^+ is not *trans* dominant to p^- since a partial diploid of the genetic structure $p^- Z^+ Y^+ / p^+ Z^- Y^+$ would be unable to grow on lactose. The presence of the Z^+ gene in a $p^- Z^+$ chromosome could be revealed by genetic recombination.

Mechanism of a *trans* Dominant Negative Mutation

Expression of the *lac* operon is regulated by the product of the *lacI* gene. Since this protein acts to turn off or turn down the expression, it is called a repressor. The protein can bind to a site partially overlapping the *lac* promoter and block transcription from the promoter into the *lac* structural genes. In the presence of inducers, the repressor's affinity for the operator is much reduced and it dissociates from the DNA. This then permits active transcription of the *lac* genes.

The repressor contains four identical subunits. Consider cells that are diploid for *lacI* where one of the *lacI* genes is *lacI*$^+$, and the other is a type we call *lacI*$^{-d}$. The d stands for "dominant." During the synthesis of the two types of repressor subunit in a cell, the probability that four newly synthesized wild-type repressor subunits will associate to form a wild-type tetramer is low. Instead, most repressor tetramers will possess both types of subunits.

If the inclusion of a single I^{-d} subunit in a tetramer interferes with the function of a tetramer, then the I^{-d} allele will be dominant and act in *trans* to nullify the activity of the good *lacI* allele, that is, it is a *trans* dominant negative mutation.

What is the physical basis for a single defective subunit inactivating the remaining three nondefective subunits in a tetramer? Lac repressor contacts the symmetrical *lac* operator with two subunits utilizing the helix-turn-helix structure that was discussed in Chapter 6. Contact with only one subunit provides far too little binding energy for the protein to bind. Therefore a subunit with a defective DNA-contacting domain may be capable of folding and oligomerizing with normal subunits, but its incorporation into a tetramer will interfere with DNA binding since two good subunits must simultaneously be involved in contacting the operator.

If a tetramer contains only one defective subunit, we might think that two nondefective subunits could still be utilized for binding to DNA. To an extent this is true. However, just as enhancers loop and complexes of proteins can contact the DNA at two or more sites, so does the *lac*

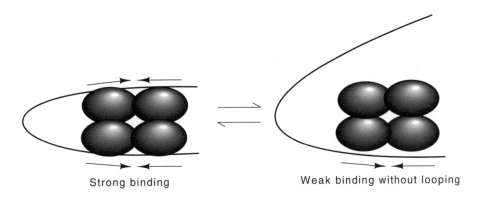

Strong binding   Weak binding without looping

Figure 8.6 When all four subunits of *lac* repressor can bind DNA and form the loop, binding to both of the DNA sites is enhanced because looping effectively increases the local concentration of the repressor at the other sites.

operon. Looping in the *lac* operon brings repressor bound at the primary operator into contact with either of two so-called pseudo-operators on either side. As will be discussed later, such looping can greatly increase the occupancy of a binding site by a protein. In the case of the tetrameric *lac* repressor, two good subunits could contact the *lac* operator or a pseudo operator, but looping could not occur, and the overall binding would be relatively weak and as a result, repression would be poor (Fig. 8.6). Thus, the inclusion of a single DNA-binding defective subunit in a tetrameric repressor can greatly interfere with repression.

Genetic Recombination

Genetic crosses between two strains, each containing a mutation in the same gene, are occasionally observed to yield nonmutant, that is, wild-type, progeny. This is the result of a crossover in which the two parental DNA molecules are precisely broken, exchanged, and rejoined.

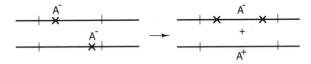

This section describes phenomenon of recombination and the next section describes experiments that make use of recombination to order genetic markers on chromosomes.

Alignment of the pairs of homologous chromosomes during meiosis has long been known through observations with light microscopes. The subsequent movement of visually identifiable sites on the chromosomes suggested that homologous chromosomes were broken and portions exchanged in a process called crossing over or recombination. Although recombination was first observed in eukaryotic cells, it appears to be almost ubiquitous. Even simple bacterial viruses can engage in genetic

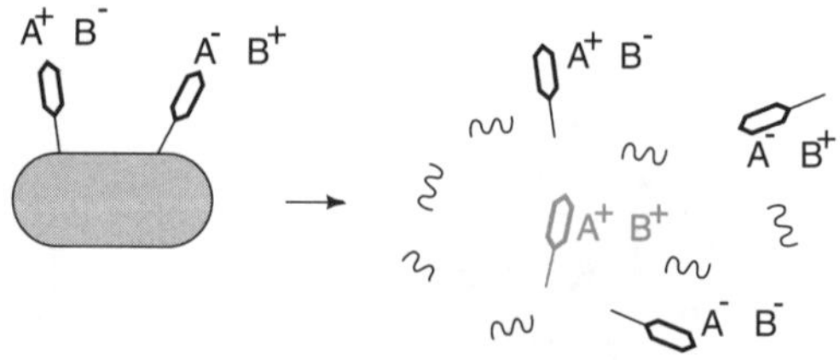

Figure 8.7 Co-infection of a single cell by phage of two different genotypes permits recombination to produce output phage of wild type in addition to output phage of the parental types.

recombination. The bacterial phage experiments were particularly important to the development of the field because phage provided a simple and small system with few variables, a high sensitivity for recombinants, and a very short generation time. These properties permitted many experiments to be done rapidly and inexpensively.

Phage mutants can easily be isolated that yield plaque morphologies different from the normal or wild-type. This is done simply by plating mutagenized phage on cells and locating the occasional different plaque. Nonsense mutations in essential phage genes may be identified as phage that grow only on nonsense-suppressing host cells. Genetic recombination between phage can be revealed by co-infection of cells with two mutants at a sufficiently high multiplicity of infection that each cell is infected with both types of phage (Fig. 8.7). Some of the progeny phage are found to carry alleles from both of the input phage! Since the phage carry only one DNA copy, such progeny have to be recombinants carrying some of the genetic information from each of the parental phage types. The discovery of genetic recombination in phage by Delbrück and Hershey opened the way for intensive study of the phenomenon of genetic recombination at the molecular level.

Mapping by Recombination Frequencies

Two or more mutations on any DNA molecule that engages in recombination can be approximately ordered along DNA molecules by measurement of the frequencies of recombination. Let us examine why this is so. Assume that the probability of a recombination or genetic crossover between two points on two almost homologous DNA molecules is a function only of the distance between the points. Up to a limit, the greater the separation, the greater the probability of crossover. This seems reasonable because increasing the separation increases the number of potential crossover sites. Recombination frequencies should be additive if the frequency is linearly proportional to marker separation. Thus, if $Rf(X,Y)$ is the recombination frequency between markers X and

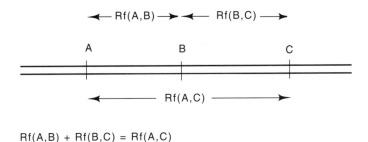

$$Rf(A,B) + Rf(B,C) = Rf(A,C)$$

Figure 8.8 Additivity relations necessary when recombination frequency is not linearly proportional to genetic distance.

Y, and if strict additivity holds, Rf(A,C) must equal Rf(A,B) + Rf(B,C) (Fig. 8.8).

Unfortunately, the assumption of a linear relation between recombination frequency and distance is not good over extremely short or long distances. At short distances the specific nucleotides involved can generate profound effects on the recombination frequencies, and additivity fails as a result. These short-distance anomalies are often called marker effects. At large distances more than one crossover is likely in the distance separating two markers. If an even number of crossovers occurs between the markers, no net recombination is observed between the markers. Therefore, as the distance between markers becomes large and multiple crossovers occur, about as many partners experience an even number of crossovers as experience an odd number. As only those with odd numbers of crossovers generate recombinants, recombination frequencies between two markers rise to 50% as an upper limit.

Given the vagaries of measuring distances by recombination frequencies, what can be done? Often the major question is merely one of marker order. The actual distances separating the markers are not of great importance. Three-factor genetic crosses are a partial solution. They permit ordering of two genetic markers, *B* and *C*, with respect to a third, *A*, that is known to lie on the outside. That is, the experiment is to determine whether the order is *A-B-C* or *A-C-B*.

Here the crosses will be described as they are performed in prokaryotes in which only one of the two participating DNA molecules is a complete chromosome, but the basic principles apply for the situation of recombination between two complete chromosomes. Mating experiments permit the introduction of a portion of the chromosomal DNA from a donor bacterial cell into a recipient cell in a process called bacterial conjugation. As a result of the inviability of such a linear DNA fragment, if a crossover occurs between the chromosome of the recipient and the incoming chromosome fragment, a second crossover between the two DNAs must also occur for the recipient chromosome not to be left open by the single crossover event (Fig. 8.9). In general, any even number of total crossovers between the incoming DNA and the recipient chromosome will yield viable recombinants.

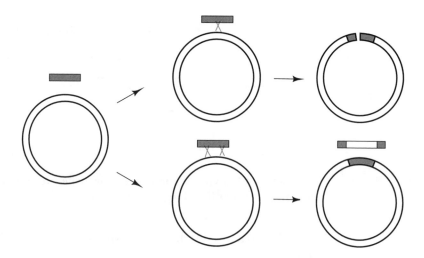

Figure 8.9 In order that a circular chromosome not be opened by genetic recombination, two crossovers between it and a fragment of linear DNA are necessary.

The genetic markers involved in the genetic cross, A, B, and C, can be different genes or different alleles within the same gene. In principle it does not matter, but in practice it could be difficult to ascertain the different varieties of recombinants if the markers all lie within the same gene.

To determine the gene order, compare the fraction of B^+C^+ recombinants among all the A^+ recombinants in the two crosses (Fig. 8.10), first between an $A^-B^-C^+$ chromosome and an $A^+B^+C^-$ DNA fragment, and then between an $A^-B^+C^-$ chromosome and an $A^+B^-C^+$ fragment. If the gene order is ABC, then generation of B^+C^+ recombinants from the first cross requires only two crossover events, whereas the second cross requires crossovers in the same two intervals as the first plus an additional two crossovers. Therefore the fraction of the B^+C^+ recombinants will be much higher from the first cross. Alternatively, if the gene order is ACB,

Figure 8.10 Three-factor crosses. The first requires only two crossovers to produce $A^+B^+C^+$ progeny whereas the second requires four.

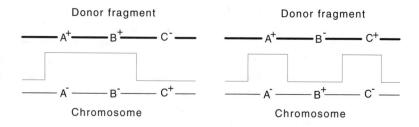

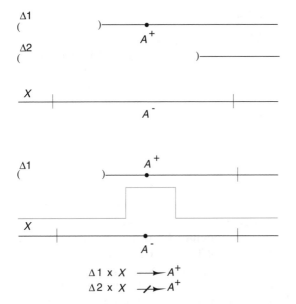

Figure 8.11 Recombination between a chromosome containing a point mutation and a DNA fragment containing a deletion. Crossovers can occur only between homologous segments, that is, outside a deleted area.

a greater fraction of B^+C^+ recombinants will result from the second cross.

Mapping by Deletions

Deletions may also be used in genetic mapping. The recombination frequency for generating functional genes is not measured in this type of mapping. Instead, all that is asked is whether or not a deletion and a point mutation can recombine to yield a functional gene. If they can, then the deletion must not have removed the nucleotide allelic to the point mutation.

Consider a series of strains each containing a deletion or a point mutation. Suppose the point mutations lie within a gene X and that the deletions all begin beyond the left end of X and extend various distances rightward into X (Fig. 8.11). If a diploid between Δ1 and point mutation A can yield an X⁺ recombinant, then A must lie to the right of the end point of Δ1. If A also fails to yield X⁺ recombinants with Δ2, then Δ2 ends to the right of A and hence to the right of Δ1. By this type of reasoning, a completely unordered set of deletions and point mutations may be ordered.

Heteroduplexes and Genetic Recombination

Having considered the existence and use of genetic recombination, we are ready to consider how it comes about. Genetic recombination yields a precise cut and splice between two DNA molecules. Even if one DNA

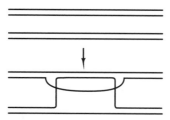

Figure 8.12 One possible mechanism for obtaining hybridization between homologous chromosomes during the process of genetic recombination.

molecule were to have been cut, it is hard to imagine how an enzyme could know where to cut the other DNA duplex so as to produce the perfect splices that genetics experiments show occur. The difficulty can be largely overcome by a mechanism utilizing the self-complementary double-stranded structure of DNA. A denatured portion of one duplex could anneal to a denatured portion of complementary sequence from the other duplex (Fig. 8.12). This would hold the two DNA molecules in register while the remainder of the recombination reaction proceeded.

The life cycle of yeast permits a direct test of the model outlined above. A diploid yeast cell undergoes recombination during meiosis, and the two meiotic cell divisions yield four haploid spores. These four spores can be isolated from one another and each can be grown into a colony or culture. In essence, the cells of each colony are identical copies of each of the original recombinants, and the cells can be tested to determine the genetic structure of the original recombinants. If one of a pair of homologous chromosomes contains a mutation and the other does not, generally two of the four resulting spores will contain the mutation and two will not.

Consider the situation resulting from melting portions of the duplexes and base pairing between complementary strands of two homologous yeast chromosomes in the process of genetic recombination. The region of pairing may include the mutation. Then a heteroduplex forms that contains the mutant sequence on one strand and the wild-type sequence on the other (Fig. 8.13). As discussed in Chapter 3, mispaired bases are subject to mismatch repair and, if it occurs, the yeast repair system in this case has no apparent reason to choose one strand to repair in preference to the other. Therefore strands may be correctly or incorrectly repaired, so the final outcome could be three copies of the wild-type or mutant sequence and one copy of the other in the meiosis from a single yeast cell. In total, a single yeast cell can produce one or three progeny spores containing the marker from one of the original chromosomes. This phenomenon is called gene conversion. It is experimentally observed and consequently it is reasonable to expect that pairing between complementary strands of recombinant partners occurs during recombination. Without heteroduplex formation and mismatch repair, there is no easy way to generate any ratio other than 2:2.

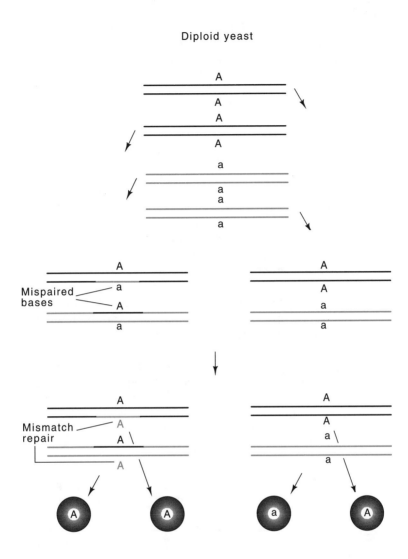

Figure 8.13 Gene conversion in yeast. A diploid (A/a) undergoes meiosis, which produces heteroduplexes A-a that are both repaired to A-A.

Branch Migration and Isomerization

Evidence mentioned in the previous section suggests that DNA duplexes engaged in recombination are likely to involve heteroduplexes consisting of one DNA strand from each parent. The problem we will address here is one way these heteroduplexes might be formed and what steps might be necessary to convert them to recombinants. As a glance at any genetics book will show, there are many schemes consisting of more or less reasonable steps that conceivably could be catalyzed by enzymes and that would ultimately lead to the generation of genetic recombination. We will outline one of these.

A single-stranded end of a DNA molecule can invade a DNA duplex in a region of homology and form a DNA heteroduplex by displacing one of the strands of the invaded parent. This can occur because

negatively supercoiled DNA is under torsion to untwist, and thereby melt a portion of duplex. Displacement of a portion of the original duplex by an invading single strand of DNA is energetically favored. Thus, once this has started, additional nucleotides from the invading strand are free to base pair because for each base pair broken in the parental duplex, a new pair forms in the heteroduplex. Then the cross-over point is free to drift in either direction along the DNA by this branch migration process.

Before proceeding with the recombination mechanism, we must consider a diversion. For simplicity we will examine a double crossover and then apply the principles to the situation described in the previous paragraph. The strands that appear to cross from one duplex over to the other are not fixed! What is the basis of this remarkable assertion? By a simple reshuffling of the DNA in the crossover region, the other strands can be made to appear to be the ones crossing over (Fig. 8.14). This reshuffling is called isomerization. To understand, consider the more dramatic transformations as shown in the Figure. These result in a change in the pair of strands that cross from one duplex to the other. In reality, however, they amount to little more than looking at the DNA

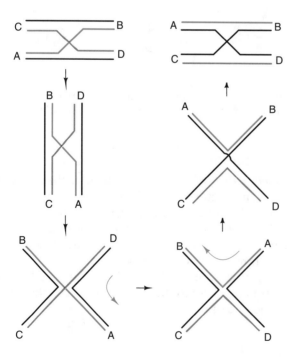

Figure 8.14 DNA crossover region during isomerization. Beginning at the upper left, a series of rotations as indicated ultimately yields the molecule at the upper right, which appears to have altered the DNA strands that cross over from one molecule to the other.

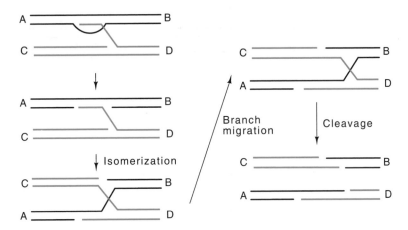

Figure 8.15 One possible pathway for genetic recombination. A nick is converted to a crossover region, which isomerizes and branch migrates, and finally the strands crossing over are cleaved.

from a different angle. Isomerization requires only minor structural shifts in the crossover region and therefore is free to occur during genetic recombination.

Now let us return to the crossover mechanism. At the one crossover stage, an isomerization yields a two strand crossover (Fig. 8.15). Branch migration followed by cleavage of these strands then produces a crossover between the two parental duplexes with a heteroduplex region near the crossover point.

Elements of Recombination in *E. coli*, RecA, RecBCD, and Chi

What types of biochemical evidence can be found in support of the mechanisms of recombination discussed in the previous section? Historically, the elucidation of metabolic pathways was often assisted by the isolation of mutations in enzymes catalyzing individual steps of the pathway. With similar objectives in mind, mutations were isolated that decreased or increased the ability of phage, *E. coli*, yeast, and some other organisms to undergo recombination. Subsequently, the enzyme products of the recombination genes from *E. coli* and yeast have been identified and purified. Of greatest importance to recombination are the RecA and RecBCD proteins. Additional enzyme activities that might be expected to play roles, such as DNA ligase, DNA polymerases, single-stranded binding protein, and proteins that wind or unwind DNA, have already been discussed and will not be further mentioned here.

RecA mutants are unable to engage in genetic recombination, and as expected, the purified RecA protein possesses a variety of activities that appear related to recombination. In the presence of ATP the protein binds to single-stranded DNA in a highly cooperative manner. Once one molecule has bound to a stretch of single-stranded DNA, other mole-

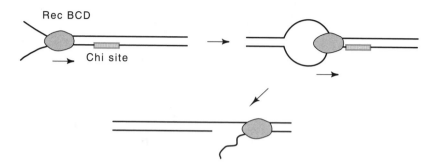

Figure 8.16 RecBCD binds to a free end of the DNA and moves at more than 200 base pairs per second. As RecBCD crosses a chi site, one of the strands is cut.

cules line up beside it. Then, with the hydrolysis of ATP the protein catalyzes the pairing of a homologous DNA strand from a double-stranded duplex to the strand covered with RecA protein.

The known activities of the RecBCD protein suggest that this protein generates the single-stranded regions that initiate the strand invasion process catalyzed by RecA. RecBCD protein binds at the free end of a double-stranded DNA duplex and moves down the DNA separating the DNA strands and leaving a single-stranded region in its wake. If it encounters a specific eight nucleotide sequence called a *chi* sequence, the enzyme cleaves one of the strands, leaving a single-stranded DNA end that can be covered with RecA protein and initiate recombination (Fig. 8.16).

Chi can stimulate recombination at distances up to 10,000 bases away from itself and appears to act only in one direction from itself, that is, it possesses a polarity. Only when *chi* is crossed by RecBCD traveling in one direction does cleavage occur. Apparently organisms other than bacteria utilize sequences that function like *chi*, since hotspots of genetic recombination are found in yeast and fungi as well. The fact that RecBCD can enter DNA only at an end restricts its stimulation of recombination to those situations in which ends are generated. In addition to random chromosome breakage, in which case genetic recombination may be a good way to attempt rescue, ends are generated during transfer of genetic information. The rate of genetic recombination between homologous DNA sequences during normal growth is very low. Recombination primarily occurs at the time of introduction of new DNA into bacteria. The analogous principle applies to meiosis of eukaryotic cells, but the recombination stimulating process there is not yet known.

Genetic Systems

One of the early steps in the investigation of biological problems is to determine roughly the complexity of the system by genetic experiments. Mutations affecting the system are isolated, their effects are determined, and in some cases the mutations are mapped. Earlier in the chapter we considered these questions in the abstract. Here we shall examine experimentally how these questions may be handled in bacteria, yeast, and the fruit fly *Drosophila melanogaster*. The same basic operations used with bacteria or yeast are also used with most other unicellular organisms or with cell cultures from multicellular organisms. Similarly, the principles and genetic operations used with *Drosophila* are similar to many used with other higher organisms, although the *Drosophila* genetic system is much more tractable than other systems.

Phage and bacteria have been important in molecular biology for many reasons. Among them is the fact that with these materials genetic experiments can easily be performed, and, most importantly, mutants can be grown and their altered genes or gene products can be isolated and definitively tested in biochemical experiments.

Yeast possess many of the virtues of bacteria. As a simple eukaryote however, many of the important questions being studied with yeast involve components or processes which are not found in bacteria or phage, for example, properties of mitochondria or messenger RNA splicing. One of the most useful properties of this organism is the ease of generating haploids and diploids. The facile generation of mutants requires use of the haploid form, but complementation studies and genetic mapping require forming diploids.

The fruit fly *Drosophila melanogaster* has been energetically studied by geneticists since about 1910. It is a eukaryote with differentiated tissues and is relatively easy and inexpensive to study. A large number of mutations as well as a wide variety of chromosome aberrations such as inversions, substitutions, and deletions have been catalogued and mapped in this organism. Fortunately *Drosophila* permits the study of many of these rearrangements with the light microscope as the chromosomes in the salivary glands are highly polytene. They contain about 1,000 parallel identical copies. This large amount of DNA and associated macromolecules generates banding patterns characteristic of each region of the chromosome. The mechanisms of tissue-specific gene expression and growth and operation of the nervous system are questions appropriate for this organism. Remarkably powerful genetics tools have been developed for investigating these questions in *Drosophila*. On the other hand, biochemical approaches to these questions are just beginning to be developed, as will be described in following chapters.

Growing Cells for Genetics Experiments

A culture should be genetically pure before one attempts to isolate a new mutant or to study the properties of an existing mutant. Being genetically pure means that all the cells of a culture are genetically identical. The easiest way to ensure the requisite purity is to grow cultures from a single cell. Then, all the cells will be descendants of the original cell, and the culture will be pure unless the spontaneous mutation rate is excessive.

The species of bacteria most widely used in molecular biology experiments is *Escherichia coli*. It can be simply purified by streaking a culture on a petri plate containing "rich" nutrient medium consisting of many nutrients such as glucose, amino acids, purines, pyrimidines, and vitamins. These plates permit growth of virtually all *Escherichia coli* nutritional mutants, and wild-type, as well as many other types of cells. Streaking is performed by sterilizing a platinum needle, poking it into a colony or culture of cells and lightly dragging it across an agar surface so that at least a few of the deposited cells are sufficiently isolated that they will grow into isolated, and therefore pure, colonies. Essentially, the only modification in the above procedure necessary for other cell-types is to use appropriate media. Often cells from higher organisms display a density-dependent growth and will not divide unless the cell density is above a critical value. Then the isolation of a culture from a single cell requires the use of tiny volumes. One method is to use microdrops suspended from glass cover slips. Evaporation from the drops is prevented by placing the cover slips upside down over small chambers filled with the growth medium. The analog of beginning bacterial genetics experiments from a single cell is beginning genetics experiments on multicellular organisms with isogenic parents. Highly inbred laboratory strains serve such a purpose.

Testing Purified Cultures, Scoring

Once a strain has been purified, its phenotype can be tested. Suppose, for example, that a desired bacterial strain is claimed to be unable to use lactose as a sole source of carbon and to be unable to synthesize arginine. What is the simplest way to test or score for Lac⁻ Arg⁻? One satisfactory growth medium for the mutant would be a medium containing K^+, Na^+, PO_4^{--}, SO_4^{--}, Mg^{++}, Ca^{++}, Fe^{++}, NH_4^+, and other trace elements which are present as chemical contaminants of these ingredients, plus glucose at a concentration of 2 gm/liter and L-arginine at 0.1 gm/liter. This is called a minimal salts medium with glucose and arginine. Each candidate colony could be spotted with sterile wooden

Table 8.1 Spot Testing for Arg⁻ and Lac⁻

Cell Type	Minimal Glucose	Minimal Glucose Arginine	Minimal Lactose Arginine
Wild Type (Lac⁺, Arg⁺)	+	+	+
Arg⁻	-	+	+
Lac⁻	+	+	-
Arg⁻, Lac⁻	-	+	-

sticks or toothpicks onto minimal salts medium containing the following additions: glucose, glucose plus arginine, and lactose plus arginine. The desired strain should grow only on glucose plus arginine medium (Table 8.1).

Isolating Auxotrophs, Use of Mutagens and Replica Plating

Consider the isolation of a bacterial strain requiring leucine as a growth supplement added to the medium. A spontaneously arising Leu$^-$ mutant might exist in a population at a frequency of about 10^{-6}. One method of finding or isolating such a leucine-requiring mutant would be to dilute cells to a concentration of about 1,000 cells/ml and spread 0.1 ml quantities on the surface of 10,000 glucose plus leucine plates. After these had grown, each of the 1,000,000 colonies could be spot tested for a leucine requirement. By this method one spontaneously occurring Leu$^-$ mutant out of 10^6 cells could be found. This is not a workable method, however, and when faced with problems like this, geneticists devised many shortcuts.

One way to decrease the work of finding a mutant is to increase its frequency of occurrence. Standard techniques for increasing the frequency of mutants in a population are treatment with chemical mutagens, exposure to UV light, or the use of mutator strains. The spontaneous mutation frequency in such strains is greatly elevated, for example due to mutations in DNA polymerase. Another shortcut is to reduce the time required to spot many colonies in the scoring steps. Replica plating allows spotting all the colonies from a plate in one operation onto a testing plate. This can be done with a circular pad of sterile velvet or paper which is first pushed against the master plate of colonies. The paper or velvet picks up many cells which can then be deposited onto a number of replica plates.

In some situations the use of a mutagen might be unwise due to the possibility of introducing more than one mutation into the strain. Therefore, a spontaneously occurring Leu$^-$ mutant might have to be found. Even with replica plating, this could entail much work and a method of selectively killing all the Leu$^+$ cells in the culture would be most valuable.

Penicillin provides a useful reverse selection for bacterial mutants. Ordinarily, mutants capable of growing in a particular medium may easily be selected. The reverse, the selection for the mutants unable to grow in a particular medium, requires a trick. Penicillin provides an answer. This antibiotic blocks the correct formation of the peptide crosslinks in the peptidoglycan layer. Walls synthesized in the absence of these crosslinks are weak. Only growing cells synthesize or try to synthesize peptido-glycan. Therefore, in the presence of penicillin, growing cells synthesize defective walls and are lysed by osmotic pressure. Non-growing cells survive. A penicillin treatment can increase the fraction of Leu$^-$ cells in the population 1000-fold. That is, a penicillin treatment can select for Leu$^-$ cells.

In practice, a culture of the desired cells is grown on minimal medium containing leucine, and then the leucine is removed either by filtration or centrifugation. The cells are resuspended in medium lacking leucine, and penicillin is added. The Leu⁺ cells continue growing and are killed by the penicillin whereas the Leu⁻ cells stop growing and remain resistant to the penicillin. The penicillin is removed and the surviving cells are plated out for spot testing to identify the Leu⁻ colonies.

Genetic Selections

In the preceding section we saw the use of penicillin for the selection of Leu⁻ mutants. Here we will further discuss genetic selections. Selective growth of mutants means using conditions in which the desired mutant will grow, but the remainder of the cells, including the wild-type parents will not. This is to be contrasted with scoring, in which all the cells grow and the desired mutant is identified by other means. Often in scoring, all the cells are plated out to form colonies. These are then spotted onto various media on which the mutant may be identified or are grown for assay of various gene products. A simple example of selecting a desired mutant would be isolating a Lac⁺ revertant from a Lac⁻ mutant by plating the cells on minimal plates containing lactose as the sole source of carbon. Up to 10^{11} cells could be spread on a single plate in search of a single Lac⁺ revertant.

A slightly more complicated mutant selection is the use of an agent whose metabolism will create a toxic compound. Cleavage of orthonitrophenyl-β-D-thiogalactoside by the enzyme β-galactosidase yields a toxic compound and cells die. Thus Lac⁺ cells expressing β-galactosidase in a medium containing glycerol and orthonitrophenyl-β-D-thiogalactoside are killed and only Lac⁻ cells survive (Fig. 8.17).

Let us examine a more complicated selection, the isolation of a nonsense mutation in the β subunit of RNA polymerase. Such a mutation would be lethal under normal circumstances since RNA polymerase is an essential enzyme and a nonsense mutation terminates translation of the elongating polypeptide chain. Tricks can be devised however, so that the desired mutant can be identified if the cells contain a temperature-sensitive nonsense suppressor, Sup⁺(ts). Such a suppressor results

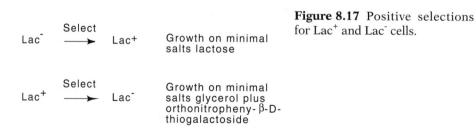

Figure 8.17 Positive selections for Lac⁺ and Lac⁻ cells.

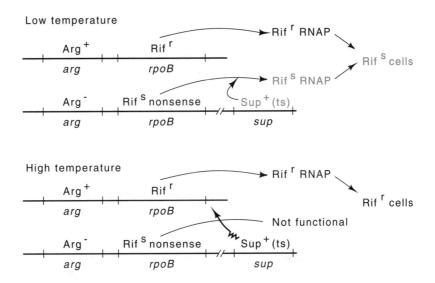

Figure 8.18 Selection for a strain containing a nonsense mutation in the β subunit of RNA polymerase.

from a mutant tRNA that suppresses nonsense codons, but loses its suppressing activity at 42°, a temperature at which *Escherichia coli* normally survives. The selection also utilizes the RNA polymerase inhibitor, rifamycin.

Two facts are key to the selection: first, that rifamycin sensitivity is dominant to rifamycin resistance, and second, that rifamycin-resistant mutants possess altered β subunits of RNA polymerase. One might have thought that resistance would be dominant to sensitivity, for it would seem that the rifamycin-resistant polymerase in a cell could function despite the presence of the sensitive polymerase. Problem 8.20 asks why sensitive polymerase is dominant to resistant polymerase. The first step in selecting the desired mutant is to mutagenize Arg⁻ Rifˢ Sup⁺(ts) Smʳ (streptomycin-resistant) cells and to grow them at the low temperature of 30°. Under these conditions, a nonsense mutation in the *rif* allele would not be lethal since the suppressor would be active and the complete β chain of polymerase would be synthesized (Fig. 8.18).

After the first step, a region of the chromosome containing the Arg⁺ Rifʳ genes could be introduced by mating with an appropriate Smˢ strain containing an Arg-Rif episome (see later section for episomes). By selecting for growth in the absence of arginine and presence of strepto-mycin, cells diploid for this region could be selected. The majority of the cells which grow will then have the genotype Arg⁺ Rifʳ/ Arg⁻ Rifˢ Sup⁺(ts) Smʳ, in which the genes before the "/" represent episomal genes. Among these will be a few of the desired genotype, Arg⁺ Rifʳ/Arg⁻ Rifˢ(amber) Sup⁺. Since rifamycin sensitivity is dominant to rifamycin resistance, the cells with the amber mutation in the β subunit will be rifamycin-resistant at 42° whereas the others will remain rifamycin-sensitive. The desired cells, but not most of the other types, would be able to grow at 42° on minimal medium lacking arginine and containing rifamycin. One of the nondesired mutant types would be those with missense mutations in the β subunit. How could these be distinguished from those with nonsense mutations in β? The answer is that the missense mutations would remain rifamycin-resistant even at 30°, whereas the amber mutants would become rifamycin-sensitive!

Mapping with Generalized Transducing Phage

As we have already discussed in Chapter 3, coliphage P1 is imperfect in its encapsulation process, and instead of packaging P1 DNA with complete fidelity, a small fraction of the phage package a segment of their host's DNA. Upon infection of sensitive cells with such a lysate, the majority of infected cells are injected with a phage DNA molecule. A small fraction of the cells however, are injected with a segment of the chromosome of the cells on which the phage lysate was prepared. This DNA segment cannot permanently survive in the cells since it lacks a DNA replication origin and furthermore, exonucleases would degrade it. It is free to engage in genetic recombination before it is degraded. This type of transfer of genetic markers via a phage is called generalized transduction.

The ability to transduce cells with a phage like P1 facilitates fine structure genetic mapping. Three factor crosses may be performed with phage P1 to determine the order of genetic markers regardless of the sex of the cells rather than being restricted to using donor and receptor cell lines for bacterial conjugation. Even more useful, however, are the consequences of the fact that the length of DNA which a P1 phage particle carries is only 1% the size of the bacterial chromosome. Therefore, the frequency with which two genetic markers are simultaneously carried in a single transducing phage particle is high if they are separated by much less than 1% of the chromosome. This frequency falls as the separation between two markers increases, and it becomes zero if they are so widely separated that a segment of DNA carrying both alleles is too large to be encapsulated. As a result, the frequency of cotransduction of two genetic markers provides a good measurement of their separation.

Cotransduction frequency often is easy to measure. For example, other experiments could have indicated that the genes for synthesis of leucine and the genes permitting utilization of the sugar arabinose were closely spaced on the chromosome. P1 mapping can measure their separation more accurately. P1 could be grown on an Ara$^+$ Leu$^+$ strain and used to transduce an Ara$^-$ Leu$^-$ strain to Ara$^+$ by spreading the P1 infected cells on agar containing minimal salts, arabinose as a carbon source, and leucine. Then, the Ara$^+$ transductants could be scored by spot testing for the state of their leucine genes.

Principles of Bacterial Sex

Lederberg and Tatum discovered that some bacterial strains can transfer DNA to recipient cells, that is, they can mate. This discovery opened the doors for two types of research. First, genetic manipulations could be used to assist other types of studies involving bacteria, and throughout the book we shall see many examples of the assistance genetics provides to biochemical, physiological, and physical studies. Second, the mechanism of bacterial mating itself was interesting and could be investigated. In this section we shall review the actual mechanism of bacterial mating.

Maleness is conferred upon cells when they contain a mating module called the F-factor. F stands for fertility. This module is a circular DNA molecule containing about 25 genes involved with conjugation. One of the F genes codes for the major protein of the F-pilus. This is an appendage essential for DNA transfer. Other genes code for additional parts of the F-pilus, its membrane attachment, and DNA replication of F-factor DNA. The system also contains at least one regulatory gene.

F-pilus contact with a suitable female activates the mating module. As a result, a break is made within the F sequences of DNA and a single strand of DNA synthesized via the rolling circle replication mode is transferred into the female (Fig. 8.19). Immediately upon entry of the single strand into the female, the complementary strand is synthesized. If no breakage occurs during transfer, all the F-factor DNA, including the portion initially left behind at the break, can be transferred into the female cell. The F-factor thus codes for transfer into female cells of itself and also any DNA to which it is connected. If the F-factor is integrated within the chromosome, then its mobilization in conjugation transfers the chromosome as well as itself. Cells which transfer their chromosomal genes to recipient cells are called Hfr for high frequency of recombination, whereas the cells which contain F-factors separate from the chromosome are called F or F', and the females are called F⁻.

A little over 100 minutes are required for transfer of the entire *Escherichia coli* chromosome to a female. Genes located anywhere on the chromosome can be mapped by determining the timing of their

Figure 8.19 Transfer of one strand of the double-stranded F factor from male cells into female cells.

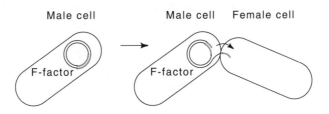

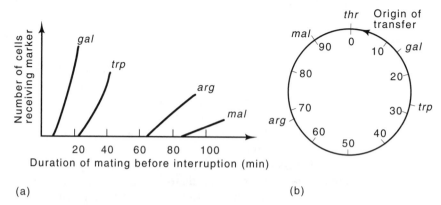

Figure 8.20 (a) The number of recombinants of the types indicated as a function of the duration of mating. (b) The approximate chromosomal locations of the markers used in the mating.

transfer to female cells. To do this mating is initiated by mixing male and female cells. Genetic markers on one side of the F-factor will be transferred to the female cells with only a few minutes of mating, while markers on the opposite side of the chromosome will not be transferred for over half an hour.

Random breakage of the DNA during mating from an Hfr to an F⁻ often interrupts the transfer so that only infrequently can the complete chromosome be transferred. This is a blessing in disguise however, for in addition to the time of transfer, the frequency of a marker's transfer also indicates its chromosomal position with respect to the integrated F-factor. More than one integration site of F has been found on the chromosome, so that a variety of origins of transfer are available and the entire chromosome can be easily mapped. Once mating has been started and been allowed to proceed for a period of time, any further transfer of markers can be stopped merely by vigorously shaking the culture to separate the mating couples.

A mating experiment might be performed in the following way. Male Arg⁺ Gal⁺ Leu⁺ Smˢ (streptomycin-sensitive) and female Arg⁻ Gal⁻ Leu⁻ Smʳ cells would be grown to densities of about 1×10^8 per ml and mixed together in equal portions. The cells would then be shaken very gently. At intervals, a sample of cells would be taken, vigorously shaken so as to separate mating couples and dilutions spread on streptomycin-containing medium so as to select for Arg⁺ or Gal⁺ or Leu⁺ Smʳ recombinants (Fig. 8.20).

F-factors facilitate genetic study with *Escherichia coli* in a second way. They need not be associated with the chromosome. Instead, they can be autonomous DNA elements existing alongside the chromosome. The transfer properties remain the same for an F-factor in this position, but the genetics are slightly altered in two ways. First, the extra-chromosomal F-factor can have some chromosomal genes associated with it, in which case it is called an F'-factor. Presumably these genes were picked up at an earlier time by an excision of the integrated F-factor from the chromosome of an Hfr cell. The second alteration of an extrachromosomal F-factor is that often the entire mating module and attached genes are transferred to a female. As a result, the female acquires a functional F-factor with its associated genes and becomes diploid for a portion of the chromosome as well as becoming capable of transferring copies of the same F′ on to other female cells.

Elements of Yeast Genetics

Yeast contain about 1.5×10^7 base pairs of DNA per cell, about five times the amount per bacterium. Baker's yeast, *Saccharomyces cerevisiae*, contains 16 chromosomes. Consequently, most are substantially smaller than the *E. coli* chromosome.

As already mentioned, yeast can be found as haploids or diploids. This property greatly facilitates mutation isolation and genetic analysis, and, combined with the small chromosome size, makes yeast a good choice for experiments which require eukaryotic cells. A diploid yeast cell can grow in culture much like a bacterium, although *Saccharomyces* divide not in half like bacteria, but generate daughter buds which enlarge and finally separate from the mother cell (Fig. 8.21). As mentioned earlier in this chapter, in contrast to *E. coli*, but like some other bacteria, yeast can sporulate. This occurs if they are starved of nitrogen in the presence of a nonfermentable carbon source like acetate. In this 24 hour process, a single diploid yeast cell undergoes meiosis and forms an ascus containing four spores. Upon incubation in rich medium, the spores germinate and grow as haploid yeast cells.

If one of the spores is isolated from an ascus and grown up separate from other spores, the resulting culture is different from the parental culture. The cells are haploid and they cannot sporulate. In fact, an ascus contains two types of spores. These are designated as mating-types, *a* and α. Both of these generate the haploid cultures and cannot sporulate.

Figure 8.21 The yeast cell cycle.

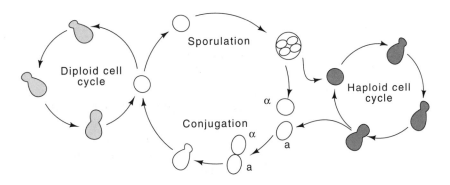

If, however, a and α cultures are mixed, cells of opposite mating-type adhere and form mating pair aggregates. First the cytoplasms of these pairs fuse, and then the nuclei fuse to generate diploid cells which grow and remain diploids. The diploids then can sporulate and regenerate the haploids of mating-types a and α.

Recombination can be used for genetic mapping in yeast much as it is used in bacteria. Since no yeast viruses are known, genetic transfer must be accomplished either by the fusion of haploids as described above or by direct DNA transfer as described in a later chapter on genetic engineering. Genetic recombination occurs during both meiosis and mitosis, but is much more frequent in meiosis.

Elements of *Drosophila* Genetics

Genetics can profitably be applied to the study of the development of tissues and tissue-specific gene expression as well as behavior, vision, muscle and nerve function in *Drosophila*. The *Drosophila* genome of about 1.65×10^8 base pairs of DNA is contained in four chromosomes. Chromosomes II, III, and IV are normally found as pairs, while in males the constitution of the fourth chromosome is XY and in females it is XX. Unlike yeast, there is no simple way to make haploid fruit flies, and therefore the genes which are located on chromosomes II, III, and IV are more difficult to study. The genes located on chromosome X may easily be studied, for males are haploid for the X chromosome and females are diploid. Therefore recessive mutations in genes located on the X chromosome will be expressed in males, and the ability of X-located genes to complement can be tested in females.

Once a mutation is generated in the DNA of a bacterium, one of the daughter cells usually is capable of expressing the mutation. The analog is also true in *Drosophila*. However, the equivalent of a cell division in bacteria is the next generation in *Drosophila*. An adult fly can be mutagenized, but many of its genes are expressed only during development. Therefore this adult will not express the mutation. Mutagenized adults must be mated and their progeny must be examined for the desired mutation. One straightforward way to mutagenize flies is to feed them a 1% sucrose solution containing ethylmethanesulfonate (EMS).

$$\text{X'Y} \quad \text{X} \quad \text{XX} \longrightarrow \left\{\begin{array}{l} \text{X'X} \\ \text{X'X} \\ \text{Y X} \\ \text{Y X} \end{array}\right.$$

Mutagenized   Female
male

If male flies are mutagenized and mated with females, four types of progeny are obtained. In the first generation only females could contain a mutagenized X chromosome. If these females are collected and mated again with unmutagenized males, then in the second generation half of the males have a chance of receiving and expressing a mutagenized X chromosome. Much tedious sorting of males and females could be required for the detection of rare mutants.

Isolating Mutations in Muscle or Nerve in *Drosophila*

How could mutations in muscle or nerve be isolated? Since these mutations could be lethal, we ought to seek conditional mutations. That is, the mutation ought to be expressed only under special conditions, for example at elevated temperature. Suzuki developed ingenious methods for the isolation of temperature-sensitive paralytic mutations. The flies we seek ought to be perfectly normal at low temperature, be paralyzed at high temperature, and recover rapidly when returned to low temperature. Undoubtedly such mutations would be exceedingly rare, and great numbers of flies would have to be screened in order to find a few candidate mutants. Such large numbers necessitated the use of tricks to eliminate the need for sorting males and females.

The first trick used an attached X chromosome. This is an inseparable pair of X chromosomes denoted as $\widehat{X}X$. Mating males with females containing an attached X chromosome, yields the expected four types of offspring (Fig. 8.22). Both the X'$\widehat{X}X$ and YY are lethal and therefore only mutagenized males or females with the attached X chromosome result from this mating. If the attached X chromosome contains a dominant temperature-sensitive lethal mutation, the females can be killed by a brief temperature pulse, leaving only the desired, mutagenized males as a pure stock. The female stocks required for the first mating also can be generated by this same technique.

The second problem was the actual selection for the temperature-sensitive paralytics. This was done by introducing up to 10^4 flies into a cubical box about two feet on a side. The temperature in the box was

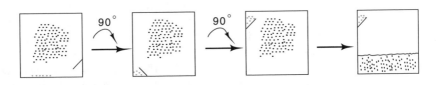

raised, and the box was given a bang on a table top to make the flies fly upwards. Any temperature-sensitive paralytics remained at the bottom of the box. These were trapped by rotating the box so that they were

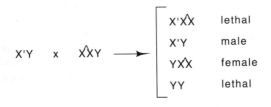

		X'$\widehat{X}X$	lethal
		X'Y	male
X'Y	x $\widehat{X}$XY $\longrightarrow$	Y$\widehat{X}X$	female
		YY	lethal

Figure 8.22 Use of the attached X chromosome to generate special populations of flies. Although an XY fly is male, an XXY fly is female.

collected on a ledge. Then the flies were anesthetized by adding carbon dioxide or ether and those flies which had been able to fly fell to the bottom of the box where they were killed by the addition of detergent and acetic acid.

Like most mutant selection schemes, several additional phenotypes were found in addition to those sought. One of these was *rex* for rapid exhaustion. After running around for few minutes a *rex* mutant shudders a bit and falls over in a paralysis which lasts a few minutes. It can then get up and is perfectly normal for about an hour. Another mutant was the *bas* for bang-sensitive. The desired mutants fell into three types: *para*ts for temperature-sensitive paralytic, *sts*ts for stoned, and *shi*ts from the Japanese word for paralyzed. The *para* mutation lies in the protein which forms the membrane sodium channel that is essential for nervous impulse transmission.

Fate Mapping and Study of Tissue-Specific Gene Expression

One obvious way to examine the tissue specificity of gene expression is to isolate the tissues and assay each for the protein or gene product in question. A slight modification of this approach is quite reasonable. The synthesis of messenger from the various tissues of a fly can be measured approximately by DNA-RNA hybridization. As we shall see in a later chapter, DNA from desired genes can be obtained and then used in such *in situ* hybridization experiments. Remarkably, genetics experiments called fate mapping can also locate the tissues in which an altered gene is expressed. This approach does not require knowledge of the gene involved and therefore it is useful in initial steps of studies. Genetic engineering has also developed techniques for examination of tissue-specific gene expression, but those approaches first require isolating DNA or RNA of the gene in question. Fate mapping is useful when the gene involved is unknown. As an example of localizing the activity of a gene, imagine a mutant fly that is unable to flap its wings. This could be because of a defective wing, a defective wing muscle, a defective nerve to the muscle, or defective neurons in the brain. Fate mapping can determine which tissue is responsible for such altered behavior.

Fate mapping relies on the developmental pathway of the fly. A fertilized *Drosophila* egg contains a single cell whose nucleus undergoes about nine divisions. These nuclei then migrate to the surface of the egg to form the blastula stage, and three more divisions occur before cell walls are laid down. At this stage, different cells on the surface ultimately

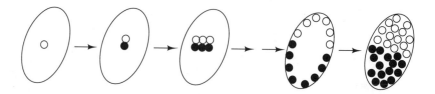

become different parts of the adult fly, but cells lying near to each other frequently develop into adjacent parts on the fly. Therefore, a map can be drawn on the egg of the parts of the adult fly that each of these cells will become. If it were possible to associate a particular phenotype in an adult with a particular location on the egg, then the tissue responsible for the adult phenotype would be determined. This is possible!

The association of tissues in the mature fly with positions on the blastula utilizes selective chromosome loss during development of the blastula. Fly development is not greatly altered in a female egg cell if one of the X chromosomes contains a defect so that it begins the first nuclear replication a little late. Consequently, this chromosome often is not segregated into one of the two resultant daughter nuclei resulting from the first nuclear division in the egg. The final result of this chromosome loss is that about half of the cells of the blastula will be diploid XX and the others will be haploid X. Since the spatial orientation of the first cleavage with respect to the egg shell is not uniform from egg to egg, and because there is little mixing of the nuclei or cells during subsequent divisions, different sets of cells will be XX and X in different blastulas. Suppose that these two types of cells can be distinguished. This can be done by placing a recessive body color marker gene, for example yellow, on the stable X chromosome. Then cells of the fly possessing the XX genotype will be black and the X genotype will be yellow. The adult fly will have a mottled appearance.

The probability that two different body parts possess different colors will be proportional to the distance that their corresponding ancestor cells in the blastula state were separated. The greater their separation, the greater the chances that the line separating the two cell-types will fall between them. If they are close together, there is little chance they will be of different cell-type and therefore there is little chance they will possess different body color.

A collage appears when the body parts of the adult fly are mapped to the blastula. The map can then be used as follows to locate the tissue in which a recessive mutation is expressed. If the mutation is located on the X chromosome which is not lost during the development, then those tissues which can express the mutant phenotype will be haploid. For example, if the mutant phenotype always appears in flies with haploid second left legs and only appears in these flies, then it can be concluded that the tissue in which the mutation is expressed is the second left leg. More generally however, the frequency of association of the mutant phenotype with a number of landmarks gives the distance on the blastula between the landmarks and the tissue in question. Transferring these distances to the blastula fate map then reveals the tissue in question.

Problems

8.1. In many eukaryotes, genetic recombination between chromosomes does not occur during mitosis and is largely restricted to the first stage of meiosis. What is the possible biological value of such a restriction?

8.2. Sperm do not transfer mitochondria to the egg. How could you recognize mitochondrial markers by their genetic behavior?

8.3. How would you select a temperature-sensitive *lacI⁻* mutation?

8.4. How would you interpret the existence of *trans* dominant negative mutations in an enzyme, and in a regulatory protein?

8.5. Show how backcrossing with eukaryotes may be used to determine the genotypes of heterozygotes. [Hint: Backcrossing means just about what it sounds like.]

8.6. Invent a scenario by which a single nucleotide change on a chromosome could lead to inactivation of all the genes of an entire chromosome. Note that one of the two X chromosomes in each cell of females of many organisms is inactivated.

8.7. Invent a scenario by which a mutation could be methylation of a particular base on DNA. Note that to qualify as a mutation, the effect must be inheritable. That is, if the base is not methylated, it remains nonmethylated, but if it somehow should become methylated, then all descendants will be methylated on the same base.

8.8. Methylation of a base of DNA in a promoter region could inactivate transcription from a gene. Invent genetic data that would lead to this conclusion. That is, the mutation might be *cis* or *trans* dominant, etc.

8.9. How could it be shown that two genetic markers were on the same chromosome in a diploid eukaryotic organism?

8.10. Trans dominant negative mutations normally are found in genes for oligomeric proteins in which multiple subunits are required for function. How could a mutation in an RNA polymerase nonessential sigma factor be *trans* dominant negative?

8.11. No experiment was described in the text that demonstrates that virtually all recombination events are precise, that is, in register, and do not delete base pairs from one DNA duplex partner and insert them into the other. Design one. [Hint: Three nonessential genes A, B, and C are in an operon and lack introns, and we have A⁻ B⁺ C⁺ and A⁺ B⁺ C⁻ strains available.]

8.12. Consider a genetic crossing experiment with yeast involving allelic markers *a* and *A*. Why, on first consideration, would you expect that the number of cells yielding one *a* type and three *A* type spores would equal the number yielding three *a* type and one *A* type? What mechanism might be responsible for a serious biasing of the numbers in one direction?

8.13. Invent genetic recombination data that would lead to the prediction that the genetic markers were arranged in a two-dimensional array rather than the one-dimensional array of markers that is actually found.

8.14. In mapping a small region by measuring the recombination frequencies between point mutations, the following recombination percentages were obtained. What is the map of the region? In addition to the most likely gene order and the relative distances between the mutations, what else can you say?

	Markers				
Markers	1	2	3	4	5
1	-	-	9	-	4
2	7	0	-	-	5
3	-	-	-	-	-
4	-	-	11	-	-
5	-	-	-	1	0

8.15. The following data set was obtained in an experiment in which deletions entering gene X from the left end and ending somewhere within X were crossed with point mutations in gene X. The ability of the indicated deletion-point mutation pair to yield X^+ recombinants is shown in the table. What is the best ordering of the markers 1-6, A-F? Discuss.

	Point Mutation					
Deletion	1	2	3	4	5	6
A	-	+	-	-	-	+
B	-	+	+	+	-	+
C	-	+	-	-	-	-
D	-	+	+	+	-	+
E	+	+	+	+	-	-
F	-	+	-	-	-	+

8.16. Why can't gene conversion easily be observed in genetic crosses between bacterial strains?

8.17. What sequences ought to be particularly prone to one- or two-base insertions or deletions?

8.18. What nucleotide would guanine base pair with in DNA replication if it were ethylated on the carboxyl of C-6?

8.19. Why is streptomycin-sensitivity dominant to streptomycin resistance? Hint, streptomycin-sensitive ribosomes synthesize toxic peptides in the presence of the drug.

8.20. Why is rifamycin sensitivity dominant to rifamycin resistance?

8.21. In the usual protocol for phage P1 transduction of most genetic markers, the P1-infected cells are immediately spread on plates selective for the marker being transduced. Why can't this protocol be used for the transduction of the streptomycin and rifamycin resistance alleles? Note that these alleles are recessive to the sensitive alleles.

8.22. In the isolation of leucine-requiring strains with the use of a penicillin selection, why are the cells grown in leucine-containing medium before the addition of penicillin?

8.23. Nitrosoguanidine mutagenizes the growing points of the bacterial chromosome. Explain how this property could be used to isolate mutations concentrated in the vicinity of a selectable genetic marker. For example, if you possessed a Leu⁻ mutation, how could you isolate many mutations near the leucine genes?

8.24. Nitrosoguanidine mutagenesis of *E. coli* growing rapidly on a rich yeast extract medium yielded double mutants that were simultaneously rifamycin-resistant and arabinose minus. The *ara* and *rif* markers lie 1/3 of the chromosome apart. Double *ara-rif* mutants were not seen if the mutagenesis was performed on cells growing on minimal medium. Why?

8.25. Expand on the following cryptic statement: "Tritium suicide was used to isolate temperature-sensitive mutations in the protein synthesis machinery..."

8.26. Physiological and *in vitro* measurements showed that the RNA polymerase from streptolydigin-resistant cells was resistant to the drug. It was then desired to map the mutation. Most likely this mutation would map near the rifamycin resistance mutation. The most straightforward way to show that the two markers were close to each other would have been to start with Stl^r Rif^s strain and to transduce it to Rif^r from Rif^r Stl^s and determine if it became Stl^s. Presumably this would show that the Rif and Stl alleles cotransduce. Why is this reasoning risky?

8.27. What are the expected effects on spontaneous mutation frequencies of hyper and hypo synthesis of the *dam* methylase? This protein is thought to mark "old" DNA strands by methylation of adenine residues in the sequence d(GATC).

8.28. Why are mutant reversion studies more sensitive than forward mutation studies in the detection of mutagenic properties of compounds? What are the limitations of determining mutagenic specificity by reversion studies?

8.29. What is a reasonable interpretation of the following? Ultraviolet light irradiation increases the frequency of mutations in phage only if *both* phage and cells are irradiated.

8.30. Look up and sketch the structural basis for the mutagenic actions of nitrous acid, hydroxylamine, ethylmethane sulfonate, 5-bromouracil, and 2-aminopurine.

8.31. What can you conclude about synthesis of the outer surface components of *Escherichia coli* from the fact that male-specific and female-specific phage exist?

8.32. In yeast how can mitotic and meiotic genetic recombination be distinguished?

8.33. Why can we neglect consideration of the behavior of mutations located on the *Drosophila* Y chromosome?

8.34. How could a stock of female *Drosophila* be generated using the properties of the attached X chromosome?

8.35. How could flies unable to detect bright light be isolated?

8.36. Could fate mapping work if the orientation of the plane of the first nuclear division with respect to the egg were not random? Could it work if there were no mixing of cells during subsequent divisions? Could it work if there were complete mixing?

8.37. In a strange experiment cells were diluted and a volume was inoculated into 1,000 tubes such that on average each tube received three cells. Upon incubation, the cells in 75% of the tubes grew, and in 25% of the tubes nothing detectable happened. What do you conclude?

8.38. In fate mapping of a behavioral mutation in *Drosophila* the following data were obtained by screening 200 flies. The mutation lies on the X chromosome which also carries a recessive mutation which leads to a yellow body color. Indicate where on the blastoderm the behavioral mutation could map to.

	Behavior			Behavior	
Eye	Normal	Mutant	Wing	Normal	Mutant
Black	41	9	Black	39	11
Yellow	11	39	Yellow	12	38

8.39. In seeking mutants defective in the transmission of nerve impulses, it was logical to hunt only for temperature-sensitive mutations. Appropriate selections yielded the desired temperature-sensitive mutations. Such a mutation turned out to be a deletion of a gene for a protein found in the cell membrane. Why is this a surprise and what could be an explanation for the temperature-sensitive phenotype?

References

Recommended Readings

Mapping of Behavior in *Drosophila* Mosaics, Y. Hotta, S. Benzer, Nature *240*, 527-534 (1972).

A General Model for Genetic Recombination, M. Meselson, C. Radding, Proc. Natl. Acad. Sci. USA *72*, 358-361 (1975).

Dominant Mutators in *Escherichia coli*, E. Cox, and D. Horner, Genetics *100*, 7-18 (1982).

Recombination and Mutation Mechanisms

A Mechanism for Gene Conversion in Fungi, R. Holliday, Genetics Res. *5*, 282-304 (1964).

Isolation and Characterization of Recombination-deficient Mutants of *E. coli* K-12, A. Clark, A. Margulies, Proc. Natl. Acad. Sci. USA *53*, 451-459 (1965).

Molecular Aspects of Genetic Exchange and Gene Conversion, R. Holliday, Genetics *78*, 273-287 (1974).

Mismatch Repair in Heteroduplex DNA, J. Wildenberg, M. Meselson, Proc. Natl. Acad. Sci. USA *72*, 2202-2206 (1975).

Repair Tracts in Mismatched DNA Heteroduplexes, R. Wagner, M. Meselson, Proc. Natl. Acad. Sci USA *73*, 4135-4139 (1976).

Kinetics of Branch Migration in Double-stranded DNA, B. Thompson, M. Camien, R. Warner, Proc. Natl. Acad. Sci. USA *73*, 2299-2303 (1976).

ATP-Dependent Renaturation of DNA Catalyzed by the *recA* Protein of *Escherichia coli*, G. Weinstock, K. McEntee, I. R. Lehman, Proc. Natl. Acad. Sci USA *76*, 126-130 (1979).

Purified *Escherichia coli recA* Protein Catalyzes Homologous Pairing of Superhelical DNA and Single-stranded Fragments, T. Shibata, C. Das-Gupta, R. Cunningham, C. Radding, Proc. Natl. Acad. Sci. USA *76*, 1638-1642 (1979).

Initiation of General Recombination Catalyzed *in vitro* by the *recA* Protein of *Escherichia coli*, K. McEntee, G. Weinstock, I. R. Lehman, Proc. Natl. Acad. Sci. USA *76*, 2615-2619 (1979).

Mutagenic Deamination of Cytosine Residues in DNA, B. Duncan, J. Miller, Nature *287*, 560-561 (1980).

RecA Protein of *Escherichia coli* Promotes Branch Migration, a Kinetically Distinct Phase of DNA Strand Exchange, M. Cox, I. R. Lehman, Proc. Natl. Acad. Sci. USA *78*, 3433-3437 (1981).

Directionality and Polarity in *recA* Protein-promoted Branch Migration, M. Cox, I. R. Lehman, Proc. Natl. Acad. Sci. USA *78*, 6018-6022 (1981).

On the Formation of Spontaneous Deletions: The Importance of Short Sequence Homologies in the Generation of Large Deletions, A. Albertini, M. Hofer, M. Calos, and J. Miller, Cell *29*, 319-328 (1982).

Substrate Specificity of the DNA Unwinding Activity of the RecBC Enzyme of *Escherichia coli*, A. Taylor, G. Smith, J. Mol. Biol. *185*, 431-443 (1985).

Mechanisms of Spontaneous Mutagenesis: An Analysis of the Spectrum of Spontaneous Mutation in the *Escherichia coli lacI* Gene, R. Schaaper, B. Danforth, B. Glickman, J. Mol. Biol. *189*, 273-284 (1986).

Spectrum of Spontaneous Frameshift Mutations, L. Ripley, A. Clark, J. deBoer, J. Mol. Biol. *191*, 601-613 (1986).

Purification and Characterization of an Activity from *Saccharomyces cerevisiae* that Catalyzes Homologous Pairing and Strand Exchange, R. Kolodner, D. Evans, P. Morrison, Proc. Nat. Acad. Sci. USA *84*, 5560-5564 (1987).

General Mechanism for RecA Protein Binding to Duplex DNA, B. Pugh, M. Cox, J. Mol. Biol. *203*, 479-493 (1988).

The Mechanics of Winding and Unwinding Helices in Recombination: Torsional Stress Associated with Strand Transfer Promoted by RecA Protein, S. Hoigberg, C. Radding, Cell *54*, 525-532 (1988).

Formation and Resolution of Recombination Intermediates by *E. coli* RecA and RuvC Proteins, H. Dunderdale, F. Benson, C. Parsons, G. Sharples, R. Lloyd, S. West, Nature *354*, 506-510 (1991).

The Structure of the *E. coli* RecA Protein Monomer and Polymer, R. Story, I. Weber, T. Steitz, Nature *355*, 318-325 (1992).

Structural Relationship of Bacterial RecA Proteins to Recombination Proteins from Bacteriophage T4 and Yeast, R. Story, D. Bishop, N. Kleckner, J. Steitz, Science *259*, 1892-1896 (1993).

Chi Recombination

Hotspots for Generalized Recombination in the *E. coli* Chromosome, R. Malone, D. Chattoraj, D. Faulds, M. Stahl, F. Stahl, J. Mol. Biol. *121*, 473-491 (1978).

Unwinding and Rewinding of DNA by the *RecBC* Enzyme, A. Taylor, G. Smith, Cell *22*, 447-457 (1980).

Chi Mutation in a Transposon and the Orientation Dependence of Chi Phenotype, E. Yagil, D. Chattoraj, M. Stahl, C. Pierson, N. Dower, F. Stahl, Genetics *96*, 43-57 (1980).

Structure of *Chi* Hotspots of Generalized Recombination, G. Smith, S. Kunes, D. Schultz, A. Taylor, K. Triman, Cell *24*, 429-436 (1981).

Identity of a Chi Site of *Escherichia coli* and Chi Recombinational Hotspots of Bacteriophage Lambda, K. Triman, D. Chattoraj, G. Smith, J. Mol. Biol. *154*, 393-398 (1982).

Orientation of Cohesive End Site *cos* Determines the Active Orientation of Chi Sequence in Stimulating *recA-recBC*-Mediated Recombination in Phage Lambda Lytic Infections, I. Kobayashi, H. Murialdo, J. Craseman, M. Stahl, F. Stahl, Proc. Natl. Acad. Sci. USA *79*, 5981-5985 (1982).

Chi-dependent DNA Strand Cleavage by RecBC Enzyme, A. Ponticelli, D. Schultz, A. Taylor, G. Smith, Cell *41*, 145-151 (1985).

RecBC Enzyme Nicking at Chi Sites During DNA Unwinding: Location and Orientation-Dependence of the Cutting, A. Taylor, D. Schultz, G. Smith, Cell *41*, 153-163 (1985).

recD: The Gene for an Essential Third Subunit of Exonuclease V, S. Amundsen, A. Taylor, A. Chaudhury, G. Smith, Proc. Nat. Acad. Sci. USA *83*, 5558-5562 (1986).

Chi-stimulated Patches are Heteroduplex, with Recombinant Information on the Phage λr Chain, S. Rosenberg, Cell *48*, 851-865 (1987).

Cutting of Chi-like Sequences by the RecBCD Enzyme of *Escherichia coli*, K. Cheng, G. Smith, J. Mol. Biol. *194*, 747-750 (1987).

Strand Specificity of DNA Unwinding by RecBCD Enzyme, G. Braedt, G. Smith, Proc. Natl. Acad. Sci. USA *86*, 871-875 (1989).

Actions of RecBCD Enzyme on Cruciform DNA, A. Taylor, G. Smith, J. Mol. Biol. *211*, 117-134 (1990).

Bacterial and Phage Genetics

Induced Mutations in Bacterial Viruses, M. Delbruck, W. Bailey Jr., Cold Spring Harbor Symposium of Quantitative Biology *11*, 33-37 (1946).

Transduction of Linked Genetic Characters of the Host by Bacteriophage P1, E. Lennox, Virology *1*, 190-206 (1955).

Selecting Bacterial Mutants by the Penicillin Method, L. Gorini, H. Kaufman, Science *131*, 604-605 (1960).

Amber Mutations of *Escherichia coli* RNA Polymerase, S. Austin, I. Tittawella, R. Hayward, J. Scaife, Nature New Biology *232*, 133-136 (1971).

Culture Medium for Enterobacteria, F. Neidhardt, P. Bloch, D. Smith, J. Bacteriol. *119*, 736-747 (1974).

Genetic Studies of the *lac* Repressor. IV. Mutagenic Specificity in the *lacI* Gene of *Escherichia coli*, C. Coulondre, J. Miller, J. Mol. Biol. *117*, 577-606 (1977).

Genetic Engineering *in vivo* using Translocatable Drug-resistance Elements, N. Kleckner, J. Roth, D. Botstein, J. Mol. Biol. *116*, 125-159 (1977).

A Genetic Analysis of F Sex Factor Cistrons Needed for Surface Exclusion in *Escherichia coli*, M. Achtman, P. Manning, B. Kusecek, S. Scwuchow, N. Willetts, J. Mol. Biol. *138*, 779-795 (1980).

DNA Transfer Occurs During a Cell Surface Contact Stage of F Sex Factor-mediated Bacterial Conjugation, M. Panicker, E. Minkley Jr., J. Bact. *162*, 584-590 (1985).

Yeast, *Drosophila,* and Mammalian Genetics

Behavioral Mutants of *Drosophila* Isolated by Countercurrent Distribution, S. Benzer, Proc. Natl. Acad. Sci. USA *58*, 1112-1119 (1967).

Temperature-sensitive Mutations in *Drosophila melanogaster*, D. Suzuki, Science *170*, 695-706 (1970).

Fate Mapping of Nervous System and Other Internal Tissues in Genetic Mosaics of *Drosophila melanogaster*, D. Kankel, J. Hall, Dev. Biol. *48*, 1-24 (1976).

Mosaic Analysis of a Drosophila Clock Mutant, R. Konopka, S. Wells, T. Lee, Mol. Gen. Genet. *190*, 284-288 (1983).

Hypervariable "Minisatellite" Regions in Human DNA, A. Jeffreys, V. Wilson, S. Thein, Nature *314*, 67-73 (1985).

Cloning and Sequence Analysis of the *para* Locus, a Sodium Channel Gene in *Drosophila*, K. Loughney, R. Kreber, G. Ganetzky, Cell *58*, 1143-1154 (1989).

A Genetic Study of the Anesthetic Response: Mutants of *Drosophila melanogaster* Altered in Sensitivity to Halothane, K. Krishnan, H. Nash, Proc. Natl. Acad. Sci. USA *87*, 8632-8636 (1990).

Expansion of an Unstable DNA Region and Phenotypic Variation in Myotonic Dystrophy, H. Harley, J. Brook. S. Rundle, S. Crow, W. Reardon, A. Buckler, P. Harper, D. Housman, D. Shaw, Nature *355*, 545-548 (1992).

Genetic Engineering and Recombinant DNA

9

The terms "genetic engineering" and "recombinant DNA" refer to techniques in which DNA may be cut, rejoined, its sequence determined, or the sequence of a segment altered to suit an intended use. For example, a DNA fragment may be isolated from one organism, spliced to other DNA fragments, and put into a bacterium or another organism. This process is called cloning because many identical copies can be made of the original DNA fragment. In another example of genetic engineering, a stretch of DNA, often an entire gene, may be isolated and its nucleotide sequence determined, or its nucleotide sequence may be altered by *in vitro* mutagenic methods. These and related activities in genetic engineering have two basic objectives: to learn more about the ways nature works and to make use of this knowledge for practical purposes.

Before 1975 the most detailed studies of biological regulatory processes were restricted to genes of small phage or bacterial genes that could be placed in the phage genome. Only by beginning with such phage could DNA or regulatory proteins be obtained in quantities adequate for biochemical study. Also, only with such phage could variant DNA sequences be easily generated for the study of altered proteins or DNA. The isolation of specialized transducing phage that carried the genes of the *lac* operon were particularly important advances in this era. These phage provided a 100-fold enrichment of the *lac* genes compared to the chromosomal DNA. They also stimulated a wide variety of important studies that greatly furthered our understanding of gene regulation as well as fostered the development of many important genetic engineering techniques. Now genetic engineering permits the same sorts of studies to be carried out on any gene from virtually any organism.

The second major reason for interest in genetic engineering is the "engineering" it makes possible. A simple application of the technology is the economical synthesis of proteins difficult or impossible to purify from their natural sources. These proteins can be antigens for use in immunization, enzymes for use in chemical processes, or specialized proteins for therapeutic purposes. Cloned DNA sequences can also be used for detection of chromosomal defects and in genetic studies. Much research in genetic engineering has also been directed at plants with the hope of improving on traditional genetic methods of crop modification. A second objective is the introduction of herbicide resistance into desired crops. This would permit spraying against weeds while the crop is growing instead of only before planting.

Genetic engineering of DNA usually involves the following steps. The DNA for study should be isolated and freed of interfering contaminants. It should be possible to cut this DNA reproducibly at specific sites so as to produce fragments containing genes or parts of genes. Next, it should be be possible to rejoin the DNA fragments to form hybrid DNA molecules. Vectors must exist to which fragments can be joined and then introduced into cells by the process called transformation. The vectors need two properties. First, they must provide for autonomous DNA replication of the vector in the cells, and second, they must permit selective growth of only those cells that have received the vectors. This chapter discusses these fundamental steps of genetic engineering as well as the crucial technique of determining the nucleotide sequence of a stretch of DNA. The next chapter discusses many of the more specialized operations that constitute genetic engineering.

The Isolation of DNA

Cellular DNA, chromosomal or nonchromosomal, is the starting point of many genetic engineering experiments. Such DNA can be extracted and purified by the traditional techniques of heating the cell extracts in the presence of detergents and removing proteins by phenol extraction. If polysaccharides or RNA contaminate the sample, they can be removed by equilibrium density gradient centrifugation in cesium chloride.

Two types of vectors are commonly used: plasmids and phage. A plasmid is a DNA element similar to an episome that replicates independently of the chromosome. Usually plasmids are small, 3,000 to 25,000 base pairs, and circular. Generally, lambda phage or closely related derivatives are used for phage vectors in *Escherichia coli*, but for cloning in other bacteria, like *Bacillus subtilis*, other phage are used. In some cases a plasmid can be developed that will replicate autonomously in more than one host organism. These "shuttle" vectors are of special value in studying genes of eukaryotes; we will consider them later.

Most often, complete purification of plasmid DNA is unnecessary and useable DNA can be obtained merely by lysing the cells, partial removal of chromosomal DNA, and the removal of most protein. Intricate DNA constructions often require highly pure DNA to avoid extraneous nucle-

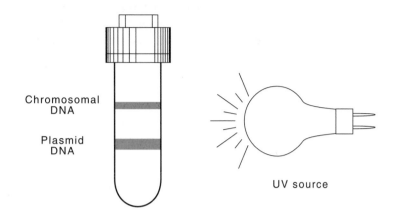

Figure 9.1 The separation of covalently closed plasmid DNA from nicked plasmid DNA and fragments of chromosomal DNA on CsCl equilibrium density gradient centrifugation in the presence of ethidium bromide. The bands of DNA fluoresce upon illumination by UV light.

ases or inhibition of sensitive enzymes. The complete purification of plasmid DNA generally requires several steps. After the cells are opened with lysozyme which digests the cell wall and detergents are added to solubilize membranes and to inactivate some proteins, most of the chromosomal DNA is removed by centrifugation. For many purposes chromatographic methods can be used to complete the purification. When the highest purity is required, however, the plasmid is purified by equilibrium density gradient centrifugation. This is done in the presence of ethidium bromide. Any chromosomal DNA remaining with the plasmid will have been fragmented and will be linear, whereas most of the plasmid DNA will be covalently closed circles. As we saw in Chapter 2, intercalating ethidium bromide untwists the DNA. For a circular molecule this untwisting generates supercoiling whereas for a linear molecule the untwisting has no major effects. Therefore a linear DNA molecule can intercalate more ethidium bromide than a circular molecule. Since ethidium bromide is less dense than DNA, the linear DNA molecules with intercalated ethidium bromide "float" relative to the circles, and therefore the two species can easily be separated. Following the centrifugation, the two bands of DNA are observed by shining UV light on the tube (Fig. 9.1). The natural fluorescence of ethidium bromide is enhanced 50 times by intercalation into DNA, and the bands glow a bright cherry red under UV light.

Lambda phage may also be partially purified by rapid techniques that remove cell debris and most contaminants. A more complete purification can be obtained by utilizing their unique density of 1.5 g/cm^3, which is halfway between the density of protein, 1.3, and the density of DNA, 1.7. The phage may be isolated by equilibrium density gradient centrifugation in which the density halfway between the top and bottom of the centrifuge tube is 1.5. They too may be easily observed in the centrifuge tube. They form a bluish band, which results from the preferential scattering of shorter wavelengths of light known as Tyndall effect. This same phenomenon is the reason the sky is blue and sunsets are red.

The Biology of Restriction Enzymes

In this section we digress into the biology of restriction enzymes and then return to their use in cutting DNA. A large number of enzymes have now been found that cut DNA at specific sites. For the most part the enzymes come from bacteria. These enzymes are called restriction enzymes because in the few cases that have been carefully studied, the DNA cleaving enzyme is part of the cell's restriction-modification system.

The phenomenon of restriction-modification in bacteria is a small-scale immune system for protection from infection by foreign DNA. In contrast to higher organisms in which identification and inactivation of invading parasites, bacteria, or viruses can be performed extracellularly, bacteria can protect themselves only after foreign DNA has entered their cytoplasm. For this protection, many bacteria specifically mark their own DNA by methylating bases on particular sequences with modifying enzymes. DNA that is recognized as foreign by its lack of methyl groups on these same sequences is cleaved by the restriction enzymes and then degraded by exonucleases to nucleotides. Less than one phage out of 10^4 wrongly methylated infecting phage is able to grow and lyse an *E. coli* protected by some restriction-modification systems. Bacteria further protect themselves from plant and animal DNA. Much plant and animal DNA is methylated on the cytosine in CpG sequences. Many strains of bacteria also contain enzymes that cleave DNA when it is methylated on specific positions.

Arber studied restriction of lambda phage in *E. coli* and found that *E. coli* strain C does not contain a restriction-modification system. Strain B has one restriction-modification system, and yet a different one recognizes and methylates a different nucleotide sequence in strain K-12. Phage P1 also specifies a restriction-modification system of its own, and this can be superimposed on the restriction-modification system of a host in which it is a lysogen.

Table 9.1 Plating Efficiencies of Phage Grown on *E. coli* C, K, and B and Plated on These Bacteria

	Plated on Strain		
Phage	C	K	B
λ-C	1	$<10^{-4}$	$<10^{-4}$
λ-K	1	1	$<10^{-4}$
λ-B	1	$<10^{-4}$	1

Recognition sequence

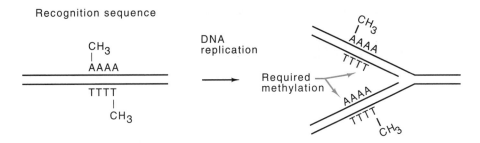

Figure 9.2 Methylation of an asymmetrical sequence necessitates recognition and methylation of two different sequences on the daughter strands on DNA replication.

Let the notation λ-C represent lambda phage that has been grown on *E. coli* strain C. Infection of strains B, K-12, and C with λ grown on the various strains yields different efficiencies of plaque formation (Table 9.1). Passage of the phage through a host of one type modifies the DNA so that it is recognized as "self" and plates at high efficiency if the phage reinfects that same strain. It is recognized as "foreign" and plates at low efficiency if the phage infects a strain with a different restriction-modification system.

Possession of a restriction-modification system introduces complexities to the process of DNA replication. Imagine that the double-stranded DNA contains methyl groups on both strands of the DNA at a recognition sequence. DNA replication creates a new duplex in which one of the strands in each of the daughter duplexes at first lacks the modification. This half-methylated DNA must not be recognized as foreign DNA and cleaved, but instead must be recognized as "self" and methylated (Fig. 9.2). Therefore, the restriction-modification system functions like a microcomputer, recognizing three different states of methylation of its recognition sequence and taking one of three different actions. If the sequence is unmethylated, the enzymes cleave. If the DNA is methylated on one of the two strands, the modification system methylates the other strand; if the DNA is methylated on both strands, the enzymes do nothing.

A palindromic recognition sequence streamlines operation of the restriction-modification system. A palindrome is a sequence that reads the same forward and backward, such as *repaper* and *radar*. Because DNA strands possess a direction, we consider a DNA sequence to be palindromic if it is identical when read 5' to 3' on the top strand and on the bottom strand (Fig. 9.3). Palindromes, of course, can be of any size, but most that are utilized as restriction-modification recognition sequences are four, five, six, and rarely, eight bases long. By virtue of the properties of palindromes, both daughter duplexes of replicated palindromic sequences are identical, and thus the modification enzyme needs to recognize and methylate only one type of substrate (Fig. 9.4). As we already saw in Fig. 9.2, the use of nonpalindromic recognition sequences would require that the modification enzyme recognize two

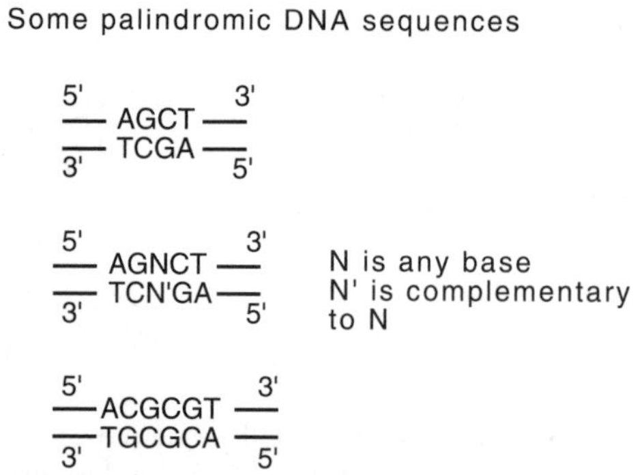

Some palindromic DNA sequences

$$\begin{array}{c} 5' 3' \\ \overline{}\text{AGCT}\overline{} \\ \overline{}\text{TCGA}\overline{} \\ 3' 5' \end{array}$$

$$\begin{array}{c} 5' 3' \\ \overline{}\text{AGNCT}\overline{} \\ \overline{}\text{TCN'GA}\overline{} \\ 3' 5' \end{array}$$ N is any base
N' is complementary
to N

$$\begin{array}{c} 5' 3' \\ \overline{}\text{ACGCGT}\overline{} \\ \overline{}\text{TGCGCA}\overline{} \\ 3' 5' \end{array}$$

Figure 9.3 Palindromic sequences. Because DNA strands possess an orientation, the reverse of the sequence is contained on the opposite strand. In palindromes with an odd number of bases, the central nucleotide is irrelevant.

different sequences. Presumably dimeric proteins are used to recognize the palindromic sequences.

Restriction enzymes are divided into three main classes. The enzymes in class I form a complex consisting of a cleaving subunit, a methylating subunit, and a sequence recognition subunit. These enzymes cleave at sites far distant from their recognition sequences and will not be further discussed here even though they were the first to be discovered. Those in class II possess their sequence recognition and cleaving activities together. They cleave in or near their recognition sequence and are of the most use in genetic engineering. The class III enzymes possess a cleavage subunit associated with a recognition and methylation subunit. These cleave near their recognition site.

A restriction enzyme within a cell is a time bomb because physical-chemical principles limit the enzyme's specificity for binding. If a restriction enzyme were to bind to a wrong sequence, and a typical bacterium contains about 4×10^6 such sequences, most likely the sequence would not be methylated and the enzyme could cleave. This would break the chromosome, and the cell would die. The experimental observation, however, is that cells containing restriction enzymes do not noticeably die any faster than cells without restriction enzymes. How, then, is the extraordinarily high specificity of the restriction enzymes generated?

Figure 9.4 Methylation of a palindromic sequence permits recognition and methylation of only one sequence during DNA replication.

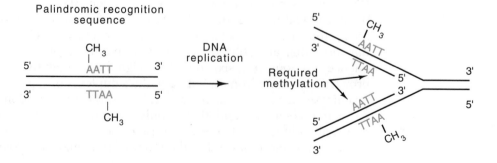

The requisite high specificity can be obtained if cutting the DNA duplex is a two-step process. An enzyme could bind to the recognition sequence, cleave one strand, wait a while, then cleave the other strand. This has the effect of utilizing the recognition sequence twice for each cleavage. If the enzyme binds at a site other than the recognition sequence, it rapidly dissociates before cleaving the second strand. Therefore, restriction enzymes are likely to produce nicks in the DNA at sites other than the recognition sequence, but these nicks can be repaired with DNA ligase and the cell will not be harmed in the process. Few restriction enzymes are likely to be found that cleave both strands of the DNA at an incorrect site in a concerted process.

Cutting DNA with Restriction Enzymes

The restriction enzymes provide a necessary tool for cutting fragments of DNA out of larger molecules. Their exquisite specificity permits very great selectivity, and because more than one hundred different restriction enzymes are known (Fig. 9.5), their wide variety permits much choice in the cleavage sites utilized. Often, fragments may be produced with end points located within 20 base pairs of any desired location.

One of the more useful properties of restriction enzymes for genetic engineering is found in the restriction-modification system produced by the *E. coli* plasmid R. The corresponding restriction enzyme is called *Eco*RI. Instead of cleaving at the center of its palindromic recognition sequence, this enzyme cleaves off-center and produces four base self-complementary ends. These "sticky" ends are most useful in recombi-

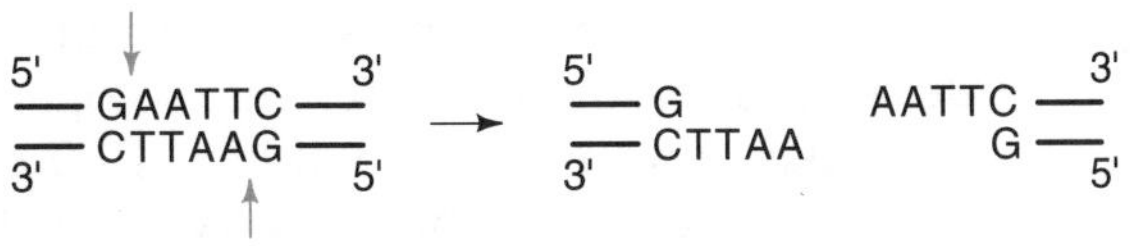

nant DNA work as they can be reannealed at low temperatures like the "sticky" ends of phage lambda. This permits efficient joining of DNA fragments during ligation steps. About half of the restriction enzymes now known generate overhanging or sticky ends. In some situations, a DNA fragment can even be arranged to have two different types of sticky

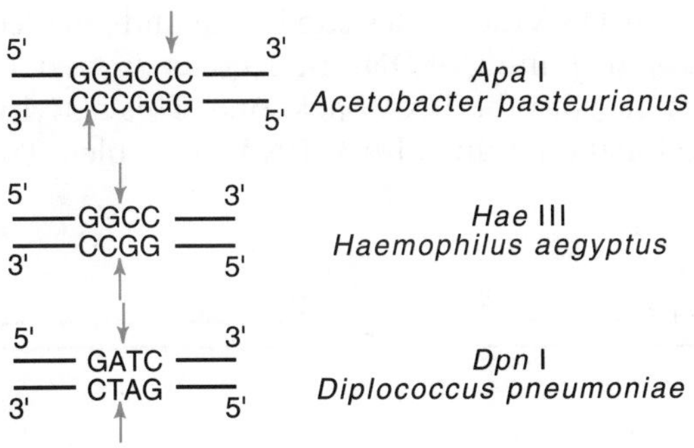

Figure 9.5 The recognition sites and cleavage sites of several restriction enzymes.

ends so that its insertion into another DNA can be forced to proceed in one particular orientation.

Isolation of DNA Fragments

Following cleavage of DNA by restriction enzymes or other manipulations to be discussed later, DNA fragments frequently must be isolated. Fortunately, fractionation according to size is particularly easy because, as discussed earlier, DNA possesses a constant charge-to-mass ratio and double-stranded DNA fragments of the same length have the same shape and therefore migrate during electrophoresis at a rate nearly independent of their sequence. Generally, the larger the DNA, the slower it migrates.

Remarkable resolution is obtainable in electrophoresis. With care, two fragments whose sizes differ by 0.5% can be separated if they lie within a range of 2 to 50,000 base pairs. No single electrophoresis run could possess such high resolution over this entire range. A typical range for adequate size separation might be 5 to 200 base pairs or 50 to 1,000 base pairs, and so on. The material through which the DNA is electrophoresed must possess special properties. It should be inexpensive, easily used, uncharged, and it should form a porous network. Two materials meet the requirements: agarose and polyacrylamide.

Following electrophoresis, bands formed by the different-sized fragments may be located by autoradiography if the DNA had been radiolabeled before the separation. Usually $^{32}PO_4$ is a convenient label because phosphate is found in RNA and DNA, ^{32}P emits particularly energetic electrons making them easily detectable, and finally ^{32}P has a short half-life so that most of the radioactive atoms in a sample will decay in a reasonable time. The isotope ^{33}P is also used. Its beta decay is weaker, and it has a half-life of 90 days. Often, sufficient DNA is present that it may be detected directly by staining with ethidium bromide. The enhanced fluorescence from ethidium bromide intercalated in the DNA compared to its fluorescence in solution permits detection of as little as 5 ng of DNA in a band. After the electrophoretic separation and detection of the DNA, the desired fragments can be isolated from the gel.

Joining DNA Fragments

Having discussed how DNA molecules can be cut and purified, it is now necessary to discuss the joining of DNA molecules. *In vivo*, the enzyme DNA ligase repairs nicks in the DNA backbone. This activity may also be utilized *in vitro* for the joining of two DNA molecules. Two require-

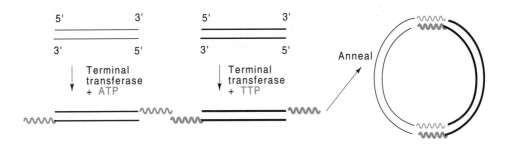

Figure 9.6 Joining two DNA fragments by poly-dA and poly-dT tails.

ments must be met. First, the molecules must be the correct substrates, that is, they must possess 3'-hydroxyl and 5'-phosphate groups. Second, the groups on the molecules to be joined must be properly positioned with respect to one another. The method for generating the proper positioning has two variations: either to hybridize the fragments together via their sticky ends or, if flush-ended fragments are to be joined, to use such high concentrations of fragments that from time to time they are spontaneously in the correct positions.

Hybridizing DNA fragments that possess self-complementary, or sticky ends, generates the required alignment of the DNA molecules. Many restriction enzymes such as *Eco*RI produce four-base sticky ends that can be ligated together after the sticky ends of the pieces to be joined have hybridized together. Because the sticky ends are usually just four base pairs, lowering the temperature during ligation to about 12°C facilitates the hybridization-ligation process.

The flush ends of DNA molecules that are generated by some restriction enzymes generate problems. One solution is to convert flush-ended molecules to sticky-ended molecules by enzyme terminal transferase. This enzyme adds nucleotides to the 3' end of DNA. Poly-dA tails can be put on one fragment and poly-dT tails can be added to the other fragment (Fig. 9.6). The two fragments can then be hybridized together by virtue of their self-complementary ends and ligated together. If the tails are long enough, the complex can be directly introduced into cells, where the gaps and nicks will be filled and sealed by the cellular enzymes. More commonly, the polymerase chain reaction as described in the next chapter would be used to generate any desired ends on the molecules.

Flush ended molecules can also be joined directly with DNA ligase. While this method is straightforward, it suffers from two drawbacks: It requires high concentrations of DNA and ligase for the reaction to proceed, and even then the ligation efficiency is low. Also, it is difficult later to excise the fragment from the vector.

Linkers can also be used to generate self-complementary single-stranded molecules (Fig. 9.7). Linkers are short, flush-ended DNA molecules containing the recognition sequence of a restriction enzyme that produces sticky ends. The ligation of linkers to DNA fragments

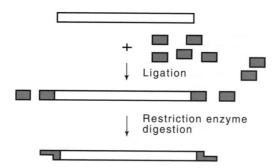

Figure 9.7 Addition of linkers by ligation and their conversion to sticky ends by restriction enzyme digestion.

proceeds with reasonably high efficiency because high molar concentrations of the linkers may easily be obtained. After the linkers have been joined to the DNA segment, the mixture is digested with the restriction enzyme, which cuts the linkers and generates the sticky ends. In this way a flush-ended DNA molecule is converted to a sticky-ended molecule that may easily be joined to other DNA molecules.

Vectors: Selection and Autonomous DNA Replication

Cloning a piece of DNA requires that it be replicated when it is put back into cells. Hence the DNA to be cloned must itself be an independent replicating unit, a replicon, or must be joined to a replicon. Additionally, since the efficiency of introduction of DNA into cells is well below 100%, cells that have taken up DNA, and are said to have been transformed, need to be readily identifiable. In fact, since only about one bacterial cell out of 10^5 is transformed, selections must usually be included to permit only the transformed cells to grow.

Vectors must fulfill the two requirements described above, replication in the host cell and selection of the cells having received the transforming DNA. As mentioned earlier, two basic types of vectors are used: plasmids and phage. Plasmids contain bacterial replicons that can coexist with the normal cellular DNA and at least one selectable gene. Usually it is a gene conferring resistance to an antibiotic. Phage, of course, carry genes for replication of their DNA. Since DNA packaged in a phage coat can enter cells effectively, selectable genes on the phage usually are unnecessary.

Plasmid Vectors

Most plasmids are small circles that contain the elements necessary for DNA replication, one or two drug-resistance genes, and a region of DNA into which foreign DNA may be inserted without damage to essential plasmid functions. One widely used plasmid, pBR322, carries genes coding for resistance to tetracycline and β-lactamase (Fig. 9.8). The latter confers resistance to penicillin and related analogs by cleaving the drugs in the lactam ring, which renders them biologically inactive. Genes conferring resistance to chloramphenicol, tetracycline, and

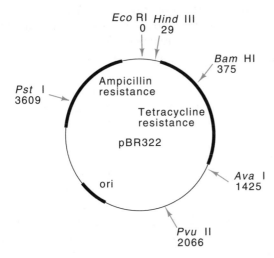

Figure 9.8 A map of the plasmid pBR322 showing the ampicillin and tetracycline resistance genes, the origin, and a few restriction enzyme cleavage sites.

kanamycin are other selectable drug-resistance markers commonly carried on plasmids.

A useful element to have on plasmids is a DNA replication origin from a single-stranded phage. When such an origin is activated by phage infection, the cell synthesizes sizeable quantities of just one strand of the plasmid. This facilitates DNA sequencing. In a typical cloning experiment, a plasmid is cut in a nonessential region with a restriction enzyme, say *Eco*RI, foreign *Eco*RI-cut DNA is added, and the single-stranded ends are hybridized together and ligated. Only a small fraction of the plasmids subjected to this treatment will contain inserted DNA. Most will have recircularized without insertion of foreign DNA. How can transformants whose plasmids contain inserted DNA be distinguished from plasmids without inserted DNA? Of course, in some conditions a genetic selection can be used to enable only transformants with the desired fragment of inserted DNA to grow. More often this is not possible and it becomes necessary to identify candidates that contain inserted DNA.

One method for identifying candidates relies on insertional inactivation of a drug-resistance gene. For example, within the ampicillin-resistance gene in pBR322 exists the only plasmid cleavage site of the restriction enzyme *Pst*I. Fortunately *Pst*I cleavage generates sticky ends and DNA may readily be ligated into this site, whereupon it inactivates the ampicillin-resistance gene. The tetracycline-resistance gene on the plasmid remains intact and can be used for selection of the cells transformed with the recombinant plasmid. The resulting colonies can be tested by spotting onto a pair of plates, one containing ampicillin, and one not containing ampicillin. Only the ampicillin-sensitive, tetracycline-resistant transformants contain foreign DNA in the plasmid. The ampicillin-resistant transformants derive from plasmid molecules that recircularized without insertion of foreign DNA.

Another way to screen for the insertion of foreign DNA utilizes the β-galactosidase gene. Insertion of foreign DNA within the gene inactivates the enzyme, which can be detected by plating transformed cells on medium that selects for the presence of the plasmid and also contains substrates of β-galactosidase that produce colored dyes when hydrolyzed. A plasmid would be unwieldy if it contained the complete 3,000 base-pair β–galactosidase gene. Therefore, only a particular N-terminal portion of the gene is put on the plasmid. The remainder of the enzyme is encoded by a segment inserted into the chromosome of the host cells. The two portions of the gene synthesize domains that bind to one another and yield active enzyme. This unusual phenomenon is called α-complementation. Cloning vectors are designed for the insertion of foreign DNA into the short, N-terminal part of β-galactosidase.

A simple technique can greatly reduce recircularization of vector molecules that lack inserted foreign DNA. If the vector DNA is treated with a phosphatase enzyme after cutting with the restriction enzyme, then recircularization is impossible because the 5'-PO$_4$ ends required by DNA ligase are absent. Foreign DNA, however, will carry 5'-PO$_4$ ends,

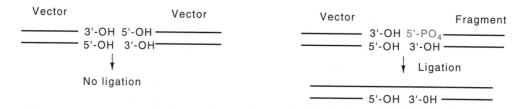

and therefore two of the four breaks flanking a fragment of foreign DNA can be ligated. This DNA is active in transformation because cells repair the nick remaining at each end of the inserted fragment.

Plasmids or phage cloning vectors can contain a short stretch of DNA containing unique cleavage sites for a number of restriction enzymes. These polylinker regions permit cleavage by two enzymes so that the resulting sticky ends are not self-complementary. The vector cannot close by itself and be religated; only when a DNA fragment containing the necessary ends hybridizes to the ends of the plasmid can the structure be ligated into a closed circle (Fig. 9.9).

Efficient genetic engineering necessitates that plasmid DNA be obtainable in high quantities. Some plasmids maintain only three or four copies per cell, whereas other plasmids have cellular copy numbers of 25 to 50. Many of the high-copy-number plasmids can be further amplified because the plasmid continues to replicate after protein synthesis and cellular DNA synthesis have ceased due to high cell densities or the presence of protein synthesis inhibitors. After amplification, a cell containing such a relaxed-control plasmid may contain as many as 3,000 plasmid copies.

Nature did not deliver plasmid vectors ready-made for genetic engineering. The most useful plasmid vectors were themselves constructed

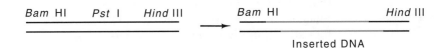

Inserted DNA

Figure 9.9 Cleavage in a polylinker region by two different restriction enzymes can create sticky ends that will not hybridize to themselves, but can only hybridize upon the entry of a foreign DNA molecule containing the appropriate sticky ends.

by genetic engineering. The starting materials were R plasmids, which are plasmids or autonomously replicating DNA elements that carry one or more drug-resistance genes. The R plasmids are the cause of serious medical problems, for various bacteria can acquire R plasmids and thereby become resistant to the normal drugs used for treatment of infections (Table 9.2). The conversion of an R plasmid into a useful vector requires elimination of extraneous DNA and removal of multiple restriction enzyme cleavage sites. In order that foreign DNA can be cloned into the plasmid, the plasmid should possess only one cleavage site for at least one restriction enzyme, and this should be in a nonessential region.

In the construction of cloning vectors, R plasmids were digested with various restriction enzymes. The resulting mixture of DNA fragments was hybridized together via the self-complementary ends and then ligated to produce many combinations of scrambled fragments. This DNA was transformed into cells. Only new plasmids containing at least the DNA segments necessary for replication and drug resistance survived and yielded colonies. The desirable plasmids containing only single cleavage sites for some restriction enzymes could be identified by amplification and purification of the DNA followed by test digestions with restriction enzymes and electrophoresis to characterize the digestion products. The plasmid pBR322 possesses single restriction enzyme cleavage sites for more than twenty enzymes. Some of the more commonly used are *Bam*HI, *Eco*RI, *Hind*III, *Pst*I, *Pvu*II, and *Sal*I.

Table 9.2 Various R-Factors Found in *E. coli*

R Plasmid	Antibiotic Resistances*	Molecular Weight $\times 10^9$
R1	Am, Cm, Km, Sm, Su	65
R6	Cm, Km, Sm, Su, Tc	65
R6K	Am, Sm	26
R28K	Am	44
R15	Su, Sm	46

* Am = ampicillin; Cm = chloramphenicol; Km = kanamycin; Sm = streptomycin; Su = sulfonamides; Tc = tetracycline

A Phage Vector for Bacteria

Phage vectors and phage-derived vectors are useful for three reasons. Phage can carry larger inserted DNA fragments than plasmids. Therefore substantially fewer transformed candidates must be examined to find a desired clone. The efficiency of infecting repackaged phage DNA into cells is considerably greater than the efficiency of transforming plasmid DNA into cells. This is an important factor when a rare clone is being sought. Finally, lambda phage permits a convenient method for screening to detect the clone carrying the desired gene. Once a desired segment of DNA has been cloned on a phage, however, the convenience of manipulating plasmids, due in part to their smaller size, dictates that the segment be subcloned to a plasmid.

Lambda was an ideal choice for a phage vector because it is well understood and easy to work with. Most importantly, the phage contains a sizable nonessential internal region flanked by *Eco*RI cleavage sites. Therefore this nonessential region could be removed and foreign DNA inserted. Before *Eco*RI-cleaved lambda DNA could be used for cloning, it was necessary to eliminate additional cleavage sites that are located in essential regions of the lambda genome. First, a lambda hybrid phage

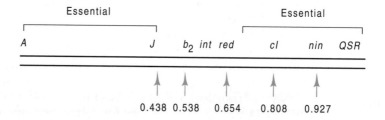

was constructed by *in vivo* genetic recombination. This lacked the three *Eco*RI cleavage sites at 0.438, 0.538, and 0.654 in the nonessential region but retained the two sites in the essential regions. Then the two remaining *Eco*RI cleavage sites were eliminated by mutation and selection. In the selection the phage were cycled between hosts lacking and containing the *Eco*RI restriction-modification systems. Any phage with cleavage sites mutated so as to be nonrecognizable by the *Eco*RI system would have a greater probability of escaping the restriction enzymes upon growth in the second host. After 10 to 20 cycles of this selection scheme, Davis found a phage that had lost one of the two R1 sites, and after an additional 9 to 10 cycles, he found a mutant that had lost the other site. This mutant phage was then made into a useful cloning vector by recombination with wild-type lambda to restore the three *Eco*RI cleavage sites located in the central portion of the genome.

The DNA isolated from lambda phage particles is linear and it can be cleaved by *Eco*RI to yield the smaller central fragments and the larger left and right arms. *Eco*RI-cleaved DNA fragments to be cloned can then be hybridized together and ligated to purified right and left arms. This DNA either can be used as it is to transfect cells made competent for its uptake or it can be packaged *in vitro* into phage heads and used to infect cells. The efficiency, per DNA molecule, of packaging and infection is much higher than transfecting with bare DNA. Therefore packaging is used when the fragment to be cloned is present in only a few copies.

Vectors for Higher Cells

Cloning DNA in higher cells poses the same problems as cloning in bacteria. The vectors must permit simple purification of sizable quantities of DNA, must permit selection of transformed cells, and must have space for inserted DNA. Shuttle vectors, which have been extensively used for cloning in yeast, are a neat solution to these requirements. In addition to containing the normal bacterial cloning-plasmid elements, they contain a yeast replicon and a genetic marker selectable in yeast (Fig. 9.10). As a result, large quantities of the vectors can be obtained by growth in *E. coli* and then transformed into yeast. The ability to shuttle between bacteria and yeast saves much time and expense in genetic engineering experiments.

Two types of yeast replication origins can be used in yeast shuttle vectors. One is a yeast chromosomal DNA replication origin, also known as an ARS element. The other is the origin from the 2 μ circles. These are plasmid-like elements with unknown function that are found in yeast. They are somewhat more stable than the ARS vectors. Nutritional markers such as uracil, histidine, leucine, and tryptophan biosynthesis have been used as selectable genes in the appropriately auxotrophic yeast.

Viruses form a basis for many vectors useful in higher plant and animal cells. For example, one of the simplest vectors for mammalian cells is the simian virus SV4O. It permits many of the same cloning operations as phage lambda.

The terminology used with mammalian cells can be confusing. "Transformation" can mean that cells have received a plasmid. It can also mean that the cells have lost their contact inhibition. In this state they continue growing past the confluent cell monolayer stage at which

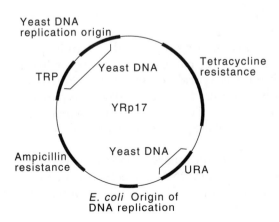

Figure 9.10 The structure of a vector for shuttling between *E. coli* and yeast. It contains genes permitting DNA replication and selection in both organisms.

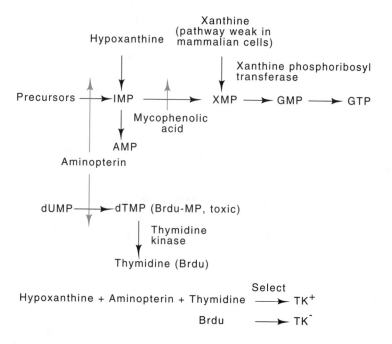

Figure 9.11 The metabolic pathways involved with some selectable genes in mammalian cells. IMP-inosine monophosphate, XMP-xanthine monophosphate, GMP-guanosine monophosphate, dUMP-deoxyuridine monophosphate, dTMP-deoxythymidine monophosphate, Brdu-bromodeoxyuridine, Brdu-MP, bromodeoxyuridine monophosphate. Aminopterin blocks tetrahydrofolate reductase, which is required for synthesis of IMP and dTMP, and mycophenolic acid blocks synthesis of XMP.

normal mammalian cells cease growth. Transformation to the uninhibited growing state can result from infection by a tumor-causing virus like SV40 or it can be a result of mutation of the genome. Although loss of contact inhibition could be useful in identifying cells that have incorporated the SV40 DNA or SV40 hybrids, this property is of limited use. Other selectable genetic markers suitable for mammalian cells are required.

One useful gene for selections in mammalian cells has been the thymidine kinase gene because TK^+ cells can be selected by growing them in medium containing hypoxanthine, aminopterin, and thymidine. Conversely, TK^- cells can be selected by growing them in medium containing bromodeoxyuridine (Fig. 9.11). Furthermore, virologists had previously discovered that the herpes simplex virus codes for its own thymidine kinase. Therefore the viral genome can be used as a concentrated source of the gene in an expressible form for initial cloning experiments.

Although the thymidine kinase gene has been useful in selecting cells that have taken up foreign DNA, a selectable gene that does not require the prior isolation of a thymidine kinase negative mutant in each cell line would also be valuable. The *E. coli* enzyme xanthine-guanine phosphoribosyl transferase gene appears to meet these requirements. The protein product of the gene functions in mammalian cells and permits selective growth of nonmutant cells that contain the enzyme (Fig. 9.11). The required growth medium contains xanthine, hypoxanthine, aminopterin, and mycophenolic acid. Other dominant genes useful for the selection of transformed cells are mutant dihydrofolate reductase that is resistant to methotrexate, a potent inhibitor of the wild-type enzyme, and kanamycin-neomycin phosphotransferase. The latter is an enzyme derived from a bacterial transposon and confers resistance to bacteria, yeast, plant, and mammalian cells to a compound called G418 . Of course, for proper expression in the higher cells the gene must be connected to an appropriate transcription unit and must contain the required translation initiation and polyadenylation signals.

Putting DNA Back into Cells

After the DNA sequence to be cloned has been joined to the appropriate vector, the hybrid must be transformed into cells for biological amplification. The phenomenon of transformation of pneumococci by DNA has been known since 1944. Once the desirable genetic properties of *E. coli* were realized, it too was tried in transformation experiments. These were unsuccessful for many years. Unexpectedly, a method for transforming *E. coli* was discovered. This occurred at a most opportune time, for the developments in the enzymology of DNA cutting and joining were almost ready to be used in a system of putting the foreign DNA into cells. Reintroducing a DNA molecule containing a replicon into a cell permits a biological amplification of that cell to greater than 10^{12}. In one day, a single molecule can be amplified to quantities sufficient for physical experiments.

The key to the initial transformation protocols of *E. coli* was treatment of the cells with calcium or rubidium ions to make them competent for the uptake of plasmid or phage DNA. The term for transformation with phage DNA that then yields an infected cell is transfection, and this term is used for infection of higher cells with nonvirus DNA as well. Yeast may be transfected after a treatment that includes incubation in the presence of lithium ions. Some types of mammalian cells, mouse L cells for example, can be transfected merely by sprinkling them with a mixture of the DNA and calcium-phosphate crystals. Here the mechanism of transfection appears to be uptake of the DNA-calcium-phosphate complex.

Direct manual injection of small volumes of DNA into cells has been highly useful for the study of cloned DNA fragments because it eliminates the need for a eukaryotic replicon or a selectable gene. Microinjection into the oocytes of the frog *Xenopus laevis* has yielded much information, and microinjection into cultured mammalian cells is also possible. DNA injected into *Xenopus* cells is transcribed for many hours and translated into easily detectable amounts of protein. As a result of these properties, a DNA segment can be cloned onto a plasmid such as pBR322, manipulated *in vitro*, and injected into the cells for examination of its new biological properties. Microinjection is also possible into a fertilized mouse embryo. The embryo can then be reimplanted to develop into a mouse. Since the fragments of injected DNA will recombine into the chromosome, the mice are transfected by the DNA. In order that all cells in a transfected mouse possess the same genetic constitution, an injected fragment must recombine into a germ line cell. Such a mouse is unlikely to be genetically homogeneous because similar fragments probably have not integrated into somatic cells. The offspring of such a mouse will be genetically homogeneous, however, and these can be profitably studied.

Another general method for incorporating DNA into cells is electroporation. Cells are subjected to a brief but intense electric field. This creates small holes in their membranes and for a short time DNA molecules present in the solution can be taken up.

Cloning from RNA

Although DNA can be extracted from cells and used in cloning steps, sometimes RNA is a better starting choice for cloning. Not only will intervening sequences be missing from mRNA, but often mRNA extracted from certain tissues is greatly enriched for specific gene sequences.

Extraction of RNA from cells yields a preponderance of ribosomal RNA. The messenger RNA can easily be separated from this ribosomal RNA since most messenger RNA from most higher organisms contains a poly-A tail at the 3' end. This tail can be used for isolation by passing a crude fraction of cellular RNA through a cellulose column to which poly-dT has been linked. At high salt concentrations, the poly-dT which is linked to the column and the poly-A tails of messenger molecules hybridize and bind the messenger RNAs to the column. The ribosomal RNA molecules flow through the column. The messenger RNAs are eluted by lowering the salt concentration so as to weaken the polyA-dT hybrids. Such a purification step frequently provides a several-hundred-fold enrichment for messenger RNA. The effort required to clone a specific gene is often greatly reduced by using this procedure in conjunction with choosing a particular tissue at a particular developmental time.

The single-stranded RNA obtained by the steps described above cannot be cloned directly. Either the RNA can be converted to DNA via a complementary strand, cDNA, or the RNA can be used to aid identification of a clone containing the complementary DNA sequence. To generate a cDNA copy of the poly-A-containing messenger, several steps are performed (Fig. 9.12). First, a poly-dT primer is hybridized to the messenger and reverse transcriptase is used to elongate the primer to yield a DNA copy. This enzyme, which is found within the free virus particle of some animal viruses, synthesizes DNA using an RNA template. At this point the sequence exists as an RNA-DNA hybrid. It is converted to a DNA duplex by the simultaneous incubation with RNAse H which cuts the RNA strand in an RNA-DNA duplex and DNA polym-

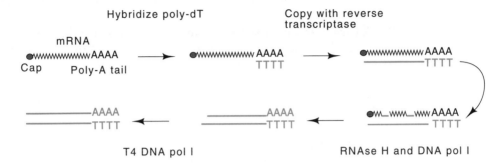

Figure 9.12 Steps in the replication of an mRNA molecule into a double-stranded DNA copy suitable for cloning.

erase pol I which synthesizes DNA using the remaining RNA as primer. DNA pol I removes the remaining RNA by nick translation. Finally, to make the ends of the DNA duplex perfectly blunt, T4 DNA polymerase pol I is added. The resulting double-stranded DNA can then be cloned by methods already discussed.

Plaque and Colony Hybridization for Clone Identification

Many techniques have been devised for the detection of the desired clones. One of the simplest is genetic selection (Fig. 9.13). Most often this simple avenue is not available, but instead one possesses a related sequence. Sometimes this related sequence is from a gene analogous to the desired gene, but from a different organism. Other times, partial amino acid sequences are known or good guesses can be made as to the amino acid sequence. From such sequences, the DNA sequences can be

Figure 9.13 Cloning DNA from *E. coli* in which a selection for X$^+$ transformants permits direct selection of the desired clones.

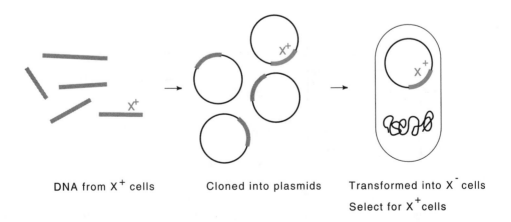

inferred. In such situations the hybridization ability of complementary DNA sequences can be used to identify cells containing the desired clone.

A variety of direct screening techniques exist for detecting a cloned gene. In these approaches, cloning the foreign DNA into a lambda vector is convenient because the phage can accommodate inserted fragments of large size and many lambda phage may be screened on a single petri plate. The candidate phage collection is called a lambda bank or library. It is plated out at thousands of plaques per plate. Then a replica copy of the phage plaques is made by pressing a filter paper onto the plate. After the paper is removed, it is immersed in alkali. By these steps the DNA is denatured and fixed onto the paper. Then radioactively labeled RNA or DNA, called a probe, is hybridized to any complementary DNA sequences in plaque images on the paper. The probe contains the known sequence derived from a previously isolated clone or a sequence deduced from amino acid sequence information. The location of the areas with bound probe are then determined by autoradiography, and viable phage containing the desired insert can be isolated from the corresponding position on the original plate. Analogous techniques exist for screening colonies containing plasmids.

Walking Along a Chromosome to Clone a Gene

Another method of cloning a desired gene is walking. Suppose a randomly isolated clone has been shown to lie within several hundred thousand bases of the DNA segment we wish to clone. Of course, demonstrating such a fact is not trivial in itself. In the case of the fruit fly *Drosophila melanogaster*, however, the demonstration is straightforward.

As explained earlier, the polytene chromosomes in cells of several *Drosophila* tissues possess characteristic bands that serve as chromoso-

Figure 9.14 (a) A gel showing DNA fragments resulting from a restriction enzyme digestion of a set of four partially overlapping DNA sequences. (b) The four DNAs and their composite.

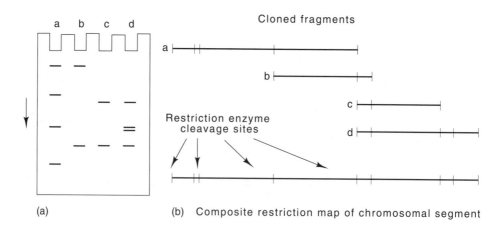

(a)

(b) Composite restriction map of chromosomal segment

mal landmarks. Therefore the DNA involved in chromosome rearrangements or deletions can readily be observed. Also, highly radioactive DNA originating from a cloned segment of *Drosophila* chromosome can be hybridized *in situ* to *Drosophila* polytene chromosomes. The position to which the fragment hybridizes can then be determined by autoradiography. Hence the chromosomal position of a cloned fragment can be determined. If it lies near a gene of interest, walking may be in order.

A chromosomal walk to clone a specific gene begins with a restriction map of the cloned piece of DNA. The right- and left-end terminal restriction fragments are used as probes of a lambda bank. Several clones are picked that hybridize to the right-hand fragment and not to the left-hand fragment. Then restriction maps of these clones are made and again the right- and left-hand fragments are used to find new clones that hybridize only to the right-hand fragments (Fig. 9.14). The successive lambda transducing phage that are identified permit walking to the right, and when a sufficient distance has been covered, *in situ* hybridization permits determination of which direction on the cloned DNA corresponds to moving toward the desired gene.

Arrest of Translation to Assay for DNA of a Gene

In vitro translation of mRNA forms the basis of one technique for identification of clones containing DNA of a particular gene. The technique requires that the enriched messenger RNA of the gene, such as the mRNA obtained from an oligo-dT column, be translatable *in vitro* to yield a detectable protein product of the desired gene. To perform the identification, DNA for a candidate clone is denatured and hybridized to the RNA used in the translation mixture. If the DNA contains sequences complementary to the mRNA, the two will hybridize together, the messenger will become unavailable for *in vitro* translation, and the gene product will not be synthesized (Fig. 9.15). DNA from a clone not possessing the DNA sequence of the gene will not interfere

Figure 9.15 Hybridization arrest of translation as an assay for DNA of a gene whose mRNA can be translated into protein.

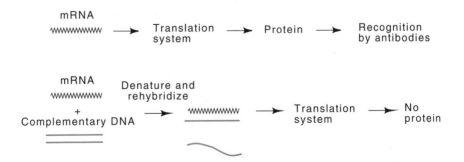

with translation of the messenger. Hence DNA carrying sequences complementary to the messenger can be detected by hybridization arrest of translation.

Chemical DNA Sequencing

Two techniques were developed for DNA sequencing: chemical and enzymatic. Initially more DNA was sequenced with the chemical method, but improvements in the enzymatic method now make it the method of choice for almost all sequencing problems. The application of the chemical methods, however, are highly useful for biochemical studies of protein-DNA interactions, and therefore also find significant use. Both the chemical and the enzymatic sequencing methods utilize electrophoresis at high temperatures in the presence of urea so as to denature the DNA. Under these conditions the single-stranded molecules migrate at rates almost independent of their sequence and dependent only on their length. In such gels two single-stranded DNA fragments differing in length by a single base can be resolved from one another if their lengths are less than 300 to 500 bases.

The basic principle of DNA sequencing by the chemical or enzymatic method is to generate a set of radioactive DNA fragments covering a region. The sizes of these fragments indicate the nucleotide sequence of the region. How is this possible? Consider a large number of single-stranded molecules (Fig. 9.16). Suppose the 5' ends of all the molecules are at the same nucleotide. Suppose also that the 3' ends of some of the molecules in the population are at the first A residue after the 5' end, some end at the second A residue and so on. That is, the population of molecules consists of some ending at each of the A's. If the first few A

Figure 9.16 DNA sequencing. Modification at a average of one A per molecule in the population followed by cleavage at the modification sites generates a population of molecules ending at the former positions of A's. Their electrophoresis generates the ladder pattern indicated.

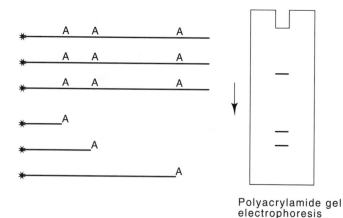

Polyacrylamide gel
electrophoresis

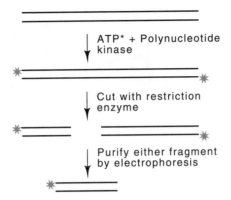

Figure 9.17 A protocol for labeling DNA on one end.

residues were located four, five, and nine nucleotides from the original 5' end, then fragments would be four, five, and nine nucleotides long. The sizes of these fragments could be determined by electrophoresis.

In the same way that A-specific ends could be used to determine the distances of A residues from the end, other base-specific terminations could be used to determine the distances of the other nucleotides from the 5' end.

Practical considerations slightly complicate the above procedure. First, the amounts of DNA that must be identified are too small to be detected by staining with ethidium bromide and observing the fluorescent DNA bands. Second, the chemical method generates extraneous fragments in addition to those originating from the one end under consideration. These other fragments could interfere with the identification of the desired fragments. Both problems are solved by radioactively labeling the DNA at its 5' end. Autoradiography of the gels then provides adequate sensitivity, and only the fragments that are radioactive and generate bands on the autoradiographs are precisely those that include the labeled end.

Obtaining a DNA fragment labeled only at one end is not difficult. Each strand in a double-stranded segment can be labeled on its 5' end with the enzyme T4 phage polynucleotide kinase. Then the two strands can be separated by denaturation of the DNA and electrophoresis under partially denaturing conditions. Frequently the two strands migrate at different rates and can be separated. If so, either can be used in chemical sequencing. A second way to obtain the DNA fragment with a radioactive label on only one end also begins with end-labeled double-stranded duplex (Fig. 9.17). This duplex is next cleaved with a restriction enzyme and if the two fragments are unequal in size, they can be separated by electrophoresis. Either of the fragments is suitable for DNA sequencing because only one of the strands in each fragment is then radioactive, and it is labeled on only one end.

To sequence a stretch of DNA, four populations of radiolabeled DNA are made, one partially cleaved at each of the four bases. These four populations are then subjected to electrophoresis in four adjacent lanes

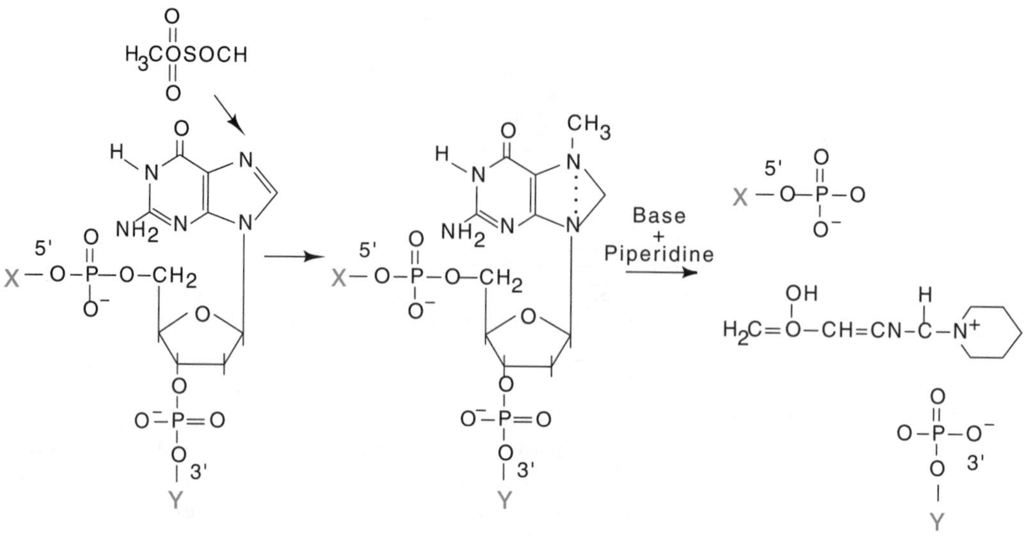

Figure 9.18 Basis of the G-specific reaction in Maxam-Gilbert sequencing. The final result is strand scission at the former position of a guanosine.

on the denaturing gels. The resulting four ladder patterns permit direct reading of the sequence of the bases.

In the chemical sequencing method the cleavages at the four bases are made by subjecting the labeled DNA to conditions that generate base-specific cleavage of the phosphodiester bonds. An average of about one cleavage per several hundred bases is optimal for most sequencing. Maxam and Gilbert discovered that to generate the requisite base specificity, the procedure needed to be broken into two parts. The first part introduces a highly base-specific chemical modification under controlled and mild conditions. Then harsh conditions are used to generate the actual cleavages at all the modified positions.

Dimethylsulfate provides a highly specific methylation of guanines (Fig. 9.18). The introduction of the methyl group permits subsequent depurination followed by cleavage of the sugar-phosphate backbone with piperidine. Slightly altered conditions yield methylation by dimethyl sulfate of guanines, and to a lesser extent, methylation of adenines. Hydrazine is used for pyrimidine-specific reactions. Both thymine and cytosine react with hydrazine unless high concentrations of salt are present. Then only cytosine reacts. These reactions yield bands for G's, G's + A's, C's + T's, and T's. These are sufficient for sequence determination.

Consider, for example, that the DNA sequence near the 5' end of the fragment was 5'-GTCAAG-3' and the fragment was labeled on its 5' end. Then electrophoresis of the four reaction products yields a ladder pattern from which the sequence may be read by proceeding upward on the gel from lane to lane (Fig. 9.19). Sequences of up to about 200 bases may be read from a single set of reactions.

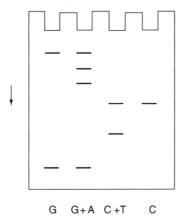

Figure 9.19 A sequencing gel in which G, G + A, C + T, and C-specific reactions were used. The sequence reads G-T-C-A-A-G.

G   G+A   C +T   C

Enzymatic DNA Sequencing

Sanger and his colleagues developed an enzymatic method for generating the set of DNA fragments necessary for sequence determination. Utilization of biological tricks now makes this method particularly efficient. When coupled with techniques to be discussed in the next chapter, thousands of nucleotides per day per person can be sequenced. The Sanger method relies on the fact that elongation of a DNA strand by DNA polymerase cannot proceed beyond an incorporated 2',3'-dideoxyribonucleotide. The absence of the 3'-OH on such an incorporated nucleotide prevents chain extension. A growing chain therefore termi-

nates with the incorporation of a dideoxyribonucleotide. Thus, elongation reactions are performed in the presence of the four deoxyribonucleotide triphosphates plus sufficient dideoxyribonucleotide triphosphate that about one of these nucleotides will be incorporated per hundred normal nucleotides. The DNA synthesis must be performed so that each strand initiates from the same nucleotide and from the same strand. This is accomplished first by using only a single-stranded template and second by hybridizing an oligonucleotide to the DNA to provide the priming 3'-OH necessary for DNA elongation by DNA polymerases.

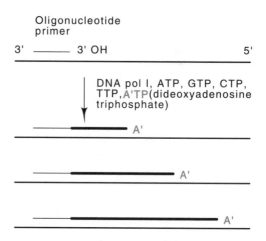

Figure 9.20 The Sanger sequencing method. Primer extension by DNA pol I in the presence of dideoxyadenosine triphosphate, A'TP, creates a nested set of oligomers ending in A'.

In sequencing, four different elongation reactions are performed, each containing one of the four dideoxyribonucleotides. Radioactive label is introduced to the fragments by the use of either labeled nucleotides or labeled primer. The fragments generated in the elongation reactions are analyzed after electrophoresis in the same manner as the fragments generated by chemical cleavage (Fig. 9.20).

Viruses like M13 and f1 that encapsidate only one strand are one source of pure single-stranded DNA. Purification of this material from as little as 2 milliliters of growth medium yields sufficient DNA for sequencing. M13 has no rigid structural limitations on the length of viral DNA that can be encapsidated because its DNA is packaged as a rod during its extrusion through the cell membrane. Therefore large and small fragments may be inserted into nonessential regions of the virus genome. Although the viral form of the virus is single-stranded, the genetic engineering of cutting and ligating is done with the double-stranded intracellular replicative form of the virus DNA. This DNA can then be transformed into cells just as plasmid DNA is.

Instead of cloning the DNA to be sequenced into the virus, it is simpler to clone the virus replication origin into a plasmid vector that contains a nonviral origin as well. Then, cloning and normal genetic engineering operations can be performed on the plasmid DNA, and when sequencing is to be done, the cells can be infected with virus. Under the influence of the virus encoded proteins, the viral replication origin on the plasmid directs the synthesis of plus strand DNA. This is then isolated and used in sequencing or purified and used in other operations.

The single-stranded DNA template necessary for Sanger sequencing may also be obtained by denaturing double-stranded DNA and preventing its rehybridization. Because partial rehybridization, often not in register, can still occur in many positions, this method works best with the use of a DNA polymerase capable of replicating through such obstacles. A DNA polymerase encoded by phage T7 is helpful in this regard. Either a chemical modification of the wild-type enzyme or a special mutant enzyme makes the T7 DNA polymerase highly processive and capable of replicating through regions containing significant amounts of secondary structure. The DNA pol I from *E. coli* or the Klenow fragment of this enzyme are adequate for sequencing from pure single-stranded DNA.

Needless to say, the vast explosion in the amount of sequenced DNA has stimulated development of highly efficient methods for analysis. Computer programs have been written for comparing homology between DNA sequences, searching for symmetrical or repeated sequences, cataloging restriction sites, writing down the amino acid sequences resulting when the DNA is transcribed and translated into protein, and performing other time-consuming operations on the sequence. The amount of information known about DNA and the ease in manipulating it has greatly extended our understanding of many biological processes and is now becoming highly important in industrial and medical areas as well.

Problems

9.1. What is a reasonable interpretation of the fact that while many different strains of bacteria possess virtually identical enzyme activities for the metabolic pathways, the nucleotide sequence specificities of their restriction-modification systems are drastically different?

9.2. What is the transformation efficiency per DNA molecule if 5×10^5 transformants are produced per microgram of 5,000-base-pair plasmid?

9.3. How could it be determined whether transformation requires the uptake of more than one DNA molecule?

9.4. Suppose you are constructing a plasmid vector from an R plasmid. The vector can be cleaved into five or so fragments by a restriction enzyme. How would you select for the elimination of a fragment containing the only cleavage site for another enzyme? How would you select for a plasmid that contains only a single copy of a particular restriction site?

9.5. How many times is the *Hae*III recognition sequence GGCC likely to occur in 50,000 base pairs of DNA with GC content 30%, 50%, and 70%?

9.6. How frequently should the recognition sequence of *Hind*II, GTpypuAC occur in 50,000 base pairs of DNA if AT ~GC ~ 50%?

9.7. How would you proceed to put a small fragment of DNA that has been generated by *Hae*III cleavage into the *Eco*RI cleavage site on a plasmid such that the fragment could be cut out of the plasmid (after cloning and growth) by *Eco*RI?

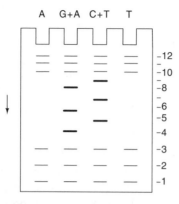

Figure 9.21 Figure for Problem 9.12.

9.8. Suppose a gene has been cloned using *Eco*RI sticky ends. Suppose also that it is necessary to remove one of the sticky ends so that the other end may be modified. How could you rather easily alter the sequence of one of the *Eco*RI ends for this purpose? [Hint: Consider partial digestions and the use of another enzyme.]

9.9. When sequencing by the Maxam-Gilbert method, how would you recognize a position that is methylated *in vivo* by a modification enzyme?

9.10. A hybrid plasmid was formed by inserting via the poly dA-poly dT method foreign DNA into the single *Eco*RI site contained on the parent plasmid. Sketch the DNA heteroduplex formed between one strand of DNA from the hybrid plasmid, opened at a random location within plasmid sequences by shear forces, and one strand from the parent plasmid, opened by *Eco*RI cleavage.

9.11. Suppose you wished to clone the gene from *Haemophilus aegyptius* coding for the restriction enzyme *Hae*. How would you select or screen for cells having received plasmids containing the genes coding for the desired restriction enzyme?

9.12. Lambda phage DNA was digested with a certain restriction enzyme that yields more than 30 fragments of the phage. These fragments were labeled on their 5' ends with polynucleotide kinase and sequenced by the Maxam-Gilbert procedure. Let us suppose that four specific base cleavage reactions were performed and it is claimed that the autoradiogram in Fig. 9.21 was obtained. What do you conclude?

9.13. Will the following method work, in principle, for DNA sequencing?

a) Label the DNA by reaction with T4 DNA kinase and γ-^{32}PO$_4$-ATP, cut into two unequal pieces, and isolate one of the pieces.

b) Lightly nick with DNAse.

c) Nick translate in four reactions with DNA pol I using all XTPs with one of each of the four dideoxytriphosphates (none radioactive).

d) Run on sequencing gels and read the sequence.

9.14. Restriction enzymes *Bam*HI and *Bcl*I cleave G$^{\downarrow}$GATCC and T$^{\downarrow}$GATCA, respectively. Why can DNA fragments produced by *Bcl*I digestion be inserted into *Bam*H1 sites? If a plasmid contains a single *Bam*HI cleavage site, how could you select for plasmids containing inserts of *Bcl*I fragments?

9.15. How would you select for a clone carrying the methylase partner of a restriction enzyme? Why is it sensible to engineer the hypersynthesis of the methylase before attempting the same for its restriction enzyme partner?

9.16. Plasmids carrying the *Eco*RI restriction-modification system may be mated into recipient cells with 100% viability. What does this tell us about the regulation of the synthesis of the *Eco*RI restriction and modification system?

9.17. How could the DNA replication origin of the *E. coli* chromosome be cloned?

9.18. Why is it unlikely that the recognition sequence of any restriction-modification system will be larger than about eight base pairs?

References

Recommended Readings

Cleavage of DNA by RI Restriction Endonuclease Generates Cohesive Ends, J. Mertz, R. Davis, Proc. Natl. Acad. Sci. USA *69*, 3370-3374 (1972).

A New Method for Sequencing DNA, A. Maxam, W. Gilbert. Proc. Natl. Acad. Sci. USA *74*, 560-564 (1977).

DNA Sequencing with Chain-terminating Inhibitors, F. Sanger, S. Nicklen, A. Coulson, Proc. Natl. Acad. Sci. USA *74*, 5463-5467 (1977).

Sequencing

Determination of a Nucleotide Sequence in Bacteriophage f1 DNA by Primed Synthesis with DNA Polymerase, F. Sanger, J. Donelson, A. Coulson, M. Kossel, D. Fischer, J. Mol. Biol. *90*, 315-333 (1974).

A Rapid Method for Determining Sequences in DNA by Primed Synthesis with DNA Polymerase, F. Sanger, A. Coulson, J. Mol. Biol. *94*, 441-448 (1975).

Mapping Adenines, Guanines, and Pyrimidines in RNA, H. Donis-Keller, A. Maxam, W. Gilbert, Nucleic Acids Res. *4*, 2527-2538 (1977).

Sequencing End-labeled DNA with Base-specific Chemical Cleavages, A. Maxam, W. Gilbert, in Methods in Enzymology, *65*, 499-559, ed. L. Grossman and K. Moldave, Academic Press, New York (1980).

A Systematic DNA Sequencing Strategy, G. Hong, J. Mol. Biol. *158*, 539-549 (1982).

Vectors

Construction of Biologically Functional Bacterial Plasmids *in vitro*, S. Cohen, A. Chang, H. Boyer, R. Helling, Proc. Natl. Acad. Sci. USA *70*, 3240-3244 (1973).

Packaging Recombinant DNA Molecules into Bacteriophage Particles *in vitro*, B. Hohn, K. Murray, Proc. Natl. Acad. Sci. USA *74*, 3259-3263 (1977).

Cosmids: A Type of Plasmid Gene-Cloning Vector that is Packageable *in vitro* in Bacteriophage Lambda Heads, J. Collins, B. Hohn, Proc. Natl. Acad. Sci. USA *75*, 4242-4246 (1978).

Construction and Characterization of New Cloning Vehicles V. Mobilization and Coding Properties of pBR322 and Several Deletion Derivatives Including pBR327 and pBR328, L. Covarrubias, L. Cervantes, A. Covarrubias, X. Soberon, I. Vichido, A. Blanco, Y. Kuparsztoch-Portnoy, F. Bolivar, Gene *13*, 25-35 (1981).

Introduction of Rat Growth Hormone Gene into Mouse Fibroblasts via a Retroviral DNA Vector: Expression and Regulation, J. Doehmov, M. Barinaga, W. Vale, M. Rosenfeld, I. Verma, R. Evans, Proc. Natl. Acad. Sci. USA *79*, 2268-2272 (1982).

Cloning of Large Segments of Exogenous DNA Into Yeast by Means of Artificial Chromosome Vectors, D. Burke, G. Carle, M. Olson, Science *236*, 806-812 (1987).

Enzymes and Enzymology of Genetic Engineering

Enzymatic Breakage and Joining of Deoxyribonucleic Acid, B. Weiss, A. Jacquemin-Soblon, T. Live, G. Fareed, C. Richardson, J. Biol. Chem. *243*, 4543-4555 (1968).

Enzymatic Joining of Polynucleotides, P. Modrich, I. Lehman, J. Biol. Chem. *245*, 3626-3631 (1970).

DNA Nucleotide Sequence Restricted by the RI Endonuclease, J. Hedgpeth, H. Goodman, H. Boyer, Proc. Natl. Acad. Sci. USA *69*, 3448-3452 (1972).

Nature of ColE1 Plasmid Replication in *Escherichia coli* in the Presence of Chloramphenicol, D. Clewell, J. Bact. *110*, 667-676 (1972).

Enzymatic Oligomerization of Bacteriophage P22 DNA and of Linear Simian Virus 40 DNA, V. Sgaramella, Proc. Natl. Acad. Sci. USA *69*, 3389-3393 (1972).

Enzymatic End-to-End Joining of DNA Molecules, P. Lobban, A. Kaiser, J. Mol. Biol. *78*, 453-471 (1973).

A Suggested Nomenclature for Bacterial Host Modification and Restriction Systems and Their Enzymes, H. Smith, D. Nathans, J. Mol. Biol. *81*, 419-423 (1973).

Ligation of *Eco*RI Endonuclease-generated DNA Fragments into Linear and Circular Structures, A. Dugaiczyk, H. Boyer, H. Goodman, J. Mol. Biol. *96*, 171-181 (1975).

*Eco*RI Endonuclease, Physical and Catalytic Properties of the Homogeneous Enzyme, P. Modrich, D. Zabel, J. Biol. Chem. *251*, 5866-5874 (1976).

The Alternate Expression of Two Restriction and Modification Systems, S. Glover, K. Firman, G. Watson, C. Price, S. Donaldson, Mol. Gen. Genet. *190*, 65-69 (1983).

Site-specific Cleavage of DNA at 8- and 10-base-pair Sequences, M. Mc-Clelland, L. Kesler, M. Bittner, Proc. Nat. Acad. Sci. USA *81*, 1480-1483 (1984).

Model for How Type I Restriction Enzymes Select Cleavage Sites in DNA, F. Studier, P. Bandyopadhyay, Proc. Natl. Acad. Sci. USA *85*, 4677-4681 (1988).

Discrimination Between DNA Sequences by the *Eco*RV Restriction Endonuclease, J. Taylor, S. Halford, Biochemistry *28*, 6198-6207 (1989).

Transformation and Selecting Transformants

The Inosinic Acid Pyrophosphorylase Activity of Mouse Fibroblasts Partially Resistant to 8-Azaguanine, J. Littlefield, Proc. Natl. Acad. Sci. USA *50*, 568-576 (1963).

Calcium-dependent Bacteriophage DNA Infection, M. Mandel, A. Higa, J. Mol. Biol. *53*, 159-162 (1970).

Injected Nuclei in Frog Oocytes Provide a Living Cell System for the Study of Transcriptional Control, J. Gurdon, E. Robertis, G. Partington, Nature *260*, 116-120 (1976).

Transfer of Purified Herpes Virus Thymidine Kinase Gene to Cultured Mouse Cells, M. Wigler, S. Silverstein, L. Lee, A. Pellicer, Y. Cheng, R. Axel, Cell *11*, 223-232 (1977).

Transformation of Yeast, A. Hinnen, J. Hicks, G. Fink, Proc. Natl. Acad. Sci. USA *75*, 1929-1933 (1978).

Selection for Animal Cells that Express the *Escherichia coli* Gene Coding for Xanthine-guanine Phosphoribosyl-transferase, R. Mulligan, P. Berg, Proc. Natl. Acad. Sci. USA *78*, 2072-2076 (1981).

Transformation of Mammalian Cells to Antibiotic Resistance with a Bacterial Gene Under Control of the SV40 Early Region Promoter, P. Southern, P. Berg, J. Molec. Appl. Genet. *1*, 327-341 (1982).

Expression of Genes Transferred into Monocot and Dicot Plant Cells by Electroporation, M. Fromm, L. Taylor, V. Walbot, Proc. Nat. Acad. Sci. USA *82*, 5824-5828 (1985).

High-velocity Microprojectiles for Delivering Nucleic Acids Into Living Cells, T. Klein, E. Wolf, R. Wu, J. Sanford, Nature *327*, 70-72 (1987).

Two Dominant-acting Selectable Markers for Gene Transfer Studies in Mammalian Cells, S. Hartman, R. Mulligan, Proc. Natl. Acad. Sci. USA *85*, 8047-8051 (1988).

DNA and RNA Isolation and Separation

A Dye-buoyant-density Method for the Detection and Isolation of Closed Circular Duplex DNA: The Closed Circular DNA in Hela Cells, R. Radloff, W. Bauer, J. Vinograd, Proc. Natl. Acad. Sci. USA *57*, 1514-1521 (1967).

Molecular Weight Estimation and Separation of Ribonucleic Acid by Electrophoresis in Agarose-acrylamide Composite Gels, A. Peacock, C. Dingman, Biochem. *7*, 668-674 (1968).

Chain Length Determination of Small Double-and-Single- Stranded DNA Molecules by Polyacrylamide Gel Electrophoresis, T. Maniatis, A. Jeffrey, H. Van de Sande, Biochemistry *14*, 3787-3794 (1975).

Colony Hybridization: A Method for the Isolation of Cloned DNAs that Contain a Specific Gene, M. Grunstein, D. Hogness, Proc. Natl. Acad. Sci. USA *72*, 3961-3965 (1975).

Sizing and Mapping of Early Adenovirus mRNAs by Gel Electrophoresis of S1 Endonuclease-Digested Hybrids, A. Berk, P. Sharp, Cell *12*, 721-732 (1977).

Interesting and Useful Techniques

Screening lambda-gt Recombinant Clones by Hybridization to Single Plaques *in situ*, W. Benton, R. Davis, Science *196*, 180-182 (1977).

The Detection of DNA-binding Proteins by Protein Blotting, B. Bowen, J. Steinberg, U. Laemmli, H. Weintraub, Nuc. Acids Res. *8*, 1-20 (1980).

Deletion of an Essential Gene in *Escherichia coli* by Site-specific Recombination with Linear DNA Fragments, M. Jasin, P. Schimmel, J. Bact. *159*, 783-786 (1984).

Synthesis of a Sequence-specific DNA-cleaving Peptide, J. Sluka, S. Horvath, M. Bruist, M. Simon, P. Dervan, Science *238*, 1129-1132 (1987).

Sequence-specific Cleavage of Double Helical DNA by Triple Helix Formation, H. Maser, P. Dervan, Science *238*, 645-650 (1987).

cDNA Cloning of Bovine Substance-K Receptor through Oocyte Expression System, Y. Masu, K. Nakayama, H. Tamaki. Y. Harada, M. Kuno, S. Nakanishi, Nature *329*, 836-838 (1987).

Detection of Specific DNA Sequences by Fluorescence Amplification: A Color Complementation Assay, F. Chehab, Y. Kan, Proc. Natl. Acad. Sci. USA *86*, 9178-9182 (1989).

Advanced Genetic
Engineering

10

The previous chapter described the fundamentals of genetic engineering: cutting, splicing, vectors, cloning, transformation, and DNA sequencing. We will continue here with descriptions of more advanced manipulations, which, for the most part, are technological aspects of genetic engineering. These are: additional cloning techniques, the polymerase chain reaction, chromosome mapping, high capacity sequencing methods, and methods for locating binding sites of proteins on DNA.

Finding Clones from a Known Amino Acid Sequence

Sometimes the protein product of a gene is available in pure form. This happy circumstance can be used to facilitate cloning of the gene. Portions of the protein can be sequenced to determine a potential DNA sequence that could have encoded this portion of the protein. An oligonucleotide with this sequence can then be used to screen a collection of clones, which is called a library, to detect those containing complementary sequences. The screening is done as described in the previous chapter. Occasionally, a clone is found in the libraries which hybridizes to the screening oligonucleotide, but which is not the correct clone. This results from the chance occurrence of a sequence complementary to the probing oligonucleotide. These incorrect positives can be detected by screening with a second oligonucleotide that should hybridize to a different part of the gene encoding the protein in question. Only the desired clones should hybridize to both oligonucleotides.

The redundancy in the genetic code prevents simple reverse translation from an amino acid sequence to a DNA sequence. The difficulty caused by the redundancy can be partially overcome by using portions

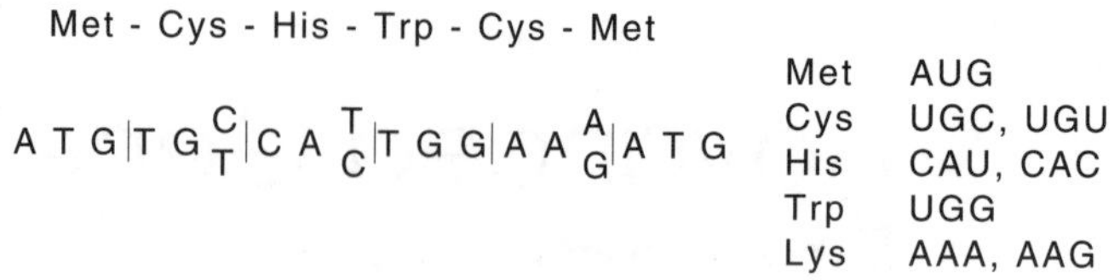

Met - Cys - His - Trp - Cys - Met

A T G|T G $_\text{T}^\text{C}$|C A $_\text{C}^\text{T}$|T G G|A A $_\text{G}^\text{A}$|A T G

Met	AUG
Cys	UGC, UGU
His	CAU, CAC
Trp	UGG
Lys	AAA, AAG

Figure 10.1 Reverse translating to obtain the sequences that could have encoded a short peptide.

of the protein's sequence containing amino acids whose codon redundancy is low. This is possible since both tryptophan and methionine have unique codons. Consider the sequence met-cys-his-trp-lys-met. Only one codon specifies an internal methionine, while the cysteine, histidine, and lysine are each specified by only two possible codons. Therefore one of only $1 \times 2 \times 2 \times 1 \times 2 \times 1 = 8$ sequences encoded the six amino acids (Fig. 10.1).

The eight necessary oligonucleotides can be synthesized simultaneously by machine by incorporating either of the two ambiguous nucleotides at the necessary positions. This is accomplished simply by supplying at the correct time a mixture of the two nucleotides to the synthesis solution.

Purification of the protein necessary for the oligonucleotide probing approach often is straightforward. Conventional purification need not be performed, however. Since all that is needed for the cloning is determination of portions of the amino acid sequence, purification and detection methods need not preserve the protein's native structure. SDS gel electrophoresis, for example, can be used as a final step in the purification of the protein. The protein in the correct band in the gel can be eluted and a portion of its amino terminal sequence determined by gas phase and mass spectrometry. As little as 10^{-12} moles of protein are sufficient for determining enough of the sequence that oligonucleotide probes can be designed to identify clones carrying the gene.

Finding Clones Using Antibodies Against a Protein

Cloning a gene becomes easier than described above if sufficient quantities of its gene product are available to permit raising antibodies against the protein. DNA from the organism is cloned into a vector designed to provide for transcription and translation of the inserted DNA. The DNA is best cloned into a site transcribed from a controllable upstream promoter and also containing an upstream ribosome binding site and protein translation initiation sequence as well (Fig. 10.2). A fragment of DNA inserted into the site and containing an open reading frame is translated if it is fused in frame with the initiation sequence. As in screening with oligonucleotides, a replica plate is made, cells on the plate are grown, and the controllable promoter is induced. The cells are lysed, the proteins are immobilized on a filter, antibody is added,

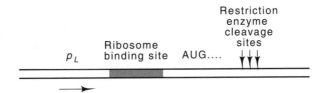

Figure 10.2 Structure of a vector suitable for antibody screening for the insertion of open reading frames of a specific protein.

and then areas with bound antibody are revealed as described below. The colony from the corresponding position on the replica plate can then be picked and studied.

Molecules of one particular antibody type bind to just one particular shape found in some other macromolecule. This is defined as their antigen. Almost any protein can be used as an antigen to elicit the synthesis of antibodies. Thus antibodies provide highly selective agents for the detection of specific proteins. The antibody selectivity for binding to the correct shape compared to binding to incorrect shapes is roughly the same as the hybridization selectivity of nucleic acids.

Although radioactive antibody could be used to detect antigen synthesized by candidate clones, it is not efficient, for different antibodies would then have to be made radioactive for the detection of different proteins. The A protein from *Staphylococcus aureus* provides a more general detection method. This protein binds to a portion of the antibody molecule so that one sample of radioactive or enzymatically tagged *Staphylococcus aureus* A protein suffices for the detection of many different antibody-protein complexes (Fig. 10.3). Another detection method is to use antibodies themselves as labels. The protein on the filter paper can be incubated with antibodies specific for the proteins that were raised in mice. Then rabbit antibodies that have been raised against mouse antibodies can be added. These will detect and bind to most mouse antibodies. The enzyme alkaline phosphatase can be linked

Figure 10.3 The use of radioactive *Staphylococcus* A protein to identify antibody-antigen complexes in Western transfers.

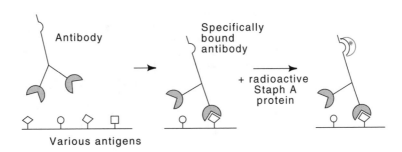

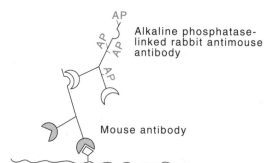

AP

Alkaline phosphatase-
linked rabbit antimouse
antibody

Mouse antibody

Figure 10.4 The use of alkaline phosphatase linked rabbit-antimouse antibody to reveal proteins to which the mouse antibody binds.

to rabbit-antimouse antibodies (Fig. 10.4). Their location can be marked by adding colorless substrate for alkaline phosphatase whose hydrolysis product is highly colored and insoluble. The product shows the location of protein to which the mouse antibody bound, to which, in turn, the rabbit antibodies containing alkaline phosphatase bound.

Southern, Northern, and Western Transfers

Here we will cover in greater detail the topic of Southern transfers that were mentioned and briefly described in Chapter 2. At the same time, since the concepts are almost the same, we will also mention the so-called Northern and Western transfers. Southern transfers once were a necessary step in chromosome mapping, but their use has been superseded by techniques based on the polymerase chain reaction. To review, DNA fragments can be separated according to size by electrophoresis through gels, denatured, transferred to a nylon or paper membrane and immobilized. Then the membrane can be immersed in a

Figure 10.5 Southern transfer methodology. After electrophoretic separation according to size, the fragments are denatured and electrophoretically transferred to a membrane before hybridization with radioactive probe.

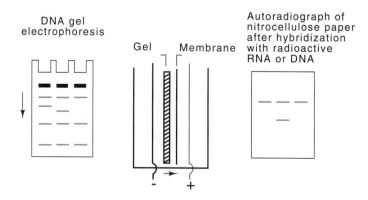

DNA gel
electrophoresis

Gel   Membrane

Autoradiograph of
nitrocellulose paper
after hybridization
with radioactive
RNA or DNA

−   +

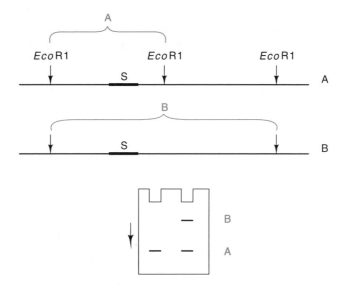

Figure 10.6 Detection of an RFLP by Southern transfer in which the radioactive probe was fragment S. The chromosome A generates band A and the chromosome B which lacks the first EcoRI cleavage site on the right generates band B. A heterozygous individual shows both bands in a Southern transfer probed with the segment S.

buffer containing a labeled oligonucleotide or DNA fragment and incubated under conditions permitting hybridization between complementary nucleic acid sequences (Fig. 10.5). The labeled fragment will therefore hybridize to its complementary sequence. The portion of the membrane carrying the fragment will then become radioactively labeled, and can be detected by autoradiography. This part of the process is analogous to plaque and colony screening described in the previous chapter. This simple technique, named a Southern transfer for Southern who first devised it, can be used in the analysis of chromosome structure.

Consider the problem of learning whether the two nearest *Eco*RI cleavage sites on either side of a segment of DNA are at the same location in two nearly homologous chromosomes. If they are not, the situation is described as a restriction fragment length polymorphism, RFLP, and the nucleotide differences producing this polymorphism can be used as a genetic marker. If the restriction fragment containing the sequence is the same size from both chromosomes, then the nearest *Eco*RI cleavage sites are likely to be in the same locations. To search for RFLPs, a DNA sample containing the two chromosomes is cut with a restriction enzyme, separated by electrophoresis, and "probed" with a radioactively labeled segment of the region (Fig. 10.6). A difference in the sizes of corresponding fragments indicates the presence of an RFLP.

Northern transfers are the converse of Southern transfers in that it is RNA rather than DNA that is separated by electrophoresis, transferred, and immobilized on membrane to preserve the original pattern.

Membrane with immobilized RNA can then be used in hybridization, just like paper with immobilized DNA.

What kinds of questions can be answered with immobilized RNA? One concerns the *in vivo* state of various RNAs. Transient precursors of a mature RNA molecule can easily be detected because they will be larger than the mature RNA and will be separated during electrophoresis. This permits tracking the maturation process of an RNA molecule. Not only can the changing sizes of the maturing species be monitored, but fates of specific regions that are removed can also be followed by probing with appropriate sequences.

Transfer-like technology can also be used to purify specific RNAs or DNAs. Either single-stranded RNA or single-stranded DNA can be bound to the paper. Then either RNA or DNA fragments complementary to the immobilized RNA or DNA can be isolated from a mixture by hybridization followed by elution. As an application, messenger RNA eluted from such immobilized DNA can be translated *in vitro* to provide a definitive identification of a candidate clone for a specific gene.

Western transfers involve proteins, not nucleic acids. The principle is the same as for Northern and Southern transfers. A pattern of proteins that have been separated by electrophoresis is transferred to paper or a membrane and then specific proteins are visualized. Some DNA- or RNA-binding proteins can easily be detected after transfer. These proteins partially renature despite being stuck to the paper. Then the paper with the immobilized proteins is incubated with the radioactive nucleic acid which binds to the immobilized protein. After washing the paper to remove unbound radioactive nucleic acid, autoradiography of the paper reveals the location of the immobilized protein. More often, the position of a specific protein is revealed by antibody probing as described in the previous section.

Polymerase Chain Reaction

A method has been devised with such remarkable sensitivity that it can detect a single molecule of a specific sequence. Furthermore, the single molecule can be detected in the presence of a 10^6 or greater excess of other sequences. This method is called the polymerase chain reaction or PCR. The polymerase chain reaction is also useful for studying specific genes or sequences. For example, it permits the sequencing in a day or two of a stretch of several hundred nucleotides with the starting point being a small sample of blood. No cloning is required for the sequencing. Such sensitivity permits the rapid characterization of the basis of mutations or of genetic defects. This extraordinary sensitivity also provides a sensitive test for the presence of a virus like HIV. Again a sample of blood can be taken and the assay can detect the presence of one copy of the single virus sequence in 100,000 cells. The polymerase chain reaction greatly facilitates generation of mutants *in vitro* and the synthesis of DNA for physical experiments.

The polymerase chain reaction is a scheme that amplifies the DNA lying between two sequences which are within several thousand base

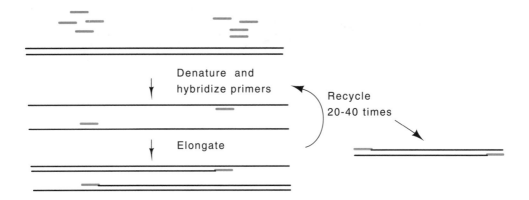

Figure 10.7 Polymerase chain reaction to amplify the sequence contained between the sites to which two primers hybridize.

pairs of one another. The amplification is accomplished by first denaturing the DNA sample, hybridizing two oligonucleotide primers to the DNA, elongating with DNA polymerase, and repeating this cycle up to 40 times (Fig. 10.7). The two oligonucleotide primers must be complementary to opposite strands of the DNA. The product of elongation primed by one oligonucleotide plus the template can then become templates for the next round of synthesis. As a result, each round of synthesis doubles the number of product DNA molecules present. The first round of synthesis produces DNA extending in one direction beyond each primer, but the DNA made in subsequent cycles from the first product DNA extends just to the ends of the primers.

Although DNA polymerase I from *E. coli* could be used in the polymerase chain reaction, its use would be inefficient because each round of denaturing the double-stranded DNA to form the single strands necessary as templates would destroy the polymerase. Therefore, these procedures use a temperature-resistant polymerase isolated from a thermophile, *Thermus aquaticus*. This polymerase withstands the 95° incubation for denaturing the DNA. Even better, after an incubation at 45° to hybridize primer to the DNA, a temperature at which the polymerase is largely inactive, the temperature can be raised to 75° to activate the polymerase. Although a little of the primer dissociates from the template at this temperature, a much greater fraction of any incorrectly hybridized primer dissociates from incorrect sites. Thus, a very great specificity is achieved for amplification of just the desired sequence of DNA.

The polymerase chain reaction can be put to a wide variety of uses. One simple example is the screening of cloning steps. Ordinarily, one must screen transformants after the simple step of inserting a DNA fragment into a plasmid. Typically 90% of the transformants contain the fragment, but since the frequency is not 100%, one must verify that

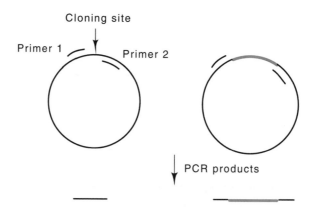

Figure 10.8 Polymerase chain reaction used to monitor the insertion of a foreign DNA between the sites to which two primers hybridize. Insertion of DNA between the primer hybridization sites increases the size of the PCR product.

the clone selected for further work is suitable. Previously such verification required growth of cultures from a collection of candidates, purification of the plasmids from each, and screening of each by restriction enzyme digestion followed by gel electrophoresis to check for the restriction fragment with properly altered size. PCR permits the same test in much less time. Transformant colonies are picked directly into tubes used for PCR. The first heat step lyses the cells. Primers are used that flank the site into which the fragment was to be cloned. The product of the amplification is run on a gel. If the fragment had been cloned between the sites, the amplified piece will have one size, but if the fragment had not been cloned, the fragment would be much smaller (Fig. 10.8).

DNA for footprinting or sequencing can be made directly from genomic DNA with the polymerase chain reaction. By labeling one of the oligonucleotide primers used in the PCR reaction, the DNA that is synthesized is already radioactive and ready for use. Such a technique eliminates the need for cloning the DNA from mutants. It also streamlines the screening of genetic diseases or of mutants isolated in the lab. Instead of cloning the DNA to determine the defect in a gene, the DNA from the organism or cells can be directly amplified and the defect determined by sequencing. PCR also greatly facilitates genetic constructions. For example, suppose a portion of a gene is to be cloned in an expression vector. The oligonucleotide primers can include not only regions homologous to the DNA to be cloned, but also additional regions necessary for cloning and expression like restriction sites, ribosome binding sites, and translation termination signals (Fig. 10.9). In the first round of amplification, the regions of homology between the primers and the template DNA hybridize. In subsequent rounds, the entire

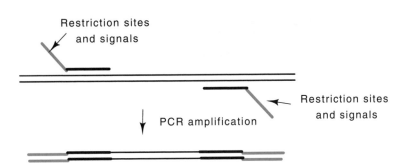

Figure 10.9 Use of PCR to both amplify a selected region of DNA as well as place specific desired sequences on the ends of the DNA.

primer base pairs and amplified DNA constitutes the fusion of the sequence plus the desired flanking sequences.

Isolation of Rare Sequences Utilizing PCR

The power of PCR can be used to simplify the cloning of genes and to isolate special rare DNA or RNA molecules out of large populations containing many different sequences. First let us consider the cloning of a gene if a tiny portion of its protein product can be isolated. The amino acid sequence of a stretch of the protein must be determined. This is used to design oligonucleotides that hybridize to the top and bottom strands of the ends of the region encoding the peptide and can be used in PCR to amplify the region from cDNA template. Of course, redundancy in the genetic code necessitates that each oligonucleotide be a mixture as discussed earlier in the chapter. For use with PCR, however, there is less need to minimize the degeneracy. Even though the proper DNA sequence is represented by only a tiny fraction of each oligonucleotide mixture, just these oligonucleotides will be functional in PCR amplification of the desired region of DNA. The other oligonucleotides may hybridize to the template cDNA, but they are unlikely to give rise to any PCR products, and it is even less likely that such products would be of the same size as the desired PCR product.

The PCR amplified product of a portion of the gene can then be used to probe a cDNA library contained on plasmid or phage clones. This will reveal the clones likely to contain an intact version of the desired gene.

PCR is also useful for the isolation of very rare sequences present in complex mixtures of sequences. During the chemical synthesis of DNA, the use of mixtures of nucleotide precursors for some synthesis cycles enables large random populations to be synthesized. Consider the determination of the optimum binding sequence of a DNA-binding protein. One strand of DNA can be synthesized with unique sequences at each end and containing a totally random interior region. First, strands complementary to the chemically synthesized molecules must be constructed. Due to the random sequences involved, the complemen-

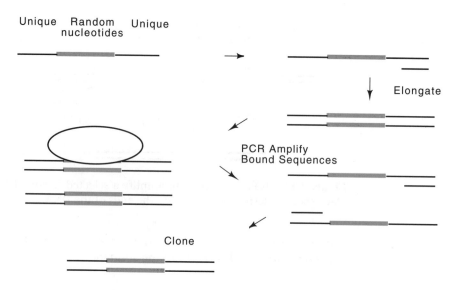

Unique   Random   Unique
          nucleotides

Elongate

PCR Amplify
Bound Sequences

Clone

Figure 10.10 Use of PCR to amplify a small amount of DNA present in a complex mixture.

tary strands cannot be synthesized chemically. They are made enzymatically by hybridizing a DNA primer to the unique sequence at one end of the chemically synthesized strand. Next, DNA pol I is used to elongate and complete the synthesis of an exact complement (Fig. 10.10). Then, the small population of DNA molecules with sequences capable of binding the protein are isolated, for example, by electrophoretic separation of protein-DNA complexes from DNA incapable of binding the protein. This tiny amount of DNA is amplified by PCR using primers complementary to the unique sequences at the ends. After amplification, the selection and amplification steps can be repeated, and finally the DNA can be cloned and sequenced.

Simple variations on the basic idea described above can be developed for the isolation of RNA capable of binding a protein or specific support on an exchange column. It is also possible to devise methods for the selection of DNA or RNA molecules that possess hydrolytic activities. Mutagenic steps can be introduced in the PCR amplification steps so that variants of the selected molecules can be created. In this way, true *in vitro* evolution experiments can be done. Living cells are not required.

Physical and Genetic Maps of Chromosomes

A genetic map contains the order and approximate recombinational distance between genetic markers whereas a physical map contains the order and physical locations of physical features of a chromosome. Most often the features used in a physical map are short segments of known sequence. Once a physical map exists, genetic markers can be placed on it. Then the mapping and cloning of the genes responsible for genetic diseases is greatly simplified. For example, the nearest physical marker can be chosen as a starting point and with chromosome walking as described in the previous chapter you can clone the gene.

Two major objectives of genetic engineering are acquiring the ability to detect genetic diseases and learning their biochemical basis. Both of these objectives are greatly aided by cloning of the mutant gene. Once the clone is available, direct screening for the mutant DNA is possible and the study of the wild type and mutant gene product becomes easier. Cloning the DNA involved with most genetic diseases is difficult, however. Frequently all that is known is an approximate map location of the defect. The usual helpful tools such as an altered enzyme, protein, or nucleic acid are missing. Here we will see how this difficulty can be circumvented.

Suppose there existed a highly detailed genetic map of the human genome. Then any new marker or genetic defect could be located by measuring its recombinational distance from the known and mapped markers. Of course, this step might require collecting data from several generations of genetic carriers. Once the map location was obtained, we could use this information in genetic counseling. Furthermore, with the integration of a physical map and a genetic map, we would also know the physical location of the genetic marker. Therefore, we could begin from the nearest physical marker and perform a chromosomal walk to clone the gene responsible for the defect.

Typically we think of the genes that code for blood type, hair color, or those genes encoding known enzymes or proteins as constituting the genetic map. While these genes can be useful, too few of them are known to permit precise mapping. Furthermore, identifying many of these markers requires the intact individual. Often this is inconvenient. Instead, we need a new kind of genetic marker, one that exists in high numbers and which is easy to score in the DNA obtained from a small number of cells.

What constitutes a suitable genetic marker? A genetic marker must be easily detected in a small sample of cells or DNA and it must exist in the population in two or more states. If the population were homozygous, then there would be no useful markers since both parents and all offspring would be genetically identical. No genetic mapping could be done. The existence of markers means that some individuals possess one allele or sequence at a position, and other individuals possess a different allele or sequence. If there are hundreds or thousands of markers at which different individuals in the population are likely to be different, then it is feasible to try to map a genetic defect with respect to these known markers.

Chromosome Mapping

It is simplest first to determine a large-scale and coarse map of a chromosome and then to build upon the initial map to increase resolution. Construction of a low-resolution physical map can begin with a

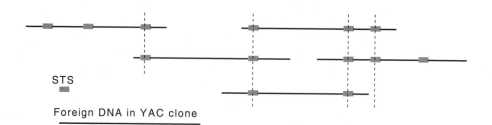

STS

Foreign DNA in YAC clone

Figure 10.11 A set of DNA fragments ordered by the STS sites they share in common.

library of overlapping clones. Mapping would then be picking and ordering a subset of these which spans an entire chromosome. The larger the clones, the fewer which need to be ordered, and the easier it is to construct of the initial map. The yeast artificial chromosome vectors, YAC vectors, are useful for chromosome mapping projects because they can accommodate very large fragments of foreign DNA, up to 10^6 base pairs. These contain autonomous replicating sequences, ARS, as origins and telomeres for the ends.

Let us first consider the physical markers which are necessary for the construction of a physical map. Unique sequences of DNA one or two hundred base pairs long can provide a good starting point for construction of a physical map. Such sequences can be obtained by sequencing random clones made from the chromosome to be mapped. The uniqueness of a sequence can be checked by hybridization. From such a unique sequence, two 20 to 30 base portions can be chosen for oligonucleotide primers that will permit PCR amplification of the sequence lying between. This amplification will occur only if the template DNA present in the PCR reaction contains the unique sequence. Such a unique

PCR

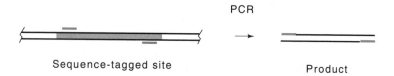

Sequence-tagged site   Product

sequence is called a sequence-tagged site, or STS. A sample of DNA is tested for the presence of an STS by PCR amplification using the two oligonucleotides followed by gel electrophoresis to assay for the amplification of the proper-sized DNA fragment.

When the number of sequence-tagged sites exceeds the number of clones necessary to span the chromosome by a factor of four or five or more, the sites can be used just like restriction enzyme cleavage sites are used in a chromosomal walk. That is, each YAC vector is assayed for the presence of each STS. If an STS site is present in two vectors, then the chromosome fragment carried by each contains this region in common (Fig. 10.11). Merely from this data, the STS sites can be ordered and an overlapping physical map of the chromosome can be constructed. Of course, for even the smallest human chromosome, tens of thousands of YAC vectors must be present in the library, and hundreds of STS sites are needed. Therefore, very many PCR amplification reactions will be required for the ordering.

Mutations, insertions, and deletions have also been used in chromosome mapping. Originally, restriction fragment length polymorphisms, RFLPs, were used as genetic and physical markers for chromosome mapping. Over the majority of the genome, sequence differences between individuals exist at a frequency of about 10^{-4}. If one of these differences occurs in the cleavage site for a restriction enzyme, an RFLP will exist there. Another source of an RFLP is an insertion or deletion or a difference in the number of repeats of a short sequence contained between the two restriction enzyme cleavage sites.

Restriction fragment length polymorphisms can be detected using Southern transfers as described in an earlier section. If individuals differ in one out of 10^4 base pairs, then the chances that a particular restriction enzyme cleavage site will be present in one chromosome but be missing in the homologous chromosome would be about 0.001. There is a reasonably high probability that either an insertion exists between two restriction enzyme cleavage sites or that two individuals differ at a particular point in the number of times that a short sequence is repeated there. One out of a hundred to one out of a thousand restriction fragments is likely to be polymorphic between some individuals in the population. Therefore, considerable work could be required to find DNA sequences useful for RFLP mapping. One method for RFLP detection is to clone random segments of human DNA and use them to probe Southern transfers of various enzyme digests from a moderate number of people. The occasional clone which reveals a polymorphism is then mapped to a specific chromosome and finally mapped at higher resolution by comparing its segregation pattern to the pattern of other markers known to lie on the same chromosome.

A more efficient method of finding and mapping genetic markers utilizes the polymerase chain reaction. Instead of using Southern transfers to detect an RFLP, PCR is used to amplify a segment of DNA that contains variable numbers of inserts. Different individuals in the popu-

Figure 10.12 A set of DNA fragments ordered by the STS sites they share in common.

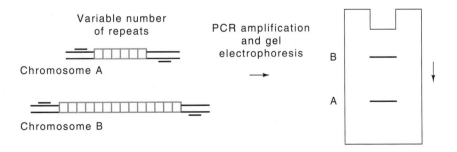

lation often possess different numbers of inserts at the same point (Fig. 10.12). The size of a fragment synthesized by PCR identifies the size of the insert. It is much easier to perform a PCR amplification and determine the size of the product than it is to perform a Southern transfer. Therefore, most chromosome mapping will shift to the use of PCR. By the approaches described above, hundreds of RFLP or PCR-based markers have already been mapped to the human genome and DNA defects of a significant number of genetic diseases have been mapped with respect to the markers.

DNA Fingerprinting—Forensics

At the other extreme from genome markers like restriction enzyme length polymorphisms would be genetic loci at which nearly every individual is unique and different. In such a case a child would inherit one or the other of the allelic states of each marker from each parent. Another child of the same parents would again inherit markers from the parents, but it would be a different set. Thus, some markers in the two children would be the same and others would be different. On the other hand, two unrelated individuals would possess virtually no markers in common.

After the above abstract description, let us consider a real situation. Scattered through the human genome are stretches containing repeats of a 32 base pair sequence. In any one of the stretches the sequence is repeated over and over again, sometimes reaching a length of thousands of nucleotides. Different individuals possess these repeated sequences, called minisatellites, in the same locations, but the number of repeats of the short sequence varies from person to person.

Consider the consequences of digesting the genomic DNA with a restriction enzyme that cleaves frequently, but lacks a cleavage site in the repeated sequence. After such cleavage the DNA is separated accord-

ing to size by electrophoresis. The fragments containing the repeated sequence are then identified by Southern transfer using a radioactive probe containing the 32 base repeated sequence.

For a typical individual, the cleaving and probing procedure resolves about 20 fragments of size greater than several thousand base pairs. Since each individual possesses different numbers of the 32 base repeats in these long stretches, each possesses a different collection of sizes for these large fragments. In short, the sizes of these fragments are DNA inheritable "fingerprints" unique to each chromosome of the individual. These DNA fingerprints are of forensic and legal value since they provide a unique linking of an individual to the DNA that can be extracted from a small amount of skin, blood, or semen, or for the determination of familial relationships.

Megabase Sequencing

In addition to mapping, sequencing often provides fundamental information for further studies on a gene or gene system. The sequencing techniques described in the previous chapter are adequate for sequencing a few genes, but one area of great interest, the immune system, possesses hundreds of genes, many of unknown function. Here hundreds of thousands of nucleotides must be sequenced. Serious effort is now also going into the early steps of determining the sequence of the entire human genome. Large sequencing projects such as these require better methods, and several have been developed. The one described below eliminates the use of radioisotopes and automates the detection of the bands on gels.

A number of steps in the standard DNA sequencing procedure seriously limit data acquisition. These are obtaining the plasmids necessary for sequencing the desired region, pouring the gels, exposing and developing the autoradiograph films, and reading the information from the films. Several of these steps can be streamlined or eliminated.

Imagine the savings in the Sanger sequencing technique if each of the four dideoxynucleotides could be tagged with a unique label. Then, instead of labeling the primer or the first nucleotides synthesized, the chain terminating nucleotide would possess the label. If this were done and each of the four labels were distinguishable, the four dideoxynucleotides could be combined in the same synthesis tube and the complex mixture of the four families of oligonucleotides could be subjected to electrophoresis in the same lane of the gel. Following electrophoresis, the four families of oligonucleotides could be distinguished and the entire sequence read just as though each one occupied a unique lane on the gel.

Figure 10.13 Excitation and emission spectra suitable for DNA sequencing. Each of the four fluorescent groups that emits at wavelengths λ_1, λ_2, λ_3, and λ_4, would be attached to a different base.

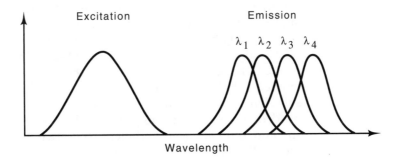

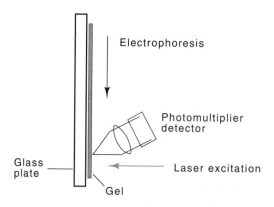

Figure 10.14 Geometry suitable for identification of fluorescent bases as they pass a point near the end of the electrophoresis gel.

Instead of using radioactive label, a fluorescent label is used. In order that this approach work well, the fluorescent adduct on the dideoxynucleotide must not interfere with the nucleotide's incorporation into DNA. Furthermore, each of the four nucleotides must be modified with a different adduct, one that fluoresces at a different wavelength from the others. In addition, it is useful if the excitation spectrum of the four fluorescent molecules substantially overlap so that only one exciting wavelength is required (Fig. 10.13).

Although the entire gel could be illuminated following electrophoresis, it is easier to monitor the passage during electrophoresis of one band after another past a point near the bottom of the gel. By measuring the color of the fluorescence passing a point near the bottom of the gel, the nucleotide terminating this particular size of oligonucleotide can be determined (Fig. 10.14). One after another, from one nucleotide to the next, the oligonucleotides pass the illumination point and the color of their fluorescence is determined, yielding the sequence of the DNA. Multiple lanes can be monitored simultaneously so that the sequence can be determined semiautomatically of many different samples simultaneously. Each lane of such a gel can provide the sequence of about 400 nucleotides of DNA.

The sensitivity of such a DNA sequencing approach is less than the radioactive techniques, but it is still sufficiently high that small DNA samples can be successfully used. A more serious problem than the sensitivity is the generation of useful samples to be sequenced. One approach is to generate many random clones from the desired DNA in a vector suitable for Sanger sequencing, to sequence at least the 300 nucleotides nearest to the vector DNA, and then to assemble the sequence of the region by virtue of the overlaps between various sequences. This shotgun approach yields the desired sequences if sufficient clones are available and sufficient time and effort are expended. In the sequencing of any sizeable amount of DNA, a pure shotgun approach is not efficient, and great effort is required to close the "statistical" gaps. When a few gaps remain, it may be easier to close them by chromosome walking than by sequencing more and more

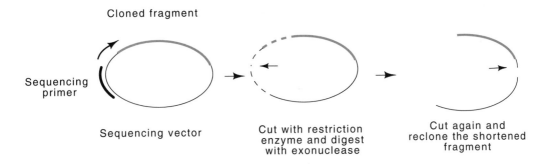

Cloned fragment

Sequencing
primer

Sequencing vector

Cut with restriction
enzyme and digest
with exonuclease

Cut again and
reclone the shortened
fragment

Figure 10.15 Generation of deletions for sequencing by exonuclease digestion. The vector is opened and digested, then a fragment is removed by cutting a second time with a restriction enzyme. This fragment is recloned and sequence is determined by using a primer to a sequence within the cloning vector adjacent to the location of the inserted fragment. By performing a series of exonuclease digestions for increasing periods, progressively larger deletions may be obtained.

randomly chosen clones, most of which will be of regions already sequenced.

Another method of generating the necessary clones for sequencing a large region is to use a nested set of overlapping deletions. By sequencing from a site within the vector sequences with the use of an oligonucleotide that hybridizes to the vector, the first 400 or so nucleotides of each of the clones can be determined. The resulting sequences can easily be assembled to yield the sequence of the entire region.

Cloned DNA

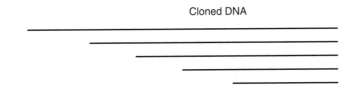

Such a set of clones can be generated by opening a plasmid containing the cloned DNA, digesting with an exonuclease for various lengths of time, and recloning so that increasing amounts of the foreign DNA inserted in the plasmid are deleted (Fig. 10.15).

Footprinting, Premodification and Missing Contact Probing

Understanding the biochemical mechanisms underlying the regulation of gene expression is a central problem in biochemistry-biology. One of the first steps in the study of a protein that binds to DNA is to determine where it binds. The use of DNAse footprinting in conjunction with Southern transfers to determine the locations of histone binding was briefly described in Chapter 2. Galas and Schmitz developed the elegant method of footprinting to solve the problem of determining the location

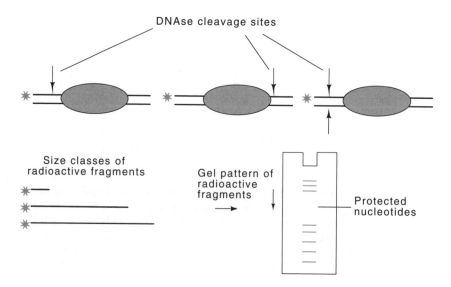

Figure 10.16 DNAse footprinting. Light digestion by DNAse I nicks each of the labeled DNA molecules on average one time. No nicking occurs on any of the molecules under the bound protein and therefore no fragments will be produced with ends in this region.

at which a protein binds to DNA. As described earlier, their method is based on the principles behind DNA sequencing. Although it was devised for mapping DNAse-sensitive sites, it can be used with any reagent that cleaves the DNA or modifies it so that later the DNA can be cleaved. The basic idea can also be used in two modes, a protection mode in which the bound protein protects the DNA from the reagent, and a prebinding interference mode in which the DNA is modified first and then the positions of modification which prevent protein binding are determined.

In the simplest form of footprinting the DNA fragment to be investigated must first be labeled on just one end of one strand. This can be done with polynucleotide kinase, with terminal transferase, which adds nucleotides to the 3'-OH end of DNA, with DNA pol I, which fills out sticky ends left by many restriction enzymes, or with PCR by amplifying a region containing the binding site and the use of one radioactively-labeled oligonucleotide primer.

After the binding of a protein to the DNA, the complex is briefly treated with DNAse. The duration of this treatment is adjusted so that about one random strand scission occurs per DNA molecule. Consequently, the population of molecules will contain examples of phosphodiester bond breakage at all positions except those covered by the protein (Fig. 10.16). Then the DNA is denatured and subjected to electrophoresis on a sequencing gel. This separates the population of molecules according to size, and the amount of DNA in any band is proportional to the amount of cleavage that occurred at the correspond-

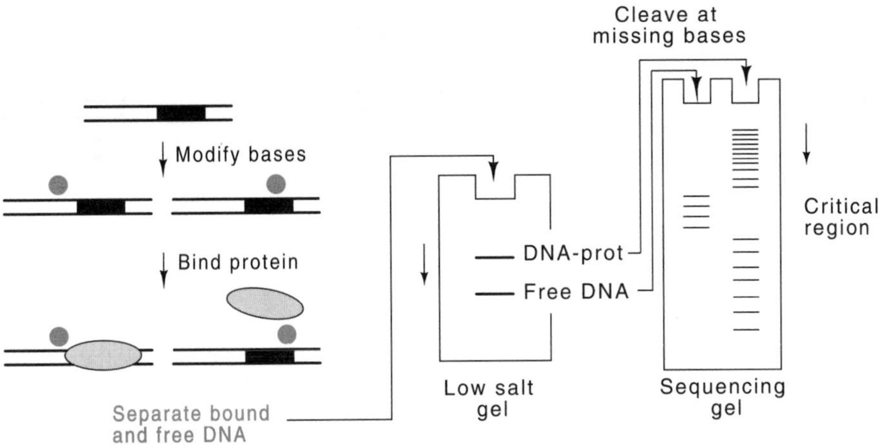

Figure 10.17 Premodification to locate contacted bases or phosphates. The intensity of the bands ultimately produced is proportional to the numbers of DNA molecules that were modified at that position. By binding protein to the population of DNA molecules, the subset modified in regions essential for protein binding are separated from the subset modified in irrelevant positions.

ing position in the DNA. No cleavages occur in the positions protected by the presence of the bound protein.

The footprinting experiments can also be performed with the protein protecting the DNA binding site from chemical attack rather than enzymatic attack. Dimethylsulfate can methylate guanine residues except some of those protected by the protein. After the methylation, the DNA can be cleaved at each of the methylated guanines, and the denatured, labeled fragments can be subjected to electrophoresis on a sequencing gel. Both DNAse I and dimethylsulfate are imperfect in that their reaction with unprotected DNA is somewhat base-specific. Consequently, both of these techniques must be performed with the control of labeled DNA free of the binding protein. The differences in the intensities of the bands between the DNA and protein plus DNA samples then identify the phosphodiester bonds protected from cleavage by the protein. The hydroxyl radical will attack and cleave the phosphodiester backbone independent of sequence. Thus, it is particularly useful for footprinting experiments.

The premodification interference mode for performing footprinting types of experiments is to modify or nick the DNA before addition of the protein whose binding site is to be mapped (Fig. 10.17). Then the protein is added. Those DNA molecules still capable of binding the protein are separated from the DNA molecules that have been modified in regions essential for binding of the protein, and as a result, do not bind the protein. These two populations of DNA molecules are separated from one another by the mobility retardation assay, in which DNA with a bound protein migrates more slowly than DNA without a bound protein. If necessary, the two populations are cleaved at the positions of modified

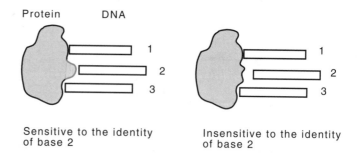

Figure 10.18 Changing a base-contacting amino acid to an alanine can make the protein insensitive to the identity of that base.

bases and the fragments are separated on DNA sequencing gels as described before.

The basic ideas of footprinting and premodification probing can even be adapted to reveal specific amino acid residue-DNA base interactions in the binding of a protein to DNA. The approach is similar to the premodification method just described. The labeled DNA is treated chemically so that each molecule, on average, has one base removed from a random position. This can be either a hydroxyl radical treatment or one of the base-specific reactions used in Maxam-Gilbert DNA sequencing. Then the entire population of molecules is permitted to bind to the protein. There is no effect on the binding if a base is missing from a position not contacted by the protein. If, however, a base is missing from a position the protein contacts, the protein will bind less tightly. Thus, if the protein and treated DNA are mixed and then diluted so that no further binding can occur, the protein will dissociate first, if it ever binds at all, from those DNA molecules that are missing bases that are contacted by the protein. The population of free DNA molecules becomes enriched with such DNA molecules. Similarly, the population of DNA molecules with bound protein becomes enriched with molecules that are missing only those bases not involved with contacts to the protein. The two DNA populations can be separated from one another with the DNA band shift assay. If necessary, the molecules are chemically cleaved at the positions of missing bases, and the positions of the cleavages are displayed after electrophoresis on DNA sequencing gels.

To demonstrate a specific residue-base interaction, the residue is modified by site-specific mutagenesis or PCR to an alanine. Because alanine is smaller than most other amino acids, most likely it will be unable to make the contact made by the amino acid it replaced (Fig. 10.18). Thus, in performing the missing contact experiment, a new base will be in the collection of bases not contacted. This is the base contacted by the amino acid at the position of the new alanine. Of course, in the execution of the experiment, allowance must be made for the fact that dissociation of the protein from the DNA is faster because of the missing contact. Either a shorter time is allowed for dissociation or buffer conditions are altered to increase the affinity of binding.

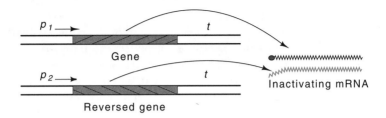

Figure 10.19 Generation of antisense RNA by inverting a gene or a segment of a gene and its subsequent hybridization *in vivo* to the mRNA from the gene, thereby preventing its translation.

Antisense RNA: Selective Gene Inactivation

Learning the *in vivo* functions of many gene products is not simple. We do learn the consequences of defects in humans of those particular proteins that have been associated with genetic diseases. As seen earlier, however, finding such proteins often involves huge amounts of work, and therefore relatively few are known. Not even this tool is available for most gene products or gene products in most organisms. Is there a general method for determining the function of genes in other organisms? Of course in bacteria, yeast, and sometimes in the fruit fly, specific genes can be intentionally inactivated, but rarely do satisfactory methods exist for inactivating genes in other organisms.

Antisense messenger RNA is one method for inactivating gene products. Massive synthesis of an RNA complementary to a messenger RNA will prevent translation of the messenger by hybridizing to it before it is translated (Fig. 10.19). Thus the synthesis of the antimessage inactivates the message. First, the gene must be cloned and then the gene or a portion of it is fused to a promoter so that its induction leads to synthesis of antimessenger. This antigene construct must then be introduced into the appropriate cells. Genes thought to be important in development have been examined with antisense messenger, and in some cases have been found to have drastic effects on the development of the organism.

Hypersynthesis of Proteins

Two reasons for cloning genes are to mutate the gene to alter the product for *in vivo* or *in vitro* studies and to increase the synthesis of a gene product either in its native organism or in bacteria. Hypersynthesis almost always seems necessary, for the more interesting the protein, the lower the levels at which it seems to be synthesized.

The most important factors limiting levels of protein synthesis in bacteria, which is where most cloned genes are expressed, are the strength of the promoter and the ribosome binding site. Secondary factors that can affect translation are the potential for folding of the mRNA and codon usage. One of the most important and most vexing problems in the hypersynthesis of proteins is proper folding. A high synthesis rate of many proteins does not yield a high concentration of the protein in the cytoplasm of the cells. Instead the protein is found in the form of granules or pellets often approaching the size of a bacterial cell. These are called inclusion bodies.

Inclusion bodies form when the protein synthesis rate so greatly exceeds its folding rate that an excess of hydrophobic areas are exposed. These bind and aggregate to form amorphous precipitates of inactive, and often insoluble protein. Several steps sometimes reduce the problem. Cells can be grown at low temperatures so that the rate of protein synthesis is greatly slowed whereas the rate of folding is only slightly depressed. The inclusion bodies can easily be purified by low speed centrifugation. They often contain essentially pure protein and occasionally the protein can be dissolved in urea or guanidine hydrochloride and renatured to yield active protein by slow removal of these denaturing agents. Sometimes the presence of chaperonin proteins during the synthesis or renaturation process increases yields of soluble protein.

Altering Cloned DNA by *in vitro* Mutagenesis

Understanding DNA-related biological mechanisms requires more than characterizing the DNA and associated proteins. It often requires alteration of the components. Not only does variation of the relevant parameters reveal more about the working mechanism, but the ability to test variants permits definitive proof of theories. Mutants have been used in molecular biology almost from its origins, first in the elucidation of biochemical pathways and now prominently in structural studies of the mechanisms by which proteins function as enzymes or recognize and bind to specific nucleotide sequences on DNA.

The efficient isolation of mutations has always posed a problem in molecular biology. Suppose mutations are desired in a particular gene or DNA sequence. If the entire organism must be mutagenized, then to obtain a reasonable number of alterations in the desired target, many more alterations will inevitably occur elsewhere on the chromosome. Often these other mutations will be lethal, so the necessary alterations in the target cannot easily be found. A method is needed for directing mutations just to the target gene. *In vitro* mutagenesis of cloned DNA fragments is a solution to the problem. Only the DNA of the target sequence is mutagenized. Just this sequence is then put back into cells.

Often random mutations need to be directed to small areas of genes or to specific nucleotides, or specific changes are desired in specific

```
——— G            GATCC ———
——— CCTAG            G ———
                 ↓
——— GGATC        GATCC ———
——— CCTAG        CTAG G ———
```

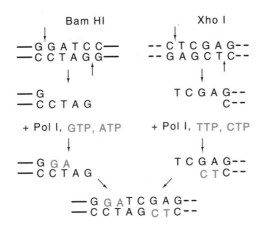

Figure 10.20 Partially filling out the overhanging ends resulting from *Bam*HI and *Xho*I digestion yields complementary ends that can easily be joined.

nucleotides. Some changes are easy to make. For example, insertions and deletions can be generated at the cleavage site of a restriction enzyme. A four-base insertion can be generated at the cleavage site of *Bam*HI by filling in the four-base single-stranded ends with DNA pol I and ligating the flush ends together.

Similarly, a four-base deletion can be generated by digesting the single-stranded ends with the single-stranded specific nuclease S1 before ligation. Variations on these themes are to use DNA pol I in the presence of only one, two, or three of the nucleotides to fill out part of the single-stranded ends before nuclease treatment and ligation (Fig. 10.20). Mixing and matching entire restriction fragments from a region under study is another closely related method of changing portions of DNA binding sites or substituting one portion of a protein for another.

More extensive deletions from the ends of DNA molecules can be generated by double-stranded exonuclease digestion. The nuclease Bal 31 from the culture medium of the bacterium *Alteromonas espejiana* is particularly useful for this purpose. With it, a set of clones with progressively larger deletions into a region can easily be isolated. The addition of linkers after Bal 31 digestion permits targeted substitution of a set of nucleotides or a change in the number of nucleotides between two sites. Deletions entering the region from both directions are isolated. Before recloning, a restriction enzyme linker is added. After these steps, a pair of deletions can be easily joined via their linkers to generate a DNA molecule identical to the wild-type except for the alteration of a stretch comprising the linker (Fig. 10.21). The use of different pairs of deletions place the linker in different locations so that the linker can be scanned through a region to determine important areas.

Bases within DNA fragments can be changed with chemical *in vitro* mutagenesis. Hydroxylamine will effectively mutagenize the cytosines in denatured DNA fragments, which can then be renatured and recloned. Alternatively, mutagenesis can be directed to particular regions. One method is to generate a single-stranded region by nicking one

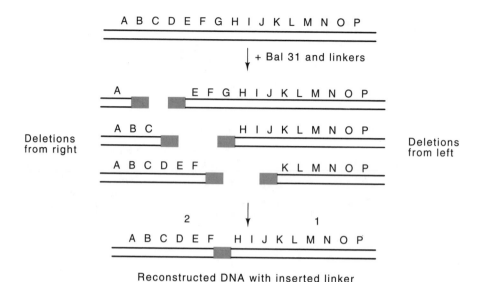

Figure 10.21 Digestion with Bal 31 from either direction and addition of linkers generates a set of molecules that can be rejoined *via* the linkers to yield a molecule like the original wild type but with a substitution of some nucleotides.

strand as a result of digestion with a restriction enzyme in the presence of ethidium bromide and then briefly digesting with exonuclease III to generate a gap. The mutagenesis is then performed with a single-stranded specific reagent such as sodium bisulfite, which mutagenizes cytosines and ultimately converts them to thymines, or by compelling misincorporation of bases during repair of a gap.

Figure 10.22 Insertional inactivation of a yeast gene.

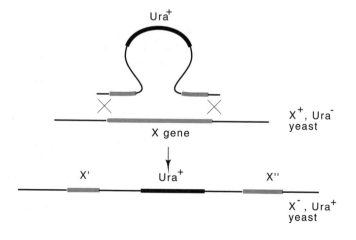

Insertional inactivation can be used to kill specific genes in yeast (Fig. 10.22). This is a prerequisite to examining the *in vivo* consequences of mutating the gene. Suppose that a cloned copy of the gene to be inactivated is available. Then the central portion of the gene can be replaced by a segment of DNA encoding one of the genes necessary for the synthesis of uracil. Uracil-requiring yeast cells are transformed with the segment of DNA containing the gene segments and the URA region, and selection is performed for cells able to grow without exogenously added uracil. Since the ends of the transforming DNA segment are highly recombinogenic, the fragment recombines into the X gene with high frequency and replaces the former intact copy of the X gene with the damaged copy. This replacement relieves the uracil requirement of the cells. That the necessary construct has been generated can be verified by Southern transfers. Restriction sites flanking the insertion are moved further apart, increasing the size of this restriction fragment.

When the steps described above are performed on diploid yeast cells, the result is one chromosome with an insertionally inactivated copy of the X gene and a second, normal copy of the X gene. To test whether the X gene is required for growth in haploid cells, haploids containing the two chromosome types can then be generated by sporulating the diploids. If the gene with the insertion is completely inviable, only two of the four spores from each tetrad will be viable.

Mutagenesis with Chemically Synthesized DNA

Khorana laid the groundwork for chemical synthesis of DNA. He developed techniques to form the phosphodiester bond between nucleotides

Figure 10.23 Outline of one method for the chemical synthesis of DNA. Note that elongation proceeds from the 3' to the 5' end in this method.

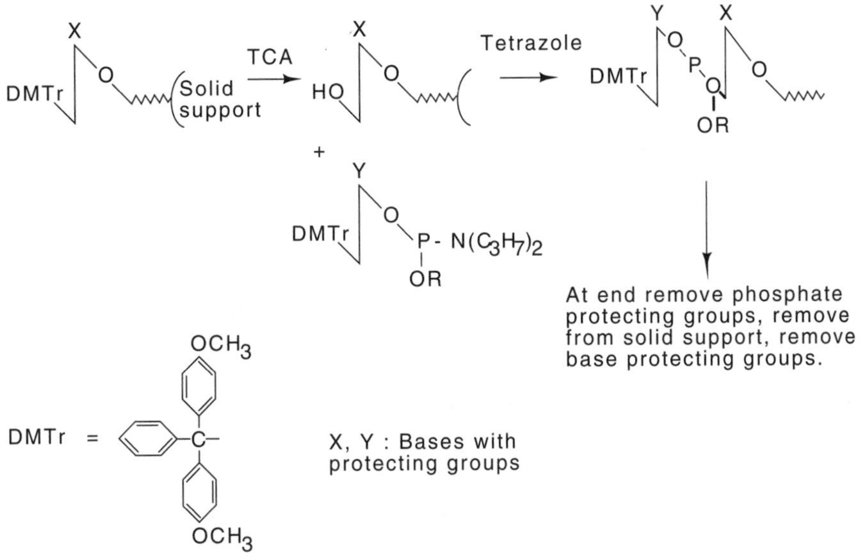

while at the same time preventing the reactive amino, hydroxyl, and other phosphorus groups from reacting. With these techniques, he and his co-workers then synthesized a complete tRNA gene. Originally many person-years were required for the synthesis of 80 nucleotide oligomers. Now, as a result of continued development by many research groups, oligonucleotide synthesis has been highly automated and as many as 100 nucleotides may be joined in specific sequence in a day.

In chemical DNA synthesis, blocking groups are placed on the reactive groups that are not to participate in the condensation to form a phosphodiester. These are then condensed to build the oligonucleotide (Fig. 10.23). After synthesis of the complete oligonucleotide, all the blocking groups are removed. If the desired oligonucleotide is particularly long, blocks of short, overlapping oligonucleotides can be synthesized, hybridized, ligated, and finally cloned.

Before 1965, no researchers would have had a good idea of what to do with the sequence of an entire chromosome if it were presented to them. We are hardly in that situation now. Similarly, before 1975 there seemed to be little reason to try to synthesize DNA chemically. Not only were relatively few interesting sequences known, but the fraction of the synthesized material that would possess the desired sequence was likely to be too small to be of use. With the development of cloning since 1975 and the overall increase in our knowledge of biological mechanisms, the picture dramatically changed. Now it is routine to synthesize a gene *de novo*. Convenient restriction sites can be placed through the gene and

Figure 10.24 Mutagenesis with chemically synthesized DNA. The oligomer hybridizes, except for the mispaired base. Extension of the primer with DNA Pol I and ligation yields heteroduplex molecules that can be transformed. Following DNA replication in the transformants, the two types of DNA molecules segregate to yield wild-type or mutant homoduplexes. Retransformation yields colonies containing entirely mutant or wild-type DNA, which can be identified by hybridization with radioactive mutant oligomer.

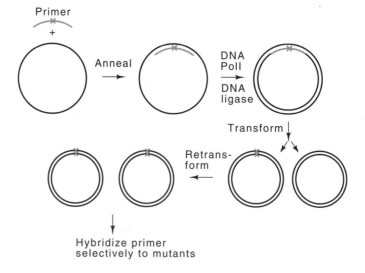

when necessary, portions of the gene can be altered by synthesis of just the region between two restriction enzyme cleavage sites.

Another method for mutating a gene is to direct mutations to a specific point. This can be done with chemically synthesized oligonucleotides in a process called oligonucleotide directed mutagenesis. An oligonucleotide containing the desired alteration, a mutation, insertion, or deletion, will hybridize to complementary, wild-type single-stranded DNA and can serve as a primer for DNA pol I (Fig. 10.24). The resulting double-stranded DNA contains one wild-type strand and one mutant strand. Upon replication in cells, one of the daughter duplexes is wild-type and the other is mutant. Sometimes it is necessary to prevent heteroduplex repair of the mutant strand. Either way, following transformation and segregation, a mutant gene can be obtained.

Problems

10.1. DNA can be injected into a mouse egg and implanted in a pseudopregnant female from which baby mice are ultimately obtained that contain copies of the injected DNA. How would you explain the finding that sometimes the resultant mice are mosaics, that is, some of the cells contain copies of the injected DNA and others do not? How could you make use of these findings to determine useful information about development of the mouse?

10.2. The homology between two sequences S1 and S2 can be compared by the dot matrix method (see, for example, Cell *18*, 865-873 [1979]) as follows: A dot is placed in the ith row and jth column of the comparison matrix (usually generated by computer) if the sequences S1 and S2 possess n out of m identical bases beginning with the ith base of S1 and the jth base of S2. Typical operating parameters would be to require identity in 8 out of 16 bases, that is, n = 8 and m = 16. Regions of homology appear as lines of dots running downward to the right. How would you use the dot matrix method to look for complementary sequences regions in an RNA molecule that might base pair with one another?

10.3. Suppose you wished to use directed mutagenesis against a portion of a gene coding for an essential enzyme in haploid yeast. How could you simultaneously inactivate the resident chromosomal gene and substitute a portion of the gene that you had modified *in vitro*?

10.4. The restriction enzyme *Xba*I cleaves the sequence TTAGA and *Hin*dIII cleaves the sequence AAGCTT. Suppose *Hin*dIII cleaves a plasmid once in a nonessential region. By what simple procedure not involving linkers or similar oligonucleotides could you clone *Xba*I-cut restriction fragments into *Hin*dIII-cut plasmid and at the same time eliminate any problems from reclosure of the plasmid without insertion of the *Xba*I-cut DNA?

10.5. List and briefly explain the important considerations when deciding on the nucleotide sequence to use when building a synthetic gene for a protein.

10.6. If the efficiency of nucleotide addition in a DNA synthesizer is 98% per cycle, what would the overall efficiency be if you were synthesizing a unique DNA sequence 150 bases long? What would you do to obtain sizeable amounts of the sequence?

10.7. How could you use PCR technology to build gene fusions if you could have the sequences for the genes involved?

10.8. Occasionally in walking along a chromosome of an organism like *Drosophila*, a region of repeated DNA of the type used for generation of DNA fingerprints is encountered. What problems does this generate for the walk, and what could be done to jump over the region?

10.9. Why does the mapping of chromosomal markers using RFLPs require statistical analysis?

10.10. How would you use the polymerase chain reaction to determine the splice site in an mRNA?

10.11. Sketch the products of PCR amplification after the first few cycles when the original DNA template extends well beyond the sites to which the primers hybridize.

10.12. How could you use PCR to introduce a restriction enzyme cleavage site at a specific position on a stretch of DNA you have cloned on a plasmid?

10.13. Recently it has been found that a human genetic disease is worsened in succeeding generations. A specific repeated triplet within a coding region of a particular gene increases from repeating several times, to repeating perhaps half a dozen times. Devise a simple assay that could detect the presence of such an expansion in a small cell sample from a fetus.

10.14. The * in the sequences below indicate positions where ethylation of a phosphate prevents protein A from binding to sequence a and protein B from binding to sequence b. What do you conclude?

(a) 5'-N-N*N*N*N*N-N-N-N-N-3'   (b) 5'-N-N-N-N-N*N*N*N*N-N-N
     3'-N-N-N-N-N-N*N*N*N*N-5'   3'-N-N*N*N*N*N-N-N-N-N-N

10.15. To perform in vitro mutagenesis, the synthetic oligonucleotides A and B were hybridized to the template strand as shown, a

Template

A   B

DNA polymerase lacking 5' to 3' exonuclease activity elongated the DNA, and DNA ligase was added. What oligonucleotides would permit PCR amplification of just the sequence represented on the bottom strand?

References

Recommended Readings

Individual-Specific "Fingerprints" of Human DNA, A. Jeffreys, V. Wilson, S. Thein, S., Nature 316, 76-79 (1985).

Searching for Pattern and Mutation in the *Drosophila* Genome with a p-*lacZ* Vector, S. Barbel, L. Ackerman, R. Crretto, T. Uemura, E. Grell, L. Jan, Y. Jan, Genes and Development 3, 1273-1287 (1989).

Cloning the Difference between Two Complex Genomes, N. Lisitsyn, N. Lisitsyn, M. Wigler, Science 259, 946-951 (1993).

Cloning and Identification Techniques

Hybridization of RNA to Double-stranded DNA: Formation of R-Loops, M. Thomas, R. White, R. Davis, Proc. Nat. Acad. Sci. USA 73, 2294-2298 (1976).

Filter Replicas and Permanent Collections of Recombinant DNA Plasmids, J. P. Gergen, R. H. Stern, P. C. Wensink, Nucleic Acids Res. 7, 2115-2136 (1979).

Total Synthesis of a Gene, H. Khorana, Science 203, 614-625 (1979).

Chemical DNA Synthesis and Recombinant DNA Studies, K. Itakura, A. Riggs, Science 209, 1401-1405 (1980).

Propagation of Foreign DNA in Plants Using Cauliflower Mosaic Virus as Vector, B. Gronenborn, R. Gardner, S. Schaefer, R. Shepherd, Nature 294, 773-776 (1981).

Efficient Isolation of Genes Using Antibody Probes, R. Young, R. Davis, Proc. Nat. Acad. Sci. USA 80, 1194-1198 (1983).

Chromosomal Walking and Jumping to Isolate DNA from *Ace* and *rosy* Loci and the Bithorax Complex in *Drosophila melanogaster*, W. Bender, P. Spierer, D. Hogness, J. Mol. Biol. 168, 17-33 (1983).

Enzymatic Amplification of β-Globin Genomic Sequences and Restriction Site Analysis for Diagnosis of Sickle Cell Anemia, R. Saiki, S. Scharf, F. Faloona, K. Mullis, G. Horn, H. Erlich, N. Arnheim, Science 230, 1350-1354 (1985).

Detection of Single Base Substitutions by Ribonuclease Cleavage at Mismatches in RNA:DNA Duplexes, R. Myers, Z. Larin, T. Maniatis, Science 230, 1242-1246 (1985).

Construction of an Ordered Cosmid Collection of the *Escherichia coli* K-12 W3110 Chromosome, S. Tabata, A. Higashitani, M. Takanami, K. Akiyama, Y. Kohara, Y. Nichimura, S. Yasuda, Y. Hirota, J. Bact. 171, 1214-1218 (1989).

By-passing Immunization, Human Antibodies from V-gene Libraries Displayed on Phage, J. Marks, H. Hoogenboom, T. Bonnert, J. McCafferty, A. Griffiths, G. Winter, J. Mol. Biol. 222, 581-597 (1991).

Screening for *in vivo* Protein-protein Interactions, F. Germino, Z. Wang, S. Weissman, Proc. Natl. Acad. Sci. USA 90, 933-937 (1993).

Sequencing and Analysis of Clones

A Simple Method for DNA Restriction Site Mapping, H. Smith, M. Birnstiel, Nucleic Acids Res. 3, 2387-2398 (1976).

Mapping Adenines, Guanines and Pyrimidines in RNA, H. Donis-Keller, A. Maxam, W. Gilbert, Nucleic Acids Res. 4, 2527-2538 (1977).

The Use of R-looping for Structural Gene Identification and mRNA Purification, J. Woolford, M. Rosbash, Nucleic Acids Res. *6*, 2483-2497 (1979).

Two-dimensional S1 Nuclease Heteroduplex Mapping: Detection of Rearrangements in Bacterial Genomes, T. Yee, M. Inouye, Proc. Nat. Acad. Sci. USA *81*, 2723-2727 (1984).

Fluorescence Detection in Automated DNA Sequence Analysis, L. Smith, J. Sanders, R. Kaiser, P. Hughes, C. Dodd, C. Connell, C. Heiner, S. Kent, L. Hood, Nature *321*, 674-679 (1986).

A Strategy for High-volume Sequencing of Cosmid DNAs: Random and Directed Priming with a Library of Oligonucleotides, F. Studier, Proc. Natl. Acad. Sci. USA *86*, 6917-6921 (1989).

DNA Sequencing by Hybridization: 100 Bases Read by a Non-gel-based Method, Z. Strezoska, T. Paunesku, D. Radosavljevic, I. Labat, R. Dramanac, Proc. Natl. Acad. Sci. USA *88*, 10089-10093 (1991).

Mapping Genes and Chromosomes

The Use of Synthetic Oligonucleotides as Hybridization Probes II. Hybridization of Oligonucleotides of Mixed Sequence to Rabbit β-Globin DNA, R. Wallace, M. Johnson, T. Hirose, T. Miyake, E. Kawashima, K. Itakura, Nuc. Acids Res. *9*, 879-894 (1981).

Chromosomal Localization of Human Leukocyte, Fibroblast, and Immune Interferon Genes by Means of *in situ* Hybridization, J. Trent, S. Olson, R. Lawn. Proc. Nat. Acad. Sci. USA *79*, 7809-7813 (1982).

A Polymorphic DNA Marker Genetically Linked to Huntington's Disease, J. Gusella, N. Wexler, P. Conneally, S. Naylor, M. Anderson, R. Tanzi, P. Walkins, K. Ottina, M. Wallace, A. Sakaguchi, A. Young, I. Shoulson, E. Bonilla, J. Martin, Nature *306*, 234-238 (1983).

DNA Polymorphic Loci Mapped to Human Chromosomes 3, 5, 9, 11, 17, 18, and 22, S. Naylor, A. Sakaguchi, D. Barker, R. White, T. Shows, Proc. Nat. Acad. Sci. USA *81*, 2447-2451 (1984).

Construction of Linkage Maps with DNA Markers for Human Chromosomes, R. White, M. Leppert, D. Bishop, D. Barker, J. Berkowitz, C. Brown, P. Callahan, T. Holm, L. Jerominski, Nature *313*, 101-105 (1985).

Forensic Application of DNA "Fingerprints," P. Gill, A. Jeffreys, D. Werrett, Nature *318*, 577-579 (1985).

Toward a Physical Map of the Genome of the Nematode *Caenorhabditis elegans*, A. Coulson, J. Sulston, S. Brenner, J. Karn, Proc. Nat. Acad. Sci. USA *83*, 7821-7825 (1986).

Random-clone Strategy for Genomic Restriction Mapping in Yeast, M. Olson, J. Dutchik, M. Graham, G. Brodeur, C. Helms, M. Frank, M. MacCollin, R. Scheinman, T. Frank, Proc. Nat. Acad. Sci. USA *83*, 7826-7830 (1986).

Localization of the Huntington's Disease Gene to a Small Segment of Chromosome 4 Flanked by D4510 and the Telomere, T. Gilliam. R. Tanzi, J. Haines, T. Bonner, A. Farniarz, W. Hobbs, M. MacDonald, S. Cheng, S. Folstein, P. Conneally, N. Wexler, J. Gusella, Cell *50*, 565-571 (1987).

The Physical Map of the Whole *E. coli* Chromosome: Application of a New Strategy for Rapid Analysis and Sorting of a Large Genomic Library, Y. Kohara, K. Akiyama, K. Isono, Cell *50*, 495-508 (1987).

Complete Cloning of the Duchenne Muscular Dystrophy (DMP) cDNA and Preliminary Genomic Organization of the DMD Gene in Normal and Affected Individuals, M. Koinig, E. Hoffman, C. Bertelson, A. Monaco, C. Ferner, L. Kunkel, Cell *50*, 509-517 (1987).

Two-Dimensional DNA Fingerprinting of Human Individuals, A. Uitterlinden, P. Slagboom, D. Knock, J. Vijg, Proc. Natl. Acad. Sci. USA *86*, 2742-2746 (1989).

Allele-specific Enzymatic Amplification of β-globin Genomic DNA for Diagnosis of Sickle Cell Anemia, D. Wu, L. Ugozzoli, B. Pal, R. Wallace, Proc. Natl. Acad. Sci. USA *86*, 2757-2760 (1989).

Completion of the Detailed Restriction Map of the *E. coli* Genome by the Isolation of Overlapping Cosmid Clones, V. Knott, D. Blake, G. Brownlee, Nucleic Acids Res. *17*, 5901-5912 (1989).

Selective Cleavage of Human DNA: RecA-assisted Restriction Endonuclease (RARE) Cleavage, L. Ferrin, R. Camerini-Otero, Science *254*, 1494-1497 (1991).

Construction of Small-insert Genomic DNA Libraries Highly Enriched for Microsatellite Repeat Sequences, E. Ostrander, P. Jong, J. Rine, G. Duyk, Proc. Natl. Acad. Sci. USA *89*, 3419-3423 (1992).

Continuum of Overlapping Clones Spanning the Entire Human Chromosome 21q, I. Chumakov, P. Rigault, S. Guillou, P. Ougen, A. Billant, G. Guasconi, P. Gervy, I. LeGall, P. Soularue, L. Grinas, L. Bougueleret, C. Bellané-Chantelot, B. Lacroix, E. Barillot, P. Gesnouin, S. Pook, G. Vaysseix, G. Frelat, A. Schmitz, J. Sambuey, A. Bosch, X. Estivill, J. Weissenbach, A. Vignal, H. Riethman, D. Cox, D. Patterson, K. Gardiner, M. Hattori, Y. Sakaki, H. Ichikawa, M. Ohki, D. LePaslier, R. Heilig, S. Antonarakis, D. Cohen, Nature *359*, 380-387 (1992).

Properties of Normal and Mutant Genes and Chromosomes

Unusual β-Globin-like Gene that has Cleanly Lost Both Globin Intervening Sequences, Y. Nishioka, A. Leder, P. Leder, Proc. Nat. Acad. Sci. USA *77*, 2806-2809 (1980).

A Mouse α-Globin-Related Pseudogene Lacking Intervening Sequences, E. Vanin, G. Goldberg, P. Tucker, O. Smithies, Nature *286*, 222-226 (1980).

Transcriptional Control Signals of a Eukaryotic Protein-coding Gene. S. McKnight, R. Kingsbury, Science *217*, 316-324 (1982).

Identification of Mutations Leading to the Lesch-Nyhan Syndrome by Automated Direct DNA Sequencing of *in vitro* Amplified cDNA, R. Gibbs, P. Nguyen. L. McBride, S. Koepf, C. Caskey, Proc. Natl. Acad. Sci. USA *86*, 1919-1923 (1989).

In vitro Mutagenesis and Engineering

Local Mutagenesis: A Method for Generating Viral Mutants with Base Substitutions in Preselected Regions of the Viral Genome, D. Shortle, D. Nathans, Proc. Nat. Acad. Sci. USA *75*, 2170-2174 (1978).

A General Method for Maximizing the Expression of a Cloned Gene, T. Roberts, R. Kacich, M. Ptashne, Proc. Nat. Acad. Sci. USA *76*, 760-764 (1979).

Targeted Deletions of Sequences from Cloned Circular DNA. C. Green, C. Tibbetts, Proc. Nat. Acad. Sci. USA *77*, 2455-2459 (1980).

Improved Methods for Maximizing Expression of a Cloned Gene: A Bacterium that Synthesizes Rabbit β-Globin, L. Guarente, G. Lauer, T. M. Roberts, M. Ptashne, Cell *20*, 543-553 (1980).

Segment-specific Mutagenesis: Extensive Mutagenesis of a *lac* Promoter/Operator Element, H. Weiher, H. Schaller, Proc. Nat. Acad. Sci. USA *79*, 1408-1412 (1982).

Gap Misrepair Mutagenesis: Efficient Site-directed Induction of Transition, Transversion, and Frameshift Mutations *in vitro*, D. Shortle, P. Grisafi, S. Benkovic, D. Botstein, Proc. Nat. Acad. Sci. USA *79*, 1588-1592 (1982).

Site-Specific Mutagenesis of *Agrobacterium* Ti Plasmids and Transfer of Genes to Plant Cells, J. Leemans, C. Shaw, R. Deblaere, H. DeGreve, J. Hernalsteens, M. Maes, M. Van Montegu, J. Schell, J. Mol. and Applied Genetics *1*, 149-164 (1982).

A Bacteriophage T7 RNA Polymerase/Promoter System for Controlled Exclusive Expression of Specific Genes, S. Tabor, C. Richardson, Proc. Nat. Acad. Sci. USA *82*, 1074-1078 (1985).

Defining the Consensus Sequences of *E. coli* Promoter Elements by Random Selection, A. Oliphant, K. Struhl, Nuc. Acids Res. *16*, 7673-7683 (1988).

An Efficient Method for Generating Proteins with Altered Enzymatic Properties: Application to β-lactamase, A. Oliphant, K. Struhl, Proc. Natl. Acad. Sci. USA *86*, 9094-9098 (1989).

Conversion of a Helix-turn-helix Motif Sequence-specific DNA Binding Protein into a Site-specific DNA Cleavage Agent, R. Ebright, Y. Ebright, P. Pendergrast, A. Gunaskera, Proc. Natl. Acad. Sci. USA *87*, 2882-2886 (1990).

Correction of the Cystic Fibrosis Defect *in vitro* by Retrovirus-mediated Gene Transfer, M. Prumm, H. Pope, W. Cliff, J. Rommens, S. Marvin, L. Tsui, F. Collins, R. Frizzell, J. Wilson, Cell *62*, 1227-1233 (1990).

Specialized Genetic Engineering Techniques

DNAse Footprinting: A Simple Method for the Detection of Protein-DNA Binding Specificity, D. Galas, A. Schmitz, Nucleic Acids Res. *5*, 3157-3170 (1978).

Genetic Transformation of *Drosophila* with Transposable Element Vectors, G. Rubin, A. Spradling, Science *218*, 348-353 (1982).

Genomic Sequencing, G. Church, W. Gilbert, Proc. Nat. Acad. Sci. USA *81*, 1991-1995 (1984).

Hydroxyl Radical "Footprinting:" High-resolution Information About DNA-Protein Contacts and Application to Lambda Repressor and Cro Protein, T. Tullius, B. Dombroski, Proc. Nat. Acad. Sci. USA *83*, 5469-5473 (1986).

Primer-directed Enzymatic Amplification of DNA with a Thermostable DNA Polymerase, R. Saiki, D. Gelfand, S. Staffel, S. Scharf, R. Higuchi, G. Horn, K. Mullis, H. Erlich, Science *239*, 487-491 (1988).

Optimized Conditions for Pulsed Field Gel Electrophoretic Separations of DNA, B. Birren, E. Lai, S. Clark, L. Hood, M. Simon, Nucleic Acids Res. *16*, 7563-7582 (1988).

Identification of the Skeletal Remains of a Murder Victim by DNA Analysis, E. Hagelberg, I. Gray, A. Jeffreys, Nature *352*, 427-429 (1991).

Assembly of Combinatorial Antibody Libraries on Phage Surfaces: The Gene III Site, C. Barbas III, A. Kang, R. Lerner, S. Benkovic, Proc. Natl. Acad. Sci. USA *88*, 7978-7982 (1991).

The Two-hybrid System: A Method to Identify and Clone Genes for Proteins that Interact with a Protein of Interest, C. Chien, P. Bartel, R. Sternglanz, S. Fields, Proc. Natl. Acad. Sci. *88*, 9578-9582 (1991).

Karyoplasmic Interaction Selection Strategy: A General Strategy to Detect Protein-protein Interactions in Mammalian Cells, F. Fearon, T. Finkel, M. Gillison, S. Kennedy, J. Casella, G. Tomaselli, J. Morrow, C. Dang, Proc. Natl. Acad. Sci. USA *89*, 7958-7962 (1992).

Repression and the *lac* Operon

11

Having discussed genetics, genetic engineering and the structure and biosynthesis of proteins and nucleic acids in the previous chapters, we are now prepared to consider biological regulatory mechanisms. What exactly is meant by "mechanism of regulation?" The term refers to the means by which the expression of a specific gene or set of genes is selectively increased or decreased, that is, induced or repressed. Thus a protein that binds only under some conditions to the promoter of one particular gene could regulate expression of this gene, whereas ATP would not be considered a regulator despite the fact that its presence is necessary for expression of the gene. Changes in ATP levels would be expected to affect expression of all genes similarly.

This section of the book describes regulation of the *lac*, *ara*, and *trp* operons, lambda phage genes, *Xenopus* 5S genes, yeast mating-type genes, and genes involved in development. The prokaryotic genes are chosen because much is known about their regulation and because each is regulated by a dramatically different mechanism. The eukaryotic systems represent clear examples of regulation and are good examples of the use of recombinant DNA technology being applied to this type of problem.

Much is known about the four prokaryotic systems, in part because intensive study was begun on them well before similar studies became possible on eukaryotes. Even before the development of genetic engineering, investigators of the bacterial systems devised means of combining genetics, physiology, and physical-chemical studies. The depth to which the studies have penetrated has revealed general principles that are likely to be operant in cells of any type. A broad diversity of regulation mechanisms are used in the systems considered here. One type is simple competition between lactose repressor and RNA polymerase for binding to DNA of the promoter region for the *lac* operon. A more complex system is seen in the *ara* operon in which a series of proteins binds to the DNA and assists RNA polymerase to initiate transcription of the *ara* operon. The *trp* operon displays a different facet of gene regulation. In it, translation is coupled to transcription. Phage lambda and the yeast mating-type genes show still more complex regulatory behavior. In lambda, a highly intricate cascade of regulatory proteins regulates gene expression. The detailed mechanisms of gene regulation are not as well known in eukaryotic systems, although many are regulated by enhancer-binding proteins. The general principles used in regulating the gene systems considered here cover a broad range, but nature's diversity in gene regulation mechanisms is certainly not exhausted by these. Other regulation systems display both minor and major differences.

Background of the *lac* Operon

The initial studies on the *lac* system, like those on most other bacterial systems, were genetic. At the Pasteur Institute in Paris during World War II, Monod began a study of the process of adaptation of *E. coli* to growth on medium containing lactose. This first led to studies on the origin of the enzymes that were induced in response to the addition of lactose to the medium and then to studies on how the induction process was regulated. Research on the *lac* operon at the Pasteur Institute flourished and spread around the world, and for many years was the most active research area in molecular biology.

By now, the essential regulatory properties of the lactose system have been characterized as a result of extensive physiological, genetic, and biochemical analyses. By no means, however, do we fully understand what happens to regulate transcription even in this simple system. Current research on the lactose system includes study of what proteins participate, which other proteins they interact with, and what they do. It also includes the more fundamental questions of how RNA polymerase recognizes promoters and initiates transcription and how proteins fold during their synthesis, recognize their substrates, and bind to other proteins or to specific sequences on DNA.

Figure 11.1 The *lac* operon of *E. coli* showing the regulatory gene *lacI*, the promoters p_{lac} and p_I, operator *O*, and the genes *lacZ*, *lacY*, and *lacA*.

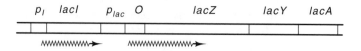

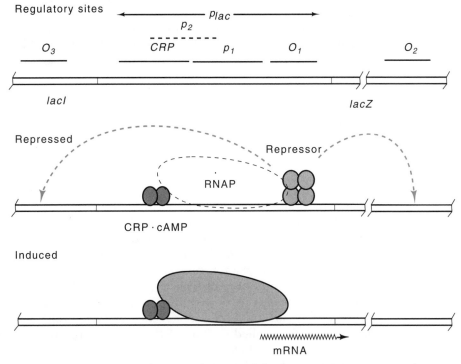

Figure 11.2 Schematic drawing of the *lac* operon in a repressed state and in an induced state in the presence of cAMP.

The lactose system consists of four genes that enable *E. coli* to grow on lactose as a source of carbon and energy (Fig. 11.1). The regulatory protein of the system is the product of the *lacI* gene. It is a repressor, and when it is bound to the DNA it prevents transcription of the *lacZ*, *lacY*, and *lacA* genes. The product of the *lacZ* gene is the enzyme β-galactosidase which cleaves lactose to yield glucose and galactose. The *lacY* product transports lactose into the cell, and the *lacA* product acetylates some toxic galactosides to detoxify them. The three genes, *lacZ*, *lacY*, and *lacA* constitute an operon by the strict definition of the word since a single promoter serves more than one gene. We will also use the word "operon" to refer to a transcriptional unit plus its related regulatory sequences even if it contains a single gene.

Figure 11.2 shows the picture that has been derived for the mechanism of regulation of the *lac* operon in *E. coli*. The *lacI* product binds to a specific site in the promoter region, termed the operator or O_1 site. While bound to this operator, and in conjunction with the two other operators located nearby, O_2 and O_3, it prevents RNA polymerase from binding to or moving from the p_1 promoter. The operators O_2 and O_3 are often called pseudo-operators, and they increase repressor binding at O_1 through loop formation in which repressor simultaneously binds to O_1 and either O_2 or O_3. Partially overlapping the p_1 promoter is a second promoter, p_2. When repressor is not bound to the operators, then

RNA polymerase has access to the two partially overlapping promoters p_1 and p_2. Little transcription results from polymerase binding to p_2 because its isomerization rate is low. Normally, p_2 plays no significant role in the system because CRP protein at its binding site blocks access of polymerase to p_2. CRP also directly stimulates transcription from p_1 promoter.

After lactose is added to a growing culture, the synthesis rate of the *lac* enzymes ultimately is increased about a thousandfold. Curiously, the induction requires activity of the very enzyme that is being induced, β-galactosidase. The low uninduced, or basal, level of β-galactosidase hydrolyzes some of the lactose that leaks into the cells or that enters via the basal levels of *lacY* protein. In a side reaction the enzyme also generates allolactose from lactose. The binding of allolactose to repressor reduces repressor's affinity for operator by about a thousandfold. As a result, repressor tends to dissociate from operator and transcription can begin. Shortly, the induced levels of β-galactosidase and lactose transporter lead to higher intracellular levels of lactose and allolactose and repression is further reduced.

The efficient initiation of transcription by RNA polymerase at the promoter of the *lac* operon requires not only the absence of bound *lac* repressor but also the presence of cyclic AMP and cyclic AMP receptor protein, which is called CAP or CRP protein. This auxiliary induction requirement is thought to result from the fact that the carbon and energy requirements of *E. coli* are most efficiently met by catabolizing glucose rather than other sugars. Consequently, cells have evolved a way to shut down the possibly inefficient use of other carbon utilization pathways if glucose is present. This phenomenon is known as the glucose effect or catabolite repression. The glucose effect is generated in two ways: by excluding inducers of some operons from the cell and by reducing the inducibility of some operons. When glucose is present and is being metabolized, the concentrations of cAMP are low, and few CRP proteins contain bound cAMP. Only when CRP has bound cAMP, which occurs when glucose is absent, can the protein specifically bind to DNA and assist RNA polymerase to initiate transcription of the CRP-dependent operons such as *lac*.

The Role of Inducer Analogs in the Study of the *lac* Operon

Compared to progress on many other problems in molecular biology, research on the *lac* operon proceeded rapidly. In part, this was due to the fact that measuring the response of the *lac* operon to changes in the cell's environment was easy. Addition of lactose induces the operon a thousandfold, and the assay of β-galactosidase is particularly simple. Another important reason for the rapid progress is that many useful analogs of lactose can readily be synthesized. These analogs facilitate the assay of the *lac* operon proteins as well as simplify the isolation of useful mutants.

The properties of several analogs are shown in Table 11.1. ONPG is particularly useful as its cleavage yields galactose and orthonitrophenol,

Table 11.1 Properties of Lactose Analogs

Analog	Inducer	Trans-ported	β-gal substrate	Comments
Lactose	-	+	+	Not an inducer
Allolactose	+	?	+	True inducer
Phenyl-β-D-galactose	-	+	+	Selects nonrepressing mutants
Orthonitro-phenyl-β-D-galactose (ONPG)	?	+	+	Produces yellow nitrophenol on hydrolysis
Thio-ONPG	-	+	+	Hydrolysis product is toxic
Isopropyl β-D-thio-galactoside (IPTG)	+	+	-	Nonmetabolized inducer
Orthonitro-phenyl-β-D-fucoside	-	-	-	Inhibits induction
X-gal, 5-bromo-4-chloro-3-indolyl-β-D-galactoside	-	-	+	Produces blue dye on hydrolysis

which when ionized at basic pH is bright yellow. This provides a simple and sensitive assay for β-galactosidase which requires only a measurement of the optical density after incubating enzyme and substrate. IPTG is useful in inducing the operon for two reasons. First, IPTG induces by binding directly to the repressor so that the presence of functional β-galactosidase in the cells to generate the true inducer is unnecessary. Second, IPTG is not metabolized. Therefore its use for induction does not disturb the metabolic pathways of the cell as might be the case for induction by lactose. Phenylgalactose is another useful analog. It is not an inducer, but it can be metabolized if β-galactosidase is present to hydrolyze it. Therefore only permanently derepressed cells grow on medium containing phenylgalactose as the sole carbon and energy source. Such permanently induced mutants are called constitutive and result either from defective repressor that is unable to bind to the operator or from mutant operator that is no longer recognized and bound by repressor.

Proving *lac* Repressor is a Protein

The phenomenon of repression in the *lac* operon was characterized by genetic experiments. These showed that a product of the *lacI* gene diffused through the cell and shut off expression of the *lac* genes. Basically these were complementation experiments that showed that *lacI* acted in *trans* to repress. A LacI⁻LacZ⁺ strain constitutively expresses β-galactosidase in the absence or presence of inducers. The *lacZ* gene in such a strain is repressed if an episome is introduced which carries a good *lacI* gene but is deleted of the *lacZ* gene. The genetic structure of the resulting strain is denoted F'*lacI*+ Δ*lacZ/lacI*⁻*lacZ*+. That is, LacI protein acts in *trans* to repress the chromosomal *lacZ* gene.

Once it was established that the LacI product repressed expression, research on the *lac* operon shifted to learning what repressor was; whether it was RNA or protein, and how it acted to block expression of the *lac* enzymes. Since the target of the repressor was a sequence that could be genetically mapped, two likely possibilities existed. The operator could be a region on the DNA to which repressor bound to reduce the intracellular levels of functional *lac* messenger. Alternatively, repressor recognized and bound to a region on the *lac* messenger and reduced translation efficiency. Either of these possibilities required recognition of a specific nucleotide sequence by the *lacI* gene product. To many investigators, such an ability seemed most reasonable for an RNA molecule. Indeed, early results obtained with RNA and protein synthesis inhibitors led to the conclusion that it was not protein. Gilbert and Müller-Hill, however, reasoned that an RNA molecule was unlikely to possess all the regulatory properties required of the *lacI* product. Therefore Müller-Hill designed a simple genetic experiment that indicated that repressor contained at least some protein.

His proof was that nonsense mutations could be isolated in the *lacI* gene. A nonsense mutation can only be isolated in a gene that encodes a protein. The starting point was a strain containing amber mutations in both the *gal* and *his* operons. This strain was then made LacI⁻ by selection of constitutives. Most of these mutants were missense, but a few were nonsense I⁻ mutants. Each of the constitutives was then reverted simultaneously to Gal⁺ and His⁺. The only way a cell could perform this double reversion was to become nonsense suppressing, Su⁺, and simultaneously suppress both nonsense mutations. These nonsense-suppressing revertants were then tested for *lac* constitutivity,

$$\text{Gal}^-_{\text{non}}\text{ His}^-_{\text{non}} \longrightarrow \text{Gal}^-\text{His}^-\text{LacI}^- \xrightarrow{\begin{array}{c}\text{Gal}^+\\\text{His}^+\end{array}} \text{Gal}^+\text{His}^+(\text{Su}^+)\left\{\begin{array}{l}\text{LacI}^+, \text{ nonsense I}^-\\\text{LacI}^-, \text{ missense I}^-\end{array}\right.$$

and a few were found to have simultaneously become I⁺. Consequently, these were nonsense I mutations that were now being suppressed and were forming functional repressor that turned off expression of the *lac* operon in the absence of inducers. This proves that the I gene product contains protein, but it does not prove that the repressor is entirely protein. The repressor had to be purified before it could be shown to be entirely protein.

An Assay for *lac* Repressor

Genetic and physiological experiments investigating properties of the *lac* operon provided information from which a number of regulatory mechanisms were proposed. These ranged from the logical mechanism of *lac* repressor binding to DNA and inhibiting transcription to complicated translational control mechanisms utilizing tRNA molecules. Clear demonstration of the regulation mechanism required purification of its components and *in vitro* reconstruction of the *lac* system.

The most important step in the reconstruction of the *lac* regulatory system was the ability to detect repressor. Lac repressor, of course, had to be highly purified from lysed cells. If regulation of the *lac* operon were efficient–and that is the main reason for the existence of regulation–then the cell should contain far fewer molecules of repressor than of the induced gene products. Furthermore, since *lac* repressor possessed no known enzymatic activity, no easy and sensitive assay for repressor was available. Without the ability to detect repressor, its purification was impossible because any fraction obtained from purification steps that was enriched in repressor could not be identified.

Repressor's only known property was that it bound inducer, IPTG being one. Therefore Gilbert and Müller-Hill developed an assay of *lac* repressor based on the protein's ability to bind to inducer molecules. Equilibrium dialysis can detect a protein that binds a particular small molecule. The protein solution to be assayed is placed in a dialysis sack and dialyzed against a buffer that contains salts to maintain the pH and ionic strength and the small molecule that binds to the protein (Fig. 11.3).

In the case of repressor, radioactive IPTG was used. After equilibrium has been attained, the concentration of free IPTG inside and outside the sack is equal, but in addition, inside the sack are the molecules of IPTG that are bound to repressor. If the concentration of repressor is sufficiently high, the increased amount of IPTG inside the sack due to the presence of repressor can be detected. Both the inside and outside concentrations of IPTG can be determined by measuring the amount of radioactivity contained in samples of known volumes taken from outside and inside the dialysis sack.

Does an equilibrium dialysis assay possess sufficient sensitivity to detect the small amounts of *lac* repressor that are likely to exist in crude

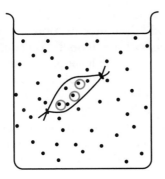

Figure 11.3 Representation of equilibrium dialysis in which the concentration of free IPTG inside and outside the sack is equal, but repressor in the sack holds additional IPTG inside the sack.

extracts of cells? The binding reaction between repressor and IPTG is closely approximated by the reaction

$$R_f + IPTG \rightleftharpoons R \cdot IPTG$$

where R_f is the concentration of free repressor, $IPTG$ is the concentration of free IPTG, and $R \cdot IPTG$ is the concentration of the complex between repressor and IPTG. A dissociation constant K_D describes the relations between the concentrations:

$$K_D = \frac{R_f \times IPTG}{R \cdot IPTG}.$$

Substituting the conservation equation, $R_f + R \cdot IPTG = R_t$, where R_t is the total amount of repressor, and rearranging yields the relation we need. Biochemists have many different names for the equivalent algebraic rearrangements of this equation but usually call the phenomenon Michaelis-Menten binding,

$$R \cdot IPTG = \frac{R_t \times IPTG}{K_D + IPTG}.$$

The ratio of radioactivity in the samples obtained from inside and outside the sack is

$$\frac{IPTG + R \cdot IPTG}{IPTG} = 1 + \frac{R_t}{K_D + IPTG}.$$

Normally in liquid scintillation, counting a 5% difference between samples with more than 100 cpm can be readily determined. Thus the quantity

$$\frac{R_t}{K_D + IPTG}$$

must be greater than 0.05 for detection of *lac* repressor by this assay.

The Difficulty of Detecting Wild-Type *lac* Repressor

Before trying to detect *lac* repressor, Gilbert and Müller-Hill estimated the signal that could be expected in the equilibrium assay and decided that they were unlikely to detect wild-type repressor. Let us examine such a calculation. Two quantities are needed: the dissociation constant of repressor for IPTG and the concentration of repressor in cell extracts.

To make a crude guess of the dissociation constant of IPTG from *lac* repressor, assume that the basal level of the *lac* operon is proportional to the fraction of *lac* operator uncomplexed with repressor. Assume also that this fraction is doubled if the effective repressor concentration is halved. Consequently, the basal level of the operon is doubled when half the total repressor in the cell is bound to IPTG and half is free of IPTG. The concentration of IPTG at which the 50-50 binding occurs and the enzyme level is twice the basal level equals the dissociation constant for

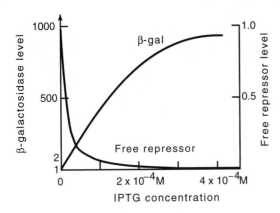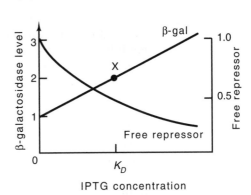

Figure 11.4 The relation between free repressor concentration and the induction level of the *lac* operon. X marks the position where [IPTG] = K_D. At this point repressor concentration has fallen to half, and the basal level has doubled.

IPTG binding to repressor. Although this concentration of IPTG could be measured in reasonably straightforward experiments, it can be calculated from data already known. The *lac* operon can be induced a thousandfold and is half-maximally induced by 2×10^{-4} M IPTG (Fig. 11.4). Roughly, then, at an IPTG concentration of $(2 \times 10^{-4})/500 = 4 \times 10^{-7}$ M IPTG, the level of expression of the *lac* operon will be twice the basal level. Consequently an estimate of the dissociation constant of *lac* repressor and IPTG is 4×10^{-7} M.

The volume of a cell is 10^{-12} cm^3 = 10^{-15} liter. Using this value, a packed cell pellet is $10^{15}/(6 \times 10^{23})$, or about 10^{-9} M in cells. If a cell contains 10 repressor molecules, the concentration of repressor in a packed cell pellet is 10^{-8} M. A cell lysate cannot easily be made at a higher concentration than that obtained by opening cells in a packed cell pellet. Hence a reasonable estimate for the concentration of repressor in the equilibrium dialysis assay is 10^{-8} M.

If high specific activity radioactive IPTG is available, it can be used in the assay at a concentration well below the K_D estimated above (see problems 11.9 and 11.10). Therefore the ratio of the radioactivities contained in samples of equal volumes taken from inside and outside the sack is

$$\frac{\text{Counts inside sack}}{\text{Counts outside sack}} = 1 + \frac{R_t}{K_D}$$

$$= 1 + \frac{10^{-8}}{4 \times 10^{-7}}$$

$$= 1 + 0.025,$$

which is less than can be reliably detected.

Detection and Purification of *lac* Repressor

The previous section showed that wild-type *lac* repressor in crude extracts of cells was not likely to produce a detectable signal in the equilibrium dialysis assay. Therefore Gilbert and Müller-Hill isolated a mutant repressor that bound IPTG more tightly than the wild-type repressor. Crude extracts made from this strain showed an excess of counts in the dialysis sack. The excess was barely detectable; nonetheless it was statistically significant, and fractionation of the extract yielded a protein sample with an easily detectable excess of counts.

Once the assay of *lac* repressor detected something, it was of great importance to prove that the origin of the signal was repressor and not something else. The proof used the tight-binding mutant. First, the tight-binding mutant was used to develop a partial purification of repressor so that a fraction could be obtained in which the signal was readily detectable. Then this same purification procedure was used to obtain a similar fraction from wild-type cells. This too generated a significant signal. The proof came with the demonstration that the apparent dissociation constants for IPTG in the fraction from the mutant and the wild-type were different. This was simply done by performing the dialysis on a series of samples at different concentrations of IPTG. The sample obtained from the mutant bound IPTG more tightly. That is, it had a smaller K_D, than the wild-type (Fig. 11.5). As the only difference between the mutant and the wild-type was a mutation in the *lacI* gene, the signal in the assay was from *lac* repressor.

The definitive detection of repressor opened the door to biochemical studies. First, with an assay, the repressor could in principle be purified and used in biochemical studies probing its mechanism of action. Second, it was possible to attempt to isolate mutants that synthesized elevated quantities of repressor so as to ease the burden of purification. With an assay, such candidates could be identified.

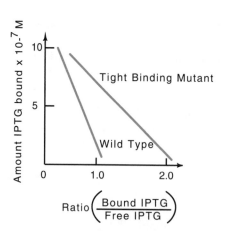

Figure 11.5 Results of equilibrium dialysis at different IPTG concentrations of wild-type repressor and a tight-binding mutant repressor. Rearranging the binding equation derived in the text yields a form convenient for plotting data, $RI = R - [K \times (RI)/I]$. In this form the value of RI when $(RI)/I = 0$ yields the concentration of R molecules capable of binding I and the slope of the binding curve gives K.

A mutation rendering *lac* repressor temperature-sensitive was used in the isolation of mutants possessing higher levels of repressor. Cells were grown at a temperature just high enough to inactivate most of the temperature-sensitive repressor. Consequently the *lac* operon was no longer repressed, and the cells expressing the *lac* operon were then killed. The survivors, which were able to repress the operon, could be of two types. Either the repressor was altered so that it could repress at

$$I^{-ts} \xrightarrow[\substack{\text{high temperature} \\ (43^0)}]{\substack{\text{t-ONPG} \\ \text{glycerol}}} \begin{array}{c} I^+ \\ Z^- \\ Y^- \end{array} \text{phenotype} \xrightarrow{\text{lactose}} I^+ \, Z^+ Y^+$$

the elevated temperature or more repressor was being synthesized. The two types of mutants could easily be distinguished with the equilibrium dialysis assay, and an overproducing mutant was identified.

The selection for the loss of constitutivity in the scheme described above used yet another lactose analog, TONPG (o-nitrophenyl-1-thio-β-D-galactoside). This inhibits growth when it is cleaved by β-galactosidase, but it is not an inducer. Mutant cells unable to cleave this compound grow in its presence. Three types of mutants have this property: the desired repressing mutants as well as *lacZ* and *lacY* mutants. Both of the undesired mutant types were easily eliminated by requiring the mutants to grow on lactose. The selection scheme was successful, and mutants were found that contained elevated amounts of *lac* repressor. The isolation of the *lacI* overproducer was the first clear example of the successful isolation of a promoter mutation and was itself a breakthrough. The resulting I^Q (Q for quantity) mutation mapped at the beginning of the I gene, as expected for a promoter mutation, and generated a 10-fold increase in the level of repressor.

Repressor Binds to DNA: The Operator is DNA

As mentioned in the introduction to this chapter, we now know that repressor acts by binding to the operator and preventing RNA polymerase from binding to the *lac* promoter. The first experiments to show that repressor binds to *lac* DNA used ultracentrifugation of radioactive repressor and the DNA from λ*lac* phage. The DNA sediments at 40S, whereas repressor sediments at 7S. If repressor bound so tightly to DNA that it did not come off during the centrifugation, it would sediment at about 40S. Indeed, DNA containing *lac* carried *lac* repressor along with it down the centrifuge tube, but only if inducers of the *lac* operon were not present (Fig. 11.6). These are the properties expected of repressor if it were to regulate by binding to DNA to prevent transcription.

A considerably easier assay of DNA binding by *lac* repressor was developed later by Riggs and Bourgeois. Pure lambda DNA passes through cellulose nitrate filters, but repressor, like many other proteins,

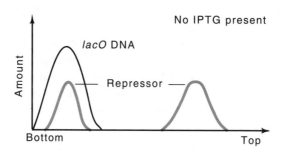

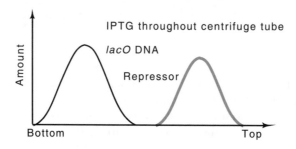

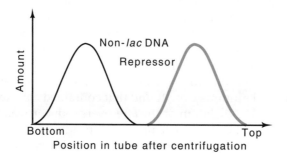

Figure 11.6 R e p r e s s o r binds to operator-containing DNA only in the absence of IPTG.

binds to the filters. It is bound by molecular interactions and not by filtration. Most surprisingly, if repressor is bound to DNA, the DNA molecule will not pass through the filter either. Thus if the DNA is made radioactive, the retention of *lac* DNA on the filter can easily be detected, providing a simple assay for repressor binding to operator (Fig. 11.7).

The filter-binding assay is very sensitive because the DNA can be made highly radioactive with $^{32}PO_4$ The long DNA molecule contains many phosphate groups, and the half-life of ^{32}P is short. This assay facilitated measurement of the rates of binding and dissociation as well as a determination of the equilibrium constant for binding.

For example, the rate of *lac* repressor dissociation from operator is easily measured by mixing repressor with radioactive DNA so that the majority of repressor is bound to operator. Then the solution is diluted and a large excess of nonradioactive *lac* DNA is added. Any repressor that subsequently dissociates from operator either remains free in solution or has a much greater chance of binding to nonradioactive DNA. At intervals after the dilution, the mixture is filtered to determine

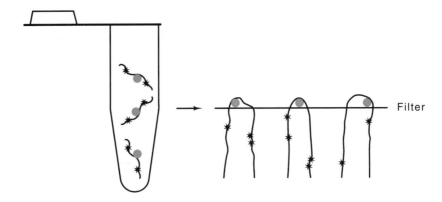

Figure 11.7 Schematic of the filter binding assay for *lac* repressor.

the fraction of repressor still bound to the radioactive *lac* operator-containing DNA. The kinetics in the reduction in the amount of radioactive DNA bound to the filters with increasing time give the dissociation rate of repressor from operator.

The Migration Retardation Assay and DNA Looping

Earlier it was mentioned that DNA can be subjected to electrophoresis under conditions compatible with protein binding. For DNA fragments in the size range of 50 to about 2,000 base pairs, the binding of a protein significantly retards the migration. Therefore, free DNA and protein-DNA complexes may easily be separated by electrophoresis and identified by staining or autoradiography. An additional virtue of the migration retardation assay is the fact that protein and DNA can be incubated in buffers containing physiological concentrations of salt, on the order of 50 mM KCl. Then, the electrophoresis can be performed at very low salt concentration. As discussed earlier, salt has a dramatic effect on the affinity with which most proteins bind to DNA. At low salt concentrations, most proteins' affinity is much higher than at physiological salt concentrations. This greatly reduces the dissociation rate of the proteins. Additionally, the presence of the gel surrounding the protein-DNA complex cages the protein and further reduces its effective dissociation rate during electrophoresis. These features make the gel migration retardation assay particularly useful for the study of protein-DNA interactions as electrophoresis "freezes" a particular solution condition.

Not surprisingly, the *lac* repressor-operator interaction can be studied with the migration retardation assay. The tetrameric *lac* repressor can bind with two of its subunits to the main operator at the promoter. Repressor's other two subunits are free to bind to either of the pseudo-operators that are located one hundred and four hundred base pairs to either side. Such a double binding by a single repressor tetramer forms

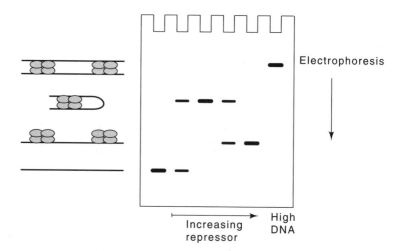

Figure 11.8 Migration retardation assay with *lac* repressor and DNA containing two operators. As repressor concentration is increased, loops, then linear structures are formed. At high DNA concentrations, sandwiches form.

a DNA loop. *In vivo* this looping reaction is facilitated by supercoiling. The same looping reaction can be facilitated *in vitro* by using a linear DNA fragment containing two strong repressor-binding sites separated by 100 base pairs. Incubation of such DNA with low concentrations of repressor permits binding of single tetramers to DNA molecules. These form DNA loops so that each tetramer contacts two operators from the same DNA molecule. Incubation at higher repressor concentrations forces a separate repressor tetramer onto each operator, and incubation at high DNA concentrations forms structures in which repressors join two DNA molecules in a sandwich structure (Fig. 11.8).

The Isolation and Structure of Operator

After the isolation of *lac* repressor and the demonstration that it bound to DNA, interest turned to the structure of operator. The first questions concerned its size and sequence. Answering such questions is now straightforward with the application of genetic engineering techniques. The questions about the *lac* operator, however, came to the fore before the era of genetic engineering. Because ingenious techniques were developed and the work directly led to the development of many of the genetic engineering techniques now used, we will review the techniques, now outdated, used to isolate and sequence the *lac* operator.

Before developing their chemical sequencing method, Gilbert and Maxam sequenced the operator, primarily by direct DNA sequencing methods, whereas somewhat later Reznikoff, Barnes, Abelson, and co-workers sequenced the entire *lac* regulatory region by purifying and

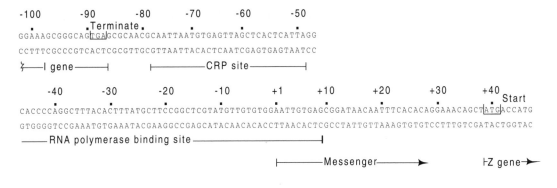

-100   -90   -80   -70   -60   -50

.Terminate.

GGAAAGCGGGCAGTGAGCGCAACGCAATTAATGTGAGTTAGCTCACTCATTAGG
CCTTTCGCCCGTCACTCGCGTTGCGTTAATTACACTCAATCGAGTGAGTAATCC

⊢—I gene——⊣   ⊢————————CRP site————⊣

-40   -30   -20   -10   +1   +10   +20   +30   +40

Start

CACCCCAGGCTTTACACTTTATGCTTCCGGCTCGTATGTTGTGTGGAATTGTGAGCGGATAACAATTTCACACAGGAAACAGCTATGACCATG
GTGGGGTCCGAAATGTGAAATACGAAGGCCGAGCATACAACACACCTTAACACTCGCCTATTGTTAAAGTGTGTCCTTTGTCGATACTGGTAC

——RNA polymerase binding site———————————————————⊣

⊢————————Messenger————⟶   ⊢Z gene⟶

⊢————Operator—————————⊣

Figure 11.9 The DNA sequence of the *lac* regulatory region and the binding sites of the proteins that bind there.

sequencing RNA copies of it (Fig. 11.9). Gilbert's method required isolation of pure operator itself. DNA was isolated from phage carrying the *lac* genes. This DNA was sonicated to fragments about 1,000 base pairs long, repressor was added, and the mixture was passed through a cellulose nitrate filter. The DNA fragments containing operator were bound by repressor and held on the filter whereas the rest passed through. Then the operator-containing fragments were specifically released by adding IPTG to the rinse buffer. Finally, repressor was again bound to these fragments, and the portions of the fragments not protected by repressor were digested by DNAse. The operator was found to be about 20 base pairs long, and it had been purified 2000-fold from the phage or 20,000-fold from the *E. coli* DNA!

The *lac* operator possesses a symmetrical sequence. This suggested that two of the repressor subunits symmetrically made contact with it (Fig. 11.10). To probe the structure of *lac*-operator-repressor complexes, Gilbert and Maxam turned to DNA protection studies. They bound repressor to operator and investigated the reaction of dimethylsulfate

Figure 11.10 The symmetry in *lac* operator and how a protein consisting of two identical subunits could bind to it.

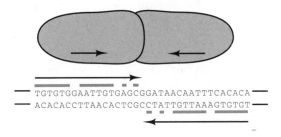

TGTGTGGAATTGTGAGCGGATAACAATTTCACACA
ACACACCTTAACACTCGCCTATTGTTAAAGTGTGT

with bases of the operator in the complex. Indeed, a symmetrical protection pattern was obtained, but, more important, they noticed that the method could easily and rapidly be adapted to yield the actual sequence of a DNA fragment. From this beginning they developed the chemical method for sequencing DNA.

In vivo Affinity of Repressor for Operator

The equilibrium dissociation constant for repressor binding to operator can be determined either from equilibrium measurements or from the ratio of the association and dissociation rates. The dissociation constant obtained by either of these methods can also be compared to *in vivo* data. A correspondence in the numbers suggests, but does not prove, that the *in vitro* reactions and conditions closely mimic the *in vivo* environment.

An estimation of the *in vivo* equilibrium dissociation constant of repressor for the operator uses the thousand-fold inducibility of the *lac* operon. Assume that the operator is completely free of repressor when the operon is fully induced and that the operator is free of repressor 1/1000 of the time when the operon is not induced. By definition,

$$K_D = \frac{R \times O}{RO}.$$

Earlier we discussed the fact that the concentration of repressor, R, is approximately 10^{-8} M. The thousand-fold inducibility means that O/RO is approximately 1/1000. Thus,

$$K_D = 10^{-8}\frac{1}{1000}$$

$$= 10^{-11}\text{M}.$$

This result closely agrees with the values found *in vitro* in buffers with salt concentrations equal to the *in vivo* value of about 0.15 M.

The DNA-binding Domain of *lac* Repressor

How could genetic experiments determine the amino acid residues that contact operator? Work by Miller and collaborators showed that nonsense mutations lying early in the *lacI* gene did not abolish expression of the entire gene. Although the ribosomes terminate translation on reaching the nonsense codon, often they do not dissociate from the mRNA before reinitiating translation at a site within the *lacI* gene. As a result, they synthesize a repressor molecule that lacks up to 61 of the amino acids normally found at the N-terminus. Most surprisingly, these truncated repressor polypeptides fold, associate as usual to form tetramers, and bind IPTG. They are, however, incapable of binding to DNA as shown *in vivo* by their inability to repress and *in vitro* by the DNA filter-binding assay. The simplest explanation of their inability to bind to DNA is that the amino acids missing from their N-terminus are the

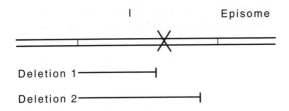

Figure 11.11 Deletion mapping of *lacI* mutations. Only the strain containing deletion 1 can recombine with the mutation indicated to reconstruct a functional *lacI* gene.

part of the wild-type repressor that recognizes and binds to operator. In addition, more detailed genetic analysis of the repressor has shown that the overwhelming majority of missense mutations affecting DNA binding by the repressor map in the region of the I gene coding for the first 60 amino acids.

A typical LacI⁻ mutant is recessive to the wild-type *lacI* gene. The amino-terminal truncated repressors, however, are dominant negatives. That is, the diploid *lacI⁺/lacI⁻ᵈ* behaves like an I⁻. As explained in Chapter 8, the dominance of Iᵈ mutations to the I⁺ allele results from the tetrameric structure of repressor and the fact that all four subunits are required for full repression.

The steps to combine to isolate a particular mutant sometimes are not apparent, so we will list and explain a series of genetics steps designed to isolate an Iᵈ mutant.

1. Isolate I⁻ mutations using phenyl-β-D-galactoside plates. Since phenyl-gal is not an inducer, but is a substrate of β-galactosidase, any cells growing on these plates must be constitutive.

2. Score to determine which are nonsense I⁻ mutations. The mutations can be isolated on an episome, F'*lac⁺ pro⁺*. To test, the episome can be mated into an F⁻ su⁺ strain. This strain should be deleted of *lac* and *pro* genes. To determine whether the episome in the suppressing strain is I⁺ or I⁻, the plates can include 5-bromo-4-chloro-3-indolyl-β–D-D-galactoside (X-gal). If X-gal is included in minimal glycerol plates, the constitutives will form deep blue colonies because the hydrolysis of this substrate by β-galactosidase produces an insoluble blue dye.

3. Map the locations of the I⁻ mutations. This can be done using the ability of strains containing the mutation on an episome and a deletion into the I gene on the chromosome to reconstruct an I⁺ gene (Fig. 11.11). A point mutation is able to recombine with deletion 1 but not deletion 2 to form I⁺ recombinants. Although this type of mapping only positions the mutation to the left or right of the end point of the deletion, an ordered set of deletions can be used to locate the mutation to within about 10 base pairs.

How can the selective plates be arranged to permit growth of only the cells that can reconstruct a functional I gene? The plates must prevent growth of each of the parent types of cells as well as mated cells that are unable to form a wild-type *lacI* gene. This can be done by using phenyl-gal and glycerol in plates. The deletion mapping can be done in a GalE⁻ strain. The resulting defective galactose epimerase gene renders the cells sensitive to galactose. The LacI⁻ cells are constitutive and cleave the phenyl-gal to produce galactose, which then prevents their growth. A LacI⁺ recombinant, however, represses β-galactosidase synthesis and does not generate galactose. Such cells can continue to grow in the presence of phenyl-gal.

4. Test the candidates for being *trans*-dominant repressor negative. This is done by mating the episome into an I⁺ strain and selecting for transfer of the *pro* marker. This strain is tested by spotting onto X-gal plates as before.

5. Physical studies of the repressor from the I⁻ᵈ mutants can test whether the repressor is a tetramer, and SDS gels can show that the mutated repressor is shorter than wild-type.

Protein sequencing can provide the final proof that the repressor synthesized in the nonsense I⁻ᵈ mutants results from translational restarts. The mutant repressor can be purified by precipitation with antibody, separated from the antibody by SDS gel electrophoresis, eluted from the gel, and the N-terminal amino acids sequenced.

A Mechanism for Induction

How do inducers reduce repressor's affinity for operator? It is possible that they bind near the operator-binding site and merely interfere with repressor's correct binding to operator. This possibility seems unlikely

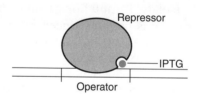

in view of the genetic data and the separability of the IPTG-binding and operator-binding substructures of repressor discussed above. It seems more likely that IPTG merely causes the subunits of repressor to alter positions slightly with respect to one another. Why should this drastically weaken the binding? Experiments with the N-terminal DNA-binding domain contain the answer.

The DNA-binding domain binds to operator with a much lower affinity than wild-type repressor. Such reduced affinity actually is expected. The wild-type tetrameric repressor molecule possesses a relatively rigid structure in which pairs of the N-terminal regions are held in positions appropriate for binding to a single operator. That is, the binding of one of the N-terminal regions to half of the symmetrical operator perforce brings a second N-terminal region into position for its binding to the other half of the operator. Therefore most of the additional interaction energy between this subunit and the DNA can be used to hold the complex onto the DNA (Fig. 11.12). This is another

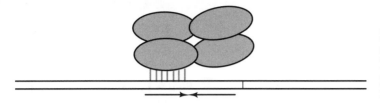

Figure 11.12 The binding of one subunit of repressor to half of the operator correctly positions the other subunit for binding to the other half of operator.

example of the chelate effect. Overall, the result is that the oligomeric repressor tightly binds to operator. The same is not true of the isolated N-terminal domains. The binding of one of these DNA-binding domains does not automatically bring another into position for binding to the other half of operator. As a result of their independent binding, the apparent affinity of a single DNA-binding domain for operator is low.

The chelate effect also streamlines the explanation of induction: binding of IPTG shifts pairs of subunits away from optimal relative positions for headpiece binding to operator. Consequently, the affinity of repressor for operator is greatly reduced, and repressor dissociates. Eventually, direct experiments may be able to test such ideas. In the meantime, two types of repressor mutations are consistent with this point of view. Repressor mutants can be found that are not located in the N-terminal region and that result in much tighter or much weaker repressor-operator binding. These mutants possess no discernible structural alterations. Most likely these types of mutation merely shift the positions of the subunits slightly with respect to one another. The tighter-binding mutant must bring the subunits into closer complementarity with operator, and the weaker-binding mutants must shift the subunits away from complementarity.

An enormous amount has been learned about the *lac* operon, and only a fraction was mentioned in this chapter. More remains to be learned about the physiology and physical chemistry underlying regulation of this set of genes and others.

Problems

11.1. Consider the repressor that is synthesized in cells containing both the I^{-d} and I^+ alleles. If subunits mix randomly during assembly of the tetrameric repressor, what is the fraction of tetrameric molecules having the composition I^+_4, $I^+_3I^{-d}_1$, $I^+_2I^{-d}_2$, $I^+_1I^{-d}_3$, I^{-d}_4 ? Consider that the synthesis of the two types of subunits is equal. Repeat, assuming that the synthesis of I^{-d} subunits is in 10-fold excess over I^+ subunits.

11.2. How can the existence of I^{-d} mutants in *lac* repressor be used to argue that the subunits of *lac* repressor exchange and that subunits are not permanently bound in the same tetrameric molecule?

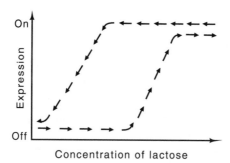

Figure 11.13 Behavior of the *lac* system as a function of the lactose concentration in the medium. Induction follows the arrows.

11.3. Define temperature-sensitive synthesis (TSS) mutations as those that lead to the protein not being able to fold at high temperatures but that allow a protein that folded at low temperature to be stable at high temperature. These are to be distinguished from TS mutations, which lead the protein to be heat-unstable regardless of the temperature at which it folded. Invent a selection for TSS Z^- mutations and for TSS I^- mutations.

11.4. Design the simplest experiment you can that could answer the question, now hypothetical, of whether repressor blocks the binding or the transcriptional progress of RNA polymerase.

11.5. Transcription of a region of DNA could alter the probability of recombination in that interval. Design an experiment to measure the effect of transcription on recombination. What precautions would be needed?

11.6. The induction curve of the *lac* operon versus IPTG concentration is approximately quadratic, that is, cooperative. The Gilbert and Müller-Hill paper shows that the IPTG binding to repressor obeys the equation derived in the text, that is, it is Michaelis-Menten. Is there a contradiction here? Why or why not?

11.7. The presence of the active transport system for the *lac* operon allows the system to behave as a bistable flip-flop by inducing at a high lactose concentration and then remaining induced at a lactose concentration well below that required for induction (Fig. 11.13). Let C_i be the concentration of IPTG inside cells and C_o its external concentration. Assume that the degree of induction of the transport system as a function of C_i and the ability of the transport system to concentrate IPTG as a function of C_o are as shown in Fig. 11.14 where 100% induction is Ind=1000. Show that the response curve *a* produces a system in which fully induced cells that are put into 10^{-6} M IPTG remain induced indefinitely. Also show that if uninduced cells are put into 10^{-6} M IPTG they do not induce despite minor fluctuations in the levels of enzymes and the active transport capabilities. Show that the curve *b* does not produce this bistable response.

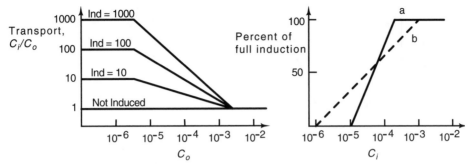

Figure 11.14 Hypothetical data giving the ability of the *lac* transport system to concentrate inducer intracellularly to a concentration C_i from an outside concentration C_o and the level of derepression of the *lac* operon as a function of the intracellular inducer concentration.

11.8. The events preceding initiation by RNA polymerase are important and need appropriate definitions. One term used in this area is "promoter." Observe the varying usages of the term in the papers J. Mol. Biol. *69*, 155-160 (1972); J. Mol. Biol. *38*, 421 (1968); Nature *217*, 825 (1968), and others these papers lead you to. Comment on the uses and make a suitable definition of your own.

11.9. (a) If cells contain 100 repressor monomers and a cell extract at 50 mg/ml made from such cells is dialyzed against 50 mCi/mM IPTG at 1×10^{-6} M, what are the dpm per 0.1 ml sample inside and outside the sack. Assume $K_D = 4 \times 10^{-6}$ M, 1 µCi = 1.2×10^6 dpm.

(b) What is the ratio if the IPTG concentration is lowered to 1×10^{-7} M?

(c) Is it of any use to use very high specific activity IPTG?

11.10. If the dissociation constant of IPTG from *lac* repressor, instead of being 2×10^{-7} M,

(a) were 1×10^{-8} M, could repressor have been detected in the original experiments?

(b) were 2×10^{-5} M, but the specific activity of IPTG had been 10 times higher, could *lac* repressor have been detected in the original experiments?

11.11. Devise a method for the isolation of a *lacI* mutation in which repressor binds IPTG less tightly than the wild-type repressor.

11.12. Like many proteins, *lac* repressor binds to nitrocellulose filters. Quite unexpectedly, however, if IPTG is bound to repressor, it is not easily released when repressor has stuck to the filters. Thus if radioactive IPTG is used, only *lac* repressor can bind radioactivity to the filters, and this constitutes a convenient assay for *lac* repressor. The assay is limited to the use of small quantities of protein, however, since about a monolayer of protein on the filters prevents further protein binding.

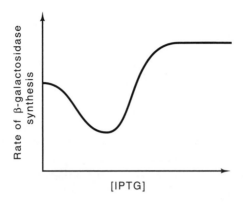

Figure 11.15 Data for Problem 11.21

What purity of repressor would be needed to bind 200 cpm of 25 mCi/mM IPTG to filters 2 cm in diameter?

11.13. What are the biological advantages of symmetrical, palindromic DNA-binding sites for proteins?

11.14. Are the CRP and repressor-binding sites the only palindromes in the *lac* regulatory region? What do you conclude about the biological significance of palindromes?

11.15. How could Müller-Hill's genetic experiment be correct but repressor be purely RNA?

11.16. What dye is produced by β-galactosidase cleavage of X-gal, and why is its insolubility important for its use as an indicator in plates?

11.17. What is the relation between the association and dissociation rates and the equilibrium binding constant for a reaction like *lac* repressor binding to operator?

11.18. After examination of the appropriate literature, describe how allolactose was found to be the true inducer of the *lac* operon.

11.19. Suppose it was observed during footprinting of *lac* repressor that the footprint of repressor was not removed by the inclusion of IPTG in the buffer during the DNAse digestion. The same experiment performed with the same components and with the same amounts of DNA and repressor, but in a volume 100-fold larger, yielded the expected result. What is the likely explanation?

11.20. How could you prove that only two of the four subunits of *lac* repressor are required for binding of the *lac* repressor to operator?

11.21. A mutant *lac* repressor shows the bizarre induction behavior graphed in Fig. 11.15. The purified repressor was found to bind to DNA, both operator and nonoperator, but with about 100 times the affinity of normal repressor. Explain these data.

11.22. What will the kinetics of β-galactosidase synthesis look like in the mated culture if F'I⁺Z⁺Y⁺ cells are mixed with F⁻Δlac cells in the absence of *lac* operon inducers? How will the kinetics be altered if the male strain is F'I^QZ^+Y^+?

References

Recommended Reading

Isolation of the *lac* Repressor, W. Gilbert, B. Müller-Hill, Proc. Natl. Acad. Sci. USA *56*, 1891-1898 (1966).

lac Repressor-operator Interaction I, Equilibrium Studies, A. Riggs, H. Suzuki, S. Bourgeois, J. Mol. Biol. *48*, 67-83 (1970).

The Nucleotide Sequence of the *lac* Operator, W. Gilbert, A. Maxam, Proc. Natl. Acad. Sci. USA *70*, 3581-3584 (1973).

Specific Destruction of the Second *lac* Operator Decreases Repression of the *lac* Operon in *Escherichia coli* Fivefold, E. Eismann, B. vonWilcken-Bergmann, B. Müller-Hill, J. Mol. Biol. *195*, 949-952 (1987).

The *lac* Operon: Physiology and Genetics

Suppressible Regulator Constitutive Mutants of the Lactose System in *Escherichia coli*, B. Müller-Hill, J. Mol. Biol. *15*, 374-379 (1966).

Transposition of the *lac* Region of *E. coli*, II. On the Role of Thiogalactoside Transacetylase in Lactose Metabolism, C. F. Fox, J. R. Beckwith, W. Epstein, E. Singer, J. Mol. Biol. *19*, 576-579(1966).

Mutational Inversion of Control of the Lactose Operon of *E. coli*, G. Meyers, J. Sadler, J. Mol. Biol. *58*, 1-28 (1971).

Distribution of Suboptimally Induced β-Galactosidase in *E. coli:* The Enzyme Content of Individual Cells, P. Maloney, B. Rotman, J. Mol. Biol. *73*, 77-91 (1973).

The Nucleotide Sequence of the Lactose Messenger Ribonucleic Acid Transcribed from UV5 Promoter Mutant of *E. coli*, N. Maizels, Proc. Natl. Acad. Sci. USA *70*, 3585-3589 (1973).

Non-specific DNA Binding of Genome Regulating Proteins as a Biological Control Mechanism, I. The *lac* Operon: Equilibrium Aspects, P. von Hippel, A. Revzin, C. Gross, A. Wang, Proc. Natl. Acad. Sci. USA *71*, 4808-4812 (1974).

Regulation of the Synthesis of the Lactose Repressor, P. Edelman, G. Edlin, J. Bact. *120*, 657-665 (1974).

Genetic Fusions Defining *trp* and *lac* Operon Regulatory Elements, D. Mitchell, W. Reznikoff, J. Beckwith, J. Mol. Biol. *93*, 331-350 (1975).

Initiation of *in vitro* mRNA Synthesis from the Wild-type *lac* Promoter, J. Majors, Proc. Natl. Acad. Sci. USA 72, 4394-4398 (1975).

Genetic Regulation: The *lac* Control Region, R. Dickson, J. Abelson, W. Barnes, W. Reznikoff, Science *187*, 27-35 (1975).

Thiogalactoside Transacetylase of the Lactose Operon as an Enzyme for Detoxification, K. Andrews, E. Lin, J. Bact. *138*, 510-513 (1976).

Genetic Studies of the *lac* Repressor, I. Correlation of Mutational Sites with Specific Amino Acid Residues: Construction of a Colinear Gene-Protein Map, J. Miller, D. Ganem, P. Lu, A. Schmitz, J. Mol. Biol. *109*, 275-301 (1977).

Genetic Studies of the *lac* Repressor, II. Fine Structure Deletion Map of the *lacI* Gene, and Its Correlation with the Physical Map, U. Schmeissner, D. Ganem, J. Miller, J. Mol. Biol. *109*, 303-326 (1977).

Genetic Studies of the *lac* Repressor, III. Additional Correlation of Mutational Sites with Specific Amino Acid Residues, C. Coulondre, J. Miller, J. Mol. Biol. *117*, 525-575 (1977).

Genetic Studies of the *lac* Repressor, IV. Mutagenic Specificity in the *lacI* Gene of *E. coli*, C. Coulondre, J. Miller, J. Mol. Biol. *117*, 577-606 (1977).

Nonspecific DNA Binding of Genome-regulating Proteins as a Biological Control Mechanism: Measurement of DNA-Bound *E. coli lac* Repressor *in vivo*, Y. Kao-Huang, A. Revzin, A. Butler, P. O'Conner, D. Noble, P. von Hippel, Proc. Natl. Acad. Sci. USA *74*, 4228-4232 (1977).

Genetic Studies of the *lac* Repressor, V. Repressors Which Bind Operator More Tightly Generated by Suppression and Reversion of Nonsense Mutations, A. Schmitz, C. Coulondre, J. Miller, J. Mol. Biol. *123*, 431-456 (1978).

Genetic Studies of the *lac* Repressor, VII. On the Molecular Nature of Spontaneous Hotspots in the *lacI* Gene of *E. coli*, P. Farabaugh, U. Schmeissner, M. Foffer, J. Miller, J. Mol. Biol. *126*, 847-863 (1978).

Sequence of the *lacI* Gene, P. Farabaugh, Nature *274*, 765-769 (1978).

Correlation of Nonsense Sites in the *lacI* Gene with Specific Codons in the Nucleotide Sequence, J. Miller, C. Coulondre, P. Farabaugh, Nature *274*, 770-775 (1978).

Molecular Basis of Base Substitution Hotspots in *E. coli*, C. Coulondre, J. Miller, P. Farabaugh, W. Gilbert, Nature *274*, 775-780 (1978).

Mapping of *I* Gene Mutations Which Lead to Repressors with Increased Affinity for *lac* Operator, M. Pfahl, J. Mol. Biol. *147*, 175-178 (1981).

Dual Promoter Control of the *Escherichia coli* Lactose Operon T. Malon, W. McClure, Cell *39*, 173-180(1984).

Detection *in vivo* of Protein-DNA Interactions within the *lac* Operon of *Escherichia coli*, H. Nick, W. Gilbert, Nature *313*, 795-798 (1985).

The Inducible *lac* Operator-repressor System is Functional in Mammalian Cells, M. Hu, N. Davidson, Cell *48*, 55-566 (1987).

Mutations in the *lac* P2 Promoter, C. Donnelly, W. Reznikoff, J. Bact. *169*, 1812-1817 (1987).

Promoters Largely Determine the Efficiency of Repressor Action, M. Lanzer, H. Bujard, Proc. Natl. Acad. Sci. USA *85*, 8973-8977 (1988).

Synergy Between *Escherichia coli* CAP Protein and RNA Polymerase in the *lac* Promoter Open Complexes, D. Straney, S. Straney, D. Crothers, J. Mol. Biol. *206*, 41-57 (1989).

Lac Repressor

Mutants That Make More Lac Repressor, B. Müller-Hill, L. Crapo, W. Gilbert, Proc. Natl. Acad. Sci. USA *59*, 1259-1264 (1968).

Translational Restarts: AUG Reinitiation of a *lac* Repressor Fragment, T. Platt, K. Weber, D. Ganem, J. Miller, Proc. Natl. Acad. Sci. USA *69*, 897-901 (1972).

Lac Repressor, Specific Proteolytic Destruction of the NH2-terminal Region, and Loss of the DNA Binding Activity, T. Platt, J. Files, K. Weber, J. Biol. Chem. *248*, 110-121 (1973).

Mutations Affecting the Quaternary Structure of the *lac* Repressor, A. Schmitz, U. Schmeissner, J. Miller, P. Lu, J. Biol. Chem. *251*, 3359-3366 (1976).

Interaction of *lac* Repressor with Inducer: Kinetic and Equilibrium Measurements, B. Friedman, J. Olson, K. Matthews, J. Mol. Biol. *111*, 27-39 (1977).

Tight-Binding Repressors of the *lac* Operon: Selection System and *in vitro* Analysis, M. Pfahl, J. Bact. *137*, 137-145 (1979).

Characteristics of Tight Binding Repressors of the *lac* Operon, M. Pfahl, J. Mol. Biol. *147*, 1-10 (1981).

Genetic Assignment of Resonances in the NMR Spectrum of a Protein: Lac Repressor, M. Jarema, P. Lu, J. Miller, Proc. Natl. Acad. Sci. USA *78*, 2707-2711 (1981).

Secondary Structure of the *lac* Repressor DNA-binding Domain by Two-dimensional ^{1}H Nuclear Magnetic Resonance in Solution, E. Zuiderweg, R. Kaptein, K. Wüthrich, Proc. Natl. Acad. Sci. USA *80*, 5837-5841 (1983).

A Chimeric Mammalian Transactivator Based on the *lac* Repressor that is Regulated by Temperature and Isopropyl-β-D-thiogalactopyranoside, S. Baim, M. Labow, A. Levine, T. Shenk, Proc. Natl. Acad. Sci. USA *88*, 5072-5076 (1991).

Repressor-DNA Interactions

The Lac Operator Is DNA, W. Gilbert, B. Müller-Hill, Proc. Natl. Acad. Sci. USA *58*, 2415-2421 (1967).

DNA Binding of the *lac* Repressor, A. Riggs, S. Bourgeois, R. Newby, M. Cohn, J. Mol. Biol. *34*, 365-368 (1968).

A Mechanism for Repressor Action, W. Reznikoff, J. Miller, J. Scaife, J. Beckwith, J. Mol. Biol. *43*, 201-213 (1969).

The *lac* Repressor-operator Interaction, III. Kinetic Studies, A. Riggs, S. Bourgeois, M. Cohn, J. Mol. Biol. *53*, 401-417 (1970).

Lac Repressor Binding to DNA Not Containing the *lac* Operator and to Synthetic Poly dAT, S. Lin, A. Riggs, Nature *228*, 1184-1186 (1970).

The Nature of Lactose Operator Constitutive Mutations, T. Smith, J. Sadler, J. Mol. Biol. *59*, 273-305 (1971).

Mapping of the Lactose Operator, J. Sadler, T. Smith, J. Mol. Biol. *62*, 139-169 (1971).

lac Repressor-operator Interaction, VI. The Natural Inducer of the *lac* Operon, A. Jobe, S. Bourgeois, J. Mol. Biol. *69*, 397-408 (1972).

lac Repressor Binding to Non-operator DNA: Detailed Studies and a Comparison of Equilibrium and Rate Competition Methods, S. Lin, A. Riggs, J. Mol. Biol. *72*, 671-690 (1972).

Photochemical Attachment of *lac* Repressor to Bromodeoxyuridine-substituted *lac* Operator by Ultra Violet Radiation, S. Lin, A. Riggs, Proc. Natl. Acad. Sci. USA *71*, 947-951 (1974).

Measurements of Unwinding of *lac* Operator by Repressor, J. Wang, M. Barkley, S. Bourgeois, Nature *251*, 247-249 (1974).

Interaction of Effecting Ligands with *lac* Repressor and Repressor-operator Complex, M. Barkley, A. Riggs, A. Jobe, S. Bourgeois, Biochem. *14*, 1700-1712 (1975).

Contacts between the *lac* Repressor and Thymines in the *lac* Operator, R. Ogata, W. Gilbert, Proc. Natl. Acad. Sci. USA *74*, 4973-4976 (1977).

Molecular Parameters Characterizing the Interaction of *E. coli lac* Repressor with Non-operator DNA and Inducer, A. Butler, A. Revzin, P. von Hippel, Biochem. *16*, 4757-4768 (1977).

Direct Measurement of Association Constants for the Binding of *E. coli lac* Repressor to Non-operator DNA, A. Revzin, P. von Hippel, Biochem. *16*, 4769-4776 (1977).

Nonspecific Interaction of Lac Repressor with DNA: An Association Reaction Driven by Counterion Release, P. deHaseth, T. Lohman, M. Record, Biochem. *16*, 4783-4791 (1977).

Minimal Length of the Lactose Operator Sequence for the Specific Recognition by the Lactose Repressor, C. Bahl, R. Wu, J. Stawinsky, S. Narang, Proc. Natl. Acad. Sci. USA *74*, 966-970 (1977).

Binding of Synthetic Lactose Operator DNAs to Lactose Repressors, D. Goeddel, D. Yansura, M. Caruthers, Proc. Natl. Acad. Sci. USA *74*, 3292-3296 (1977).

Salt Dependence of the Kinetics of the *lac* Repressor-operator Interaction: Role of Nonoperator DNA in the Association Reaction, M. Barkley, Biochem. *20*, 3833-3842 (1981).

lac Repressor: A Proton Magnetic Resonance Look at the Deoxyribonucleic Acid Binding Fragment, K. Arndt, F. Boschelli, P. Lu, J. Miller, Biochem. *20*, 6109-6118 (1981).

lac Repressor-*lac* Operator Interaction: NMR Observations, H. Nick, K. Arndt, F. Boschelli, M. Jarema, M. Lillis, J. Sadler, M. Caruthers, P. Lu, Proc. Natl. Acad. Sci. USA *79*, 218-222 (1982).

Repressor-operator Interaction in the *lac* Operon, II. Observations at the Tyrosines and Tryptophans, H. Nick. K. Arndt, F. Boschelli, M. Jarema, M. Lillis, H. Sommer, P. Lu, J.Sadler, J. Mol. Biol. *161*, 417-438 (1982).

A Perfectly Symmetric *lac* Operator Binds *lac* Repressor Very Tightly, J. Sadler, H. Sasmor, J.Betz, Proc. Natl. Acad. Sci. USA *80*, 6785-6789 (1983).

Thermodynamic Origins of Specificity in the *lac* Repressor-Operator Interaction, M. Mossing, M. Record, Jr., J. Mol. Biol. *186*, 295-305 (1985).

MalI, A Novel Protein Involved in Regulation of the Maltose System of *Escherichia coli*, Is Highly Homologous to the Repressor Proteins GalR, CytR, and LacI, J. Reidl, K. Römisch, M. Ehrmann, W. Boos, J. Bact. *171*, 4888-4899 (1989).

lac Repressor Acts by Modifying the Initial Transcribing Complex So that it Cannot Leave the Promoter, J. Lee, A. Goldfarb, Cell *66*, 793-798 (1991).

Orientation of the *lac* Repressor DNA Binding Domain in Complex with the Left Operator Half-site Characterized by Affinity Cleaving, J. Shin, R. Ebright, P. Dervan, Nuc. Acids Res. *19*, 5233-5236 (1991).

DNA Looping

lac Repressor Forms Loops with Linear DNA Carrying Two Suitably Spaced *lac* Operators, H. Krämer, M. Niemöller, M. Amouyal, B. Revet, B. von Wilcken-Bergmann, B. Müller-Hill, EMBO Journal *6*, 1481-1491 (1987).

DNA Supercoiling Promotes Formation of a Bent Repression Loop in *lac* DNA, J. Borowiec, L. Zhang, S. Sasse-Dwight, J. Gralla, J. Mol. Biol. *196*, 101-111 (1987).

lac Repressor Forms Stable Loops *in vitro* with Supercoiled Wild-type DNA Containing all Three Natural *lac* Operators, E. Eismann, B. Müller-Hill, J. Mol. Biol. *213*, 763-775 (1990).

Induction, Repression, and the *araBAD* Operon

12

Molecular biologists were greatly surprised to discover that nature regulates gene activity in more than one way. Many of the early molecular biologists had been trained as physicists and perhaps this conservatism was a carryover from physics where relatively few principles govern a broad range of phenomena. As we have already seen in previous chapters, biological systems like the mechanism of DNA replication show a surprisingly large variability between organisms. This variability first became apparent during examination of the mechanism of gene regulation in the arabinose operon. Data began to accumulate indicating that this set of genes is regulated by a protein that turned on their expression rather than turned off their expression as had been found in the lactose operon. Because nature's variability at this level was not yet appreciated, once the lactose operon was characterized as negatively regulated, evidence that a different gene system might not be regulated similarly was received by some with hostility. Now, of course, we understand that, at this level, nature uses an almost endless set of variations. The challenge now is discerning and understanding the underlying principles.

Genetic study of the arabinose operon of *Escherichia coli* was begun by Gross and Englesberg and pursued for many years by Englesberg. The work started as a straightforward genetic mapping exercise using phage P1. As data began to accumulate that indicated that the regulation mechanism might not be a simple variation of that used in the *lac* operon, Englesberg began a more intensive characterization of the system. The mechanisms regulating the arabinose operon are yet to be fully understood. This chapter will explain how the genetic data indicated that the operon is positively regulated, and then consider the *in vitro* experiments that proved that the system is positively regulated. Next, the unexpected findings that indicated that the *ara* system is also negatively regulated will be discussed. These led to the discovery of the "action at a distance" phenomenon, which was first explained in the *ara* system as the result of looping the DNA so as to bring together two sites that are separated by 200 base pairs. Subsequently, looping has been found in many other systems and is a general mechanism to permit proteins bound at some distance from a site to influence what happens at the site.

The Sugar Arabinose and Arabinose Metabolism

The pentose L-arabinose occurs naturally in the walls of plant cells. The bacterium *E. coli*, but not humans, can use this sugar as a source of carbon and energy. Therefore arabinose is a free meal to intestinal flora when we eat a meal containing vegetables. Before arabinose can be metabolized by the bacterial intracellular arabinose enzymes, it must be transported from the growth medium through the inner membrane to the cytoplasm. This is performed by two independent arabinose transport systems, the products of the *araE* and *araFGH* genes. The *araE* system possesses a low affinity for arabinose and therefore is most effective in the presence of high concentrations of arabinose. The *araFGH* system possesses high affinity for arabinose uptake and therefore may be most valuable when arabinose concentrations are very low, on the order of 10^{-7} M.

Arabinose, like many sugars, exists in solution predominantly in a ring form in either of two conformations. These interconvert with a half-time of about 10 minutes (Fig. 12.1). Therefore it is possible that one of the transport systems is specific for the α anomeric form of arabinose and the other transports the β form. The β form is a substrate for the first enzyme of the catabolic pathway, but it is not known whether the other ring form or the linear form, which exists in only trace

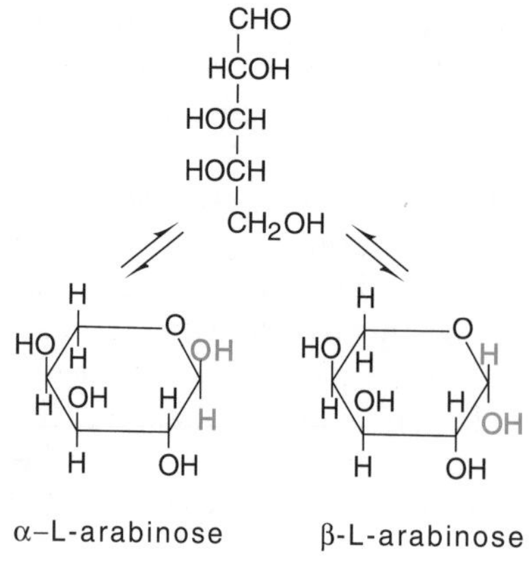

α–L-arabinose   β-L-arabinose

Figure 12.1 The open chain and pyranose ring forms of L-arabinose.

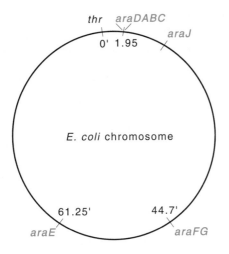

Figure 12.2 The locations on the *E. coli* chromosome of the genes induced by L-arabinose. Genes *araE* and *araFGH* are for uptake, *araBAD*, for catabolism, and *araJ* is of unknown function. The numbers indicate the gene positions on the genetic map.

concentrations, might be the true inducer and play a role analogous to that of the true inducer, allolactose, of the *lac* operon.

The duplication of arabinose uptake systems hindered their genetic and physiological study because mutants are then difficult to identify. A defect in either system is masked by the activity of the other system. The genes coding for the enzymes required for arabinose catabolism, however, have been easier to map and study. They are located near the top of the genetic map as it is usually drawn (Fig. 12.2). Arabinose is first converted to L-ribulose by arabinose isomerase, the *araA* gene product. Then ribulose is phosphorylated by ribulokinase, the product of the *araB* gene, to yield L-ribulose-5-phosphate (Fig. 12.3). The ribulose phosphate is next converted to D-xylulose-5-phosphate by the *araD* gene product, ribulose phosphate epimerase. Xylulose phosphate enters the pentose phosphate shunt, and the enzymes subsequently involved are not induced specifically by the presence of arabinose.

The genes of the two arabinose active transport systems plus an additional arabinose-inducible gene of unknown function map to three different regions of the chromosome. These are different from the map locations of the genes required for the catabolism of arabinose. There-

Figure 12.3 The path of catabolism of L-arabinose and the gene products catalyzing the conversion.

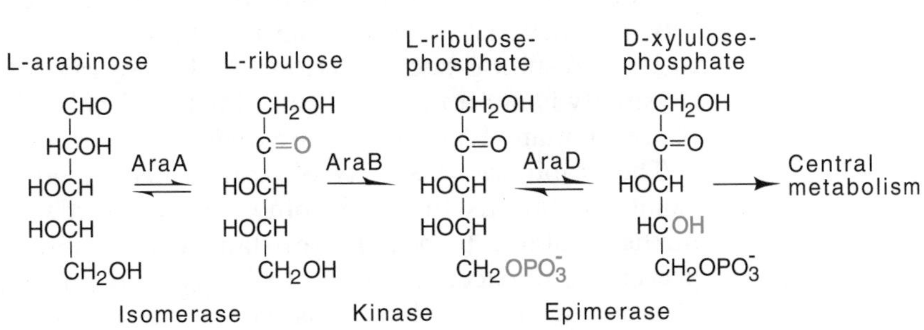

L-arabinose	L-ribulose	L-ribulose-phosphate	D-xylulose-phosphate
CHO	CH_2OH	CH_2OH	CH_2OH
HCOH	C=O	C=O	C=O
HOCH	HOCH	HOCH	HOCH
HOCH	HOCH	HOCH	HCOH
CH_2OH	CH_2OH	$CH_2 OPO_3^-$	$CH_2OPO_3^-$

AraA (Isomerase) → AraB (Kinase) → AraD (Epimerase) → Central metabolism

fore a total of four sets of genes have been discovered whose activities are regulated by arabinose. Why should this be? Why not have all the genes in one large operon? Two possible reasons will be discussed below.

At low arabinose concentrations, a high-affinity, but perhaps energy-inefficient or low-capacity, uptake system might be necessary for the cells to be able to utilize any arabinose at all. On the other hand, in the presence of high concentrations of arabinose, a different uptake system might be more useful. One with a high transport rate would be necessary. This need not have a particularly high affinity for arabinose. These different needs would necessitate splitting the corresponding genes into separate operons so that they could be differentially regulated. Alternatively, suppose arabinose were suddenly presented to bacteria, as it might be in the gut. It is to the cell's great advantage to begin metabolizing this new nutrient just as soon as possible. If, however, all the genes for the uptake and metabolism of arabinose were in a single long operon, then the interval from induction until RNA polymerase could transcribe to the end of the operon would be about three minutes. Any strain that divided its arabinose operon into two or three separately transcribed units could induce all its *ara* enzymes more quickly and begin appreciable arabinose metabolism a minute or two sooner than cells with an undivided arabinose operon. This time saved each time the operon was induced could be of enormous selective value over evolutionary time scales.

Induction of the arabinose operon also requires the presence of cyclic AMP and the cyclic AMP receptor protein CRP. The main role of this protein is to enable induction of the arabinose operon only in the absence of glucose. This prevents the cell's attempting to utilize arabinose when glucose, which is a better carbon source, is present. The general phenomenon of being able to induce an operon well only in the absence of glucose is called catabolite repression. A significant number of bacterial operons display catabolite repression.

Genetics of the Arabinose System

Most mutations in the arabinose operon generate the expected phenotypes. The *araB*, *araA*, and *araD* genes lie in one transcriptional unit served by promoter p_{BAD} (Fig. 12.4), and mutations in these genes abolish the activity of the enzyme in question as well as leave the cells arabinose-negative. Mutations in *araD*, however, have the useful property that they make the cells sensitive to the presence of arabinose due to their resulting accumulation of ribulose phosphate. This type of sensitivity is not an isolated case, for high levels of many sugar phosphates in many types of cells are toxic or growth inhibitory.

The arabinose-sensitivity of AraD⁻ cells provides a simple way to isolate mutations in the arabinose genes. AraD⁻ mutants that have further mutated to become resistant to arabinose contain mutations preventing the accumulation of ribulose phosphate. Each cell capable of growing into a colony must contain a secondary mutation in the

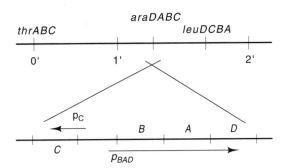

Figure 12.4 Fine structure of the *E. coli* genome in the region of the *araBAD* genes, the *ara* promoters, and the directions of transcription of the genes. In the top, the genes are drawn in the conventional orientation of the *E. coli* genetic map, and in the bottom, the arabinose genes have been inverted to conform with the conventional usage, in which the major transcription is rightward.

arabinose operon in addition to the AraD⁻ mutation. Such mutants are isolated by spreading large numbers of AraD⁻ cells onto plates containing arabinose and some other source of carbon and energy, such as glycerol or yeast extract. The few cells that grow into colonies contain additional mutations in *araC, B,* or *A*. This scheme does not yield transport-negative mutants since both transport systems would have to be inactivated before the cells would be arabinose-resistant, and such double mutants are too rare to be detected.

The behavior of mutations in one of the arabinose genes was paradoxical. The cells became arabinose-negative, but no enzymatic activity could be associated with the gene. In fact, none seemed necessary because the other arabinose gene products performed all the required metabolic conversions. Additionally, cells with mutations in this gene, *araC*, had the strange property of not possessing any of the arabinose-induced enzymes or active transport systems (Table 12.1). If the gene product were not any of the logical proteins, perhaps it wasn't a protein. This wasn't possible, however, because nonsense mutations existed in the *araC* gene. Only genes encoding proteins can possess nonsense mutations. Therefore, the *araC* gene product had to be a protein.

Formally, the behavior of the *araC* mutants was consistent with several regulatory mechanisms. First, contrary to expectations prevalent at the time of this work, *araC* could code for a positive regulator,

Table 12.1 Induction of the Arabinose Operon

Genotype	Isomerase Levels		Transport Genes	
	Basal	Induced	Basal	Induced
C⁺	1	300	1	150
C⁻	1	1	1	1
C⁺/C⁻	1	300	1	150
F'C⁺A⁻/C⁻A⁺	1	300	1	150

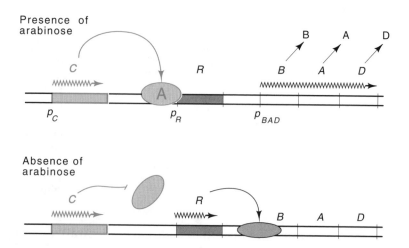

Figure 12.5 A double-negative regulation system in which C looks like an activator in genetic and physiological tests. C is a repressor of R synthesis, R is a repressor of the *BAD* operon.

one whose presence is necessary for synthesis of the other *ara* gene products. Second, *araC* could code for a component directly involved with the uptake of arabinose into cells. Its absence would mean that the intracellular level of arabinose would never become high enough to permit derepression of the arabinose genes. Third, *araC* could be part of a double-negative regulation system (Fig. 12.5). That is, AraC protein in the presence of arabinose could repress the synthesis of the true repressor of the arabinose operon. In this case, if *araC* were inactive or arabinose were absent, then the repressor would be synthesized and the arabinose enzymes would not be synthesized.

Detection and Isolation of AraC Protein

However ingenious, genetic and physiological or cloning and mapping experiments suffer from the defect that they are rarely able to provide rigorous proofs of mechanisms of action. Proof of a model usually requires purification of the system's individual components and *in vitro* reconstruction of the system. How then could biologically active AraC protein be purified for biochemical studies? Detection of the *lac* repressor, which had been accomplished earlier, was difficult enough. Its isolation capitalized upon its tight-binding to an inducer of the *lac* operon, IPTG. Not even this handle was available for detection of AraC protein. *In vivo* experiments measuring the induction level of arabinose enzymes in cells as a function of the intracellular arabinose concentration showed that the affinity of AraC protein for arabinose was too low to permit its detection by the equilibrium dialysis that was used to isolate *lac* repressor.

Work on the isolation of AraC was performed well before genetic engineering permitted facile isolation of many proteins. Now proteins

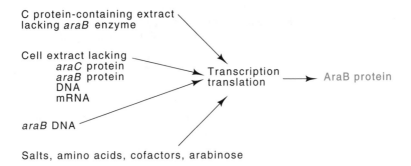

Figure 12.6 Schematic of the coupled transcription-translation system.

often can be detected and purified merely by engineering such a great hypersynthesis that cell lysates possess a prominent new band upon SDS gel electrophoresis. Such gels then provide a way to follow the purification of the overproduced protein, but they do not ensure that the protein is biologically active. How can an assay be devised for the biological activity of AraC protein? The only activity that possesses sufficient sensitivity was based on the ability of AraC protein to induce synthesis of the *araBAD* genes. Working with the *lac* operon, Zubay had developed a means of preparing a partially fractionated cell lysate in which added DNA could be transcribed and translated. When a concentrated source of *lac*Z DNA is added to this system, *lac* mRNA is synthesized and translated into active β-galactosidase. This system was adapted to the arabinose genes to search for the AraC protein.

The necessary coupled transcription-translation system consists of a cell extract from which much cellular DNA and all mRNA has been removed (Fig. 12.6). Then DNA template, salts, amino acids, and enzyme cofactors are added. The cell extracts must be made from *araB* mutants so that the small quantities of the *araB* enzyme, ribulokinase, which will be synthesized *in vitro*, can be detected. Also, the extracts must be free of AraC protein so that it can be detected on its addition to the extracts. Of course, it is sensible that the source of AraC protein be as concentrated as possible as well as lack ribulokinase. Finally, a highly concentrated source of *araB* DNA must be added to the synthesis extract.

It was possible to meet the various requirements of the *in vitro* system. The extract for *in vitro* synthesis was prepared from cells deleted for the *araCBAD* genes. The source of AraC protein was the *araCBAD* deleted cells infected with a phage carrying the *araC* gene but not an active *araB* gene. The source of *araB* DNA was a phage carrying the arabinose genes. Of course, similar experiments done at the present time would utilize plasmids carrying the desired genes. When the ingredients were mixed, AraB protein was synthesized only if both arabinose and AraC-containing extract were added to the reaction. This result rigorously excludes the possibility that AraC protein only appears to regulate positively because its actual role is transporting arabinose into cells. It also makes the possibility of a double repressor regulation system exceedingly remote.

Of course, the last word on the positive nature of the regulation system came with a completely pure *in vitro* transcription system consisting of just cyclic AMP receptor protein, RNA polymerase, *ara* DNA, and AraC protein plus arabinose, cAMP and triphosphates. Ara-specific messenger is synthesized if and only if these components and arabinose are present in the reaction. The fact that this system works also shows that the regulation is exerted at the level of initiation of transcription rather than via degradation of mRNA or even at a translational level.

Although the assay was cumbersome, the coupled transcription-translation system permitted quantitation of AraC protein in fractions obtained from purification steps. Micrograms of the protein could be purified. When the *araC* gene was fused to the *lacZ* promoter and placed on a high-copy-number plasmid the level of AraC protein reached values permitting straightforward purification. Now it is possible in just a few days to purify 20 milligrams of the protein. This facilitates physical experiments designed to examine the basis of the protein's activities.

Repression by AraC

In the presence of arabinose, AraC protein induces expression of the metabolic and active transport genes. That is, it is a positive regulator. Most unexpectedly, AraC protein also appears to repress expression of these genes. Three types of experiments demonstrate the repression exerted by AraC protein. The simplest utilizes Ara$^+$ revertants that are isolated in strains deleted of the *araC* gene. These mutations are called I^c. They lie in the p_{BAD} RNA polymerase-binding site and they permit a low rate of polymerase binding and initiation in the absence of AraC protein. Repression is revealed by the fact that the constitutive promoter activity of the I^c mutants is reduced by the presence of AraC protein. The protein can act in *trans* to repress just like the *lac* repressor.

Experiments with the I^c mutations also show that at least part of the site required for repression of p_{BAD} lies upstream from all of the sites required for induction. This is shown by the properties of strains containing the two deletions, $\Delta 1$ or $\Delta 2$. $\Delta 1$ extends just to the end of the *araC* gene and $\Delta 2$ ends beyond the *araC* gene, so that half of the regulatory region between the *araC* and *araBAD* genes containing p_{BAD} has been deleted. The promoter p_{BAD} in both of the strains is undamaged

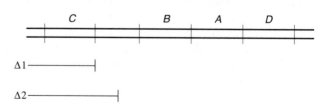

Table 12.2 Repression of I^c in the Absence of
Arabinose Depends on an Upstream Site

Episome	Expression of chromosomal *araA*	
	$\Delta 1 I^c (BAD)^+$	$\Delta 2 I^c (BAD)^+$
No C gene	30	30
F'C$^+\Delta$BAD	1	30

by the deletions because it remains fully inducible when AraC protein
is provided in *trans*. Repression of the I^c mutation by AraC protein
occurs only in the strain containing $\Delta 1$ (Table 12.2). Therefore at least
part of the site required for repression has been deleted by $\Delta 2$.

Conceivably, the repression exerted by AraC occurs only in the I^c
mutants. Therefore experiments will be described that do not use the I^c
mutations. The experiments use the same two deletions as described
above, $\Delta 1$ and $\Delta 2$. AraC$^+\Delta$(BAD) episomes are introduced into the
deletion strains. When these cells are grown in the absence of arabinose,
the basal level in the strain containing $\Delta 1$ is normal but the basal level
in the strain containing $\Delta 2$ is 10 to 30 times normal. This result confirms
the repression effect and also shows that in the absence of arabinose, a
little of the AraC protein is in the inducing state and can weakly induce
p_{BAD} if the region defined by $\Delta 2$ has been deleted.

A third demonstration of repression in the *ara* system utilizes *araC*
constitutive mutations, *araC^c*. This type of mutation causes the arabinose enzymes to be induced even in the absence of the normal inducer,
L-arabinose. Diploids containing both *araC^c* and *araC$^+$* mutations are
surprising, for the C$^+$ allele is almost completely dominant to the C^c
allele (Table 12.3). C$^+$/C^c diploids possess nearly the normal uninduced
level of arabinose enzymes in the absence of arabinose and nearly the
fully induced level of enzymes in the presence of arabinose. In light of
the other experiments showing repression, these results are most simply
explained as resulting from repression by the C$^+$ protein despite the
presence of the C^c protein. These results, however, are also consistent
with AraC protein being an oligomer in which the *in vivo* dominance of
C$^+$ results from subunit mixing.

Table 12.3 Dominance of C$^+$ to C^c in the
Absence of Arabinose

Genotype	Expression of BAD Genes
C$^+$	1
C^c	300
C^c/C$^+$	20

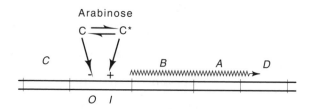

Figure 12.7 The presence of arabinose drives AraC protein into the inducing state, which acts positively at the *I* site. Before arabinose addition, the protein was acting negatively through the *O* site.

The *araC^c* mutants mentioned above are easily isolated with the aid of the arabinose analog 5-methyl-L-arabinose, which is also known as D-fucose. D-fucose cannot be metabolized by *E. coli*, but it does interact with AraC protein to inhibit the normal induction by arabinose. Mutants able to grow on arabinose in the presence of fucose are called *araC^c*

The results described above indicate that AraC protein can induce or repress the initiation of transcription of the arabinose operons in *E. coli*. AraC protein must therefore exist in at least two states, a repressing and an inducing state, and arabinose must drive the population of AraC protein molecules in a cell toward the inducing state (Fig. 12.7). The deletions show that the most upstream site required for repression lies further upstream than the most upstream site required for induction. By analogy to the *lac* operon, the site required for repression is called an operator, *O*. The *I* site stands for induction.

Regulating AraC Synthesis

The *araC* gene is oriented oppositely to the *araBAD* genes, and its promoter, p_C, is adjacent to the regulatory sites for the BAD operon. This location raises the possibility that the activity of p_C might be regulated by the same proteins that regulate the activity of p_{BAD}.

Since the measurement of AraC protein is difficult, lengthy, and imprecise, Casadaban chose not to study regulation of the p_C promoter by measuring AraC protein itself but instead to fuse the promoter to the

Figure 12.8 Deletion of intervening DNA fuses p_C to a transposed *lac*Z gene.

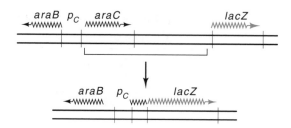

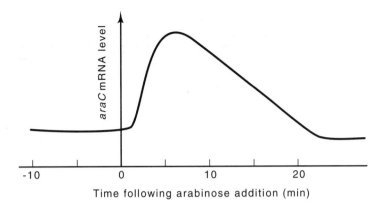

Figure 12.9 Transient derepression of *araC* messenger level following the addition of arabinose to growing cells.

β-galactosidase gene(Fig. 12.8). The assay of this protein is simple. The general strategy of fusing β-galactosidase to promoters has become a widespread tool. It is used not only in the study of bacterial genes, but also in higher cells. Its use in *Drosophila*, for example, permits facile study of the spatial and temporal specificity of gene regulation. Also, the Berk-Sharp S1 nuclease mapping method which was discussed in Chapter 5 now permits a straightforward characterization of the p_C promoter activity under a variety of conditions. The original p_C-*lacZ* fusion was constructed in a series of intricate genetic operations. First the β-galactosidase gene was brought near the *araC* gene and then intervening DNA was deleted so that the structural gene of β-galactosidase was fused to p_C. Hence the measurement of β-galactosidase became a measurement of the activity of p_C.

The first finding was that p_C is stimulated about threefold by the presence of cyclic AMP-CRP. A more surprising finding was that p_C is about six times more active in the absence of AraC protein than in its presence. That is, AraC protein represses its own synthesis. A third finding was that on the addition of arabinose to cells, the level of *araC* messenger increases about fourfold in several minutes and then slowly falls back to its prior level (Fig. 12.9). Whether this transient derepression is of any physiological value is unknown. One might expect that the resulting elevated level of AraC protein could facilitate induction of the *ara* operons not already saturated with the protein.

Binding Sites of the *ara* Regulatory Proteins

Now that the regulatory phenomena of the *ara* operon have been laid out, what can be said about its mechanism of regulation? The nucleotide sequences of the promoters p_{BAD}, p_E, p_{FGH}, show three distinct regions of homology among themselves. One of these is the RNA polymerase-binding site. Another is the cyclic AMP receptor protein-binding site, and the third is the AraC protein-binding site. Inexplicably, the *araBAD*

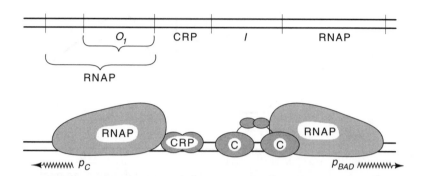

Figure 12.10 Schematic of the binding sites of AraC protein, CRP, and RNA polymerase on the *araBAD* regulatory region. The *araO₁* site is the operator for p_C and the *araI* site is required for induction.

RNA polymerase-binding site is quite similar to the RNA polymerase sites of promoters that do not require auxiliary proteins for activity. Since p_{BAD} requires the auxiliary proteins AraC and CRP for its activity, the polymerase-binding sequence might have been expected to be markedly different from the consensus RNA polymerase-binding sequence.

The identity of the protein-binding sites was established by DNAse footprinting. In *araBAD*, the AraC protein-binding site is just upstream from RNA polymerase, and the CRP-binding site is just upstream from that of *araC* (Fig. 12.10). The AraC-binding site is called *araI* for induction. These three sites are required for induction of p_{BAD}.

A second AraC protein-binding site called *araO₁* lies another 60 nucleotides upstream of *araI*. This site overlaps the RNA polymerase-binding site of p_C, and since occupancy of *araO₁* by AraC protein sterically interferes with the binding of RNA polymerase at p_C, *araO₁* is an operator of p_C. *AraO₁* is not directly involved with repression of p_{BAD}. A simple demonstration of this is the behavior of p_C and p_{BAD} as the level of AraC protein is increased in a series of strains containing plasmids with *araC* fused to promoters of different strengths. As the level of AraC

Table 12.4 p_c Activity as a Function of AraC Level

	Promoter Activity	
Level of C Protein	p_c	p_{BAD}
No C	50	1
Wild type, C⁺	10	300
2 x Wild Type	5	300
60 x Wild Type	1	300

protein is increased, p_C shuts down but p_{BAD} remains fully inducible (Table 12.4).

Deletion analysis showed that one of the sites required for repression of p_{BAD} lies beyond *araO₁*. It is called *araO₂*, and it lies 270 base pairs upstream from the p_{BAD} transcription start point. Footprinting has confirmed that AraC protein binds to this site, and mutation analysis

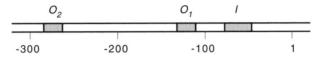

Position with respect to p_{BAD} transcription start

has shown that a single nucleotide change in this region can eliminate repression of p_{BAD}. Deletion of this site is the reason that the original deletion that was isolated by Englesberg could no longer repress p_{BAD} but could induce. His deletion extended through *araO₂* into *araO₁*, and during the interval of several years between the discovery of O_1 and the discovery of O_2, *araO₁* was thought to be required for repression of p_{BAD}.

The location of the CRP site required for stimulation of the p_{BAD} promoter is most surprising. In the *lac* operon, the CRP site lies at positions -48 to -78 with respect to the start of transcription, just next to RNA polymerase. At the *ara* promoter, AraC protein occupies this site and CRP lies further away, binding to positions -80 to -110. CRP makes specific contacts with RNA polymerase in activating transcription of the *lac* promoter. Does it make a different set of specific contacts with AraC protein, and does AraC protein make yet another set of contacts with RNA polymerase? Alternatively, both proteins might simultaneously contact RNA polymerase to activate transcription, but it seems more likely that CRP activates through a DNA bending mechanism and that AraC activates through direct protein-protein contacts.

DNA Looping and Repression of *araBAD*

How can repression of p_{BAD} be generated from the *araO₂* site that is located more than 200 base pairs upstream? There are three possibilities (Fig. 12.11). A signal could be transmitted through the DNA, for example, by changing the angle of the base tilt; something could polymerize along the DNA starting at *araO₂* and finally cover *araI*; or the DNA could bend back and bring the *araO₂* region near p_{BAD}. This latter possibility was shown to be the case by a series of experiments in which the spacing between *araO₂* and the promoter region was varied. If five base pairs, which is half a helical turn, are added between these two sites, the ability to repress is greatly diminished, but if 11 base pairs are added, the ability to repress is restored. An addition of 15 base pairs eliminates repression, and a longer addition of 31 base pairs restores repression.

These results show that the absolute distance between the two sites is not greatly important but that the angular orientation between the

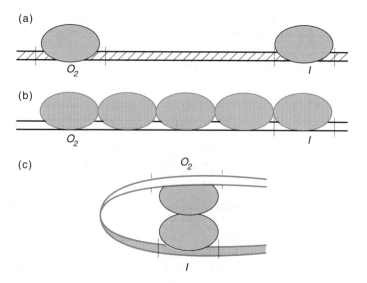

Figure 12.11 Three possible mechanisms for repression from a site several hundred nucleotides from a promoter (a) properties of the DNA can be altered, (b) protein can polymerize along the DNA, or (c) loops can form.

two is. This implies that the DNA loops back to form contacts between AraC protein bound at $araO_2$ and something in the promoter region. If the protein at $araO_2$ is on the opposite side of the DNA (Fig. 12.12), the additional energy required to twist the DNA half a turn so the sites again face one another is too great and the loop required for repression does not form. The isolation of repression deficient point mutations located the sites necessary for repression. As expected, some were in $araO_2$. Additional repression negative mutations were in *araI*, thereby identifying the other end of the loop.

Figure 12.12 How a five-base-pair insertion can introduce half a helical turn and prevent formation of a loop.

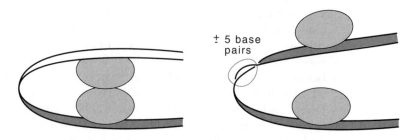

In vivo Footprinting Demonstration of Looping

The *lac* operon was the first well-understood gene system. The genes of the *lac* operon are turned on when *lac* repressor dissociates from the *lac* operator and the genes are turned off when the repressor binds. Everyone came to believe that the turning on or off of gene activity was accomplished by the acts of binding or dissociation of proteins from DNA. In the *ara* operon the repression that occurs in the absence of arabinose involves the *araI* site, for mutations in *araI* can interfere with repression. In the presence of arabinose the *araI* site is also required for induction. These facts require that at least part of the *araI* site be occupied by AraC protein both in the presence and absence of arabinose. Induction results from changing the state or conformation of AraC protein, and not the *de novo* binding or dissociation of the protein. A crucial test of the looping model that made these predictions was that *araI* be occupied by AraC protein in the absence of arabinose.

How can one test for the binding of a specific protein to a specific site in growing cells? Dimethylsulfate can be used for footprinting in addition to DNAse. In contrast to DNAse, dimethylsulfate easily enters cells. Its rate of methylation of guanines at a protein's binding site can be either increased or decreased by the presence of the protein. Following a brief treatment with dimethylsulfate, analogous to light nicking by DNAse in DNA footprinting, the frequency of methylation at different guanines can be measured by isolating, labeling, cleaving at the positions of methylations and separation of fragments on a sequencing gel.

In vivo footprinting showed that AraC protein occupies *araI* both in the absence and presence of arabinose, thus fulfilling an essential requirement of the looping model. The footprinting experiments showed a second fact. The $araO_2$ site, to which AraC protein binds only weakly *in vitro*, is also occupied *in vivo*. This is as required by the looping model. It is not, however, occupied when the *araI* and $araO_1$ sites are deleted. That is, its occupancy depends on the presence of sites located more than a hundred nucleotides away. This is, of course, what happens in looping. It is the cooperativity generated by looping that leads to the occupancy of $araO_2$. The binding of AraC protein to *araI* increases its concentration in the vicinity of $araO_2$ to such an extent that this second site is occupied.

The degree of cooperativity generated by the looping can be roughly estimated. AraC protein is present at about 20 molecules per cell. This is a concentration of about 2×10^{-8} M. The concentration of $araO_2$ in the presence of *araI* can be estimated to be at least 10^{-6} M. Thus, by virtue of looping, the concentration of AraC protein near $araO_2$ can be increased more than 100-fold.

How AraC Protein Loops and Unloops

Two lines of *in vivo* evidence suggest that the loop between *araI* and $araO_2$ is broken upon the addition of arabinose. First, the operon is fully inducible if $araO_2$ is deleted. This implies that the loop is not involved with induction. Second, the occupancy of *araO₂* falls immediately after arabinose addition. One good way to study such regulated looping is to perform *in vitro* experiments. AraC loops sufficiently weakly, however, that looping of linear DNA just does not occur. Instead, looping had to be studied using small supercoiled circles of about 400 base pairs. These are sufficiently small, and the supercoiled DNA is sufficiently wound upon itself, that AraC protein can easily bridge the gap between *araO₂* and *araI*. These supercoiled, looped circles migrate upon electrophore-

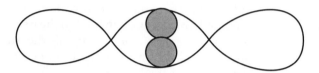

sis at a rate different from free circles or circles with AraC bound at a single point. Therefore binding, looping, and unlooping can all be assayed.

Study of the looped supercoils revealed an unexpected property. As AraC is a dimer in solution, and a dimer binds to linear DNA containing the *araI* site, it had seemed likely that the looped species would be formed from a dimer of AraC bound to *araI* and a dimer bound to *araO₂*. This proved not to be the case. The looped species contains only one dimer! In the absence of arabinose one of the monomers of an AraC dimer binds to the left half of *araI*, and the other monomer binds to *araO₂* (Fig. 12.13). The left and right halves of *araI* are called *araI₁* and *araI₂*. Upon the addition of arabinose the protein reorients, and the subunit contacting *araO₂* lets go, and contacts *araI₂*. The subunit rearrangement occurs in the absence of free protein and is largely independent of the precise sequences at the sites involved. In the absence of arabinose the protein prefers to loop, that is, contact to nonlocal sites. In the presence of arabinose, the protein prefers to contact local sites. The contacting of the *araI₂* site by AraC protein provides the inducing signal to RNA polymerase. Only if this site is properly positioned so as

Figure 12.13 Reorientation of the two subunits of AraC protein shifts the most favored conformation from a looped state to the unlooped state.

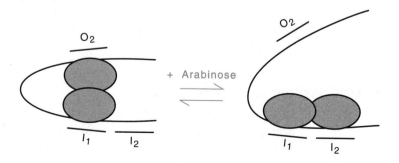

Absence of arabinose

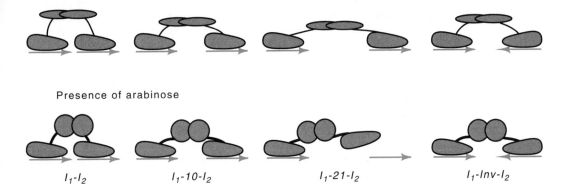

Presence of arabinose

I_1-I_2 I_1-10-I_2 I_1-21-I_2 I_1-Inv-I_2

Figure 12.14 Binding of AraC protein to sequences with different distances separating the half-sites or with one half-site inverted. In the presence of arabinose, the reach of AraC is reduced.

to overlap partially the -35 sequence of p_{BAD} does AraC activate transcription.

One simple mechanism that could generate the observed behavior of AraC protein is a subunit reorientation. Suppose that in the absence of arabinose the subunits are oriented such that binding to two half-sites in a looping structure is energetically easier than binding to two adjacent half-sites. Then the protein would prefer to loop in the absence of arabinose. If the presence of arabinose causes the subunits to reorient so that binding to the two half-sites of *araI* is favored, then the protein would unloop, bind to both halves of *araI*, and induce p_{BAD}.

The data supports a slight modification of the reorientation model outlined above. AraC protein monomers consist of a dimerization domain flexibly connected to a DNA-binding domain. The half-sites to which AraC binds normally are oriented in a direct repeat orientation. AraC still binds with high affinity when one half-site is inverted so that the total site possesses inverted repeat symmetry. This shows that the dimerization and DNA-binding domains are flexibly connected. Further, the half-sites of *araI* can be separated by an additional 10 or 21 base pairs, or even inverted, without greatly affecting the affinity of AraC. When, however, arabinose is added, the affinity of AraC for the wild-type *araI* site, the "+10" site, and the inverted site is increased, but that for the "+21" site is decreased. Because arabinose did not increase the affinity of AraC for all the sites, arabinose cannot be increasing the intrinsic affinity of a DNA-binding domain for a half-site. Instead, arabinose alters the relative positions of the DNA-binding domains or changes the flexibility or length of the connection between the dimerization and DNA-binding domains (Fig. 12.14).

Why Looping is Biologically Sensible

DNA looping solves three regulation problems involving DNA. This regulation can be of gene expression, DNA recombination, or DNA replication. The discussion below is couched in terms of gene regulation, but the same arguments apply to the other looping situations. The first problem solved is that of space. Most gene regulatory proteins must bind to specific DNA sequences to be able to direct their activities to specific promoters. There may not be room immediately adjacent to a bound RNA polymerase for all the proteins that need to affect transcription from that promoter. DNA looping provides a solution. A regulatory protein can bind at some distance from an initiation complex and still directly affect transcription by means of DNA looping.

The second problem that DNA looping helps solve is concentrations. This is particularly a problem in eukaryotic cells. Many cell-types have the capability of inducing very large numbers of genes. This means that the relevant regulatory proteins must all be present in the nuclei of these cells. The concentration of any single regulatory protein cannot be high since all the proteins must share the same volume. Therefore the system might try to design the binding sites so that the regulatory proteins bind particularly tightly to the DNA. Unfortunately, such tight-binding could interfere with other cellular operations like replication, repair, and recombination. Nonetheless, systems can be built so that the binding affinity of the proteins for the sites is not too high and yet the binding sites are well occupied. The general method of doing this is to increase the concentration of the protein in the immediate vicinity of its binding site. DNA looping accomplishes this.

The third problem DNA looping solves concerns time. Due to the low concentrations of the individual regulatory proteins that must be maintained within cells, significant time could elapse until a regulatory protein could find its site and bind. If the protein is already on its site, but kept from activating transcription by looping, then induction can be extremely rapid. It is merely the time required to unloop.

Why Positive Regulators are a Good Idea

One of the reasons for positive regulatory systems is even more important for eukaryotic cells. Let us use the *lac* regulatory system as the basis for comparison. At any instant, more than half of the *lac* repressor in a bacterial cell is nonspecifically bound at nonoperator sites. If the cell contained a thousand times as much DNA, then at least a thousand times as much *lac* repressor would be required to maintain full repression of the *lac* operon. If the cell contained 20,000 operons regulated by repressors with similar characteristics, then the nucleus could not accommodate the total amount of regulatory proteins required to regulate these operons. Of course, it may be possible for a repressor to possess greater selectivity than *lac* repressor has for its operator site, but there has to be an upper limit.

Positive regulatory systems do not require high levels of repressor to maintain low basal expression rates since a positively regulated promoter is naturally off. For many genes, the occupancy of only a fraction of the promoter regions by positive regulators may be sufficient to provide sufficient activity to the promoter when it must be turned on.

Problems

12.1. How could subunit mixing explain the dominance of C^+ to C^c without requiring that C^+ protein be a true repressor?

12.2. Frequently it is claimed that phosphorylation of sugars prevents their leakage out of cells. If this were the case, it would seem reasonable to phosphorylate arabinose before the other enzymatic conversions. Is there a biochemical reason for not phosphorylating arabinose before the isomerization step?

12.3. Suppose that two different missense mutations in the arabinose isomerase were found to complement one another. How could this be possible? Suppose that a point mutation in the *araC* gene could nonetheless yield Ara^+ cells when a particular CRP mutation was introduced into the cell. What could you infer about the functioning of the system, and what could you conclude definitely?

12.4. What experiment would be the best to determine whether α- or β-L-arabinose is actively transported by cells? What value would it be to further research to know which it is?

12.5. Why is it of interest to know whether $AraI^c$ promoters can be repressed in strains lacking active CRP?

12.6. Suppose that dominant Ara^- mutations are very rare but do exist. Devise an efficient selection method for their isolation.

12.7. Why do you expect that placing a *lac* operator between $araO_1$ and $araO_2$ would somewhat increase *araBAD* inducibility in $LacI^+$ but not in $LacI^-$ cells?

12.8. The PaJaMa experiment, named for Pardee, Jacob, and Monod, consists of mating $F'I^+Z^+$ episomes into F^- cells deleted of the *lac* region. Immediately after entry of the episome into the female, synthesis of LacZ protein begins, but its synthesis is turned off after about 30 minutes because of the accumulation of *lac* repressor. What could a PaJaMa experiment tell about the arabinose operon and what strains would be needed?

12.9. Devise an efficient method for recombinational mapping of C^c mutations.

12.10. Why is it a paradox if experiments show that a deletion from p_C into the CRP-binding site in the p_{BAD} promoter of the arabinose operon leaves the promoter p_{BAD} partially inducible, but is completely uninducible in strains unable to synthesize cyclic AMP? What is the interpretation of this finding? Note, arabinose is able to leak into such $cAMP^-$ cells perfectly well.

12.11. Two variants, A and B, were made with insertions in nonessential sites between $araO_1$ and CRP and between CRP and $araI$. One can imagine two possible sets of results for the inducibility of these strains. What would each data set suggest as to the mechanism of CRP protein function in induction of the arabinose operon?

Inducibility of p_{BAD}

	Possibility A	Possibility B
Wild Type	1.0	1.0
Insertion between O_1 and CRP	.05	1
Insertion between CRP and I	.05	.05

12.12. How could you test the possibility that the role of CRP in induction of p_{BAD} is to help break the repression loop? Does this make biological sense?

12.13. The claim has been made that the existence of temperature-sensitive mutations in the *araC* gene proves that the gene product is a protein. Why is such reasoning faulty?

12.14. Design an experiment to show whether or not the turn down in the activity of p_C 10 minutes after the addition of arabinose is due to catabolite repression generated by the metabolism of arabinose.

12.15. Cells growing on either a rich broth medium or on glucose minimal medium were treated with nitrosoguanidine or UV. Each of the resulting two groups of rifamycin-resistant mutants was then tested for its Ara phenotype. What is an explanation for the appearance of Rif^r-Ara^- double mutants only when cells were treated with nitrosoguanidine and then only when they had grown on rich medium?

12.16. Lac UV5 is a mutation that makes activity of the *lac* promoter independent of cyclic AMP receptor protein. Suppose promoter minus mutants of *lac* UV5 occur at a frequency of 10^{-9}. Suppose also that promoter minus mutations in the arabinose operon occurred only at a frequency of 10^{-10}. Not only that, but on analysis, all of these turn out to be little insertions or deletions, whereas the *lac* mutations were primarily base changes. What would be the best interpretation of these data?

12.17. What would you conclude if a small fraction of AraC$^-$ mutations could be suppressed by compensating mutations in RNA polymerase, but that no mutations exist in AraC that can be corrected by mutations in CRP protein?

12.18. Estimate the effective concentration in molarity of the $araO_2$ site in the vicinity of the $araI$ site.

References

Recommended Readings

The L-arabinose Operon in *Escherichia coli* B/r: A Genetic Demonstration of Two Functional States of the Product of a Regulator Gene, E. Englesberg, C. Squires, F. Meronk, Proc. Natl. Acad. Sci. USA *62*, 1100-1107 (1969).

Arabinose C Protein: Regulation of the Arabinose Operon *in vitro*, J. Greenblatt, R. Schleif, Nature New Biology *233*, 166-170 (1971).

DNA Looping and Unlooping by AraC Protein, R. Lobell, R. Schleif, Science *250*, 528-532 (1990).

Genetics of the *ara* System

Determination of the Order of Mutational Sites Governing L-arabinose Utilization in *E. coli* B/r by Transduction with Phage P1bt, J. Gross, E. Englesberg, Virology *9*, 314-331 (1959).

Direct Selection of L-arabinose Negative Mutants of *E. coli* Strain B/r, H. Boyer, E. Englesberg, R. Weinberg, Genetics *47*, 417-425 (1962).

An Analysis of "Revertants" of a Deletion Mutant in the *C* Gene of the L. Arabinose Complex in *E. coli*, B/r: Isolation of Initiator Constitutive Mutants (I^c), E. Englesberg, D. Sheppard, C. Squires, F. Meronk, J. Mol. Biol. *43*, 281-298 (1969).

Directed Transposition of the Arabinose Operon: A Technique for the Isolation of Specialized Transducing Bacteriophages for any *E. coli* Gene, S. Gottesman, J. Beckwith, J. Mol. Biol. *44*, 117-127 (1969).

Arabinose-leucine Deletion Mutants of *E. coli* B/r, D. Kessler, E. Englesberg, J. Bact. *98*, 1159-1169 (1969).

Selection of *araB* and *araC* Mutants of *E. coli* B/r by Resistance to Ribitol, L. Katz, J. Bact. *102*, 593-595 (1970).

Nonsense Mutants in the Regulator Gene *araC* of the L-arabinose System of *E. coli* B/r, J. Irr, E. Englesberg, Genetics *65*, 27-39 (1970).

Initiator Constitutive Mutants of the L-arabinose Operon (0IBAD) of *E. coli* B/r, L. Gielow, M. Largen, E. Englesberg, Genetics *69*, 289-302 (1971).

Fine Structure Deletion Map of the *E. coli* L-arabinose Operon, R. Schleif, Proc. Natl. Acad. Sci. USA *69*, 3479-3484 (1972).

Novel Mutation to Dominant Fucose Resistance in the L-arabinose Operon of *E. coli*, N. Nathanson, R. Schleif, J. Bact. *115*, 711-713 (1973).

Fusion of the *E. coli lac* Genes to the *ara* Promoter: A General Technique Using Bacteriophage Mu-l Insertions, M. Casadaban, Proc. Natl. Acad. Sci. USA *72*, 809-813 (1975).

The Isolation and Characterization of Plaque-forming Arabinose Transducing Bacteriophage Lambda, J. Lis, R. Schleif, J. Mol. Biol. *95*, 395-407 (1975).

Paucity of Sites Mutable to Constitutivity in the *araC* Activator Gene of the L-arabinose Operon of *E. coli*, N. Nathanson, R. Schleif, J. Mol. Biol. *96*, 185-199 (1975).

Deletion Analysis of the *Escherichia coli* P$_C$ and P$_{BAD}$ Promoters, T. Dunn, R. Schleif, J. Mol. Biol. *180*, 201-204 (1984).

Dominance Relationships Among Mutant Alleles of Regulatory Gene *araC* in the *Escherichia coli* B/r L-arabinose Operon, D. Sheppard, J. Bact. *168*, 999-1001 (1986).

Physical Map Location of the *araFGH* Operon of *Escherichia coli*, W. Hendrickson, K. Rudd, J. Bact. *174*, 3836-3837 (1992).

Physiology of the Regulation of the *ara* system

Positive Control of Enzyme Synthesis by Gene C in the L-arabinose System, E. Englesberg, J. Irr, J. Power, N. Lee, J. Bact. *90*, 946-957 (1965).

The L-arabinose Permease System in *E. coli* B/r, C. Novotny, E. Englesberg, Biochim. et Biophys. Acta *117*, 217-230 (1966).

Further Evidence for Positive Control of the L-arabinose System by Gene *araC*, D. Sheppard, E. Englesberg, J. Mol. Biol. *25*, 443-454 (1967).

Induction of the L-arabinose Operon, R. Schleif, J. Mol. Biol. *46*, 197-199 (1969).

Dual Control of Arabinose Genes on Transducing Phage λd*ara*, R. Schleif, J. Greenblatt, R. Davis, J. Mol. Biol. *59*, 127-150 (1971).

Hyperinducibility as a Result of Mutation in Structural Genes and Self-catabolite Repression in the *ara* Operon, L. Katz, E. Englesberg, J. Bact. *107*, 34-52 (1971).

Induction of the *ara* Operon of *E. coli* B/r, M. Doyle, C. Brown, R. Hogg, R. Helling, J. Bact. *110*, 56-65 (1972).

Different Cyclic AMP Requirements for Induction of the Arabinose and Lactose Operons of *E. coli*, J. Lis, R. Schleif, J. Mol. Biol. *79*, 149-162 (1973).

In vivo Experiments on the Mechanism of Action of L-arabinose C Gene Activator and Lactose Repressor, J. Hirsh, R. Schleif, J. Mol. Biol. *80*, 433-444 (1973).

Induction Kinetics of the L-arabinose Operon of *E. coli*, R. Schleif, W. Hess, S. Finkelstein, D. Ellis, J. Bact. *115*, 9-14 (1973).

Regulation of the Regulatory Gene for the Arabinose Pathway *araC*, M. Casadaban, J. Mol. Biol. *104*, 557-566 (1976).

Hypersensitivity to Catabolite Repression in the L-arabinose Operon of *E. coli* B/r is *trans* acting, D. Sheppard, M. Eleuterio, J. Bact. *126*, 1014-1016 (1976).

The *araC* Promoters; Transcription, Mapping and Interaction with the *araBAD* promoter, J. Hirsh, R. Schleif, Cell *11*, 545-550 (1977).

The *E. coli* L-arabinose Operon: Binding Sites of the Regulatory Proteins and a Mechanism of Positive and Negative Regulation, S. Ogden, D. Haggerty, C. M. Stoner, D. Kolodrubetz, R. Schleif, Proc. Natl. Acad. Sci. USA 77, 3346-3350 (1980).

Regulation of the L-arabinose Transport Operons in *Escherichia coli*, D. Kolodrubetz, R. Schleif, J. Mol. Biol. *151*, 215-227 (1981).

L-arabinose Transport Systems in *Escherichia coli* K-12, D. Kolodrubetz, R. Schleif, J. Bact. *148*, 472-479 (1981).

Hyperproduction of *araC* Protein from *Escherichia coli*, R. Schleif, M. Favreau, Biochem. *21*, 778-782 (1982).

Transcription Start Site and Induction Kinetics of the *araC* Regulatory Gene in *Escherichia coli* K-12, C. Stoner, R. Schleif, J. Mol. Biol. *170*, 1049-1053 (1983).

The *araE* Low Affinity L-arabinose Transport Promoter, Cloning, Sequence, Transcription Start Site and DNA Binding Sites of Regulatory Proteins, C. Stoner, R. Schleif, J. Mol. Biol. *171*, 369-381 (1983).

In vivo Regulation of the *Escherichia coli araC* Promoter, S. Hahn, R. Schleif, J. Bact. *155*, 593-600 (1983).

Characterization of the *Escherichia coli araFGH* and *araJ* Promoters, W. Hendrickson, C. Stoner, R. Schleif, J. Mol. Biol. *215*, 497-510 (1990).

Mapping, Sequence, and Apparent Lack of Function of *araJ*, a Gene of the *Escherichia coli* Arabinose Regulon, T. Reeder, R. Schleif, J. Bact. *173*, 7765-7771 (1991).

Characterization and Study of Parts of the *ara* System

L-arabinose Operon Messenger of *E. coli*, R. Schleif, J. Mol. Biol. *61*, 275-279 (1971).

Anomeric Specificity and Mechanism of Two Pentose Isomerases, K. Schray, I. Rose, Biochemistry *10*, 1058-1062 (1971).

Direction of Transcription of the Regulatory Gene *araC* in *E. coli* B/r, G. Wilcox, J. Boulter, N. Lee, Proc. Natl. Acad. Sci. USA *71*, 3635-3639 (1974).

The Arabinose C Gene Product of *E. coli* B/r is Hyperlabile in a Cell-Free Protein Synthesis System, D. Steffen, R. Schleif, Mol. and Gen. Genet. *128*, 93-94 (1974).

Overproducing *araC* Protein with Lambda-arabinose Transducing Phage, D. Steffen, R. Schleif, Mol. and Gen. Genet. *157*, 333-339 (1977).

In vitro Construction of Plasmids which Result in Overproduction of the Protein Product of the *araC* gene of *E. coli*, D. Steffen, R. Schleif, Mol. and Gen. Genet. *157*, 341-344 (1977).

Nucleotide Sequence of the L-arabinose Regulatory Region of *E. coli* K-12, B. Smith, R. Schleif, J. Biol. Chem. *253*, 6931-6933 (1978).

DNA Sequence of *araBAD* Promoter Mutants of *E. coli*, A. Horowitz, C. Morandi, G. Wilcox, J. Bact. *142*, 659-667 (1980).

Identification of *araC* Protein on Two-dimensional Gels, Its *in vivo* Instability and Normal Level, D. Kolodrubetz, R. Schleif, J. Mol. Biol. *149*, 133-139 (1981).

Is the Amino Acid Sequence but Not the Nucleotide Sequence of the *Escherichia coli araC* Gene Conserved?, C. Stoner, R. Schleif, J. Mol. Biol. *154*, 649-652 (1982).

Mutations in the Sigma Subunit of *E. coli* RNA Polymerase Which Affect Positive Control of Transcription, J. Hu, C. Gross, Mol. Gen. Genetics *199*, 7-13 (1985).

AraB Gene and Nucleotide Sequence of the AraC Gene of *Erwinia carotovora*, L. Lei, H. Lin, L. Heffernan, G. Wilcox, J. Bact. *164*, 717-722 (1985).

High-affinity L-arabinose Transport Operon, Nucleotide Sequence and Analysis of Gene Products, J. Scripture, C. Voelker, S. Miller, R. O'Donnell, L. Polgar, J. Rade, B. Horazdovsky, R. Hogg, J. Mol. Biol. *197*, 37-46 (1987).

High-affinity L-arabinose Transport Operon, B. Horazdovsky, R. Hogg, J. Mol. Biol. *197*, 27-35 (1987).

Effect of Mutations in the Cyclic AMP Receptor Protein-binding Site on *araBAD* and *araC* Expression, L. Stoltzfus, G. Wilcox, J. Bact. *171*, 1178-1184 (1989).

Genetic Reconstitution of the High-affinity L-arabinose Transport System, B. Horazdovsky, R. Hogg, J. Bact. *171*, 3053-3059 (1989).

Activation of *ara* Operons by a Truncated AraC Protein Does Not Require Inducer, K. Menon, N. Lee, Proc. Natl. Acad. Sci. USA *87*, 3708-3712 (1990).

Physical Studies

The Regulatory Region of the L-arabinose Operon: Its Isolation on a 1000 Base-pair Fragment from DNA Heteroduplexes, J. Lis, R. Schleif, J. Mol. Biol. *95*, 409-416 (1975).

The Regulatory Region of the L-arabinose Operon: A Physical, Genetic and Physiological Study, R. Schleif, J. Lis, J. Mol. Biol. *95*, 417-431 (1975).

The General Affinity of *lac* Repressor for *E. coli* DNA: Implications for Gene Regulation in Procaryotes and Eucaryotes, S. Lin, A. Riggs, Cell *4*, 107-111 (1975).

Electron Microscopy of Gene Regulation: The L-arabinose Operon, J. Hirsh, R. Schleif, Proc. Natl. Acad. Sci. USA *73*, 1518-1522 (1976).

A Gel Electrophoresis Method for Quantifying the Binding of Proteins to Specific DNA Regions—Application to Components of the *Escherichia coli* Lactose Operon Regulatory System, M. Garner, A. Revzin, Nuc. Acids. Res. *9*, 3047-3060 (1981).

Equilibria and Kinetics of *lac* Repressor Operator Interactions by Polyacrylamide Gel Electrophoresis, M. Fried, D. Crothers, Nuc. Acids Res. *9*, 6505-6525 (1981).

Arabinose-inducible Promoter from *Escherichia coli*, Its Cloning from Chromosomal DNA, Identification as the *araFG* Promoter and Sequence, B. Kosiba, R. Schleif, J. Mol. Biol. *156*, 53-66 (1982).

Regulation of the *Escherichia coli* L-arabinose Operon Studied by Gel Electrophoresis DNA Binding Assay, W. Hendrickson, R. Schleif, J. Mol. Biol. *174*, 611-628 (1984).

A Dimer of AraC Protein Contacts Three Adjacent Major Groove Regions of the *araI* Site, W. Hendrickson, R. Schleif, Proc. Nat. Acad. Sci. USA *82*, 3129-3133 (1985).

Altered DNA Contacts Made by a Mutant AraC Protein, A. Brunelle, W. Hendrickson, R. Schleif, Nuc. Acids. Res. *13*, 5019-5026 (1985).

Equilibrium DNA-binding of AraC Protein, Compensation for Displaced Ions, K. Martin, R. Schleif, J. Mol. Biol. *195*, 741-744 (1987).

Determining Residue-base Interactions Between AraC Protein and *araI* DNA, A. Brunelle, R. Schleif, J. Mol. Biol. *209*, 607-622 (1989).

Variation of Half-site Organization and DNA Looping by AraC Protein, J. Carra, R. Schleif, EMBO Journal *12*, 35-44 (1993).

DNA Looping

An *araBAD* Operator at -280 Base Pairs that is Required for P$_{BAD}$ Repression—Addition of DNA Helical Turns between the Operator and Promoter Cyclically Hinders Repression, T. Dunn, S. Hahn, S. Ogden, R. Schleif, Proc. Nat. Acad. Sci. USA *81*, 5017-5020 (1984).

Upstream Repression and CRP Stimulation of the *Escherichia coli* L-arabinose Operon, S. Hahn, T. Dunn, R. Schleif, J. Mol. Biol. *180*, 61-72 (1984).

The DNA Loop Model for *ara* Repression: AraC Protein Occupies the Proposed Loop Sites *in vivo* and Repression-negative Mutations Lie in these Same Sites, K. Martin, L. Huo, R. Schleif, Proc. Nat. Acad. Sci. USA *83*, 3654-3658 (1986).

Transcription of the *Escherichia coli ara* Operon *in vitro*, The Cyclic AMP Receptor Protein Requirement for P_{BAD} Induction that Depends on the Presence and Orientation of the *araO$_2$* Site, S. Hahn, W. Hendrickson, R. Schleif, J. Mol. Biol. *188*, 355-367 (1986).

Why Should DNA Loop?, R. Schleif, Nature *327*, 369-370 (1987).

Three Binding Sites for AraC are Required for Autoregulation of AraC in *Escherichia coli*, E. Hamilton, N. Lee, Proc. Natl. Acad. Sci. USA *85*, 1749-1753 (1988).

Alternative DNA Loops Regulate the Arabinose Operon in *Escherichia coli*, L. Huo, K. Martin, R. Schleif, Proc. Natl. Acad. Sci. USA *85*, 5444-5448 (1988).

Novel Activation of AraC Expression and a DNA Site Required for *araC* Autoregulation in *Escherichia coli* B/r, L. Cass, G. Wilcox, J. Bact. *170*, 4179-4180 (1988).

In vivo DNA Loops in *araCBAD*: Size Limit and Helical Repeat, D. Lee, R. Schleif, Proc. Nat. Acad. Sci. USA *86*, 476-480 (1989).

AraC-DNA Looping: Orientation and Distance-dependent Loop Breaking by the Cyclic AMP Receptor Protein, R. Lobell, R. Schleif, J. Mol. Biol. *218*, 45-54 (1991).

Attenuation and the *trp* Operon

13

In the previous chapter we saw an example of regulation of gene expression by control of RNA polymerase initiation frequency. Many of the mechanisms appearing in regulation of the arabinose operon also apply to eukaryotic systems as well. In this chapter we shall consider a different mechanism of gene regulation, one that first appeared to be confined to prokaryotic cells, but which now is known to function in eukaryotic cells as well. This is controlling transcription termination at a special point after the mRNA has been initiated but before the RNA polymerase has completed transcription of the operon's structural genes. Each transcript that is initiated has two possible fates. Either its synthesis terminates before the completion of the entire messenger, or no premature termination occurs, and useable messenger is synthesized. Regulation alters the ratio of incomplete to complete messenger. That is, the effectiveness of termination is regulated. Often this type of regulation mechanism is termed attenuation. Ribosomal RNA, ribosomal proteins, amino acid biosynthetic genes, and other genes whose expression rate is related to growth rate of the cells use attenuation mechanisms. In this chapter most of our attention will be directed to the *trp* operon. In it we shall see that the major regulator of the effectiveness of *trp* messenger RNA transcription termination is the secondary structure of the RNA itself.

No cells are continuously exposed to ideal growth conditions. Indeed, the majority of the time *Escherichia coli* cells are exposed to adverse conditions, and growth is slowed or stopped. From time to time however, spurts of growth are possible when nutrients suddenly appear or when populations of cells at high concentration are diluted. Therefore cells must possess regulatory mechanisms that turn on and turn off at appropriate times the synthesis of enzymes like those of the tryptophan biosynthetic pathway. Three important cellular states should be considered with respect to synthesis of the tryptophan biosynthetic enzymes:

1. Tryptophan is absent, but otherwise cells are capable of synthesizing protein;

2. Tryptophan is absent, and in addition, no protein synthesis is possible for additional reasons;

3. Tryptophan is present;

Only in the first state is it sensible for cells to synthesize *trp* messenger RNA. In both the second and third states it is energy-efficient for cells not to synthesize *trp* messenger RNA. A simple repressor mechanism like that of the *lac* operon that blocks synthesis of messenger whenever tryptophan is present would permit synthesis of *trp* messenger whether or not protein synthesis is possible. This is insufficient. In addition, cells need a way to tell when tryptophan is absent but protein synthesis is possible. The tryptophan operon has evolved a clever way to do this.

The Aromatic Amino Acid Synthetic Pathway and its Regulation

The first chemical reaction common to the synthesis of tryptophan, tyrosine, and phenylalanine is the condensation of erythrose-4-P and phosphoenolpyruvate to form 3-deoxy-D-arabino-heptulosonate-7-P, DAHP, a reaction catalyzed by DAHP synthetase (Fig. 13.1). Due to its position at the head of the aromatic amino acid biosynthetic pathway and to the fact that this reaction is irreversible, it is logical that this reaction should be a key point of regulation. Indeed, this expectation is met. *Escherichia coli* regulates both the amount and activity per molecule of DAHP synthetase as a function of the intracellular levels of the aromatic amino acids.

A double regulation of total DAHP synthetase activity is logical. While regulation of the enzyme level minimizes unnecessary consumption of

Figure 13.1 Outline of the aromatic amino acid biosynthetic pathway in *Escherichia coli*. Each arrow represents an enzymatic step. Tryptophan feedback inhibits one of the DAHP synthetases, AroH, as well as the first enzyme of the pathway committed to tryptophan synthesis.

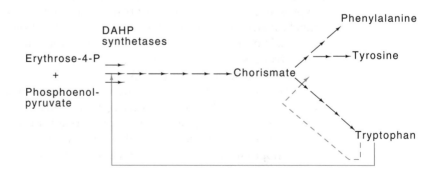

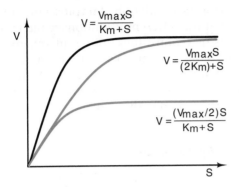

Figure 13.2 Two possibilities for feedback inhibition of an enzyme. The top curve is enzyme velocity as a function of substrate concentration, and the equation is the activity of an uninhibited enzyme as a function of its K_m, V_{max}, and substrate concentration S.

amino acids and energy in the synthesis of the enzyme, this type of regulation is incapable of producing appreciable changes in the enzyme levels or of total enzyme activity on time scales shorter than minutes. A much more rapidly-responding regulation mechanism is also necessary to adjust the synthesis rates of tryptophan, tyrosine, and phenylalanine on time scales of seconds.

In addition to stabilizing the activity of the synthetic pathway against random fluctuations, a rapidly responding regulation would fine-tune the biosynthetic flow rates of the aromatic amino acids and would be able to respond rapidly to growth rate changes generated by changes in the nutrient medium. Feedback inhibition of an enzyme's activity meets the requirements, as this mechanism can alter an enzyme's activity in milliseconds.

Feedback inhibition is an example of an allosteric interaction in which accumulation of the product of the pathway leads to inhibition of the activity of an enzyme in that pathway. This is a specific example of an allosteric interaction in which a molecule dissimilar in shape to the substrates of an enzyme can bind to the enzyme, usually at a site on the enzyme far from the active site, and can generate conformational changes that alter the catalytic activity of the enzyme. Feedback inhibition can reduce an enzyme's activity in either of two fundamental ways (Fig. 13.2). The tryptophan-sensitive DAHP synthetase is feedback-inhibited largely as a result of a change in its V_{max}, whereas the first enzyme of the pathway used solely for tryptophan synthesis, anthranilate synthetase, is an example of the other possibility. It is feedback-inhibited by tryptophan via a change in its K_m.

Bacillus subtilis possesses a single DAHP synthetase whose synthesis and activity is regulated by the three aromatic amino acids. *Escherichia coli*, however, possesses three different DAHP synthetases. The activity of one, the AroH protein, is feedback-inhibited by tryptophan, another is feedback-inhibited by tyrosine, and the third is feedback-inhibited by phenylalanine. Only if the cell's growth medium possesses all three amino acids is all DAHP synthetase activity within an *E. coli* cell fully inhibited.

This is an example of the fact that different microorganisms possess different overall schemes for the regulation of tryptophan synthesis. On one hand, it is possible that the different evolutionary niches occupied by different microorganisms require these different schemes. On the other hand, perhaps one scheme is no better than another, and the different ones just happened to have evolved that way. Either way, this diversity means that the overall scheme for regulation of tryptophan synthesis in *E. coli* is not the only one that works.

Rapid Induction Capabilities of the *trp* Operon

The tryptophan operon consists of five genes that code for the enzymes unique to the synthesis of tryptophan, *trpE, D, C, B,* and *A* (Fig. 13.3). In addition, the *trpR* gene codes for a repressor that helps regulate expression of these genes. TrpR repressor protein bound to the *trp* operator blocks access of RNA polymerase to the promoter. Since this repressor binds to *trp* operator far better in the presence of its corepressor tryptophan than in its absence, the transcription of the *trp* genes is repressed in the presence of excess tryptophan and derepressed during times of tryptophan deficiency.

A classical repression mechanism, in which the repressor protein blocks binding or initiation by RNA polymerase, is adequate for meeting part of the regulatory requirements of the tryptophan operon. Such a regulation mechanism, however, cannot monitor the overall capability of the cell to synthesize protein, and it would permit *trp* mRNA synthesis whenever tryptophan was in short supply, even when protein synthesis was impossible. Such a mechanism is not adequate for regulating an

Figure 13.3 The *trp* operon that codes for the enzymes necessary for the conversion of chorismate to tryptophan. Shown is the operator, *o*; promoter, *p*; leader region, *trpL*; and the positions within the operon of the five *trp* structural genes.

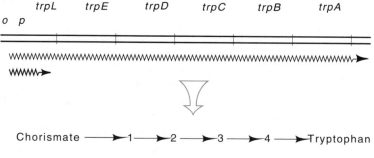

1. anthranilate
2. phosphoribosyl anthranilate
3. carboxphenylamino-deoxyribulose-phosphate
4. indoleglycerol phosphate

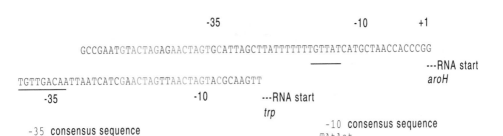

```
                    -35                              -10        +1
        GCCGAATGTACTAGAGAACTAGTGCATTAGCTTATTTTTTTGTTATCATGCTAACCACCCGG
                                                ─────
                                                                    ---RNA start
TGTTGACAATTAATCATCGAACTAGTTAACTAGTACGCAAGTT                          aroH
    ─────                          ────
    -35                            -10        ---RNA start
                                                 trp
                                                        -10 consensus sequence
      -35 consensus sequence                            TAtAat
      tcTTGACat
```

Figure 13.4 DNA sequences of the *aroH* and *trp* operon RNA polymerase binding sites aligned according to the homologies between the TrpR binding sites. Identical bases are shown in red. The -35 and -10 RNA polymerase recognition sequences are underlined.

amino acid biosynthetic operon. In the next section we shall discuss how the cell can link protein synthetic ability to mRNA synthesis. Before that, however, we shall examine an amusing consequence of repression in *trp* as well as look at one possible reason for the existence of repression in the *trp* operon.

TrpR represses synthesis of messenger for *aroH*, which encodes DAHP synthetase messenger as well as the *trp* operon messenger. For both, in the presence of tryptophan, it binds to DNA and blocks RNA polymerase from binding to the promoter. An interesting variation is shown between the two promoters. In the *trp* operon, the operator is centered around the -10 region of the RNA polymerase binding site, whereas in the *aroH* operon, the repressor-binding site is located around the -35 region. Since both of these operators are similar, the -10 region of the *trp* promoter and the -35 region of the *aroH* promoter can only weakly resemble the sequences typical of these regions in active promoters (Fig. 13.4). Apparently to compensate for this drastic alteration in part of the RNA polymerase-binding site, the other portions of the RNA polymerase-binding sites of these two promoters are homologous to highly active consensus promoter sequences.

In addition to regulating the *trp* operon and *aroH*, TrpR repressor in the presence of tryptophan also represses its own synthesis. As we will see, the consequences of this self-repression are that the cellular levels of the *trp* enzymes can rapidly increase to optimal levels following tryptophan starvation. The *ara* operon uses a positive-acting regulation mechanism to generate a rapid induction response followed by a lower steady-state response as catabolite repression turns down induction once catabolism of arabinose begins. The, *trp* operon achieves a rapid response using only negative-acting elements.

How is the rapid enzyme induction accomplished? Consider cells growing in the presence of excess tryptophan. In such a case the *trp* operon, *trpR*, and *aroH* genes are all repressed, but not fully off. A balance is maintained such that the level of TrpR repressor represses the trpR gene so as to maintain that level. Upon tryptophan starvation, these genes are all derepressed, and the gene products are synthesized

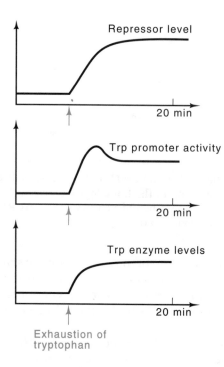

Figure 13.5 Repressor level, *trp* promoter activity, and *trp* enzyme levels following exhaustion of tryptophan from the medium of growing cells.

at a high rate (Fig. 13.5). As the intracellular concentration of TrpR repressor and tryptophan itself both increase, appreciable repression can set in, and transcription of the three sets of genes decreases. Finally, when steady state has been reached in minimal medium lacking tryptophan, the *trp* operon is 90% repressed, in part because the level of *trp* repressor is much higher than it is in the presence of tryptophan.

The Serendipitous Discovery of *trp* Enzyme Hypersynthesis

We now return to the main topic of this chapter. The discovery of polarity is an illustration of the fact that frequently the most important discoveries result from attempting to solve some other problem. Polarity is the decrease in the expression of a gene downstream in an operon from a nonsense mutation. This phenomenon commanded the attention of many molecular biologists from 1970 to about 1976. Experiments by Yanofsky and Jackson at this time were designed to study the phenomenon of polarity by locating within the *trp* operon any elements that affected polarity. They sought to isolate deletions that eliminated the polar effects of a mutation near the beginning of the operon on the expression of a gene near the end of the operon. As deletions are rare to begin with, special selections and scorings must be used to identify cells containing the desired deletion. The basic selection method was to use conditions in which polarity reduced expression of a promoter-distal *trp* gene to the extent that cells could not grow and then to select for secondary mutants that could grow. Among them would be deletions

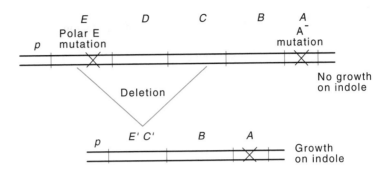

Figure 13.6 How an internal *trp* deletion can relieve polarity generated by a polar *trpE* mutation and increase expression of the *trpB* gene.

removing any elements that generated polarity. Scoring by replica plating onto petri plates spread with lawns of phage carrying parts of the *trp* operon then identified the candidates containing deletions internal to the *trp* operon.

The last two genes of the tryptophan operon, *trpA* and *trpB*, code for the α and β subunits of tryptophan synthetase. This enzyme normally has an $\alpha_2\beta_2$ structure. The synthetase from a *trpA* mutant has a β_2 structure. This dimer possesses part of the enzyme's normal activity and catalyzes the conversion of indole and serine to tryptophan and water at about 3% the rate that the wild-type enzyme catalyzes the conversion of serine and indole-3-glycerol phosphate to tryptophan and water. The activity of the β_2 complex is such that if indole is present in the growth medium, tryptophan can be synthesized at a rate sufficient to supply the cells' tryptophan requirements. The reduction in expression of the *trpB* gene in a *trpA* mutant strain containing a polar *trpE* mutation leaves the cells unable to convert indole to tryptophan at a rate adequate for growth (Fig. 13.6). Under these conditions, a deletion of the *trpE* mutation that does not include the *trp* promoter and the *trpB* gene relieves the polarity on *trpB* expression and permits cells to satisfy their tryptophan requirements with indole. Amongst the indole$^+$ colonies will be deletions of various sizes. One test was determining their sizes.

Candidates were scored for loss of the *trpD* and *trpC* genes with special phage that carried copies of these genes, but with mutations in the genes. The phage can infect cells, but many cells are not lysed. Often the phage either recombine into the chromosome or simply remain in the cell for several generations until diluted away by cell growth. While present in the cells, the phage can recombine with homologous DNA on the host chromosome if it is present. Thus, cells that retain *trpD* or *trpC* can reconstruct a functional *trpD* or *trpC* genes on some copies of the phage. These can subsequently be detected by replica plating onto another lawn of *trpC* or *trpD* mutant cells. By these means Jackson and Yanofsky identified many internal deletions in the *trp* operon that relieved polarity. We might have expected that fusions of the *trpB* gene

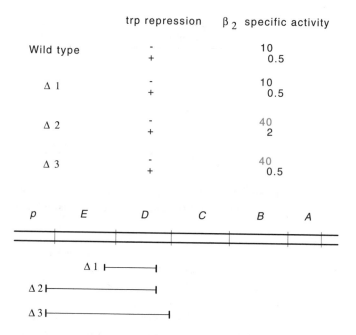

Figure 13.7 The specific activity of the β2 protein of the *trp* operon in the presence and absence of *trp* repressor in wild-type cells and in cells containing the three different deletions indicated.

to other genes of the chromosome might also have occurred, but, curiously, none were found.

Out of 34 internal deletions that had removed polarity on the expression of the *trpB* gene, all were still regulated by the Trp repressor. Most unexpectedly, two deletion strains hypersynthesized the TrpB protein by a factor of three to ten; each of these was deleted of all of the *trpE* gene (Fig. 13.7). The source of this hypersynthesis was then examined more carefully, as described in the next section.

Early Explorations of the Hypersynthesis

The experiments described in this section were performed before the development of modern genetic engineering. It is interesting to see how Yanofsky and his collaborators cleverly utilized the technology available at the time to learn the source of the hypersynthesis and thereby discover the phenomenon of attenuation. The first question in their careful investigation was whether the hypersynthesis in the internal deletion strains mentioned above resulted from an alteration in the levels of messenger. Either more messenger was present in the deletion strains or messenger was being translated more efficiently. The path of subsequent experiments depended on which it was.

Messenger RNA in cells was briefly labeled with tritiated uridine, extracted, and hybridized to complementary *trp* DNA immobilized on

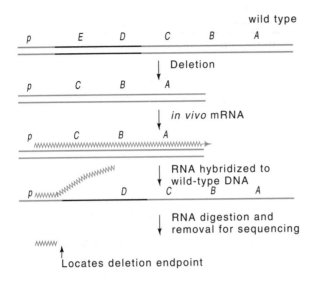

Figure 13.8 RNA synthesized from a deletion, hybridized to wild-type DNA, and digested with RNAse can reveal the deletion end point.

paper. *Trp*-containing DNA was available in the form of φ80 and λ phage in which part of the phage genome was substituted with *trp* sequences. Use of such phage containing all or part of the *trp* operon substituted for the use of fragments that today would be isolated by restriction enzyme cleavage of an appropriate plasmid or by PCR amplification using appropriate primers and crude *E. coli* DNA as template. The hybridization revealed that in all strains, the levels of *trpB* messenger paralleled the levels of TrpB enzyme. Therefore the cause of *trp* enzyme hypersynthesis in some of the internal deletion strains was elevated messenger levels.

Some of the internal deletion strains hypersynthesized the TrpB protein and messenger and some did not. Hence it was important to learn whether the effect could be correlated with the extent of the deletion. This was particularly important because sequencing of *trp* messenger had revealed that the *trp* operon contained an unexpectedly long leader of 162 bases between the start of transcription and the translation start of the first *trp* enzyme.

The deletion end points were determined by isolating *trp* messenger synthesized *in vivo* by the deletion strains. Total RNA from the deletion strains was isolated and hybridized to wild-type *trp* DNA (Fig. 13.8). The region of complementarity between the RNA and DNA was resistant to RNAse. After the digestion, this region was then melted off the DNA and the exact extent of trp-specific sequence obtained from each deletion was apparent on RNA sequencing. RNA sequencing was used since DNA sequencing techniques had not yet been invented. It was found that only those deletions that removed a site located 20 nucleotides before the *trpE* gene hypersynthesized TrpB protein.

Conceivably, the deletions caused hypersynthesis by altering the activity of the *trp* promoter, although this possibility seemed unlikely because some of the deletions in the hypersynthesizing strains ended as

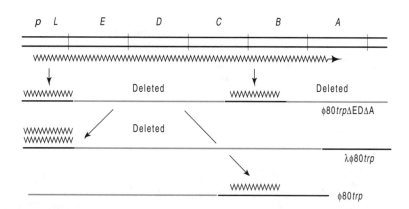

Figure 13.9 Two-step hybridization for determination of the relative amounts of *trpL* and *trpB* mRNA in cells. Messenger from the *trp* operon as indicated on the top line is hybridized to phage DNA carrying part of the *trp* operon. RNA eluted from the first phage was then hybridized to phage containing sequences homologous to either the leader region or the *trpB* region of the operon.

far as 100 nucleotides downstream from the promoter. More likely, the deletions removed a site located 20 bases ahead of the start of the *trpE* gene that terminated most transcription before it entered the *trp* structural genes. This idea was tested by comparing the amount of *trp* messenger specified by a sequence lying upstream of the *trpE* gene and upstream of the potential termination site to the amount of messenger specified by sequences downstream from the site (Fig. 13.9). The downstream messenger contained the *trpB* region of the operon. A two-step hybridization permitted quantitation of messenger RNA in cells from each of these regions. This experiment provided the telling clue. Wild-type repressed cells as well as *trpR⁻* cells contained an eight-to-ten-fold molar excess of messenger from the leader region preceding the *trpE* gene over messenger from the *trpB* region. In cells deleted of only the critical region just ahead of the *trpE* gene, the amount of *trpB* messenger was elevated and was nearly comparable to the amount of *trpL* RNA. These results prove that, indeed, termination occurs at the site 20 bases ahead of the *trpE* gene and that under normal conditions a majority of the transcription does not enter the *trp* structural genes. The site at which termination occurs is called *trp a*.

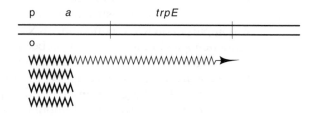

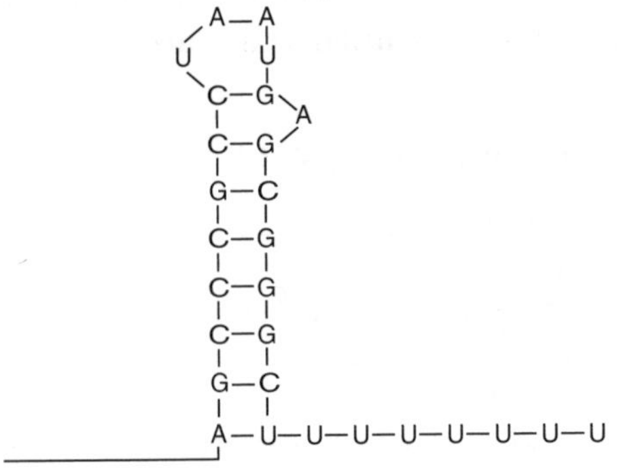

Figure 13.10 The termination site of *trpL* RNA.

The obvious and most important question to ask at this point was whether transcriptional termination, which ordinarily permitted only one in eight to ten polymerases to transcribe into the *trp* structural genes, was regulated or whether the efficiency of termination at *trp a* was independent of growth conditions. This was tested by starving for tryptophan and examining the attenuation level. Suitable starvation was generated by the addition of indolylacrylic acid that inhibits both tryptophan synthesis and tryptophanyl-tRNA synthetase. It did decrease termination, and thus termination at *trp a* is regulated, and is called attenuation. Starvation for other amino acids did not decrease termination at *trp a*, thereby proving that termination does play a regulatory role in the *trp* operon.

In vitro transcription of the *trp* operon DNA was the logical next step in these studies. The *trp* RNA from such experiments could be quantitated by its synthesis from φ80 and λ-*trp* phage and appropriate RNA-DNA hybridizations. These showed that virtually all transcripts ended at the point identified in the *in vivo* experiments as responsible for termination. The sequence of this region showed that it resembles other sequences at which RNA transcription terminates (Fig. 13.10). It contains a G-C-rich region that can form a hairpin followed by a string of eight U's.

The leader region also contains an AUG codon and can bind ribosomes just like the beginning of authentic genes. This was shown by the straightforward but technically difficult experiment of isolating radioactive leader region RNA from cells, binding ribosomes to it in the presence of initiation factors and f-met-tRNA^{fmet}, lightly digesting with RNAse, and sequencing the RNA that was protected from digestion. Following the AUG codon are 13 more codons before a translation termination codon. Clearly it was of great interest to determine whether this leader peptide is synthesized *in vivo*. Vigorous attempts to isolate it failed, however. Therefore an indirect experiment was used to prove that the leader is translated. The leader peptide region was fused by deletion to the *lac* repressor gene, and the corresponding fusion protein was found to be synthesized at a high rate. Since the leader peptide contains two tryptophans, termination efficiency appeared to be coupled to translation. This suspicion was deepened by the findings that *trp* regulation is altered in some *trp* synthetase mutants.

trp Multiple Secondary Structures in *trp* Leader RNA

Once the hairpin secondary structure in the RNA immediately preceding the *trp* attenuation site was associated with transcription termination, several critical experiments were apparent. One was to examine the consequences of changing the sequence of the region. Mutations changing the stability of the hairpin might be expected to alter termination efficiency. A second and easier experiment was to examine naturally occurring sequence variants. Because related bacterial strains must have evolved from a common predecessor, they probably share the same basic mechanism of *trp* regulation. If the leader sequences of their *trp* operons possess structural features in common, these features are likely to be important in the regulatory mechanisms. Indeed, this was found. The *trp* operon in *Salmonella typhimurium* and *Serratia marcescens* also contain leader regions. Both code for a leader peptide containing several tryptophans, and both mRNAs are capable of forming a hairpin just before a string of U's located about 20 bases ahead of the *trpE* gene.

Close examination of the leader region of the *trp* operon of *E. coli*, the *trp* operons from other strains of bacteria, and other amino acid biosynthetic operons has revealed two additional facts. First, the peptides encoded by the leader regions always contain one to seven of the amino acid residues synthesized by that operon. Second, the leader mRNAs possess at least four regions that can form intramolecular base-paired hairpins (Fig. 13.11). Here they are numbered 1, 2, 3, and

Figure 13.11 The two possible structures that the second half of the *trp* leader region is capable of assuming. Above is indicated the leader region with the complementary regions. Opposed arrows at the same height above the leader are homologous and can base pair. The wavy line below the leader mRNA is the region encoding the leader peptide.

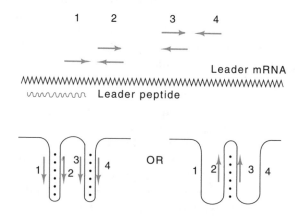

4. The pairing of 3 and 4 forms the hairpin discussed above that is required for termination of transcription. Additionally, region 2 can base pair either with region 3 or with region 1. Likewise, region 3 can base pair with region 2 or 4. The particular hybrids that form on a leader molecule determine whether transcription of that molecule will terminate as RNA polymerase passes the attenuation region. As we will see, the pairing choice is regulated by the position on the leader of a ribosome translating the leader peptide.

Coupling Translation to Termination

The location of the leader peptide with respect to the four different regions capable of hybrid formation in the *trp* leader region provides a simple mechanism for regulating termination. If the 3-4 hybrid forms during transcription of this region, termination is possible because this hybrid is the "termination" loop. Conversely, if the 2-3 hybrid forms during transcription of leader, then formation of the "termination" loop is prevented. Finally, if the 1-2 hybrid forms, then region 2 is not available for formation of the 2-3 hybrid, but as RNA polymerase transcribes regions 3 and 4, they are free to base pair, and transcription terminates.

How can the presence or absence of tryptophan affect formation of the 1-2 or 2-3 hybrid? Ribosomes translating the leader region in the absence of charged tRNAs will stall at the *trp* codons and, owing to their location, they will block the formation of the 1-2 hybrid (Fig. 13.12). Formation of the 2-3 hybrid is not blocked, and as soon as these regions of the mRNA are synthesized, this hybrid forms. Consequently, the 3-4 hybrid does not form in time to terminate transcription, and termination does not occur.

In the absence of any protein synthesis, ribosomes cannot bind to the leader region, and the 1-2 hybrid forms. In turn, this permits formation of the 3-4 hybrid. Termination follows. This happens in an *in vitro*

Figure 13.12 Possible structures of the leader mRNA in the presence of ribosomes under the three conditions: tryptophan starvation, no protein synthesis, and tryptophan excess.

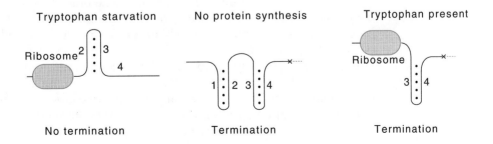

Table 13.1 Free Energy of Leader Hairpins

Hairpin	ΔG
1-2	-11.2 Kcal/mol
2-3	-11.7 Kcal/mol
3-4	-20 Kcal/mol

transcription system, which, as we saw above, terminates transcription with a high probability at the attenuation site.

Finally, what happens in the presence of excess tryptophan? In cells containing adequate levels of tryptophan, much initiation, of course, will be blocked by the *trp* repressor. The transcription that is initiated however, will be largely terminated, for ribosomes either complete translation of the leader peptide and permit hybrids 1-2 and 3-4 to form, or ribosomes remain awhile at the termination codon. From this position they block formation of the 1-2 hybrid. In either case, the 3-4 hybrids form, and termination at the attenuator occurs.

It is necessary to note that the relative thermodynamic stability of the various leader hybrids is not important to attenuation. The factor determining whether termination will occur is which structures are not blocked from forming. The kinetics of formation of the base-paired structures should be on time scales less than milliseconds so that if they are not blocked from forming, they should form while the RNA is being synthesized. Then several seconds after initiating transcription, RNA polymerase reaches the attenuation site and terminates or not depending on whether or not the 3-4 hybrid has formed. The interval between initiation and termination is much shorter than the interconversion time of many hybrid structures, and they therefore will not necessarily have had time to adopt their lowest-energy conformation (Table 13.1). The factor determining termination is which of the hybrid structures exists at the time RNA polymerase transcribes past the potential termination site.

How does the system ensure that ribosomes initiate translation as soon as the leader has been synthesized? If ribosomes do not promptly initiate translation of the leader, premature termination at the attenuator will result. At the typical rates of ribosome binding to messenger, transcription could easily extend beyond the termination site before the first ribosome had a chance to bind to messenger and affect loop formation. The problem of forcing a ribosome onto the mRNA just as it emerges from the polymerase is solved in a simple way. The *trp* leader sequence possesses regions at which transcription is slowed due to pausing by polymerase. Most likely these pauses result from hairpin structures in the newly synthesized RNA. The duration of the pauses could be random, but if the average is long enough, most messengers could then have a ribosome bind and initiate translation. When polymerase resumes transcription, the ribosome following immediately behind is properly situated to regulate attenuation.

RNA Secondary Structure and the Attenuation Mechanism

Point mutations provide excellent support for the attenuation mechanism as outlined above. Changing the AUG codon of the leader to an AUA has the expected effect (see Problem 13.2). A base change abolishing a base pair in the 2-3 hybrid, but one having no effect on the base pairs of hybrids 1-2 or 3-4 prevents the leader from forming the 2-3 hybrid. Consequently, the 3-4 hybrid always forms, and termination inevitably results. Mutations that reduce the stability of the 3-4 hybrid reduce transcription termination, both *in vitro* and *in vivo*.

Direct experiments also provide evidence for the existence of secondary structure in the leader RNA. Light digestion of the RNA with RNAse T1 cleaves only at those guanosines postulated not to be in the hybrids (Fig. 13.13). A second experiment also suggests that base pairing in the leader region is important. As mentioned above, in an *in vitro* transcription system with purified DNA and RNA polymerase, virtually all transcripts are terminated at the attenuation site. The substitution of inosine triphosphate (ITP) for GTP however, eliminates the premature termi-

Cytosine-guanine

Cytosine-inosine

nation. While it is possible that specific contacts between the inosine and RNA polymerase or between the inosine and the DNA are responsible for the lack of termination, it seems more likely that the failure of the substituted RNA to form its 3-4 hybrid is the reason. Whereas G-C base pairs possess three hydrogen bonds, I-C base pairs have only two

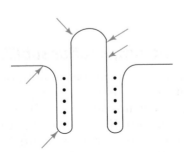

Figure 13.13 The points of RNAse T$_1$ cleavage of the leader RNA with respect to the 1-2 and 3-4 hybrid structure that likely forms on naked *trp* leader RNA *in vitro*.

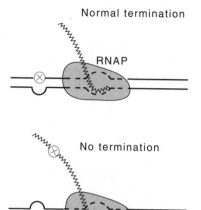

Normal termination

No termination

Figure 13.14 The use of hetero-duplex template DNAs for demonstrating that conformation of the product RNA determines whether termination occurs. When the mutation is in the strand such that the product RNA is altered, no termination occurs.

because inosine lacks an amino group that is present on guanosine. The two hydrogen bonds do not provide sufficient energy to form the usual hybrids, and thus termination does not occur. Analogs incorporated into the DNA generate much smaller perturbations on termination, also suggesting that it is the base-paired structure of the transcribed RNA that is necessary for termination at the attenuation site. The use of DNA molecules heteroduplex in the leader region also yields data consistent with the attenuation mechanism. DNA molecules can be constructed that are wild-type on one strand and mutant in a single base in the leader region on the other (Fig. 13.14). Only when the mutation is on the strand copied by RNA polymerase and therefore producing an altered RNA product molecule does the mutant nucleotide alter attenuation. If the RNA product is wild-type, so is transcription and termination.

Finally, direct physical evidence for the attenuation model has also been obtained. In an *in vitro* transcription system, the usual termination at the attenuator can be blocked by including a large excess of an oligonucleotide complementary to region 1. Apparently this oligonucleotide anneals to region 1 as it is synthesized and prevents its pairing with region 2. Consequently, the 2-3 hairpin forms, and the 3-4 hairpin, which is required for termination, cannot form. Therefore, termination does not occur.

Other Attenuated Systems: Operons, *Bacillus subtilis* and HIV

Study of the sequences of a number of amino acid biosynthetic operons has revealed that they are likely to be regulated by attenuation. Their leader peptides contain dramatic runs of the amino acid whose synthesis the operon codes for (Fig. 13.15). In summary, the attenuation mechanism seems to be an exceptionally efficient method regulating the amino acid biosynthetic operons because the necessary regulation is

```
his    met-thr-arg-val-gln-phe-lys-HIS-HIS-HIS-HIS-HIS-HIS-HIS-pro

ile    met-thr-ala-LEU-LEU-arg-VAL-ILE-ser-LEU-VAL-ILE-ser-VAL-VAL
       pro-pro-cys-gly-ala-ala-leu-gly-arg-gly-lys-ala

leu    met-ser-his-ile-val-arg-phe-thr-gly-LEU-LEU-LEU-LEU-asn-ala
       gly-arg-pro-val-gly-gly-ile-gln-his

phe    met-lys-his-ile-pro-PHE-PHE-PHE-ala-PHE-PHE-PHE-thr-PHE-pro

thr    met-lys-arg-ILE-ser-THR-THR-ILE-THR-THR-THR-ILE-THR
       ala-gly

trp    met-lys-ala-ile-phe-val-leu-lys-gly-TRP-TRP-arg-thr-ser
```

Figure 13.15 Sequences of leader peptides from some amino acid operons.

obtained by the properties of just 160 nucleotides of RNA. In the case of the *trp* operon, the double-barreled regulation provided by *trp* repressor and attenuation provides up to a 700-fold regulation range, 70-fold coming from repression and another 10-fold from attenuation. The autoregulation of *trpR* permits rapid accumulation of optimal enzyme level in cells on tryptophan starvation followed by a slower rate of enzyme synthesis when steady-state conditions have been reached.

It is not just amino acid biosynthetic operons that are regulated by attenuation. The synthesis of aspartic transcarbamoylase is also regulated by such a mechanism. This enzyme ultimately leads to the synthesis of uracil and hence UTP, and therefore we can imagine a coupling between UTP levels and the speed of transcription in a leader region, and indeed such has been found.

Escherichia *coli* and a number of closely related bacteria regulate their *trp* operons similarly. In *Bacillus subtilis,* however, a less closely related bacterium, has evolved a significant variation of the attenuation mechanism. Introduction of multiple copies of *trp* operon DNA into *B. subtilis* leads to the deregulation of the chromosomal gene copy. This becomes active whether or not tryptophan is present. One might first guess that the multiple gene copies merely bind a limited number of repressor molecules so that overall, any copy of the operon is derepressed because on average it is repressed only a small part of the time.

The actual situation on the *trp* operon was more interesting than mere repressor titration. Expressing *trp* messenger under control of the *lac* promoter provided a simple demonstration that it is multiple copies of *trp* messenger RNA and not multiple copies of DNA that lead to deregulation. While the *lac* promoter is repressed and little *trp* messenger is present, the cellular *trp* promoter is regulated normally. When *trp* messenger is synthesized at high rates under control of the *lac* promoter, the chromosomal *trp* operon loses regulation and becomes constitutive. The simplest explanation of this finding is that the *trp* RNA itself sequesters a molecule present in the cell in limited amounts. This molecule could be required for repression of the *trp* operon.

Simply by deleting portions of the *trp* messenger, the region necessary for titration of the presumed protein could be mapped. Not surprisingly, it lies ahead of the *trp* structural genes. The messenger ahead of the genes also has the potential to form multiple hairpin structures. One of the structures contains the typical transcription termination signal of a G-C rich hairpin followed by a string of U's. The alternative structure to the RNA precludes formation of the termination hairpin. In the presence of tryptophan the regulatory protein binds to the structure that leads to termination at the attenuator.

The human immunodeficiency virus HIV-1 regulates synthesis of its RNA via an attenuation mechanism. Of course, since the transcription occurs in the nucleus, it cannot be directly coupled to translation as it is for the *trp* operon. In the absence of the HIV-1 Tat protein, RNA polymerase begins transcription at the HIV promoter, but pauses after synthesis of about 60 nucleotides. These transcripts usually terminate. When Tat protein is provided, the transcripts elongate to completion. Although Tat may interact with the promoter to affect its activity as well, the primary target of Tat action is the elongating mRNA, and Tat binds to a site on this RNA called TAR.

Problems

13.1. If you wished to gather evidence for the attenuation mechanism of the *trp* operon by hybridizing an oligonucleotide to RNA *in vitro* as it is being synthesized, what would you attempt?

13.2. What would the effect be of changing the AUG codon of the *trp* leader peptide to AUA or of damaging the Shine-Dalgarno sequence?

13.3. If the leader region of the *trp* operon of the messenger RNA has signaled the RNA polymerase passing the attenuation region whether or not it is useful to terminate, it would be sensible for the attentuation region to no longer be involved in binding ribosomes and synthesizing a short polypeptide. How could this hypothesis be tested?

13.4. Suppose the enzymes of the *trp* operon were highly unstable, that is, had a lifetime of about five minutes. What types of regulation would then be necessary or unnecessary?

13.5. Pretend the *trp* repressor is not self-regulating, that is, that it is synthesized at a low constant rate from a weak promoter. Invent a scheme, perhaps requiring some genetic engineering and based on colony color, which would reveal a clone carrying *trp* operator.

13.6. What is likely to happen to regulation of the *trp* operon in a mutant *E. coli* in which RNA polymerase elongates at twice the normal rate?

13.7. Suppose *E. coli* were being modified to permit maximum synthesis of tryptophan while retaining the basic regulatory mechanism. Suppose also that the signal that has been chosen to induce this maximal synthesis is starvation or semistarvation for leucine. What cellular modifications would be necessary?

13.8. In view of the rates discussed in earlier chapters, estimate the average time between initiations by ribosomes on the *trp* leader sequence. What would be the effect on *trp* attenuation if the ribosome initiation frequency were limited only by the peptide elongation rate?

13.9. How could an attenuation mechanism be used for regulation of synthesis of purines or pyrimidines?

13.10. The data presented in this chapter on the TrpB enzyme levels in the different deletions are anomalous. What is unusual and what is a likely explanation?

13.11. Considering that a DNA sequence that encodes transcription termination contains a region rich in G's and C's that can base pair followed by a string of about six U's, devise several mechanisms that could use this information to signal the RNA polymerase to terminate transcription. For example, one might be that the separated DNA strands at the transcription point themselves form hairpins in addition to the RNA, and the three hairpins bind to sites within the polymerase to cause it to terminate.

13.12. Suppose, in the days before definitive experiments could be performed with purified components, you wished to determine whether tryptophan, trp-tRNA^{trp}, or tryptophanyl-tRNA-synthetase was involved in repression of the *trp* operon. Invent data you could obtain with crude extracts or partially purified extracts, perhaps using appropriate mutants or tryptophan analogs, that would prove that tryptophan was the corepressor of the *trp* operon.

13.13. In describing the experiments of Yanofsky, no mention was made in this chapter of testing whether the hypersynthesizing deletions might be the effect of a diffusible product in the cells. Design an experiment and invent results that would prove that the hypersynthesis was a *cis* effect.

13.14. How do you reconcile the existence of suppressors of polarity that function on polar mutants in *lac* and *trp* with the fact that *in vitro* transcription of *trp* terminates nearly 100% at the attenuator even in the absence of rho?

13.15. Consider the use of hybridization to quantitate *trp* leader and distal mRNA resulting from *in vitro* transcription. What would the results look like and what mistaken conclusion would be drawn if the DNA used for hybridization were not present in molar excess over the RNA?

13.16. The paper, "Transcription Termination at the *trp* Operon Attenuators of *E. coli* and *S. typhimurium*: RNA Secondary Structure and Regulation of Termination," F. Lee, C. Yanofsky, Proc. Natl. Acad. Sci. USA 75, 4365-4369 (1977), did not present the complete attenuation model. What critical information was not yet available?

References

Recommended Readings

The Region Between the Operator and First Structural Gene of the Tryptophan Operon of *E. coli* May Have a Regulatory Function, E. Jackson, C. Yanofsky, J. Mol. Biol. *76*, 89-101 (1973).

Transcription Termination *in vivo* in the Leader Region of the Tryptophan Operon of *E. coli*, K. Bertrand, C. Squires, C. Yanofsky, J. Mol. Biol. *103*, 319-337 (1976).

Novel Form of Transcription Attenuation Regulates Expression of the *Bacillus subtilis* Tryptophan Operon, H. Shimotsu, M. Kuroda, C. Yanofsky, D. Henner, J. Bact. *166*, 461-471 (1986).

Genetics and Physiological Studies

Nonsense Codons and Polarity in the Tryptophan Operon, C. Yanofsky, J. Ito, J. Mol. Biol. *21*, 313-334 (1966).

Transcription of the Operator Proximal and Distal Ends of the Tryptophan Operon: Evidence that *trpE* and *trpA* are the Delimiting Structural Genes, J. Rose, C. Yanofsky, J. Bact. *108*, 615-618 (1971).

Expression of the Tryptophan Operon in Merodiploids of *E. coli*, H. Stetson, R. Somerville, Mol. Gen. Gen. *111*, 342-351 (1971).

Internal Promoter of the Tryptophan Operon of *E. coli* is Located in a Structure Gene, E. Jackson, C. Yanofsky, J. Mol. Biol. *69*, 307-313 (1972).

Internal Deletions in the Tryptophan Operon of *E. coli*, E. Jackson, C. Yanofsky, J. Mol. Biol. *71*, 149-161 (1972).

Structure and Evolution of a Bifunctional Enzyme of the Tryptophan Operon, M. Grieshaber, R. Bauerle, Nature New Biology *236*, 232-235 (1972).

Isolating Tryptophan Regulatory Mutants in *E. coli* by Using a *trp-lac* Fusion Strain, W. Reznikoff, K. Thornton. J. Bact. *109*, 526-532 (1972).

Regulation of Transcription Termination in the Leader Region of the Tryptophan Operon of *Escherichia coli* Involves Tryptophan or its Metabolic Product, K. Bertrand, C. Yanofsky, J. Mol. Biol. *103*, 339-349 (1976).

Polarity Suppressors Defective in Transcription Termination at the Attenuator of the Tryptophan Operon of *E. coli* have Altered Rho Factor, L. Korn, C. Yanofsky, J. Mol. Biol. *106*, 231-241 (1976).

Tandem Termination Sites in the Tryptophan Operon of *E. coli*, A. Wu, G. Christie, T. Platt, Proc. Natl. Acad. Sci. USA *78*, 2913-2917 (1981).

trp Aporepressor Production is Controlled by Autogenous Regulation and Inefficient Translation, R. Kelley, C. Yanofsky, Proc. Natl. Acad. Sci. USA *79*, 3120-3124 (1982).

Repression is Relieved before Attenuation in the *trp* Operon of *Escherichia coli* as Tryptophan Starvation Becomes Increasingly Severe, C. Yanofsky, R. Kelley, V. Horn, J. Bact. *158*, 1018-1024 (1984).

Genetic Analysis of the Tryptophan Operon Regulatory Region Using Site-directed Mutagenesis, R. Kolter, C. Yanofsky, J. Mol. Biol. *175*, 299-312 (1984).

Transcription, Termination, and mRNA Structure

In vitro Synthesis of Enzymes of the Tryptophan Operon of *Escherichia coli*, P. Pouwels, J. Van Rotterdam, Proc. Natl. Acad. Sci. USA *69*, 1786-1790 (1972).

Punctuation of Transcription *in vitro* of the Tryptophan Operon of *E. coli*, H. Pannekoek, W. Brammer, P. Pouwels, Mol. Gen. Gen. *136*, 199-214 (1975).

Nucleotide Sequence of the 5' End of Tryptophan Messenger RNA of *Escherichia coli*, C. Squires, F. Lee, K. Bertrand, C. Squires, M. Bronson, C. Yanofsky, J. Mol. Biol. *103*, 351-381 (1976).

Termination of Transcription *in vitro* in the *Escherichia coli* Tryptophan Operon Leader Region, F. Lee, C. Squires, C. Squires, C. Yanofsky, J. Mol. Biol. *103*, 383-393 (1976).

Nucleotide Sequence of Region Preceding *trp* mRNA Initiation Site and its Role in Promoter and Operator Function, G. Bennett, M. Schweingruber, K. Brown, C. Squires, C. Yanofsky, Proc. Natl. Acad. Sci. USA *73*, 2351-2355 (1976).

Transcription Termination at the *trp* Operon Attenuators of *E. coli* and *S. typhimurium*: RNA Secondary Structure and Regulation of Termination, F. Lee, C. Yanofsky, Proc. Natl. Acad. Sci. USA *74*, 4365-4369 (1977).

Nucleotide Sequence of the Promoter-operator Region of the Tryptophan Operon of *E. coli*, G. Bennett, M. Schweingruber, K. Brown, C. Squires, C. Yanofsky, J. Mol. Biol. *121*, 113-137 (1978).

RNA Polymerase Interaction at the Promoter-operator Region of the Tryptophan Operon of *Escherichia coli* and *Salmonella typhimurium*, K. Brown, G. Bennett, F. Lee, M. Schweingruber, C. Yanofsky, J. Mol. Biol. *121*, 153-177 (1978).

Attenuation in the *E. coli* Tryptophan Operon: Role of RNA Secondary Structure Involving the Tryptophan Codon Region, D. Oxender, G. Zurawski, C. Yanofsky, Proc. Natl. Acad. Sci. USA *76*, 5524-5528 (1979).

E. coli Tryptophan Operon Leader Mutations, which Relieve Transcription Terminators are cis-dominant to *trp* Leader Mutations which Increase Transcription Termination, G. Zurawski, C. Yanofsky, J. Mol. Biol. *142*, 123-129 (1980).

Structure and Regulation of *aroH*, the Structural Gene for the Tryptophan-repressible 3-Deoxy-D-arabinose-heptulosonic acid-7-phosphate Synthetase of *E. coli*, G. Zurawski, R. Gunsalus, K. Brown, C. Yanofsky, J. Mol. Biol. *145*, 47-73 (1981).

Pausing of RNA Polymerase During *in vitro* Transcription of the Tryptophan Operon Leader Region, M. Winkler, C. Yanofsky, Biochem. *20*, 3738-3744 (1981).

Effects of DNA Base Analogs on Transcription Termination at the Tryptophan Operon Attenuator of *Escherichia coli*, P. Farnham, T. Platt, Proc. Natl. Acad. Sci. USA *79*, 998-1002 (1982).

Transcription Termination at the Tryptophan Operon Attenuator is Decreased *in vitro* by an Oligomer Complementary to a Segment of the Leader Transcript, M. Winkler, K. Mullis, J. Barnett, I. Stroynowski, C. Yanofsky, Proc. Natl. Acad. Sci. USA *79*, 2181-2185 (1982).

Transcript Secondary Structures Regulate Transcription Termination at the Attenuator of *S. marcescens* Tryptophan Operon, I. Stroynowski, C. Yanofsky, Nature *298*, 34-38 (1982).

Transcription Analyses with Heteroduplex *trp* Attenuator Templates Indicate that the Transcript Stem and Loop Structure Serves as the Termination Signal, T. Ryan, M. Chamberlin, J. Biol. Chem. *258*, 4690-4693 (1983).

Use of Complementary DNA Oligomers to Probe *trp* Leader Transcript Secondary Structures Involved in Transcription Pausing and Termination, R. Fisher, C. Yanofsky, Nucleic Acids Research *12*, 3295-3302 (1984).

Analysis of the Requirements for Transcription Pausing in the Tryptophan Operon, R. Fisher, A. Das, R. Kolter, M. Winkler, C. Yanofsky, J. Mol. Biol. *182*, 397-409 (1985).

Rho-dependent Transcription Termination in the Tryptophanase Operon Leader Region of *Escherichia coli* K-12, V. Stewart, R. Landick, C. Yanofsky, J. Bact. *166*, 217-223 (1986).

Isolation and Structural Analysis of the *Escherichia coli trp* Leader Paused Transcription Complex, R. Landick, C. Yanofsky, J. Mol. Biol. *196*, 363-377 (1987).

Detection of Transcription-pausing *in vivo* in the *trp* Operon Leader Region, R. Landick, J. Carey, C. Yanofsky, Proc. Nat. Acad. Sci. USA *84*, 1507-1511 (1987).

tRNA, Ribosomes, and Translation Termination

Tryptophanyl-tRNA and Tryptophanyl-tRNA Synthetase are not Required for *in vitro* Repression of the Tryptophan Operon, C. Squires, J. Rose, C. Yanofsky, H. Yang, G. Zubay, Nature New Biol. *245*, 131-133 (1973).

An Intercistronic Region and Ribosome-binding Site in Bacterial Messenger RNA, T. Platt, C. Yanofsky, Proc. Natl. Acad. Sci. USA *72*, 2399-2403 (1975).

Ribosome-protected Regions in the Leader-trpE Sequence of *Escherichia coli* Tryptophan Operon Messenger RNA, T. Platt, C. Squires, C. Yanofsky, J. Mol. Biol. *103*, 411-420 (1976).

Mutations Affecting tRNA[Trp] and its Charging and their Effect on Regulation of Transcription Termination at the Attenuator of the Tryptophan Operon, C. Yanofsky, L. Soll, J. Mol. Biol. *113*, 663-677 (1977).

Translation of the Leader Region of the *E. coli* Tryptophan Operon, G. Miozzari, C. Yanofsky, J. Bact. *133*, 1457-1466 (1978).

The Effect of an *E. coli* Regulatory Mutation on tRNA Structure, S. Eisenberg, M. Yarus, L. Soll, J. Mol. Biol. *135*, 111-126 (1979).

Translation Activates the Paused Transcription Complex and Restores Transcription of the *trp* Operon Leader Region, R. Landick, J. Carey, C. Yanofsky, Proc. Nat. Acad. Sci. USA *82*, 4663-4667 (1985).

Control of Phenylalanyl-tRNA Synthetase Genetic Expression, J. Mayaux, G. Fayat, M. Panvert, M. Springer, M. Grunberg-Manago, S. Balquet, J. Mol. Biol. *184*, 31-44 (1985).

trp Repressor and Repressor-operator Interactions

Regulation of *in vitro* Transcription of the Tryptophan Operon by Purified RNA Polymerase in the Presence of Partially Purified Repressor and Tryptophan, J. Rose, C. Squires, C. Yanofsky, H. Yang, G. Zubay, Nature New Biol. *245*, 133-137 (1973).

Interaction of the Operator of the Tryptophan Operon with Repressor, J. Rose, C. Yanofsky, Proc. Natl. Acad. Sci. USA *71*, 3134-3138 (1974).

Interaction of the *trp* Repressor and RNA Polymerase with the *trp* Operon, C. Squires, F. Lee, C. Yanofsky, J. Mol. Biol. *92*, 93-111 (1975).

Sequence Analysis of Operator Constitutive Mutants of the Tryptophan Operon of *E. coli*, G. Bennett, C. Yanofsky, J. Mol. Biol. *121*, 179-192 (1978).

Purification and Characterization of *trp* Aporepressor, A. Joachimiak, R. Kelley, R. Gunsalus, C. Yanofsky, Proc. Natl. Acad. Sci. USA *80*, 668-672 (1983).

trp Repressor Interactions with the *trp*, *aroH*, and *trpR* Operators, L. Klig, J. Carey, C. Yanofsky, J. Mol. Biol. *202*, 769-777 (1988).

Attenuation Systems Other than *trp* in *E. coli*

Nucleotide Sequence of the Promoter-operator Region of the Tryptophan Operon of *Salmonella typhimurium*, G. Bennett, K. Brown, C. Yanofsky, J. Mol. Biol. *121*, 139-152 (1978).

Nucleotide Sequence of the Attenuator Region of the Histidine Operon of *Escherichia coli* K-12, P. DiNocera, F. Blasi, R. Frunzio, C. Bruni, Proc. Natl. Acad. Sci. USA *75*, 4276-4282 (1978).

DNA Sequence from the Histidine Operon Control Region: Seven Histidine Codons in a Row, W. Barnes, Proc. Natl. Acad. Sci. USA *75*, 4281-4285 (1978).

The Regulatory Region of the *trp* Operon of *Serratia marcescens*, G. Miozzari, C. Yanofsky, Nature *276*, 684-689 (1978).

Regulation of the Threonine Operon: Tandem Threonine and Isoleucine Codons in the Control Region and Translational Control of Transcription Termination, J. Gardner, Proc. Natl. Acad. Sci. USA *76*, 1706-1710 (1979).

leu Operon in *Salmonella typhimurium* is Controlled by an Attenuation Mechanism, R. Gemmil, S. Wessler, E. Keller, J. Calvo, Proc. Natl. Acad. Sci. USA *76*, 4941-4945 (1979).

Alternative Secondary Structures of Leader RNAs and the Regulation of the *trp*, *phe*, *his*, *thr*, and *leu* Operons, E. Keller, J. Calvo, Proc. Natl. Acad. Sci. USA *76*, 6186-6190 (1979).

E. coli RNA Polymerase and *trp* Repressor Interaction with the Promoter-Operator Region of the Tryptophan Operon of *Salmonella typhimurium*, D. Oppenheim, G. Bennet, C. Yanofsky, J. Mol. Biol. *144*, 133-142 (1980).

Functional Analysis of Wild-type and Altered Tryptophan Operon Promoters of *Salmonella Typhimurium* in *E. coli*, D. Oppenheim, C. Yanofsky, J. Mol. Biol. *144*, 143-161 (1980).

Model for Regulation of the Histidine Operon of *Salmonella*, H. Johnson, W. Barnes, F. Chumley, L. Bossi, J. Roth, Proc. Natl. Acad. Sci. USA *77*, 508-512 (1980).

Nucleotide Sequence of *ilvGEDA* Operon Attenuator Region of *Escherichia coli*, F. Nargang, C. Subrahmanyam, H. Umbarger, Proc. Natl. Acad. Sci. USA *77*, 1823-1827 (1980).

Multivalent Translational Control of Transcription Termination at Attenuator of *ilvGEDA* Operon of *Escherichia coli* K-12, R. Lawther, G. Hatfield, Proc. Natl. Acad. Sci. USA *77*, 1862-1866 (1980).

DNA Sequence Changes of Mutations Altering Attenuation Control of the Histidine Operon of *Salmonella typhimurium*, H. Johnston, J. Roth, J. Mol. Biol. *145*, 735-756 (1981).

Antitermination Regulation of *ampC*, Ribosome Binding at High Growth Rate, B. Jaurin, T. Grundstrom, T. Edlund, S. Normark, Nature *290*, 221-225 (1981).

Superattenuation in the Tryptophan Operon of *Serratia marcescens*, I. Stroynowski, M. van Cleemput, C. Yanofsky, Nature *298*, 38-41 (1982).

Nucleotide Sequence of Yeast *LEU2* Shows 5'- Noncoding Region Has Sequences Cognate to Leucine, A. Andreadis, Y. Hsu, G. Kohlhaw, P. Schimmel, Cell *31*, 319-325 (1982).

Attenuation Control of *purB1* Operon Expression in *Escherichia coli* K-12, C. Turnbough, K. Hicks, J. Donahue, Proc. Natl. Acad. Sci. USA *80*, 368-372 (1983).

Attenuation of the *ilvB* Operon by Amino Acids Reflecting Substrates or Products of the *ilvB* Gene Product, C. Hauser, G. Hatfield, Proc. Natl. Acad. Sci. USA *81*, 76-79 (1984).

Characterization of the *Bacillus subtilis* Tryptophan Promoter Region, H. Shimatsu, D. Henner, Proc. Natl. Acad. Sci. USA, *81*, 6315-6319 (1984).

Role of an Upstream Regulatory Element in Leucine Repression of the *Saccharomyces cerevisiae leu2* Gene, A. Martinez-Arias, H. Yast, M. Casadaban, Nature *307*, 740-742 (1984).

Regulation of Aspartate Transcarbamylase Synthesis in *Escherichia coli*: Analysis of Deletion Mutations in the Promoter Region of the *purBI* Operon, H. Levin, H. Schachman, Proc. Nat. Acad. Sci. USA *82*, 4643-4647 (1985).

Comparison of the Regulatory Regions of *ilvGEDA* Operons from Several Enteric Organisms, E. Harms, J. Hsa, C. Subrahmanyam, H. Umbarger, J. Bact. *164*, 207-216 (1985).

Evidence for Transcription Antitermination Control of Tryptophanase Operon Expression in *Escherichia coli* K-12, V. Stewart, C. Yanofsky, J. Bact. *164*, 731-740 (1985).

Antitermination of Transcription within the Long Terminal Repeat of HIV-1 by *tat* Gene Product, S. Kao, A. Calman, P. Luciw, B. Peterlin, Nature *330*, 489-493 (1987).

cis-Acting Sites in the Transcript of the *Bacillus subtilis trp* Operon Regulate Expression of the Operon, M. Kuroda, D. Henner, C. Yanofsky, J. Bact. *170*, 3080-3088 (1988).

The Role of Tat in the Human Immunodeficiency Virus Life Cycle Indicates a Primary Effect on Transcriptional Elongation, M. Feinberg, D. Baltimore, A. Frankel, Proc. Natl. Acad. Sci. USA *88*, 4045-4049 (1991).

MtrB from *Bacillus subtilis* Binds Specifically to *trp* Leader RNA in a Tryptophan-dependent Manner, J. Otridge, P. Gollnick, Proc. Natl. Acad. Sci. USA *90*, 128-132 (1993).

Reconstitution of *Bacillus subtilis trp* Attenuation *in vitro* with TRAP, the *trp* RNA-binding Attenuation Protein, P. Babitzke, C. Yanofsky, Proc. Natl. Acad. Sci. USA *90*, 133-137 (1993).

Lambda Phage Genes and Regulatory Circuitry

14

As a result of its convenient size and interesting biological properties, lambda phage has been intensively studied for many years. Lambda is neither so small that each of its genes must play multiple roles in phage development nor so large that there are too many genes to study or understand. Furthermore, lambda has an interesting dual mode of existence. On one hand, a lambda phage can infect a cell, grow vegetatively to produce a hundred copies of itself, and lyse the cell. On the other hand, a lambda phage can infect a cell and enter a quiescent phase. In this lysogenic state only three phage genes are expressed, and both daughters of cell division by a lysogenic cell are similarly lysogenic. Although it is highly stable and can be passed for many generations to descendant cells, lysogeny need not be permanent. The lambda within a lysogenic cell can be induced to enter its vegetative mode and will then multiply and lyse the cell.

Lambda raises two fundamental questions. First, how does the virus regulate its growth? Lytic growth itself requires regulation because genes must be expressed at the right time and to the correct extent to maximize viable phage yield. A second aspect to lambda's regulation problems is lysogeny. The phage existing as a lysogen must keep most of its genes off and switch them on upon induction. The regulation system controlling the two states must be reasonably stable because lambda in a lysogenic cell only rarely induces spontaneously to begin vegetative growth.

The second fundamental question raised by the existence of the lysogenic state is entering and excising from the chromosome. When lambda infects a cell and generates a lysogen, its DNA must integrate into the host chromosome. When the integrated lambda induces, its DNA is excised from the chromosome. How are these steps accomplished? What enzymes and DNA substrates are required for these reactions? How is the reaction forced to proceed in the direction of integration following infection, and how is it forced to go in the direction of excision upon phage induction?

The research that has been concentrated on lambda has told us much about this phage, particularly the research that combined genetic analysis and physiological studies. One unexpected practical result of these studies was the ability to use lambda as a vector in genetic engineering. Now most research on lambda is concentrated on the more difficult questions of how things happen rather than what happens, or where the genes are, or what the different genes accomplish for the phage. This chapter explains the structure and scheme of gene regulation in phage lambda, and a later chapter describes the integration and excision process of the phage.

A. The Structure and Biology of Lambda

The Physical Structure of Lambda

The lambda phage that has been studied in the laboratory possesses an isometric head of diameter 650 Å and a tail 1,500 Å long and 170 Å wide. A tail fiber extends an additional 200 Å from the tail. This tail fiber makes a specific contact with a porin protein found in the outer membrane of the host cell, and the phage DNA is injected through the tail into the cell. Apparently in an early isolation of lambda, investigators chose a mutant form of the virus that lacked tail fibers. The natural form possesses a number of tail fibers that attach the phage to something other than the

Figure 14.1 Cleavage at two *cos* sites between the *R* and *A* genes generates lambda monomers from oligomers during encapsidation of lambda DNA. Because each *cos* site is cut with a 12 base staggered cleavage, complementary end sequences are generated.

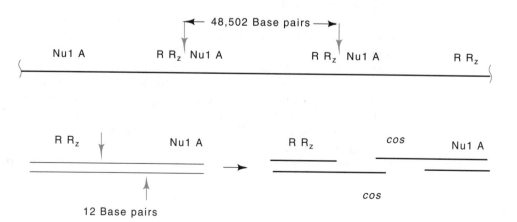

porin protein. The DNA within the phage particle is double-stranded and linear, with a length of 48,502 base pairs. This is roughly 1% the size of the *Escherichia coli* chromosome. Intracellularly, lambda DNA exists as monomeric circles or polymeric circular forms, but during encapsidation, unit-length lambda linear genomes are cut out of the polymeric forms (Fig. 14.1). The cuts of the two DNA strands are offset from one another by 12 bases. Thus the ends of the encapsidated phage DNA are single-stranded and complementary.

Roughly half the mass of a lambda particle is DNA and half is protein. Consequently, the particle has a density in CsCl halfway between the density of DNA, 1.7 gm/cm^3, and the density of protein, 1.3 gm/cm^3. This density facilitates purification of the phage since isopycnic banding in CsCl density gradients easily separates the phage from most other cellular components.

The Genetic Structure of Lambda

A turbid or opaque lawn of cells is formed when 10^5 or more cells are spread on the agar surface in a petri plate and allowed to grow until limited by nutrient availability. If a few of the original cells are infected with a virus, several cycles of lysis of the infected cells and infection of adjacent cells during growth of the lawn leaves a clear hole in the bacterial lawn. These holes are called plaques. The first mutations isolated and mapped in phage lambda were those that changed the morphology of its plaques. Ordinarily, lambda plaques are turbid or even contain a minicolony of cells in the plaque center. Both result from growth of cells that have become immune to lambda infection. Consequently, lambda mutants that do not permit cells to become immune produce clear plaques. These may easily be identified amid many turbid plaques. Kaiser isolated and mapped such clear plaque mutants of lambda. These fell into three complementation groups, which Kaiser called *CI*, *CII*, and *CIII*, with the C standing for clear.

One particular mutation in the *CI* gene is especially useful. It is known as CI$_{857}$, and the mutant *CI* product is temperature-sensitive. At temperatures below 37° the phage forms normal turbid plaques, but at temperatures above 37° the mutant forms clear plaques. Furthermore, lysogens of lambda CI$_{857}$ can be induced to switch from lysogenic to lytic mode by shifting their temperature above 37°. They then excise from the chromosome and grow vegetatively.

Many additional lambda mutations were isolated and mapped by Campbell. These could be in essential genes, as he used conditional mutations. Such mutants are isolated by plating mutagenized phage on a nonsense-suppressing strain. Plaques deriving from phage containing nonsense mutations may be identified by their inability to grow after being spotted onto a nonsuppressing strain.

Phage with nonsense mutations in various genes may then be studied by first preparing phage stocks of the nonsense mutants on suppressing hosts. The phage can then be used in a variety of studies. For example, pairs of mutants can be crossed against one another and the frequency

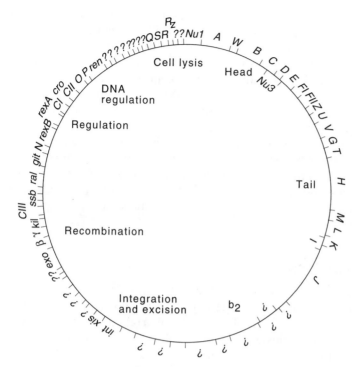

Figure 14.2 The complete physical map of lambda determined from its DNA sequence. The sizes of genes are indicated along with the functions of various classes of proteins. The genetic map is similar except that the circle is opened between genes R_z and $Nu1$.

of generation of wild-type phage by recombination between the two mutations can be quantitated by plating on nonsuppressing strains. In this way a genetic map can be constructed. The nonsense mutants also facilitate study of phage gene function. Nonsuppressing cells infected with a nonsense mutant phage stock progress only partway through an infective cycle. The step of phage development and maturation that is blocked by the mutation can be determined with radioactive isotopes to quantitate protein, RNA, and DNA synthesis, or electron microscopy to determine which phage macromolecules or structures are synthesized.

Campbell named the genes he found and mapped A through R, left to right on the genetic map. The genes identified and mapped after his work are identified by the remaining letters of the alphabet, by three-

A		J	b2	int	β	γ	CIII		N		CI		O		P	Q		S	R	R_z

letter names, and by other symbols.

The sequence of the entire lambda DNA molecule has been determined. A few of the known genes could be identified in the sequence by the amino acid sequences of their products or by mutations that changed the DNA sequence. Many others could be identified with a high degree of confidence by the correspondence between genetic and physical map location (Fig. 14.2). The identification of such open reading frames was greatly assisted by examination of codon usage. An open reading frame that is not translated into protein usually contains all sense codons at about the same frequency, whereas the reading frames that are translated into protein tend to use a subset of the codons. That is, many proteins use certain codons with a substantially higher frequency than other codons. Unexpectedly, applying these criteria to the DNA sequence of phage lambda revealed more than ten possible genes not previously identified genetically or biochemically. It remains to be shown how many of these actually play a role in phage growth and development.

The DNA sequencing revealed a second unexpected property of the lambda genes. Many are partially overlapped. The function of this overlap could be to conserve coding material, although it may also play

a role in translation. Since ribosomes drift phaselessly forward and backward after encountering a nonsense codon and before dissociating from the mRNA, this overlapping may help adjust the translation efficiency of the downstream gene product.

Genes of related function are clustered in the lambda genome. The genes *A* through *F* are required for head formation, and *Z* through *J* for tail formation. The genes in the b_2 region are not essential for phage growth under normal laboratory conditions and may be deleted without material effect. Since such deletion phage still possess the normal protein coat but lack about 10% of their DNA, they are less dense than the wild-type phage; that is, they are buoyant density mutants. The genes *int* and *xis* code for proteins that are involved with integration and excision and will be discussed in a later chapter. The genes *exo*, β, and γ are involved with recombination. The proteins from genes *CIII*, *N*, *CI*, *cro*, and *CII* are expressed early in the growth cycle and regulate the expression of phage genes expressed early. The genes *O* and *P* code for proteins required for initiation of lambda DNA replication. The *Q* gene product regulates expression of late genes. These genes include *S*, *R*, and R_z, which are required for lysis of the cell as well as the genes that encode proteins comprising the head and tail of the phage particle.

Lysogeny and Immunity

When lambda lysogenizes a cell, it inserts its DNA into the DNA of the host. There, the lambda DNA is passively replicated by the host just as though it were host DNA. Such a lambda lysogen is highly stable, and lysogeny can be passed on for hundreds of generations. Upon induction of the phage, which may be either spontaneous or caused by exposure to inducing agents such as UV light, lambda excises itself from the chromosome and enters a lytic cycle. About one hundred phage are produced, and the cell lyses.

The advantages to the phage and cell of lysogeny would largely be lost if a lysogenic cell could be lytically infected by another lambda phage. Indeed, lambda lysogenic cells cannot be lytically infected with more phage. The cells are said to possess lambda immunity. The superinfecting phage can adsorb to immune cells and inject their DNA into the cell, but the superinfecting phage DNA does little else. This DNA is not replicated or integrated into the host chromosome and therefore is diluted away by cell growth since it is passed to only one of the two daughter cells on cell division.

Lambda repressor protein, the product of the *CI* gene, confers the immunity. Repressor encoded by the lysogenic phage diffuses throughout the cytoplasm of the cell. When another lambda phage injects its DNA into this cell, the repressor binds to specific sites on the superinfecting DNA and inactivates the promoters necessary for the first steps of vegetative phage growth. This same repressor activity also prevents the lysogenic phage from initiating its growth. Upon induction of the lysogen however, the repressor is destroyed and the phage can begin a lytic growth cycle.

Lambda's Relatives and Lambda Hybrids

Lambda does not uniquely occupy its ecological niche. A variety of both close and distant relatives of lambda are known. The near relatives all possess the same sticky ends of the chromosome and are of almost the same size and genetic structure as lambda. Some possess the same immunity as lambda, which means that they cannot grow in lambda lysogens, but others possess different immunities and can grow in lambda lysogens (Table 14.1). That is, they are heteroimmune. Remarkably, lambda relatives can form recombinants between one another to form hybrids. DNA heteroduplexes between them show that their genes tend to be either closely homologous or quite dissimilar. It is as though nature possesses a few fundamental lambda-type phage and can interchange their parts to produce the large number of different lambdoid phage that are observed.

Table 14.1 Properties of a Number of Lambdoid Phage

Phage	Immunity	Position When Integrated	Sticky Ends
λ	λ	*gal*	λ
21	21	*trp*	λ
φ80	φ80	*trp*	λ
φ81	φ81	*gal*	λ
82	82	*gal*	
434	434	*gal*	λ

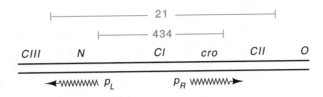

Figure 14.3 The regions of the phage λ genome that are replaced by segments of 21 and 434 in hybrid λimm^{21} and λimm^{434}.

A study of the similarities and differences between lambda's relatives has been fruitful in understanding lambda. Phage 21, φ80, and 434 have been particularly useful in the research on phage lambda. They have highlighted the crucial invariants of lambdoid growth and regulation. Recombinant hybrids have also been constructed between lambda and both 21 and 434 (Fig. 14.3). These were constructed so as to retain the immunity of the heteroimmune phage but to contain as much of the rest of lambda as possible. These lambda hybrids are called λimm^{434}, or λi^{434}, and λi^{21}. They permit recombination and complementation studies that otherwise would be impossible because of the existence of immunity.

B. Chronology of a Lytic Infective Cycle

Lambda Adsorption to Cells

Phage particles ought not to inject their DNA wantonly into anything. For the highest survival value, they should release their encapsidated DNA only after they have made stable contact with a suitable host cell. Not unexpectedly, then, lambdoid phage make specific contacts with structures on the outer surface of *E. coli*. Only when the proper receptor is present can the phage adsorb and inject.

Many different membrane structures are used by different phage in the adsorption and injection process. The laboratory form of lambda that has been studied for the last two decades uses a protein specified by the maltose operon. It is named the LamB protein because its first known function was lambda adsorption and only later was it discovered to be part of the maltose operon. The normal cellular function of the LamB protein is to create a pore through the outer membrane somewhat larger than the pores usually found there. This maltose-inducible pore is necessary for the diffusion of maltodextrins, to the periplasmic space. The pore is also necessary if maltose is present at low concentrations. Induction of the maltose operon increases the levels of this protein and speeds the adsorption process; eliminating synthesis of this protein or mutationally altering it makes cells resistant to lambda and can leave them Mal⁻.

Early Transcription of Genes *N* and *Cro*

Within a minute after injection into the host, the complementary ends of the lambda DNA anneal to each other and are covalently joined by the DNA ligase of the host. Supercoiling of the circle by DNA gyrase is then possible, and the DNA rapidly acquires the superhelical density necessary for maximum activity of the phage early promoters p_L and p_R. These are located on either side of *CI*.

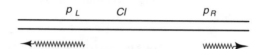

As on all promoters, initiation of transcription from the promoters p_L and p_R is followed by the dissociation of the sigma subunit of RNA polymerase after the polynucleotide chains have reached a length of 6 to 15 nucleotides. Then a host protein, the NusA product, binds to RNA polymerase. Nus stands for N utilization substance. Most likely it binds in the site previously occupied by the sigma subunit because only one or the other of these proteins can bind to RNA polymerase at once.

The RNA polymerase molecules elongate past the point where the sigma and NusA subunits exchange and they transcribe through the *N*

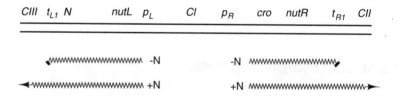

gene on the left and the *cro* gene on the right. This is all that happens at this stage of the lambda infection, for just beyond these genes the RNA

Figure 14.4 The sigma subunit dissociates from RNA polymerase and is replaced by the NusA protein. At the terminator, the rho protein interacts with the complex and releases RNA and RNA polymerase.

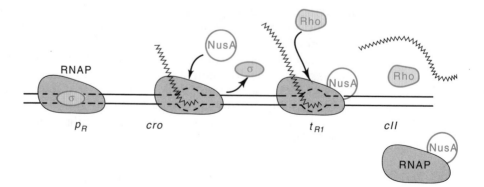

polymerase pauses, assisted by the NusA protein, and the host transcription terminator, rho, releases the RNA chain and RNA polymerase from the DNA (Fig. 14.4). This pausing and termination occurs at the first transcription termination sites in the p_L and p_R operons, t_{L1} and t_{R1}. Additional terminators with slightly different properties, t_{L2} and t_{R2}, lie farther downstream in both operons. As a result of t_{L1} and t_{R1}, the first stage of lambda vegetative growth is confined to the accumulation of N and Cro proteins.

N Protein and Antitermination of Early Gene Transcription

Several minutes after infection, N protein reaches sufficiently high concentrations in the cytoplasm to function. It binds to RNA polymerase and to a special sequence in the mRNA whose corresponding sequence in the DNA is called the *nut*, for N utilization, site. Without a functional *nut* sequence, or without the NusA protein already bound to RNA polymerase, the N protein does not bind to the polymerase. RNA polymerase with N protein bound to it no longer terminates transcription at the sites t_{L1} and t_{R1} and instead continues across these sites to synthesize messenger for genes *CII* and *CIII* as well as the more distal genes in these operons (Fig. 14.5).

It should be mentioned at this time that a number of *E. coli* mutants exist in which wild-type lambda does not grow. Some of these behave as though the lambda possessed a mutation in the N protein. These are called *nus* mutants, and the NusA protein is the product of one such gene. Presumably, NusA protein can be altered so that it still fulfills its normal cellular function, but lambda N protein does not properly interact with it. Consequently, termination always occurs at t_{L1} and t_{R1} in *nus* mutants. Other *nus* mutations, *nusB*, *nusE*, and *nusG*, lie in RNA polymerase and in ribosomal genes. These other proteins stabilize the N protein-RNA polymerase interaction.

Figure 14.5 A representation of how lambda N protein could antagonize the action of rho protein at the terminators.

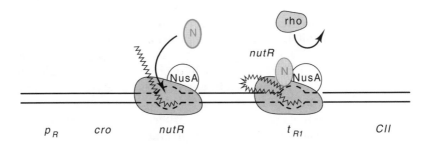

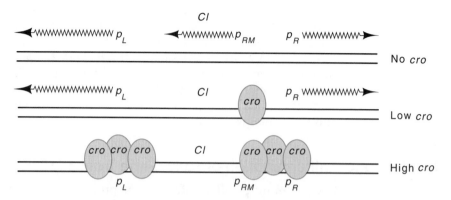

Figure 14.6 Cro protein first inactivates p_{RM} and at higher concentrations reduces activity of p_L and p_R.

The Role of Cro Protein

Once lambda has started down the lytic pathway of development, it should not waver in its efforts to maximize efficiency. The three regulatory genes we discussed earlier, *CI*, *CII*, and *CIII* all act to turn off genes used in the lytic cycle. Their presence would be deleterious to lytic growth. Not surprisingly, then, the phage makes a protein to turn down excessive synthesis of these three proteins. It is called the Cro protein, for control of repressor synthesis. First, Cro binds to an operator adjacent to the promoter for the maintenance synthesis of CI repressor in lysogens, p_{RM} (Fig. 14.6). Thus, no CI is made via this route. Later, when Cro reaches higher concentrations, it binds to the operators adjacent to p_L and p_R and represses transcription from them as well. This secondary effect reduces the synthesis of the phage CII and CIII proteins.

Initiating DNA Synthesis with the O and P Proteins

Transcription originating from p_R leads to the accumulation of the O and P proteins. These, plus transcription into or near the *ori* site, initiate a cascade of assembly and disassembly whose function is to denature the origin region. Denatured origin is the target of DNA primase and its availability permits replication to begin.

When the O protein has reached appropriate levels, it binds to a series of four repeats of a palindromic DNA sequence located at *ori* (Fig. 14.7). After the O protein has bound, the P protein with bound host protein DnaB binds to O protein. The host chaperonin proteins DnaJ and DnaK, that were mentioned earlier for their roles in renaturing denatured proteins, release P protein and the helicase activity of DnaB plus torsion on the DNA generated by transcription nearby helps separate the strands at the origin. With the additional assistance of the host proteins, DnaE, DnaG, DnaZ, RNA polymerase, and DNA gyrase, DNA synthesis initiates and proceeds outward in both directions from *ori*. Ordinarily

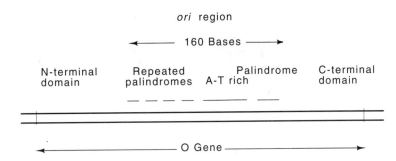

Figure 14.7 The structure and location of the *ori* region within the lambda *O* gene.

the requirement for transcriptional activation of *ori* is met by transcription originating from p_R. Other lambda mutants may be isolated that create other promoters that activated *ori*. Their transcription does not have to cross *ori*. One mutant promoter that activates lambda phage DNA replication was 95 base pairs away, and its transcription is directed away from *ori*.

Curiously, *ori* is located within the lambda *O* gene itself! The coding regions of the *O* gene on either side of *ori* specify well-defined domains. The amino acids encoded by the portion of *O* containing *ori*–the four repeats, an A-T-rich region, and a palindrome–are not predicted to possess much secondary structure and are very sensitive to proteases. Such a picture is consistent with the experimental findings that the N-terminal portion of the O protein contains the phage-specific DNA determinants and that the carboxy-terminus of O binds the lambda P protein. The O protein itself is highly unstable *in vivo*.

Mutations in the *ori* site itself support the structure described above. *Ori* mutants are recognized because they are *cis*-dominant mutations affecting DNA replication and they generate very tiny plaques. The sequence of several such mutations has shown that they lie in the *ori* section of *O*. Some are small deletions within the *ori* region, but they all preserve the reading frame; that is, they are multiples of three bases. Therefore, the amino acids encoded by the *ori* portion of the *O* gene are unimportant to the functioning of O protein.

Proteins Kil, γ, β, and Exo

Leftward transcription beyond *CIII* leads to the accumulation of the Kil, γ, β, and Exo proteins in addition to others whose functions are not known but that are nonessential. Kil stops cell division. The γ protein

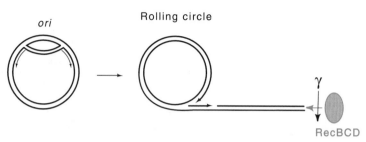

Figure 14.8 The rolling circle replicative forms are sensitive to RecBCD, but γ will inactive RecBCD, thus sparing the replicative forms.

shuts off the host recombination pathway, and the Exo and β proteins open a new pathway for genetic recombination.

Turning off the host recombination pathway is essential to the phage. As mentioned above, lambda DNA is first replicated bidirectionally from the *ori* region. Later in its growth cycle, however, lambda's replication shifts to a rolling circle mode. This generates the oligomers of lambda DNA that are obligatory for encapsidation. Few, if any, rolling circles survive in the presence of active host RecBCD enzyme. Apparently this enzyme binds at the free end of DNA in a rolling circle and moves toward the circle. To prevent the cell from inactivating the rolling circles, lambda inactivates RecBCD with a protein of its own, γ (Fig. 14.8). Lambda still has need for recombination, and it therefore potentiates another pathway for genetic recombination by synthesizing the Exo and β proteins.

Q Protein and Late Protein Synthesis

The final genes in the early operons are the *int* and *xis* genes on the left and the *Q* gene on the right. The *int* and *xis* genes are important in the lysogenization process and are discussed in a later chapter. The Q protein functions as an antiterminator of a promoter located immediately to its right. Without Q protein, polymerase initiates at this promoter and terminates 190 bases later (Fig. 14.9). When Q protein is

Figure 14.9 Action of Q protein in preventing premature termination at $t_{R'}$.

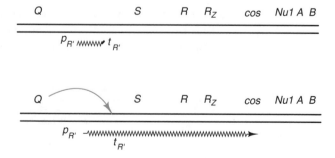

present, this termination is abolished, and transcription proceeds across the S and R genes, across the ligated sticky ends, and into the remaining late genes. These code for the head and tail structures of the phage. Since transcription of the late genes occurs at a time when about 40 copies of the DNA exist in the cell, large quantities of late mRNA are synthesized.

Lysis

One problem the phage must solve is lysing the cells at the right moment. Most likely this time is a compromise, worked out over the eons, between maximizing the number of completed phage particles released and minimizing the interval after infection or induction until some completed phage are released. Although infection can begin by slipping lambda's DNA molecule into the cell through a small hole, the release of newly assembled phage particles requires something more drastic. At the least, holes large enough for the phage head must be punched in the rigid peptidoglycan layer, and the inner and outer membranes must be ruptured as well.

Three lambda proteins are known to participate in the lysis process. They are all late proteins synthesized under control of the Q gene product. The first to be identified is the product of the lambda R gene. Originally this was called the lysozyme for its ability to lyse cells, then for a while it was mistakenly called an endopeptidase or endolysin for the specific bonds in the peptidoglycan the enzyme was thought to cleave, and now it is correctly known to be a transglycosylase (Fig. 14.10) that cleaves between N-acetylglucosamine and N-acetylmuramic acid. Another lambda-encoded protein also degrades the peptidoglycan layer of the cell. It is the product of the R_Z gene, and it is an endopeptidase. The third protein required for lysis is the product of the S gene. Experiments indicate that this protein forms a pore through the inner membrane so that the R and R_Z products, which are cytoplasmic proteins, are provided access to their peptidoglycan substrate.

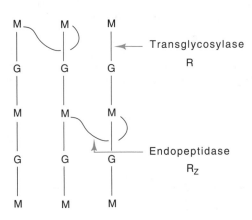

Figure 14.10 The structure of the peptidoglycan layer and the bonds cleaved by the R and R_Z proteins.

The behavior of cells infected with S⁻ phage is consistent with the idea that the S gene codes for a pore. As expected, the R and R$_Z$ products accumulate in the cytoplasm of such cells, and lysis does not occur unless the inner membrane is damaged. Chloroform treatment or freezing and thawing are two methods for disrupting the integrity of the inner membrane in S⁻ mutants. Protein synthesis, DNA synthesis, and respiration do not shut off in cells infected with S⁻ phage as they normally do 40 minutes after infection with S⁺ phage. The shutoff of macromolecular synthesis that normally occurs 40 minutes after infection results from leakage of crucial intracellular components, as at this time the cell loses its ability to concentrate small molecules intracellularly.

S⁻ mutants facilitate work with lambda. First, since macromolecular synthesis does not shut off at 40 minutes, phage continue to be made and the phage yield per infected or induced cell is raised from about 100 to about 500. Second, phage may easily be harvested from the cell growth medium by centrifuging cells full of phage into pellets, resuspending in small volumes, and lysing by the addition of chloroform. As a result, large quantities of highly concentrated phage are easily obtained.

C. The Lysogenic Infective Cycle and Induction of Lysogens

Chronology of Becoming a Lysogen

We now examine how an infecting lambda phage eventually shuts down gene activity of all but three of its genes when it lysogenizes a cell. The beginning steps are the same as in a lytic infection: DNA is injected into the cell, and the N and Cro proteins, which are sometimes called the immediate early genes, begin to accumulate. Then the remainder of the proteins synthesized under control of the p_L and p_R promoters, including CII and CIII, begin to be synthesized. These latter proteins are crucial to the lysogenic response in three ways. First, they are necessary for the initial synthesis of CI; second, they are necessary for adequate synthesis of the phage protein required for integration of phage DNA into the host DNA, the Int protein; and third, they delay or reduce the expression of the Q protein that turns on late protein synthesis.

The CII protein activates transcription from three phage promoters p_{RE}, p_I, and p_{AQ}. The CIII protein works with CII somehow to protect CII from the action of proteases within the cell. The sensitivity of CII to proteases and the protection provided by CIII generate a response highly

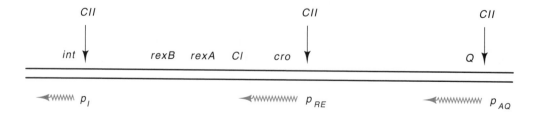

sensitive to the physiological state of the cell. It is this response that determines whether or not a cell will proceed down the lytic or the lysogenic developmental pathways. The promoter p_{RE} is located to the right of *cro* and is oriented in the direction opposite to p_R. Thus the initial part of the RNA synthesized under control of this promoter is anti-*cro* messenger and likely reduces Cro protein synthesis. The remainder of the messenger encodes repressor protein, the *CI* gene product, as well as two other proteins in this operon, RexA and RexB. The p_I promoter lies partly in the *xis* gene and is just in front of the *int* gene. The p_{AQ} promoter is located within the Q gene and oriented in the opposite direction. Thus, the p_{AQ} promoter specifies the synthesis of antimessenger that reduces Q protein synthesis until CII is gone.

In infected cells that will become lysogens, repressor soon reaches a level sufficient to bind to the operators that overlap p_L and p_R. This binding shuts off synthesis of N, Cro, and the other early gene products. Hence, although a little Exo, P, γ, O, P, and Q have accumulated, their levels are insufficient to sustain growth. Lambda destined to become a lysogen may even undergo several rounds of DNA replication, but then the early genes are shut off and the instability of O prevents further development. The Int protein, which also has been synthesized at high rates under control of p_I, participates with a number of host proteins in the integration of one copy of the lambda DNA into the host.

After high levels of lambda repressor have effectively shut off the phage early promoters, four phage promoters can still produce functional messenger, p_{RM}, p_{RE}, p_I, and p_{AQ}. However as CII and CIII lose activity and are diluted away with cell growth, $p_{RE}, p_I,$ and p_{AQ} shut down and leave only the promoter for the maintenance of repressor synthesis, p_{RM}, active. The program of shut-down occurs whether or not lambda has been successful in integrating itself into the host chromosome. If integration has occurred, then the lambda DNA is replicated by the host machinery upon each synthesis cycle and each daughter cell inherits a copy of the lambda genome and is lysogenic. If integration has not occurred, then lambda is passed to only one daughter and soon is effectively lost by dilution.

Site for Cro Repression and CI Activation

Regulation of the early promoters is more complicated than has been indicated. At both p_L and p_R are three binding sites for both Cro and repressor, O_{L1}, O_{L2}, and O_{L3} on the left, and O_{R1}, O_{R2}, and O_{R3} on the right. Repressor bound at O_{L1} is sufficient to inactivate p_L. Repressor

and Cro binding on the right is more interesting and complicated because there these proteins regulate the activities of both p_R and p_{RM}.

Table 14.2 Effects of Increasing Concentrations of Lambda Repressor, its Binding to Operators, and its Effects on the Promoters p_{RM} and p_R

Repressor Levels	Activity of p_{RM}	Operator Occupied by Repressor			Activity of p_R
		O_{R3}	O_{R2}	O_{R1}	
None	Weak	-	-	-	On
Low	Activated	-	+	+	Repressed
High	Repressed	+	+	+	Repressed

Repressor binding to the operators on the right acts as a repressor for p_R and a stimulator or, at higher concentrations, as a repressor for p_{RM} (Table 14.2). These different activities are accomplished by the following means. Repressor bound at O_{R1} inactivates p_R, but repressor bound at O_{R1} and O_{R2} simultaneously represses p_R and activates p_{RM}. Repressor bound at O_{R3} inactivates p_{RM}. Cro protein bound at O_{R3} represses p_{RM}, and Cro bound to O_{R1} or O_{R2} represses p_R (Table 14.3).

Despite the fact that Cro and repressor bind to virtually the same sequences, at least as assayed by DNAse protection and the behavior of mutations lying in the three operator sites on the right, their binding is not the same. As the level of Cro begins to rise in cells, for example during a lytic infective cycle, Cro binds first to O_{R3} and shuts off the synthesis of CI repressor. Only later, after the level of Cro has risen still higher, does it bind to O_{R2} and O_{R1} and shut down the activity of p_R. On the other hand, during a phage developmental cycle that will result in the production of a lysogen, as repressor begins to accumulate it first binds to O_{R1} and O_{R2} and shuts off p_R and turns on p_{RM}. At still higher concentrations, repressor binds to O_{R3} and shuts off p_{RM}. The basis for the differential affinity of repressor and Cro for the three operators lies in the slight sequence differences among the sites and the structural

Table 14.3 Effects of Increasing Concentrations of Cro Protein, its Binding to Operators, and its Effects on the Promoters p_{RM} and p_R

Cro Level	Activity of p_{RM}	Operator Occupied by Cro			Activity of p_R
		O_{R3}	O_{R2}	O_{R1}	
None	Weak	-	-	-	On
Low	Repressed	+	-	-	On
Medium	Repressed	+	+	-	Repressed
High	Repressed	+	+	+	Repressed

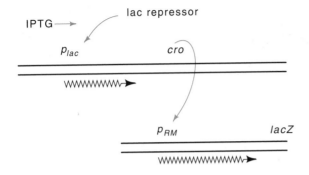

Figure 14.11 Decoupling of Cro and Repressor from their promoters and reconstruction to facilitate examination of their interactions via the inducibility of *lac* promoter and the assay of β-galactosidase.

differences in the proteins. Repressor and Cro "read" the sequences with different emphasis on different bases.

We owe the preceding picture to a series of clever *in vivo* and *in vitro* experiments by Ptashne and his collaborators. Two operations had to be performed to examine the *in vivo* effects of Cro and repressor on the activities of p_R and p_{RM}. The synthesis of the two proteins had to be decoupled and a means had to be found of varying the level of one protein while examining the activity of each of the promoters. Genetic engineering came to the rescue. In one case Cro protein synthesis was placed under control of the *lac* promoter via a p_{lac}-*cro* fusion (Fig. 14.11). Since the *lac* promoter was still regulated by *lac* repressor, the level of Cro in cells could be varied by varying the concentration of *lac* inducer, IPTG, added to the culture medium. Quantitating the activities of p_R and p_{RM} was facilitated by fusing either promoter to the β-galactosidase gene. Hence, although IPTG was added to cells and β-galactosidase was measured, the results elucidated behavior not of the *lac* operon, but of lambda phage.

As the intracellular concentration of Cro was increased, first p_{RM} and then p_R was repressed (Fig. 14.12). These promoters showed much

Figure 14.12 Response of the p_R and p_{RM} promoters to increasing concentrations of Cro protein.

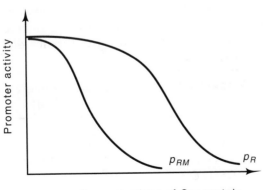

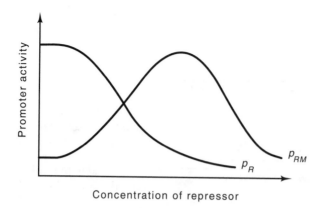

Figure 14.13 Response of the p_R and p_{RM} promoters to increasing concentrations of repressor protein.

different responses to repressor (Fig. 14.13). At one concentration, p_R was shut off and p_{RM} was turned on. Then, at higher concentrations, p_{RM} was shut off as well. *In vitro* transcription experiments using DNA fragments several hundred base pairs long yielded the same results.

Quantitation of the binding of Cro and repressor to the three operators was done with DNAse footprinting. At the lowest Cro concentrations, only O_{R3} is occupied, but as Cro levels are increased, O_{R2} and O_{R1} become occupied as well. The results with repressor are the reverse. O_{R1} has the highest apparent affinity for repressor, being 50% occupied at a concentration of 3 nM *in vitro*. At twice this concentration, O_{R2} is 50% occupied, but 25 times this concentration is required for O_{R3} to be 50% occupied.

Cooperativity in Repressor Binding and its Measurement

As seen in the preceding section, repressor does not bind with equal affinity to the three operators on the right. More detailed experiments also reveal that repressor molecules do not bind independently to the

Table 14.4 The Binding Energies of Repressor Dimers to the Various Operators and the Interaction Energies that Generate Cooperative Binding of Repressor Dimers

Components	Binding Energy, ΔG (Kcal)
R-O_{R1}	-12
R-O_{R2}	-10
R-O_{R3}	-10
R-O_{R1}-R-O_{R2}	-2
R-O_{R2}-R-O_{R3}	-2

three operators (Table 14.4). Not surprisingly, repressors bound at the operators interact with one another and change the overall binding energy of repressor for an operator. Most unexpectedly, however, the results show that a repressor molecule bound at the middle operator can interact either with a repressor bound at O_{R1} or with a repressor bound at O_{R3}, but it does not simultaneously interact with both. This is termed a pairwise interaction.

Operator mutants were used to measure the intrinsic affinity of repressor for the three operator sites as well as to reveal interactions between adjacently bound repressor molecules. By eliminating repressor binding to O_{R2}, the binding energy of repressor to O_{R1} and to O_{R3} could be measured without complications caused by interactions with adjacent repressors. Similarly, eliminating binding at O_{R1} or O_{R3} permitted determination of the other binding energies. Ultimately, this series of measurements provided enough data to permit calculation of the interaction energies of adjacent repressor molecules. The measurements showed that repressor binding at O_{R2} has a dramatic effect on the binding of adjacent repressors.

The Need for and the Realization of Hair-Trigger Induction

Why does lambda possess regulation systems as complicated as those described above? Part of the answer is that lambda must switch between two states, lytic and lysogenic. Even so, this does not explain why a simple repressor like that found in the *lac* operon could not suffice for lambda's needs. In part the answer lies in the fact that lambda must regulate its gene activities over a much greater range than *lac*. The basal level of the *lac* operon is 0.1% of the fully induced level, and there is no apparent reason why a lower basal level would serve any useful purpose. A similar basal level of repression of lambda early genes would be disastrous, for lambda would then have a high rate of spontaneous induction from the lysogenic state. Experimentally the spontaneous induction rate is low. Typically in a doubling time, fewer than one cell out of 10^6 spontaneously cures of lambda or induces. This low rate means that the basal level of expression of the lambda early genes is very low indeed.

Suppose the basal level of expression of a repressible operon similar to the *lac* operon is to be generated by increasing the concentration of repressor. To reduce the basal level to 1/1000 of normal, the repressor level would have to be raised 1000-fold. Not only would this be wasteful of repressor, but even worse are the implications for induction. With 10,000 molecules of repressor per cell, inducing the early genes to greater than 50% of maximal would require inactivating approximately of 9,998 repressor molecules per cell. Although lambda phage may like to hitch a free ride in healthy cells, once viability of the cells is in question because of damage to the host DNA, lambda bails out and induces. The lambda in very few cells indeed would be capable of inducing if 99.995% of the repressor within a cell had to be inactivated before early genes could be turned on efficiently.

Table 14.5 Derepression of the Lambda Promoter p_R and the *lac* Promoter as a Function of the Amount of Repressor Remaining Active

Concentration of Repressor	Activity	
	p_R	p_{lac}
100%	<0.01	<0.01
10%	0.5	0.03
1%	-	0.5

One of lambda's solutions to the problem of being either repressed or induced has been to evolve a nonlinear induction response. At the normal levels of repressor, of about 100 dimers per cell, lambda is fully repressed. However, if 90% of repressor has been inactivated, the p_R promoter of lambda is 50% of fully induced. For comparison, inactivation of 90% of *lac* repressor induces the *lac* operon only 3% (Table 14.5). The highly cooperative binding of repressor to O_{RI} and O_{R2} is largely responsible for lambda's nonlinear response. This can be understood quantitatively as follows. Since the high cooperativity in repressor binding means that most often O_{RI} and O_{R2} are either unoccupied or simultaneously occupied, it is a good approximation to assume that two repressors bind to operator at the same time. Combining the binding equation that defines the dissociation constant with the conservation equation and solving yields the ratio of free operator O to total operator O_T, that is, the relative amount of derepression as a function of repressor concentration:

$$O + 2R \leftrightarrows O{\cdot}R_2,$$

$$K_D = \frac{O \times R^2}{O{\cdot}R_2},$$

$$O + O{\cdot}R_2 = O_T$$

$$\frac{O}{O_T} = \frac{K_D}{K_D + R^2}.$$

As the concentration of free repressor falls from a value a bit above $K_D^{1/2}$ to a value below $K_D^{1/2}$, promoter activity, which is proportional to the fraction of operator unoccupied by repressor, O/O_T, increases rapidly due to the R^2 term. Hence very sharp changes in operator occupancy can be produced with relatively small changes in the concentration of repressor (Fig. 14.14).

Lambda repressor uses another mechanism in addition to cooperative binding to the operators to increase the nonlinearity in its response to repressor concentration. Only dimers of the repressor polypeptide are able to bind to lambda operators under physiological conditions. The dissociation constant governing the repressor monomer-dimer

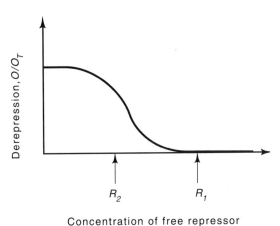

Figure 14.14 Exaggerated example of the nonlinearity in the repression of p_R as a function of repressor concentration. The operator is fully occupied and off at concentration R_1; at a repressor concentration about half of R_1, the operator is unoccupied and the promoter is largely derepressed.

reaction is such that a small reduction from the normal intracellular concentration of repressor sharply reduces the concentration of active repressor dimers. The combination of repressor monomer association-dissociation, cooperativity in repressor binding, and the need for sustained synthesis of N protein for lambda induction and development produces a very nonlinear induction response to changes in repressor level. At one concentration of repressor, lambda is highly stable as a lysogen, but at a lower concentration, lambda almost fully derepresses p_L and p_R.

Induction from the Lysogenic State

In this section we discuss how lambda escapes from the lysogenic state and enters a lytic cycle. One time at which this occurs follows host DNA damage and repair. The signal a phage uses for detection of this damage is the extensive binding of RecA protein to single-stranded DNA that accumulates at a replication site when extensive DNA damage is present. Apparently when polymerized along the single-stranded DNA, RecA is held in a shape that interacts with LexA and lambda repressor. The interaction stimulates the inherent self-cleavage activity of these repressors (Fig. 14.15). Once cleaved, these repressors are no longer able to repress.

In a nonlysogenic cell the genes that are normally repressed by LexA protein are the *lexA* gene itself, the *recA* gene, and a set of about 20 others that are part of the SOS system. Known functions of this system are to repair damaged DNA and to postpone cell division until repair is completed. Once repair is completed, the single-stranded DNA no longer available to activate RecA. Consequently, newly synthesized LexA protein is no longer cleaved and it therefore represses synthesis of RecA and the other proteins of the SOS system. In a normal nonlysogenic cell, the SOS system switches off when it is no longer needed.

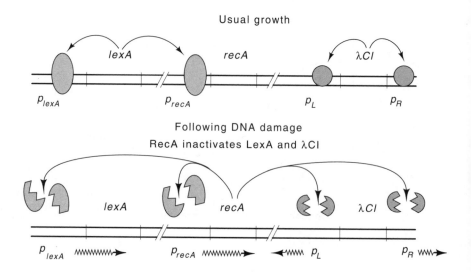

Figure 14.15 LexA protein represses its own synthesis as well as that of RecA protein. Cleavage of LexA and lambda repressor activated by RecA derepresses the *lexA*, *recA*, and lambda operons.

The proteolytic activity of LexA and CI which is stimulated by RecA protein disconnects the N-terminal domain of the lambda repressor from the C-terminal domain (Fig. 14.16). A number of physical experiments with protease-digested repressor or with N-terminal fragments of repressor produced from nonsense mutations have revealed that the N-terminal half of the protein folds up to form a compact domain that can bind to operator. The C-terminus also folds to form a compact domain, but this domain is primarily responsible for the dimerization of repressor. The C-terminal domains lacking N-terminal domain still dimerize, whereas the N-terminal fragments do not.

Figure 14.16 Representation of the structure of lambda repressor and the effects of cleavage by RecA protease. The DNA-binding domains are shown contacting one another while on DNA because a small bit of the dimerization energy derives from such an interaction.

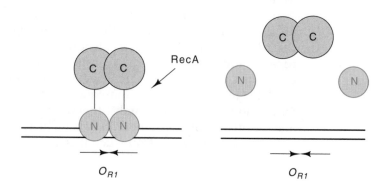

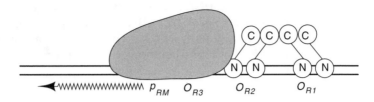

Figure 14.17 The activation of RNA polymerase at p_{RM} by two repressor dimers bound at O_{R1} and O_{R2}.

As required for induction, the affinity of the monomeric N-terminal domain for operator is much lower than that of the intact dimeric repressor. Furthermore, the N-terminal domains do not show strong cooperativity in their binding to adjacent operators. In the next section we will see why the proteolytic cleavage and elimination of dimeric DNA-binding domains greatly reduce the affinity of repressor for operator. As a result of this reduced affinity, repressor comes off the DNA and the phage induces.

The N-terminal domain of repressor makes protein-protein contacts with RNA polymerase when it stimulates transcription from the p_{RM} promoter (Fig. 14.17). The stimulation provided by this positive-acting factor is similar to the stimulation provided by CRP protein. The repressor accelerates the isomerization rate by RNA polymerase after it has bound to p_{RM}. CRP bound near position -41 on promoters also stimulates isomerization.

Two types of evidence indicate that it is the N-terminal domain of repressor that contacts RNA polymerase. The first is the existence of mutations in lambda repressor that eliminate the positive stimulation of p_{RM} by repressor when it occupies O_{R1} and O_{R2}. These lie in the portion of the protein immediately adjacent to the helices of the N-terminal domain that contact operator. The second is that high levels of the N-terminal domain are capable of stimulating p_{RM}.

Entropy, a Basis for Lambda Repressor Inactivation

The binding of dimeric repressor to operator can be described by a dissociation constant that is related to the change in the free energy via the standard thermodynamic relationship:

$$K_D = e^{-\frac{\Delta G}{RT}}$$

$$= e^{-\frac{(\Delta H - T\Delta S)}{RT}},$$

where K_D is the dissociation constant of the dimer from operator, ΔG is the change in free energy, R is the gas constant, T is the temperature, ΔS is the change in entropy, and ΔH is the change in enthalpy.

It is natural to assume that if monomeric repressor were binding to operator, then roughly half as many contacts would be formed between repressor and DNA and roughly half as many water molecules and ions would be displaced from DNA and the protein as the repressor bound. Therefore we might write for the dissociation constant of the monomer where K_M is the dissociation constant for monomer,

$$K_M = e^{-\frac{[\Delta H - T\Delta S]}{2RT}}$$

or $K_M = K_D{}^{1/2}$. This is not correct, however. To make this clear, let ΔS be written as the sum of the entropy changes involved with the contacts and displacement of water, $\Delta S_{\text{interaction dimer}}$ plus the change in entropy involved with immobilization and orientation of the dimeric repressor, $\Delta S_{\text{intrinsic dimer}}$:

$$\Delta S_{\text{dimer}} = \Delta S_{\text{interaction dimer}} + \Delta S_{\text{intrinsic dimer}}.$$

The same type of equation can be written for the monomer:

$$\Delta S_{\text{monomer}} = \Delta S_{\text{interaction monomer}} + \Delta S_{\text{intrinsic monomer}}.$$

Roughly the monomer makes half as many interactions as the dimer and displaces half as many water molecules and ions as it binds. Therefore

$$\Delta S_{\text{interaction monomer}} = \Delta S_{\text{interaction dimer}}/2.$$

The same is not true of the intrinsic entropies. The entropy change associated with bringing repressor monomer to rest on operator by correctly positioning and orienting it is nearly the same as the change associated with the dimeric repressor. Thus

$$\Delta S_{\text{monomer}} \neq \Delta S_{\text{dimer}}/2,$$

Figure 14.18 Why a dimer can bind much more tightly than a momomer. Almost all the additional ΔH provided by the second monomer's binding can go into increasing ΔG, whereas for binding of a monomer, almost no ΔG is left over to contribute to binding.

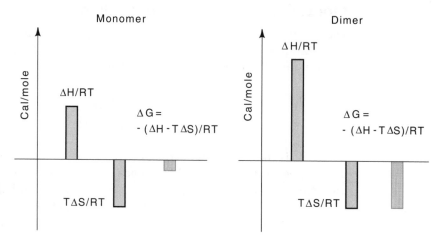

and detailed calculations based on statistical mechanics or experiments show that the inequality can be severe. This is another example of the chelate effect discussed earlier. In other words, a sizable fraction of the total entropy change involved with repressor binding to DNA is associated with its correct positioning. Roughly the same entropy is required to orient a monomer or dimer, but in the case of lambda repressor, the presence of the second subunit of the dimer adds twice as much to the binding energy. Most of this additional energy can go into holding the dimer on the DNA, and hence K_D is much less than K_M^2 (Fig. 14.18).

Problems

14.1. How would you perform genetic crosses between lambda and phage 434 to produce a λimm^{434} phage?

14.2. The *rex* genes of lambda are in the CI operon, and Rex protein gives lambda lysogens immunity to a class of T4 mutants possessing defects in the r_I or r_{II} genes. Wild-type T4 grows well on lambda lysogens. How would you isolate *rex* mutants of lambda?

14.3. An interesting cloverleaf structure was proposed for the *ori* region with bound *O* protein in "DNA Sequences and Structural Homologies of the Replication Origins of Lambdoid Bacteriophages," R. Grosschedl, H. Hobom, Nature *277*, 621-627 (1979). What precisely does the DNAse footprinting of O bound to *ori* say about this model? See "Purified Bacteriophage Lambda O Protein Binds to Four Repeating Sequences at the Lambda Replication Origin," T. Tsurimoto, K. Matsubara, Nucleic Acids Res. *9*, 1789-1799 (1981).

14.4. How would you isolate nonsense mutations in the *S* gene of lambda?

14.5. Why does lambda possess a linear genetic map with the *A* and *R* genes at opposite ends when in fact the DNA molecules undergoing recombination are circular and the *A* and *R* genes are adjacent?

14.6. What would you do to make a bacterial strain that would maximize the frequency of lysogenization upon infection by phage lambda?

14.7. How could one show genetically that v_1v_3, a double-point mutation in the rightward operators, makes p_R insensitive to repression? Similarly, how could one show genetically that v_2, a mutation in the leftward operators, makes p_L repressor insensitive?

14.8. C_{17} is a repressor-insensitive new promoter for genes *CII*, *O*, and *P*. Why might lambda $C_{17}CI^-$ be virulent (capable of forming plaques on a lambda lysogen) but lambda $C_{17}CI^+$ not be virulent?

14.9. What is a reasonable interpretation of the fact that cells infected with S^+ phage and treated with metabolic inhibitors, such as cyanide, or protein synthesis inhibitors, such as chloramphenicol, later than 15 minutes after infection promptly lyse?

14.10. The lambda R gene product was once identified as an endopeptidase in "Endopeptidase Activity of phage Lambda-Endolysin," Nature New Biol. *234*, 144-145 (1971). What is a likely reason for the misidentification and what control experiments would you have performed?

14.11. A lysogen of lambda $CI_{857}N^-O^-P^-$ is grown at 42° for many generations and then grown at 35°. Infection of the cells grown in this manner with lambda yielded clear plaques and a lysogenization frequency much lower than seen on infection of a similarly treated nonlysogenic strain. Why?

14.12. Simultaneous infection of a lambda lysogen with λ and $λi^{434}$ produced a burst containing only $λi^{434}$. Why? How would you isolate lambda mutants that would grow in a lambda lysogen when coinfected with $λi^{434}$, and where would you expect these to map?

14.13. Write a few insightful comments about the locations of genes and their sites of action in phage lambda.

14.14. A culture of a lambda lysogen was mutagenized, grown ten doublings, and plated out. Why is it at first paradoxical that some plaques should be clear, and what is the explanation?

14.15. Conceivably a *nus* mutation might act by causing a host protein to inactivate the N protein of phage lambda. Although a complicated biochemical experiment could be used to test this idea, what simple genetic experiment could disprove the proposal?

14.16. Devise a selection or scoring method for the isolation of *CI* mutations that repress normally but do not stimulate p_{RM}.

14.17. If the LamB porin possesses larger holes than the two porins normally present in the outer membrane, why not just use the LamB porin and forget the others? Note that none of the porins is ion-selective.

14.18. Given the genes in which *nus* mutations have been found, invent a mechanism of antitermination.

14.19. How would you isolate lambda mutants that have bypassed their requirement for transcriptional activation from p_R for activation of DNA replication?

14.20. You discover an *E. coli* strain that is lysogenic for lambda and that this lambda induces when the cultures are heated above 37°. The resulting phage form clear plaques when plated on all strains, including a cured version of the strain from which the mystery lambda phage was obtained. What mutation(s) does the lambda carry?

14.21. What are the colonies resulting from the following treatments likely to be or to contain? A. Infect with lambda CI_{857} and grow at 35°. B. Like A., but grow at 40°. C. Like A., but spread on plates containing 10^9 lambda CI^-. D. Like A., but spread on plates containing 10^9 lambda CI^- and 10^9 λvir.

14.22. (a) What happens to lambda induction in response to DNA damage if the intracellular repressor levels become too high? (b) Contrast the wisdom of maintaining a low intracellular level of repressor via making repressor messenger infrequently but translating it many times versus making messenger often but translating each molecule only a few times. (c) With respect to the answer in part b, comment on the fact that the 5' ribonucleotide of *CI* messenger is the A of the initiation AUG codon.

14.23. In the lambdoid family there exist at least ten different Q gene specificities. Based on this, what do you predict about the mechanism of Q action?

14.24. Suppose someone has made a variant lambda phage in which the coding regions for repressor and cro protein are interchanged. Your objective is to try to make lystates and lysogens of this phage. How could you do it? For example, you might infect cells full of CII and CIII protein. If it is not possible, what is the reason?

14.25. Predict the consequences to phage lambda if the dimerization of CI repressor were made ten-fold tighter.

References

Recommended Readings

Control of Lambda Repressor Synthesis, L. Reichardt, A. Kaiser, Proc. Nat. Acad. Sci. USA *68*, 2185-2189 (1971).

Positive Control of Endolysin Synthesis *in vitro* by the Gene N Protein of Phage Lambda, J. Greenblatt, Proc. Nat. Acad. Sci. USA *69*, 3606-3610 (1972).

Interactions between DNA-bound Repressors Govern Regulation by the Lambda Phage Repressor, A. Johnson, B. Meyer, M. Ptashne, Proc. Nat. Acad. Sci. USA *76*, 5061-5065 (1979).

Phage Lambda Gene Q Antiterminator Recognizes RNA Polymerase Near the Promoter and Accelerates It Through a Pause Site, E. Grayhack, X. Yang, L. Lau, J. Roberts, Cell *42*, 259-269 (1985).

Genetics, Physical Mapping, DNA Structure and Sequence

Cohesive Sites on the DNA from Several Temperate Coliphages, R. Baldwin, P. Barrand, A. Fritsch, D. Goldthwait, F. Jacob, J. Mol. Biol. *17*, 343-357 (1966).

Electron-Microscopic Visualization of Deletion Mutations, R. Davis, N. Davidson, Proc. Nat. Acad. Sci. USA *60*, 243-250 (1968).

Note on the Structure of Prophage Lambda, P. Sharp, M. Hsu, N. Davidson, J. Mol. Biol. *71*, 499-501 (1972).

The Topography of Lambda DNA: Insertion of Ordered Fragments and the Physical Mapping of Point Mutations, J. Egan, D. Hogness, J. Mol. Biol. *71*, 363-381 (1972).

Nucleotide Sequence of *cro*, *cII* and Part of the *O* Gene in Phage Lambda DNA, E. Schwarz, G. Scherer, G. Hobom, H. Kossel, Nature *272*, 410-414 (1978).

Nucleotide Sequence of the *O* Gene and of the Origin of Replication in Bacteriophage Lambda DNA, G. Scherer, Nucleic Acid Res. *5*, 3141-3156 (1978).

DNA Sequence of the *att* Region of Coliphage 434, D. Mascarenhas, R. Kelley, A. Cambell, Gene *15*, 151-156 (1981).

Nucleotide Sequence of Bacteriophage Lambda DNA, F. Sanger, A. Coulson, G. Hong, D. Hill, G. Petersen, J. Mol. Biol. *162*, 729-773 (1982).

The *Rex* Gene of Bacteriophage Lambda is Really Two Genes, K. Matz, M. Schandt, G. Gussin, Genetics *102*, 319-327 (1982).

N and Q Function

A Site Essential for Expression of all Late Genes in Bacteriophage Lambda, I. Herskowitz, E. Signer, J. Mol. Biol. *47*, 545-556 (1970).

Bypassing A Positive Regulator: Isolation of a Lambda Mutant that Does not Require N Product to Grow, N. Hopkins, Virology *40*, 223-229 (1970).

The Specificity of Lamboid Phage Late Gene Induction, R. Schleif, Virology *50*, 610-612 (1972).

Regulation of the Expression of the N Gene of Bacteriophage Lambda, J. Greenblatt, Proc. Nat. Acad. Sci. USA *70*, 421-424 (1973).

Gene N Regulator Function of Phage Lambda imm2l: Evidence that a Site of N Action Differs from a Site of N Recognition, D. Friedman, G. Wilgus, R. Mural, J. Mol. Biol. *81*, 505-516 (1973).

Release of Polarity in *Escherichia coli* by Gene N of Phage Lambda: Termination and Antitermination of Transcription, S. Adhya, M. Gottesman, B. deCrombrugghe, Proc. Nat. Acad. Sci. USA *71*, 2534-2538 (1974).

Altered Reading of Genetic Signals Fused to the N Operon of Bacteriophage Lambda: Genetic Evidence for Modification of Polymerase by the Protein Product of the N Gene, N. Franklin, J. Mol. Biol. *89*, 33-48 (1974).

Transcription Termination and Late Control in Phage Lambda, J. Roberts, Proc. Nat. Acad. Sci. USA *72*, 3300-3304 (1975).

Coliphage Lambda nutL⁻: A Unique Class of Mutants Defective in the Site of Gene N Product Utilization for Antitermination of Leftward Transcription, J. Salstrom, W. Szybalski, J. Mol. Biol. *124*, 195-221 (1978).

N-independent Leftward Transcription in Coliphage Lambda: Deletions, Insertions and New Promoters Bypassing Termination Functions, J. Salstrom, M. Fiandt, W. Szybalski, Mol. Gen. Gen. *168*, 211-230 (1979).

L Factor that is Required for β-galactosidase Synthesis is the *nusA* Gene Product Involved in Transcription Termination, J. Greenblatt, J. Li, S. Adhya, D. Friedman, L. Baron, B. Redfield, H. Kung, H. Weissbach, Proc. Nat. Acad. Sci. USA 77, 1991-1994 (1980).

The *nusA* Gene Protein of *Escherichia coli*, Its Identification and a Demonstration that it Interacts with the Gene N Transcription Anti-terminating Protein of Bacteriophage Lambda, J. Greenblatt, J. Li, J. Mol. Biol. *147*, 11-23 (1981).

Interaction of the Sigma Factor and the *nusA* Gene Protein of *E. coli* with RNA Polymerase in the Initiation-termination Cycle of Transcription, J. Greenblatt, J. Li, Cell *24*, 421-428 (1981).

Analysis of *nutR*: A Region of Phage Lambda Required for Antitermination of Transcription, E. Olson, E. Flamm, D. Friedman, Cell *31*, 61-70 (1982).

Evidence that a Nucleotide Sequence, "boxA," is Involved in the Action of the *NusA* Protein, D. Friedman, E. Olson, Cell *34*, 143-149 (1983).

Conservation of Genome Form but Not Sequence in the Transcription Antitermination Determinants of Bacteriophages Lambda, φ21, and P22, N. Franklin, J. Mol. Biol. *181*, 75-84 (1985).

The *nut* Site of Bacteriophage λ is Made of RNA and is Bound by Transcription Antitermination Factors on the Surface of RNA Polymerase, J. Nodwell, J. Greenblatt, Genes and Development *5*, 2141-2151 (1991).

Cro Structure and Function

Role of the *cro* Gene in Bacteriophage Lambda Development, H. Echols, L. Green, B. Oppenheim, A. Oppenheim, A. Honigman, J. Mol. Biol. *80*, 203-216 (1973).

Sequence of *cro* Gene of Bacteriophage Lambda, T. Roberts, H. Shimatake, C. Brady, M. Rosenberg, Nature *270*, 274-275 (1977).

The Essential Role of the *Cro* Gene in Lytic Development by Bacteriophage Lambda, A. Folkmanis, W. Maltzman, P. Mellon, A. Skalka, H. Echols, Virology *81*, 352-362 (1977).

Mechanism of Action of *Cro* Protein of Bacteriophage Lambda, A. Johnson, B. Meyer, M. Ptashne, Proc. Nat. Acad. Sci. USA *75*, 1783-1787 (1978).

Structure of the *Cro* Repressor from Bacteriophage Lambda and its Interaction with DNA, W. Anderson, D. Ohlendorf, Y. Takeda, B. Mathews, Nature *290*, 754-758 (1981).

Lambda Phage *cro* Repressor, DNA Sequence-dependent Interactions Seen by Tyrosine Fluorescence, F. Boschelli, K. Arndt, H. Nick, Q. Zhang, P. Lu, J. Mol. Biol. *162*, 251-266 (1982).

Repressor Structure and Function

Specific Binding of the Lambda Phage Repressor to Lambda DNA, M. Ptashne, Nature *214*, 232-234 (1967).

Isolation of the Lambda Phage Repressor, M. Ptashne, Proc. Nat. Acad. Sci. USA *57*, 306-313 (1967).

Structure of the Lambda Operators, T. Maniatis, M. Ptashne, Nature *246*, 133-136 (1973).

Control of Bacteriophage Lambda Repressor Synthesis after Phage Infection: The Role of *N, cII, cIII* and *cro* Products, L. Reichardt, J. Mol. Biol. *93*, 267-288 (1975).

Control of Bacteriophage Lambda Repressor Synthesis: Regulation of the Maintenance Pathway by the *cro* and *cI* Products, L. Reichardt, J. Mol. Biol. *93*, 289-309 (1975).

The Kil Gene of Bacteriophage Lambda, H. Greer, Virology *66*, 589-604 (1975).

Inactivation and Proteolytic Cleavage of Phage Lambda Repressor *in vitro* in an ATP-dependent Reaction, J. Roberts, C. Roberts, D. Mount, Proc. Nat. Acad. Sci. USA *74*, 2283-2287 (1977).

Sites of Contact between Lambda Operators and Lambda Repressor, Z. Humayun, D. Kleid, M. Ptashne, Nucleic Acids Research *4*, 1595-1607 (1977).

Regulatory Functions of the Lambda Repressor Reside in the Amino-terminal Domain, R. Sauer, C. Pabo, B. Meyer, M. Ptashne, K. Backman, Nature *279*, 396-400 (1979).

The Lambda Repressor Contains Two Domains, C. Pabo, R. Sauer, J. Sturtevant, M. Ptashne, Proc. Nat. Acad. Sci. USA *76*, 1608-1612 (1979).

Gene Regulation at the Right Operator (O_R) of Bacteriophage Lambda I. O_R3 and Autogenous Negative Control by Repressor, R. Maurer, B. Meyer, M. Ptashne, J. Mol. Biol. *139*, 147-161 (1980).

Gene Regulation at the Right Operator (O_R) of Bacteriophage Lambda II. O_R1, O_R2 and O_R3: Their Roles in Mediating the Effects of Repressor and *cro*, B. Meyer, R. Maurer, M. Ptashne, J. Mol. Biol. *139*, 163-194 (1980).

Gene Regulation at the Right Operator (O_R) of Bacteriophage Lambda III. Lambda Repressor Directly Activates Gene Transcription, B. Mayer, M. Ptashne, J. Mol. Biol. *139*, 195-205 (1980).

A Fine Structure Map of Spontaneous and Induced Mutations in the Lambda Repressor Gene, Including Insertions of IS Elements, M. Leib, Mol. and Gen. Genet. *184*, 364-371 (1981).

The N-Terminal Arms of Lambda Repressor Wrap Around the Operator DNA, C. Pabo, W. Krovatin, A. Jeffrey, R. Sauer, Nature *298*, 441-443 (1982).

The Operator-binding Domain of Lambda Repressor: Structure and DNA Recognition, C. Pabo, M. Lewis, Nature *298*, 443-447 (1982).

Quantitative Model for Gene Regulation by Lambda Phage Repressor, G. Ackers, A. Johnson, M. Shea, Proc. Nat. Acad. Sci. USA *79*, 1129-1133 (1982).

Mutant Lambda Phage Repressor with a Specific Defect in its Positive Control Function, L. Guarente, J. Nye, A. Hochschild, M. Ptashne, Proc. Nat. Acad. Sci. USA *79*, 2236-2239 (1982).

Control of Transcription by Bacteriophage P22 Repressor, A. Poteete, M. Ptashne, J. Mol. Biol. *157*, 21-48 (1982).

Repressor Structure and the Mechanism of Positive Control, A. Hochschild, N. Irwin, M. Ptashne, Cell *32*, 319-325 (1982).

Autodigestion of *lexA* and Phage Lambda Repressors, J. Little, Proc. Nat. Acad. Sci. USA *81*, 1375-1379 (1984).

The O_R Control System of Bacteriophage Lambda, A Physical-Chemical Model for Gene Regulation, M. Shea, G. Ackers, J. Mol. Biol. *181*, 211-230 (1985).

NH_2-terminal Arm of Phage Lambda Repressor Contributes Energy and Specificity to Repressor Binding and Determines the Effects of Operator Mutations, J. Eliason, M. Weiss, M. Ptashne, Proc. Nat. Acad. Sci. USA *82*, 2339-2343 (1985).

Activation of Transcription by the Bacteriophage 434 Repressor, F. Bushman, M. Ptashne, Proc. Nat. Acad. Sci. USA *83*, 9353-9357 (1986).

Feeling for the Bumps, R. Schleif, Nature *325*, 14-15 (1987).

Lambda Repressor Mutants that are Better Substrates for RecA-mediated Cleavage, F. Gimble, R. Sauer, J. Mol. Biol. *206*, 29-39 (1989).

Lambda DNA Replication

Regulation of Bacteriophage Lambda DNA Replication, M. Green, B. Gotchel, J. Hendershott, S. Kennel, Proc. Nat. Acad. Sci. USA *58*, 2343-2350 (1967).

Intracellular Pools of Bacteriophage Lambda Deoxyribonucleic Acid, B. Carter, M. Smith, J. Mol. Biol. *50*, 713-718 (1970).

Replication of Bacteriophage Lambda DNA Dependent on the Function of Host and Viral Genes, I. Interaction of *red*, *gam*, and *rec*, L. Enquist, A. Skalka, J. Mol. Biol. *75*, 185-212 (1973).

Recognition Sequences of Repressor and Polymerase in the Operators of Bacteriophage Lambda, T. Maniatis, M. Ptashne, H. Backman, D. Kleid, S. Flashman, A. Jeffrey, R. Mauer, Cell *5*, 109-113 (1975).

Specificity Determinants for Bacteriophage Lambda DNA Replication II. Structure of *O* Proteins of Lambda-Φ80 and Lambda-82 Hybrid Phages and of a Lambda Mutant Defective in the Origin of Replication, M. Furth, J. Yates, J. Mol. Biol. *126*, 227-240 (1978).

Purified Bacteriophage Lambda O Protein Binds to Four Repeating Sequences at the Lambda Replication Origin, T. Tsurimoto, K. Matsubara, Nuc. Acids. Res. *9*, 1789-1799 (1981).

Specificity Determinants for Bacteriophage Lambda DNA Replication III. Activation of Replication in Lambda ri^c Mutants by Transcription Outside of *ori*, M. Furth, W. Done, B. Meyer, J. Mol. Biol. *154*, 65-83 (1982).

Specialized Nucleoprotein Structures at the Origins of Replication of Bacteriophage Lambda: Complexes with Lambda O Protein and with Lambda O, Lambda P, and *Escherichia coli* DnaB Proteins, M. Dodson, J. Roberts, R. McMacken, H. Echols, Proc. Nat. Acad. Sci. USA *82*, 4678-4682 (1985).

Overproduction of an Antisense RNA Containing the *oop* RNA Sequence of Bacteriophage Lambda Induces Clear Plaque Formation, K. Takayama, N. Houba-Herin, M. Inouye, Mol. Gen. Genetics *210*, 180-186 (1988).

Regulation of Repressor Synthesis and Destruction

Genetic Characterization of a prm⁻ Mutant of Bacteriophage Lambda, K. Yen, G. Gussin, Virology *56*, 300-312 (1973).

Location of the Regulatory Site for Establishment of Repression by Bacteriophage Lambda, M. Jones, R. Fischer, I. Herskowitz, H. Echols, Proc. Nat. Acad. Sci. USA *76*, 150-154 (1979).

Site Specific Recombination Functions of Bacteriophage Lambda: DNA Sequence of Regulatory Regions and Overlapping Structural Genes for *Int* and *Xis*, R. Hoess, C. Foeller, K. Bidwell, A. Landy, Proc. Nat. Acad. Sci. USA *77*, 2482-2486 (1980).

Promoter for the Establishment of Repressor Synthesis in Bacteriophage Lambda, U. Schmeissner, D. Court, H. Shimatake, M. Rosenberg, Proc. Nat. Acad. Sci. USA *77*, 3191-3195 (1980).

Purified Lambda Regulatory Protein *cII* Positively Activates Promoters for Lysogenic Development, H. Shimatake, M. Rosenberg, Nature *292*, 128-132 (1981).

Purified *lexA* Protein is a Repressor of the *recA* and *lexA* Genes, J. Little, D. Mount, C. Yanisch-Perron, Proc. Nat. Acad. Sci. *78*, 4199-4203 (1981).

Mechanism of Action of the *lexA* Gene Product, R. Brent, M. Ptashne, Proc. Nat. Acad. Sci. USA *78*, 4204-4208 (1981).

SOS Induction and Autoregulation of the *himA* Gene for the Site-specific Recombination in *Escherichia coli*, H. Miller, M. Kirk, H. Echols, Proc. Nat. Acad. Sci. USA *78*, 6754-6758 (1981).

Multilevel Regulation of Bacteriophage Lambda Lysogeny by *E. coli himA* Gene, H. Miller, Cell *25*, 269-276 (1981).

Control of Phage Lambda Development by Stability and Synthesis of *cII* Protein: Role of the Viral *cIII* and *hflA*, *himA* and *himD* Genes, A. Hoyt, D. Knight, A. Das, H. Miller, H. Echols, Cell *31*, 565-573 (1982).

Bacteriophage Lambda Protein *cII* Binds Promoters on the Opposite Face of the DNA Helix from RNA Polymerase, Y. Ho, D. Wulff, M. Rosenberg, Nature *304*, 702-708 (1983).

A C_{II}-dependent Promoter is Located within the Q Gene of Bacteriophage Lambda, B. Hoopes, W. McClure, Proc. Nat. Acad. Sci. USA *82*, 3134-3138 (1985).

Cleavage of the CII Protein of Phage Lambda by Purified HflA Protease: Control of the Switch Between Lysis and Lysogeny, H. Cheng, P. Mahlrad, M. Hoyt, H. Echols, Proc. Natl. Acad. Sci. USA *85*, 7882-7886 (1988).

Regulation of Delayed Early Genes Like *int*

Transcription of the *int* Gene of Bacteriophage Lambda, New RNAP Binding Site and RNA Start Generated by int-Constitutive Mutations, R. Fischer, Y. Takeda, H. Echols, J. Mol. Biol. *129*, 509-514 (1979).

Posttranscriptional Control of Bacteriophage Lambda *int* Gene Expression from a Site Distal to the Gene, G. Guarneros, C. Montanez, T. Hernandez, D. Court, Proc. Nat. Acad. Sci. USA *79*, 238-242 (1982).

Regulation of Bacteriophage Lambda *int* Gene Expression, A. Oppenheim, S. Gottesman, M. Gottesman, J. Mol. Biol. *158*, 327-346 (1982).

Lysis and Regulation of Late Genes

Mutations in Bacteriophage Lambda Affecting Host Cell Lysis, A. Harris, D. Mount, C. Fuerst, L. Siminovitch, Virology *32*, 553-569 (1967).

Control of Development in Temperate Bacteriophages, II. Control of Lysozyme Synthesis, C. Dambly, M. Couturier, R. Thomas, J. Mol. Biol. *32*, 67-81 (1968).

New Mutations in the S Cistron of Bacteriophage Lambda Affecting Host Cell Lysis, A. Goldberg, M. Howe, Virology *38*, 200-202 (1969).

Cell Lysis by Induction of Cloned Lambda Lysis Genes, J. Garrett, R. Fusselman, J. Hise, L. Chiou, D. Smith-Grillo, J. Schulz, R. Young, Mol. and Gen. Genet. *182*, 326-331 (1981).

The R Gene Product of Bacteriophage Lambda is the Murein Transglycosylase, K. Bienkowska-Szewczyk, B. Lipinska, A. Taylor, Mol. and Gen. Genet. *184*, 111-114 (1981).

S Gene Product: Identification and Membrane Localization of a Lysis Control Protein, E. Altman, R. Altman, J. Garrett, R. Grimaila, R. Young, J. Bact. *155*, 1130-1137 (1983).

Dominance in Lambda S Mutations and Evidence for Translational Control, R. Raab, G. Neal, C. Sohaskey, J. Smith, R. Young, J. Mol. Biol. *199*, 95-105 (1988).

Other

Entropic Contributions to Rate Accelerations in Enzymic and Intramolecular Reactions and the Chelate Effect, M. Page, W. Jencks, Proc. Nat. Acad. Sci. USA *68*, 1678-1683 (1971).

Escape Synthesis of the Biotin Operon in Induced Lambda *b2* Lysogens, K. Krell, M. Gottesman, J. Parks, J. Mol. Biol. *68*, 69-82 (1972).

Isolation and Properties of rex⁻ Mutants of Bacteriophage Lambda, G. Gussin, V. Peterson, J. Virology *10*, 760-765 (1972).

Isolation of the Bacteriophage Lambda Receptor from *E. coli*, L. Randall-Hazelbauer, M. Schwartz, J. Bact. *116*, 1436-1446 (1973).

An Endonuclease Induced by Bacteriophage Lambda, M. Rhoades, M. Meselson, J. Biol. Chem. *248*, 521-527 (1973).

Studies on the Late Replication of Phage Lambda: Rolling-circle Replication of the Wild-type and a Partially Suppressed Strain *O*am29 *P*am80, D. Bastia, N. Sueoka, E. Cox, J. Mol. Biol. *98*, 305-320 (1975).

Receptor for Bacteriophage Lambda of *E. coli* Forms Larger Pores in Black Lipid Membranes than the Matrix Protein, B. Boehler-Kohler, W. Boos, R. Dieterle, R. Benz, J. Bact. *138*, 33-39 (1979).

Lambda Encodes an Outer Membrane Protein: The *lom* Gene, J. Reeve, J. Shaw, Mol. Gen. Gen. *172*, 243-248 (1979).

Downstream Regulation of *int* Gene Expression by the *b2* Region in Phage Lambda, C. Epp, M. Pearson, Gene *13*, 327-337 (1981).

Effect of Bacteriophage Lambda Infection on Synthesis of *groE* Protein and Other *Escherichia coli* Proteins, D. Drahos, R. Hendrix, J. Bact. *149*, 1050-1063 (1982).

Alternative mRNA Structures of the *CIII* Gene of Bacteriophage Lambda Determine the Rate of Its Translation Initiation, S. Altuvia, D. Kornitzer, D. Teff, A. Oppenheim, J. Mol. Biol. *210*, 265-280 (1989).

Xenopus 5S RNA Synthesis

15

The cells resulting from successive divisions of a fertilized egg cell ultimately become specialized for different functions. How do they do this? The answer is equivalent to understanding regulation of the relevant genes. The genes involved in growth and development must respond to internal signals generated by the organism as it develops. This is in contrast to the externally generated and easily controlled signals that regulate the activities of many of the gene systems discussed up to this point. Important developmental biology questions revolve around the signals: how they are timed, how they set up spatial patterns, and how they regulate expression of the appropriate genes. In this chapter we shall be concerned with a relatively simple developmental problem: ribosomal 5S RNA synthesis, but in a relatively complex organism, the South African clawed toad *Xenopus*. Later we shall consider pattern development and the regulation of genes in the fruit fly *Drosophila*.

Biology of 5S RNA Synthesis in *Xenopus*

Oocytes develop in the ovaries of female frogs in a development process that takes from three months to several years. Eventually the oocytes mature as eggs ready to be fertilized. These huge cells contain large amounts of nutrients and protein synthesis machinery so that later, the fertilized egg can undergo many cell divisions without significant intake of nutrients and without additional ribosome synthesis.

Curiously, during oogenesis components of the ribosomes are not synthesized in parallel. In the ovaries the 5S RNA components of ribosomes are synthesized first, for about two months. As they are synthesized, they are stored away in two types of RNA-protein particles.

Origin

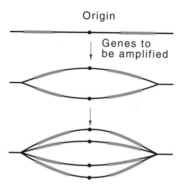

Genes to
be amplified

Figure 15.1 Two rounds of initiation from an origin can yield four copies of genes immediately surrounding that origin.

These are known as 7S and 42S particles from their rates of sedimentation in centrifugation. Even though 5S RNA synthesis proceeds for a long time, the enormous number of RNA molecules needed in an egg necessitates that multiple genes encode the RNA. The *Xenopus* genome contains about 100,000 copies of the 5S gene. A special RNA polymerase transcribes these genes as well as genes coding for tRNAs. This is RNA polymerase III. Although it is different from the two other cellular RNA polymerases, it shares a number of subunits in common with them.

After synthesis of the 5S RNA, synthesis of the 18S, 28S, and 5.8S ribosomal RNAs and the ribosomal proteins begins. We might expect the frog genome also to contain around 100,000 copies of the genes coding for 18S, 28S and 5.8S ribosomal RNAs. It doesn't, however. Apparently the load of carrying large numbers of these much bigger genes would be too great. Instead, *Xenopus* generates the required number of DNA templates for the other ribosomal RNAs by selective replication of the 450 ribosomal RNA genes before the ribosomal RNA is to be synthesized.

Once, a key to success in developmental biology was finding an organism or cell-type devoted to the activity you wished to study. In such cells the specialization in activity reduces the level of other and interfering activities. This simplifies the experimenter's measurements. Also, the purification for biochemical study of the components involved is simpler since they are present in sizeable concentrations in these cells. For this reason *Xenopus* permitted careful examination of several important developmental biology questions before the appearance of genetic engineering techniques.

The strategy of selective gene amplification used by *Xenopus* on its ribosomal RNA genes is not unique. Other organisms also utilize it for genes whose products are needed in high level. In terminally differentiated tissues where no further cell division occurs, the gene amplification can occur right on the chromosome by the selective activity of a replication origin (Fig. 15.1). Since an egg cell, however, must later divide, the mechanism of gene amplification must not irreparably damage the genome. In *Xenopus* a copy of the rRNA gene is replicated

rDNA cluster of ~600 genes

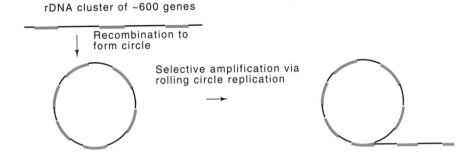

Figure 15.2 Selective replication of rRNA genes yields as many as 4,000,000 gene copies per cell.

free of the chromosome to generate high numbers of templates necessary for the synthesis of the RNA (Fig. 15.2). As the last components of a ribosome are synthesized, mature ribosomes form. Some begin translating maternal mRNA while others remain inactive until fertilization triggers a more vigorous rate of protein synthesis.

Why the genes coding for 5S RNA cannot be amplified along with the 18S, 28S, and 5.8S rRNA genes is not apparent. *Xenopus*, and most other organisms, have gone to considerable trouble to separate the synthesis of 5S and the other ribosomal RNAs. Even the RNA polymerases used are different: 5S uses RNA polymerase III whereas 18S, 28S, and 5.8S use polymerase I.

As the oocyte matures, ribosome synthesis and most other activity ceases until fertilization activates protein synthesis. Then, after the developing embryo has reached the 4,000 cell stage, 5S ribosomal RNA synthesis begins again. A small fraction of the RNA synthesized at this time is the RNA species which was synthesized in the oocyte. Most, however, possesses a slightly different sequence. It is called somatic 5S RNA. The oocyte 5S RNA synthesis soon ceases altogether, leaving just the somatic 5S RNA synthesis. The haploid genome of *Xenopus* contains about 450 copies of the somatic variety of 5S gene in contrast to about 100,000 copies of the variety expressed only in oocytes.

Study of 5S RNA synthesis is attractive because the 5S genes show developmental regulation in that the oocyte 5S genes are expressed only in the developing oocyte whereas the somatic 5S genes are expressed both in the oocyte and in the developing embryo. In addition to the simple timing pattern of 5S RNAs, the product of the synthesis is simple. The 5S RNA is not processed by capping, polyadenylation, or splicing, nor is there a tissue specificity in the expression. Hence this seems like an ideal material for study of developmental questions.

In vitro 5S RNA Synthesis

What is required to reproduce *in vitro* the regulation of 5S RNA synthesis that is seen *in vivo*? A conservative approach in attempting to answer this type of question is to begin as near to the biological situation as possible and to work backward towards a system containing only known and purified components, each performing a well-understood function. During each of the steps in moving from the native biological system to the completely defined simple system, one's criterion of a properly working system is that correct regulation be maintained.

A logical first step in reproducing the biological regulation of *Xenopus* 5S RNA synthesis is to try injecting DNA coding for 5S RNA into oocytes. Such injections are of medium technical difficulty. Due to the large size of later stage oocytes, they can be isolated free of ovarian tissue, injected, incubated, and synthesis products from individual or pooled oocytes analyzed. Gurdon and Brown found that microinjecting the nuclei of oocytes with either DNA purified from frog erythrocytes or with pure plasmid DNA containing one or more 5S genes yielded synthesis of 5S RNA. This synthesis could continue many hours or even for days after the injection.

The next step toward a defined system also determined whether the "living" oocyte possessed any mystical properties. That is, could 5S RNA synthesis be achieved by an extract of nuclei purified from oocytes, or were intact oocytes required? The experiments showed that nuclei were capable of synthesizing the RNA. A major drawback to the experiments, however, was the labor required to obtain the oocyte nuclei from the ovaries. Not surprisingly, then, a simple cell extract of the entire oocyte was tried, and found to be almost as effective in synthesizing 5S RNA as the nuclear extracts. Generally chromatin, that is DNA plus associated proteins and histones, that is obtained by gently lysing cells and taking whatever sediments with DNA, is a good template DNA.

Oocyte nuclear extracts using either DNA or chromatin were active in synthesizing 5S RNA. When RNA polymerase III was tried instead of the extracts, polymerase plus DNA was inactive, but polymerase with chromatin as a template was active. These experiments demonstrate that at least one factor, presumably a protein, in addition to DNA and RNA polymerase III, is necessary for 5S RNA synthesis and that the factor is associated with chromatin. The question then became one of determining how many factors were required for synthesis of 5S RNA, how many of the factors were associated with chromatin, and whether properties of these factors explained either the shutoff of oocyte and somatic 5S RNA synthesis in the oocyte or the later resumption of only somatic 5S RNA synthesis.

Extracts prepared from unfertilized eggs do not support the synthesis of either type of 5S RNA from a pure DNA template. If these extracts are supplemented with extracts prepared from oocytes that were actively synthesizing 5S RNA, they do synthesize 5S RNA. Thus the unfertilized eggs provide an assay and the immature oocytes provide a source for one of the required factors. Through this sort of approach three factors have been found that stimulate synthesis of 5S RNA,

TFIIIA, TFIIIB, and TFIIIC. Other genes that are transcribed by RNA polymerase III, like some tRNA genes, the gene encoding the U6 RNA which is involved in splicing, and the adenovirus VA gene, utilize TFIIIB and TFIIIC only. TFIIIA was the easiest to purify and study, and therefore the most is known about it. This protein is monomeric and has a molecular weight 37,000.

TFIIIA Binding to the Middle of its Gene as Well as to RNA

Prokaryotic promoters are located upstream from their genes. Therefore it was logical to look in front of 5S genes for the sequences necessary for synthesis of 5S RNA. Finding these sequences could be approached with the tools of molecular genetics. Nuclear extracts capable of accurately synthesizing 5S RNA were used with pure DNA templates. A series of templates could be generated with progressively larger deletions approaching, and even entering the 5' end of the 5S gene. After deletion of the natural sequences, the plasmids were ligated closed and transformed into bacteria for biological synthesis of large quantities of the DNA. The product of the constructions was a substitution of plasmid sequence for what had been native *Xenopus* sequence ahead of the 5S gene (Fig. 15.3).

The deletion experiments were remarkable. Deleting all of the natural sequence ahead of the 5S sequence did not reduce synthesis of the 5S RNA to less than 50% of normal. Even when the first 40 nucleotides of the sequence were deleted, an RNA product of about the 120 nucleotide size of 5S RNA was synthesized. In this case the first 20 nucleotides synthesized were specified by plasmid DNA sequence (Fig. 15.4). Only when the deletions extended beyond about position 50 inside the 5S gene did they block synthesis.

The location of the downstream edge of the sequences necessary for 5S RNA synthesis is also of interest. The experiments to determine this position required a technical trick. In the experiments probing for the

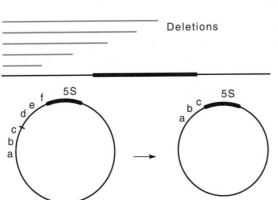

Deletions

Figure 15.3 Deletions upstream of a gene do not eliminate DNA, they substitute new sequences like *abc* for the deleted DNA *def*.

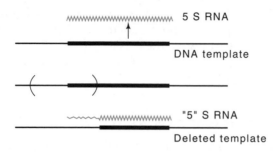

5 S RNA

DNA template

"5" S RNA

Deleted template

Figure 15.4 Deletion of part of the 5S gene and all upstream sequences still yields transcription of the gene. Transcription begins from approximately the same location as that used when the normal sequence is present.

sequences necessary for transcription initiation, the natural termination sequence of the 5S gene remained intact. Therefore, whatever the sequence of the RNA at its 5′ end, transcription still terminated normally and yielded an RNA molecule of about 120 nucleotides. The presence of this RNA could be determined simply by arranging that the RNA synthesized *in vitro* be radioactive and then running the resultant products on an acrylamide gel. When 5S RNA had been synthesized in a reaction, it formed a prominent band in the gel. Such a simple assay technique could not work if the sequences at the 3′ end of the gene that are required for termination were deleted.

Hybridization assays could be devised to look for RNA sequences from the remaining 5′ end of the deleted 5S gene. These assays however, would also detect any readthrough transcription resulting from incorrect initiation events occurring upstream of the 5S gene (Fig. 15.5). A better assay for 5S RNA synthesis was to force the 5S RNA to terminate prematurely, before it reached sequences that might be altered due to deletion of the 3′ end of the gene. These terminations could be generated by introducing a transcription terminator in the region between the start site of transcription and the beginning of the sequence already determined to be required for transcription (Fig. 15.6). A simpler approach,

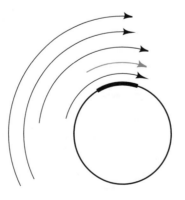

Figure 15.5 Mere total RNA including 5S sequences is not a good assay for activity of specific transcription of the 5S RNA gene due to nonspecific transcription which begins upstream of the gene.

5S

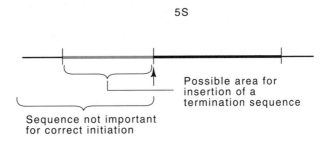

Possible area for
insertion of a
termination sequence

Sequence not important
for correct initiation

Figure 15.6 A region in which a transcription terminator could be placed for assay of specific transcription of the 5S gene.

but one based on this basic idea, was taken. This used the same trick as is used in DNA sequencing. In the triphosphates used for transcription was included a low concentration of a 3'-dideoxy-triphosphate. When a molecule of this analog was incorporated into the growing RNA chain, no further elongation was possible. At appropriately low concentrations of a 3'deoxy-triphosphate, a characteristic spectrum of sizes of prematurely terminated RNA molecules results (Fig. 15.7). Transcription initiating at other points on the chromosome generate other spectra of prematurely terminated molecules. Despite the presence of these groups of RNA molecules, the spectrum indicative of initiation from the 5S RNA promoter could be identified and therefore the activity of this promoter could easily be assayed.

The experiments probing for the location of the 3' end of the sequences necessary for normal initiation of the 5S gene were successful. The downstream edge of the sequences required was between positions 80 and 83. That is, within the gene, between positions 50 and 80 lies a stretch of 30 nucleotides which are required for correct synthesis of the 5S RNA.

Does TFIIIA bind within the interior of the 5S gene? Indeed it does. DNAse footprinting experiments show that it protects the region from about +50 to +80. Similarly, pre-ethylation experiments in which those phosphates are determined whose ethylation prevents or substantially

Figure 15.7 A spectrum of specific RNA sizes generated by transcription beginning at the 5' end of the 5S RNA gene and terminating at particular nucleotides.

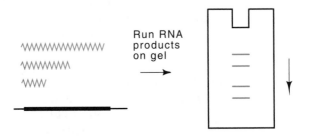

Run RNA
products
on gel

weakens binding of the protein also revealed the binding site of the protein to be entirely within the gene. These deletion and biochemical experiments show that TFIIIA determines where RNA polymerase III initiates on the 5S gene.

The TFIIIA protein revealed an additional unexpected property. It was noticed that an excess of 5S RNA in the *in vitro* transcription reactions inhibited transcription of the 5S gene. This type of result suggests that the RNA inhibits RNA polymerase. This possibility could be simply tested since RNA polymerase III also transcribes tRNA genes. If the 5S RNA inhibited RNA polymerase, it should block synthesis of both 5S RNA and tRNA. It did not. It blocked only 5S RNA synthesis. Therefore, 5S RNA must bind TFIIIA, thereby titrating TFIIIA from the DNA. Indeed, this is true and the dual binding ability of TFIIIA appears to be of biological importance as one of the proteins in the 7S complex of 5S RNA is TFIIIA.

Switching from Oocyte to Somatic 5S Synthesis

In vitro transcription experiments revealed an unexpected property of 5S transcription. Brown and co-workers discovered that once TFIIIA had bound to a DNA molecule, it did not easily or rapidly dissociate and bind to a different molecule. Even multiple rounds of transcription could not displace a bound TFIIIA molecule. Demonstrating this property is rather simple. Consider two different kinds of DNA templates for 5S RNA, one containing an insertion of a few bases so that the resulting product is distinguishably larger when run on polyacrylamide gels. When the two types of templates were added to a reaction together, both sizes of RNA were synthesized, but under conditions of limiting TFIIIA; whichever template was first added encoded all the resulting RNA. The TFIIIA remained associated with its DNA upon dilution or even through 40 rounds of transcription.

Complexes of TFIIIA formed *in vivo* also possess high stability. These experiments can be done in the following way. As increasing quantities of DNA containing the 5S gene are injected into oocytes, eventually the synthesis of 5S RNA saturates. This can be shown to be due to the titration of TFIIIA factor. Injection of TFIIIA factor relieves the saturation and permits synthesis of still more 5S RNA. Secondly, if template for tRNA, which does not require TFIIIA, is included in the second injection of 5S template along with template for 5S RNA, only the tRNA template is active. This shows that RNAP III has not been titrated out. Once it is known that TFIIIA is limiting, the inactivity of the 5S template added to an already template saturated oocyte means that no significant amount of TFIIIA dissociates from the first template during the experiment.

The somatic 5S genes, but not the oocyte 5S genes in chromatin isolated from somatic cells are active in some types of extracts. What maintains the set state of these two types of genes? Not surprisingly, the somatic genes contain bound TFIIIA. Not only do the oocyte genes lack TFIIIA, but provision of extra TFIIIA does not activate them. One explanation could be that histones on the DNA block access of TFIIIA to its binding site. Indeed, various histones can be stripped from chromatin by extraction in buffers containing high concentrations of salt. Extraction with 0.6 M NaCl removes histone H1 and at the same time permits TFIIIA activation of oocyte 5S genes. Similarly, stripping the chromatin of histones by high salt, addition of TFIIIA, and then reformation of chromatin by addition of the histones and slow removal of the salt leave the oocyte genes active.

We can generate a simple picture to explain the histone results. After replication, the DNA is bare behind the replication fork. Under these conditions both TFIIIA and histones should be able to bind to the 5S gene. If TFIIIA binds first, a complete transcription complex forms. This possesses sufficient stability to remain in place for long periods. Once it has the complex is present, nucleosomes are free to form on either side. The presence of the TFIIIA and the absence of histone bound at the 5S gene means that transcription can begin despite the presence of histones elsewhere on the template. Conversely, if histones bind first to the gene, they remain bound and prevent formation of a transcription complex. In this case the 5S gene cannot be activated, even by the subsequent addition of large amounts of TFIIIA. Hence the long-term transcriptional abilities of the 5S genes could be determined at the time of their replication.

Just as a transcription complex must be present to prevent histones from binding to the critical part of the 5S gene, those genes transcribed by RNAP II must also be prevented from being locked in an off state. General or specific transcription factors must perform this role for mRNA genes. One attractive candidate for a general transcription factor that would often serve this role is TFIID, which binds to the TATA box of the promoter region of genes transcribed by RNA Pol II.

The high stability of TFIIIA binding raises the possibility of a simple explanation for the switch between oocyte and somatic 5S RNA synthesis. In principle, the developmental switch could be simply accomplished if TFIIIA itself had a substantial binding preference for somatic genes. Then, in the developing oocyte with its vast excess of TFIIIA, both somatic and oocyte genes would be active. However, as TFIIIA is titrated out into the 7S and 42S particles, the functional excess would vanish and the genes would be occupied by TFIIIA in proportion to the protein's binding preference for the two slightly different sequences. The main problem with this explanation is that, experimentally, TFIIIA has too small a preference for the somatic genes to explain the thousand-fold change in the relative activities of the two types of genes. If the oocyte sequence of the 5S gene itself is changed nucleotide by nucleotide to that of the somatic 5S, there is some effect in the correct direction. The changes at positions 47, 53, 55, and 56 of the gene do make the oocyte gene behave more like the somatic genes, in oocyte extracts in which the activity of TFIIIA is being monitored, but the change is insufficient to explain the dramatic difference between the two genes *in vivo*. The conclusion has to be that the nucleotide differences between the somatic and oocyte 5S RNA genes are sensed by the entire transcription complex consisting of TFIIIA, B, C, and RNA polymerase III.

Structure and Function of TFIIIA

One of the motivations for studying regulation of the 5S genes was the hope that phenomena totally unheard of in prokaryotes would be found in eukaryotes, and the first well-studied eukaryotic system seemed likely to yield a rich harvest. For the most part, what has been found has turned out to be phenomena that are not unique either to the 5S genes or to eukaryotes. For example, the stable state that is determined by a protein or complex of proteins that remain bound for long periods of time has proven to be a valuable concept, but it is not unique either to eukaryotes or to the 5S system. Even the startling discovery of an internal control region turns out not to be unique, either to eukaryotes or to genes transcribed by RNA polymerase III. The zinc fingers, which were first found in TFIIIA, are a common structure in eukaryotes, but they also exist in prokaryotes.

The discovery of the zinc fingers is illustrative of the unexpected source for many discoveries. While exploring the cause for substantial losses during the purification of TFIIIA from 7S particles, workers noticed that steps like gel filtration that could separate the protein from small molecules led to substantial loss of TFIIIA activity. Also, addition of metal ion chelators increased losses. Prior addition of Zn^{++}, but not other ions, prevented the losses. With this strong clue that zinc ion was involved, they performed atomic adsorption spectroscopy and found indeed that each molecule of the protein contained five to ten zinc ions.

The presence of multiple ions in the protein suggested that an ion-binding substructure or domain might be repeated within the protein. This suspicion was reinforced by the finding that upon extended proteolytic digestion of the protein there remained multiple species of a 3,000-4,000 dalton peptide. Upon examining the sequence of the protein for a repeated sequence of about 30 amino acids, one was found. The marvelous property of these repeating units was that they possessed

Figure 15.8 Locations of cysteines and histidines in zinc fingers.

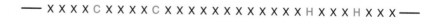

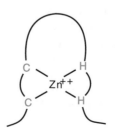

two cysteines and two histidines spaced just as required for these four amino acids to chelate zinc (Fig. 15.8), as was discussed in Chapter 6.

Since TFIIIA possesses nine zinc fingers, and each can contact three bases of of the DNA, the protein can contact part of the 40 bases of the essential internal control region in the 5S gene. The portion of the protein that does not contact DNA likely contacts one or both of the other proteins required for most active transcription of 5S genes, TFIIIB or IIIC. There is a 10,000 dalton portion of the protein at the N-terminal end that probably is involved in this function since it does not possess zinc fingers and its removal eliminates transcriptional activation by TFIIIA.

The structure of a zinc finger was predicted by comparison with other proteins containing liganded metal ions and close examination of the sequences of zinc fingers. Subsequent determination of the structure of artificial zinc fingers by nuclear magnetic resonance methods and X-ray crystallography confirmed the predicted structure.

Problems

15.1. If the unfertilized egg trick had not worked in developing the assay for 5S transcription factors, how could you proceed to assay and purify the factors?

15.2. How many ribosomes does a mature *Xenopus* egg contain?

15.3. Which is most relevant to the developmental regulation shown by 5S RNA—the rate of binding of TFIIIA to the 5S genes or the equilibrium constant for its binding?

15.4. How was the existence of two different 5S RNA sequences discovered?

15.5. How could you use PCR technology to probe for synthesis of RNA from a modified 5S gene lacking its normal transcription terminator?

15.6. Comment on the absence of free 5S RNA in developing eggs and the significance of "stable states."

15.7. Suppose you could perform genetics operations with *Xenopus* at the same time scale and with the same numbers of individuals as you can with *E. coli*. How would you proceed to determine the basis for shutoff of oocyte 5S RNA synthesis in somatic cells?

15.8. What experiments would you like to do to investigate why nature has gone to so much trouble to separate the synthesis of 5S RNA from that of the other ribosomal components?

15.9. How would you prove or disprove the possibility that in the determined state transcription experiments, TFIIIA actually is displaced from its DNA molecule by RNA polymerase III; but, at the DNA concentrations used in the experiments, the most probable DNA molecule for TFIIIA to bind to a second time is the very molecule from which it was displaced?

15.10. How do we know that RNAP III transcribes 5S genes *in vivo*?

References

Recommended Readings

Repetitive Zinc-binding Domains in the Protein Transcription Factor from *Xenopus* Oocytes, J. Miller, A. McLachlan, A. Klug, EMBO Journal *4*, 1609-1614 (1985).

The Primary Structure of Transcription Factor TFIIIA Has 12 Consecutive Repeats, R. Brown, C. Sander, P. Argos, Federation of European Biochemical Societies Letters *186*, 271-274 (1985).

Transcription Factor TFIIIC Can Regulate Differential *Xenopus* 5S RNA Gene Transcription *In Vitro*, A. Wolffe, Eur. J. Mol. Biol. *7*, 1071-1079 (1988).

Developmental Regulation of Two 5S Ribosomal Genes,, A. Wolffe, D. Brown, Science *241*, 1626-1632 (1988).

5S Genes, Structure, and *in vitro* Regulation

Non-coordinated Accumulation and Synthesis of 5S Ribonucleic Acid by Ovaries of *Xenopus laevis*, P. Ford, Nature *233*, 561-564 (1971).

Different Sequences for 5S RNA in Kidney Cells and Ovaries of *Xenopus laevis*, P. Ford, E. Southern, Nature New Biology *241*, 7-12 (1973).

Isolation of a 7S Particle from *Xenopus laevis* Oocytes: a 5S RNA-protein Complex, Proc. Natl. Acad. Sci. USA *76*, 241-245 (1979).

A Control Region in the Center of the 5S RNA Gene Directs Specific Initiation of Transcription: I. The 5' Border of the Region, S. Sakonju, D. Bogenhagen, D. Brown, Cell *19*, 13-25 (1980).

Onset of 5S RNA Gene Regulation During *Xenopus* Embryogenesis, W. Wormington, D. Brown, Dev. Biol. *99*, 248-257 (1983).

Transcription Complexes that Program *Xenopus* 5S RNA Genes Are Stable *in vivo*, M. Darby, M. Andrews, D. Brown, Proc. Natl. Acad. Sci. USA *85*, 5516-5520 (1988).

Whole Genome PCR: Application to the Identification of Sequences Bound by Gene Regulatory Proteins, K. Kinzler, B. Vogelstein, Nucleic Acids Res. *17*, 3645-3653 (1989).

Transcription Factors

A Specific Transcription Factor That Can Bind Either the 5S RNA Gene or 5S RNA, H. Pelham, D. Brown, Proc. Natl. Acad. Sci. USA 77, 4170-4174 (1980).

Multiple Factors are Required for the Accurate Transcription of Purified Genes by RNA Polymerase III, J. Segall, T. Matsui, R. Roeder, J. Biol. Chem. *255*, 11986-11991 (1980).

Specific Interaction of a Purified Transcription Factor with an Internal Control Region of 5S RNA Genes, D. Engelke, S. Ng, B. Shastry, R. Roeder, Cell *19*, 717-728 (1980).

Related 5S RNA Transcription Factors in *Xenopus* Oocytes and Somatic Cells, H. Pelham, W. Wormington, D. Brown, Proc. Natl. Acad. Sci. USA *78*, 1760-1764 (1981).

The Binding of a Transcription Factor to Deletion Mutants of a 5S Ribosomal RNA Gene, S. Sakonju, D. Brown, D. Engelke, S. Ng, B. Shastry, R. Roeder, Cell 23, 665-669 (1981).

Contact Points Between a Positive Transcription Factor and the Xenopus 5S Gene, S. Sakonju, D. Brown, Cell 31, 395-405 (1982).

Domains of the Positive Transcription Factor Specific for the Xenopus 5S RNA Gene, D. Smith, I. Jackson, D. Brown, Cell 37, 645-652 (1984).

A Positive Transcription Factor Controls the Differential Expression of Two 5S RNA Genes, D. Brown, M. Schlissel, Cell 42, 759-767 (1985).

Structural Homology of the Product of the Drosophila krüppel Gene with Xenopus Transcription Factor IIIA, U. Rosenberg, C. Schroder, A. Preiss, A. Kienlin, S. Cote, I. Riede, H. Jackle, Nature 319, 336-339 (1986).

An Elongated Model of the Xenopus laevis Transcription Factor IIIA-5S Ribosomal RNA Complex Derived from Neutron Scattering and Hydrodynamic Measurements, P. Timmins, J. Langowski, R. Brown, Nucleic Acids. Res. 16, 8633-8644 (1988).

The Role of Highly Conserved Single-stranded Nucleotides of Xenopus 5S RNA in the Binding of Transcription Factor IIIA. P. Romaniuk, Biochemistry 38, 1388-1395 (1989).

Transcription Experiments

Multiple Forms of Deoxyribonucleic Acid-dependent Ribonucleic Acid Polymerase in Xenopus laevis, R. Roeder, J. Biol. Chem. 249, 249-256 (1974).

Selective and Accurate Transcription of the Xenopus laevis 5S RNA Genes in Isolated Chromatin by Purified RNA Polymerase III, C. Parker, R. Roeder, Proc. Natl. Acad. Sci. USA 74, 44-48 (1977).

A Nuclear Extract of Xenopus laevis Oocytes That Accurately Transcribes 5S RNA Genes, E. Berkenmeier, D. Brown, E. Jordan, Cell 15, 1077-1086 (1978).

Transcription of Cloned Xenopus 5S RNA Genes by X. laevis RNA Polymerase III in Reconstituted Systems, S. Ng, C. Parker, R. Roeder, Proc. Natl. Acad. Sci. USA 76, 136-140 (1979).

A Quantitative Assay for Xenopus 5S RNA Gene Transcription, W. Wormington, D. Bogenhagen, E. Jordan, D. Brown, Cell 24, 809-817 (1981).

Stable Transcription Complexes of Xenopus 5S RNA Genes: A Means to Maintain the Differentiated State, D. Bogenhagen, W. Wormington, D. Brown, Cell 28, 413-421 (1982).

Purified RNA Polymerase III Accurately and Efficiently Terminates Transcription of 5S RNA Genes, N. Cozzarelli, S. Gerrard, M. Schlissel, D. Brown, D. Bogenhagen, Cell 34, 829-835 (1983).

The Transcriptional Regulation of Xenopus 5S RNA Genes in Chromatin: The Roles of Active Stable Transcription Complexes and Histone H1, Cell 37, 903-913 (1984).

DNA Replication in vitro Erases a Xenopus 5S RNA Gene Transcription Complex, A. Wolffe, D. Brown, Cell 47, 217-227 (1986).

Differential 5S RNA Gene Expression in vitro, A. Wolffe, D. Brown, Cell 51, 733-740 (1987).

Effect of Sequence Differences Between Somatic and Oocyte 5S RNA Genes on Transcriptional Efficiency in an Oocyte S150 Extract, W. Reynolds, Mol. Cell. Biol. 8, 5056-5058 (1988).

Kinetic Control of 5S RNA Gene Transcription, C. Seidel, L. Peck, J. Mol. Biol. *227*, 1009-1018 (1992).

Zinc Fingers

EXAFS Study of the Zinc-binding Sites in the Protein Transcription Factor IIIA, G. Diakun, L. Fairall, A. Klug, Nature *324*, 698-699 (1986).

Potential Metal-binding Domains in Nucleic Acid Binding Proteins, J. Berg, Science *232*, 485-487 (1986).

Proposed Structure for the Zinc-binding Domains from Transcription Factor IIIA and Related Proteins, J. Berg, Proc. Natl. Acad. Sci. USA *85*, 99-102 (1988).

Mapping Functional Regions of Transcription Factor TFIIIA, K. Vrana, M. Churchill, T. Tullius, D. Brown, Molecular and Cellular Biology *8*, 1684-1696 (1988).

Three-dimensional Solution Structure of a Single Zinc Finger DNA-binding Domain, M. Lee, G. Gippert, K. Soman, D. Case, P. Wright, Science *245*, 635-637 (1989).

Determination of the Base Recognition Positions of Zinc Fingers from Sequence Analysis, G. Jacobs, EMBO Journal *11*, 4507-4517 (1992).

Regulation of Mating Type in Yeast

16

Rapid progress has been made in learning the basic structure and properties of a great many regulated genes and gene families of eukaryotic cells. Much of the progress is attributable to the power of genetic engineering, which has made the task of cloning and sequencing many eukaryotic genes relatively straightforward. Once a gene has been cloned, its own sequences, and sequences governing its regulation can be determined. In the most amenable cases, sequences of the gene and of its promoter and regulatory elements, and any sequences the gene product interacts with, may all be deleted or altered. In some cases the DNA can be reintroduced into the original organism to examine the consequences of the alterations.

After the effects of DNA modifications on the expression of a gene have been explored, the effect on the organism of the altered regulation can be investigated and the biochemical mechanisms underlying the regulation can be more easily studied. In a few systems the depth of our understanding of eukaryotic gene regulation approaches or exceeds that of prokaryotic genes. One example of a well-studied eukaryotic system is the mating-type genes of baker's yeast *Saccharomyces cerevisiae*.

A haploid yeast cell contains both *a*- and α-mating type genetic information. Although normally only one or the other is expressed, a cell can switch from one to the other. This appears to be a more complicated example of differentiation than that considered in the previous chapter, but one that is still relatively simple. Nonetheless, it appears to contain many of the elements found in the much more complicated differentiation systems of higher eukaryotes. One of the virtues of this system is that both genetic and biochemical approaches may be applied to it.

The Yeast Cell Cycle

Yeast come in three cell-types, the haploid *MATa* and *MATα* mating types, and diploid *MATa/α*. Haploid cells of opposite cell types can conjugate to form a diploid *a/α* cell which is incapable of further mating. Under starvation conditions the *a/α* diploid, but not haploid cells, undergo meiosis to generate two *a* and two α mating-type spores in a structure called an ascus.

The status of the mating type of a clone or colony can easily be

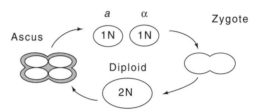

determined by mating with tester strains of known mating type (Fig. 16.1). Colonies to be checked should possess at least one nutritional requirement for growth. These colonies can be grown on rich plates containing all necessary nutrients and then replica plated onto lawns of tester cells of mating type *MATa* and of mating type *MATα* that possess different nutritional requirements. After several hours of mating on rich medium, the plates are replicated onto minimal medium plates lacking the nutritional requirements of both the colonies to be checked and the tester strains. Combinations of parental cells that can mate will yield diploid conjugants capable of growing on the minimal medium through complementation of the defective alleles.

Figure 16.1 Scheme for the detection of colonies capable of mating with *a* and with α mating type cells.

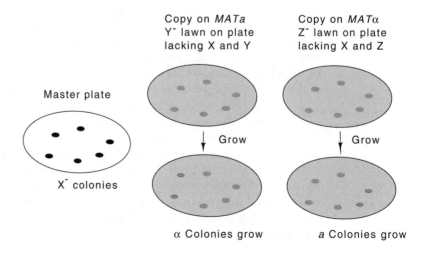

Most yeasts isolated from nature change their mating type nearly every generation. A culture of such yeast will therefore contain cells of *a* and α mating type that can conjugate to produce *a*/α diploids which grow, but which do not mate further or change mating type. These switching cells are therefore called homothallic for their capability of mating with sister cells. To assay the current mating type status of such homothallic yeast, the behavior of individual cells must be examined. Fortunately this is straightforward since cells of opposite mating type expand towards one another in a characteristic shape called a shmoo which is easily recognized in the microscope.

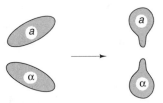

Yeast that are to conjugate must induce and repress appropriate genes so as to synchronize their cell cycles, grow and adhere together, and finally fuse both cytoplasm and nucleus. Cells of one mating type sense the nearby presence of the opposite mating type by pheromone peptides of 12 and 13 amino acids long. Each cell-type secretes one pheromone and responds to the pheromone secreted by the opposite cell-type. The pheromone receptors span the membrane and are similar to rhodopsin and β-adrenergic receptors. On their cytoplasmic side the receptors interact with signaling proteins known as G proteins and through cascades of regulatory proteins, ultimately induce and repress the genes necessary for conjugation.

Mating Type Conversion in *Saccharomyces cerevisiae*

Yeast possessing the *ho* allele switch from mating type *MATa*, to *MATα* or vice versa at a frequency of about 10^{-6} per doubling time, whereas yeast possessing the *HO* allele, the form most commonly found in nature, switch mating types as frequently as once each generation.

As in the study of many biological phenomena, the study of mutants revealed hidden details of the mating-type system. *Saccharomyces* mating-type mutants showed an astonishing fact. If an *ho* strain which is *MATα*⁻, making it defective in mating, is switched to *MATa* and then switched back to *MATα*, the original defect, the *MATα*⁻ mutation, disap-

$$\alpha^- \longrightarrow a^+ \longrightarrow \alpha^+ \longrightarrow a^+ \longrightarrow \alpha^+ \ldots$$

pears! The original *MATα*⁻ allele cannot be found, no matter how many additional times the *a*-α conversion is performed. New *MATα*⁻ mutations can be isolated in the healed strain, and these too heal upon mating-type switching. The only reasonable conclusion is the one first postulated by

Chromosome III

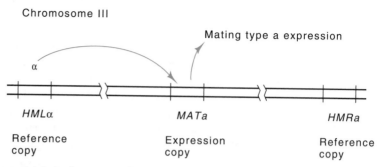

Figure 16.2 Reference and expression copies of yeast mating-type genes. In this example α mating type information is being copied from *HMLα* to *MAT* to switch the *a* mating type cell to α mating type.

Oshima and Takano–that the cells contain nonexpressed reference copies of the *MATa* and *MATα* information on the chromosome in addition to the expressible forms of *MATa* or *MATα* information. The reference sites are named *HML* and *HMR*, and the expression site is named *MAT* (Fig. 16.2). Usually, but not always, *HML* contains α mating-type information, and *HMR* contains *a* type information. As explained below, direct physical data have confirmed this conclusion.

Cloning the Mating Type Loci in Yeast

How could the existence of reference copies of mating-type information at *HML* and *HMR* and an expression copy at *MAT* be proven physically? The most straightforward proof is simply to clone DNA copies from the three loci and use Southern transfers to show that the DNA occupying the *MAT* locus comes from *HML* or *HMR* and changes with a shift in mating type.

Cloning the mating-type loci required both a convenient genetic selection for the presence of mating-type genes and a suitable vector. Due to the low efficiency of yeast transformation, unreasonably large

Figure 16.3 Scheme for cloning yeast mating type genes.

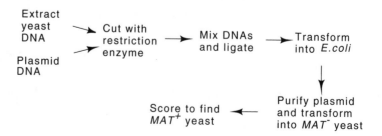

quantities of yeast DNA, restriction enzymes, and ligase would have been required if investigators had tried directly to transform yeast. Instead, the yeast DNA was ligated into a bacterial-yeast shuttle vector, transformed into bacteria, amplified to suitable levels, extracted, and this was used to transform the yeast (Fig. 16.3).

The genetic selection for mating-type DNA was straightforward. The source DNA derived from yeast capable of mating, that is, having a functional *MAT* locus. The amplified plasmid DNA was transformed into a yeast strain possessing a defect in the *MAT* locus causing sterility. Therefore, after the transformation, the yeast having received a plasmid carrying the *MAT* locus would become capable of mating and could easily be detected. The plasmid could then be isolated from the yeast and transferred into *E. coli* for amplification.

Transfer of Mating Type Gene Copies to an Expression Site

Once the *MATα* mating-type clone had been obtained by Hicks, Strathern, and Klar, and by Nasmyth and Tatchell, Southern transfer experiments testing DNA transposition to the *MAT* locus were possible. Yeast DNA was cut with various restriction enzymes and was separated according to size by electrophoresis. The locations of the α-specific or *a*-specific sequences were determined by hybridization using the cloned α or *a* sequences as probe.

We might expect that the α mating-type locus would have no homology to the *a* mating-type locus. Then, in the DNA from a *MATα* cell there would be two different DNA fragments with homology to the *MATα* locus, one from *MATα* itself, and one from *HML* or *HMR*. From a *MATa* cell we would expect only one fragment with homology to the α mating type sequence. Surprisingly, a different answer emerged. The α sequence also possesses partial homology to *a* mating-type sequences. Therefore three different sizes of restriction fragments were observed in the DNA from *MATa* or *MATα* cells probed with either *a* or α sequences (Fig. 16.4). These are sequences at *HML*, *HMR*, and *MAT*.

The fact that the *a*-specific and α-specific sequences were partially homologous, but of different size permitted a direct and simple demon-

Figure 16.4 Cleavage of DNA containing the *HML*, *HMR*, and *MAT* loci by a restriction enzyme which does not cleave within the loci..

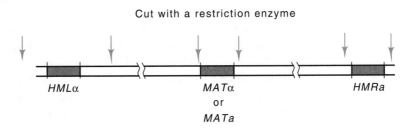

Cut with a restriction enzyme

HMLα   *MATα*   *HMRa*
                            or
                          MATa

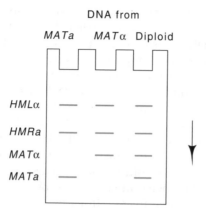

DNA from

MATa   *MATα* Diploid

HMLα

HMRa

MATα

MATa

Figure 16.5 Southern transfer showing that *MATa* sequences are about 100 base pairs shorter than *MATα* sequences.

stration of the basis of mating-type conversion. The genetic data indicated that sequences from *HML* or *HMR* were copied into *MAT*, where they were expressed. Because the *a* mating-type sequence is about 100 nucleotides shorter than the α mating type, the identity of the sequence occupying the *MAT* locus can be directly determined on Southern transfers (Fig. 16.5). The experiments showed, as expected, that mating type *a* cells possessed the *a* sequences at *MAT*, and that mating type α cells possessed the α sequences at *MAT*. The Southern transfer of a diploid yeast that possesses a *MATa*/MATα genotype yields two bands originating from the *MAT* locus that differ in size by 100 base pairs. The identity of the shorter and longer mating-type segments was determined by isolating spores, growing, and testing both genetically as well as performing a transfer. The *a*-type cultures possess only the shorter *MAT* sequence and the α-type cultures possess only the longer *MAT* sequence.

Structure of the Mating Type Loci

Cloning the mating-type sequences permitted direct determination of their structures. DNA-DNA heteroduplexes revealed the overall structural similarities and differences, while DNA sequencing was used to examine the detailed aspects of the structures. The *a*- and α-specific sequences are denoted by *Ya* and *Yα* are relatively small, only about 800 base pairs long. In addition to these regions, the *HML*, *MAT*, and *HMR* loci are flanked by common sequences *W, X, Y, Z1* and *Z2*.

The RNA transcripts from either of the *MAT* regions originate from near the center and extend outward in both directions, giving transcripts a1 and a2 or α1 and α2. These transcripts were identified by S1 mapping by extracting RNA from cells, hybridizing it to end-labeled DNA fragments, and digesting the remaining single-stranded RNA and DNA. Measurement of the size of the DNA that was protected from digestion by the RNA and knowledge of the locations of the labeled end of the DNA fragment give the transcription start points. On the basis of genetic complementation tests, the α1, α2, and a1 transcripts are translated into protein, but the a2 is not translated. Consistent with this conclusion is

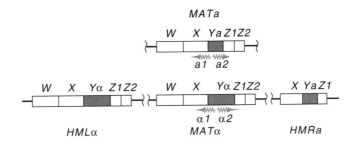

the fact that no mutations have been found in the a2 region and the a2 transcript lacks a good open reading frame preceded by an AUG codon.

The Expression and Recombination Paradoxes

Two important questions are raised by the structure of the mating-type loci and the locations of transcripts in *MATa* and *MATα*. First, how is transposition always forced to proceed from *HML* or *HMR* to *MAT* and never the reverse, and second, how is expression of *a* or *α* from *HML* or *HMR* prevented?

Relatively little is known about the biochemistry of the transposition reaction. The available data suggest that a direct transfer of DNA through a free diffusible intermediate does not occur, but instead that mating-type shift is a result of gene conversion (See Chapter 8) type of substitution reaction in which the appropriate information from *HML* or *HMR* is copied into *MAT* (Fig. 16.6). In part, this conclusion is based on the failure to find free DNA copies of sequences from *HML* or *HMR* in yeast.

One simple method for expressing sequences at *MAT* and not expressing the sequences at *HML* or *HMR* would have been for the region around *MAT* to provide a promoter that specifies transcription across whatever sequence has been inserted at *MAT*. This is not the explanation, however, for transcription of α and *a* genes begins from within the mating type-specific *Y* sequences. Another possibility is that an enhancer is near the *MAT* region so that whatever is inserted into *MAT* is expressed. It is also possible that repression occurs at *HML* and *HMR*.

Figure 16.6 Folding the chromosome to bring *HMR* near *MAT* so copied sequence information can be passed directly from *HMR* to *MAT*.

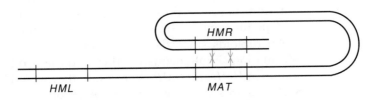

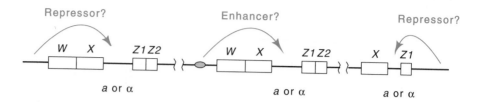

Figure 16.7 Something has to act at a distance to regulate the promoters of mating type genes. Either repression acts from afar at *HML* and *HMR* or activation acts over a distance at *MAT*.

Since it is hundreds of nucleotides from the mating type-specific promoter to sequences unique to the *HML, HMR* or *MAT* loci, activation or repression would have to occur over a substantial distance (Fig. 16.7). While such "action at a distance" events are frequently found for activation of transcription, they are less often seen for repression of transcription. Nonetheless, genetics experiments revealed that it is repression at *HML* and *HMR* that actually occurs. This same repression or silencing mechanism of the mating-type promoters also blocks transfer of the donor information at *HML* and *HMR* to *MAT*.

Silencing *HML* and *HMR*

The roles of the mating-type genes have been defined for cell-type determination. α1 is an activator of α mating-type genes, α2 is a repressor of *a* mating-type genes, and *a1* plays no role in haploid cells (Fig. 16.8). To generate the *a* and α mating types it would suffice if α1 product activated α type genes, which otherwise would be off, and if α2 product repressed *a* type genes, which otherwise would be on. This, however, would not provide for correct expression in diploid cells where the haploid-specific genes must be repressed and diploid-specific genes must be turned on. Genetic experiments examining the expression of mating type-specific genes have revealed that in addition to the pattern mentioned above, in diploid cells the *a1* product and α2 product combine to form a repressor of haploid-specific genes and activator of diploid-specific genes. Without such a repressor, α type genes would be expressed in diploids. One way to start to work out the control circuitry is to isolate mutations that disrupt the normal regulation patterns.

If proteins repress expression of genes located in *HML* and *HMR*, then mutations in such repressor proteins should permit expression of mating-type information from these loci. Such mutant cells could behave as *a*/α diploids since they would express both *a* and α mating-type information simultaneously if they had the two types of mating information at *HML* and *HMR*. As a result, they would be sterile for mating. Such mutants have been found. A second approach to isolation of repression negative mutants is to begin with cells that are *HMLα-MATa⁻*

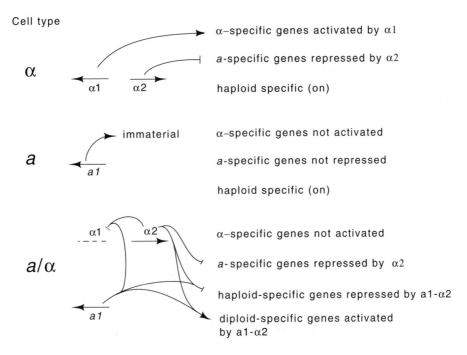

Cell type

Figure 16.8 The various gene activating and repressing activities necessary to regulate *a*- and α-specific, haploid-specific, and diploid-specific genes in the three cell types of yeast. Note that α1 itself is a haploid-specific gene.

-*HMR*α (Fig. 16.9). Even though such cells express no active mating-type information, they are of *a* mating type by default. Now, if a mutant loses

Figure 16.9 Cells can be made to switch mating type with the loss of repression at *HM* loci if they possess an a1⁻ allele at *MAT* and cannot transfer copies from *HML* or *HMR* to *MAT*. As a result they switch from constitutively expressing *a* type genes to repressing these genes and expressing genes activated by α1.

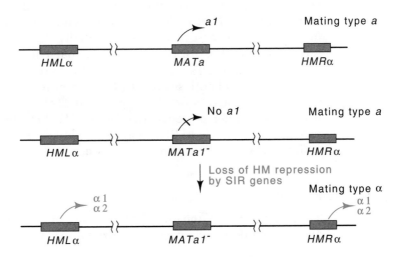

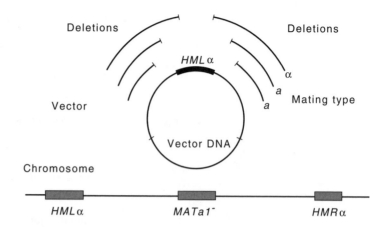

Figure 16.10 Yeast of the genotype shown containing a vector carrying *HMLα* are *a* mating type unless a deletion enters the region necessary for repression of expression from *HMLα*.

the ability to repress at the *HML* or *HMR* loci, then that cell becomes α mating type.

Cells possessing the necessary genotype were mutagenized and colonies were screened for a shift from *a* to α mating type. This was a simple replica plating assay for mating ability. Amongst the large number of candidates screened, four repressor genes turned up. These are called the *MAR* or *SIR* genes. The existence of these genes and the behavior of strains mutant in these genes proves that repression prevents expression of mating-type genes at the *HML* and *HMR* loci.

The next question that can be asked about repression of the mating-type information is where are the DNA sites that are required to generate the repression? These must lie in the DNA sequences unique to the *HML* and *HMR* loci, for otherwise the repressor *SIR* genes would repress mating-type information at *MAT* as well. Precisely where they lie could be found by deletion mapping. This was most easily done by placing an entire *HML* or *HMR* region on a yeast shuttle vector which could be introduced into yeast possessing essentially the same genotype as described above (Fig. 16.10). A deletion that eliminates repression of *HMLα* shifts the cell's mating type from *a* to α. Such sites lie on both sides of *HML* and both sides of *HMR*. Currently the best guess as to the mechanism of action of silencing is that *MAR* or *SIR* gene products, histone, and other proteins organize the *HM* loci chromatin such that they are inaccessible to RNA polymerase but accessible for interaction with the *MAT* locus for gene conversion.

Isolation of α2 Protein

The combination of the low synthesis levels of most regulatory proteins and the fact that few possess any enzymatic activity makes their detection difficult. The *lac* repressor was first detected by virtue of its binding to iso-propyl-thio-galactoside, IPTG, an inducer of the system that is not cleaved by β-galactosidase. No small molecule effector of the mating-type factors is known, and as will be seen below, none is necessary. Some other method had to be used to identify the α protein.

One approach to the isolation of mating type encoded factors could be the DNA migration retardation assay using short DNA fragments containing an appropriate binding site. If the protein binds, the electrophoretic migration rate of the fragment is greatly reduced. This assay is sufficiently selective that it can detect a specific DNA-binding protein present at only a tiny fraction of total protein. That is, the assay often can detect a specific DNA-binding protein in crude extracts of the cells. Nowadays this assay is the method of choice for initial attempts to detect proteins that bind to DNA.

Since α2 protein does not come equipped with a convenient enzyme assay and the DNA retardation assay was not then in wide use, Herskowitz and Johnson used a clever trick for the isolation of α2. They

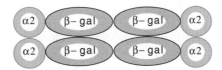

fused the entire α2 gene to β-galactosidase. The fusion protein functions *in vivo* to repress *a*-specific genes normally, and the β-galactosidase also retains its activity. Therefore, the enzyme can be used both to assist the purification by its known behavior in purification steps, and to assay protein fractions for the hybrid protein during the purification. Despite susceptibility to proteolytic cleavage in the cell extracts, the fusion protein could be partially purified by conventional techniques by tracking the β-galactosidase activity through conventional fractionation steps.

Two assays of DNA binding by α2 fusion protein could be used. The first links a radioactive DNA fragment containing an α2-binding site to *Staphlococcus aureus* cells via antibody and α2-β-galactosidase in a DNA–α2-β-gal-antibody-cell sandwich (Fig. 16.11). The bacteria and their associated complex can be sedimented with a low speed centrifugation and separated from other proteins and from control DNA fragments not containing an α2-binding site. DNA isolated from the complex was then subjected to electrophoresis to verify that only the correct DNA fragment was sedimented in the complex. Another assay for specific binding of α2 protein was DNAse footprinting.

The binding of α2 to its recognition sequence represses expression of *a* mating-type genes. Might α2 be able to repress other genes if its binding sequence were placed in the regulatory region? The answer was yes, but the results were not simple. We can imagine repression as

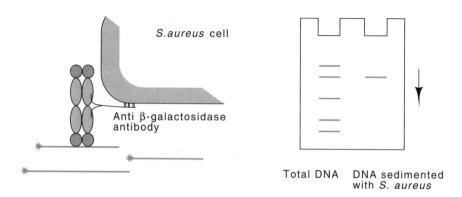

Figure 16.11 Antibody against β-galactosidase that binds to *Staphlococcus aureus* cells can couple α2 protein bound to DNA carrying the α2 specific sequence. The specific DNA can then easily be separated from other DNA fragments. The selectivity can be displayed by electrophoresis and autoradiography of the radioactive DNA fragments.

resulting from steric exclusion of the binding of another protein. This was not the case for α2 repressor. Introducing its binding site anywhere upstream of the cytochrome c gene promoter generated repression. The binding sequence could be between the enhancer and the TATA box, or even somewhat upstream of the enhancer element. It did not have to overlap another protein-binding site. Thus, α2 seems to repress by a different mechanism than by simple steric exclusion.

α2 and MCM1

DNAse footprinting revealed an unexpectedly large binding site for α2. It had two protected regions around a central unprotected region. Since the central protected region possessed a symmetric sequence, it seemed possible that an additional protein bound there.

Two kinds of experiments could be used to look for a possible second protein. The first was to ask whether the central region of the α2-binding site is involved in repression, and the second was to dissect α2 protein into functional domains. It is relatively straightforward to construct and insert altered α2-binding sites in front of genes. When this was done, it was found that if their sites were altered in the central region, repression did not occur despite the binding of α2. This means that either the central region affects the activity of the bound protein without affecting its binding, or that an additional protein binds in this region by making use of the central region DNA sequence.

The DNA migration retardation assay was used to look for a protein that binds with α2 to repress genes. One was found. It could be purified, and its binding to the site studied. It binds cooperatively with α2 protein.

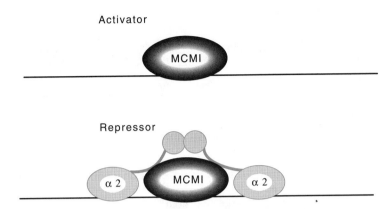

Figure 16.12 The MCM1 protein functions as an activator when alone, and as a repressor when it is flanked by the α2 protein.

The sequence to which it binds is found in front of some other genes whose activity is regulated by the mating type of the cells. In some cases the protein is an activator, and in others it is a repressor. When the α2 sequences flank the site, the complex acts as a repressor (Fig. 16.12). This protein, which has finally come to be called MCM1 has also been found in other types of cells. Its close relative is found in human cells where it is a transcription factor that activates genes in response to the presence of serum and is known as the serum response factor.

The α2 protein could also be dissected by deleting DNA segments from a plasmid containing the α2-β-galactosidase fusion gene. The resulting proteins lost their DNA-binding activities only when a 60 amino acid portion near the carboxyl terminus of the protein was deleted. These proteins lost their ability to repress, however, when short stretches of amino acids near the N-terminus were deleted. The resulting proteins could still bind to DNA. Thus they were synthesized, stable, capable of binding to DNA, but had lost a domain required for repression.

Sterile Mutants, Membrane Receptors and G Factors

One method for exploring the complexity of the mating-type system is to determine the range of defective mutant types that can be isolated. Additionally, mutants greatly facilitate biochemical studies, as they often permit associating a particular biochemical defect with a response of the system. Sterile mutants are particularly easy to isolate. Wild-type haploid cells cease growth in the presence of the opposite mating type sex pheromone. Therefore, any cells that continue growth must be defective in their detection of or response to the sex pheromones. Suitable pheromone for this selection can be obtained from the medium by growing α mating type cells to a high density, although chemically synthesized α type pheromone is now cheaper and more pure. The *a*

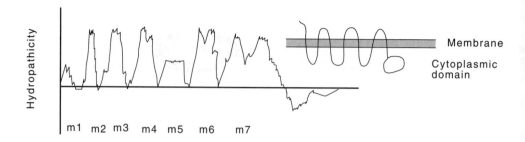

Figure 16.13 A representation of the hydropathicity plot of mating factor receptor. These plots are running averages of the hydropathicity of seven amino acids. A large hydropathicity indicates a membrane spanning region of the protein.

pheromone cannot easily be synthesized because it is only active when when post-translationally modified to a form not easily synthesized *in vitro*.

Amongst the sterile mutants which can be found by resistance to growth factor inhibition are those lacking the receptor for mating-type pheromone. These mutants can be identified by their lack of binding of radioactive factor to intact cells. Of course, once such defective mutants are found, the gene responsible can be isolated. This has been done, and the sequences of the *a* and α receptors turn out to be similar to a class of membrane receptors including both rhodopsin and β-adrenergic receptor. These proteins possess seven hydrophobic regions that cross the membrane seven times (Fig. 16.13). In general these proteins couple extracellular events to intracellular actions.

In higher cells the next protein in the chain from membrane receptors to alterations in gene expression is called G protein because it normally binds GDP. In the absence of stimulation, one of the three subunits of G binds GDP. Upon stimulation, the α subunit of G binds GTP and

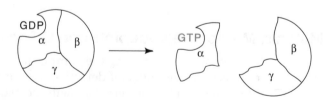

dissociates from the β and γ subunits. It has been difficult in higher cells to determine whether it is the α subunit that activates or the β-γ dimer that activates the next step in a signaling pathway. In yeast, however, it is quite clear that the β-γ pair play an important role in regulation, for an absence of the α subunit or an excess of β and γ generates the same response as exposure to mating factor. Judged on the similarity of their amino acid sequences to proteins with known activities, the later proteins in the signal chain are likely to be protein kinases. The final protein in the signaling chain is a transcription factor which binds and activates transcription of the appropriate genes.

DNA Cleavage at the *MAT* Locus

A key step in the shifting of mating type is the transfer of the genetic information from *HML* and *HMR* to *MAT*. Apparently the transfer is initiated by cleaving the DNA at the *MAT* locus. A specific endonuclease, much like a restriction enzyme, could generate a double-stranded break in DNA at the *MAT* locus. How could such an enzyme be sought? Restriction enzymes can be detected by their cleavage of DNA, such as lambda phage or plasmid DNA. If the mating-type endonuclease is specific, then only DNA carrying the *MAT* locus and possibly the *HML* and *HMR* loci would be substrates. Also, the enzyme generating the cleavages should be found in haploid *HO* yeast but probably would not be found in *ho* or in diploid yeast since these latter types do not switch mating type.

A particularly sensitive assay was used to search for the predicted cleavage (Fig. 16.14). A plasmid containing the mating-type region was cut and end-labeled with $^{32}PO_4$. Incubation of this DNA with extracts prepared from various yeast strains should cut the DNA somewhere within the *MAT* locus. Such cleavage would generate a smaller radioactive fragment which could be detected on electrophoresis of the digestion mixture and autoradiography. As expected, *HO* haploids contain a cleaving activity, and, furthermore, this activity is present at only one part of the cell cycle, just before the period of DNA synthesis. Using this cleavage assay, the enzyme has been partially purified, and biochemical studies on it are now possible.

The site of DNA cleavage lies just within the *MAT* locus. Some of the mutations that act in *cis* to prevent the normal high frequency of

Figure 16.14 A sensitive scheme for detecting an endonuclease that cleaves specifically in *MAT* sequences.

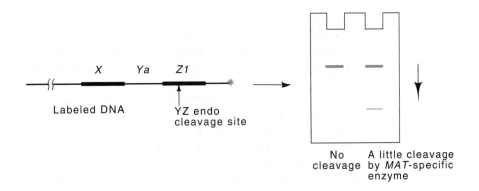

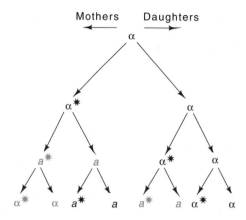

Figure 16.15 The pattern of mating type switching in *HO S. cerevisiae.* The star indicates that the mother cell will switch mating type in the next generation. Red indicates a switch.

mating-type conversion do not generate the double-stranded break *in vivo.* Most likely they prevent binding or cleavage by the HO nuclease, and, as expected, they prevent cleavage *in vitro.* As is typical for most restriction enzyme cleavage sites, these changes lie within 10 base pairs of the cleavage site.

The HO endonuclease is particularly unstable. Presumably this results from the requirement that any cleavage that is to generate a mating-type switch must be made in a narrow window of time. The regulation of the synthesis of HO messenger RNA also must be tight. Apparently a sizeable number of proteins are involved in this regulation. A region of DNA preceding the *HO* gene of over a thousand nucleotides and about fifteen different proteins seem to be involved.

In HO cells the pattern of mating-type conversion is fixed. Only in a cell that has produced at least one daughter, called a mother cell, does a mating-type conversion occur (Fig. 16.15). At this time both the mother and daughter convert. The complex set of proteins that regulate HO protein expression also compute this cell lineage and permit expression of the *HO* gene only in the mother and not in first-born daughters.

DNA Strand Inheritance and Switching in Fission Yeast

A yeast other than *Saccharomyces cerevisiae* is also widely studied. This is *Schizosaccharomyces pombe, schizo* meaning divide, and *pombe* an African word for beer. These are rod-shaped cells that elongate and divide by fission into two equal-sized cells, in contrast to budding. Some of the processes like mRNA splicing in *S. pombe* are more similar to those found in higher eukaryotic cells than the corresponding functions in *S. cerevisiae.*

The two mating types in *S. pombe* are called *P* for plus, and *M,* for minus. The mating-type conversion occurs by transposition of copies of genetic information derived from reference locations, *mat2* and *mat3,* into the expression locus *mat1,* much like that found in *S. cerevisiae.*

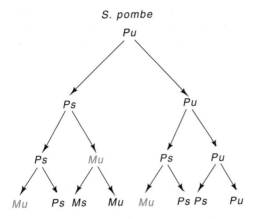

Figure 16.16 Pattern of mating type switching observed in the fission yeast *S. pombe*.

The pedigree of switching is much different between the two yeasts, however. Instead of mother and daughter switching together as in *S. cerevisiae*, only a single sister switches in *pombe*. That is, only one of four granddaughters of an originally unswitchable *Pu* cell switches (Fig. 16.16).

Figure 16.17 A strand marking mechanism of determining the developmental fate of daughter cells. The two cell types are denoted by *M* and *P* with curved and angular symbols. Switchable and unswitchable are indicated by *s* and *u*.

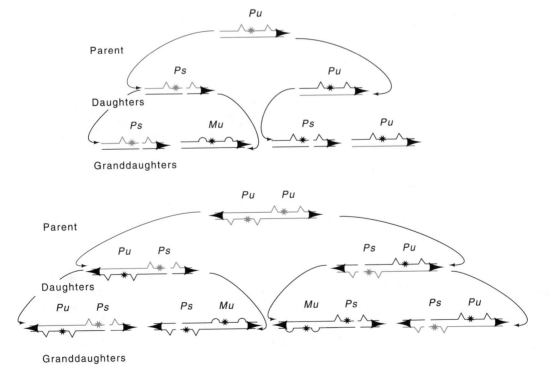

What mechanism directs just one of the two sisters of a *Ps* cell to switch? We might expect that some molecule in the cytoplasm or nucleus is unequally expressed, distributed, or stabilized in the two daughters, and therefore controls their different behavior. Instead, the marking system appears to use strand-specific marking of the DNA. One of the strands in a *Pu* is marked. After replication, only the chromatid inheriting the marked strand is cleaved. This cell is then competent to switch and is denoted *Ps*, and it produces one switched sister in the next generation.

An elegant experiment led to the proposal of the strand marking mechanism. A strain was constructed to contain an inverted duplication of *mat1* (Fig. 16.17). Thus, each of the two strands of the DNA duplex contain a copy of the site that is marked for eventual cutting. Hence, in the *Pu* cell, both strands are marked, and both sister cells derived from the cell should be switchable, and indeed, both are. Southern blot analysis to quantitate the level of cleavage also fits expectations. Thus, for *S. pombe*, inheritance of parental DNA strands, which are complementary and not identical, confers developmental asymmetry to daughter cells.

Problems

16.1. What can we infer about the lifetimes of the biological molecules involved in mating-type determination and function in light of the fact that mating type can be detected to change nearly every generation in *HO* cells?

16.2. Describe how the yeast transformant carrying a mating-type gene could be detected.

16.3. How can dominance and recessivity of *SIR* genes be tested since mating type can ordinarily be determined for haploid yeast only?

16.4. A mutation known as *rad*52 is lethal in haploid *HO* strains, apparently because of the double-stranded break that is generated at the *MAT YZ* site. How could the properties of this mutation be used to infer that the mating system generates no similar breaks elsewhere in the yeast chromosomes?

16.5. How would you test whether the absence of the mating-specific endonuclease in diploid and *ho* yeast results from its failure to be synthesized or from the synthesis of an inactivator of the enzyme?

16.6. Draw a scheme by which a double-stranded cleavage near *MAT* could initiate a gene conversion in which an adjacent region is altered to that of *HML* or *HMR*.

16.7. In seeking the enzyme that makes the double-stranded cut in the *MAT* locus, why would it be wise to use *HO* haploid and not diploid cells? What problem would this raise and how would you overcome it?

16.8. Why is it logical that the mating-type conversion be initiated by a cut in the DNA at *MAT* rather than a cut at *HML* or *HMR*?

16.9. Suppose a yeast *ho* changes mating type at a frequency of 10^{-6} whereas an *HO* mutant changes with almost each cell division. What is a good way to try to isolate an *HO* mutant from the *ho* parent?

16.10. A diploid yeast cell was sporulated and dissected. All of the resulting four spores from each ascus were fertile while germinating and while at the single cell stage. Upon separation of the cells and further growth, fertility was lost. What could be going on?

16.11. Null mutations in cholesterol biosynthesis in yeast leave *MATa*, but not *MATα* cells almost unable to mate. What single-day experiment would you do first to begin to learn the basis for this behavior?

16.12. Devise a mechanism of chromosome marking that would permit only one of the eight greatgranddaughters of a cell to switch its mating type.

16.13. Normally α mating-type information is at *HML* and *a* mating-type information is at *HMR*. Of cells which could switch, starred in Fig. 16.15, about 94% do switch. When *a* mating-type information is at *HML* and α mating-type information is at *HMR*, only 6% of the cells competent to switch actually do switch. What does this mean?

References

Recommended Readings

Transposable Mating Type Genes in *Saccharomyces cerevisiae*, J. Hicks, J. Strathern, A. Klar, Nature *282*, 478-483 (1979).

A Repressor (MATα2 Product) and Its Operator Control Expression of a Set of Cell Type Specific Genes in Yeast, A. Johnson, I. Herskowitz, Cell *42*, 237-247 (1985).

Transcriptional Regulation in the Yeast Life Cycle, K. Nasmyth, D. Shore, Science *237*, 1162-1170 (1987).

A Regulatory Hierarchy for Cell Specialization in Yeast, I. Herskowitz, Nature *342*, 749-757 (1988).

Yeast Repressor alpha2 Binds to its Operator Cooperatively with Yeast Protein MCM1, C. Keleher, S. Passmore, A. Johnson, Molecular and Cellular Biology *9*, 5228-5230 (1990).

Yeast Physiology and Genetics

A New Method for Hybridizing Yeast, C. Lindegren, G. Lindegren, Proc. Nat. Acad. Sci. USA *29*, 306-308 (1943).

Mating Types in *Saccharomyces*: Their Convertibility and Homothallism, Y. Oshima, I. Takano, Genetics *67*, 327-335 (1971).

Mapping of the Homothallism Genes *HMα* and *HMa* in *Saccharomyces* Yeasts, S. Harashima, Y. Oshima, Genetics *84*, 437-451 (1976).

Asymmetry and Directionality in Production of New Cell Types During Clonal Growth: The Switching Pattern of Homothallic Yeast, J. Strathern, I. Herskowitz, Cell *17*, 371-381 (1979).

Control of Cell Type in Yeast by the Mating Type Locus: The α1-α2 Hypothesis, J. Strathern, J. Hicks, I. Herskowitz, J. Mol. Biol. *147*, 357-372 (1981).

Regulation of Transcription in Expressed and Unexpressed Mating Type Cassettes of Yeast, A. Klar, J. Strathern, J. Broach, J. Hicks, Nature *289*, 239-244 (1981).

Molecular Analysis of a Cell Lineage, K. Nasmyth, Nature *302*, 670-676 (1983).

Identification and Comparison of Two Sequence Elements that Confer Cell-type Specific Transcription in Yeast, A. Miller, V. Mackay, K. Nasmyth, Nature *314*, 598-602 (1985).

Nucleotide Sequences of *STE2* and *STE3*, Cell Type-specific Sterile Genes from *Saccharomyces cerevisiae*, N. Nakayama, A. Miyajima, K. Arai, EMBO Journal *4*, 2643-2648 (1985).

Activation of Meiosis and Sporulation by Repression of the RME1 Product in Yeast, A. Mitchell, I. Herskowitz, Nature *319*, 738-742 (1986).

Translation and a 42-Nucleotide Segment within the Coding Region of the mRNA Encoded by the MATα1 Gene are Involved in Promoting Rapid mRNA Decay in Yeast, R. Parker, A. Jacobson, Proc. Natl. Acad. Sci. USA *87*, 2780-2784 (1990).

Degradation of a-Factor by a *Saccharomyces cerevisiae* α-Mating-Type-Specific Endopeptidase: Evidence for a Role in Recovery of Cells from G1 Arrest, S. Marcus, C. Xue, F. Naider, J. Becker, Mol. Cell. Biol. *11*, 1030-1039 (1991).

Structure and Action of a1, α1, α2, and MCM1

A Conserved DNA Sequence in Homoeotic Genes of the *Drosophila* Antennapedia and Bithorax Complexes, W. McGinnis, M. Levine, E. Hafen, A. Kuroiwa, W. Gehring, Nature *308*, 428-433 (1984).

The Yeast α-Factor Receptor: Structural Properties Deduced from the Sequence of the *STE2* Gene, A. Burkholder, L. Hartwell, Nuc. Acids Res. *13*, 8463-8475 (1985).

Homeo-domain Homology in Yeast MATα2 is Essential for Repressor Activity, S. Porter, M. Smith, Nature *320*, 766-768 (1986).

Yeast Peptide Pheromones, a-Factor and α-Factor, Activate a Common Response Mechanism in their Target Cells, A. Bender, G. Sprague, Jr., Cell *47*, 929-937 (1986).

Homeo Domain of the Yeast Repressor α2 Is a Sequence-specific DNA-Binding Domain But is Not Sufficient for Repression, M. Hall, A. Johnson, Science *237*, 1007-1012 (1987).

Matα1 Protein, a Yeast Transcription Activator, Binds Synergistically with a Second Protein to a Set of Cell-type Specific Genes, A. Bender, G. Sprague, Cell *50*, 681-691 (1987).

a1 Protein Alters the DNA Binding Specificity of α2 Repressor, C. Goutte, A. Johnson, Cell *52*, 875-882 (1988).

The Yeast Cell-type-specific Repressor α2 Acts Cooperatively with a Non-cell-type-specific Protein, C. Kelcher, C. Goutte, A. Johnson, Cell *53*, 927-936 (1988).

The Short-lived MATα2 Transcriptional Regulator is Ubiquitinated *in vivo*, M. Hochstrasser, M. Ellison, V. Chau, A. Varshavsky, Proc. Natl. Acad. Sci. USA *88*, 4606-4610 (1991).

The *Saccharomyces cerevisiae STE14* Gene Encodes a Methyltransferase that Mediates C-terminal Methylation of a-Factor and RAS Proteins, C.

Hrycyna, S. Sapperstein, S. Clark, S. Michaelis, EMBO J. *10*, 1699-1709 (1991).

Significance of C-terminal Cysteine Modification to the Biological Activity of the *Saccharomyces cerevisiae* a-Factor Mating Pheromone, S. Marcus, G. Caldwell, D. Miller, C. Xue, F. Naider, J. Becker, Mol. Cell Biol. *11*, 3603-3612 (1991).

HO Action and Regulation

Homothallic Switching of Yeast Mating Type Cassettes is Initiated by a Double-stranded Cut in the *MAT* Locus, J. Strathern, A. Klar, J. Hicks, J. Abraham, J. Ivy, K. Nasmyth, C. McGill, Cell *31*, 183-192 (1982).

Deletions and Single Base Pair Changes in the Yeast Mating-type Locus that Prevent Homothallic Mating-type Conversions, B. Weiffenbach, D. Rogers, J. Haber, M. Zoller, D. Russell, M. Smith, Proc. Nat. Acad. Sci. USA *80*, 3401-3405 (1983).

A Site-specific Endonuclease Essential for Mating-type Switching in *Saccharomyces cerevisiae*, R. Kostriken, J. Strathern, A. Klar, J. Hicks, F. Heffron, Cell *35*, 167-174 (1983).

Activation of the Yeast *HO* Gene by Release from Multiple Negative Controls, P. Sternberg, M. Stern, I. Clark, I. Herskowitz, Cell *48*, 567-577 (1987).

Both Positive and Negative Regulators of *HO* Transcription are Required for Mother-cell-specific Mating-type Switching in Yeast, K. Nasmyth, D. Stillman, D. Kipling, Cell *48*, 579-587 (1987).

Cell Cycle Control of the Yeast *HO* Gene: *Cis* and *Trans*-acting Factors, L. Breeden, K. Nasmyth, Cell *48*, 389-397 (1987).

Signal Transduction Pathways

Binding of α-Factor Pheromone to Yeast a Cells: Chemical and Genetic Evidence for an α-Factor Receptor, D. Jenness, A. Burkholder, L. Hartwell, Cell *35*, 521-529 (1983).

Evidence the Yeast *STE3* Gene Encodes a Receptor for the Peptide Pheromone *a* Factor: Gene Sequence and Implications for the Structure of the Presumed Receptor, D. Hagen, G. McCaffrey, G. Sprague, Jr., Proc. Natl. Acad. Sci. USA *83*, 1418-1422 (1986).

Occurrence in *Saccharomyces cerevisiae* of a Gene Homologous to the cDNA Coding for the α Subunit of Mammalian G Proteins, M. Nakafuku, H. Itoh, S. Nakamura, Y. Kaziro, Proc. Natl. Acad. Sci. USA *84*, 2140-2144 (1987).

STE2 Protein of *Saccharomyces kluyveri* is a Member of the Rhodopsin/β-adrenergic Receptor Family and is Responsible for Recognition of the Peptide Ligand α Factor, L. Marsh, I. Herskowitz, Proc. Nat. Acad. Sci. USA *85*, 3855-3859 (1988).

The Yeast STE12 Protein Binds to the DNA Sequence Mediating Pheromone Induction, J. Dolan, C. Kirkman, S. Fields, Proc. Natl. Acad. Sci. USA *86*, 5703-5707 (1989).

The *STE4* and *STE18* Genes of Yeast Encode Potential β and γ Subunits of the Mating Factor Receptor-coupled G Protein, M. Whiteway, L. Hougan, D. Dignard, D. Thomas, L. Bell, G. Saari, F. Grant, P. O'Hara, V. MacKay, Cell *56*, 467-477 (1989).

Constitutive Mutants in the Yeast Pheromone Response: Ordered Function of the Gene Products, D. Blinder, S. Bouvier, D. Jenness, Cell *56*, 479-486 (1989).

Regulation of the Yeast Pheromone Response Pathway by G Protein Subunits, S. Nomoto, N. Kakayama, K. Arai, K. Matsumoto, EMBO J. *9*, 691-696 (1990).

The Carboxyl Terminus of Scg1, the Gα Subunit Involved in Yeast Mating is Implicated in Interactins with the Pheromone Receptors, J. Hirsch, C. Dietzel, J. Kurjan, Genes and Dev. *5*, 467-474 (1991).

Mutations in the Guanine Nucleotide-binding Domains of a Yeast Gα Protein Confer a Constitutive or Uninducible State to the Pheromone Response Pathway, J. Kurjan, J. Hirsch, C. Dietzel, Genes and Dev. *5*, 475-483 (1991).

MAT, HM, and Mating Type Switching

MAR1-A Regulator of *HMa* and *HMα* Loci in *Saccharomyces cerevisiae*, A. Klar, S. Fogel, K. MacLeod, Genetics *93*, 37-50 (1979).

A Suppressor of Mating Type Locus Mutations in *Saccharomyces cerevisiae*: Evidence for and Identification of Cryptic Mating Type Loci, J. Rine, J. Strathern, J. Hicks, I. Herskowitz, Genetics *98*, 837-901 (1979).

Interconversion of Yeast Mating Type by Transposable Genes, A. Klar, Genetics *95*, 631-638 (1980).

The Structure and Organization of Transposable Mating Type Cassettes in *Saccharomyces* Yeast, J. Strathern, E. Spatola, C. McGill, J. Hicks, Proc. Nat. Acad. Sci. USA *77*, 2839-2843 (1980).

The Structure of Transposable Yeast Mating Type Loci, K. Nasmyth, K. Tatchell, Cell *19*, 753-764 (1980).

The Sequence of the DNAs Coding for the Mating-type Loci in *Saccharomyces cerevisiae*, C. Astell, L. Ahlstrom-Jonasson, M. Smith, K. Tatchell, K. Nasmyth, B. Hall, Cell *27*, 16-23 (1981).

The Regulation of Yeast Mating-type Chromatin Structure by SIR: An Action at a Distance Affecting Both Transcription and Transposition, K. Nasmyth, Cell *30*, 567-578 (1982).

Molecular Cloning of Hormone-responsive Genes from Yeast *Saccharomyces cerevisiae*, G. Stetler, J. Thorner, Proc. Nat. Acad. Sci. USA *81*, 1144-1148 (1984).

Characterization of a "Silencer" in Yeast: A DNA Sequence with Properties Opposite to Those of a Transcriptional Enhancer, A. Brand, L. Breeden, J. Abraham, R. Sternglanz, K. Nasmyth, Cell *41*, 41-48 (1985).

Cloning and Characterization of Four SIR Genes of *Saccharomyces cerevisiae*, J. Ivy, A. Klar, J. Hicks, Mol. and Cell Biol. *6*, 688-702 (1986).

Identification of the DNA Sequences Controlling the Expression of the MATα Locus of Yeast, P. Siliciano, K. Tatchell, Proc. Nat. Acad. Sci. USA *83*, 2320-2324 (1986).

RAP-1 Factor is Necessary for DNA Loop Formation *in vitro* at the Silent Mating Type Locus HML, J. Hofmann, T. Laroache, A. Brand, S. Gasser, Cell *57*, 725-737 (1989).

The Developmental Fate of Fission Yeast Cells is Determined by the Pattern of Inheritance of Parental and Grandparental DNA Strands, A. Klar, EMBO J. *9*, 1407-1415 (1990).

Genes Regulating Development

17

A developing organism must regulate genes spatially and temporally. In contrast to genes of the arabinose operon where the inducing signal comes from the environment of the cell, a developing embryo generates its own signals for regulating genes. Such signals might be as simple as cells keeping track of their ancestors, or as complicated as schemes involving signaling between groups of cells. Electrical signals might be used, but likely their range would be short. The simplest form of general signaling would seem to be to use chemicals, which we shall call morphogens, whose concentrations can indicate positions. This would be much like specifying locations on the earth with latitude and longitude.

General Considerations on Signaling

In principle, three chemicals whose concentrations varied in the x, y, and z directions would be sufficient to specify every important location in a developing organism (Fig. 17.1). After determining its position, any cell could induce or repress the genes appropriate to its position.

Generating a simple coordinate system in which the locations of points are determined by the concentrations of three chemicals presents several problems. The first is simply creating the gradients. How are gradients to be built? The embryo cannot leave it to chance that an appropriate cell will start off the process of building a gradient, perhaps by synthesizing and secreting some compound. Therefore, either certain cells are special as a result of their lineage and they will set up the gradients, or an external influence directs some of the cells in the embryo to behave differently from the rest. In either case, the embryo

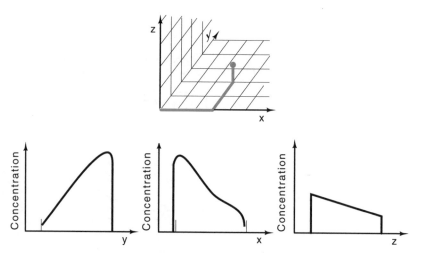

Figure 17.1 Three morphogens whose concentrations vary with position can uniquely locate a point in three dimensional space.

develops an asymmetry, and this asymmetry defines an initial direction or directions.

Commonly, at least one external influence or definition of an axis of the developmental coordinate system comes from the mother. The maternal developmental signals must be special. They cannot be introduced into the egg during its development if the signal is a freely diffusing molecule. Any gradient in such a molecule would diffuse away before fertilization and embryo development. Therefore maternal effects either must be generated during development or they must be placed in the egg in a way that diffusion cannot alter. Another useable source of asymmetry for an embryo's coordinate system is the entry point of the fertilizing sperm.

To reach all cells, the chemicals used to specify location in an embryo must be freely diffusible. If the gradients are set up in the egg after it has been subdivided into cells, the chemicals must enter and leave cells freely. This requires that the molecule be small. On the other hand, position might be determined before a fertilized egg has divided into cells. *Drosophila* operates this way, and the embryo reaches about 4,000 nuclei before cell walls are synthesized. Thus, the morphogens can be proteins.

If many different coordinate positions are to be distinguished along one concentration gradient, precise measurements of the morphogen concentration must be made. The standard biochemical means for measuring concentration are simply measuring the amount of binding of a chemical to a receptor with appropriate affinity (Fig. 17.2). This method is incapable of discerning small differences in the concentration of a substance, but detecting such differences would be necessary if many different developmental cues are to be derived from one gradient. Instead of using a few gradients to determine everything about a

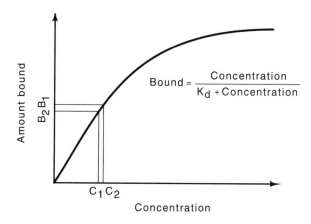

Figure 17.2 As the concentration of a molecule increases, the amount bound to receptors increases, at first linearly with the concentration, and then decreasing. Small changes in concentration lead to small changes in the amount bound.

developing embryo, it is more logical that gross patterns are first developed from simple gradients. These can be subdivided as many times as necessary to produce as many different states as are required for the developing embryo (Fig. 17.3).

Not only must cells determine where they are in the coordinate system of morphogens, but they must do so crisply. The division lines between different tissues must be sharp. Interpenetration of one tissue into another would generate problems. Therefore we can expect developing organisms to use special techniques to make the division lines sharp. One simple technique is to make decisions when there are as few cells as possible. Another is to use a small embryo, for the smaller the embryo, the steeper the gradients, and therefore the easier it is to make decisions. Once an individual cell has determined what tissue or body part it is to become, it can go through multiple cell divisions to generate as much tissue as necessary. A second general way to make sharp dividing lines in differentiating tissue is to make the processes non-linear. First, let us examine the linear situation. Suppose the decision between becoming tissue of type A or type B depends on the amount of morphogen bound to a protein, and that the binding is described by the

Figure 17.3 Repeated subdivision of an embryo can generate many areas.

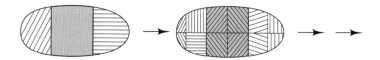

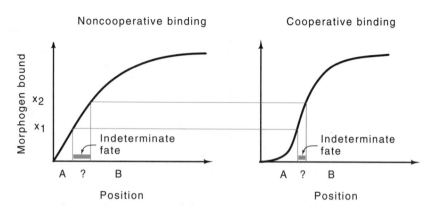

Figure 17.4 The lengths of the red boxes are proportional to the numbers of cells with indeterminate developmental fates.

standard Michaelis-Menten equation. Further, assume that all cells whose receptors bind less than x_1 of the morphogen become cell-type A, and all cells whose receptors bind more than x_2 of the morphogen become type B. Cells that bind an intermediate amount of morphogen can become type A or B (Fig. 17.4). A moderate number of cells could exist in this indeterminate position. Suppose instead, that cells use a more complex method of measuring concentration. They could use two receptors and respond to two gradients or respond to a single morphogen by requiring that two binding events occur. Then the response of the receptor system could be made to be the product of two Michaelis-Menten curves. This is nonlinear, and the transition between the critical concentrations becomes much shorter. Therefore fewer cells are subjected to indeterminate concentrations of morphogen. This principle of making responses depend on multiple sensing systems can be used to make particularly sharp responses. It is analogous to that used by phage lambda in deciding between the lysogenic and lytic responses.

As a result of the general principles mentioned above, we have the following questions concerning developmental systems. What kinds of genes and gene products are involved? How is signaling accomplished? Are there cascades of regulation so that subdivisions are generated within an embryo? Are multiple gradients used in some decision processes so as to make sharp boundaries? Once a cell knows where it is, what happens next to determine its developmental fate?

Outline of Early *Drosophila* Development

As in the development of spores of yeast, a single diploid cell in the ovary of a female fly gives rise to a number of haploid cells. In *Drosophila* the precursor cell undergoes four divisions to yield sixteen cells (Fig. 17.5). These are not separated, but retain communication with one another via cytoplasmic bridges. Eventually, one of the two cells with four

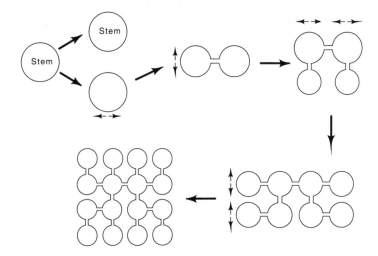

Figure 17.5 Creation of nurse and egg cells from a precursor stem cell in the ovary of a *Drosophila*. The first division of the stem cell yields one cell which remains stem cell type and a daughter cell that will yield nurse and egg cells. Four additional divisions yield a network of interconnected cells.

connections to other cells moves to the end of the group and begins the development process of generating an egg. The fifteen connected cells assist the process and are called nurse cells. While it is developing, the future egg cell is called an oocyte. The nurse cells absorb nutrients from the surrounding fluid and synthesize rRNA, mRNA, and other components. Additionally, maternal diploid cells also influence the developing oocyte. (Fig. 17.6). These are called the follicle cells, and they surround the nurse cells and developing oocyte. Partway through the oogenesis pathway, the nurse cells empty their contents into the developing oocyte. When the maturation process is complete, the synthetic activities in the follicle cells almost ceases.

Upon fertilization, DNA synthesis begins at a high rate in the egg. Many replication origins are utilized per chromosome so that a chro-

Figure 17.6 Later stage of egg development showing an egg cell plus nurse cells surrounded by follicle cells. Although each cell possesses a nucleus and 16 are present, not all nuclei are in the plane of the cross section. The future egg cell is at the far right. At a still later stage of development the egg is enlarging and the nurse cells are found at the left.

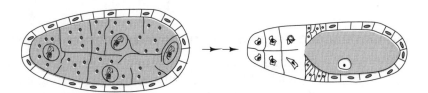

mosome doubling can take as little as ten minutes. The original diploid chromosome doubles ten times, then a few nuclei migrate to the posterior pole of the egg where they ultimately become the germ line cells. The remainder of the nuclei migrate to the surface of the egg and divide four more times, yielding a total of about 4,000 nuclei. At this point nuclear membranes and cell walls develop, creating isolated cells. This stage consists of a single layer of cells and is called a blastula.

Up to the time of wall formation, the nuclei are not committed to any developmental fate, for they can be moved to different locations without altering the development of the embryo. After the walls form, the cells are partially committed, and relocating them interferes with normal development. About six hours after fertilization the embryo develops about 14 distinguishable segments. Six of these ultimately fuse to yield the head, three become the thorax, and about eight develop into the abdomen. By 24 hours after fertilization the larva is fully developed.

Classical Embryology

Early experiments in embryology indicated that long-range influences help determine pattern formation in embryos. Sander examined the consequences of isolating parts of the embryo by tying off portions so as to prevent chemical communication between two groups of cells. Similar experiments showed that the anterior portion of the egg exerted an influence over the pattern development in the remainder of the egg. Material removed from the anterior pole and microinjected elsewhere behaved as though it contained the source of the pattern maker. Also, destroying material at the anterior end of the egg eliminated head structures and yielded an embryo with a posterior replacing the head structures.

Removal of about 10% of the cytoplasm from the anterior end of an egg also affects the resulting pattern development. Usually the abdominal region of the resulting organism is defective. Removal of an equal amount of cytoplasm from other parts of the egg has relatively little effect on the developing embryo. Dominance experiments can even be done with microinjection. Injecting cytoplasm from the posterior region into the anterior region suppresses head development. Similarly, anterior cytoplasm taken from an embryo somewhat later in development represses posterior development if it is injected into the posterior region. In either case, removal of cytoplasm from one end reduces the tendency to develop structures characteristic of that end, and injection of cytoplasm from the other end of the embryo can reverse the identity of structures near the end. In this way two-headed or two-tailed embryos can be formed.

Using Genetics to Begin Study of Developmental Systems

Although the classical experiments with insect embryos indicated the existence of long-range effects, further progress has required new techniques. Genetic approaches are one method for proceeding with a deeper analysis of development. Mutants have the potential for indicating the complexity of a system by revealing the approximate number of genes or gene products involved and the ways in which the system can fail. Recently, molecular genetics has greatly streamlined the process of obtaining and studying mutants and genes involved in the development process.

We might expect the existence of two easily distinguishable classes of developmental mutations, maternal and embryonic. That is, some genes and gene products necessary for spatial development likely are expressed only in the nurse or follicle cells and are required for egg development. Other mutations likely are embryonic and expressed only in the developing embryo.

How can maternal lethal or embryonic lethal mutations be isolated and studied? Certainly the technology is not as simple as mutagenizing bacteria and providing leucine as they grow, and then identifying those colonies that require leucine for growth. Nonetheless, the operations are not particularly complicated either. Two problems must be solved. The first is handling the diploid chromosomes. Most likely defects will be lethal. Since dominant lethal mutants cannot be propagated, the mutants which can be propagated and studied must be recessive. Recessive lethals can be generated as long as the homologous chromosome does not carry the mutation. Whenever the mutation is to be detected or to be studied, both chromosomes must carry the mutation. The second problem is preventing recombination between homologous

Figure 17.7 Scheme for isolating and studying maternal and embryonic lethal mutations. A mutagenized male is mutated with a female carrying a genetically marked balancer chromosome, indicated by B. The genotypes of subsequent offspring can be easily identified and ultimately, a homozygous female can be generated.

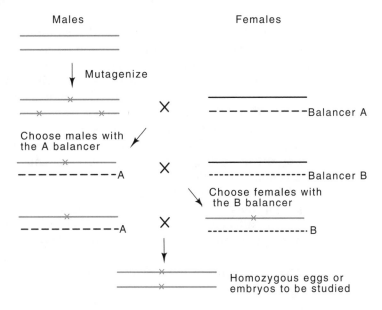

chromosomes. Due to the absence of the necessary mutants, this cannot be done simply by using a recombination deficient *Drosophila*.

Ordinarily at meiosis, extensive pairing and crossing over occurs between homologous chromosomes. Classical geneticists noticed, however, that if one of the chromosome pairs carries a chromosome inversion, then pairing between the two chromosomes is impeded in the vicinity of the inversion. Therefore recombination between a particular pair of chromosomes can be greatly reduced by combining multiple inversions and rearrangements into one of the chromosome pairs. Such a chromosome is called a balancer chromosome.

Suppose we wish to isolate mutations in chromosome III. The process can be started by mutagenizing a male and mating with a female carrying a normal chromosome III and a balancer chromosome III with a dominant, easily identified marker like wing shape or bristle pattern (Fig. 17.7). Progeny carrying the balancer will have inherited the other copy of chromosome III from the mutagenized male. Males can be taken from this point and used to generate females carrying the mutagenized chromosome III. Finally, the resultant males and females carrying the mutagenized chromosome III can be mated, and the resulting eggs or embryos lacking the dominant markers on the balancer chromosomes can be examined for lethality or the presence of developmental abnormalities. In one such screen of over 5,000 mutagenized chromosomes, about sixty maternal effect mutations were found. On average four alleles of each of the fifteen genes were found. If the mutations were randomly distributed among the relevant genes, these statistics imply that most of the maternal effect genes had received at least one mutation. That is, that there are on the order of 20 such genes. This is not far off. Ultimately, about 30 were found.

The general conclusions from these genetic experiments are that a relatively small number of genes are required for spatial development in *Drosophila*. Many of the genes exert a strong effect over a sizeable portion of the embryo. Three groups of maternal effect mutations have been found: anterior, which lack head and thoracic structures; terminal, which lack structures at both ends; and posterior, which lack posterior structures. Among the mutations affecting anterior structures was the mutation known as *bicoid*. Embryos carrying this mutation behave similarly to embryos with their anterior cytoplasm removed. They often lack head and thorax structures and replace them with a mirror image of the posterior region. These defects can be complemented by microinjection with anterior cytoplasm from wild-type embryos. Other mutations affecting head and thorax development are the *swallow* and *exuperantia* mutations.

Additional classes of mutations that affect embryonic pattern development independent of maternal effects have also been found. Some of these affect segmentation in the embryo. Mutations in the gene named *fushi tarazu* have too few segments. Some mutations remove the even or the odd numbered segments, or lack a sizeable block of segments or have segments or regions duplicated.

Developmental genes also act after segmentation. The identities of the segments are determined by homeotic genes. For example, the *Antennapedia* complex of genes specifies development of part of the thorax, and mutants in this complex may grow legs from the head instead of antennae. The bithorax cluster of homeotic genes studied by Lewis specifies identity of the posterior portion of the thorax and abdomen.

Cloning Developmental Genes

Clones of developmental genes are required for many studies. As in any research, the techniques devised for the studies and the results obtained greatly aided later and similar efforts. The only information available to assist cloning a developmental gene the first time was the known chromosomal location of the gene. Such genes could be mapped both genetically and cytologically. The cytological mapping utilized the characteristic banding pattern seen in the polytene chromosomes from *Drosophila* salivary glands and the existence of deletions and inversions affecting a developmental complex like bithorax.

Cloning of bithorax by Bender began with a collection of random *Drosophila* clones. These could be located approximately on the *Drosophila* chromosomes by *in situ* hybridization. Then a chromosomal walk was done to reach the gene complex. One of the major problems with walking along a chromosome is knowing when you get to your gene. Sometimes deletions will delineate a gene. Inversions that inactivate the desired gene are better, however. The end points are easy to locate in a walk and they must lie within or very near the gene.

One approach to complete the identification of a clone is to seek RNAs to which the cloned DNA hybridizes. Hybridizations of radioactive DNA from candidate clones to mRNA can be done *in situ* to slices of embryos or even adult flies to identify any RNAs with expression patterns suggestive of developmental genes. Once a candidate RNA can be detected, RNA extracted at the appropriate time from embryos of the the right tissue can be used to make a cDNA library that is then screened with the cloned DNA. In the case of *Antennapedia*, the probe to the gene hybridized not only to the gene, but also to another region of the same chromosome, a region containing the bithorax developmental gene complex.

Enhancer Traps for Detecting and Cloning Developmental Genes

Molecular genetics provides a streamlined method for the detection of developmentally regulated genes in *Drosophila*. This regulation can be either with respect to time or with respect to position of the cell. A segment of DNA containing an enhancerless promoter connected to a β-galactosidase gene is inserted into the genome at random locations. Only if the inserted DNA falls near an enhancer will the promoter be activated and will β-galactosidase be synthesized. If the enhancer regu-

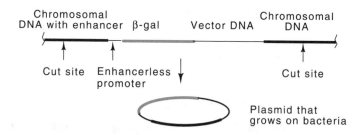

Figure 17.8 An enhancer trap vector inserted into chromosomal DNA places the β-galactosidase gene under control of a nearby enhancer so that the adjacent DNA can be easily cloned.

lates a developmental gene, then the pattern of β-galactosidase synthesis will similarly be developmentally regulated. As a result, the pattern in space or time of β-galactosidase synthesis will be the same as that normally followed by the gene under influence of the enhancer. This pattern can be visualized by immersing the embryos in a substrate like X-gal whose product after cleavage by β-galactosidase is insoluble and highly colored.

Once an insertion of interest has been found, subsequent cloning steps can proceed efficiently. If the inserted DNA is a complete *E. coli* vector and lacks a particular restriction enzyme cleavage site, then a clone can simply be made containing DNA from either side of the insertion. DNA from the insertion strain is cleaved with the enzyme, circularized, and transformed into *E. coli*. Since the cleavage will occur only in the *Drosophila* DNA, such DNA will flank the vector sequences, and the recircularized vector will contain some surrounding DNA (Fig. 17.8). Once this has been accomplished, a cDNA library can be screened for sequences also contained on the clone from the enhancer trap. Any such sequences that are expressed with the same pattern as the original β-galactosidase expression pattern are good candidates for developmentally regulated genes.

Expression Patterns of Developmental Genes

How can it be demonstrated that a developmental gene actually does encode a protein and that the protein is involved in development? Many developmental genes look like they code for proteins. The bithorax gene complex of *Drosophila* helps specify identities of segments from the second thoracic to the eighth abdominal segment. The *Ultrabithorax* region within the bithorax complex generates a number of transcripts from a very large region containing a few exons and large introns (Fig.

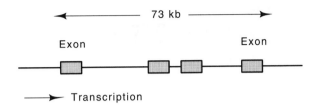

Figure 17.9 The *Ubx* locus showing the distance between the exons in the transcript.

17.9). The 5′ exon contains a substantial open reading frame and looks like it codes for a protein.

Antibodies against the presumed ultrabithorax protein, if they could be made, would provide proof that the gene encodes a protein. They would also provide the basis of an assay for the presence of such a protein so that it could be purified. The traditional approach of purifying a protein and injecting it into a mouse or rabbit could hardly be used for a protein that cannot be assayed and hasn't been purified. Molecular genetics techniques provided inexpensive solutions to the problem of purifying the protein, making useful antibodies, and locating the protein within the developing larva. A cloning vector was used from which β-galactosidase could be synthesized in *E. coli* only if DNA containing an open reading frame were inserted upstream of the enzyme (Fig. 17.10). Thus Lac⁺ colonies were selected from cells transformed with a plasmid into which fragments of the 5′ exon had been ligated. These

Figure 17.10 A vector for cloning stretches of DNA containing open reading frames. Protein 1 and β-galactosidase are fused out of frame. The correct reading frame can be restored and cells containing the vector will be Lac⁺ if a third segment of DNA is added which restores the correct reading frame and which contains no translation termination codons.

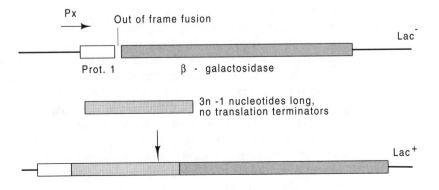

Figure 17.11 Representation of the pattern of expression of a homeotic gene in the *Drosophila* embryo.

would synthesize the *Ubx*-β-galactosidase fusion protein, and furthermore, because of the size of the protein, it could easily be isolated from SDS protein gels. β-galactosidase is one of the largest polypeptides synthesized in *E. coli*, and the fusion protein likely would be even larger. Hence the fusion protein could be separated from other proteins simply by electrophoresis. The region of the gel containing the fusion protein was cut out, protein eluted, and injected into rabbits or mice for direct induction of antibodies or for production of monoclonal antibodies.

The resulting antibodies against a portion of the homeotic protein permitted assay of the developing fly for the protein's location and timing of synthesis. Not surprisingly, immunofluorescence shows that the homeotic protein is found in the nuclei of cells, suggesting that it is a DNA-binding protein. It is found in restricted parts of the embryo.

In general, in studying developmental genes, two variables must be examined, the RNA and the protein product. We would like to know when the mRNA is synthesized and where, and when the RNA is translated and where the protein product goes. Questions about the RNA product can be answered with *in situ* RNA-DNA hybridization between slices of embryo and radioactive DNA. The protein products can be located with similar *in situ* probing by using antibodies directed against the protein in question. The antibodies can be obtained as outlined above or by injecting rabbits with synthetic polypeptides based on DNA sequences in the developmental genes.

As might be expected, the patterns of RNA and protein synthesis from different developmental genes roughly parallel their effects on embryos. For example, the products of genes effecting segmentation are often expressed in stripes across the embryo (Fig. 17.11). Most of the proteins are found in the nuclei, implying that many of the proteins regulate gene expression directly by binding to DNA. Of course for developmental proteins expressed after cell walls have formed, we would expect to find developmental genes coding for secreted proteins, membrane-bound receptors, and proteins involved in regulation cascades linking membrane receptors to gene activity in the nucleus. An example of such a system is the G-protein protein kinase pathway.

Not all proteins used early in development look like they bind to DNA. One such example is the product of the *vasa* gene. This is a maternal effect gene that is expressed only in nurse cells. It has a sequence with about 24% amino acid identity to the translation initiation factor eIF-4A, which in combination with 4B and 4F recognizes the cap structures on mRNA.

The bicoid mRNA and protein appears in a particularly pleasing pattern. Bicoid mRNA is synthesized in the nurse cells and transferred to the oocyte during oogenesis. This RNA binds just inside the oocyte at the extreme anterior end by means of the *swallow* and *exuperantia* gene products. The bicoid mRNA is not translated until the egg is fertilized. This is a marvelous way for the egg to generate a gradient in the protein just at the time it is needed. The egg can sit for a long time and diffusion cannot destroy the gradient. These results are also consistent with the microinjection results.

Similarities Among Developmental Genes

The sequences of the *fushi tarazu* and *Antennapedia* genes revealed that they encoded protein, and more importantly, that over a stretch of about 60 amino acid residues, the sequences of the two proteins were even more highly conserved than the corresponding coding DNA sequences. A portion of another homeotic developmental gene complex, bithorax, also possessed a closely related sequence. This sequence is now called a homeo box. With the conservation of the homeo box among three proteins, it seemed likely to be present in other developmental regulators from the fly as well as in closely related organisms. This proved to be true. Not only do other species of *Drosophila* contain homeo boxes; so do all higher organisms, including humans. Typically a Southern transfer of genomic DNA hybridized to a homeo sequence reveals five to ten similar sequences. If hybridization at low stringency is performed, less closely related sequences are revealed. These have been cloned and sequenced, yielding the expected more variable homeo-like sequences. Such sequences can be used again to find still less closely related genes. Many organisms possess multiple clusters of four or five homeo box proteins.

Homeo boxes were also noticed to be similar to the alpha mating-type factors of yeast which were known to bind to DNA. Therefore it seemed likely that proteins containing a homeo-box sequence bind to DNA. This is true. Homeo-box domains bind DNA, and the structure of several has been determined. The homeo-box fits into the major groove of the DNA, but its positioning is notably different from that of the helix-turn-helix motif. Not all developmental genes possess homeo boxes. A number of other DNA-binding motifs are utilized, including zinc finger, zinc domains, helix-loop-helix proteins and others.

Overall Model of *Drosophila* Early Development

From knowledge of developmental genes and their expression patterns, we can form a rough picture of *Drosophila* development. Initially two gradients are set up. The anterior gradient in the *bicoid* product is generated by activating *bicoid* mRNA for translation at the time of fertilization. A posterior morphogen is also utilized. This is the *nanos* product, which is located in the rear half of the egg. Next the developmental genes responsible for sizeable portions of the embryo are activated. *Hunchback*, *krüppel*, and *knirps* are examples of such genes. Defects in these genes lead to sizeable gaps in the shapes of developing embryos. The concentrations of *bicoid* and *nanos* regulate the expression of these proteins. Likely they are synthesized only in regions in which proteins lie in certain concentration ranges. Next, the pair rule genes function. These eight or so genes further subdivide the developing embryo into six to eight stripes. A gene like *engrailed* helps sharpen the boundaries between the segments. After this, the *runt* and *hairy* genes are expressed in alternate stripes. The homeotic genes then function to specify the actual identity of the different areas.

Problems

17.1 How would you seek maternal lethal mutations, that is, mutations rendering females unable to lay fertile eggs?

17.2. What is the likely appearance of embryos containing the *swallow* or *exuperantia* mutations?

17.3. Suppose a gradient from one end of a *Drosophila* egg to the other exists in the concentration of a protein of 50,000 molecular weight. How long would be required for diffusion to flatten significantly, say by a factor of two or three, the gradient in concentration?

17.4. Illustrate how an appropriate deletion will permit a time-saving jump to be made when cloning a gene by walking along a chromosome.

17.5. What sort of sequence drift between two genes could indicate that evolution was conserving the protein sequence and not the DNA sequence?

17.6. In the study of embryonic and maternal lethals, why couldn't both males and females be taken from the cross between the mutagenized males and females carrying the marked balancer chromosome? It would seem that males and females taken at this point could be crossed, thereby eliminating one of the crosses necessary in generating homozygous individuals.

17.7. If you were seeking developmental mutants in *Drosophila* located on an autosomal chromosome, why would it be a poor idea to design the search such that the mutations you preserved would be located on the balancer chromosome?

References

Recommended Readings

Mutations Affecting Segment Number and Polarity in *Drosophila*, C. Nüsslein-Volhard, E. Wieschaus, Nature *287*, 795-801 (1980).

Coordinate Expression of the Murine Hox-5 Complex Homoeobox-Containing Genes During Limb Pattern Formation, P. Dolle, J. Izpisua-Belmonte, H. Falkenstein, A. Renucci, D. Duboule, Nature *342*, 767-772 (1989).

Regulation of a Segmentation Stripe by Overlapping Activators and Repressors in the *Drosophila* Embryo, D. Stanojevic, S. Small, M. Levine, Science *254*, 1385-1387 (1991).

Techniques for Development Study and Other Organisms

The Embryonic Cell Lineage of the Nematode *Caenorhabditis elegans*, J. Sulston, E. Schierenberg, J. White, J. Thomson, Developmental Biology *100*, 64-119 (1983).

Phenocopies Induced with Antisense RNA Identify the *wingless* Gene, C. Cabrera, M. Alonso, P. Johnston, R. Phillips, P. Lawrence, Cell *50*, 659-663 (1987).

Identification of a Novel Retinoic Acid Receptor in Regenerative Tissues of the Newt, C. Ragsdale, Jr., M. Petkovich, P. Gates, P. Chambon, J. Brockes, Nature *341*, 654-657 (1989).

Searching for Pattern and Mutation in the *Drosophila* Genome with a P-*lacZ* Vector, E. Bier, H. Vaessin, S. Shepherd, K. Lee, K. McCall, S. Barbel, L. Ackerman, R. Carretto, T. Uemura, E. Grell, L. Jan, Y. Jan, Genes and Development *3*, 1273-1287 (1989).

P-element-mediated Enhancer Detection: A Versatile Method to Study Development in *Drosophila*, H. Bellen, C. O'Kane, C, Wilson, U. Grossniklaus, R. Pearson, W. Gehring, Genes and Development *3*, 1288-1300 (1989).

Manipulation of Flower Structure in Transgenic Tobacco, M. Mandel, J. Bowman, S. Kempin, H. Ma, E. Meyerowitz, M. Yanofsky, Cell *71*, 133-143 (1992).

Initial Gradients and Early Events in Differentiation

Analysis of *Krüppel* Protein Distribution During Early *Drosophila* Development Reveals Posttranscriptional Regulation, U. Gaul, E. Seifert, R. Schuh, H. Jäckle, Cell *50*, 639-647 (1987).

The *Toll* Gene of *Drosophila*, Required for Dorsal-ventral Embryonic Polarity, Appears to Encode a Transmembrane Protein, C. Hashimoto, K. Hudson, K. Anderson, Cell *52*, 269-279 (1988).

A Gradient of *bicoid* Protein in *Drosophila* Embryos, W. Driever, C. Nüsslein-Volhard, Cell *54*, 83-93 (1988).

The *bicoid* Protein Determines Position in the *Drosophila* Embryo in a Concentration-dependent Manner, W. Driever, C. Nüsslein-Volhard, Cell *54*, 95-104 (1988).

Drosophila Nurse Cells Produce a Posterior Signal Required for Embryonic Segmentation and Polarity, K. Sander, R. Lehmann, Nature *335*, 68-70 (1988).

Function of *torso* in Determining the Terminal Anlagen of the *Drosophila* Embryo, M. Klinger, M. Erde'lyi, J. Szabad, C. Nüsslein-Volhard, Nature *335*, 275-277 (1988).

The Product of the *Drosophila* Gene Vasa is Very Similar to Eukaryotic Initiation Factor-4A, P. Lasko, M. Ashburner, Nature *335*, 611-617 (1988).

The Bicoid Protein is a Positive Regulator of *hunchback* Transcription in the Early *Drosophila* Embryo, W. Driever, C. Nüsslein-Volhard, Nature *337*, 138-143 (1989).

The *Drosophila* Gene *torso* Encodes a Putative Receptor Tyrosine Kinase, F. Sprenger, L. Stevens, C. Nüsslein-Volhard, Nature *338*, 478-483 (1989).

Specific Proteolysis of the c-mos Proto-oncogene Product by Calpain on Fertilization of *Xenopus* Eggs, N. Watanabe, G. Woude, Y. Ikawa, N. Sagata, Nature *342*, 505-511 (1989).

A Gradient of Nuclear Localization of the *dorsal* Protein Determines Dorsoventral Pattern in the Drosophila Embryo, S. Roth, D. Stein, C. Nüsslein-Volhard, Cell *59*, 1189-1202 (1989).

Interactions between Peptide Growth Factors and Homeobox Genes in the Establishment of Antero-posterior Polarity in Frog Embryos, R. Altaba, D. Melton, Nature *341*, 33-38 (1989).

Positive Autoregulation of Sex-lethal by Alternative Splicing Maintains the Female Determined State in *Drosophila*, L. Bell, J. Horabin, P. Schedl, T. Cline, Cell *65*, 229-239 (1991).

The Polarity of the Dorsoventral Axis in the *Drosophila* Embryo is Defined by an Extracellular Signal, D. Stein, S. Roth, E. Vogelsang, C. Nüsslein-Volhard, Cell *65*, 725-735 (1991).

Segmentation

A Gene Complex Controlling Segmentation in *Drosophila*, E. Lewis, Nature *276*, 265-270 (1978).

A Gene Product Required for Correct Initiation of Segmental Determination in *Drosophila*, G. Struhl, Nature *293*, 36-41 (1981).

Sequence of a *Drosophila* Segmentation Gene: Protein Structure Homology with DNA Binding Proteins, A. Laughon, M. Scott, Nature *310*, 25-31 (1984).

Structure of the Segmentation Gene *paired* and the *Drosophila* PRD Gene Set of a Gene Network, G. Frigerio, M. Burri, D. Bopp, S. Baumgartner, M. Noll, Cell *47*, 735-746 (1986).

Control Elements of the *Drosophila* Segmentation Gene *fushi tarazu*, Y. Hiromi, A. Kuroiwa, W. Gehring, Cell *43*, 603-613 (1987).

Finger Protein of Novel Structure Encoded by *hunchback*, a Second Member of the Gap Class of *Drosophila* Segmentation Genes, D. Tautz, R. Lehmannn, H. Schürch, E. Seifert, A. Kienlin, K. Jones, H. Jäckle, Nature *327*, 383-389 (1987).

The *Drosophila* Homolog of the Mouse Mammary Oncogene *int-1* Is Identical to the Segment Polarity Gene *wingless*, F. Rijsewijk, M. Schuermann, E. Wagenaar, P. Parren, D. Weigel, R. Nusse, Cell *50*, 649-657 (1987).

Regulation of Segment Polarity Genes in the *Drosophila* Blastoderm by *fushi tarazu* and *even skipped*, P. Ingham, N. Baker, A. Martinez-Arias, Nature *331*, 73-75 (1987).

Two-tiered Regulation of Spatially Patterned *Engrailed* Gene Expression During *Drosophila* Embryogenesis, S. DiNardo, E. Sher, J. Hoemskerk-Jongens, J. Kassis, P. O'Farrell, Nature *332*, 604-609 (1987).

Regulation of the *Drosophila* Segmentation Gene *hunchback* by Two Maternal Morphogenetic Centres, D. Tautz, Nature *332*, 281-284 (1988).

The *GLI-Krüppel* Family of Human Genes, J. Ruppert, K. Kinzer, A. Wong, S. Bigner, F. Kao, M. Law, H. Seuanez, S. O'Brien, B. Vogelstein, Mol. and Cell. Biol. *8*, 3104-3223 (1988).

Homeotic and Late Steps in Differentiation

Molecular Genetics of the *bithorax* Complex in *Drosophila melanogaster*, W. Bender, M. Akam, F. Karch, P. Beachy, M. Peifer, P. Spierer, E. Lewis, D. Hogness, Science *221*, 23-29 (1983).

The Elements of the *Bithorax* Complex, P. Lawrence, G. Morata, Cell *35*, 595-601 (1983).

The Location of *Ultrabithorax* Transcripts in *Drosophila* Tissue Sections, M. Akam, EMBO Journal *2*, 2075-2084 (1983).

Homoeosis in *Drosophila*: The *Ultrabithorax* Larval Syndrome, P. Hayes, T. Sato, R. Denell, Proc. Natl. Acad. Sci. USA *81*, 545-549 (1984).

Structural Relationships Among Genes that Control Development: Sequence Homology Between the *Antennapedia*, *Ultrabithorax*, and *fushi tarazu* Loci of *Drosophila*, M. Scott, A. Weiner, Proc. Natl. Acad. Sci. USA *81*, 4115-4119 (1984).

A Conserved DNA Sequence Found in Homeotic Genes of *Drosophila Antennapedia* and *Bithorax* Complexes, W. McGinnis, M. Levine, E. Hafen, A. Kuroiwa, W. Gehring, Nature *308*, 428-433 (1984).

Fly and Frog Homeo Domains Show Homologies with Yeast Mating Type Regulatory Proteins, J. Shepherd, W. McGinnis, A. Carrasco, E. De Robertis, W. Gehering, Nature *310*, 70-71 (1984).

A Homologous Protein-coding Sequence in *Drosophila* Homeotic Genes and its Conservation in Other Metazoans, W. McGinnis, R. Garber, J. Wirz, A. Kiroiwa, W. Gehring, Cell *37*, 303-308 (1984).

Protein Products of the Bithorax Complex in *Drosophila*, R. White, M. Wilcox, Cell *39*, 163-171 (1984).

The *Drosophila* Developmental Gene *snail* Encodes a Protein with Nucleic Acid Binding Fingers, J. Boulay, C. Dennefeld, A. Alberga, Nature *330*, 395-398 (1987).

The Structural and Functional Organization of the Murine HOX Gene Family Resembles that of Drosophila Homeotic Genes, D. Duboule, P. Dollé, EMBO Journal *8*, 1497-1505 (1989).

Involvement of the *Xenopus* Homeobox Gene *Xhox3* in Pattern Formation Along the Anterior-posterior Axis, A. Ruiz-Altaba, D. Melton, Cell *57*, 317-326 (1989).

Structure and Activity of the Sevenless Protein: A Protein Tyrosine Kinase Receptor Required for Photoreceptor Development in *Drosophila*, M. Simon, D. Bowtell, G. Rubin, Proc. Natl. Acad. Sci. USA *86*, 8333-8337 (1989).

The *Drosophila* Seven-up Gene, a Member of the Steroid Receptor Gene Superfamily, Controls Photoreceptor Cell Fates, M. Miodzik, Y. Hiromi, U. Weber, C. Goodman, G. Rubun, Cell *60*, 211-224 (1990)

A Human HOX4B Regulatory Element Provides Head-specific Expression in *Drosophila* Embryos, J. Malicki, L. Cianetti, C. Peschle, W. McGinnis, Nature *358*, 345-347 (1992).

Lambda Phage Integration and Excision

18

This chapter discusses lambda phage DNA integration into and excision from the bacterial chromosome. Integrated phage are often called prophage because, although they are not expressing most of their genes, they have the capability of producing phage upon induction. Early studies had suggested that the DNA of a prophage might be associated with the chromosome in a specific location, but the idea of integration rather than some other form of attachment was hard to accept because it seemed too dangerous biologically, and too difficult biochemically for specific pieces of DNA to be cut and rejoined. Now, of course, having seen the typical reactions utilized in genetic engineering, these concepts are familiar, and we can focus on the aspects yet to be understood. In retrospect, we might have expected that chromosomes were not unchanging since long before study of phage lambda began, classical geneticists had observed that DNA segments of the *Drosophila* chromosome generated by X-rays reintegrated so as to generate chromosomal rearrangements.

In addition to this chapter on lambda phage integration, two other chapters also cover topics related to site-specific genetic recombination. The recombination processes to be discussed require special proteins or enzymes and usually involve a special DNA site on at least one of the participating DNA molecules. Later we will cover other DNA segments that integrate into or excise from the chromosome. One of the most striking findings about these integrating sequences, now called transposons, is that they are ubiquitous. They have been found in all organisms that have been carefully examined. The final chapter of the three treats the generation of diversity of binding specificities in antibody molecules. In part this is accomplished by shuffling DNA sequences within the chromosomes of cells in the immune system. Hence, in contrast to transposons, which can be viewed as parasites, the moveable DNA sequences may also be of great positive value to an organism.

Mapping Integrated Lambda

How do we know lambda phage associates with the chromosome when it lysogenizes a cell? Lambda lysogens could possess copies of the phage DNA freely floating in the cytoplasm. A problem generated by this mode of existence is that the molecules would be randomly inherited by daughter cells on division. As a result, the lower the average number of lambda molecules per cell, the higher the fraction of daughters that might fail to inherit any and therefore become nonlysogenic. To explain lambda's low rate of spontaneously becoming nonlysogenic, which is about 10^{-6} per doubling, cells would have to possess unacceptably large numbers of the phage genomes. A second method to ensure proper inheritance of lysogeny would be for a nonintegrated lambda to possess a special segregation mechanism. Several genetic elements use this mechanism for stable inheritance. These include the phage P1 and the F factor, both of which have been discussed earlier in other contexts. This mechanism does not require large numbers of the lambda genome to be present. The third and simplest segregation mechanism would be for lambda to attach to or integrate into the host chromosome. Then it would be replicated and segregated into daughter cells with the host chromosomes.

Elegant genetic experiments showed not only that lambda associates with the chromosome but that it associates with a specific site on the chromosome. Now, of course, a simple Southern transfer experiment could settle the issue. The original data suggesting that the lambda DNA was associated with a particular region of the chromosome came from

Figure 18.1 The transfer frequency of markers from lambda lysogens or non-lysogens. Marker X^+ is located near the origin of transfer, and markers Y^+ and Z^+ are transferred after the point of lambda insertion.

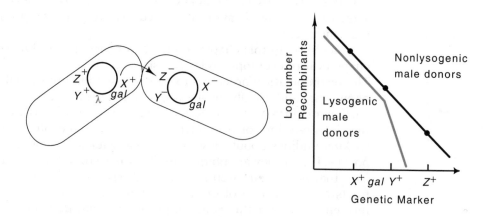

mating experiments between lysogenic male bacteria and nonlysogenic female bacteria.

Transfer of genetic markers (see Chapter 8) lying between the origin of transfer and the *gal* region showed no differences between lysogenic and nonlysogenic males. The frequency of female recombinants incorporating genetic markers transferred after the *gal* genes, however, revealed a discrepancy (Fig. 18.1). Their frequency was much lower if the DNA was transferred from lysogenic males than if the DNA came from nonlysogenic males. This can be understood as follows: if lambda is attached to the host genome in the vicinity of *gal*, then when it enters a nonlysogenic female, it finds itself in an environment lacking lambda repressor. As a result, the phage induces and proceeds through a lytic cycle analogous to infecting a cell. This process of a phage inducing upon its transfer into a nonlysogenic female is called zygotic induction. Its parallel for the *lac* operon leads to temporary induction of the *lacZYA* genes upon their introduction into cells deleted of the operon. Later as *lac* repressor accumulates, transcription of the genes shuts off.

A minor paradox is raised by the lambda results discussed above. The DNA is transferred to females in a linear fashion beginning at a particular point. Therefore it would seem that if a genetic marker lying ahead of the integrated lambda were transferred to females, the lambda itself eventually would also be transferred. Why are such cells not killed by zygotic induction?

The resolution to the paradox is that the majority of recipient cells that receive a *gal*-proximal marker also do not also receive *gal*. The DNA strand being transferred frequently breaks and conjugation stops. Even though genetic markers lying near the origin are transferred at high efficiency, markers lying farther away, and therefore transferred later in the mating process, are transferred with substantially lower efficiencies. Consequently, most cells receiving a marker proximal to lambda or *gal* never receive lambda or *gal*; so zygotic induction occurs in only a tiny fraction of the cells that receive markers transferred before *gal*.

Simultaneous Deletion of Chromosomal and Lambda DNA

The simplest method for attaching lambda to the host chromosome would be to insert it directly into the DNA. The simplest ideas are not always correct, however, and in the mid-60s when the question was being considered, chromosomal DNA seemed sacrosanct. Experiments were designed to determine whether lambda did integrate into the chromosome or whether it merely was stuck onto a special place on the chromosome. The genetic demonstration of lambda's insertion into the host chromosome utilized deletions. The conclusion would be that lambda is integrated into the chromosome if a deletion of host genes extends into the lambda and removes some, but not all, lambda genes.

Often the identification of deletions is difficult. In this case, however, their identification and isolation were easy. The nitrate reductase complex permits *E. coli* to use nitrate as an electron sink in the absence of oxygen. This enzyme complex is not entirely specific, and it will reduce

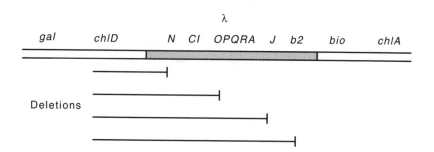

Figure 18.2 Deletions from *chlD* into lambda genes that are consistent with lambda being directly integrated into the host chromosome.

chlorate as well. The product, chlorite, however, is toxic. Such a situation is a geneticist's dream, for then mutants that are not killed by the chlorate can easily be isolated. These mutants have lost the ability to reduce chlorate, and many have been deleted of part or all of the nitrate reductase genes. Although it would be appropriate to name the genes involved after nitrate, the genetic locus is frequently called *chl* for chlorate resistance.

The deletions in the various *chl* loci frequently extend into adjacent genes. If chlorate-resistant mutants are isolated in a lambda lysogen, a few are found to be deletions and some of these are missing some but not all lambda genes. The pattern of lambda genes remaining or deleted is always consistent with the lambda genome being linearly incorporated into the chromosome in a specific orientation with respect to the flanking genes (Fig. 18.2). If some of the lambda genes have been deleted, how does one show that some lambda genes remain? The cell certainly cannot produce infective lambda. A method called marker rescue answers the question. Suppose we wish to test whether a lysogen with a deletion still has an intact *J* gene. A lawn of these cells is poured in soft agar on a plate and a small volume of a $\lambda imm^{434}susJ$ (nonsense mutation in *J*) stock is spotted onto the lawn. Being heteroimmune, the superinfecting phage is not repressed even if the deleted lysogen still possesses immunity. A few of these infecting phage exchange their defective *J* gene by recombination for the good one from what remains of the partially deleted lysogenic lambda. Sufficient phage make this replacement, that cycles of growth, lysis, and infection of surrounding cells produces a turbid spot on the opaque lawn. If no *J* gene is present on the partially deleted chromosome, the infecting $\lambda imm^{434}susJ$ cannot proceed through multiple growth cycles, and the lawn is unperturbed. In some cases, a gene on the partially deleted lysogen will be able to complement the superinfecting phage as well as recombine with it, but the mapping results are unaltered. Only if the gene is present in the host cells can the superinfecting phage grow.

Marker rescue experiments on *chl* deletions revealed many deletions that extended only part way into the lambda genes. Certainly the simplest explanation for this result is that the DNA of lambda is contiguous with the host DNA.

DNA Heteroduplexes Prove that Lambda Integrates

Davidson and Sharp directly demonstrated the insertion of lambda DNA into the genome using electron microscopy. They used a procedure developed to examine sequence homologies between strands from different DNA molecules. After denaturation to separate strands of the two input DNA populations, the strands are allowed to anneal. Then the DNA is mixed with a cytochrome C solution to coat the molecules and increase their diameter as well as improve their staining properties. To spread out the long, snarled DNA molecules, a small volume of the DNA plus cytochrome C is layered on top of a buffer. As the protein spreads and forms a monolayer, the DNA molecules are stretched out and can then be picked up on an electron-transparent support and stained. Single- and double-stranded DNA visualized by this procedure are easily distinguished because single-stranded DNA is more flexible and therefore curlier than double-stranded DNA.

When two homologous DNA strands reanneal, a simple double-stranded molecule results. A heteroduplex between a strand deleted of a stretch of sequence generates a single-stranded "bush" on the other strand, and if the two strands possess a stretch of sequence without homology, a "bubble" is formed. The heteroduplexes formed between an F'-factor containing an integrated lambda genome and lambda DNA showed the structure expected if lambda were to insert itself into the chromosome (see problem 18.4).

Gene Order Permutation and the Campbell Model

Determination of the lambda prophage gene order in lysogens by deletion mapping exposed an anomaly. The order is a particular circular permutation of the order found by recombination frequencies between genes for vegetatively growing lambda (Fig. 18.3). On the basis of a small

Figure 18.3 The gene orders of lambda phage determined for lysogens by deletion mapping and for lytically infected cells.

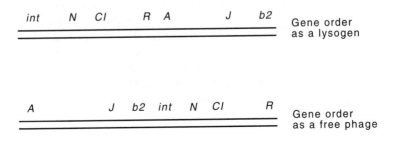

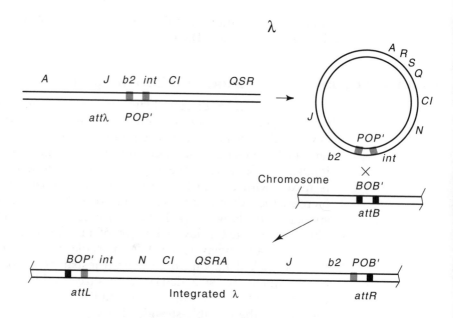

Figure 18.4 The Campbell model for lambda integration. Linear phage DNA circularizes soon after infection by association of the *A* and *R* ends. Integration occurs by means of a site-specific recombination event in the *O* regions of the phage and bacterial DNA to generate the integrated phage. *P* and *P'*, and *B* and *B'* are the phage and bacterial DNA sequences required for this process.

number of genetic experiments Campbell proposed a brilliant explanation for the gene order problem and also made several other predictions
concerning problems on integration. The Campbell model was that the
lambda DNA, which is linear within the virion, circularizes upon entering a cell and then opens its genome at another point as it reciprocally
recombines via a site-specific recombination event mediated by a specialized recombination machinery into the chromosome of the host
(Fig. 18.4). A simple notation has developed for the DNA elements
involved in integration and excision. The region of the host DNA into
which lambda inserts is labeled *BOB'*. *B* denotes the left portion of the
bacterial sequence into which lambda inserts, *O* is the crossover point
or region of the recombination between phage and chromosomal DNA,
and *B'* is the right portion of the bacterial sequence. The entire region
BOB' is called *attB*. Similarly, the phage regions are called *POP'* and *attP*.
The products of integration are *BOP'*, which is also called *attL*, and *POB'*,
which is called *attR*. Note that this notation does not prejudice the
discussion by assuming that *B* and *P* are homologous or that *B* is the
same as *B'*. In fact, we can predict that the system would not depend on
lengthy homologies between *BOB'* and *POP'* because if it were, the
general recombination system could then recombine to insert or excise
a phage genome.

Isolation of Integration-Defective Mutants

A regulated *BOB'-POP'*-specific recombination system would solve lambda's problem of restricting integration and excision reactions to the right time and right sites on the phage and host genomes. That lambda possesses such a recombination system could be proven by the isolation of lambda mutants unable to lysogenize due to the absence of an integration enzyme. The main difficulty in such a proof is a common problem in genetics—that of identifying the desired mutant.

At the time that integration mutants were being sought, a lambda mutant was already known that permitted good guesses to be made about the properties of the desired integration-defective mutants. The lambda mutation deletes the phage *b2* region. It extends into the *POP'* region and leaves the phage deficient in its integration abilities. Most of the λ*b2* phage that infect cells without lysing them establish repression but do not integrate into the chromosome. They therefore form turbid plaques similar to those of wild-type lambda. If cells from the turbid center of such a plaque are streaked on solidified medium and allowed to proceed through additional generations of growth in the absence of superinfecting phage, the λ*b2* are diluted away since they do not replicate by their own *ori*, and they are not integrated into the chromosome to be replicated by the host. Therefore the resultant cells are nonlysogenic.

When λ*b2* phage and wild-type λ coinfect cells, the wild-type λ cannot complement the defect in the λ*b2*. This shows that the defect in λ*b2* is in a site required for the integration, and not in a protein that can act in *trans*. Even so, the λ*b2* provide a good indication of the behavior to be expected of mutants defective in any phage-encoded proteins required for integration. Mutants defective in the ability to integrate should form turbid plaques but not stably integrate. They could have mutations in phage proteins required for the process of integration or mutations in the DNA sites directly used in integration. Because the number of nucleotides in the DNA sites ought to be much smaller than the coding region for the integration proteins, the majority of integration-defective mutants isolated ought to be in the integration proteins.

The two steps of growing cells to dilute away the nonintegrated lambda and testing for lambda immunity can be combined. The candidate cells are streaked on plates containing the pH indicators eosine yellow and methylene blue as well as having about 10^7 λ*CI*⁻ mutants spread on the surface. The cells spread on the plate grow, divide, and occasionally encounter a λ*CI*⁻. If a cell infected by a λ*CI*⁻ particle does not possess lambda immunity, the infecting phage is not repressed, and it grows and lyses the cell. This releases some acid and changes the color of the pH indicators. Adjacent cells may also be infected and lysed, but because the colony stops growing before all cells can be lysed, some cells remain in the colony. A colony of nonlysogenic cells spotted on these plates therefore grows into a ragged, purple colony, whereas colonies of immune cells yield smooth, pink colonies.

A brute-force pick-and-spot technique based on the knowledge of the behavior of $\lambda b2$ mutants on the indicating plates permitted the isolation of lambda mutants unable to lysogenize. These are called *int* mutants. Some of them were nonsense mutants, proving that the phage encodes a protein that is required for integration.

Isolation of Excision-Deficient Mutants

As expected, the *int* mutants can be helped to lysogenize by complementation. Not surprisingly, these mutants are also found to excise poorly without assistance. Hence Int protein is also required for excision. Although it would seem that the process of integration should be readily reversible with the same components that are used for integration, surprisingly, a phage-encoded protein is required for excision but not integration.

A phenomenon known as heteroimmune curing forms the basis of a simple demonstration that excision requires a protein in addition to Int protein. If lambda lysogens are infected with a heteroimmune phage like λimm^{434}, most cells are lysed, but many of the survivors are found

$$\lambda\,imm^{434} \quad + \quad \text{Lysogen} \quad \longrightarrow \quad \text{Some nonlysogens}$$

to have been cured of the lambda. This is called superinfection curing. Apparently the superinfecting heteroimmune phage provides diffusible products that facilitate excision of the prophage. Some Int$^+$ deletion phage, but not others, can promote curing when they superinfect lysogens of different immunity. Thus something in addition to Int protein must be involved in excision.

The isolation of a nonsense mutation in the gene required for excision proved that it coded for a protein. This isolation used heat-pulse curing. If a lysogen of λCI_{857} growing at 32° is heated to 42° for five minutes and then grown at 32°, the heat-sensitive CI_{857} repressor is first denatured, and phage growth begins. Then, after the cells are cooled to 32°, the repressor renatures and further phage development ceases before sufficient phage products have accumulated to kill the cells. In the five minutes of derepression however, sufficient phage proteins are synthesized that lambda can excise from the chromosome. After further cell growth, the excised lambda genome is diluted away, and daughter cells that are cured of lambda appear at high frequency.

Heat-pulse curing was used to isolate excision-defective mutants in the following way (Fig. 18.5). A mutagenized stock of λCI_{857} was used to lysogenize cells. This step selected for phage retaining the ability to lysogenize. The lysogenic cells resulting from this step were grown at 32° and then replicated onto three petri plates. One was incubated at 32°, one was incubated at 42°, and the third was incubated at 42° for five minutes, at 32° for two hours, and then at 42° overnight.

The first plate kept viable copies of all the colonies. The second plate showed which colonies were infected with lambda, as growth at 42°

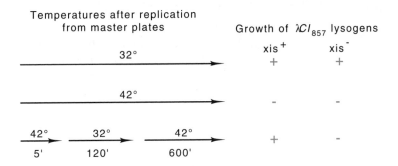

Figure 18.5 Temperature protocols for the identification of *xis*⁺ and *xis*⁻ by the inability of the latter to excise following a short heat induction. Each of the three temperature time lines represents the conditions the three petri plates were exposed to.

would induce the phage and kill lysogens. Growth on the third plate indicated which colonies could heat-pulse cure. On this plate, the lambda lysogens would heat-pulse cure, and the cured cells in such a colony would be capable of growing into colonies during the subsequent growth at 42°. Any colony whose cells possessed excision-defective phage could not heat-pulse cure and would therefore be killed by the attempted growth of the unexcised prophage during the subsequent extended exposure to 42°.

Later, a simpler method for detecting excision-defective mutants was devised. This scheme uses cells in which lambda has mistakenly integrated into a site within the *gal* genes that resembles the authentic *attB* site. The cells are Gal⁻ as a result. Infection of these cells by heteroimmune phage able to provide Int and Xis proteins in *trans* catalyzes excision of the phage from the *gal* genes, and the cells become Gal⁺. If such infected cells are plated on galactose plates, plaques with Gal⁺ revertants are red on medium containing galactose indicator dye and Gal⁻ plaques are white. With this convenient assay, the excision abilities of many different phage can be assayed on a single galactose indicator plate.

Using the plate assay for excision, Enquist and Weisberg performed a thorough genetic analysis of the *att-int-xis* region of the lambda chromosome. They isolated and characterized hundreds of *int* and *xis* point mutations. The mutations fell into two complementation groups indicating that two genes were involved. One codes for the Int protein and one for the Xis protein. The mutations were mapped with a set of deletions ending in the region. The large number of mutants studied permitted a reasonable estimation of the sizes of the genes. The *int* gene appeared to be about 1,240 base pairs long, and the *xis* gene was very small, only 110 base pairs long. No additional phage genes acting in *trans* and directly involved in the integration or excision process were discovered.

Properties of the *int* and *xis* Gene Products

How many phage-encoded proteins are required for integration and excision? If Int and Xis proteins will function in *trans*, then complementation tests can be performed to determine the number of *int* genes and the number of *xis* genes carried on the phage. Experiments showed that the Int protein of one phage will indeed help another phage integrate, for cells coinfected with a λ*int* mutant and a heteroimmune λ*imm*⁴³⁴ phage yield some lambda lysogens in addition to λ*imm*⁴³⁴ lysogens and

$$\lambda \, imm^{434} \; + \; \lambda int^-_x \quad \xrightarrow{\text{Infect}} \quad \lambda \text{ and } \lambda imm^{434} \text{ lysogens}$$

double lysogens. In contrast, coinfection of cells with two different λ*int* mutants does not yield lysogens at the frequencies normally observed for complementation. All the *int* mutations tested therefore must lie in

$$\lambda \, int^-_x \; + \; \lambda int^-_y \quad \xrightarrow{\text{Infect}} \quad \text{No \; lysogens}$$

the same gene. Analogous experiments with *xis* mutants show that they too are confined to a single gene.

Int and *xis* mutations are not complemented by most phage other than lambda. Although φ21 possesses its own Int and Xis proteins, these do not complement lambda *int* or *xis* mutants. Thus, *int* and *xis* can be phage-specific.

$$\lambda \, int^-_x \; + \; \phi 21 \quad \xrightarrow{\text{Infect}} \quad \phi 21 \text{ Lysogens only}$$

$$\lambda \, int^-_x \; + \; \phi 80 \quad \xrightarrow{\text{Infect}} \quad \phi 80 \text{ Lysogens only}$$

Incorrect Excision and *gal* and *bio* Transducing Phage

Lambda phage can act as a specialized transducer of certain host genes. This means that each phage particle in some lambda lines carries a host gene in addition to essential phage genes. The host gene on the phage DNA can be brought into an infected cell and either complement a defective copy of a gene in the host or can recombine with the host copy of the gene. The process of carrying the DNA across and incorporating it into the chromosome of the infected cell is called transduction. Because only certain genes can be transduced by lambda, and because each phage particle in a lysate of a transducing phage carries a copy of the host gene, the process is called specialized transduction. Specialized transduction by lambda should be contrasted to generalized transduction by P1 in which only a small fraction of the phage particles contain nonphage DNA, only a small fraction of these contain DNA of any particular bacterial gene, and the DNA in a transducing particle contains no phage DNA.

Transducing phage once were necessary for the enrichment and ultimate isolation of regions of DNA involved in regulation of bacterial genes. Also, cells infected with transducing phage often hypersynthesize the proteins they encode owing to the presence of multiple phage genomes during much of the phage growth cycle. Without this gene dosage effect, the proteins could be present in quantities too low for biochemical study. The enormous value of transducing phage in research on bacterial gene regulation stimulated efforts to develop genetic engineering techniques so that DNA, proteins, and gene regulation systems from other organisms could also be studied.

Upon induction of a lysogen, most prophage excise uneventfully. At a frequency of 10^{-5} to 10^{-7}, however, a prophage excises incorrectly from the host chromosome and picks up bacterial DNA bordering the phage integration site. These incorrect excision events produce transducing phage. One of the joys and powers of genetics research is that exceptionally rare events such as these may be captured and perpetuated for further study.

Figure 18.6 Production of *gal* and *bio* transducing phage by recombination events located to the left or to the right of the end of an integrated lambda.

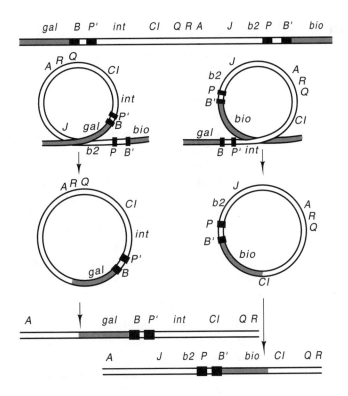

When lambda excises incorrectly from a point other than *BOP'* and *POB'*, bacterial DNA on one side or the other of the integrated lambda DNA can be picked up by the phage (Fig. 18.6). These incorrect excision events could utilize the *int* and *xis* products, but more likely use cellular enzymes that break and rejoin DNA. If the crossover sites are not too far apart for the intervening DNA to fit within the phage head, the resulting DNA can be encapsidated to form a viable phage particle. Usually, for packaging to occur some phage genes must be deleted from the resulting excision product. Looping out to the left produces a phage with *gal* genes substituting for the *b2* region. That is, *gal* DNA in the phage is flanked on both sides by lambda genes. Also, *attL*, or *BOP'*, replaces the normal phage attachment region *POP'*.

The *gal* genes are sufficiently far from *attλ* that phage having picked up *gal* genes must, of necessity, have left at least the *J* gene behind in order that the transducing phage genome be small enough to be packaged in a lambda coat. These phage are defective because they do not yield viable phage upon infection of cells. They are propagated by coinfecting them with phage that provide their missing *J* gene product. Such defective *gal* transducing phage are called λ*dgal*.

Phage that excise the other way pick up the biotin genes. These substitute for the nonessential *int*, *xis*, and other early genes under control of p_L. Usually these phage are not defective and hence are called λ*pbio* for their ability to form plaques and transduce biotin genes. They possess *attR*, *POB'*.

Transducing Phage Carrying Genes Other than *gal* and *bio*

Isolation of *gal* or *bio* transducing phage is straightforward because these genes flank the lambda attachment site. For genes located farther away, a simple improper excision event cannot produce transducing phage. Genes located perhaps as far as a minute away on the genetic map, about 40,000 base pairs, from an integrated phage can be picked up by a combination of a deletion of the intervening DNA followed by an improper excision event. In this case the combined probabilities for such a double event are so low that large cell cultures must be used to find the transducing phage. What can be done to isolate phage that carry genes located more than a minute from *attB* of lambda?

Two approaches were used to bring an integrated phage and a particular gene close enough together to permit generation of transducing phage. Either the gene was moved close to the integrated phage genome or the phage was moved close to the gene. The phage can be brought near the gene by selecting for lysogens following infection of a host deleted of the phage attachment region *attB*. At very low but usable frequencies, lambda will then integrate into sites that possess some similarity to the normal chromosomal integration site. These secondary integration sites are so widely scattered over the chromosome that lambda can be forced, albeit at very low frequencies, to integrate into or adjacent to most genes. Either the desired integrant can be selected by genetic means or the entire population of cells with phage integrated at many different sites can be used as a source of phage. Some of the phage will excise incorrectly from their abnormal positions and transduce the desired adjacent bacterial genes. The second method of forcing an integrated phage to be near a target gene is to select for the illegitimate recombinational insertion of an episome carrying the gene into a site adjacent to the prophage.

Use of Transducing Phage to Study Integration and Excision

One use of transducing phage is to demonstrate that lambda normally integrates and excises at precisely the same point. Of course, lambda can be integrated and excised many times from the bacterial *att* region, and the region apparently suffers no harm. Nonetheless, how do we know, without sequencing, that bases are not inserted or deleted in the process? The integration of lambda into secondary *att* sites provided the proof that, as far as the sequence of the host chromosome is concerned, excision is the exact opposite of integration.

When lambda integrates into a secondary *att* site, the gene into which lambda has inserted is disrupted and therefore inactivated, but when the lambda is induced and excises from these sites, the majority of the surviving cells possess a perfectly normal gene. Few of the lambda improperly excise and produce transducing phage as described in the previous section. That is, except for the products of the rare improper excision events, no nucleotides are inserted or deleted at the pseudo *att* site. Therefore it is reasonable to infer that the integration and excision cycle at the normal *att* site similarly does not alter its sequence. For example, one site for secondary lambda integration is a gene coding for a protein required for proline synthesis. The insertion of lambda makes the cells Pro⁻, but heat-pulse curing leaves the cells Pro⁺ (Fig. 18.7).

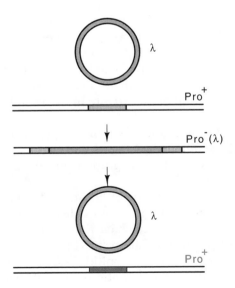

Figure 18.7 Integration of lambda phage into a pseudo *att* site in a gene inactivates that gene, but excision restores the original nucleotide sequence and the gene is reactivated.

A second use of transducing phage is in the study of the biochemistry of the integration and excision reactions themselves. These site-specific recombination events take place at the *att* regions, but the partners need not be confined to a phage and host chromosome. For example, the enzymatic equivalent of an excision reaction can be performed between λ*dgal* and a λ*pbio* to form a λ and a λ*dgal-bio*. To detect these recombination products, the input phage must be genetically marked. This can be done by using nonsense mutations located in the *A* and *R* genes, which are at opposite ends of the lambda DNA. The cross might therefore be between λ*dgalA⁻R⁺* and λ*pbioA⁺R⁻*; the frequency of generation of wild-type lambda, those able to form plaques on su⁻ cells, could

$$BOP' \;+\; POB' \;\longrightarrow\; BOB' \;+\; POP'$$
$$(\lambda A^-dgal) \quad (\lambda R^-pbio) \qquad (\lambda A^-R^-dgalbio) \quad (\lambda)$$

be measured. Crosses between all combinations of *att* regions can be performed by similar approaches. The main results of such studies examining the site-specific recombination events catalyzed by Int and Xis proteins are that all combinations of *att* regions will recombine, but at different rates, and that the Xis protein is required only for the excision type of reaction.

The Double *att* Phage, *att²*

The preceding sections have described a little of what has been learned from genetic experiments about lambda's ability to integrate and excise. One way to study the biochemistry of integration and excision is to construct an *in vitro* system that mimics the *in vivo* reaction. The first requirement for integration or excision is to bring the two participating molecules close together. *In vivo*, this requirement is partially met as a simple consequence of the fact that both DNA molecules are confined to the volume of the cell and therefore are held in close proximity. The *in vitro* integration or excision reaction should be greatly speeded if the two *att* regions similarly can be forced close to one another. One way to accomplish this is to place both *att* sites on the same DNA molecule. Then the concentration of one *att* site in the vicinity of the other is, of necessity, high.

This section describes the isolation and properties of such a double *att* phage, and the following sections describe its use and the use of a similar phage to study the integration and excision reactions more deeply both *in vivo* and *in vitro*.

Some distance to either side of an integrated lambda are two sites that the host recombination system will recombine at a reasonable frequency. This event excises a phage that has picked up host DNA from both sides of the phage integration site (Fig. 18.8). Ordinarily this event cannot be detected because the resulting DNA is too long to be packaged in a lambda coat, but if several deletions have first been put into lambda,

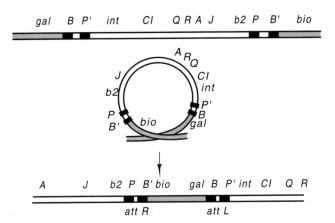

Figure 18.8 Recombination between two points beyond the ends of an integrated lambda phage generated the λ*att²* phage containing *attR* and *attL*.

this particular excision product will fit into the head, and viable phage are formed.

The structure of the phage produced by the recombination between two sites flanking lambda is most interesting. The phage contains two *att* regions, *attL* and *attR*, and is called *att²*. Such an *att²* phage can lose the extra bacterial DNA by a reaction analogous to normal excision:

$$BOP' + POB' \longrightarrow POP' + BOB'$$

This process requires the Int and Xis proteins and produces a viable lambda phage genome and a minicircle (Fig. 18.9).

Study of the excision reaction requires quantitating the input phage λ*att²* as well as the product phage. Fortunately, the two may be readily distinguished. The concentrations of λ*att²* and the λ generated by excision can be assayed in a mixture of the two by first separating the phage on the basis of density in equilibrium centrifugation. The λ*att²* phage is more dense than lambda as a result of its additional DNA.

A simpler assay of the two types of phage makes use of the sensitivity of lambda to heat and chelators of Mg⁺⁺. Removal of Mg⁺⁺ ion from the phage reduces the charge neutralization between phosphates of the

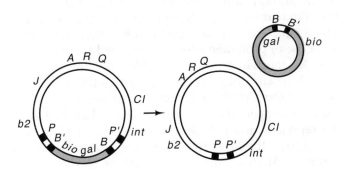

Figure 18.9 An excision reaction between the two *att* regions on *att²* generates a viable phage with a single *att* region and a minicircle.

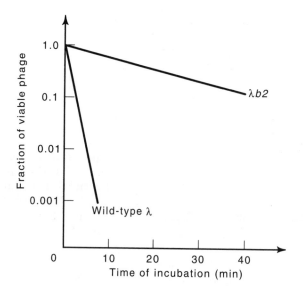

Figure 18.10 Killing of λ*b2* and wild-type λ by heating in the presence of a Mg^{++} chelator.

DNA backbone, and as a result, the DNA expands and can burst the phage coat. Phage particles possessing less than the usual amount of DNA are more resistant to heating and removal of Mg^{++}. Combinations of Mg^{++} chelators and elevated temperatures can be found in which wild-type lambda is killed by factors of 10^3 to 10^4, but the λ*b2* deletion mutant is unharmed (Fig. 18.10). The same principle can be used to titer λ in the presence of λ*att²*. If the mixture is titered on plates containing pyrophosphate, which chelates magnesium, only the λ can proceed through multiple infective cycles and generate plaques. By titering also on normal plates, the total number of both phage types present can be determined since both form plaques.

Demonstrating Xis is Unstable

When lambda infects a cell and proceeds down the lysogenic pathway, ultimately it must integrate stably into the chromosome. If both Int and Xis proteins are synthesized, the phage DNA would repeatedly integrate and then excise from the chromosome until these proteins have been diluted away by growth or have become inactive. What ensures that the last event is integration rather than excision? The phage utilizes two mechanisms to maximize the probability that integration activity predominates under these conditions. One is the preferential synthesis of Int protein via the p_I promoter which is activated by CII protein. The second is for Xis protein to be less stable than Int. Weisberg and Gottesman used the *att²* phage to simplify testing this hypothesis with an *in vivo* assay of the stability of Int and Xis proteins.

The test used a lysogen of λ*CI$_{857}$* briefly heat-pulsed to denature the thermolabile repressor and induce a burst of Int and Xis protein synthesis. The repressor then renatured, and the phage-encoded protein synthesis stopped. At times thereafter, portions of the culture were

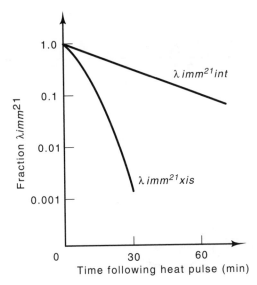

Figure 18.11 The fraction of $\lambda imm^{21}att^2int^-$ or xis^- as a function of time that is converted to λimm^{21} by the Int and Xis proteins induced at zero time.

additionally infected with $\lambda imm^{21}att^2xis^-$ or $\lambda imm^{21}att^2int^-$. Recall that an excision-type of reaction on λatt^2 removes the extra bacterial DNA from the phage. At early times after the heat pulsing, the Int and Xis protein already present from the first phage complemented the *int* or *xis* defect of the att^2 and catalyzed the excision reaction to produce the smaller single *att* λimm^{21} from the $\lambda imm^{21}att^2$. At later times, however, the Int or Xis synthesized by the first phage had decayed, and therefore fewer wild-type λimm^{21} were produced (Fig. 18.11). Simply infecting at various times after the heat pulse and assaying for the number of non-att^2 lambda that were produced in the resulting phage burst allowed the kinetics of decay of the Int and Xis protein to be measured. The half-life of Int was greater than an hour, whereas the half-life of Xis was less than half an hour.

Inhibition By a Downstream Element

A lambda phage not integrated into the host chromosome and proceeding down the lytic pathway would do well to minimize its synthesis of Int protein. On the other hand, an induced lambda located in the chromosome has great need for Int protein to assist in excision. How could the phage know whether or not it is integrated into the chromosome and how could it therefore adjust Int protein synthesis? In both situations, the *int*-specific promoter p_I is off and the synthesis and regulation of Int and Xis proteins derives from the p_L promoter.

The location of the *b2* region of the phage signals whether it is integrated in the chromosome or is free in the cytoplasm. When the phage has excised it is circular, and the *b2* region is adjacent to the *int* gene (Fig. 18.12). On the other hand, when the phage is integrated, host DNA is adjacent to the *int* gene. These two situations have different

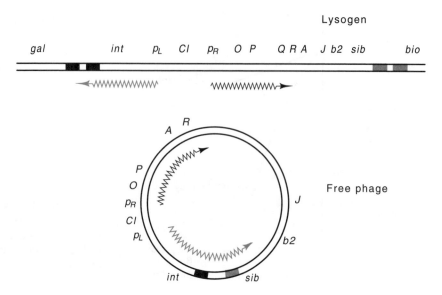

Figure 18.12 *Sib* is not part of the *int* messenger when lambda is transcribed as a lysogen.

effects on the translation efficiency of *int* messenger deriving from the p_L promoter. A sequence called *sib* lying in the *b2* region is transcribed and becomes part of the *int* messenger only when the phage is not integrated. This sequence inhibits translation of the *int* message.

The synthesis of Int and Xis proteins was characterized in cells that had been exposed with a low dose of UV light to block host-encoded protein synthesis. These cells were then infected with phage, and radioactive amino acids were added. At various times, samples were taken and subjected to SDS polyacrylamide gel electrophoresis. The Int and Xis proteins were easily detected and quantitated by autoradiography. Alternatively, no UV light was used, and instead the Int or Xis proteins were precipitated with antibody, further separated from the other proteins by the electrophoresis, and finally quantitated from autoradiographs.

Infection by wild-type lambda yields only a low synthesis of Int protein, but infection by λ*b2* yields high synthesis of Int protein. Point mutations and deletions have been isolated that yield high levels of Int protein following infection. These narrowly define a region in the *b2* region called *sib* which is responsible for the altered synthesis rate of Int. The deletions and point mutations have a strictly *cis* effect on Int protein synthesis. Therefore the effect is called downstream or retro-inhibition because it acts backward on expression of the preceding gene.

The *sib* region contains a typical transcription termination sequence, and the *sib* mutations alter the structure of the RNAse III-sensitive structure that can be formed in messenger that initiates from p_L (Fig. 18.13). Transcripts initiated at p_I normally terminate near the end of this region, but the transcripts initiated at p_L ignore termination signals because of the phage N protein bound to RNA polymerase, and the

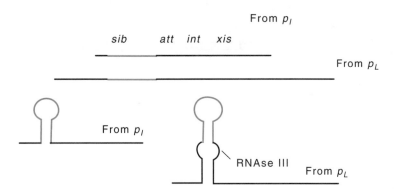

Figure 18.13 Secondary structures of the *sib* region in transcripts originating from p_I and p_L. The promoter p_I lies just ahead of *xis*, so the transcription start point does not include all of *xis*.

transcription continues past this region. The p_L-initiated transcripts are capable of forming an additional base-paired region that is sensitive to RNAse III. This enzyme that cleaves special double-stranded RNA structures. The transcripts initiated at p_I do not form the RNAse III-sensitive structure. Most likely, then, cleavage by RNAse III leads to degradation of the adjacent *int* gene messenger. As expected, the *sib* effect is absent in RNAse III-defective mutants.

In vitro Assay of Integration and Excision

Although ingenious genetic experiments can reveal much about biological systems, ultimately the details must be studied biochemically. Therefore it was of great interest to isolate the Int and Xis proteins. Such an isolation requires an assay for the proteins. The most reliable assay for Int protein activity is to seek a protein that carries out the entire integration reaction *in vitro*. As mentioned above, one likely requirement for such a reaction to proceed *in vitro* is a high concentration of *attB* and *attP* regions. The easiest way to obtain these is to place both on the same DNA molecule.

Can *attB* and *attP* be put on the same molecule, just as *attL* and *attR* could be put on λatt^2? Nash constructed such phage by selecting the product of a rare recombination event in a region of nonhomology between a $\lambda dgal\text{-}bio$ containing *attB* and a λ (Fig. 18.14). Similar to the excision-type of reaction that occurs on λatt^2, on $\lambda attB\text{-}attP$ an integration-type of reaction can occur. This removes the *bio* region and leaves the phage considerably more resistant to Mg^{++} chelation or heat. Hence, the parental phage and derivatives that have undergone an integration reaction and have become smaller can easily be distinguished.

Initial tests with the $\lambda attB\text{-}attP$ phage showed that an *in vitro* integration reaction catalyzed by an extract from cells would work. The

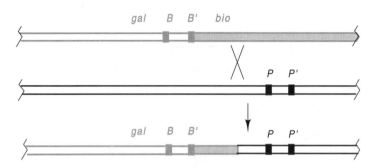

Figure 18.14 The illegitimate recombination by which the λattB-attP phage was generated.

experiment was performed by incubating the λattB-attP DNA with a cell extract prepared from heat-pulsed lambda lysogens. Then the DNA was extracted from the mix and used to transfect cells that had been made capable of taking up naked lambda DNA. The phage from the transfected spheroplasts were spread on a lawn of sensitive cells on a plate containing pyrophosphate. Under these conditions, only products from the integrative reaction would produce plaques. This provided an extremely sensitive assay for the integration reaction. Later, as the assay conditions were improved, the integration reaction could be assayed merely by physical quantitation of the DNA products. The locations of restriction enzyme cleavage sites in the DNA are rearranged by the integration reaction. This creates new sizes of restriction fragments that can be detected by separating the resulting fragments by electrophoresis (Fig. 18.15).

The *in vitro* integration reaction required Mg^{++}, ATP, and spermidine. More careful examination showed that the use of supercoiled DNA eliminated the requirement for ATP and Mg^{++}. The assays permitted the purification and characterization of biologically active Int protein, and analogous experiments have been done with the λatt^2 for the purification of Xis protein.

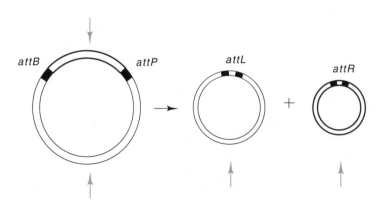

Figure 18.15 Digestion with a restriction enzyme generates different fragments from the λattB-attP phage and its integration "product," lambda and the minicircle.

Host Proteins Involved in Integration and Excision

As mentioned above, the integration reaction requires supercoiled DNA or relaxed circular DNA plus ATP and Mg^{++}. This suggests that the cell extracts contain a protein that can supercoil DNA using ATP for the required energy. Indeed, this proved to be so. By following the ability of an extract either to promote the integration reaction with relaxed circular substrates, or more simply, to produce supercoiled DNA from relaxed circular DNA, Gellert and coworkers identified and purified DNA gyrase. As explained in Chapter 2, this enzyme introduces negative superhelical twists in covalently closed DNA circles.

Further study of the *in vitro* integration reaction revealed the requirement for yet another host protein. This protein is directly involved in the integration and excision reactions. Extensive purification and characterization of this protein, which is called IHF, for integration host factor, showed it to be a dimer of rather small subunits, 11,000 and 9,500 molecular weight. The genes coding for these peptides, *himA* and *himD*, had been identified genetically from host mutations that block phage integration. It is surprising that cells should possess a nonessential protein that is required for integration of lambda. The IHF protein assists lambda phage integration and excision by helping the DNA bend into the complicated intasome structure that will be discussed below. This protein plays a similar role in other systems in which DNA bending is required.

Another host protein, called FIS, is also involved in the excision process. As the intracellular levels of this protein vary with the growth condition of the cells, it likely plays an important role in directing the phage into either the lytic or the lysogenic mode of existence, but it regulates only when Xis is present in limited amounts.

Structure of the *att* Regions

Historically the ability to sequence DNA came after geneticists had isolated the *att* mutants and had determined much about the integration

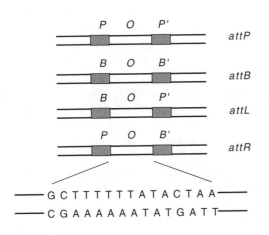

Figure 18.16 The *att* regions and nucleotide sequence of the O region of *att*.

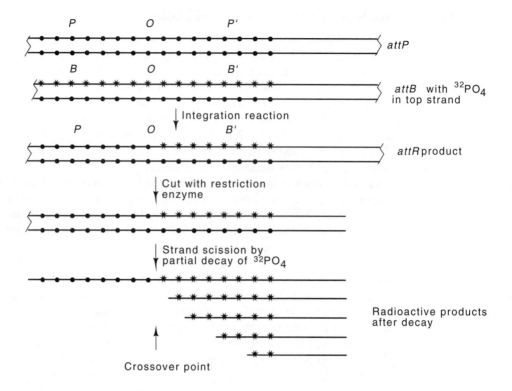

Figure 18.17 Scheme for determining the site of strand transfer in the integration reaction. Small circles represent nonradioactive phosphates and stars represent radioactive phosphates. More precision is gained by placing radioactive phosphates adjacent to only one of the four bases by using the appropriate α-labeled triphosphates when the strand is labeled.

and excision reactions. Therefore, obtaining the sequence of *att* would permit checking the genetic inferences. Sequencing the *att* region was straightforward because phage carrying the four possible *att* regions—*POP'*, *BOB'*, *BOP'*, and *POB'*—were available. Landy identified the appropriate restriction fragments from these phage, isolated and sequenced them. As expected from the genetic experiments, all four contained an identical central or core sequence corresponding to the *O* part of the *att* region. The size of the core could not have been determined from the genetic experiments, but it turned out to be 15 base pairs long (Fig. 18.16).

The recognition regions flanking the cores are of different lengths. By deletion analysis, the sequences on either side of *O* at the bacterial *att* region that are necessary for integration have been shown to be just large enough to specify an Int protein binding site. By contrast, *attP* is about 250 base pairs long.

The sites of Int-promoted cleavage of the *O* sites have been determined exactly using radioactive phosphate both as a label and as an agent to cleave the DNA. An *in vitro* integration reaction was performed between an *attP* and a short DNA fragment containing *attB* that had

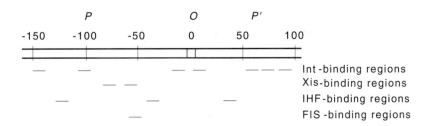

Figure 18.18 The regions in *attP* to which Int, Xis, FIS, and IHF proteins bind.

been extensively labeled with radioactive phosphate on one and then the other strand. This generated *attL* and *attR* sites. In the *attR* site, all phosphates that lie to the right of the crossover point derive from *attB* and are therefore both radioactive and subject to strand scission by radioactive decay. The positions of these radioactive phosphates were then determined by permitting some to decay. The decay event cleaves the phosphate backbones; subsequent electrophoresis of the denatured fragments on a sequencing gel yields bands corresponding to each position that had a radioactive phosphate and no bands of other sizes (Fig. 18.17). The exchange points are deduced to be at fixed positions on the top and bottom strands, but separated by seven base pairs.

Structure of the Intasome

One approach to the study of the integration and excision reactions is to determine where purified Int, Xis, FIS, and IHF proteins bind in the *att* regions. The proteins have a complex binding pattern in *attP* (Fig. 18.18). Int protein binds to seven sites in *attP*: two in the common core, two on the *P* arm, and three on the *P'* arm. Interestingly, the core sequences for Int protein binding are different from the arm sequences. Not surprisingly then, Int protein possesses two domains, an N-terminal domain which binds the arm sequences and the C-terminal domain, which binds the core sequence at the crossover point. IHF binds to three sites, Xis protein binds to two sites, and FIS binds to one site partially overlapping an Xis site. Together these binding sites cover the entire region from -150 to +100 with respect to the center of the common core.

Several facts suggest that the protein-*attP* complex is folded rather than being extended in a line. When the complex of the various *att* regions and the Int, Xis, FIS, and IHF proteins is examined in the electron microscope, a compact and topologically complicated structure is seen. This would make us fear that it could be an artifact except that a complex of Int plus Xis proteins bound at *attR* appears to pair specifically with a similar complex formed at *attL* on a different DNA molecule. Since the bacteria and phage *att* regions possess no sequence homology except for their common core sequences, either it is the bound proteins or it is recombined DNA that holds the pair of DNA

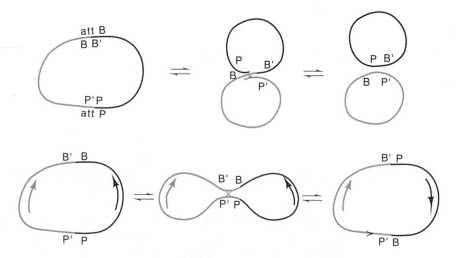

Figure 18.19 An integrative reaction between the *PP'* and *BB'* sites in opposite orientations on two different plasmids either generates two smaller circles, or it inverts one of the segments between the sites with respect to the other site.

molecules together. The *attB* site with which *attP* interacts is much more compact, possessing only the two core binding sites of Int protein. The *attB* site probably does not bind Int protein by itself under normal physiological conditions, but the bare DNA merely collides with the intasome complex at *attP* and the strand exchange process then begins.

One way to estimate the shape of *attP* might be to consider the bending in the DNA that is generated by each of the proteins which bind in the region. Another approach for learning about the structure of the intasome is to examine the topology of the DNA that engages in recombination. Consider the crossover reaction when performed on a small circular plasmid lacking supercoiled turns. When the *att* sites are in one orientation, the crossover reaction generates two smaller circles from the original, but when one of the *att* sites is reversed, the product of a crossover reaction does not generate two circles (Fig. 18.19). Instead, the segment of DNA between the sites is reversed in orientation. Further, depending on the topology of the DNA at the time of the crossover, either a simple circle or a knotted product called a trefoil is generated (Fig. 18.20). When one of the *att* sites participating in the crossover reaction is wrapped around a core of proteins as indicated, a superhelical turn is generated. This can be trapped in the reaction so that the product is a trefoil rather than a simple circle.

Nash and colleagues found two important results when examining the products of Int protein-catalyzed segment reversal. First, that half the products were trefoils, and second, that all the trefoils were topologically identical, that is, they possessed crossover points of the same sense. The presence of only one kind of crossover points or nodes in the trefoils means that the polarities of the wrapping in all the original PP' sites

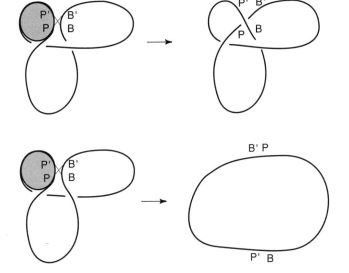

Figure 18.20 An integrative reaction between *POP'* sites wrapped around a protein and *BOB'* sites on a circle can generate a trefoil if the DNA strand is trapped by the reaction and if the correct strand, in this case the *P'B* product, is placed above the *PB'* product.

were the same. Had any been wrapped the other way any resultant trefoils would have possessed nodes of the opposite topological sign and would be the mirror image of the one shown. Second, the fact that the fraction of trefoils was as large as one half means that almost all of the *attP* substrate was wrapped around protein at the time of recombination.

Holliday Structures and Branch Migration in Integration

The integration-excision reaction of phage lambda could proceed by concerted double-stranded cleavages of both of the parental molecules. Alternatively, one strand from each parental molecule could be cut and exchanged, then the strand exchange point could be shifted by seven base pairs to the location of the exchange for the other DNA strand. This type of shifting is called branch migration, and normally it costs little or no energy since for each base pair whose hydrogen bonds are broken, a new hydrogen bonded base pair is formed (Fig. 18.21). The second alternative is similar to the pathway for general recombination that utilizes a Holliday structure and was discussed earlier. Several lines of evidence lead to the conclusion that it is the second pathway that is followed.

The first evidence favoring utilization of a Holliday structure is the experimental finding that Int protein will catalyze the conversion of an artificially constructed *att* site Holliday structure to the products expected from a complete integration reaction. Part of the interest in this important result is the trick used to generate the substrate DNA. Restriction enzyme fragments containing the four types of *att* sites were denatured and reannealed. In addition to regenerating the original duplexes, cross hybridization creates a Holliday-type "chi" structure

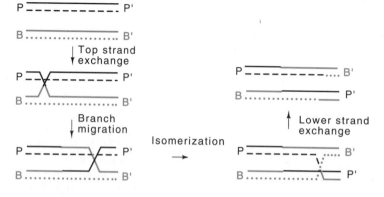

Figure 18.21 Single-strand exchange between top strands of two DNA molecules carrying PP' and BB' sites will generate a Holliday structure.

(Fig. 18.22). Int protein can cleave this to generate the expected products.

A second line of experiments also leads to the conclusion that the integration reaction proceeds in discrete cutting steps. In this experiment an *attP* site on a supercoiled circle is reacted with a linear *attB* site. If the integration reaction is permitted to proceed to completion, a long linear product results. If, however, the reaction is stopped early, then the "alpha" structures sketched in Fig. 18.23 are formed. These could form only if a single strand is cut and exchanged. By these means one has an assay for completion of only the first strand exchange, or both strand exchanges.

The first cut occurs in the upper strand at the left end of the seven base cohesive region. Placing a nick at this site in the DNA substrate could deprive Int of the energy from the phosphodiester bond. Indeed, such a nick blocks initiation of the integration reaction. The next step in the reaction is the branch migration or branch displacement that shifts the strand crossover point seven base pairs to the right. Genetic

Figure 18.22 Denaturing and reannealing four DNA duplexes, two carrying the PP' and BB' sites, and two carrying BP' and PB' sites will generate two types of Holliday structures as well as the original duplexes.

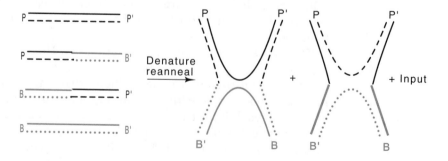

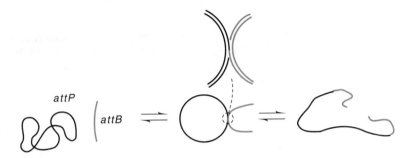

Figure 18.23 The integration reaction between a supercoiled PP' site and a linear BB' site will generate first the "alpha" structure after one strand has exchanged, and then the linear structure after both strands have exchanged.

experiments showed that mispaired bases in the central region greatly reduced the rate of strand exchange. Similar experiments done *in vitro*, or crosslinking from one DNA strand to the other in this region similarly block completion of the exchange reaction. Finally, placing a nick at the site of the lower strand crossover point blocks completion of the reaction.

Problems

18.1. Suppose lambda DNA molecules freely float in the cell cytoplasm and are randomly segregated into daughter cells. How many DNA molecules per cell would be required to generate a spontaneous curing rate of 10^{-6} per generation?

18.2. Breakage of a DNA strand during conjugation prevents transfer of genetic markers lying after the breakage point. If the probability of breakage per unit length of chromosome transferred to female cells is a constant b, show that the probability of transfer of a marker located at position l from the origin of transfer is e^{-bl}.

18.3. Using Southern transfer technology, how would you show that lambda DNA in a lysogenic strain is directly inserted into the *E. coli* genome?

18.4. Sketch the expected structure of the heteroduplex between F'(λ) and λ DNA.

18.5. If λ and $\lambda b2$ coinfect cells, the only $b2$ lysogens found are double lysogens simultaneously possessing λ and $\lambda b2$ integrated into their chromosomes. Why?

18.6. Design a simple experiment in which the frequency of integration into an episome and chromosome would be capable of proving that $attB \neq attP$ and/or $attB' \neq attP'$.

18.7. Phage Mu is much like lambda except that it appears to be almost indiscriminate with respect to the site into which it integrates.

How would you show that Mu phage could insert in either direction within a gene?

18.8. Suppose that Int⁻ lambda cannot lysogenize at all. It is found, however, that Mu phage can help lambda integrate. It is also found that the Mu-assisted integration of lambda is not altered if the *POP'* of lambda is deleted. Predict the genetic structure of Mu-assisted lambda lysogens.

18.9. The integration-excision reaction is asymmetric in that Int is required for integration and both Int and Xis are required for excision. How can this be true without violating the fundamental physical law of microscopic reversibility?

18.10. How would the following data be interpreted?

Infection by	Gives Bursts of
$\lambda N^- + \lambda imm^{434}$	$50\ \lambda N^- + 50\ \lambda imm^{434}$
$\lambda N^- + \lambda imm^{21}$	$50\ \lambda N^- + 50\ \lambda imm^{21}$

Induction of the Lysogen	Plus Infection by	Gives Bursts of
λN	λimm^{434}	$50\ \lambda N^- +\ \ 50\ \lambda imm^{434}$
λN^-	λimm^{21}	$0\ \lambda N^- + 100\ \lambda imm^{21}$

18.11. Consider excision from a lambda lysogen with one copy of the phage in the chromosome and one copy in an episome. What would be the consequences of a reaction between *attL* on the chromosome and *attR* on the phage? What might prevent this reaction if it is deleterious?

18.12. Lambda's requirement of supercoiled DNA for integration may merely be a way to detect whether its DNA is circular. Why would integration of linear lambda DNA into the chromosome be lethal? Which substrate would be more likely to possess a requirement for supercoiling, the bacterial or the phage DNA?

18.13. The superhelical twists required for the lambda excision reaction could be required to place the DNA in a conformation that is recognized by the Int and Xis proteins, or the strain put into the DNA could assist part of the biochemistry, for example, by melting out the crossover region. What experiments could you devise that could distinguish between these two possibilities?

18.14. How does the ability to cross *att* mutations from the phage onto the chromosome suggest that lambda makes staggered cuts at *O* when it integrates and excises?

18.15. Why would λimm^{434} but not λimm^{21} be expected to heteroim-mune-cure a wild-type lambda lysogen?

18.16. A $\lambda para$ phage was produced by insertion of lambda into the *araB* gene followed by excision to produce a phage carrying the *araC* gene and part of the *araB* gene. What would be the interpretation of the fact that induction of a single lysogen of this phage yielded only a few phage per cell unless arabinose was also present in the medium? Then it would produce normal quantities of phage.

18.17. How could you make a lambda lysogen of a Gal⁻ strain that is also deleted of the maltose gene activator so that it is λ^r? You might try infecting with lambda phage containing $\phi80$ tail fibers, but the cells are $\phi80^r$ as well as resistant to all other lambdoid phage and their host range mutants. The conditions of the problem are to make the lysogen in one step. Transducing the cells to make them Mal⁺ and then lysogenizing them is forbidden.

18.18. When lambda is forced to integrate into the chromosome at sites other than the normal lambda attachment site, it does so at low frequencies that vary widely from site to site. What correlation would you expect between integration frequency at a site and the fraction of excised phage from sites that transduce adjacent bacterial genes?

18.19. In studying integration or excision events from att^2 types of substrates, the two *att* regions can be in parallel or opposite orientations on the DNA molecule. How do the products of the reaction differ in these two cases? What use might be made of either of these?

18.20. Look up and describe the basis of the genetic selection Nash used in the construction of the $\lambda attB$-$attP$.

18.21. Predict the consequences of introducing a gap of one base in the phosphodiester backbone of the DNA in the middle of the seven base region between the two sites at which single-strand cuts are made during the integration or excision reactions.

References

Recommended Readings

Use of Site-specific Recombination as a Probe of DNA Structure and Metabolism *in vivo*, J. Bliska, N. Cozzarelli, J. Mol. Biol. *194*, 205-218 (1987).

Site-specific Recombination Intermediates Trapped with Suicide Substrates, S. Nunes-Düby, L. Matsumoto, A. Landy, Cell *50*, 779-788 (1987).

Functional Replacement of a Protein-induced Bend in a DNA Recombination-Site, S. Goodman, H. Nash, Nature *341*, 251-254 (1989).

Lambda Int Protein Bridges Between Higher Order Complexes at Two Distant Chromosomal Loci *attL* and *attR* S. Kim, A. Landy, Science *256*, 198-203 (1992).

Int, Xis and *att* Mutants

Mutants of Bacteriophage Lambda Unable to Integrate into the Host Chromosome, R. Gingery, H. Echols, Proc. Nat. Acad. Sci. USA *58*, 1507-1514 (1967).

A Deletion Analysis of Prophage Lambda and Adjacent Genetic Regions, S. Adhya, P. Cleary, A. Campbell, Proc. Nat. Acad. Sci. USA *61*, 956-962 (1968).

Integration-Negative Mutants of Bacteriophage Lambda, M. Gottesman, M. Yarmolinsky, J. Mol. Biol. *31*, 487-505 (1968).

Evidence for a Prophage Excision Gene in Lambda, A. Kaiser, T. Masuda, J. Mol. Biol. *47*, 557-564 (1970).

New Mutants of Bacteriophage Lambda with a Specific Defect in Excision from the Host Chromosome, G. Guarneros, H. Echols, J. Mol. Biol. *47*, 565-574 (1970).

Deletion Mutants of Bacteriophage Lambda, I. Isolation and Initial Characterization, J. Parkinson, R. Huskey, J. Mol. Biol. *56*, 369-384 (1971).

Deletion Mutants of Bacteriophage Lambda, II. Genetic Properties of *att*-defective Mutants, J. Parkinson, J. Mol. Biol. *56*, 385-401 (1971).

Deletion Mutants of Bacteriophage Lambda, III. Physical Structure of *att*, R. Davis, J. Parkinson, J. Mol. Biol. *56*, 403-423 (1971).

Attachment Site Mutants of Bacteriophage Lambda, M. Shulman, M. Gottesman, J. Mol. Biol. *81*, 461-482 (1973).

A Genetic Analysis of the *att-int-xis* Region of Coliphage Lambda, L. Enquist, R. Weisberg, J. Mol. Biol. *111*, 97-120 (1977).

Biochemical Analysis of *att*-Defective Mutants of the Phage Lambda Site-specific Recombination System, W. Ross, M. Shulman, A. Landy, J. Mol. Biol. *156*, 505-529 (1982).

Mutational Analysis of Integrase Arm-type Binding Sites of Bacteriophage Lambda, C. Aayer, S. Hesse, R. Gumport, J. Gardner, J. Mol. Biol. *192*, 513-527 (1986).

Structure of the *att* Regions

Viral Integration and Excision, Structure of the Lambda *att* Sites, A. Landy, W. Ross, Science *197*, 1147-1160 (1977).

The Lambda Phage *att* Site: Functional Limits and Interaction with *int* Protein, P. Hsu, W. Ross, A. Landy, Nature *285*, 85-91 (1980).

Structure and Function of the Phage Lambda *att* Site: Size, Int-binding Sites, and Location of the Crossover Point, K. Mizuuchi, R. Weisberg, L. Enquist, M. Mizuuchi, M. Buraczynska, C. Foeller, P. Hsu, W. Ross, A. Landy, Cold Spring Harbor Laboratory Symposium of Quantitative Biology *45*, 429-437 (1981).

DNA Sequence of the *att* Region of Coliphage 434, D. Mascarenhas, R. Kelley, A. Campbell, Gene *5*, 151-156 (1981).

Regulation of Int and Xis Synthesis and Activity

Regulation of Lambda Prophage Excision by the Transcriptional State of DNA, R. Davis, W. Dove, H. Inokuchi, J. Lehman, R. Roehrdanz, Nature New Biology *238*, 43-45 (1972).

Int-constitutive Mutants of Bacteriophage Lambda, K. Shimada, A. Campbell, Proc. Nat. Acad. Sci. USA *71*, 237-241 (1974).

Constitutive Integrative Recombination by Bacteriophage Lambda, H. Echols, Virology *64*, 557-559 (1975).

Retroregulation of the *int* Gene of Bacteriophage Lambda: Control of Translation Completion, D. Schindler, H. Echols, Proc. Nat. Acad. Sci. *78*, 4475-4479 (1981).

Downstream Regulation of *int* Gene Expression by the *b2* Region in Phage Lambda, C. Epp, M. Pearson, Gene *13*, 327-337 (1981).

Posttranscriptional Control of Bacteriophage Lambda *int* Gene Expression from a Site Distal to the Gene, G. Guarneros, C. Montanez, T. Hernandez, D. Court, Proc. Nat. Acad. Sci. USA *79*, 238-242 (1982).

Regulation of Bacteriophage Lambda *int* Gene Expression, A. Oppenheim, S. Gottesman, M. Gottesman, J. Mol. Biol. *158*, 327-346 (1982).

Deletion Analysis of the Retroregulatory Site for the Lambda *int* Gene, D. Court, T. Huang, A. Oppenheim, J. Mol. Biol. *166*, 233-240 (1983).

Removal of a Terminator Structure by RNA Processing Regulates *int* Gene Expression, U. Schmeissner, K. McKenney, M. Rosenberg, D. Court, J. Mol. Bio. *176*, 39-53 (1984).

Interactions between Lambda Int Molecules Bound to Sites in the Region of Strand Exchange are Required for Efficient Holliday Junction Resolution, B. Franz, A. Landy, J. Mol. Biol. *215*, 523-535 (1990).

A Switch in the Formation of Alternative DNA Loops Modulates λ Site-specific Recombination, L. Moitoso, A. Landy, Proc. Natl. Acad. Sci. USA *88*, 588-592 (1991).

The Integration-Excision Reaction, *in vitro* Reactions

Lambda *attB-attP*, a Lambda Derivative Containing both Sites Involved in Integrative Recombination, H. Nash, Virology *57*, 207-216 (1974).

Elements Involved in Site Specific Recombination in Bacteriophage Lambda, S. Gottesman, M. Gottesman, J. Mol. Biol. *91*, 489-499 (1975).

Integrative Recombination in Bacteriophage Lambda: Analysis of Recombinant DNA, H. Nash, J. Mol. Biol. *91*, 501-514 (1975).

Integrative Recombination of Bacteriophage Lambda DNA *in vitro*, H. Nash, Proc. Nat. Acad. Sci. USA *72*, 1072-1076 (1975).

Restriction Assay for Integrative Recombination of Bacteriophage Lambda DNA *in vitro*: Requirement for Closed Circular DNA Substrate, K. Mizuuchi, H, Nash. Proc. Nat. Acad. Sci. USA *73*, 3524-3528 (1976).

DNA Gyrase: An Enzyme that Introduces Superhelical Turns into DNA, M. Gellert, K. Mizuuchi, M. O'Dea, H. Nash, Proc. Nat. Acad. Sci. USA *73*, 3872-3876 (1976).

Interaction of Int Protein with Specific Sites on Lambda *att* DNA, W. Ross, A. Landy, Y. Kikuchi, H. Nash, Cell *18*, 297-307 (1979).

Role of DNA Homology in Site-specific Recombination, R. Weisberg, L. Enquist, C. Foeller, A. Landy, J. Mol. Biol. *170*, 319-342 (1983).

Role of the *xis* Protein of Bacteriophage Lambda in a Specific Reactive Complex at the *attR* Prophage Attachment Site, M. Better, W. Wickner, J. Auerbach, H. Echols, Cell *32*, 161-168 (1983).

Resolution of Synthetic *att*-site Holliday Structures by the Integrase Protein of Bacteriophage Lambda, P. Hsu, A. Landy, Nature *311*, 721-726 (1984).

Role of Homology in Site-specific Recombination of Bacteriophage Lambda: Evidence Against Joining of Cohesive Ends, H. Nash, C. Bauer, J. Gardner, Proc. Nat. Acad. Sci. USA *84*, 4049-4053 (1987).

Protein-Protein Interactions in a Higher-order Structure Direct Lambda Site-specific Recombination, J. Thompson, L. Mitoso de Vargas, S. Skinner, A. Landy, J. Mol. Biol. *195*, 481-493 (1987).

Homology-dependent Interactions in Phage Lambda Site-specific Recombination, P. Kitts, H. Nash, Nature *329*, 346-348 (1987).

Synapsis of Attachment Sites During Lambda Integrative Recombination Involves Capture of a Naked DNA by a Protein-DNA Complex, E. Richet, P. Abcarian, H. Nash, Cell *52*, 9-17 (1988).

Empirical Estimation of Protein-induced Bending Angles: Applications to Lambda Site-specific Recombination Complexes, J. Thompson, A. Landy, Nucleic Acids Res. *16*, 9687-9705 (1988).

An Intermediate in the Phage Lambda Site-Specific Recombination Reaction is Revealed by Phosphorothioate Substitution in DNA, P. Kitts, H. Nash, Nuc. Acids Res. *16*, 6839-6856 (1988).

Suicide Recombination Substrates Yield Covalent Lambda Integrase-DNA Complexes and Lead to Identification of the Active Site Tyrosine, C. Pargellis, S. Nunes-Düby, L. de Vargas, A. Landy, J. Biol. Chem. *263*, 7678-7685 (1988).

Helical-repeat Dependence of Integrative Recombination of Bacteriophage Lambda: Role of the PI and HI Protein Binding Sites, J. Thompson, U. Snyder, A. Landy, Proc. Nat. Acad. Sci. USA *85*, 6323-6327 (1988).

Phasing of Protein-induced DNA Bends in a Recombination Complex, U. Snyder, J. Thompson, A. Landy, Nature *341*, 255-257 (1989).

Half-*att* Site Substrates Reveal the Homology Independence and Minimal Protein Requirements for Productive Synapsis in Lambda Excisive Recombination, S. Nunes-Düby, L. Matsumoto, A. Landy, Cell *59*, 197-206 (1989).

Mapping of a Higher Order Protein-DNA Complex: Two Kinds of Long-range Interactions in λ*attL*, S. Kim, L. de Vargas, S. Nunes-Düby, A. Landy, Cell *63*, 773-781 (1990).

Symmetry in the Mechanism of Bacteriophage λ Integrative Recombination, A. Burgin, Jr., H. Nash, Proc. Natl. Acad. Sci. USA *89*, 9642-9646 (1992).

Deformation of DNA During Site-specific Recombination of Bacteriophage λ: Replacement of IHF Protein by HU Protein or Sequence-directed Bends, S. Goodman, S. Nicholson, H. Nash, Proc. Natl. Acad. Sci. USA *89*, 11910-11914 (1992).

Topology and the Integration-Excision Reaction

Involvement of Supertwisted DNA in Integrative Recombination of Bacteriophage Lambda, K. Mizuuchi, M. Gellert, H. Nash, J. Mol. Biol. *121*, 375-392 (1978).

Knotting of DNA Caused by a Genetic Rearrangement, Evidence for a Nucleosome-like Structure in Site Specific Recombination of Bacteriophage Lambda, T. Pollock, H. Nash, J. Mol. Biol. *170*, 1-18 (1983).

Site-specific Recombination of Bacteriophage Lambda, The Change in Topological Linking Number Associated with Exchange of DNA Strands, H. Nash, T. Pollock, J. Mol. Biol. *170*, 19-38 (1983).

Genetic Rearrangements of DNA Induces Knots with a Unique Topology: Implications for the Mechanism of Synapsis, J. Griffith, H. Nash, Proc. Nat. Acad. Sci. USA *82*, 3124-3128 (1985).

Host Involvement in the Integration-Excision Reaction

An *Escherichia coli* Gene Product Required for Lambda Site-specific Recombination, H. Miller, D. Friedman, Cell *20*, 711-719 (1980).

Direct Role of the *himA* Gene Product in Phage Lambda Integration, H. Miller, H. Nash, Nature *290*, 523-526 (1981).

Multilevel Regulation of Bacteriophage Lambda Lysogeny by *E. coli himA* Gene, H. Miller, Cell *25*, 269-276 (1981).

Purification and Properties of the *E. coli* Protein Factor Required for Lambda Integrative Recombination, H. Nash, C. Robertson, J. Biol. Chem. *256*, 9246-9253 (1981).

Control of Phage Lambda Development by Stability and Synthesis of cII Protein: Role of the Viral *CIII* and *hflA* and *himD* Genes, A. Hoyt, D. Knight, A. Das, H. Miller, H. Echols, Cell *31*, 565-573 (1982).

E. coli Integration Host Factor Binds to Specific Sites in DNA, N. Craig, H. Nash, Cell *39*, 707-716 (1984).

Cellular Factors Couple Recombination with Growth Phase: Characterization of a New Component in the Lambda Site-specific Recombination Pathway, J. Thompson, L. de Vargas, C. Koch, R. Kahmann, A. Landy, Cell *50*, 901-908 (1987).

Transducing Phage

Specialized Transduction of Galactose by Lambda Phage from a Deletion Lysogen, K. Sato, A. Campbell, Virology *41*, 474-487 (1970).

Prophage Lambda at Unusual Chromosomal Locations, I. Location of the Secondary Attachment Sites and the Properties of the Lysogens, K. Shimada, R. Weisberg, M. Gottesman, J. Mol. Biol. *63*, 483-503 (1972).

Isolation of Plaque-forming, Galactose-transducing Strains of Phage Lambda, M. Feiss, S. Adhya, D. Court, Genetics *71*, 189-206 (1972).

Prophage Lambda at Unusual Chromosomal Locations, II. Mutations Induced by Bacteriophage Lambda in *E. coli* K12. K. Shimada, R. Weisberg, M. Gottesman, J. Mol. Biol. *80*, 297-314 (1973).

Prophage Lambda at Unusual Chromosome Locations, III. The Components of the Secondary Attachments Sites, K. Shimada, R. Weisberg, M. Gottesman, J. Mol. Biol. *93*, 415-429 (1975).

Int and Xis Proteins

Purification of Bacteriophage Lambda Int Protein, H. Nash, Nature *247*, 543-545 (1974).

Site-Specific DNA Condensation and Pairing Mediated by the *int* Protein of Bacteriophage Lambda, M. Better, C. Lu, R. Williams, H. Echols, Proc. Nat. Acad. Sci. USA *79*, 5837-5841 (1982).

Bacteriophage Lambda *int* Protein Recognizes Two Classes of Sequence in the Phage *att* Site: Characterization of Arm-Type Sites, W. Ross, A. Landy, Proc. Nat. Acad. Sci. USA *79*, 7724-7728 (1982).

Purification of the Bacteriophage Lambda *xis* Gene Product Required for Lambda Excisive Recombination, K. Abremski, S. Gottesman, J. Biol. Chem. *257*, 9658-9662 (1982).

The Mechanism of Lambda Site-specific Recombination: Site-Specific Breakage of DNA by *Int* Topoisomerase, N. Craig, H. Nash, Cell *35*, 795-803 (1983).

Patterns of Lambda *Int* Recognition in the Regions of Strand Exchange, W. Ross, A. Landy, Cell *33*, 261-272 (1983).

Interaction of the Lambda Site-specific Recombination Protein Xis with Attachment Site DNA, S. Yin, W. Bushman, A. Landy, Proc. Nat. Acad. Sci. USA *82*, 1040-1044 (1985).

Autonomous DNA Binding Domains of Lambda Integrase Recognize Two Different Sequence Families, L. de Vargas, C. Pargellis, N. Hasan, E. Bushman, A. Landy, Cell *54*, 923-929 (1988).

Related Topics

Ter, a Function Which Generates the Ends of the Mature Lambda Chromosome, S. Mousset, R. Thomas, Nature *221*, 242-244 (1969).

Packaging of Prophage and Host DNA by Coliphage Lambda, N. Sternberg, R. Weisberg, Nature *256*, 97-103 (1975).

The Terminase of Bacteriophage Lambda, Functional Domains for *cosB* Binding and Multimer Assembly, S. Frackman, D. Siegele, M. Feiss, J. Mol. Biol. *183*, 225-238 (1985).

Transposable Genetic Elements

19

Transposable elements are special DNA sequences ranging upwards in length from several hundred base pairs. These sequences can spread within a cell by being copied into new DNA locations as well as within a population through infection, transformation, transduction, or conjugation.

The phenomenon of transposing elements was first described in maize in the elegant genetic studies of McClintock. General interest in the subject was then stimulated by genetic and physical studies with MIEscherichia *coli*. The development of genetic engineering made the detection, characterization, and study of transposing elements much easier than before. Such elements have now been found in all organisms that have been carefully examined. This chapter considers the discovery, structure, properties, and use of transposable elements that are found in the bacteria *Escherichia coli* and *Salmonella typhimurium*, the yeast *Saccharomyces*, the fruit fly *Drosophila*, and humans.

The cell is an ideal home for a parasitic sequence of nucleotides. There it should enjoy almost the same treatment as an integrated lambda phage genome. Furthermore, the copying of such a sequence into a new chromosomal location likely requires biochemical activities already present or easily synthesized in the host cell. Given such an environment, it is not surprising that parasitic sequences have evolved and now exist. A deeper question is why such sequences do not constitute most of the cell's DNA. Most likely, the tendency of such sequences to proliferate is countermanded by evolution on a grander scale. A cell line or organism that is weighted down by an excessive number of unused DNA sequences would be at a survival disadvantage compared to others containing fewer such sequences. Consequently, evolution will constantly select for cells without too many transposable sequences, and, overall, an unhappy balance will exist between the sequence and the host cell line. This is not to say that transposable elements are without value to their hosts. As we will see, the presence of repeated DNA sequences can facilitate chromosome rearrangements. This reshuffling of genetic material may greatly speed evolution and aid cells that contain at least a few repeated sequences.

In addition to being a burden upon DNA replication, the insertion of parasitic DNA sequences into the genome will inevitably damage crucial genes. We might therefore expect sophisticated parasitic sequences to devise ways to avoid killing genes or proteins by their insertion. Two methods could be used. A sequence might arrange to splice itself out of RNA when the region has been transcribed, or the sequence might code for a protein that splices itself out of translated protein. If the host RNA or protein product is rejoined, effects upon the host of the parasitic sequence will be minimal. Transposable sequences have been identified that perform either of these functions.

IS Elements in Bacteria

Nonsense mutations in many operons have polar effects, that is, they reduce the expression of downstream genes. In efforts to study these effects more carefully, in the late 1960s Malamy, Shapiro, and Starlinger each isolated and characterized strongly polar mutations in the *lac* and *gal* operons. In addition to the expected classes of nonsense mutations, another type was also found. This new type was highly polar but not suppressible by known nonsense suppressors. Since the mutations reverted, albeit at very low rates, it seemed likely that they were insertions and not deletions. Therefore it became of great interest to find out just what these mutations were. Nowadays this question would probably be answered by cloning a portion of the DNA containing the mutation

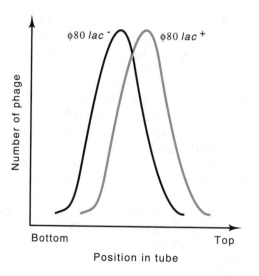

Figure 19.1 The positions of Lac$^+$ and strongly polar Lac$^-$ phage particles in the centrifuge tube after CsCl equilibrium gradient centrifugation. The polar mutation makes the phage more dense and it bands closer to the bottom of the tube.

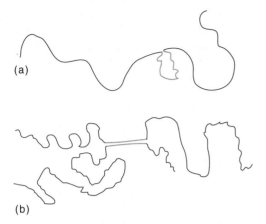

Figure 19.2 (a) DNA heteroduplex between lambda and lambda containing IS2. The single-stranded bubble is the DNA of IS2. (b) DNA heteroduplex formed between two r strands of lambda, each containing IS2 but in opposite orientations.

and sequencing it to determine the exact nature of the mutation. Such a procedure was impossible at the time and less direct experiments were performed.

One likely cause of the polarity was the insertion of extraneous nucleotides into the *lac* and *gal* genes. Fortunately, this possibility could be tested by the techniques then available. If the insertions were larger than a few hundred nucleotides, they could be detected directly because they would increase the density of *lac* or *gal* transducing phage. Therefore the mutations were crossed onto λ*dgal* and φ80*plac* phage and the densities of phage both with and without the mutations were carefully compared by equilibrium gradient centrifugation (Fig. 19.1). The results showed that transducing phage carrying the polar mutations were denser than the corresponding transducing phage not carrying the mutations. Therefore the mutations were caused by the insertion of DNA elements. These later came to be called IS, for insertion sequences.

Table 19.1 Sizes of Various Bacterial IS Sequences and the Number of Base Pairs of Host DNA they Duplicate Upon Insertion

IS Element	Size of Element	Size of Host Duplication
IS1	768	9
IS2	1,327	5
IS3	1,300	3, 4
IS4	1,426	11, 12
IS5	1,195	4
IS10-R	1,329	9
IS50-R	1,531	9

At first, the insertions were suspected of being merely random sequences that somehow had been inserted into the genes being studied. The mechanism of their generation is not so easy to imagine, however. Point mutations, insertions and deletions of a nucleotide or two, and sizable deletions may all be created by simple reactions likely to occur in cells. Not so for sizable insertions.

Additional study of the IS sequences began to clarify their origin. One particularly valuable technique for their study was the formation of DNA heteroduplexes (Fig. 19.2). DNA isolated from the *gal* and *lac* transducing phage with and without the IS sequences was used in these experiments. When necessary, the strands of these phage could even be separated and purified. The lengths of IS sequences fell into discrete size classes, and any particular IS sequence was found to insert into a gene in either of the two possible orientations. The existence of the length classes suggested that a small number of IS sequences might be responsible for all the insertions. Indeed, fewer than ten different IS sequences have been found in *Escherichia coli* (Table 19.1). The demonstration that IS elements were specific sequences did not explain how they were inserted into different sites in the cell's chromosome, but by analogy to lambda phage, the elements were expected to code for one or more proteins involved in the process. By now the IS elements have been obtained on phage and plasmids and they have been sequenced and characterized using *in vitro* mutagenesis as described below. From these studies a comprehensive picture of their structure and function has emerged.

Structure and Properties of IS Elements

Most of the bacterial IS sequences contain two characteristic elements. First, a sequence of 10 to 40 base pairs at one end is repeated or nearly repeated in an inverse orientation at the other end. This is called IR, for inverted repeat. Second, the IS sequences contain one or two regions that appear to code for protein because they possess promoters, ribosome-binding sequences, ATG (AUG) codons, and sense codons fol-

Figure 19.3 The organization of IS5. The large and small proteins are in the same codon register.

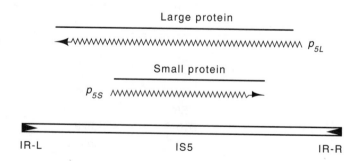

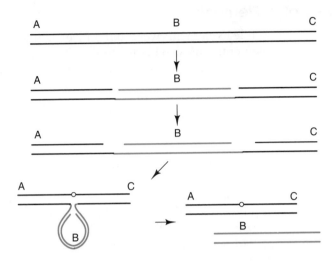

Figure 19.4 One mechanism by which nicking DNA can stimulate generation of deletions.

lowed by peptide chain termination codons. If proteins were made under control of these sequences, they would likely be involved with the process of copying the IS into a new position in the process of transposition. Indeed, one of the gene products is generally called the transposase. As only extremely low quantities of a transposase are needed, its transcription and translation are specially designed to be inefficient. Additionally, some premium is placed on possessing a minimum size, and genes of a transposon frequently overlap. For example, in IS5 two oppositely oriented reading frames overlap. The shorter is contained entirely within the longer (Fig. 19.3).

A high expression of the IS genes most likely would stimulate a high rate of transposition into new chromosomal sites. Since IS transposition rates appear low, something must protect IS genes from readthrough transcription and translation if the sequence has inserted into an active gene of the host cell. IS5 possesses six peptide chain termination codons flanking the protein genes—one at each end of the transposon, one in each reading frame. Additionally, the left end of the IS5 contains a transcription termination sequence oriented so as to prevent transcription from adjacent bacterial genes from entering the IS in the reading direction of the transposase.

Not only do some IS sequences copy themselves or jump into new DNA locations, but they also stimulate the creation of deletions located near the IS elements. These deletions have one end precisely located at the end of an IS element and apparently arise in abortive transposition attempts that begin with transposase nicking the DNA at one of the boundaries of the IS element (Fig. 19.4). Such a facile creation of deletions in any selected region is highly useful to geneticists, for, as we have seen, a set of such deletions can be used for fine structure mapping.

Discovery of Tn Elements

While IS elements were under investigation, another category of genetic elements was also being studied. These were plasmids that carry genes encoding proteins that confer resistance to antibiotics. They are called R-factors.

Not long after antibiotics began to be widely used, many bacterial isolates from human infections were found to be resistant to one or more of the drugs. These isolates were not resistant by virtue of a mutation altering the cellular target of the antibiotics; instead, they synthesized specific proteins or enzymes, which conferred resistance to the cells either by detoxifying the antibiotics or by blocking their entry into the cell. For example, penicillin-resistant strains were found to synthesize a β-lactamase that opens the lactam ring of penicillin and renders it harmless to the cells (Fig. 19.5).

Although many R-factors were fertile and could transfer themselves to other cells by conjugation, the speed with which different drug resistances appeared on R-factors was astonishingly rapid. Soon after the introduction of a new antibiotic, R-factors that carried genes conferring resistance to the new antibiotic would be found in many geographic locations. Unfortunately this has created great problems in the treatment of infections. The drug-resistance factors rapidly pick up genes encoding resistance to the most commonly used drugs, and soon treatment of infections becomes difficult. Now physicians use restraint in prescribing antibiotics to slow the spread of drug-resistance genes.

The ability of drug-resistance genes to spread rapidly suggested that they could hop from one R-factor to another or to other DNA sequences such as phage or the bacterial chromosome. One demonstration of this property came from an attempt to construct a lambda phage carrying the kanamycin-resistance gene from an R-factor.

The objective of the experiment was to generate a λkan transducing phage by forcing lambda to insert at random locations, and then excise. A small fraction of the excised phage would have picked up the kanamycin genes and be able to transduce cells to Kan^r. Cells deleted of attB and carrying a kanamycin-resistance R-factor were infected with lambda. This lambda phage was deleted of part of the b2 region so that the DNA of the resulting kanamycin-transducing phage would not be

Figure 19.5 The structure of the lactam ring of penicillins and related antibiotics and their structure following ring opening by β-lactamase. Different penicillins possess different R groups.

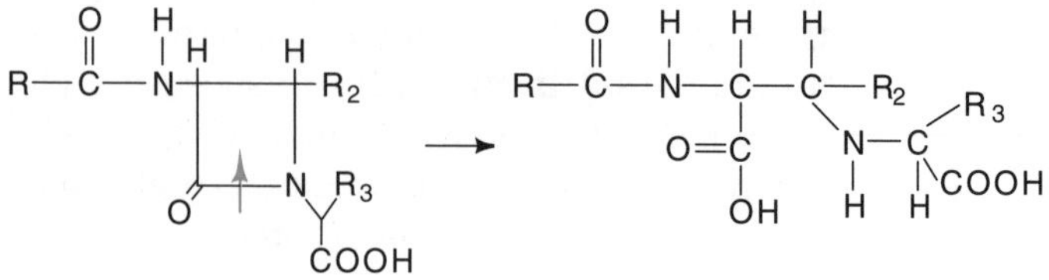

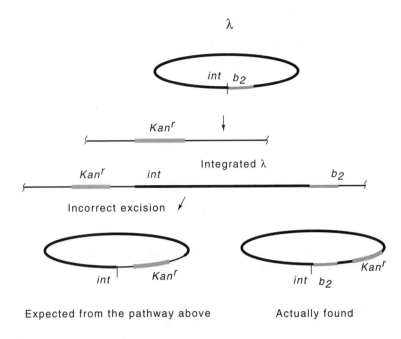

Figure 19.6 The structure of λ-*kan*^r expected if the transducing particles were generated by the same mechanism by which λ-*gal* and λ-*bio* transducing phages are produced, and the structure of λ-*kan* actually found.

too large to be packaged. To permit the phage to integrate via the Int pathway, the deletions chosen did not extend into the *att* region. As seen in the previous chapter, a small fraction of the resulting lysogens should then contain a lambda inserted into DNA adjacent to the kanamycin-resistance gene(s) by virtue of integrating into sites that weakly resemble lambda's normal integration site (Fig. 19.6). Upon induction, a small fraction of the phage were expected to excise incorrectly, pick up the adjacent kanamycin-resistance genes, and be capable of transducing kanamycin resistance to other cells.

Indeed, kanamycin-transducing phage were isolated. The phage possessed several unexpected properties, however. First, the additional DNA in these phage was not located immediately adjacent to *attP*. Second, kanamycin-resistant transformants obtained by infecting cells with the λ*kan* phage were lysogenic for lambda only if they contained an *attB*. These properties can be understood as follows. The DNA coding for kanamycin resistance originally hopped or was copied onto lambda by a recombination event rather than arriving there after an improper phage excision of the type that produces *gal* and *bio* transducing particles. Therefore the foreign DNA could be located anywhere on the phage and need not be immediately adjacent to *attP*. Second, a copy of the sequence encoding the kanamycin resistance could also hop or be copied from the phage onto the chromosome.

The DNA responsible for conferring the kanamycin resistance was examined by DNA heteroduplexes. It was found to be 5,400 base pairs

Table 19.2 Properties of Some Selected Transposons

Transposon	Genes Carried	Size (kilobases)	IS Element	Repeats
Tn1	Ampicillin^R	5	-	-
Tn3	Ampicillin^R	4.957	-	-
Tn4	Ampicillin^R, Streptomycin^R, Mercuric^R	20.5	IS50	-
Tn5	Kanamycin^R	5.7	IS50	Inverted
Tn6	Kanamycin^R	4.2	-	-
Tn7	Trimethoprim^R	13.5	-	-
Tn9	Chloramphenicol^R	2.5	IS1	Direct
Tn10	Tetracycline^R	9.3	IS10	Inverted

long and flanked by inverted repeated sequences 1,500 base pairs long. Some of the drug-resistance genes studied on the R-factors themselves

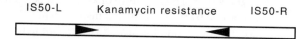

IS50-L   Kanamycin resistance   IS50-R

also possessed similar flanking sequences. Further study showed that the repeated sequences flanking the drug-resistance genes were transferred when the drug-resistance gene hopped. Finally, it was found that the repeated elements at the ends of many of the drug-resistance genes possessed DNA sequences essentially identical to known IS elements. Thus it appears that two IS elements can flank a gene and then facilitate copying the composite element into other DNA locations by the same transposition reactions as are used by IS elements themselves. Sequences with these general properties and structure are called transposons, or Tn elements (Table 19.2). The element described in this section is called Tn5, and the IS sequences associated with it are called IS50 so as not to be confused with IS5.

Structure and Properties of Tn Elements

Bacterial transposons can be grouped into four broad structural classes. Members of the class containing Tn5 and Tn10 possess a pair of IS elements on either side of a DNA sequence that is transposed. Although the right- and left-hand IS elements are nearly identical, only one of the IS elements encodes the transposase that is required for transposition. The difference between the left element of Tn5, which is called IS50-L, and the right element, IS50-R, is a single nucleotide change. This creates an ochre mutation in the genes both for the transposase and for a second protein that is translated in the same reading frame, and at the same time, creates a promoter pointed to the right.

Transposition by members of the Tn5-Tn10 class of transposons appears to require at least one protein encoded by the transposon and only DNA sites located very near the ends of the element. If internal sequences are deleted and if the necessary proteins are provided in *trans*, the deleted element will transpose, although at a very low rate, for the transposase from Tn10 is exceedingly unstable. This protein functions far better in *cis*, and even has trouble translocating longer versions of the transposon because of the distance separating its ends. Transposons in this class cut themselves out of one location in the chromosome and insert themselves into another. Their main problem then becomes one of arranging that such a repositioning ultimately results in a net gain in the number of transposon copies.

Tn3 belongs to a second class of transposons. Members of this group do not contain flanking IS sequences and possess only short inverted repeats at their ends. Their transposition requires two transposon-encoded proteins—a transposase and another protein. Moreover, transposition by these elements requires more than the two proteins and the ends of the transposon. A DNA site in the middle of the transposon is

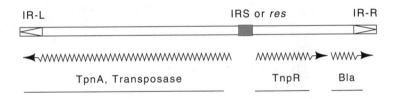

also required. Without this site, the transposon does not complete transposition events; it copies itself into new locations but connects the old location to the new (Fig. 19.7). Consequently, a transposon lacking the internal site fuses the donor and target DNA replicons when it attempts transposition. These properties demonstrate that transposition by these transposons occurs by copying rather than by excision and reintegration.

Figure 19.7 Incomplete transposition can fuse two replicons with the duplication of the transposition.

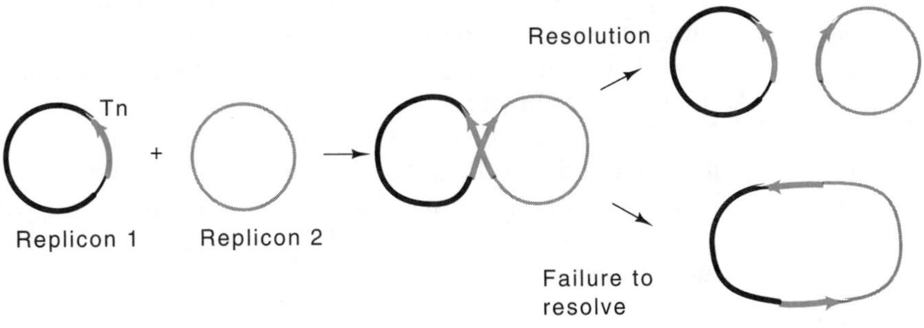

The third class of transposon is represented by phage Mu. Superficially this can be thought of as a transposon that happens to have picked up a phage. Therefore it replicates not by freeing itself from the chromosome but by copying itself into multiple locations. This property is particularly useful for study of the process because instead of transposing at a frequency of perhaps 10^{-4} per generation like many transposons, when induced, it transposes at a frequency of 10^2 per cell doubling time.

Tn7 constitutes a fourth class of transposon. It transposes into a unique site on the *E. coli* chromosome with high specificity, but if this site is absent, the transposon integrates nearly at random throughout the chromosome.

Inverting DNA Segments by Recombination, Flagellin Synthesis

Inversion of a stretch of DNA and excision of a stretch of DNA, which will be discussed later, are closely related. Both involve crossover events at the boundaries of the segment. As we saw in the previous chapter, (Fig. 18.19), site-specific recombination between two recognition regions on a DNA molecule can invert the intervening DNA. If one of the recognition regions is reversed, however, the same excision reaction will excise the DNA.

Here we will be concerned with a form of regulation that involves inverting a DNA segment. Later, a full chapter will be devoted to chemotaxis, the ability of cells to sense and move toward attractants or away from repellents. The movement is generated by rotation of flagella. In *Salmonella* two different genes encode flagellin. Depending on the

orientation of an invertible controlling DNA element, only one or the other of the flagellin genes is expressed at one time.

In the habitat of *Salmonella*, the gut, considerable selective advantage accrues to the strain that can express more than one flagellin type. Antibody secreted by patches of special cells on the walls of the intestine can bind to a flagellum and block rotation of the flagella and therefore stop chemotaxis. Hence, to survive and swim in the presence of antiflagellin antibody, it is to a cell's advantage to be able to synthesize a second type of flagellin against which the host does not yet possess antibody. In these circumstances, the simultaneous synthesis of two types of flagellin within a cell would be harmful. If both types of flagellin were incorporated into flagella, then either of two types of antibody could block chemotaxis rather than just one. Consequently, if two flagellin genes are present, one should be completely inactive while the other is functioning.

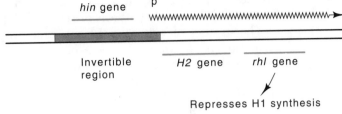

Figure 19.8 The structure of the DNA at the flagellin controlling sequence in *Salmonella*. The *hin* gene codes for a protein that can invert the controlling region. The region also contains a promoter for synthesis of *H2* and *rhl* messenger when the region is in one orientation.

How can an appropriate flagellin synthesis flip-flop circuit be constructed using regulatory elements available to a cell? Only one of the flagellin genes should possess any activity at one time, but reversing the activity pattern should be readily possible. These requirements are not easily met using regulatory elements of the type we have thus far considered. In general, the regulation systems we have considered in previous chapters do not permit construction of simple circuits possessing two states of dramatically different activities and at the same time permit simple switching on time scales of hours between the two states. For example, although the lambda phage early gene system possesses the necessary selectivity in having two states, fully off and fully on, this system is not reversible. Alternatively, genes regulated like *lac*, *trp*, and *ara* either do not possess large differences between their on and off states, or they flip states on time scales short compared to a doubling time of the cells.

An invertible controlling segment of DNA solves *Salmonella's* problem. In one orientation, only one type of flagellin is synthesized, and in the other orientation, only the other type is synthesized. The apparatus responsible for flipping the DNA segment can be completely independent of flagellin or expression of the flagellin genes. That is, the protein or proteins necessary for inversion of the DNA segment can be synthesized independently at a rate appropriate for optimum survival of the population.

The structure of the invertible segment *hin* in *Salmonella* is similar to IS elements and also to the portion of Mu phage called the G segment. The *hin* segment is 995 base pairs long. It has inverted repeats of 14 bases at its ends, and its central region encodes a protein that is necessary for inversion. At its right end the element contains a promoter oriented to the right. Just to the right of the invertible segment lies one of the flagellin genes, *H2*, and another gene called *rhl*. The Rhl protein is a repressor of the second flagellin gene, *H1*, located elsewhere on the chromosome. As a result of this structure, one orientation of the invertible element provides synthesis of H2 protein and Rhl repressor. In this state the Rhl repressor blocks synthesis of H1 flagellin (Fig. 19.8). When the controlling element is in the other orientation, H2 and Rhl protein are not synthesized because they have been separated from their promoter, and therefore synthesis of H1 is not repressed.

The strategy of switching the antigenic properties of crucial surface structures is not unique to bacteria. African trypanosomes, unicellular flagellated cells, can evade the immune systems of vertebrates by switching their synthesis of surface proteins. These protozoa have hundreds of different genes for the same basic surface protein. The process of switching on the synthesis of any one of them, however, resembles more closely the mating-type switch of yeast than the orientation switching involved with the switching of *Salmonella* flagellin synthesis.

Mu Phage As a Giant Transposable Element

A transposon could carry DNA with phage-like functions rather than a simple antibiotic resistance gene. Such a transposon would be able to travel freely from cell to cell autonomously as a virus. Phage Mu first drew attention by its ability to generate mutations in infected cell populations. These mutations were generated by its insertion into various bacterial genes. Careful mapping of many Mu phage insertions into the β-galactosidase gene showed that Mu inserted without high specificity in its target sequence. Thus, Mu behaved like a lambda phage with little sequence specificity in its choice of the bacterial *att* site.

Two facts, however, indicate that Mu is more like a transposon than like phage lambda. First, Mu duplicates five bases of chromosomal DNA upon its insertion. Most transposons generate such duplications. These arise as a result of staggered nicking of the target sequence followed by insertion of the transposon and replication (Fig. 19.9). Second, Mu does not excise from the chromosome and replicate in the cytoplasm. Even though an induced cell may yield a hundred Mu phage upon lysis, never during the lytic cycle are any free Mu DNA molecules observed. The Mu DNA is replicated only by transposition. Packaging of the Mu takes place directly on the integrated DNA. A dramatic demonstration of replication by transposition is provided by Southern transfers of DNA taken from a Mu lysogen before and after induction of the phage. Before induction, only a single restriction fragment of the host DNA contains sequences

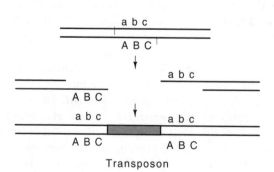

Figure 19.9 Insertion of a transposon into a site with staggered nicks followed by filling in of the gaps generates a duplication of the sequence included between the nicks.

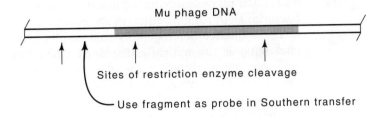

Sites of restriction enzyme cleavage

Use fragment as probe in Southern transfer

Chromosomal DNA cut by the restriction enzyme

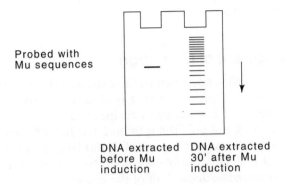

Probed with
Mu sequences

DNA extracted   DNA extracted
before Mu   30' after Mu
induction   induction

Figure 19.10 DNA taken from a Mu lysogen before and after phage induction used in a Southern transfer.

homologous to a restriction fragment including the end of Mu, but half an hour after induction of Mu growth, many restriction fragments contain these sequences (Fig. 19.10).

Mu packages its DNA right out of the chromosomal insertions. Part of the head structure recognizes a sequence near the left end of the DNA, reaches out beyond the phage 50 to 150 base pairs, and begins packaging. A headfull of DNA is packaged, and the remaining DNA is then

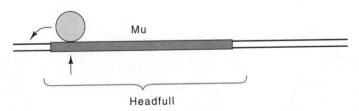

cleaved and the tail is attached. Since a Mu phage headfull of DNA is slightly larger than the genome of the phage, the packaged DNA usually extends beyond the right end of the phage and includes about 3,000 base pairs of bacterial DNA as well. The headfull hypothesis explains the fact that a Mu phage carrying an insertion packages a smaller amount of bacterial DNA roughly equal in amount to the size of the insertion.

The sequence heterogeneity that results from the headfull mode of packaging is dramatically displayed when heteroduplexes of phage genomes are analyzed in the electron microscope. Since it is highly unlikely that the two strands from the same phage will rehybridize upon formation of the heteroduplex, two unrelated strands will associate. The phage DNA portions of the strands will, of course, be complementary and will form a duplex. The strands of bacterial DNA on the right end, however, are unlikely to be complementary, and these will remain single-stranded and are observed as "split ends." The similar split ends from the left end are too small to be clearly observed.

An Invertible Segment of Mu Phage

The split-ended heteroduplexes described in the preceding section are observed with DNA that is obtained from Mu-infected cells. Mu DNA obtained instead from an induced lysogen yields heteroduplexes containing not only the split end, but half of them also contain an internal single-stranded region of 3,000 base pairs that forms a bubble (Fig. 19.11). This bubble results from the fact that the region, called G, is inverted in about half of the phage.

The ability of Mu to invert G can be abolished by a small deletion at the end of G or by a point mutation in the adjacent region of the DNA called *gin*. This point mutation can be complemented, thereby proving that *gin* encodes a protein that participates in inversion of G. Curiously, phage P1 also contains the same G loop, and P1 can complement Mu *gin* mutants.

The G-loop segment codes for proteins that determine the host range of Mu. Phage induced from lysogens with G in the plus orientation are able to adsorb to *Escherichia coli* K-12, but not to *Citrobacter freundii* or *Escherichia coli* C. When G is in the minus orientation, the abilities of the resultant phage to adsorb to these strains are reversed. This phenomenon provides an explanation of the heteroduplexing experiments. Those Mu able to adsorb to *Escherichia coli* K-12 must be in the plus orientation. Since the rate of inversion is low, the G region of most

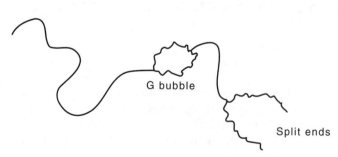

G bubble

Split ends

Figure 19.11 Representation of an electron micrograph of a portion of a denatured and reannealed Mu phage DNA duplex showing the split ends and G loop.

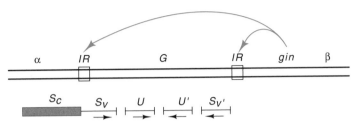

Figure 19.12 Structure of the *G* region of phage Mu. Transcription of the S and U or S' and U' genes initiates at a promoter in the α region. The *gin* gene product acts at the inverted repeats, IR, to invert the *G* region at a low rate.

of the phage in the resulting lysate will still be in the plus orientation, and heteroduplexes prepared from this DNA will not contain *G* bubbles. Lysogens are different. Although the phage originally entered cells with the *G* loop in the plus orientation, during the many generations of growth from a single lysogen, *G* can invert, and eventually about half the cells in the population will contain Mu with *G* in the plus orientation and half will contain Mu with *G* in the minus orientation. Induction of the phage in this cell population and preparation of DNA heteroduplexes will yield half the molecules with *G* bubbles.

The mutants with *G* stuck in one orientation because of the deletion at the end of *G* or the point mutation in *gin* permit a particularly simple demonstration of the infectivity properties. Mu lysogens with *G* stuck in the minus orientation yield phage particles, but these are not infective on *Escherichia coli* K-12. Mu lysogens containing *G* stuck in the plus orientation are fully infective when plated on *Escherichia coli* K-12.

Genetic analysis of the *G* region of the phage shows that two sets of genes are involved: *S* and *U*, and *S'* and *U'*. *S* and *U* are expressed when *G* is in the plus orientation and *S'* and *U'* are expressed when *G* is in the minus orientation (Fig. 19.12). The promoter for these genes and the initial portion of the *S* gene is contained in the α region adjacent to G. Thus, *G* contains a variable portion of *S* and *S'* as well as the intact *U* and *U'* genes. This is an interesting example of a situation in which a given DNA sequence can be used to specify different products.

In vitro Transposition, Threading or Global Topology?

Like most systems, deeper understanding of transposition requires learning the biochemical details of the transposition reactions. This, in turn, requires assay of the reaction and the use of purified components. One convenient assay for transposition utilizes a plasmid containing the DNA sites involved in a transposition reaction. Rearrangement of segments of the plasmid resulting from transposition will change the sizes

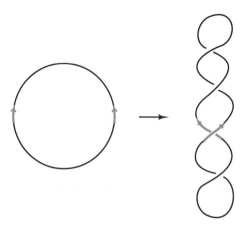

Figure 19.13 Well-separated inverted repeats of a sequence on a circle can be adjacent to one another and parallel when the circle is supercoiled.

of DNA fragments resulting from subsequent restriction enzyme cleavage.

Experiments with a plasmid containing two *res* sites of Tn3 showed that resolution would proceed *in vivo* in the presence of a Tn3 with a functional resolvase gene. The reaction would also take place *in vitro* upon the addition of purified resolvase. This reaction occurred, however, only if the plasmid DNA was supercoiled and if the two *res* sites on the plasmid were correctly oriented. Reversing one blocked the reaction, in contrast to what is seen with lambda phage where the excision reaction will proceed with the *att* sites in either orientation with respect to one another.

Why should an orientation requirement exist? We might expect that the two sites encounter each other by random flailings about of the DNA, in which case, there could be no orientation requirement. One attractive hypothesis to explain the orientation requirement is threading. Once the protein binds one site, it reaches the other by crawling along the DNA, thus preserving its knowledge of the orientation of the first site. Threading the DNA through the protein until the second site arrives at the protein accomplishes the same end.

The requirement for a specific orientation for the transposition sites can be explained in a different way. Suppose that the sites had to interact in a specific orientation at the time of transposition. Due to negative supercoiling, two sites which are in inverted repeat orientation on a circular DNA molecule will interact in a parallel orientation (Fig. 19.13). Interacting in an antiparallel orientation would require massive rearrangement of the supercoils and would be highly unlikely to occur.

The two possible explanations for the orientation requirement of the interacting sites of the Mu-Tn3 class of transposons can be tested by making use of the lambda phage integration enzymes. After a plasmid containing Mu sites which are flanked by lambda phage attachment sites has supercoiled itself, then reaction between the lambda *att* sites can be performed. Such an interaction should have little effect upon the overall supercoiled structure of the plasmid. Yet, if the sites are in one

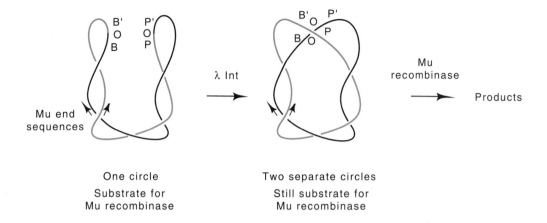

Figure 19.14 Action of lambda phage Int protein on a circle containing the Mu ends can separate it into two chemically separate circles without dramatically altering the wrapping structure in the vicinity of the Mu ends.

orientation, the reaction will invert the Mu sites, and if the lambda *att* sites are in the opposite orientation, the reaction will generate two separate DNA molecules, each containing one of the Mu sites (Fig. 19.14). Neither orientation of the lambda *att* sites had any effect on the Mu transposition reaction. This result shows that tracking is not used by the Mu enzymes to determine the transposition sites' orientations, but rather it is the orientation of the sites with respect to one another in the supercoiled DNA that matters.

Hopping by Tn10

An intermediate in the transposition process of Tn3 contains both the source and target molecules joined together in a cointegrate. Then a recombination event in the *res* site resolves these structures. The existence of cointegrates proves that Tn3 transposes by replication. Transposon Tn10 is different. No cointegrate structures were ever observed. This fact suggests that Tn10 might transpose without replication.

A conceptually simple experiment can test whether Tn10 replicates as it transposes. Kleckner and Bender constructed a lambda phage that could not replicate upon entering a normal host. The phage contained Tn10 that carried both a tetracycline resistance gene and the *lacZ* gene. A second phage was also made that was identical to the first except that it contained a nonsense mutation in *lacZ*.

Consider the consequences of infecting a LacZ⁻ cell with a phage that is heteroduplex for the point mutation (Fig. 19.15). Since the phage does not replicate, the only way for cells to become tetracycline resistant is for the transposon to enter the cell's chromosome. If the transposon replicates as it transposes, then the *lacZ* that is inserted into the chro-

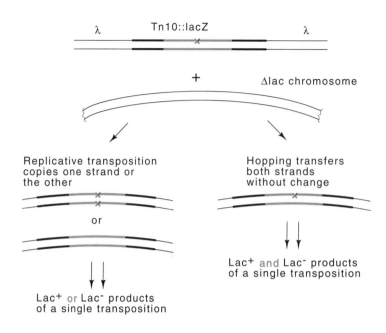

Figure 19.15 How a λ phage with a single base pair mismatch in *lacZ* within a Tn10 can generate homozygous replicative transposition products or heterozygous nonreplicative products.

mosome will be homozygous. It will either be wild type or mutant on both strands. If, however, the transposon excises itself from the phage and inserts itself into the chromosome before replicating, then the *lacZ* gene will remain heterozygous. If the cells are repair deficient, only after the first chromosome replication will each DNA molecule within the cells become homozygous. This means, however, that the descendants of the original cell will be of two types, Lac⁺ and Lac⁻. Because cells forming a colony do not mix, one half the colony will be Lac⁺ and one half Lac⁻. On Lac indicating plates a sector comprising about half the colony will indicate Lac⁺ and the other half will indicate Lac⁻.

A simple control experiment was possible by infecting cells lysogenic for lambda phage. The infecting lambda phage could recombine into the prophage in regions of homology at considerably higher frequency that the transposon relocates on its own. Therefore, tetracycline-resistant cells would have obtained a heterozygous copy of the *lacZ* gene, and should show sectoring on the indicating plates. They did. Therefore Tn10 hops.

In order for Tn10 to survive, it must hop infrequently, and shortly after passage of a replication fork (see problem 19.19). The infrequent hopping results from a low synthesis rate and extreme instability of its transposase. The transposase is so unstable that the larger the Tn10 element, the less frequently it transposes, and transposase barely functions in *trans* at all. The low synthesis rate of transposase results from two factors. One is a low activity of the promoter serving transposase, p_{IN} on IS10-L, and the second is antisense messenger deriving from p_{OUT} (Fig. 19.16). This second promoter is active, and its product hybridizes with the beginning of transposase messenger to block translation. This

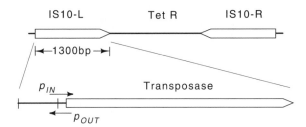

Figure 19.16 The locations of p_{IN} and p_{OUT} in IS10-L

is one of several examples of antisense messenger playing an important biological role.

A simple experiment showed that antisense messenger from p_{OUT} reduces translation of the p_{IN} messenger rather than interfering with its transcription. Two fusions of β-galactosidase to the transposase gene were made (Fig. 19.17). One was a protein fusion in which the N-terminal portion of transposase substituted for the N-terminal of β-galactosidase. In this, the levels of transposase messenger as well as any regulation of translation would affect the synthesis of β-galactosidase. In the second fusion, β-galactosidase was placed further downstream so that only the synthesis level of transposase messenger would affect the synthesis of β-galactosidase. A plasmid with a copy of the p_{OUT} region was introduced to these two cell lines. The extra copies of the p_{OUT} RNA decreased β-galactosidase synthesis from the protein fusion, but not

Figure 19.17 Hybridization of transcripts from p_{OUT} to p_{IN} interferes with translation of the protein fusion between transposase and β-galactosidase, but not the gene fusion.

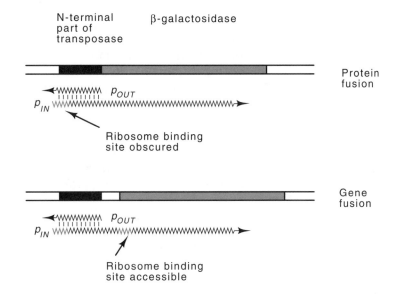

from the gene fusion, thus showing that the antisense RNA affects translation of transposase messenger.

Another type of control of transposition also exists. It has been found that damaging the host cell *dam* methylase system increases transposition frequency. The *dam* methylase is the system that methylates GATC sequences to permit repair enzymes to identify newly replicated DNA. There are two GATC sequences in IS10, one in the -10 region of p_{IN}, and the other in the binding site of transposase. Methylating both of these sites greatly decreases the transposition frequency by reducing both the promoter activity and the transposase activity. What about half methylated, or hemimethylated DNA, as is generated after replication? *In vitro* experiments showed that DNA methylated on one strand is significantly more active than fully methylated DNA, both in the promoter activity and in the transposase activity.

The activity of hemimethylated DNA could also be examined *in vivo*. Conjugating male cells transfer only one strand. The recipients synthesize the complementary strand as it enters. Donor Dam$^+$ cells were mated with Dam$^-$ recipient cells to generate hemimethylated DNA. This DNA was prevented from recombining into the female by the presence of a RecA$^-$ mutation. Thus, the only way recipient cells could become tetracycline resistant would be for the transposon to hop. The frequency of transposition hopping could be measured by determining the frequency of tetracycline-resistant cells. Hemimethylated Tn10 transposed much more frequently than fully methylated Tn10. These results mean than Tn10 is much more likely to transpose just after the replication fork has passed.

Retrotransposons in Higher Cells

The genomes of higher cells contain substantial numbers of repeated sequences. For example, the *Alu* sequence of 300 base pairs constitutes about 5% of the human genome. Among these repeated sequences are two main classes of transposable elements. One is similar to Tn10 in its DNA transposition mode. This includes the Ac element in maize, Tc1 in nematodes, and the P element in *Drosophila*. Members in the other class transpose by means of an RNA intermediate, and they include the Ty1 factor in yeast, the copia-like elements in *Drosphila* and the long interspersed, LI elements in mammals. These elements are retrotransposons and are closely related to retroviruses, if not identical to retroviruses in some cases.

Hybridization and sequencing of Ty1 and the regions into which it inserts have revealed its structure. It duplicates five bases upon insertion and consists of two flanking regions, called delta, of 330 base pairs oriented as direct repeats around a 5,600-base-pair central region that contains considerable homology to retroviruses. Not all delta elements found in yeast are identical, nor are the central regions identical, for some Ty elements are able to block expression of nearby yeast genes while others stimulate expression of genes near the point of integration.

A recombination event between the two delta elements deletes the central region and one delta element. Hence it is not surprising that the yeast genome contains about 100 of these solo delta elements. Recombination between different Ty elements can create various chromosomal rearrangements. Although the consequences of recombination can be determined easily in yeast, similar chromosome rearrangements catalyzed by recombination between repeated sequences must also occur in other organisms. Consequently transposons may be of positive value to an organism because they may speed chromosome rearrangements that may directly generate new proteins and new schemes of gene regulation.

Retroviruses, whose study began long before their discovery as a part of the yeast Ty1 factors, are particles that contain single-stranded RNA. This is both translated upon infection as well as converted via an RNA-DNA hybrid to a DNA-DNA duplex that is often integrated into the genome of the infected cell and is called a provirus. This form of the virus and defective variants can be considered the retrotransposon. The generation of the RNA-DNA duplex is catalyzed by reverse transcriptase. This enzyme is packaged within the virus particle. Upon their insertion into the chromosome, retroviruses duplicate a small number of bases as a result of generating staggered nicks in the target sequence. At each end of the retrovirus sequence is an inverted repeat of about 10 base pairs that is part of a few-hundred-base-pair direct repeat. Between the direct repeats, which are named long terminal repeats or LTRs, are sequences of about 5,000 base pairs that code for viral coat protein and other proteins. Transcription begins near the end of one LTR, proceeds through the central region, and ends within the other LTR.

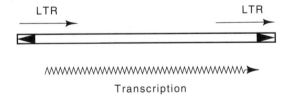

Transcription

Some of the DNA that is found in repeated sequences in *Drosophila* has been analyzed, and its properties suggest that it too is related to proviruses. One such family is *copia*. Small virus-like particles containing RNA homologous to *copia* DNA sequences can even be isolated from the nucleus of *Drosophila* cells. This RNA is translatable into one of the proteins that coats the RNA.

Retroviruses have been much harder to demonstrate in humans than in other animals. Nonetheless, they have been found. One was found to have inserted a copy of itself into the gene encoding factor VIII which is necessary for blood coagulation.

A more easily observed element in the human genome is the *Alu* family of sequences. Humans contain 100,000 to 500,000 copies of this sequence. The name derives from the fact that the restriction enzyme *Alu* cleaves more than once within the sequence. Consequently, digestion of human DNA with *Alu* yields 100,000 identical fragments, which upon electrophoresis generate a unique band in addition to the faint smear generated by the heterogeneity of the remainder of the DNA. The *Alu* sequences look like direct DNA copies of mRNA molecules because they contain a stretch of poly deoxyadenosine at their 3' ends. Like transposons, the *Alu* sequences also are flanked by direct repeats of chromosomal sequences of 7 to 20 base pairs. It is not possible to tell from their structure whether retroviruses and *Alu* sequences evolved from transposable elements or the reverse.

An RNA Transposition Intermediate

With the similarity of the yeast Ty1 sequence to retroviruses, it was of considerable interest to determine whether the element transposes using an RNA intermediate or a DNA intermediate. One simple way to test whether Ty1 proceeds through an RNA intermediate was to make use of the fact that in eukaryotic cells intervening sequences are removed from mRNA. Fink and co-workers placed a strong and controllable promoter upstream of a Ty element (Fig. 19.18). In the middle of the element they placed a segment of a yeast ribosomal protein gene containing an intervening sequence. Inducing the promoter led to a high frequency of transposition. Most importantly, the transposed elements lacked the intervening sequence, thereby proving they had transposed via an RNA intermediate, and while in the RNA state, had been subjected to the cellular processing that removes intervening sequences.

Figure 19.18 When Ty1 passes through a mRNA state, the intervening sequence in the ribosomal protein gene is spliced out and the resulting transposon, which drives His expression, lacks the sequence.

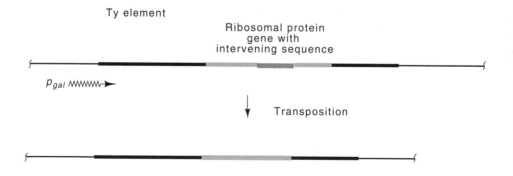

P Elements and Transformation

Transposons should form ideal vectors for genetic engineering. DNA to be transposed into a chromosome merely has to be inserted at a nonessential site within a transposon active in the desired cells, and then the transposon DNA has to be injected into the cells. Indeed, such constructs do work, and were first made with bacterial transposons Tn3, Tn5 and Tn10. The P element from *Drosophila* has many characteristics of a transposable sequence capable of regulating its transposition, and it too can be engineered for transformation of DNA into cells.

When a male P+ line is mated with a female non-P line, a short interval of intense transposition in the developing embryo follows, and many genes are inactivated by the insertion of P elements. Afterwards, the transposition activity ceases, and the resultant strains are genetically stable. The reverse–mating a P+ female with a non-P male–does not stimulate transposition. By analogy to the bacterial transposons or to phage lambda, the P elements behave as though they encode a repressor and enzymes necessary for transposition. Upon entry of a P element into a new cytoplasm, the temporary absence of repressor permits the transposition enzymes to be synthesized, and transposition likely occurs. Additionally, the *Drosophila* P elements appear to be stimulated to excise from their former locations as well as to copy themselves into new locations because many of the mutations generated by insertions of P elements can be induced to revert by the same mating process that stimulates P element mutations.

A P element would form a useful vector for inserting DNA into a chromosome of a fruit fly. By analogy with the bacterial transposons, nearly any sequence of foreign DNA could be inserted into the middle of the P element. If this P element plus a P element providing transposase were introduced into a non-P egg cell, transposition enzymes should be synthesized and move or copy the altered P element into the cellular DNA (Fig. 19.19).

Figure 19.19 Using a plasmid with a functional *Drosophila* P element to provide transposase so that the P element containing foreign DNA can transpose into the *Drosophila* genome.

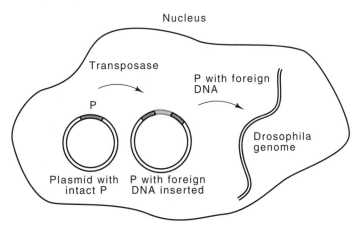

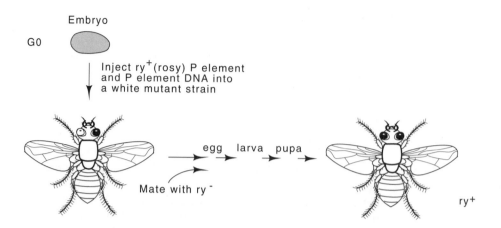

Figure 19.20 Schematic of the transformation process in *Drosophila*. Some cells deriving from one end of the egg including the germ line cells of the GO generation may be transformed. Only in the G1 generation can all the cells in a fly contain copies of the transforming DNA.

Spradling and Rubin found that all these expectations were correct and that P elements do provide a basis for efficient transformation of DNA into *Drosophila*. They inserted the *rosy* gene into the middle of an internally deleted P element. Then they injected both plasmid DNA containing the P element with inserted DNA and plasmid DNA containing an intact P into developing *rosy* negative *Drosophila* embryos. The injection was into a portion of the embryo in which the germ line cells develop. It was done while the embryo was multinucleate but before cell walls had developed. Many of the injected embryos survived and yielded mosaic adults whose progeny would then be homogeneous. Some possessed a functional rosy gene in all the fly's tissues (Fig. 19.20). These flies were easily detected because their eye color was different from that of the untransformed *rosy* negative mutants.

The sequence of events in transformation is as follows. The injected P element DNA is transcribed in the nucleus, and the resulting RNA is transported to the cytoplasm where it is translated into transposition enzymes that return to the nucleus. There they catalyze the transposition of the P element containing the *rosy* gene into a *Drosophila* chromosome. In the meantime, cell walls form, and the embryo develops to a larva and finally to an adult. Some of the cells of its germ line contain the functional *rosy* gene carried in the P element. When this fly matures and mates with a mutant *rosy* strain, those progeny that receive egg or sperm containing the *rosy* gene on the P element will then express functional xanthine dehydrogenase that is encoded by the *rosy* gene.

P element transformation is a valuable technique for the introduction of DNA into *Drosophila*. It is used routinely in the study of gene regulation and development. P element transformation possesses several drawbacks, however. First, the size of DNA that can be inserted into the chromosome is limited to what can be placed within a P element and handled on a plasmid carrying a P element. Some interesting genes are much larger than this size. A second problem is that P elements insert randomly. Because the expression level of a gene in *Drosophila* depends upon the position into which it has been placed, comparisons of gene activity in a collection of transformants cannot be closely compared.

P Element Hopping by Chromosome Rescue

The above-mentioned difficulties with P element transformation can be overcome by utilizing a property of the cells. Not surprisingly, parasites have capitalized on the cell's need for mechanisms to preserve fragments of broken chromosomes. The P elements appear to use this machinery by leaving a gap behind when they excise themselves from one chromosomal location and insert themselves into another. The cell's gap repair or chromosome rescue enzymes then replace the DNA missing from the gap by making use of a copy of the chromosome that still contains a P element.

Ordinarily, the ends of chromosomes are not recognized by this repair pathway due to the presence of the telomere structures. The gaps left by P element excision, however, lack telomeres and are acted upon by the rescue enzymes.

In the rescue process a small amount of DNA is digested from the ends by an exonuclease. Then, at each end, one of the two DNA strands can be further digested so as to leave single-stranded tails. As a result of RecA-like activity, these single-stranded tails can invade double-stranded DNA possessing complementary sequences. Subsequent elongation by DNA polymerase and recombination completes the process that removes the gap. If the P element excision and repair process occurs shortly after chromosome replication, then the sister chromatids are adjacent to one another and the centromere has not yet divided. Hence, the most probable template for repair of a P element excision event is simply the sister chromatid (Fig. 19.21). Thus, the gap is repaired with another copy of the P element. *In toto*, a P element has been released with no net change to the chromosome. Sometimes, however, a P element can excise from the chromosome before chromosome replication. In this case the template for repair of the gap can be the homologous chromosome. This other chromosome need not carry a P element in this position, and therefore the healed chromosome may not contain a P element.

Consider a P element near or in a region we wish to modify. If it can be induced to excise, then the repair mechanisms in the cell can use either the sister chromatid, or the homologous chromosome, or even just a homologous sequence located on a different chromosome. Therefore, we can modify the sequence as we desire, and insert it without a means of expression into the cells. Then we can induce excision of the P element inactivating our gene. Repair of the resulting gap can use the specially doctored DNA, and a fraction of the surviving cells will have replaced a portion of their original gene with our modified sequence.

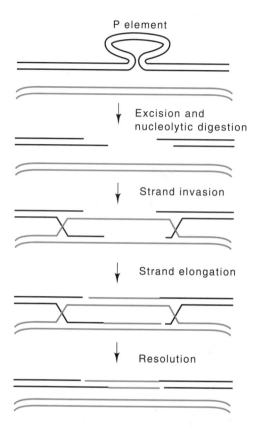

P element

Excision and
nucleolytic digestion

Strand invasion

Strand elongation

Resolution

Figure 19.21 Gap replacement in which DNA adjacent to a P element that excises is replaced by a DNA sequence from elsewhere.

Hence, a directed mutagenesis step has been performed. This replacement does not require that the P element have carried the entire modified gene, and further, the DNA sequence we have introduced by this method goes exactly to the desired chromosome site.

In summary, P elements are useful in genetic engineering in three basic types of experiments. First, they can be used as mutagens. They can transpose into genes and inactivate them. Because it is straightforward to clone the DNA flanking such an insertion event, it is possible to search for P element transformation events that have inactivated a desired gene, and then to clone the inactivated gene. A second use of the P elements is to carry DNA into *Drosophila*. Either intact genes can be carried into whatever location a P element hops into, or fragments of genes can be engineered to replace exactly a region of DNA already in the chromosome. The third use of P elements is the enhancer trap experiments that were described in Chapter 17. In those experiments, a P element transposes a reporter gene into the influence region of an enhancer.

Problems

19.1. Describe how R-factors or R-factors and F'-factors, instead of lambda phage as was described in the text, could be used to demonstrate transposition by drug-resistance transposons.

19.2. Consider the problem of comparing the densities of lambdoid transducing phage carrying polar mutations to the densities of phage without the mutations. What sequence of steps could be used to cross a mutation on the chromosome onto a transducing phage such that the only difference between the progeny and parent phage is the mutation?

19.3. Demonstration of the identity of insertion sequences in different lambda phage required the separation of the DNA strands of the phage. Look up how the DNA strands, sometimes called r for right and l for left or Watson and Crick, of lambda can be separated.

19.4. Why are oppositely oriented overlapping genes likely to be in the same codon register?

19.5. DNA from a bacterium was denatured, permitted to rehybridize to the extent of 1%, then digested to completion with S1 nuclease. The DNA was run on an acrylamide gel. Describe the resulting gel.

19.6. Show how deletions and inversions can be generated by the insertion of a Tn factor into its own replicon. A replicon is an autonomously replicating element like a chromosome or plasmid.

19.7. How could you determine the genetic locations of IS sequences that are naturally resident on the chromosome of *Escherichia coli*?

19.8. What is a straightforward genetic method of looking for replicon fusion generated by a transposon?

19.9. The left split ends of DNA heteroduplexes of phage Mu are too short to be be easily seen in electron microscopy of DNA heteroduplexes. How could you precisely measure the distribution of sizes of host DNA located at the left end?

19.10. Show how denaturing and reannealing DNA containing two inverted repeated sequences separated by some other DNA will create a stem and loop structure, a "lollipop."

19.11. What problem, unaddressed in the text, is raised by the following fact? Density labeling Mu phage and infecting show that the same DNA molecule that was packaged in the phage coat, not a copy of this DNA molecule, is integrated into the chromosome of a newly formed Mu lysogen.

19.12. Consider a transposon in the Tn5-Tn10 class that is flanked by inverted IS sequences. During transposition, the outside ends of the IS elements function as ends. Using a plasmid, how could you test whether the opposite ends could be used instead? That is, instead of transposing with the outside ends of the IS sequences, perhaps the inside ends could be used.

19.13. To make use of its broad host range, Mu has a great need to circumvent an infected cell's restriction-modification system. How might it do this and how would you test your idea?

19.14. Without the use of mutants unable to invert the *G* segment, stuck or *gin* mutants, how could you demonstrate the variable host range specificities of Mu phage?

19.15. Based on the reasons for shifting flagellin type in *Salmonella*, estimate sensible limits for the frequency of inverting the flagellin controlling segment.

19.16. Look up and describe how the invertible segment of *Salmonella* that controls flagellin synthesis was discovered.

19.17. Consider sequences such as transposable elements in the chromosome of a cell. Show how recombination events between these portable regions of homology can create deletions, duplications, inversions, and, in cells with more than one chromosome, translocations.

19.18. Give a plausible sequence of events by which a transposable element such as Tn10 could evolve to become a retrovirus.

19.19. Why does transposition of Tn10 just after passage of a replication fork replicate the transposon faster, on average, than the chromosome of the cell?

19.20. In testing segregation of Lac$^+$ and Lac$^-$ cells following transposition of the heterozygous Tn10, how are Lac$^-$ transformants distinguished from untransformed cells? Why is this necessary?

References

Recommended Readings

Genetic Evidence that Tn10 Transposes by a Nonreplicative Mechanism, J. Bender, N. Kleckner, Cell *45*, 801-815 (1986).

Interaction of Proteins Located at a Distance Along DNA: Mechanism of Target Immunity in the Mu DNA Strand-Transfer Reaction, K. Adzuma, K. Mizuuchi, Cell *57*, 41-47 (1989).

Targeted Gene Replacement in *Drosophila* via P Element-induced Gap Repair, G. Gloor, N. Nassif, D. Johnson-Schlitz, C. Preston, W. Engels, Science *254*, 1110-1117 (1991).

Isolation of an Active Human Transposable Element, B. Dombroski, S. Mathias, E. Nonthakumar, A. Scott, H. Kazazian, Jr., Science *254*, 1805-1808 (1991).

Is, Tn, and P Elements

The Origin and Behavior of Mutable Loci in Maize, B. McClintock, Proc. Nat. Acad. Sci. *36*, 344-355 (1950).

Frameshift Mutations in the Lactose Operon of *E. coli*, M. Malamy, Cold Spring Harbor Symp. Quant. Biol. *31*, 189-201 (1966).

Strong Polar Mutations in the Transferase Gene of the Galactose Operon of *E. coli*, E. Jordan, H. Saedler, P. Starlinger, Mol. Gen. Genet. *100*, 296-306 (1967).

Mutations Caused by the Insertion of Genetic Material into the Galactose Operon of *Escherichia coli*, J. Shapiro, J. Mol. Biol. *40*, 93-105 (1969).

Insertion Mutations in the Control Region of the Galactose Operon of *E. coli*, II. Physical Characterization of the Mutations, H. Hirsch, H. Saedler, P. Starlinger, Mol. Gen. Genet. *115*, 266-276 (1972).

Polar Mutations in *lac*, *gal*, and Phage lambda Consist of a Few IS-DNA Sequences Inserted with Either Orientation, M. Fiandt, W. Szybalski, M. Malamy, Mol. Gen. Genet. *119*, 223-231 (1972).

Mutagenesis by Insertion of a Drug-resistance Element Carrying an Inverted Repetition, N. Kleckner, R. Chan, B. Tye, D. Botstein, J. Mol. Biol. *97*, 561-575 (1975).

Transposition of R Factor Genes to Bacteriophage lambda, D. Berg, J. Davies, B. Allet, J. Rochaix, Proc. Nat. Acad. Sci. USA *72*, 3628-3632 (1975).

IS1 is Involved in Deletion Formation in the *gal* Region of *E. coli* K-12, H. Reif, H. Saedler, Mol. Gen. Genet. *137*, 17-28 (1975).

Genetic Engineering *in vivo* Using Translocatable Drug-resistance Elements, N. Kleckner, J. Roth, D. Botstein, J. Mol. Biol. *116*, 125-159 (1977).

DNA Sequence at the Integration Sites of the Insertion Element IS1, M. Calos, L. Johnsrud, J. Miller, Cell *13*, 411-418 (1978).

In vitro Mutagenesis of a Circular DNA Molecule by Using Synthetic Restriction Sites, F. Heffron, M. So, B. McCarthy, Proc. Nat. Acad. Sci. USA *75*, 6012-6016 (1978).

Physical Structures of Tn10-promoted Deletions and Inversions: Role of 1400 bp Inverted Repetitions, D. Ross, J. Swan, N. Kleckner, Cell *16*, 721-731 (1979).

DNA Sequence Analysis of the Transposon Tn3: Three Genes and Three Sites Involved in Transposition of Tn3, F. Heffron, B. McCarthy, H. Ohtsubo, E. Ohtsubo, Cell *18*, 1153-1163 (1979).

Identification of the Protein Encoded by the Transposable Element Tn3 which is Required for its Transposition, R. Gill, F. Heffron, S. Falkow, Nature *282*, 797-801 (1979).

Transposition Protein of Tn3: Identification and Characterization of an Essential Repressor-controlled Gene Product, J. Chou, P. Lemaux, M. Casadaban, S. Cohen, Nature *282*, 801-806 (1979).

Regulation of Tn5 Transposition in *Salmonella typhimurium*, D. Biek, J. Roth, Proc. Nat. Acad. Sci. USA 77, 6047-6051 (1980).

Resolution of Cointegrates between Transposons Gamma-Delta and Tn3 Defines the Recombination Site, R. Reed, Proc. Nat. Acad. Sci. USA *78*, 3428-3433 (1981).

Transposon Tn3 Encodes a Site-specific Recombination System: Identification of Essential Sequences, Genes, and Actual Site of Recombination, R. Kostriken, C. Morita, F. Heffron, Proc. Nat. Acad. Sci. USA *78*, 4041-4045 (1981).

Nucleotide Sequence of Terminal Repeats of 412 Transposable Elements of *Drosophila melanogaster*, B. Will, A. Bayev, D. Finnegan, J. Mol. Biol. *153*, 897-915 (1981).

Genetic Organization of Transposon Tn10, T. Foster, M. Davis, D. Roberts, K. Takeshita, N. Kleckner, Cell *23*, 201-213 (1981).

DNA Sequence Organization of IS10-Right of Tn10 and Comparison with IS10-Left, S. Halling, R. Simons, J. Way, R. Walsh, N. Kleckner, Proc. Nat. Acad. Sci. USA *79*, 2608-2612 (1982).

Genetic Transformation of *Drosophila* with Transposable Element Vectors, G. Rubin, A. Spradling, Science *218*, 348-353 (1982).

Inverted Repeats of Tn5 are Transposable Elements, D. Berg, L. Johnsrud, L. McDivitt, R. Ramabhadran, B. Hirschel, Proc. Nat. Acad. Sci. USA *79*, 2632-2635 (1982).

Control of Tn5 Transposition in *Escherichia coli* Is Mediated by Protein from the Right Repeat, J. Johnson, J. Yin, W. Reznikoff, Cell *30*, 873-882 (1982).

Transposition of Cloned P Elements into *Drosophila* Germ Line Chromosomes, A. Spradling, G. Rubin, Science *218*, 341-347 (1982).

Identification of a Transposon Tn3 Sequence Required for Transposition Immunity, C. Lee, A. Bhagwat, F. Heffron, Proc. Nat. Acad. Sci. USA *80*, 6765-6769 (1983).

DNA Sequence of the Maize Transposable Element *Dissociation*, H. Doring, E. Tillman, P. Starlinger, Nature *307*, 127-130 (1983).

Tn10 Transposase Acts Preferentially on Nearby Transposon Ends *in vivo*, D. Morisato, J. Way, H. Kim, N. Kleckner, Cell *32*, 799-807 (1983).

Site-specific Relaxation and Recombination by the Tn3 Resolvase: Recognition of the DNA Path between Oriented *res* Sites, M. Krasnow, N. Cozzarelli, Cell *32*, 1313-1324 (1983).

Three Promoters Near the Termini of IS10: pIN, pOUT, and PIII, R. Simons, D. Hoopes, W. McClure, N. Kleckner, Cell *34*, 673-682 (1983).

Translational Control of IS10 Transposition, R. Simons, N. Kleckner, Cell *34*, 683-691 (1983).

Isolation of the Transposable Maize Controlling Elements Ac and Ds, N. Fedoroff, S. Wessler, M. Shure, Cell *35*, 235-242 (1983).

IS10 Transposition is Regulated by DNA Adenine Methylation, D. Roberts, B. Hoopes, W. McClure, N. Kleckner, Cell *43*, 117-130 (1985).

The Maize Transposable Element *Ds* Is Spliced from RNA, S. Wessler, G. Baran, M. Varagona, Science *327*, 916-918 (1987).

Recombination by Resolvase is Inhibited by *lac* Repressor Simultaneously Binding Operators Between *res* Sites, R. Saldanha, P. Flanagan, M. Fennewald, J. Mol. Biol. *196*, 505-516 (1987).

Site-Specific Recombination by Tn3 Resolvase: Topological Changes in the Forward and Reverse Reactions. W. Stark, D. Sherratt, M. Boocock, Cell *58*, 779-790 (1989).

The Two Functional Domains of γδ Resolvase Act on the Same Recombination Site: Implications for the Mechanism of Strand Exchange, P. Dröge, G. Hatfull, N. Grindley, N. Cozzarelli, Proc. Natl. Acad. Sci. USA *87*, 5336-5340 (1990).

The Hin Invertasome: Protein-mediated Joining of Distant Recombination Sites at the Enhancer, K. Heichman, R. Johnson, Science *249*, 511-517 (1990).

Bacterial Transposon Tn7 Utilizes Two Different Classes of Target Sites, K. Kubo, N. Craig, J. Bact. *172*, 2774-2778 (1990).

High-Frequency P Element Loss in *Drosophila* is Homology Dependent, W. Engels, D. Johnson-Schlitz, W. Eggleston, J. Sved, Cell *62*, 515-525 (1990).

A Tn3 Derivative that Can be Used to Make Short in-frame Insertions Within Genes, M. Hoekstra, D. Burbee, J. Singer, E. Mull, E. Chiao, F. Heffron, Proc. Natl. Acad. Sci. USA *88*, 5457-5461 (1991).

Artificial Mobile DNA Element Constructed from the *EcoRI* Endonuclease Gene, S. Eddy, L. Gold, Proc. Natl. Acad. Sci. USA *89*, 1544-1547 (1992).

Mu Phage

Bacteriophage-induced Mutation in *Escherichia coli*, A. Taylor, Proc. Nat. Acad. Sci. USA *50*, 1043-1051 (1963).

Mutations in the Lactose Operon Caused by Bacteriophage Mu, E. Daniell, R. Roberts, J. Abelson, J. Mol. Biol. *69*, 1-8 (1972).

Electron Microscope Heteroduplex Study of the Heterogeneity of Mu Phage and Prophage DNA. M. Hsu, N. Davidson, Virology *58*, 229-239 (1974).

State of Prophage Mu DNA upon Induction, E. Ljunquist, A. Bukhari, Proc. Nat. Acad. Sci. USA *74*, 3143-3147 (1977).

Molecular Model for the Transposition and Replication of Bacteriophage Mu and other Transposable Elements, J. Shapiro, Proc. Nat. Acad. Sci. USA *76*, 1933-1937 (1979).

Heterogeneous Host DNA Attached to the Left End of Mature Bacteriophage Mu DNA, M. George, A. Bukhari, Nature *292*, 175-176 (1981).

Conservative Integration of Bacteriophage Mu DNA into pBR322 Plasmid, J. Liebart, P. Ghelardini, L. Paolozzi, Proc. Nat. Acad. Sci. USA *79*, 4362-4366 (1982).

G Inversion in Bacteriophage Mu: A Novel Way of Gene Splicing, M. Gilphart-Gassler, R. Plasterk, P. Van de Putte, Nature *297*, 339-342 (1982).

Transcription Initiation of Mu *mom* Depends on Methylation of the Promoter Region and a Phage-coded Transactivator, R. Plasterk, H. Vrieling, P. Van de Putte, Nature *301*, 344-347 (1983).

Mechanism of Transposition of Bacteriophage Mu: Structure of a Transposition Intermediate, R. Craigie, K. Mizuuchi, Cell *41*, 867-876 (1985).

Role of DNA Topology in Mu Transposition: Mechanism of Sensing the Relative Orientation of Two DNA Segments, R. Craigie, K. Mizuuchi, Cell *45*, 793-800 (1986).

Inverting Elements, Hin, G

Role of the G Segment in the Growth of Phage Mu, N. Symonds, A. Coelho, Nature *271*, 573-574 (1978).

Recombinational Switch for Gene Expression, J. Zieg, M. Silverman, M. Hilmen, M. Simon, Science *196*, 170-172 (1978).

The Invertible Segment of Bacteriophage Mu DNA Determines the Adsorption Properties of Mu Particles, A. Bukhari, L. Ambrosio, Nature *271*, 575-577 (1978).

Inversion of the G DNA Segment of Phage Mu Controls Phage Infectivity, D. Kamp, R. Kahmann, D. Zipser, T. Broker, L. Chow, Nature *271*, 577-580 (1978).

Phase Variation in *Salmonella*: Genetic Analysis of a Recombinational Switch, M. Silverman, J. Zieg, M. Hilmen, M. Simon, Proc. Nat. Acad. Sci. USA *76*, 391-395 (1979).

Analysis of the Nucleotide Sequence of an Invertible Controlling Element, J. Zieg, M. Simon, Proc. Nat. Acad. Sci. USA *77*, 4196-4200 (1980).

Invertible DNA Determines Host Specificity of Bacteriophage Mu, P. van de Putte, S. Cramer, M. Giphart-Gassler, Nature *286*, 218-222 (1980).

Phase Variation: Genetic Analysis of Switching Mutants, M. Silverman, M. Simon, Cell *19*, 845-854 (1980).

Homology between the Invertible Deoxyribonucleic Acid Sequence that Controls Flagellar-phase Variation in *Salmonella sp.* and Deoxyribonucleic Acid Sequences in Other Organisms, E. Szekely, M. Simon, J. Bact. *148*, 829-836 (1981).

Mapping of the *pin* Locus Coding for a Site-specific Recombinase that Causes Flagellar-phase Variation in *Escherichia coli* K-12, M. Enomoto, K. Oosawa, H. Momota, J. Bact. *156*, 663-668 (1983).

Hin-mediated Site-specific Recombination Requires Two 26 bp Recombination Sites and a 60 bp Recombinational Enhancer, R. Johnson, M. Simon, Cell *41*, 781-791 (1985).

RNA and Protein Transposition-related Activities

Evidence for Transposition of Dispersed Repetitive DNA Families in Yeast, J. Cameron, E. Loh, R. Davis, Cell *16*, 739-751 (1979).

DNA Rearrangements Associated with a Transposable Element in Yeast, G. Roeder, G. Fink, Cell *21*, 239-249 (1980).

Insertion of the *Drosophila* Transposable Element *copia* Generates a 5 Base Pair Duplication, P. Dunsmuir, W. Brorein, Jr., A. Simon, G. Rubin, Cell *21*, 575-579 (1980).

Retrovirus-like Particles Containing RNA Homologous to the Transposable Element *copia* in *Drosophila melanogaster*, T. Shiba, K. Saigo, Nature *302*, 119-124 (1983).

Ty Elements Transpose Through an RNA Intermediate, J. Boeke, D. Garfinkel, C. Styles, G. Fink, Cell *40*, 491-500 (1985).

Nucleotide Sequence of a Yeast Ty Element: Evidence for an Unusual Mechanism of Gene Expression, J. Clare, P. Farabaugh, Proc. Nat. Acad. Sci. USA *82*, 2829-2833 (1985).

Ty Element Transposition: Reverse Transcriptase and Virus-like Particles, D. Garfinkel, J. Boeke, G. Fink, Cell *42*, 507-517 (1985).

A Copia-like Transposable Element Family in *Arabidopsis thaliana*, D. Voytas, F. Ausubel, Nature *336*, 242-244 (1988).

Protein Splicing Converts the Yeast *TFP1* Gene Product to the 69-kD Subunit of the Vacuolar H^+-Adenosine Triphosphatase, P. Kane, C. Yamashiro, D. Wolczyk, N. Neff, M. Goebl, T. Stevens, Science *250*, 651-657 (1990).

Reverse Transcriptase Encoded by a Human Transposable Element, S. Mathias, A. Scott, H. Kazazian, Jr., J. Boeke, A. Gabriel, Science *254*, 1808-1810 (1991).

Protein Splicing in the Maturation of *M. tuberculosis* RecA Protein: A Mechanism for Tolerating a Novel Class of Intervening Sequences, E. Davis, P. Jenner, P. Brooks, M. Colston, S. Sedgwick, Cell *71*, 201-210 (1992).

Homing of a DNA Endonuclease Gene by Meiotic Gene Conversion in *Saccharomyces cerevisiae*, F. Gimble, J. Thorner, Nature *357*, 301-306 (1992).

Generating Genetic Diversity: Antibodies

20

The restriction-modification systems discussed in earlier chapters provide bacteria with the ability to identify and destroy foreign DNA. Multicellular organisms also must be able to identify and destroy foreign invaders. Such invaders include not only viruses but also bacteria and yeasts as well as multicelled parasites. Additionally, higher organisms must protect themselves against their own uncontrolled growth. From time to time, a few of their cells lose proper regulation and begin uncontrolled growth. Often these runaway cells are stopped by the immune system, but when they are not, the result is cancer.

Similar to a bacterium's "immunity" to infection by phage lambda, an organism can be resistant to infection by a parasite through the action of a specific protein-mediated interaction. The specific portion of the immune response of vertebrates can be considered to consist of two steps: first, recognizing a macromolecule that is foreign to the organism; and second, doing something with the recognized macromolecule. Because of the complexity of the immune system, this chapter will be concerned primarily with only a portion of the first part of the system, that involving adaptive and specific recognition of foreign molecules. The core of this response is a set of rearrangements of DNA segments and alterations in DNA sequences that generate a multitude of genes, each slightly different from the others. These code for the many different proteins that recognize the foreign molecules.

The Basic Adaptive Immune Response

Vertebrates can detect and dispose of most foreign macromolecules. Foreign macromolecules that can stimulate the immune system to produce a response are called antigens. These can include protein, nucleic acids, polysaccharides, and lipids that originate in a vertebrate from bacterial, viral, or parasitic infections, from injections of foreign materials, or from the transformation of a normal cell into a cancer cell. The transformed cells usually possess altered cell surfaces that often are recognized by the immune system.

Evolution has chosen a compromise between the ability of an animal to respond immediately to any one of millions of foreign antigens and the enormous burden that would result from being able to respond quickly. Young vertebrates possess the ability to recognize and respond to most foreign invaders, but this ability begins at a very low level for most antigens. Only some time after exposure to an antigen does the animal's ability to respond to that antigen begin to be significant. Thereafter, the animal can respond more quickly and more vigorously to the reappearance of a particular antigen. As a result of such exposure animals acquire immunity.

In most cases, the immune system can mobilize to deal with a new antigen before the presence of its source can harm the animal. Speeding and amplifying the immune response is the objective of medical immunization. In such immunizations, antigens are injected to prime the immune system. To avoid transmitting the disease associated with the antigen, antigenic material extracted from the organism is injected, or the organism is killed before injection, or a relatively harmless variant is injected. Therefore the animal generates an immune response without suffering the actual dangerous infection. Often, a second or third injection is made after several weeks to boost the animal's response.

Two responses begin to be apparent within a week of the introduction of a new antigen into an animal. New proteins called antibodies appear in the serum. They are able to bind specifically to the antigen and not to other molecules. Also, a subpopulation of the cell type called T lymphocytes appears. T lymphocytes, which ordinarily are found in the blood, spleen, and lymph nodes, have the ability to bind parts of the antigen to their surface.

In some cases, the mere binding of antibodies or T cells to an antigen is enough to protect the animal from the source of the antigen. As an example, the binding of antibody to the flagella of *Salmonella*, discussed in the previous chapter, may sufficiently immobilize the cell that it cannot survive in the intestinal tract. Usually, however, an animal's immune response includes more than just antibody molecules or T cells binding to the antigen. Additional proteins and cell types may participate in elimination of antigens. These recognize complexes of antibody bound to antigen and act to kill, digest, or eliminate the foreign material as well as to modulate the additional synthesis of antibodies and specific T cells. One such array of proteins is called the complement system.

In addition to the T lymphocytes, B lymphocytes also play an important role in the immune response. They are found in bone marrow, blood, spleen, and lymph nodes where they synthesize and secrete antibody molecules. All the antibody molecules synthesized by a particular B cell are of a single type. These bind to best to antigens possessing the same shape which elicited their synthesis and less well to molecules possessing similar shapes. Analogously, each T cell has a single binding specificity. One of the problems of immunology is understanding how the different B or T cells acquire the ability to synthesize proteins with recognition specificities. As we will see in more detail later, different immature lymphocytes possess the ability to recognize different antigens. The presence of one of these antigens can trigger one of the pre-B cell lymphocytes to begin growth and division. This growth requires both the antigen binding to the surface of the B cell and stimulation from a class of T cells that is defined by a surface antigen of its own called CD4. After about twelve divisions of B cells, this expanded clone of identical cells can synthesize and secrete appreciable quantities of antibody. This chapter will concern itself primarily with B cells. The same gene structures and gene rearrangements found in B cells also exist in T cells, and will not be discussed in detail here.

Telling the Difference Between Foreign and Self

What keeps a vertebrate from synthesizing antibodies against its own macromolecules? In principle, one solution to this problem would be to have the entire repertoire of an animal's antibody specificities explicitly encoded in the genome and expressed during development. Then the problem of preventing synthesis of antibodies against itself would be simple. None would be coded in the collection of specificities. This would be inefficient for at least two reasons. First, as we will see in the next section, the very large number of specificities needed by the immune system would use an inordinate amount of the coding potential of the genome. Second, such an inflexible system would be highly dangerous, for the moment an invader mutated and developed surface molecules not neutralized by the host cells' immune system, the invader could freely grow in the host animal, and the entire population of host animals could be killed. It would be as though the animals didn't possess any immune system at all, for the mutated invader would encounter no opposition.

For these reasons we expect considerable randomness, even in genetically identical animals, in the generation of antibody-binding specificities. Therefore, at least a few animals in any population will be capable of recognizing any antigen. It is this randomness, however, that generates the problem of responding only to foreign macromolecules. If the generation of the ability to recognize diverse antigens is random and not explicitly encoded in the DNA, the same randomness will generate antibodies against the organism itself, and these self-recognizing antibodies must be eliminated if the animal is to survive.

The problem of preventing self-recognition is largely solved by killing the lymphocytes that recognize macromolecules of the organism itself. This, of course, must be done after development of the immune system but before the organism is exposed to foreign macromolecules. For purposes of the molecular biology of the immune system, it is useful to consider the following simplified picture of the development of an

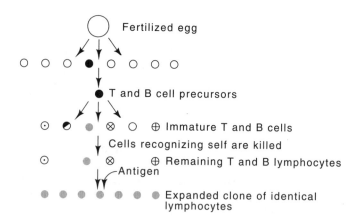

Figure 20.1 Growth and differentiation of a fertilized egg to yield the immature T cells, some of which recognize self and are killed. The remainder recognize other antigens and reside in the animal until the binding of antigen to their surface triggers them to begin growth, after which "clonal expansion" yields a population of size sufficient to synthesize appreciable quantities of immunoglobulin.

animal's lymphocytes (Fig. 20.1). Eventually in the cells deriving from a fertilized egg, there arises a single cell that can be considered the precursor of the T and B cells. Subsequent divisions of this cell produce only T and B cells. Each of the T or B cells generally develops the ability to recognize only one antigen. Some of these antigens are macromolecules present in the animal itself. The lymphocytes that recognize these "self" antigens are either killed or made unresponsive. The time of such deletion or inactivation varies with the animal. In mice this occurs at about the time of birth, whereas in humans it is much earlier. The lymphocytes capable of synthesizing antibodies to antigens not present at this time survive and form the pool of immature T and B lymphocytes that are capable of responding to foreign antigens.

The consequences of incomplete killing of lymphocytes that recognize "self" or the subsequent alteration of specificity so that "self" is recognized are predictable. In either case the immune system turns against the animal, and autoimmune conditions result. Autoimmunity is not at all uncommon. Forms of diabetes mellitus result from improper recognition of pancreatic cells. Myasthenia gravis is a condition in which antibodies are directed against muscle-nerve connections, and systemic lupus erythematosus is a disease in which antibodies are directed against many cellular components. Many other examples are also known. Autoimmune conditions also appear to result from loss of the ability to repress activation of T or B cells.

The Number of Different Antibodies Produced

How many antibodies with different specificities can be generated by a single mouse or human? If the number is small, then all the information for their synthesis could be explicitly stored in the genome, however unwise this would be, but if it is very large, some other mechanism must be used for generation of the diversity. A straightforward physical measurement made with mice provides an estimation of the lower limit in the number of different antibodies animals are capable of synthesizing.

After injection of mice with phosphocholine linked to a macromolecule, antibodies are induced. Isoelectric focusing shows that the serum then contains more than 100 different antibodies capable of binding phosphocholine. Each of these has a somewhat different binding specificity. Furthermore, the levels of each of these 100 antibodies have been increased by at least a factor of 100 compared to their level before injection. Together these induced phosphocholine-binding antibodies constitute about 1/200 of the total antibody protein in the serum of the mice.

These numbers lead to the estimate that mice can synthesize antibodies with more than 2×10^6 different specificities. The reasoning is as follows. Let N be the number of different antibodies present in a mouse. If these are all synthesized in nearly the same levels, each constitutes about 1/N of the total antibody protein in the serum. If one particular antibody is induced by a factor of 100, that one then constitutes 100/N of the total serum. The 100 different induced phosphocholine-specific antibodies then constitute $100 \times (100/N)$ of the total antibody protein in the serum, which is about 1/200 of the total. That is, $100 \times (100/N) = 1/200$, or $N = 2 \times 10^6$. If the mechanisms of antibody synthesis required one gene for each different antibody specificity, these numbers would necessitate that a sizable portion of the genome be used to code for the immune system,

$$2 \times 10^6 \text{ genes} \times 10^3 \frac{\text{nucleotides}}{\text{gene}} = 2 \times 10^9 \text{ nucleotides.}$$

Myelomas and Monoclonal Antibodies

The tremendous diversity in antibody specificities present in the serum of an animal precludes any simple study of homogeneous antibody. Although purifications based on binding to an antigen might be attempted for obtaining antibody with one specificity, the resultant antibodies would still be highly heterogeneous, for many different antibodies normally are induced by a single antigen. Even a molecule as simple as phosphocholine can induce synthesis of hundreds of different antibodies. Instead of pursuing the purification of a single antibody out of the thousands that are induced, researchers have made use of a cancer of the immune system to obtain homogeneous antibodies.

A lymphocyte with unregulated cell division generates a cancer called myeloma. The serum of patients with myeloma contains high concentrations of one single antibody. Myelomas can also be induced in mice, and the cancerous cells can be transferred from mouse to mouse. The

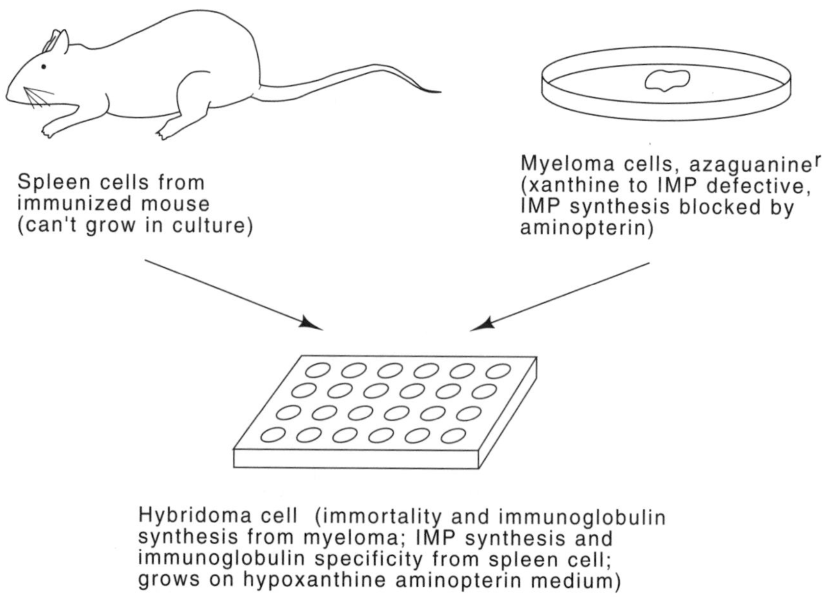

Figure 20.2 Scheme for the production of cell lines synthesizing monoclonal antibodies. Neither of the parental cell types can grow in the hypoxanthine aminopterin medium.

ascites fluid collected from such mice is a solution of almost pure, homogeneous antibody. A particular myeloma can be indefinitely propagated, and its corresponding antibody can be easily purified and studied.

The major problem with the use of myelomas is that their antigens are unknown. Of course, it is possible to screen hundreds of myeloma antibodies against thousands of potential antigens to find a molecule the antibodies bind to, but this does not prove to be highly successful. It would be ideal to be able to induce a myeloma for any desired antigen. This can be done. Milstein and Kohler discovered that the immortality of a myeloma cell line can be combined with the selectable antigen specificity of antibodies that are obtained from B cells of immunized animals. If a variant myeloma line has been used that itself does not produce antibody, the result of the fusion is a cell line that synthesizes a single type of antibody. This has the desired binding specificity. The cell line can be grown *in vitro* or injected into mice, giving them myeloma but providing a convenient source of large quantities of their antibody product. The antibodies produced by these means are called monoclonal because they derive from a single cell or clone.

Mouse cell lines synthesizing monoclonal antibodies are isolated by fusing myeloma cells with lymphocytes extracted from the spleen of a mouse about five days after its immunization with the desired antigen (Fig. 20.2). The cell fusion is accomplished merely by mixing the two cell types in the presence of polyethylene glycol. This agent induces fusion of the cell membranes; after an interval of unstable growth in which a number of chromosomes are lost, relatively stable chromosome numbers, called karyotypes, are established in the population. To reduce the background deriving from growth of unfused parental cell types, a genetic selection against both of the parental cell types is used so that only a fusion product can grow. The fused cells are then diluted to a concentration of a few cells per well of a culture tray, and after they have had time to grow and secrete antibody to the growth medium, the individual cultures from the wells are screened for the presence of antibody with the desired specificity.

Monoclonal antibodies find wide use in diagnostic and therapeutic medicine and in many types of basic research. In addition to providing an almost limitless supply of antibody with unchanging characteristics, monoclonal technology permits antibodies to be made against materials that cannot be completely purified. For example, monoclonal antibodies can be made against the cell walls of neurons from an animal. The neurons are dissected out, cell walls isolated and injected into a mouse, and the fusion of spleen cells to the myeloma line performed. Even unfractionated brain could be used for the initial injections. The antibodies synthesized by the fused cells can be screened for desired properties. Some of the resulting monoclonal antibodies have been found to bind to all cells, some bind only to neurons, and others bind to just a few neurons. Presumably neurons of the latter class are closely related to one another in lineage or function.

The Structure of Antibodies

Determining the amino acid sequences and three-dimensional structures of antibodies has greatly helped our understanding of how they are encoded by a reasonable number of genes and how they function. Although eight major classes of human antibody are known, this chapter deals primarily with the IgG group which contains IgG$_3$, IgG$_1$, IgG$_{2b}$, and IgG$_{2a}$. This is the most prevalent antibody in serum. The other antibodies are IgM, the first detectable antibody synthesized in response to an antigen, IgD, IgE, and IgA. These have slightly different structures and apparently are specialized for slightly different biological roles. IgG contains four polypeptide chains, two identical light, or L, chains of 22,500 molecular weight, and two identical heavy, or H, chains of about 50,000 molecular weight, giving a structure H$_2$L$_2$. Two subclasses of light chains are known in mouse and humans, kappa and lambda, whereas the eight major immunoglobulin classes are determined by the eight types of heavy chain, M, D, G3, G1, G2b, G2a, E, and A.

The chains of IgG are linked by inter- and intrachain disulfide bonds into a Y-shaped structure (Fig. 20.3). Papain digestion of one mole of IgG yields two moles of Fab fragments and one mole of Fc fragment. The "ab" indicates that these fragments possess the same binding specificity as the intact IgG antibody, and the "c" indicates that these fragments are readily crystallizable. Analysis of fragments of IgG has

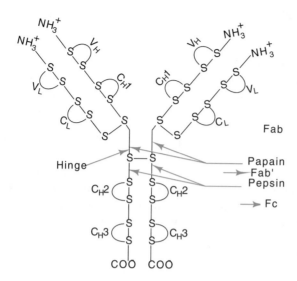

Figure 20.3 The structure of IgG. The antigen-binding site is at the amino terminal end. V and C indicate variable and constant regions, respectively. H and L indicate heavy and light chains, respectively, and the numbering refers to domains.

been highly important because fragments are much more easily studied than the intact IgG molecule.

The amino acid sequences of both the L and the H chains of IgG isolated from myelomas revealed an astonishing fact. About the first 100 amino acid residues of each of these chains are highly variable from antibody to antibody, whereas the remainder of the chains is constant (Fig. 20.4). Even more remarkable, however, is the finding that three or four regions within the variable portion of the protein are hypervariable. That is, most of the variability of the proteins lies in these regions. In view of their variability, such regions are likely to constitute the antigen-binding site of the antibody molecule. This suspicion was confirmed by affinity labeling of antibodies with special small-molecule antigens that chemically link to nearby amino acid residues. They were found to link to amino acids of the hypervariable portions of the L and H chains, just as expected.

Proteins from myelomas also assisted the X-ray crystallography of antibodies. Without the homogeneity provided by monoclonal antibodies, the crystal structures of antibodies could never have been deter-

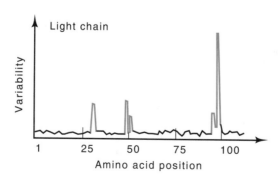

Figure 20.4 The variability of the light chains as a function of residue position. The hypervariable regions show as regions with significantly greater variability.

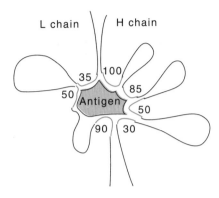

Figure 20.5 Schematic of the structure of the antigen-binding region of the V domain showing the confluence of the hypervariable regions to form the antigen-binding pocket.

mined. The structure of Fab fragments confirms most of the postulates that the amino acid sequence raised. The hypervariable portions of the L and H chains together form the antigen-binding site at the end of each arm of the antibody molecule in a domain formed by the variable regions (Fig. 20.5). In addition, the constant region of the L chain and the first constant region, C_{H1}, of the H chains form another domain (shown earlier in Fig. 20.3). The constant regions C_{H2} and C_{H3} of the two heavy chains form two additional domains of similar structure.

The fact that the variable portions of the L and H chains together form the antigen-binding site greatly reduces the number of genes required to specify antibodies. If any one of 1,000 L chains could combine with any one of 1,000 different H chains, then a total of 2,000 different genes could code for antibodies, with a total of 1×10^6 different binding specificities. The next sections describe how B lymphocytes combine small numbers of segments of the L and H genes to generate large numbers of different L and H genes, which then generate still larger numbers of antibodies because an L chain can function with most H chains. Techniques similar to those used for study of the antibody genes from B cells have also been applied to the identification and analysis of clones containing the genes encoding the T cell antigen receptors. These proteins also contain two polypeptides with variable and constant regions, and DNA rearrangements during the ontogeny of the organism also generate diversity.

Many Copies of V Genes and Only a Few C Genes

The patterns of sequence variability of the first 100 residues of the L and H chains and the constancy of the remainder of these chains suggested that the genes for L and H proteins were split into two parts. The V regions could then be encoded by any one of a set of different V region segments; upon differentiation into lymphocytes, one particular V region segment would be connected to the appropriate constant region segment to form a gene coding for an entire L or H chain. Such a proposal should be readily verifiable by the techniques of genetic engineering.

Cloning DNA coding for the V and C regions was not difficult. Since antibodies are secreted, they are synthesized by ribosomes attached to the endoplasmic reticulum of lymphocytes. Therefore, isolation of poly-A RNA from membranes of mouse myeloma cells yielded an RNA fraction greatly enriched for antibody messenger. This was then converted to cDNA with reverse transcriptase and cloned. Sequencing of the clones provided positive identification of those that contained the immunoglobulin genes. While this approach provided DNA copies of the mature RNA that is translated to form immunoglobulin, it did not directly indicate the status of the portion of the genome that encoded the RNA.

Southern transfer experiments done by Tonegawa and by Leder using fragments of cloned cDNA as a probe showed three facts about the genes involved with immunoglobulin synthesis. First, any single V region sequence possesses detectable but variable homology with up to about 30 different V region segments in the genome. The total number of V region coding segments in the genome can then be estimated from the numbers of such segments sharing homology to more than one V region probe. Results of such experiments indicate that the genome of the mouse contains from 100 to 1,000 kappa V region genes, with the most likely number being about 300. The second result from the Southern transfer experiments was that the genome contains much smaller numbers of the C region genes. The mouse genome contains four of the lambda subclass C region L genes and only one of the kappa subclass. The astounding third result obtained from Southern transfers was that the genome in the vicinity of a V and C region gene undergoes a rearrangement during differentiation from a germ line cell to a lymphocyte. That is, the size of DNA fragment generated by digestion with a restriction enzyme and containing homology to a V region differed between DNA extracted from embryos and DNA extracted from the myeloma source of the cDNA V region clone.

Answering deeper questions about the structure and rearrangements of the genes involved with antibody synthesis required additional cloning. Copies of undifferentiated V and C region genes could be obtained by cloning DNA isolated from mouse embryos. The differentiated copies could be obtained by cloning from myelomas. DNA extracted from mouse embryos and myelomas was cloned in lambda phage, and the plaques were screened for those containing DNA homologous to the V or C region from the cDNA clone. The results of these experiments in conjunction with the Southern transfer experiments showed that in the germ line cells, the C and V region genes are situated far from one another. In a myeloma cell, however, a DNA rearrangement has placed the expressed V and C region genes beside one another. After the rearrangement, they are separated by only about 1,000 base pairs. An R-loop between light chain mRNA and the rearranged DNA shows the V and C region genes separated by an intervening sequence plus an additional portion of V region gene called V'. This codes for the leader sequence that is required for secretion of the polypeptide.

The J Regions

The segments of embryonic DNA encoding the V regions of both the lambda and kappa light chains of mouse IgG do not code for about 13 amino acids normally considered to be the last part of the variable region of the protein. The sequence of the embryonic C region DNA does not solve the problem, for it contains just the amino acids that are constant from antibody to antibody. The missing amino acids have been found to be encoded by a segment of DNA lying from 1 to 3 kilobases (Kb) upstream from the 5' end of the C region gene. This region is called the J region because it is joined to the V region in the process of DNA rearrangement that connects a V and C region pair.

Of course, the most direct demonstration of the existence of a J region was provided by sequencing. But Tonegawa had first performed R-looping to locate the J region approximately. Light chain messenger from myelomas was hybridized to C region DNA cloned from embryos (Fig. 20.6). The R-looped structures of the lambda chain located a short region with the necessary coding properties about 1,000 to 3,000 nucleotides upstream from the C region. By contrast, kappa light chain mRNAs extracted from a series of myelomas indicated the existence of four different J regions ahead of the kappa C segment. Thus the presence of these J regions adds another means for generation of antibody

Figure 20.6 Genesis of an R-loop demonstrating the existence and location of a J region. In the embryonic DNA no V region is near a J-C region.

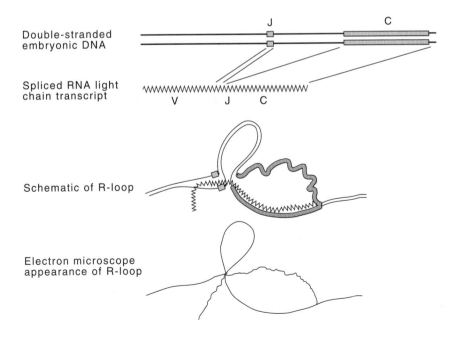

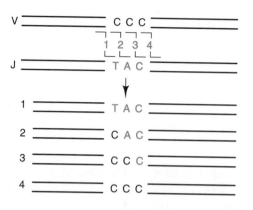

Figure 20.7 The variable position of the crossover joining the V and J regions can create several different codons at the join region.

diversity. Any of the several hundred V segments can be linked to the C segment by any one of the four J regions. Since the J regions of the light chains lie just beyond the third hypervariable region of the antibody, variability in this position has a direct effect on the structure of the antigen-binding site of the resultant antibody and the affinity of the antibody-antigen binding.

While nucleotide sequencing revealed a total of five kappa chain J regions, sequencing of myeloma proteins showed that one of the J regions is not functional. This finding is in keeping with the absence of a GT sequence at the beginning of a splice site on this potential J region gene.

Nucleotide sequencing by Leder demonstrated that additional diversity was also generated in connection with the J regions. The DNA

Figure 20.8 The process of genome rearrangement and transcript splicing that finally yields a processed mRNA molecule capable of being translated into an intact immunoglobulin light chain.

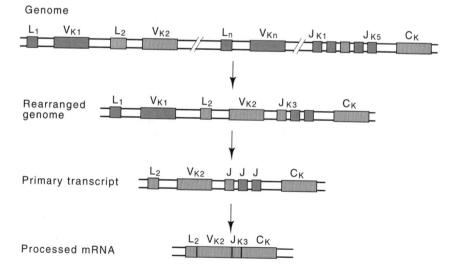

crossover site that connects the V segment to one of the J regions is not fixed. Slight variability in its position creates codons not present on either the V or the J segment being joined (Fig. 20.7).

The J region and the surrounding sequences are special. Immediately upstream of the J region itself lie signals that are used in the DNA splicing of a V region to the J region (Fig. 20.8). Then, immediately after the J region are the nucleotides that signal the mRNA processing machinery so that the intervening sequence of about 2,500 bases between the J region and the beginning of the C region can be removed and the messenger can be made continuous for translation. Later sections explain two additional functions of this region: activating transcription and serving as one end of a deletion that changes the C region joined to the V-J region.

The D Regions in H Chains

The heavy chains of mice and humans have also been examined to determine whether they are spliced together from segments as are the light chains. Indeed, the heavy chains are. About 300 variable H chain segments were found as well as four J regions in mice and ten in humans. There also exists a stretch of C regions, one gene for each of the constant region classes. These segments, however, do not encode all the amino acids found in the heavy chains. Thus it seemed likely that

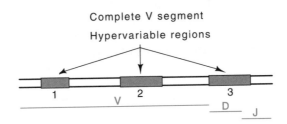

an additional segment of DNA is spliced in to generate the complete immunoglobulin heavy chain. These segments are called D, for diversity, because they lie in the third hypervariable region of the heavy chain.

Almost the same strategy as was used to find the J regions worked in finding the missing segments of the heavy chains. Tonegawa had observed that the J regions in some myelomas were rearranged even though they were not connected to V segments. Most likely the rearrangements involved the D segments. This hypothesis proved to be true. By cloning a rearranged J region and using the D region it contained, the D regions were located on the embryonic DNA ahead of the J regions.

One of the interesting features of the D regions is that they possess flanking sequences of seven and nine bases separated by 12 or 23 base pairs (Fig. 20.9) as do the sequences and spacers found alongside the V and J regions of the light chain. These flanking sequences are recognized by the DNA rearranging system. Cutting and joining is between a pair of sequences, one possessing a 12 base spacer, and one containing a 23 base spacer.

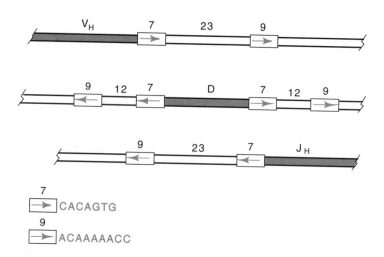

Figure 20.9 The V, D, and J regions and the positions, sizes, and sequences of the flanking recognition sequences used in the processing to produce an intact immunoglobulin gene.

Not only does introduction of J and D regions generate diversity, but imperfections in the rearranging process itself also generate diversity. That is, the cutting and joining are not perfect. Often deletions or insertions of several bases are made as the D segment is spliced into position (Fig. 20.10).

The variable region of the H chain that is generated by the fusions and rearrangements of different DNA regions is joined to the C region.

The C region of the heavy chain itself contains three intervening sequences. One lies between the coding region for each stretch of 100

Figure 20.10 An insertion of three bases generated upon splicing a D and J segment.

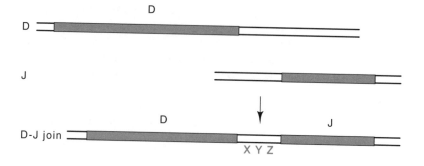

amino acids. These groups of 100 amino acids form the domains of the protein that interact with the portions of the immune system that dispose of foreign molecules bound to antibody molecules. The fact that intervening sequences separate well-defined structural domains of a protein led to the prediction mentioned earlier that most intervening sequences in other proteins would also be found to separate the nucleotides coding for functional and structural domains of the protein. Such a spacing increases the chances that sequences coding for intact domains could be shuffled between proteins in the course of evolution.

Induced Mutations and Antibody Diversity

We have seen that lymphocytes use reasonably small numbers of V, J, and D segments to generate a large number of different genes coding for antibody polypeptide chains. This, plus the fact that antigenic specificity is generated by the combination of two unrelated antibody chains, permits an animal to synthesize millions of different antibodies. As mentioned earlier, however, survivability of a species dictates that antibody specificities be randomly determined. Furthermore, it is likely that within the population of animals of one species, the genes involved with antibody diversity should themselves be considerably more variable than other genes. Such variability at a genetic locus is known as a genetic polymorphism and indeed is observed in the antibody genes.

The evidence for variability beyond that introduced at the ends of D segments and the variable crossover points in the splicing of J segments is direct. In one study involving a special line of mice, myelomas whose IgG products bound phosphocholine were examined. In contrast to most similar situations, the light chain variable region RNAs from these myelomas was found to hybridize to only one genomic V region. Thus, only one V region gene encoded all the different V region RNAs. Some of the V regions cloned from mouse myelomas were found to possess exactly the same V region sequence as the homologous V region cloned from embryonic mouse DNA. Others, however, differed from the embryonic sequence in one or more locations. Since all the V region genes originated from the same germ line V region and the corresponding sequence in nonlymphocyte cells is unaltered, the slight sequence differences represent mutations that were selectively introduced into the gene. Some of the differences encoded different amino acids, whereas others were neutral. Furthermore, the changes were localized to within about 2 Kb of the V region gene. These results mean that the mechanism responsible for generating the mutations makes mutations randomly but confines the changes to the immediate vicinity of the V region gene.

Class Switching of Heavy Chains

As mentioned above, the first detectable antibody synthesized in response to an antigen is IgM. Later, the other classes, IgM, IgD, IgG, IgE, and IgA can be detected. Remarkably, myelomas in culture also can switch the class of antibody they are synthesizing, but without changing the antigen specificity. Since the class of an antibody is determined solely by the constant region of the heavy chain, class switching must be generated by a change in the C segment that is connected to the V, D, and J segments.

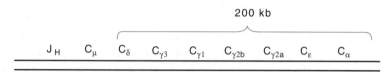

As mentioned above, following the J segment, the chromosome contains a series of C region segments beginning with the one coding for the M class of immunoglobulin. This array of eight C region segments is spread over 200 Kb of DNA and possesses an order consistent with the sequence of class switching that has been observed. Class switching replaces one of the downstream C regions for the M constant region by deleting intervening DNA. Hence, the intervening sequence between the J and C region and sites ahead of the other C region genes contain signals used in the generation of the deletions that produce class switching.

Enhancers and Expression of Immunoglobulin Genes

Up to now in this chapter, we have ignored the problem of regulation of expression of the immunoglobulin genes. Consider the 300 or so kappa-type V regions. The cell would waste considerable resources on

Figure 20.11 An enhancer in the region between the J and C regions stimulates a promoter lying in front of a V region segment after the V region has been joined to the C region.

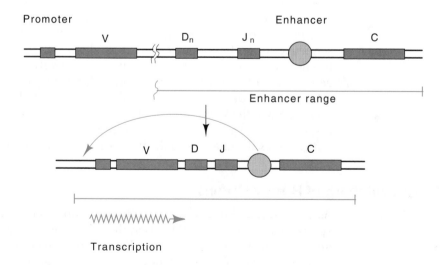

the synthesis of messenger, and perhaps its translation, if each of these gene segments were transcribed. How then can the fusion of a V segment to a C segment activate transcription of the spliced gene and only the spliced gene? The explanation, of course, is that each V region contains a promoter, but lacks an enhancer for activation. The necessary enhancers are located near the C genes (Fig. 20.11). Therefore, not only does gene splicing generate a substantial portion of the diversity in antibodies, but it also activates transcription of a spliced gene. Only after a V region has been placed near the C region is the enhancer sufficiently close to the promoter to activate it.

Like many enhancers, those serving the immunoglobulin genes are highly complex. Many different proteins bind to the enhancer. Some of these function in other tissues and must serve in a general way to activate. Others must be specific for cells of the immune system. Overall, however, the combination of enhancer proteins present in the cells of the immune system distinguishes these cells, and genes of the immune system are turned on only in the appropriate tissues.

The AIDS Virus

The AIDS virus is also known as the human immunodeficiency virus, or HIV, because of its effects on the immune system. It is a retrovirus that ultimately proves lethal to humans. The lethal effects of this virus derive from its crippling of the immune system by its killing of the CD4 helper T cells. Without these cells, the B cells cannot multiply in response to newly present antigens, and ultimately the infected person dies from infections that normally are of little consequence. The virus binds the CD4 cells with a viral coat protein specific for a transmembrane cellular protein unique to these cells and a few other cell types in the human body. After binding the membrane coat of the virus fuses with the membrane of the cell and the RNA is released within the cell

Figure 20.12 Gene products of HIV.

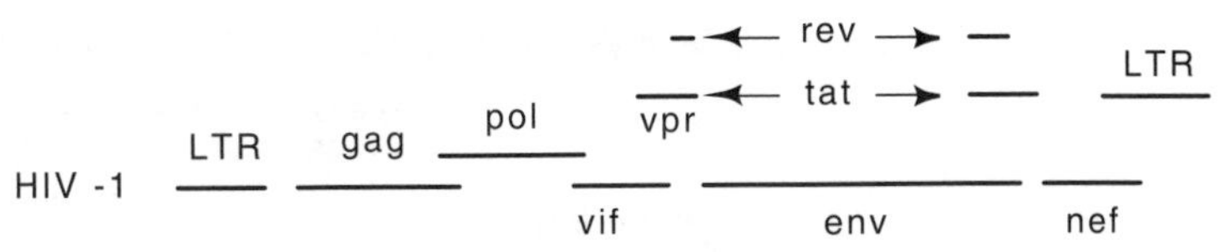

tat - transactivator

rev - regulates expression

vif - virion infectivity

vpr - ?

nef - negative factor

vpx - ?

where it frequently integrates into the chromosome after its conversion to double-stranded DNA. A protein encoded by the virus or provirus (Fig. 20.12) enables the infected cell to fuse with other uninfected T4 cells, at least in cell culture. Both infected cells and cells in fusion complexes ultimately are killed.

The AIDS virus is considerably more complex than the more simple retroviruses. In addition to the *gag*, *pol*, and *env* genes, the virus contains five additional open reading frames that encode proteins that activate or repress viral protein synthesis or play other roles in the viral infection.

Engineering Antibody Synthesis in Bacteria

There are limitations to polyclonal and monoclonal antibodies that can be found. Antibodies desired for special purposes, for example investigation of enzyme catalysis, could require antibodies of such selectivity that few are likely to be found by conventional methods. While animals may synthesize hundreds of antibodies against a large antigen, the number is limited by the size of the B cell population. Even somatic mutation can be imagined not to generate all desired structures in antigen-binding sites of antibodies. How could vast numbers of variant antibodies be generated for the study of antibody-antigen interactions?

One method for increasing the number of antibody specificities available for study would be to express a large library of light and heavy chains in bacteria. If these large libraries could be combined, then the same multiplicative increase in diversity as occurs in nature could also be available. Furthermore, expressing the proteins in bacteria simplifies

Figure 20.13 Amplification of an mRNA sequence coding for an immunoglobulin gene. Reverse transcriptase is used at the first step to make DNA. Thereafter the thermoresistant *Taq* polymerase is used in normal PCR reactions. The sequences encoding the restriction enzyme cleavage sites do not hybridize to the template on the first cycle. Thereafter they do.

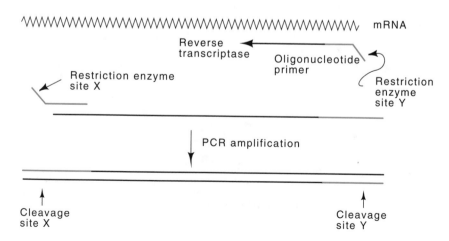

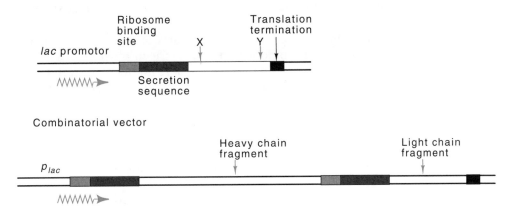

Figure 20.14 Design of vectors suitable for expression of light chain or heavy chain fragments. The combinatorial vector expresses both the light chain and the fragment of the heavy chain.

altering specific amino acids or mutagenizing regions of the protein. Time and cost scales would also become much more reasonable if the work is done in bacteria.

How might one begin to express a wide collection of light or heavy chains in bacteria? We cannot simply take a collection of V region genes, for most in the genome are not connected to the D or J regions and no diversification has occurred at the splicing step. We need to clone after rearrangement has taken place, and it would be best to clone only genes that have rearranged. If, however, we clone from DNA, the intervening sequences still interrupt the genes. The best approach is to clone from the mRNA using reverse transcriptase to make DNA copies. Then the polymerase chain reaction can be used to amplify the DNA. Due to the high fraction of conserved amino acids in the variable and light chain regions, it is possible to design oligonucleotides that will specifically hybridize to regions of the genes. The oligonucleotides can also be made with restriction enzyme cleavage sites so that later the resulting DNA can be easily inserted into a cloning vector (Fig. 20.13).

Cleverness in designing the cloning vector also is helpful. The incoming DNA must be expressed; therefore the plasmid must contain an active promoter. It should be possible to regulate this promoter in case the gene products are lethal to the cells. The *lac* promoter is a good one to use. Downstream from the promoter must be a good ribosome-binding site. Beyond this it is helpful to have a sequence that will direct the resulting protein to be secreted from the cells. Then, at last, comes the cloning site which is flanked by the two restriction enzyme cleavage sites. The same sites were engineered into the DNA that was amplified by PCR. Finally, a translation termination sequence is needed. Additional factors are that after generating libraries of light and heavy chains, combinations of the two must be formed. Resulting vectors must carry one light and one heavy chain with both being expressed under control of the *lac* promoter (Fig. 20.14).

Screening of the recombined library for vectors carrying a light and heavy chain that bind a particular antigen must be easy. Techniques have been developed for screening large numbers of lambda plaques on plates, and thus the best vector for initial screening would be lambda phage rather than a plasmid. Later, once a desired recombinant has been found, the light and heavy chains should be transferred to a plasmid vector. Finally, rather than attempting to synthesize the entire antibody molecule, just the Fab' fragment should be sufficient for many studies, and the probes for PCR amplification can be designed accordingly.

The technology described above has worked well and libraries have been made containing up to 10^6 different light chains and 10^6 different heavy chain fragments. These have been combined to yield libraries with truly gargantuan numbers of different antibody specificities.

Assaying for Sequence Requirements of Gene Rearrangements

It is not easy to assay for the sequence requirements of the joining signals used in the assembly of the immunoglobulins. Many copies of these sequences exist in normal cells and it is impossible to change all of them and then measure the new rates of recombination. An additional difficulty in the study of the recombination process is the low rate of recombination. Such low rates are necessary so that two recombination events do not often simultaneously occur in one cell.

To assay for recombination we need a method that will permit alteration of the relevant sequences and permit us to look at just the genes that are recombined. One way is to design a vector that can be transformed into cell lines established from different cell types of the immune system. The DNA could be left a while in the cells, then removed and transformed into *E. coli*, where we want to be able to detect easily just those rare DNA molecules that have been acted upon by the

Figure 20.15 Recombinational joining between signal sequences generates a coding joint and a signal joint.

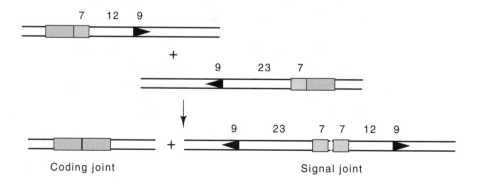

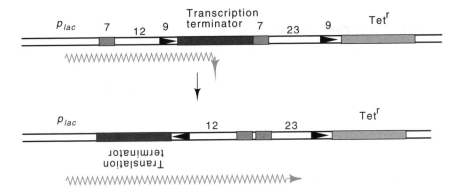

Figure 20.16 Recombination between the signal sequences inverts the region between the two, thereby inactivating the terminator by turning it backwards and upside-down. This activates expression of tetracycline resistance.

site-specific recombination machinery that assembles the immuno-globulin genes.

Sequence analysis has shown that the recombinant junction signals always recombine a signal with a 12 base spacing with a signal containing a 23 base spacing (Fig. 20.15). Two products result, a coding joint and a signal joint. Part of the diversity of the immune system is the inaccuracy that is generated at the coding joint. Nucleotides are deleted and inserted at this point. Very few such inaccuracies occur at the signal junction. If one of the junction signals were inverted, then, just like integration and excision of phage lambda, the DNA lying between the coding regions would not be deleted, but would be inverted.

The inversion of a DNA segment can be utilized in the design of a vector to assay for recombinational joining. By designing the vector to generate inversions rather than deletions, the products of both the coding joint and the signal joint can be preserved for further study. This can be done by arranging that a promoter on the vector be prevented from driving expression of a drug resistance gene by the presence of a transcription terminator between the two (Fig. 20.16). Flanking the terminator are a pair of recognition sequences oriented so that recombination inverts the DNA, thereby nullifying the terminator and permitting transcription of the drug resistance gene.

Samples of the plasmid DNA with the promoter, terminator, and recognition sequences, can be transformed into cells, allowed to rest there for a long period, extracted, and then transformed into bacteria. The total number of transformed bacteria can be assayed by the number of colonies receiving the plasmid, and the fraction which are also tetracycline resistant provides a measure of the recombination which occurred while the DNA was within the animal cells. The results of such experiments show that bases near the coding splice site are most important, as are the spacing distances.

Cloning the Recombinase

Direct assay of the immune system recombinase is not easy, and unsuccessful attempts to detect the enzyme were made for years. An indirect approach was to seek the DNA that either codes for the recombinase or that induces the synthesis of the recombinase. This was possible with the same trick described in the previous section. Recombinational joining was used to invert a sequence and permit expression of a selectable marker. In this case, however, since the join was to occur within animal cells not normally expressing the recombinase, cells were transformed with vectors containing fragments of chromosomal DNA. Any vectors that expressed the immune system recombinase would flip the DNA sequence and express the selectable marker.

One might wonder why this experiment had a chance of working. Why should DNA transformed into cells be any more likely to express the recombinase than the recombinase gene resident on the chromosome? One might picture that DNA in the chromosome was prevented from expressing by mechanisms similar to those that repress the yeast mating-type genes at the *HML* and *HMR* loci. Alternatively, newly arriving DNA might be able to express the recombinase gene until it is repressed just like zygotic induction of phage lambda or the P element transposition events that occur in *Drosophila* just after mating. Nonetheless, the assay did work, and it has been possible to detect and clone two clustered and divergently transcribed genes that code for the recombinase or the recombinase activator gene RAG.

Problems

20.1. Why are some antigens precipitated by IgG, but never by F_{ab} fragments?

20.2. With the introduction of as few additional postulates as possible, explain why the new antibodies elicited by repeated boostings of the same animal often possess higher and higher affinities for the antigen.

20.3. Estimate a theoretical maximum number of different kappa antibody specificities an animal could synthesize merely by splicing, and without mutation, deletion, or insertion, if the animal possessed 300 V_L, 10 J_L, 300 V_H, 10 D, and 10 J_H regions. Again, in the absence of diversity generated by variable position of splicing, deletions or insertions, and introduction of mutations, why could this theoretical maximum not be achieved?

20.4. Invent data using DNA probes homologous to just 30 different V regions that would indicate that the total mouse genome contains 200 different V region segments.

20.5. How would you interpret data obtained from sequencing a large number of V regions of the L and H chains that showed that about 10 amino acid residues of each chain were absolutely invariant?

20.6. Sketch the structure of an R-loop formed by annealing IgG heavy chain mRNA and its genomic DNA.

20.7. Suppose you have a cDNA clone containing a V region and you wish to identify a clone containing its embryonic V region gene. Considering that a number of genomic V genes will possess appreciable homology to the V region on the cDNA clone, how could a clone containing the desired V region be identified without sequencing?

20.8. Devise a method using known enzyme activities by which lymphocytes could hypermutate DNA in the vicinity of the appropriate recombined V(D)JC gene. The mutations must be concentrated in the area, and restricted to only the one recombined gene region.

20.9. The somatic mutations in a V region gene are confined to the one linked to the C region gene. Why is it logical that IgM not possess these mutations and that only the classes that are generated after IgM possess them?

20.10. Hypothesize a mechanism for splicing the V, J, D, and C regions together based on the sequences and spacings between the flanking conserved nucleotides of these regions.

20.11. What is the importance of knowing the orientation of the V segments on the chromosome relative to the D, J, and C segments?

20.12. No mention was made of the fact that mouse and human cells are diploid. What is a likely reason and explanation for the fact that lymphocytes synthesize only one antibody type despite the fact that with two chromosomes they could synthesize two or more types?

20.13. How would you look for the DNA that encodes the carboxy terminal amino acids that anchor the membrane-bound forms of antibody?

20.14. Without sequencing or cloning, how could you test the idea that class switching results from RNA processing and not from DNA rearrangements?

20.15. What techniques might one use to increase the chances that a collection of "in vitro" monoclonal antibodies might contain the desired specificity?

20.16. About what fraction of the human or mouse genome would be required to code for 2×10^6 average-sized genes?

20.17. On the vector used to examine recombinational joining why would it be sensible to place a DNA sequence that would lead to replication of the DNA while it was in the cultured animal cells?

20.18. Beyond the gene encoding the immunoglobulin recombinase are seven AUUUA sequences (See Cell 46, 659-667 [1986]). Why might this be?

References

Recommended Readings

Continuous Cultures of Fused Cells Secreting Antibody of Predefined Specificity, G. Kohler, C. Milstein, Nature 256, 495-497 (1975).

Somatic Mutation of Immunoglobulin Light-chain Variable-region Genes, E. Selsing, U. Storb, Cell 25, 47-58 (1981).

Identification of D Segments of Immunoglobulin Heavy Chain Genes and their Rearrangement in T Lymphocytes, Y. Kurosawa, H. von Boehmer, W. Haas, H. Sakano, A. Trauneker, S. Tonegawa, Nature 290, 565-570 (1981).

RAG-1 and RAG-2, Adjacent Genes That Synergistically Activate V(D)J Recombination, M. Oettinger, D. Schatz, C. Gorka, D. Baltimore, Science 248, 1517-1523 (1990).

Structure of Antibodies and T Cell Receptors

The Participation of A and B Polypeptide Chains in the Active Sites of Antibody Molecules, H. Metzger, L. Wofsy, S. Singer, Proc. Nat. Acad. Sci. USA 51, 612-618 (1964).

The Molecular Basis of Antibody Formation: A Paradox, W. Dreyer, J. Bennett, Proc. Nat. Acad. Sci. USA 54, 864-869 (1965).

The Three-dimensional Structure of the F_{ab}' Fragment of a Human Myeloma Immunoglobulin at 2.0-Å Resolution, R. Poljak, L. Amzel, B. Chen, R. Phizackerley, F. Saul, Proc. Nat. Acad. Sci. USA 71, 3440-3444 (1974).

Binding of 2,4 Dinitrophenyl Compounds and Other Small Molecules to a Crystalline Lambda-type Bence-Jones Dimer, A. Edmundson, K. Ely, R. Girling, E. Abola, M. Schiffer, F. Westholm, M. Fausch, H. Deutsch, Biochem. 13, 3816-3827 (1974).

A Human T Cell-specific cDNA Clone Encodes a Protein Having Extensive Homology to Immunoglobulin Chains, Y. Yanagi, Y. Yoshikai, K. Leggett, S. Clark, I. Aleksander, T. Mak, Nature 308, 145-149 (1984).

Isolation of cDNA Clones Encoding T Cell-specific Membrane-associated Proteins, S. Hedrick, D. Cohen, E. Nielsen, M. Davis, Nature 308, 149-153 (1984).

Complete Primary Structure of a Heterodimeric T-cell Receptor Deduced from cDNA Sequences, H. Saito, D. Kranz, Y. Takagaki, A. Hayday, H. Eisen, S. Tonegawa, Nature 309, 757-762 (1984).

Regulatory Pathways Governing HIV-1 Replication, R. Cullen, W. Greene, Cell 58, 423-426 (1989).

Human Immunodeficiency Virus Integration Protein Expressed in Escherichia coli Possesses Selective DNA Cleaving Activity, P. Sherman, J. Fyfe, Proc. Natl. Acad. Sci. USA 87, 5119-5123 (1990).

Cloning, Sequencing, and Structure of the V, D, J, and C Genes

An Analysis of the Sequences of the Variable Regions of Bence Jones Proteins and Myeloma Light Chains and their Implications for Antibody Complementarity, T. Wu, E. Kabat, J. Exp. Med. 132, 211-250 (1970).

Sequences of Five Potential Recombination Sites Encoded Close to an Immunoglobulin kappa Constant Region Gene, E. Max, J. Seidman, P. Leder, Proc. Nat. Acad. Sci. USA 76, 3450-3454 (1979).

Domains and the Hinge Region of an Immunoglobulin Heavy Chain are Encoded in Separate DNA Segments, H. Sakano, J. Rogers, K. Huppi, C. Brack, A. Traunecker, R. Maki, R. Wall, S. Tonegawa, Nature 277, 627-633 (1979).

An Immunoglobulin Heavy Chain Variable Region Gene is Generated from Three Segments of DNA: V_H, D, J_H, P. Early, H. Huang, M. Davis, K. Calame, L. Hood, Cell 19, 981-992 (1980).

Two mRNAs Can be Produced from a Single Immunoglobulin μ Gene by Alternative RNA Processing Pathways, P. Early, J. Rogers, M. Davis, K. Calame, M. Bond, R. Wall, L. Hood, Cell 20, 313-319 (1980).

Complete Nucleotide Sequence of Immunoglobulin γ2b Chain Gene Cloned from Newborn Mouse DNA, Y. Yamawaki-Kataoka, T. Kataoka, N. Takahashi, M. Obata, T. Honjo, Nature 283, 786-789 (1980).

Gene Segments Encoding Transmembrane Carboxyl Termini of Immunoglobulin γ Chains, J. Rogers, E. Choi, L. Souza, C. Carter, C. Word, M. Kuehl, D. Eisenberg, R. Wall, Cell 26, 19-27 (1981).

Structure of the Human Immunoglobulin μ Locus: Characterization of Embryonic and Rearranged J and D Genes, J. Ravetch, U. Siebenlist, S. Korsmeyer, T. Waldmann, P. Leder, Cell 27, 583-591 (1981).

Identification and Nucleotide Sequence of a Diversity DNA Segment (D) of Immunoglobulin Heavy-chain Genes, H. Sakano, Y. Kurosawa, M. Weigert, S. Tonegawa, Nature 290, 562-565 (1981).

Organization, Structure, and Assembly of Immunoglobulin Heavy Chain Diversity DNA Segments, Y. Kurosawa, S. Tonegawa, J. Exp. Med. 155, 201-218 (1982).

Structure of Genes for Membrane and Secreted Murine IgD Heavy Chains, H. Cheng, F. Blattner, L. Fitzmaurice, J. Mushinski, P. Tucker, Nature 296, 410-415 (1982).

Genomic Organization and Sequence of T-cell Receptor β-chain Constant and Joining-region Genes, N. Gascoigne, Y. Chien, D. Becker, J. Kavaler, M. Davis, Nature 310, 387-391 (1984).

A Third Rearranged and Expressed Gene in a Clone of Cytotoxic T Lymphocytes, H. Saito, D. Kranz, Y. Takagaki, A. Hayday, H. Eisen, S. Tonegawa, Nature 312, 36-40 (1984).

Organization and Sequences of the Diversity, Joining and Constant Region Genes of the Human T-cell Receptor β-chain, B. Toyonaga, Y. Yoshikai, V. Vadasz, B. Chin, T. Mak, Proc. Nat. Acad. Sci. USA 82, 8624-8628 (1985).

Complete Physical Map of the Human Immunoglobulin Heavy Chain Constant Region Gene Complex, M. Hofker, M. Walter, D. Cox, Proc. Natl. Acad. Sci. USA 86, 5567-5571 (1989).

Making Antibody Fragments Using Phage Display Libraries, T. Clackson, H. Hoogenboom, A. Griffiths, G. Winter, Nature 352, 624-628 (1991).

Semisynthetic Combinatorial Antibody Libraries: A Chemical Solution to the Diversity Problem, C. Barbas III, J. Bain, D. Hoekstra, R. Lerner, Proc. Natl. Acad. Sci. USA 89, 4457-4461 (1992).

V, D, J, and C Joining and Class Switching

A Complete Immunoglobulin Gene is Created by Somatic Recombination, C. Brack, M. Hirama, R. Lenhard-Schuller, S. Tonegawa, Cell *15*, 1-14 (1978).

The Arrangement and Rearrangement of Antibody Genes, J. Seidman, P. Leder, Nature *276*, 790-795 (1978).

Sequences at the Somatic Recombination Sites of Immunoglobulin Light-chain Genes, H. Sakano, K. Huppi, G. Heinrich, S. Tonegawa, Nature *280*, 288-294 (1979).

Amino Acid Sequence of Homogeneous Antibodies to Dextran and DNA Rearrangements in Heavy Chain V-region Gene Segments, J. Schilling, B. Clevinger, J. Davie, L. Hood, Nature *283*, 35-40 (1980).

The Joining of V and J Gene Segments Creates Antibody Diversity, M. Weigert, R. Perry, D. Kelly, T. Hunkapiller, J. Schilling, L. Hood, Nature *283*, 497-499 (1980).

Deletions in the Constant Region Locus Can Account for Switches in Immunoglobulin Heavy Chain Expression, S. Cory, J. Jackson, J. Adams, Nature *285*, 450-456 (1980).

DNA Sequences Mediating Class Switching in α-Immunoglobulins, M. Davis, S. Kim. L. Hood, Science *209*, 1360-1365 (1980).

Repetitive Sequences in Class-switch Recombination Regions of Immunoglobulin Heavy Chain Genes, T. Kataoka, T. Miyata, T. Honjo, Cell *23*, 357-368 (1981).

Two Pairs of Recombination Signals are Sufficient to Cause Immunoglobulin V-(D)-J Joining, S. Akira, K. Okazaki, H. Sakano, Science *238*, 1134-1138 (1987).

Stable Expression of Immunoglobulin Gene V(D)J Recombinase Activity by Gene Transfer into 3T3 Fibroblasts, D. Schatz, D. Baltimore, Cell *53*, 107-115 (1988).

The scid Defect Affects the Final Step of the Immunoglobulin VDJ Recombinase Mechanism, B. Malynn, T. Blackwell, G. Fulop, G. Rathbun, A. Farley, P. Ferrier, L. Heinke, R. Phillips, G. Yancopoulos, F. Alt, Cell *54*, 453-460 (1988).

V(D)J Recombination: A Functional Definition of the Joining Signals, J. Hesse, M. Lieber, K. Mizuuchi, M. Gellert, Genes and Dev. *3*, 1053-1061 (1989).

The V(D)J Recombination Activating Gene, RAG-1, D. Schatz, M. Oettinger, D. Baltimore, Cell *59*, 1035-1048 (1989).

Circular DNA is a Product of the Immunoglobulin Class of Switch Rearrangement, U. von Schwedler, H. Jäck, M. Wabl, Nature *345*, 453-456 (1990)

Impairment of V(D)J Recombination in Double-strand Break Repair Mutants, G. Taccioli, G. Rathbun, E. Oltz, T. Stamato, P. Jeggo, F. Alt, Science *260*, 207-210 (1993).

Somatic Variation

Two Types of Somatic Recombination are Necessary for the Generation of Complete Immunoglobulin Heavy-chain Genes, H. Sakano, R. Maki, Y. Kurosawa, W. Roeder, S. Tonegawa, Nature *286*, 676-683 (1980).

A Single V_H Gene Segment Encodes the Immune Response to Phosphorylcholine: Somatic Mutation is Correlated with the Class of the Antibody, S. Crews, J. Griffin, H. Huang, K. Calame, L. Hood, Cell *25*, 59-66 (1981).

Clusters of Point Mutations are Found Exclusively Around Rearranged Antibody Variable Genes, P. Gearhart, D. Bogenhagen, Proc. Nat. Acad. Sci. USA *80*, 3439-3443 (1983).

Somatic Mutations of Immunoglobulin Genes are Restricted to the Rearranged V Gene, J. Gorski, P. Rollini, B. Mach, Science *220*, 1179-1181 (1983).

Somatic Diversification of the Chicken Immunoglobulin Light Chain Gene is Limited to the Rearranged Variable Gene Segment, C. Thompson, P. Neiman, Cell *48*, 369-378 (1987).

A Hyperconversion Mechanism Generates the Chicken Light Chain Preimmune Repertoire, C. Reynaud, V. Anquez, H. Grimal, J. Weill, Cell *48*, 379-388 (1987).

Monoclonal and Artifical Antibodies

A Monoclonal Antibody for Large-scale Purification of Human Leukocyte Interferon, D. Secher, D. Burke, Nature *285*, 446-450 (1980).

Monoclonal Antibodies Distinguish Identifiable Neurons in the Leech, B. Zipser, R. McKay, Nature *289*, 549-554 (1981).

Monoclonal Antibody 18B8 Detects Gangliosides Associated with Neuronal Differentiation and Synapse Formation, G. Grunwald, P. Fredman, T. Magnani, D. Trisler, V. Ginsberg, M. Nirenberg, Proc. Nat. Acad. Sci. USA *82*, 4008-4012 (1985).

Generation of a Large Combinatorial Library of Immunoglobulin Repertoire in Phage Lambda, W. Huse, L. Sastry, S. Iverson, A. Kang, M. Alting-Mees, D. Burton, S. Benkovic, R. Lerner, Science *246*, 1275-1281 (1989).

Cloning of the Immunological Repertoire in *Escherichia coli* for Generation of Monoclonal Catalytic Antibodies: Construction of a Heavy Chain Variable Region-specific cDNA Library, L. Sastry, M. Alting-Mees, W. Huse, J. Short, J. Sorge, B. Hay, K. Janda, S. Benkovic, R. Lerner, Proc. Natl. Acad. Sci. USA *86*, 5728-5732 (1989).

Regulation of Antibody Synthesis

A Tissue-specific Transcription Enhancer Element is Located in the Major Intron of a Rearranged Immunoglobulin Heavy Chain Gene, S. Gillies, S. Morrison, V. Oi, S. Tonegawa, Cell *33*, 717-728 (1983).

A Lymphocyte-specific Cellular Enhancer is Located Downstream of the Joining Region in Immunoglobulin Heavy Chain Genes, J. Banerji, L. Olson, W. Schaffner, Cell *33*, 729-740 (1983).

Protein-Binding Sites in Ig Gene Enhancers Determine Transcriptional Activity and Inducibility, M. Lenardo, J. Pierce, D. Baltimore, Science *236*, 1573-1577 (1987).

Identification of a DNA Binding Protein that Recognizes the Nonamer Recombinational Signal Sequence of Immunoglobulin Genes, B. Halligan, S. Desidero, Proc. Nat. Acad. Sci. USA *84*, 7019-7023 (1987).

A Protein Binding to the J_k Recombination Sequence of Immunoglobulin Genes Contains a Sequence Related to the Integrase Motif, N. Matsunami, Y. Hamaguchi, Y. Yamamoto, K. Kuze, K. Kangawa, H. Matsuo, M. Kawaichi, T. Honjo, Nature *342*, 934-937 (1989).

A Second B Cell-specific Enhancer 3' of the Immunoglobulin Heavy-chain Locus. S. Pettersson, G. Cook. M. Brüggemann, G. Williams, M. Neuberger, Nature *344*, 165-168 (1990).

Molecular Requirements for the μ-Induced Light Chain Gene Rearrangement in pre-B Cells, A. Iglesias, M. Kopf, G. Williams, B. Bühler, G. Köhler, EMBO J. *10*, 2147-2156 (1991).

Biological Assembly, Ribosomes and Lambda Phage

21

We have already dealt with biological assembly in the folding of polypeptide chains. In that case, although we do not understand the process well enough to predict how a sequence of amino acids will fold, we have some of the basic principles well in hand. For most proteins, folding in the environment of the cell or in physiological buffers appears to be completely specified by the polypeptide itself. No external folding engine or scaffolding is required although some of the heat shock proteins assist folding by binding to proteins in unfolded states and preventing their irreversible aggregation. Is it also true that large biological structures themselves contain all the information for their formation, or are scaffoldings and blueprints somehow needed?

Two fundamentally different types of larger structures are considered in this chapter: the structure and assembly of ribosomes and the structure and assembly of lambda phage. As we saw in Chapter 7, ribosomes consist of two nonidentical subunits, each containing one or two RNA molecules and many different ribosomal proteins. The great majority of these proteins are present in each ribosome as a single copy. Thus, ribosomes are irregular and asymmetric. On the other hand, lambda phage, and most other viruses, possess a highly regular coat that covers DNA or RNA molecules. The coat of lambda consists of many copies of a few proteins that form an icosahedral head in addition to a number of other proteins that form a tail and tail fibers for absorption to cells. Both in the case of ribosome assembly and in the case of virus assembly, one important fundamental problem is determining the assembly order.

A. Ribosome Assembly

RNAse and Ribosomes

Originally, RNAse I contaminating most preparations of ribosomes caused much trouble in the study of ribosomes and their constituents. One might ask how the ribosomes could function *in vivo* without degradation if they are so quickly degraded *in vitro*. The answer is that RNAse I is located in the cell's periplasmic space. There it does no harm to the ribosomes until the cells are lysed and it is released, whereupon it adventitiously binds tightly to ribosomes and degrades their RNA. The degradation is rapid at room temperature or above, but even occurs at temperatures near 0°.

Molecular biologists used two sensible approaches to solve the RNAse problem. The first was a classic case of applying genetics to solve a biochemical problem—isolate an RNAse I⁻ mutant. This was not a trivial task because no genetic selection was apparent for permitting the growth of just RNAse I⁻ mutants, nor was any physiological trait likely to reveal the desired mutants. The only known characteristic of the desired mutants would be their lack of RNAse I in the periplasmic space. The obvious solution to the problem of isolating the desired mutant under the circumstances, but apparently not one used before, was merely to use a brute-force approach and score several thousand candidate colonies from a heavily mutagenized culture for absence of the enzyme.

To minimize the work of scoring, the mutagen had to be highly effective. Fortunately, nitrosoguanidine can induce multiple mutations into each cell. As a result, any mutant lacking a nonessential gene activity can be found in a population of a few thousand candidates from nitrosoguanidine-mutagenized cultures. The work required to perform conventional RNAse I assays on several thousand different cultures is large. Therefore Gesteland devised two simple scoring methods. In one, the whole cells from individual colonies grown from a mutagenized culture were resuspended at 42° in buffer containing radioactive ribosomal RNA and EDTA. The high temperature and the EDTA released the RNAse from the cells without lysing them. Then, after an incubation in the presence of radioactive RNA, the undegraded RNA was precipitated by addition of acid, and its radioactivity was determined. Several hundred colonies per day could be assayed for the ability of their RNAse I to degrade the RNA. The second scoring method used a clever plate technique in which duplicate plates contained the colonies to be tested. One was overlaid with several milliliters of agar containing a high concentration of tRNA and EDTA and was incubated at 42°. During a few hours of incubation, those colonies containing RNAse I digested the tRNA to short oligonucleotides. The mutant colonies lacking RNAse I could not digest the tRNA in their immediate vicinity. Then the plate was flooded with concentrated HCl. The acid precipitated the tRNA, leaving the plate opaque except in the areas surrounding colonies containing RNAse I. The desired RNAse I⁻ colonies lacked cleared halos and could easily be detected.

The second approach for elimination of RNAse I problems was biological. A number of different bacterial strains were examined for this enzyme. Strain MRE600 was found to lack the enzyme. Therefore this strain has been used as a source of ribosomes in some structural and assembly studies.

The Global Structure of Ribosomes

The topic of *in vivo* assembly or *in vitro* reassembly cannot be studied without some knowledge of the structure being assembled. Learning the structure of ribosomes has been a tantalizing problem. Most of the interesting details of ribosomes are too small to be seen by electron microscopy, but ribosomes are too large for application of the conventional physical techniques such as nuclear magnetic resonance or other spectroscopic methods. Although crystals of ribosomes have been formed for X-ray crystallography, it will be a very long time until these yield detailed structural information.

One of the first questions to resolve about ribosome structure is the folding of rRNA. It could be locally folded, which means that all nucleotides near each other in the primary sequence are near each other in the tertiary structure (Fig. 21.1), or it could be nonlocally folded. Recall that many proteins, IgG for instance, are locally folded and contain localized domains, each consisting of a group of amino acids contiguous in the primary sequence. Ribosomal RNA possesses a sequence that permits extensive folding, most of which is local but a part of which is nonlocal. Many stretches of the RNA can form double-helical hairpins. The remainder is single-stranded in the form of small or medium sized regions separating double-stranded regions of complementary sequences (Fig. 21.2). Several lines of evidence support the secondary structure which is predicted primarily on the basis of sequence. One of the most compelling is interspecies or phylogenetic sequence comparisons. Frequently, a stretch A_1, which is postulated to base pair with stretch B_1, is altered between species. In such a case, the sequences of A_2 and B_2 in the second species are both found to be altered

Figure 21.1 Examples of local and nonlocal folding.

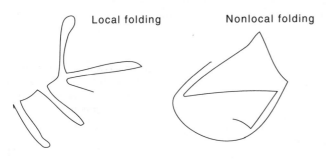

Local folding   Nonlocal folding

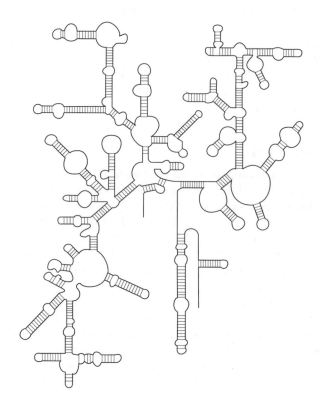

Figure 21.2 Secondary folding of 16S rRNA based on sequence, RNAse susceptibility, and interspecies comparisons.

so as to preserve the complementarity. Such compensating sequence changes are a strong sign that the two stretches of nucleotides are base paired in both organisms (Fig. 21.3). Additional evidence supporting the secondary structure predictions is the locations of sites susceptible to cleavage by RNAses or to crosslinking. The cleavage sites of RNAse A lie in loops or in unpaired regions. Crosslinking portions of the rRNA with psoralen, a chemical which will intercalate into helical DNA or RNA and covalently link to nucleotides in the presence of UV light, joins regions postulated to be paired and also connects nucleotides that are in close proximity because of the tertiary structure of the ribosome.

One of the most obvious methods for obtaining structure information on ribosomes is to look at them in the electron microscope. Due to its small size, a ribosomal subunit looks like an indistinct blob. One can,

Figure 21.3 How base differences in two regions of RNA strengthen proposals for base pairing between the regions involved.

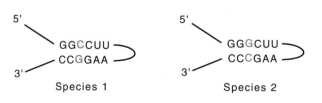

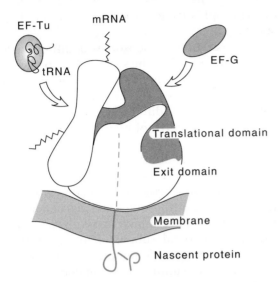

EF-Tu

mRNA

tRNA

EF-G

Translational domain

Exit domain

Membrane

Nascent protein

Figure 21.4 Diagrammatic representation of the exit and translational domains of the ribosome and their orientations with respect to the membrane binding site. The nascent protein is shown as an unfolded, extended chain during its passage through the ribosome. (Adapted from Bernabeu and Lake, Proc. Nat. Acad. Sci. USA 79, 3111-3115 [1982].)

however, see several discernible shapes of blob, suggesting that the subunit is not symmetrical and that it binds to the support in only a few discrete orientations, much like a book resting on a table. As a result, it is possible to average a number of images using electronic image processing. It is even possible to subtract the averaged image of a ribosome from the averaged image of a ribosome plus tRNA to reveal the binding location of a tRNA molecule.

Individual ribosomal proteins can also be marked with antibodies

and located by microscopy. IgG is bivalent, so it will join two ribosomal subunits. Owing to the asymmetry of the subunits, the position at which the two are linked by the antibody can often be ascertained. Not only can many ribosomal proteins be located, but the exit site of nascent polypeptide chains has been found through the use of antibodies to a nonribosomal protein whose synthesis can be highly induced (Fig. 21.4). Other features located by this method are the 5' and 3' ends of the rRNAs, dimethyl A residues, translation factor-binding sites, and tRNAs.

Assembly of Ribosomes

How are ribosomes formed? We would like to have an answer to this question concerning *in vivo* assembly because we want to know how things work. We also would like to be able to carry out *in vitro* assembly so that we can perform sophisticated experiments on the structure and function of ribosomes.

Study of the *in vivo* assembly of ribosomes is difficult because the experimenter can do so little to alter the system. One or two ribosomal precursors can be distinguished from mature ribosomes on the basis of their sedimentation velocities. They contain only a subset of the complement or ribosomal proteins. Additionally, since RNA is methylated in a number of positions, the degree of methylation of RNA extracted from the precursors can be measured. Not surprisingly, such studies have not greatly illuminated the subject of ribosome assembly.

Nomura made pivotal contributions to our understanding of *in vitro* ribosome assembly. As a first step in the attack on this difficult problem, he removed a few proteins from ribosomes and learned how to replace them and restore the ribosome's ability to synthesize protein. The proteins of the smaller ribosomal subunit were separated into two classes: those few that split off the 30S subunit when it is centrifuged to equilibrium in CsCl and the remaining cores that contain the 16S rRNA and the majority of the smaller subunit's proteins. He then tried to reassemble the ribosomes from these two fractions. The assay of assembly was sedimentation velocity of the products or their ability to function in an *in vitro* translation assay.

The next step in the study of ribosome assembly was the reconstitution from isolated rRNA and ribosomal proteins. The RNA was purified by phenol extraction to remove the proteins and then extensively dialyzed to remove the phenol. The ribosomal proteins were obtained by extracting purified 30S subunits with urea and lithium chloride. This harsh treatment denatures the proteins but solves one of the harder technical problems in the study of ribosomal proteins, that of their insolubility. Much time was invested by many investigators to find ways of solubilizing these proteins. Why the proteins should be poorly soluble under some conditions is not at all clear; perhaps they cannot achieve their correct conformation in the absence of the correct rRNA-binding site and, as a result, the proteins easily aggregate.

To reconstitute ribosomes, a solution containing ribosomal proteins is slowly added to a stirred buffer containing the rRNA. In the solution it is a race between reconstitution and aggregation. A substantial number of intact 30S functional ribosomal subunits are formed, however. Once functional 30S subunits could be formed, optimal reconstitution conditions could be found. Not surprisingly, they were similar to *in vivo* conditions. They are about 0.3 M KCl, pH 6 to 8, and Mg^{++} greater than 10^{-3} M. Soon it became possible to reconstitute ribosomes from purified 16S rRNA and the individually purified ribosomal proteins. These results definitively proved that ribosome assembly requires no additional scaffolding or assembly proteins that are not found in the mature particle. The remarkable self-assembly process can be likened to assembling watches by shaking their parts together in a paper bag.

One major reason Nomura succeeded where many others failed was that he tried to reconstitute the ribosomes in buffers closely resembling conditions found *in vivo* and at temperatures at which cells grow well, 37°. Most other workers had attempted reconstitution at temperatures near freezing to minimize the effects of proteases and nucleases on the ribosomes.

Experiments with *in vitro* Ribosome Assembly

Sophisticated experiments on ribosome structure and assembly became possible with the ability to isolate individual ribosomal proteins and to reconstitute ribosomes from them. For example, the concentration of each protein could be varied in the reconstitution experiments. This allowed determination of the kinetic order of the reaction, that is, the number of components that had to interact simultaneously for the reaction to proceed. It is certainly to be expected that the assembly process would be sequential, with proteins being added one after another; the alternative case—all 21 of the 30S ribosomal subunit proteins and the rRNA coming together simultaneously in a reaction involving 22 components—would have far too low an assembly rate. Indeed, the probability that even a few of the components come together at once is low. Quite surprisingly, however, the most rate-limiting step in ribosome assembly is not the coming together of even two components; it is the intramolecular rearrangement of a structure already formed (Fig. 21.5). That is, the rate-limiting reaction is unimolecular, and its rate relative to the amount of components present cannot be speeded by increasing the concentration of any components. Such a unimolecular step is analogous to the isomerization of RNA polymerase to form an open complex at a promoter.

The rate of *in vitro* ribosome assembly is strongly temperature-dependent. From this dependence, the activation energy can be calculated to have the unusually high value of 38 Kcal/mole and is indicative of extensive structural rearrangements. This high value also suggested that *in vivo* ribosome formation also requires high energy. If so, the activation energy might be increased in some ribosome mutants. This has been found. A sizable fraction of cold-sensitive mutants in *E. coli* are defective in ribosome assembly. At low temperatures, such mutants accumulate precursors of ribosomes that contain some, but not all, of

Figure 21.5 Assembly of 30S subunits. A subset of 30S proteins and 16S rRNA form an RI complex that then must be activated by heat to form RI*, to which the remainder of the ribosomal proteins bind to form the complete 30S subunit.

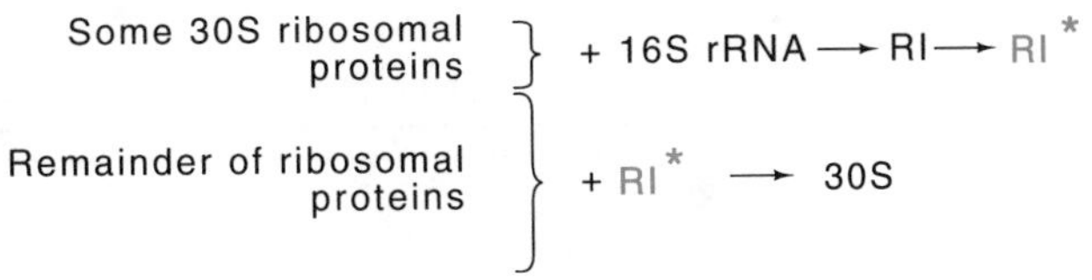

$$\left.\begin{array}{l} \text{Some 30S ribosomal proteins} \\ \\ \text{Remainder of ribosomal proteins} \end{array}\right\} \quad \begin{array}{l} + \text{16S rRNA} \longrightarrow \text{RI} \longrightarrow \text{RI}^* \\ \\ + \text{RI}^* \longrightarrow \text{30S} \end{array}$$

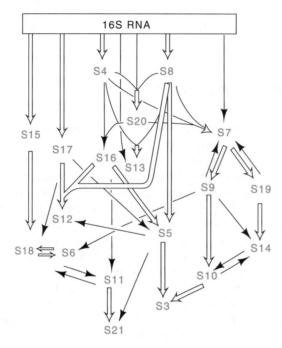

Figure 21.6 Assembly map of the 30S ribosomal subunit. (From Assembly Mapping of 30S Ribosomal Proteins from *E. coli*, W. Held, B. Ballow, S. Nizushima, M. Nomura, J. Biol. Chem. *249*, 3103-3111 [1974].)

the ribosomal proteins. Raising the temperature allows these precursors to mature.

Reconstitution of ribosomal subunits from cold-sensitive mutants permitted pinpointing the target of the mutations. They were in the ribosomal proteins. A similar approach can also be used to locate the altered component in other types of ribosome mutations. For example, the protein that is altered in streptomycin-resistant mutants is a protein designated S21. The ribosomal proteins are designated by S for proteins of the smaller, 30S subunit and L for proteins of the larger, 50S subunit. They are numbered in the order of their locations on a two-dimensional gel in which they all are separated from one another.

The order that individual ribosomal proteins bind to the maturing ribosome can also be determined by *in vitro* assembly experiments (Fig.

Figure 21.7 Gene structure of the ribosomal protein operon containing S13, S11, and S4. This operon also contains the gene for the sigma subunit of RNA polymerase and the ribosomal protein L17.

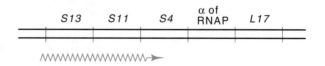

21.6). The binding of the isolated radioactive ribosomal proteins to rRNA can be measured by sedimentation of the RNA after incubation. Only a subset of the 30S proteins are capable of binding to the naked 16S rRNA. The complexes thus formed can then be used to determine which proteins can bind next. Building in this fashion, a complete assembly map of the ribosomal subunits has been constructed. In general it shows that any of several proteins can bind to the structure at any stage in its assembly. Also, a few of the proteins that interact strongly during assembly are encoded by genes in the same ribosomal protein operon (Fig. 21.7).

Determining Details of Local Ribosomal Structure

Consider the fundamental question of determining which proteins are close neighbors in the ribosome. One direct approach to this question is to crosslink two proteins on the intact ribosome with bifunctional crosslinking reagents. If two ribosomal proteins are connected by the reagent when they are in a ribosome, but not when they are free in solution, it can be concluded that the proteins are near one another in the ribosome. This technique is fraught with artifacts, however, and results from different laboratories frequently do not agree, leading some to believe only those crosslinking results that have been duplicated in more than two laboratories.

Many of the proteins that are crosslinked to each other are proteins that depend on one another during assembly of the ribosomal subunit. A few of these proteins are encoded in the same operons. In one case, proteins that are adjacent to one another in the ribosome derive from adjacent genes in the chromosome. For example, ribosomal proteins S4, S11, and S13 lie in the same operon, S13-S11-S4. S4 and S13 and S13 and S11 crosslink, S4 and S13 interact during assembly, and together they interact with S11 during assembly.

The ability to reassemble ribosomes from their isolated components greatly facilitates structural studies. A ribosome can be partially assembled, for example, and then antibody against a component in the immature ribosome can be added. If the presence of the antibody blocks the subsequent association of a ribosomal protein added later, it is reasonable to expect that the antibody directly blocks access of the protein to its site.

If all ribosomal proteins were spherical, their complete spatial arrangement would be determined by knowing the distances between the centers of proteins. Some of the requisite measurements can be made with fluorescence techniques or slow neutron scattering. Fluorescent molecules possess an absorption spectrum such that illumination by photons within this wavelength band excites the molecule, which then emits a photon of longer wavelength within what is called the emission spectrum of the molecule (Fig. 21.8).

In vitro assembly of ribosomes can be used to construct a ribosome in which two of the proteins contain the fluorescent probes. By illuminating the rebuilt ribosomes with light in the excitation spectrum of the

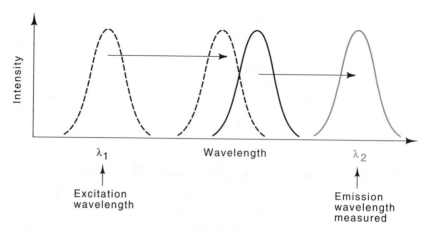

Figure 21.8 Spectra used in measuring distances separating ribosomal proteins. Dotted line is the excitation and emission spectrum of fluorescent molecule 1 and the solid line is the excitation and emission spectra for molecule 2.

first molecule and measuring the strength of the fluorescence in the wavelength of the emission spectrum of the second molecule, the distance between the two fluorescent molecules can be determined. The amount of light in the second emission spectrum varies as the sixth power of the distance separating the molecules:

$$E = \frac{R_0^6}{R_0^6 + R^6}$$

where R is the distance between the fluorescent molecules and R_o is a constant that depends on the orientations of the molecules, the spectral overlap of the fluorescent emission and excitation spectra, and the index of refraction of the medium separating the molecules. The method yields the most reliable data for proteins separated by 25 to 75 Å; that is, the method is best at determining the distances of nearest neighbors in the ribosome.

Neutron diffraction is another method of measuring distances between ribosomal proteins. This method has yielded the most information and the most reliable information on ribosome structure. It too relies on reassembly of ribosomal subunits. Two proteins in the ribosome are replaced by their deuterated equivalents. These proteins are obtained from cells grown on deuterated medium. Since the neutron scattering properties of hydrogen and deuterium are different, an interference pattern is generated by the presence in the ribosome of the two proteins with different scattering properties. The angular separation in the peaks of the interference pattern can be related to the distance separating the two altered proteins in the reconstituted ribosome. Overall, the results of crosslinking, assembly cooperativity, immune

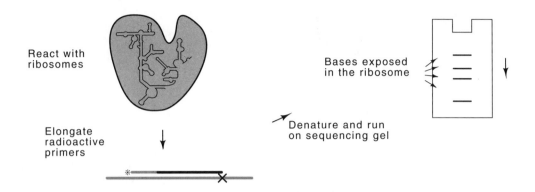

Figure 21.9 Technique for footprinting rRNA in the intact ribosome. The enhanced bands correspond to exposed bases. Normally a control would be done reacting denatured rRNA and rRNA that is in an intact ribosome. Then the bases protected and not protected are revealed by comparing the band intensities from the free RNA and the RNA from the ribosomes.

microscopy, fluorescence transfer, and neutron scattering give a consistent picture for the locations within the ribosomal subunits of the ribosomal proteins.

Ribosomal RNA can also be footprinted like DNA. Either bare RNA, RNA with a few proteins bound, or even an intact ribosome with or without a bound protein synthesis inhibitor like streptomycin can be used. The RNA is treated with chemicals like dimethylsulfate or kethoxal that react with unprotected nucleotides. Then the RNA is purified and a DNA oligonucleotide that will serve as a primer for reverse transcriptase is hybridized. The elongation by reverse transcriptase ends at the modified bases, and the locations of protected bases can be determined (Fig. 21.9).

B. Lambda Phage Assembly

General Aspects

The presence of a coat on phage lambda, and most other virus particles as well, can be broken down into three basic problems. How is the coat assembled? How is the nucleic acid placed inside the coat? How is the nucleic acid released from the virus particle into the appropriate cell? At present, partial answers are known to these questions, but it is not yet possible even to imagine the shapes of a set of folded proteins that would self-assemble into a structure the size of a phage, encapsidate DNA, and when properly triggered, release the DNA.

Not only is the structure of lambda phage dramatically different from a ribosome, but its assembly process also is notably different. First, as in the case of many larger viruses, one virus-encoded protein is used

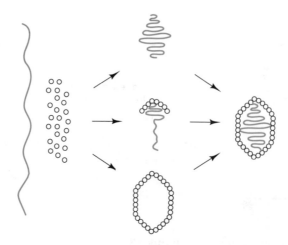

Figure 21.10 Three schemes for maturation of head subunits and the lambda DNA molecule into an assembled head with the DNA inside.

during assembly but is not present in the final particle. Second, a host protein participates enzymatically in the assembly process. Third, a number of proteins are cleaved or covalently joined during assembly.

An interesting problem involved in phage maturation is how the DNA is packaged in the head. There are three possible models for encapsidation of the DNA in the head (Fig. 21.10). The first is condensation of the DNA followed by assembly of the protein capsid around it. The second is a concerted condensation of DNA and capsid components, and the third is insertion of the DNA into a preformed capsid. Of these three possibilities, phage lambda matures by the third.

The Geometry of Capsids

Previously, we saw that lambda's genes are crowded together and even overlapping in places. Other bacterial virus genomes are similarly squeezed, and it is likely that survival of many types of viruses depends on their packing as much information as possible into as short a genome as possible. This being the case, it is likely that as few genes as possible will be used to code for DNA encapsidation. What general principles might then apply to the structure of such viruses?

A virus coat could be constructed like a regular polyhedron. For example, 24 identical subunits could form a cube with one subunit at

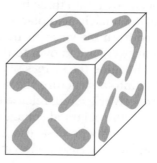

Figure 21.11 A cube with subunits at each vertex.

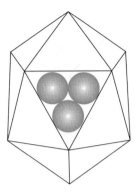

Figure 21.12 A regular icosahedron.

each corner of each face (Fig. 21.11). Note that each of the 24 subunits would make identical contacts with its neighbors. The maximum volume that could be enclosed by a set of subunits, each making identical contact with its neighbors, is based on the 20-sided icosahedron (Fig. 21.12). Each face could consist of three subunits. Thus the maximum number of subunits that can be utilized to enclose a volume in which each subunit is exactly equivalent to any other is 60. Experimentally, however, many viruses, lambda included, are found to be approximately icosahedral but to possess more than 60 subunits in their coats.

Caspar and Klug have investigated the structures that can be constructed when the constraint that each subunit be exactly equivalent to any other is weakened to permit them to be only quasi-equivalent. That is, each subunit will have nearly the same shape, but will still make homologous contacts with its neighbors. Using the same contact points over and over necessitates that the final structure be symmetrical. Maximizing the similarity of contacts with neighbors necessitates that the structure be helical or a variation on an icosahedron. We shall consider only those based on icosahedra.

The restriction to icosahedral symmetry can be most easily understood by considering why some other symmetry is less favored. For example, why is a large cubic structure unlikely? Suppose a large planar network of subunits is to be converted to a cube. One of the eight vertices of a cube can be generated by converting a point of fourfold symmetry to a point of threefold symmetry by removing a semi-infinite quadrant of subunits (Fig. 21.13). Repeating this process at the other seven vertices generates the cube. In such a structure there are three kinds of subunit interactions; between subunits on the same face, between subunits across an edge, and between subunits at a vertex. When the cube contains 24 subunits, each subunit engages in each of these three types of interactions and is equivalent to any other subunit. When the cube possesses more subunits, however, not all subunits engage in these three types of interactions, and the different subunits are non-equivalent and therefore distorted with respect to one another.

A similar analysis can be performed if a plane is covered with equilateral triangles and some points of sixfold symmetry are converted to points of fivefold symmetry. Of course, if the fivefold symmetric

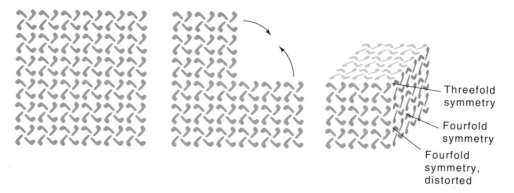

Figure 21.13 Conversion of a planar net of subunits to a cube by removing a sector of the plane and bringing the edges together.

vertices are adjacent to one another, a regular icosahedron is generated. If the fivefold vertices are not adjacent, and a larger volume is enclosed, then again three types of interactions are possible, but the differences amongst them are smaller, and hence the required distortions are smaller than in the cubic case. Hence, nature frequently will choose to construct virus coats based on the five and six neighbor structure. Often the units possessing the fivefold or sixfold symmetry can be isolated intact or observed as a unit in the electron microscope. These capsomeric polymers are called pentons and hexons.

One way to view some, but not all, of the higher-order structures is that they subdivide the equilateral triangles of the regular icosahedron. The triangles can be subdivided into 4, 9, 16, 25, and so on, subtriangles. Other icosahedra, however, are also possible. Consider a plane net of equilateral triangles and unit vectors **a** and **b** (Fig. 21.14). Satisfactory icosahedra can be constructed if one of the fivefold vertices is placed at an origin and the second at position (n**a**, m**b**), where n and m are integers. If we define $T = (n^2 + nm + m^2)$, then the regular icosahedron has $T = 1$ since $n = 1$ and $m = 0$. The number of subunits, S, in all the

Figure 21.14 Coordinates for locating vertices of an icosahedron with quasi-equivalent subunits. X is one vertex and Y the second, which is located 2 units in the **a** direction and 1 unit in the **b** direction.

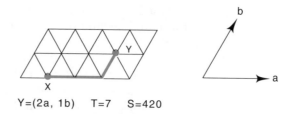

Y=(2a, 1b)   T=7   S=420

icosahedra is S = 60T. Although the coat of lambda phage is based on a T = 7 icosahedron, it uses two types of coat protein. The two bear a fixed spatial relationship to one another as though they derived from a single

polypeptide chain with a segment removed. In addition to these proteins, smaller numbers of other proteins are found in the head. Some of these form the connection between the head and the tail.

The Structure of the Lambda Particle

The physical structure of lambda phage and the organization of the genetic map of lambda show the same relationship as the physical structure of the ribosome and the organization of the genetic map of ribosomal proteins. The genes for proteins that lie near one another in the particles often lie near one another in the genetic map (Fig. 21.15). Even though a single species of a polycistronic messenger is synthesized that includes all the phage late genes, they are translated at vastly different efficiencies closely paralleling the numbers of the different types of protein molecules that are found in the mature phage particle.

The A protein, often denoted pA, in conjunction with pNu1 is responsible for producing the cohesive ends of the DNA. Proteins pD and pE are the major capsid proteins. Protein pE plus Nu3 followed by pB and pC is able to form a precursor of the head, pNu3 is then degraded and pD is added. In the final structure, protein pD is interspersed with equal numbers of protein pE, that is, pD and pE are the two major capsid proteins. Protein pF$_{II}$ as well as protein pW are involved with steps preparing the head assembly for receipt of the tail. pF$_{II}$ provides a specificity for the type of tail that is attached to the head. Protein pV is the major structural protein of the tail. Proteins pH, pM, pL, and pK are

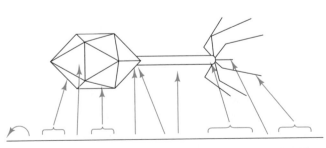

A W B C Nu3 D E F$_1$ F$_{11}$ Z U V G T H M L K I J Stf Tfa

Figure 21.15 The genetic map of some genes involved with forming the head and tail of lambda and the locations of the gene products in the physical structure of the head and tail.

all involved in the transition from the tail tube to the tail fibers at the end, which are made from pJ, pTfa, and pStf.

The Head Assembly Sequence and Host Proteins

Head precursors of lambda can be isolated by centrifugation and can thereby be studied. The first consists of pE, pNu3, pB and pC (Fig. 21.16). Without pE, nothing forms, and without pNu3, only large amorphous shapes appear. The formation of this structure also requires that about 12 molecules of the pB protein be polymerized. This is done with the assistance of the host proteins GroEL and GroES. The pB protein ultimately forms the connector between head and tail.

In the head maturation process there is both cleavage of polypeptide chains and fusion of chains. For example, some molecules of pB are cleaved and some molecules of pE are fused to protein pC. The final head is formed by the addition of lambda DNA, its cleavage to unit-length molecules containing the 12-base single-stranded cos ends, and the addition of pD on an equimolar basis to pE. Proteins pW and pF$_{II}$ prepare the head for attachment of the tail. One of the functions of such a protein is to act as an adapter between the hexagonal tail and the pentagonal vertex to which the tail is attached.

The protein pNu3 occupies a role predicted for many other proteins but thus far found rather infrequently. It appears to be a structural protein and is used during the assembly of the phage particle but is not present in the assembled phage particle. It is synthesized in high quantities and forms a part of the scaffolding during formation of the particle. In the case of lambda, this protein is cleaved after a single use, but in some other phage the analogous protein is used more than once.

The GroEL-GroES protein complex is a member of the class of proteins that assists the formation of complex protein structures. Often these proteins possess ATPase activities. Some disaggregate oligomers that have formed from hydrophobic interactions or prevent their formation. Others assist the molten globule state to perform final rearrangements of secondary structure elements. The protein DnaK that functions in assembling the replication complex of lambda along with the phage P protein and the host DnaB protein, also participates in the lambda capsid maturation reactions. Mutants with altered GroE protein

Figure 21.16 Maturation scheme of lambda heads.

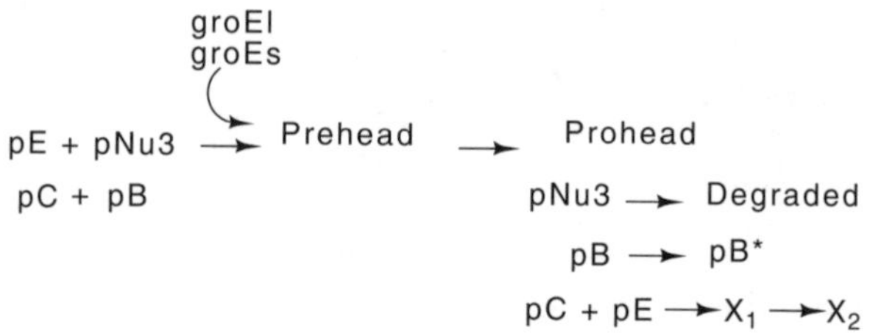

frequently fail to support the growth of not only phage lambda but also related phage and even phage T4. In addition to failing to cleave pB, they do not package DNA. Mutants of lambda that overcome the *groE* defect can be isolated. These are usually found to be altered in the lambda *E* gene. Some of the rarer *groE* mutations can be overcome by compensating mutations in the lambda *B* gene. These results strongly suggest that GroE protein touches the pE and pB proteins during the maturation of the phage.

The *groE* mutation joins the list of other cellular mutations that may block maturation of the phage. In addition to *groE*, there are also the *groP* mutations in the host cell, which block DNA replication of the phage, and *groN* mutations which make the phage appear to be N defective. The *groP* mutations lie within a subunit of DNA polymerase. Some *groN* mutations lie in the RNA polymerase and can be overcome by a compensating mutation in the phage N gene.

Packaging the DNA and Formation of the *cos* Ends

Up to this point, we have not considered the system that converts circular lambda DNA, which is found in cells, into the sticky-ended linear molecules that are found in phage particles. The termini-producing protein, the A gene product, generates the required staggered nicks while packaging the DNA. The generation of these nicks can easily be assayed by using a double lysogen in which both prophage are Int⁻ or Xis⁻. Normal phage excision is not possible from such a double lysogen, but the *ter* system can clip an intact lambda genome out of the middle of the two excision-defective lambda genomes (Fig. 21.17).

In experiments designed to study the effects on lambda of duplicating portions of its DNA, a gratuitous duplication of the cohesive end occurred. A study of this mutant has yielded appreciable insight into the packaging of phage DNA. The starting phage for generation of the duplications was lambda deleted of a sizable fraction of the *b2* region of the phage and containing an amber mutation in the *red* gene. These were grown, and selection was made for phage containing duplications by selecting denser phage after separation according to density on equilibrium centrifugation in CsCl. As expected, most of the resulting dense phage contained duplications as shown by the following criteria:

Figure 21.17 Cutting a mature full-length lambda out of a tandem double lysogen in which each individual phage is excision-defective.

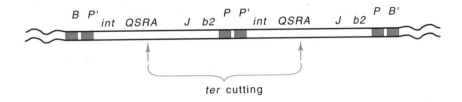

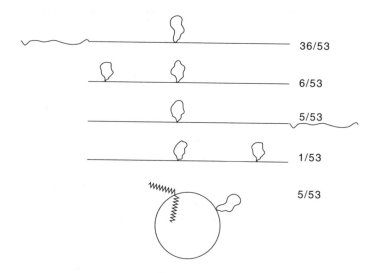

Figure 21.18 The structures of heteroduplexes formed between a lambda containing a duplication on the left arm and a lambda containing a duplication of the *cos* site. The numbers on the right indicate the number of occurrences of the various structures.

heteroduplexes between strands from these phage and strands from wild-type phage contained a bubble; the phage were denser than the parental-type; and, upon growth in Rec⁺ cells or in Su⁺ cells to suppress the *red* mutation, unequal crossing over around the duplication yielded both triplications and phage lacking the duplication altogether.

If the duplication phage were grown on cells unable to recombine and unable to suppress the phage *red* mutation, no change occurred in most of the duplication phage. One duplication phage, however, was unstable under these conditions. At an appreciable frequency it segregated phage lacking the duplication. Upon heteroduplex formation, an even more startling result was found. The duplication could appear at either end of the phage. To distinguish the ends in these experiments, the heteroduplexes were formed between this strange phage and another phage containing a duplication in the left arm. In all cases, one duplication bubble was formed in the left half of the molecule and an additional bubble, that from the duplication in the strange phage, was found near the right or left end.

A number of types of heteroduplexes were found (Fig. 21.18). Two important conclusions can be drawn from these experiments: the phage contains a duplicated *cos* site, but both *cos* sites need not be cleaved for the DNA to be packaged, and packaging of the DNA proceeds from left

to right. The reasoning for this is left as a problem at the end of the chapter.

The finding of a polarized left-to-right packaging of the lambda DNA is consistent with the *in vitro* packaging experiments investigating the nature of the DNA required for packaging. These packaging experiments found that a molecule containing only a single genome's equivalent of lambda DNA, a monomer, was not capable of being packaged. Only polymers of the lambda genome were capable of being packaged. *In vivo*, the necessary polymers could derive from the rolling circle mode of DNA replication or from recombination between circular molecules. The obvious experiments were designed to discover the minimum DNA capable of being packaged. The minimum is a lambda monomer containing a sticky left end and a right *cos* end covalently joined to a left *cos* end. These results all strongly suggest that the left end of lambda is packaged first and the right end enters last. Since the tail is put on after packaging of the DNA, it is natural to expect that the right end of lambda would be contained either within the tail or just at the union of the tail to the head; indeed, when tailless lambda phage that contain lambda DNA are isolated and are lightly treated with DNAse, it is the right end of the DNA that is attacked. All of this strongly suggests that upon infection, the right end of lambda should be injected first into the cells.

Formation of the Tail

The tail of lambda is formed independently of the head and is then attached to the head. The most striking property of tail maturation is a mechanism that permits the tail to grow to a fixed length. First, proteins pG, pH, pI, pJ, pK, pL, and pM interact to form an initiator of polymerization of the major tail protein, pV. The pV protein polymerizes on this initiator to form the tail tube. Finally, the terminator of polymerization, pU, is added. After this, pZ protein functions, and the tail is attached to the head. The tail fibers formed from the pTfa and pStf are added somewhere in this sequence. Unbeknownst until recently, the lambda phage chosen for laboratory work in the 70s and 80s lacked these fibers and adventitiously adsorbed, albeit slowly, via the pJ fiber to the maltose porin in the outer cell wall. As a result of this slow adsorption, the tail fiber-deficient lambda form large plaques. Perhaps for the good of science, the resulting large-plaque morphology made it possible to distinguish clear and turbid plaques, and hence, permitted the study of lysogeny and its regulation.

If mutations in the tail genes are present, abnormal tails may be formed. For example, if the terminator of polymerization, pU, is absent, then normal tails are formed, but the pZ protein cannot function and the phage is largely inviable. Upon prolonged incubation of U⁻ extracts, *in vivo* or in an *in vitro* reconstitution system, the tails will extend beyond their normal length to form what is called a polytail. This structure is sufficiently normal that the pZ protein can function on it, and the polytail is attached to the head to yield a phage particle, but one of very low infectivity. Of course, if the pV protein is absent, no tail polymerization occurs. Similarly, mutations in any of the genes *G* through *M* block initiator formation and therefore no tail is formed. The pH protein of the tail undergoes proteolytic cleavage during maturation to yield a protein called H*. Apparently the pH protein acts much like a tape measure in determining the length of the tail. Internal deletions in *H* yield phage with shorter tails.

In vitro Packaging

In vitro packaging of lambda DNA is possible and, indeed, is widely used to package restructured DNA in genetic engineering operations. Such packaging is done by preparing and mixing two highly concentrated extracts of cells, each infected with lambda mutants incapable of forming complete lambda particles or synthesizing DNA. Together these extracts contain all the required proteins for packaging. *In vitro* packing proceeds well if one of the extracts contains a head precursor formed as a result of mutations in genes *D* or *F*. If such an extract is mixed with an extract prepared from bacteria growing lambda-defective in the *E* gene and lambda DNA is added, mature lambda are formed. This *in vitro* packaging requires ATP, as would be expected from the fact that DNA somehow is stuffed into the head.

In this chapter we have seen two extremes for biological assembly. The "unique" structure of the ribosome is generated with ribosomal RNA and, except for one duplicated protein, with single copies of each of the ribosomal proteins. On the other extreme, many phage coats are virtually crystalline in that they contain mainly one protein used over and over in a regular array. Although we know many facts about assembly, in reality we understand very little about the actual process. We cannot look at the structure of a protein like the lambda coat protein and predict that it is capable of forming a regular icosahedron. We are also quite in the dark about understanding how nucleic acid is inserted into phage coats. Far more remains to be learned in this area than is already known.

Problems

21.1. (a) How would you prove that an antibiotic that blocked protein synthesis did so by interfering with the functioning of ribosomes?

(b) How would you show that its site of action is restricted to the 30S or 50S subunits?

(c) If the effect were on the 30S subunit, how would you tell whether a mutation providing resistance to the antibiotic altered the RNA or protein?

(d) Suppose parts a, b, and c above showed that the target was 16S RNA. Suppose sequencing showed that all rRNA from strains resistant to the drug were altered identically. How could this be possible in light of the fact that *E. coli* has seven genes for ribosomal RNA?

21.2. If the 30S ribosomal subunit were to be characterized by a set of distances between the centers of mass of the individual proteins, what is a minimum number of distances generally required to specify the topography of 21 proteins? How many different arrangements are compatible with this set of distances?

21.3. Devise a method for the isolation of nonsense mutations in the gene coding protein S12. You might wish to use the facts that some changes in this protein produce streptomycin resistance and that streptomycin sensitivity is dominant to resistance.

21.4. What would you conclude if it were determined that an icosahedral phage hydrolyzes 30.0 ATP molecules per 10.5 base pairs of DNA packaged?

21.5. To demonstrate the improbability that ribosome assembly requires the simultaneous occurrence of all 21 proteins in a small volume, estimate the probability that in a solution at 10 micrograms per ml in each of the proteins whose weight may be taken as 20,000, all 21 are present simultaneously in a sphere of radius twice that of a ribosome.

21.6. Suppose antibodies were made against a "purified" ribosomal protein and used in electron microscopy to locate the ribosomal protein on the ribosome. The microscopy located two sites 100 Å apart to which the antibodies bound. We are left concluding either that the protein has a highly asymmetric shape in the ribosome, or . . . ? What simple experiments could resolve the issue?

21.7. How would you determine which proteins lie on the interface between the 30S and 50S subunits?

21.8. What would you conclude, and what experiments would you perform if the nucleotide sequence at the end of the gene for ribosomal protein S11 and the beginning of the gene for S4 were XYZUGAXXXXXAUG?

21.9. Of what evolutionary value is it for proteins that are adjacent in the ribosome to derive from genes that are adjacent on the chromosome?

21.10. How could an experiment be done to test for ter-mediated interchange of *cos* sites?

21.11. The geodesic dome (Fig. 21.19) was designed according to the same principles as those that govern icosahedral virus assembly. How many triangular facets would the entire sphere contain?

21.12. Where does trypsin cut peptides? What fraction of peptides cleaved from random protein by trypsin could be expected to contain sulphur? If cells are labeled at 10 mCi/μM in sulphur, how many counts per minute would exist in a typical tryptic spot of lambda phage E protein derived from 10 micrograms of phage?

21.13. It was once proposed that a protein found in lambda tails was a product of the A gene. How, with labeling but with no use of peptide

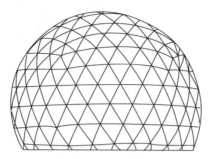

Figure 21.19 A geodesic dome constructed according to the principles of quasi-equivalence.

maps, could this be disproved? Remember that an A⁻ mutant makes no phage.

21.14. If the binding of lambda phage to *E. coli* triggers the injection of lambda DNA, there could exist *E. coli* mutants capable of absorbing lambda but incapable of triggering DNA injection. How would you isolate such a mutant?

21.15. What is the origin of the circular structures with single-stranded tails observed in the heteroduplexes containing a duplication of the *cos* site?

21.16. How do the data presented on the heteroduplexes of the *cos* duplication lead to the conclusion that packaging proceeds from left to right?

21.17. There exist bacterial *groE* mutants that are able to grow but that do not support the growth of lambda phage. How could a lambda transducing phage be isolated carrying the *groE* gene?

21.18. The single-stranded regions in heteroduplexes between lambda and lambda containing small duplications are not found at unique sites; rather, they "wander." Why? How can such a wandering bubble be forced to appear at a unique location with another duplication?

21.19. The products of genes F$_{II}$ and W prepare lambda heads for attachment of the tails. How would you determine whether these proteins must act in a particular order, and if they do, how would you determine which acts first?

21.20. In a λ*para* transducing phage, the *E. coli* DNA replaced the *int* gene of lambda. A double lysogen of this phage and wild-type lambda was made, but the two phage types could not be separated by equilibrium density gradient centrifugation because they were the same density. Therefore a double lysogen was made with the *ara* transducing phage and λb2. Induction of this double lysogen yielded transducing phage of *b2* density and plaque forming nontransducing phage of wild-type density. How did the density interchange of the two phage come about?

21.21. Show that the number of subunits in the icosahedral viruses is proportional to $n^2 + nm + m^2$, where n and m are as described in the text relating to Fig. 21.14.

21.22. Consider the construction of a cube from a piece of paper by successively cutting out quadrants as illustrated in Fig. 21.13. Define the angular deficit as the angle between the two lines defining the quadrant. In the case of a cube, the angular deficit required to form each vertex is $90°$. Therefore, the total angular deficit of the cube is $720°$. Generalize to the other regular polyhedra, irregular convex polyhedra, convex and concave polyhedra.

21.23. For the smaller prokaryotic ribosomal subunit, draw to the same scale a line representing the average diameter of the subunit, a line representing the length of the 16S rRNA, and a line representing the average diameter of a ribosomal protein of molecular weight 20,000.

References

Recommended Readings

Assembly Mapping of 30S Ribosomal Proteins from *Escherichia coli*, S. Mizushima, M. Nomura, Nature *226*, 1214-1218 (1970).

Bacteriophage Lambda Derivatives Carrying Two Copies of the Cohesive End-Site, S. Emmons, J. Mol. Biol. *83*, 511-525 (1974).

Complete Mapping of the Proteins in the Small Ribosomal Subunit of *Escherichia coli*, M. Capel, D. Engelman, B. Freeborn, M. Kjeldgaard, J. Langer, V. Ramakrishnan, D. Schindler, B. Schoenborn, I-Y. Sillers, S. Yabuki, P. Moore, Science *238*, 1403-1406 (1988).

Ribosomal RNA

Isolation and Characterization of Ribonuclease I Mutants of *Escherichia coli*, R. Gesteland, J. Mol. Biol. *16*, 67-84 (1966).

Mechanism of Kasugamycin Resistance in *E. coli*, T. Helser, J. Davies, J. Dahlberg, Nature New Biology *235*, 6-9 (1972).

Localization of 3' Ends of 5S and 23S rRNAs in Reconstituted Subunits of *Escherichia coli* Ribosomes, M. Stoffler-Meilicke, G. Stoffler, O. Odom, A. Zinn, G. Kramer, B. Hardesty, Proc. Nat. Acad. Sci. USA *78*, 5538-5542 (1981).

Topography of 16S RNA in 30S Subunits and 70S Ribosomes, Accessibility to Cobra Venom Ribonuclease, S. Vassilenko, P. Carbon, J. Ebel, C. Ehresman, J. Mol. Biol. *152*, 699-721 (1981).

Structure of *E. coli* 16s RNA Elucidated by Psoralen Crosslinking, J. Thompson, J. Hearst, Cell *32*, 1355-1365 (1983).

Structure of the *Escherichia coli* 16S ribosomal RNA, Psoralen Crosslinks and N-acetyl-N'-(p-glyoxylylbenzoyl) cystamine Crosslinks Detected by Electron Microscopy, P. Wollenzien, R. Murphy, C. Cantor, J. Mol. Biol. *184*, 67-80 (1985).

Rapid Chemical Probing of Conformation in 16S Ribosomal RNA and 30S Ribosomal Subunits Using Primer Extension, D. Moazed, S. Stern, H. Noller, J. Mol. Biol *187*, 399-416 (1986).

Direct Localization of the tRNA-Anticodon Interaction Site on the *Escherichia coli* 30S Ribosomal Subunit by Electron Microscopy and Computerized Image Averaging, T. Wagenknecht, J. Frank, M. Boublik, K. Nurse, J. Ofengand, J. Mol. Biol. *203*, 753-760 (1988).

Scrambled Ribosomal RNA Gene Pieces in *Chlamydomonas reinhardtii* Mitochondrial DNA, P.Boer, M. Gray, Cell *55*, 399-411 (1988).

Model for the Three-dimensional Folding of 16S Ribosomal RNA, S. Stern, B. Weiser, H. Noller, J. Mol. Biol. *204*, 447-481 (1988).

Mutation at Position 791 in *Escherichia coli* 16S Ribosomal RNA Affects Processes Involved in the Initiation of Protein Synthesis, W. Tapprich, D. Gross, A. Dahlberg, Proc. Natl. Acad. Sci. USA *86*, 4927-4931 (1989).

Binding of tRNA to the Ribosomal A and P Sites Protects Two Distinct Sets of Nucleotides in 16 S rRNA, D. Moazed, H. Noller, J. Mol. Biol. *211*, 135-145 (1990).

DNA Hybridization Electron Microscopy, Localization of Five Regions of 16S rRNA on the Surface of 30S Ribosomal Subunits, M. Oakes, J. Lake, J. Mol. Biol. *211*, 897-906 (1990).

The Excision of Intervening Sequences from *Salmonella* 23S Ribosomal RNA, A. Burgin, K. Parodos, D. Lane, N. Pace, Cell *60*, 405-414 (1990).

Computer Modeling 16S Ribosomal RNA, J. Hubbard, J. Hearst, J. Mol. Biol. *221*, 889-907 (1991).

Ribosome Structure and Function

Identity of the Ribosomal Proteins Involved in the Interaction with Elongation Factor G, J. Highland, J. Bodley, J. Gordon, R. Hasenbank, G. Stoffler, Proc. Nat. Acad. Sci. USA *70*, 147-150 (1973).

Mapping *E. coli* Ribosomal Components Involved in Peptidyl Transferase Activity, N. Sonenberg, M. Wilchek, A. Zamir, Proc. Nat. Acad. Sci. USA *70*, 1423-1426 (1973).

Determination of the Location of Proteins L14, L17, L18, L19, L22 and L23 on the Surface of the 50S Ribosomal Subunit of *Escherichia coli* by Immune Electron Microscopy, G. Tischendorf, H. Zeichhardt, G. Stoffler, Mol. Gen. Genet. *134*, 187-208 (1974).

Singlet Energy Transfer Studies of the Arrangement of Proteins in the 30S *Escherichia coli* Ribosome, K. Huang, R. Fairclough, C. Cantor, J. Mol. Biol. *97*, 443-470 (1975).

Identification by Diagonal Gel Electrophoresis of Nine Neighboring Protein Pairs in the *Escherichia coli* 30S Ribosome Crosslinked with Methyl-4-Mercaptobutyrimidate, A. Sommer, R. Traut, J. Mol. Biol. *97*, 471-481 (1975).

Ribosomal Proteins S5, S11, S13 and S19 Localized by Electron Microscopy of Antibody-labeled Subunits, J. Lake, L. Kahan, J. Mol. Biol. *99*, 631-644 (1975).

Neutron Scattering Measurements of Separation and Shape of Proteins in 30S Ribosomal Subunit of *Escherichia coli*: S2-S5, S5-S8, S3-S7, D. Engelman, P. Moore, B. Schoenborn, Proc. Nat. Acad. Sci. USA *72*, 3888-3892 (1975).

Ribosome Structure Determined by Electron Microscopy of *Escherichia coli* Small Subunits, Large Subunits and Monomeric Ribosomes, J. Lake, J. Mol. Biol. *105*, 131-159 (1976).

Triangulation of Proteins in the 30S Ribosomal Subunit of *E. coli*, P. Moore, J. Langer, B. Schoenborn, D. Engelman, J. Mol. Biol. *112*, 199-234 (1977).

Ribosomal Proteins S3, S6, S8, and S10 of *E. coli* Localized on the External Surface of the Small Subunit by Immune Electron Microscopy, L. Kahan, D. Winkelmann, J. Lake, J. Mol. Biol. *145*, 193-214 (1981).

Topography of the *E. coli* 5S RNA-Protein Complex as Determined by Crosslinking with Dimethylsuberimidate and Dimethyl-3, 3' dithiobis-propionimidate, T. Fanning, R. Traut, Nuc. Acids Res. *9*, 993-1004 (1981).

Nascent Polypeptide Chains Emerge from the Exit Domain of the Large Ribosomal Subunit: Immune Mapping of the Nascent Chain, C. Bernabeu, J. Lake, Proc. Nat. Acad. Sci. USA *79*, 3111-3115 (1982).

Ribosomal Protein S4 is an Internal Protein: Localization by Immunoelectron Microscopy on Protein-deficient Subribosomal Particles, D. Winkelmann, L. Kahan, J. Lake, Proc. Nat. Acad. Sci. USA *79*, 5184-5188 (1982).

Three-dimensional Reconstruction and Averaging of 50S Ribosomal Subunits of *Escherichia coli* from Electron Micrographs, H. Oettl, R. Hegerl, W. Hoppe, J. Mol. Biol. *163*, 431-450 (1983).

Positions of Proteins S14, S18, S20 in the 30 S Ribosomal Subunit of *Escherichia coli*, V. Ramakrishnan, M. Capel, M Kjeldgaard, D. Engelman, P. Moore, J. Mol. Biol. *174*, 265-284 (1984).

Interaction of Ribosomal Proteins S6, S8, S15, and S18 with the Central Domain of 16S Ribosomal RNA from *Escherichia coli*, R. Gregory, M. Zeller, D. Thurlow, R. Gourse, M. Stark, A. Dahlberg, R. Zimmerman, J. Mol. Biol. *175*, 287-302 (1984).

Three-dimensional Reconstruction of the 30 S Ribosomal Subunit from Randomly Oriented Particles, A. Verschoor, J. Frank, M. Radermacher, T. Wagenknecht, J. Mol. Bio. *178*, 677-698 (1984).

Elongation Factor Tu Localized on the Exterior Surface of the Small Ribosomal Subunit, J. Langer, J. Lake, J. Mol. Biol. *187*, 617-621 (1986).

Localization of the Binding Site for Protein S4 on 16S Ribosomal RNA by Chemical and Enzymatic Probing and Primer Extension, S. Stern, R. Wilson, H. Noller, J. Mol. Biol. *192*, 101-110 (1986).

Positions of S2, S13, S16, S17, S19, and S21 in the 30S Ribosomal Subunit of *Escherichia coli*, M. Capel, M. Kjeldgaard, D. Engelman, P. Moore, J. Mol. Biol. *200*, 65-87 (1987).

Labeling the Peptidyltransferase Center of the *Escherichia coli* Ribosome with Photoreactive tRNA[Phe] Derivatives Containing Azidoadenosine at the 3' End of the Acceptor Arm: A Model of the tRNA-Ribosome Complex, J. Wower, S. Hixson, R. Zimmerman, Proc. Natl. Acad. Sci. USA *86*, 5232-5236 (1989).

Localization of the Release Factor-2 Binding Site on 70S Ribosomes by Immuno-electron Microscopy, B. Kastner, C. Trotman, W. Tate, J. Mol. Biol. *212*, 241-245 (1990).

Ribosome Assembly

Assembly Mapping of 30S Ribosomal Proteins from *Escherichia coli*, W. Held, B. Ballou, S. Mizushima, M. Nomura, J. Biol. Chem. *249*, 3103-3111 (1974).

Assembly Map of the Large Subunit (50S) of *Escherichia coli* Ribosomes, R. Rohl, K. Nierhaus, Proc. Nat. Acad. Sci. USA *79*, 729-733 (1982).

Interaction of Ribosomal Protein S5, S6, S11, S12, S18, and S21 with 16S rRNA, S. Stern, T. Powers, L. Changchien, H. Noller, J. Mol. Biol. *201*, 683-695 (1988).

Virus Structure

Physical Principles in the Construction of Regular Viruses, D. Caspar, A. Klug, Cold Spring Harbor Symposium on Quantitative Biology *27*, 1-24 (1962).

The Capsid Structure of Bacteriophage Lambda, M. Bayer, A. Bocharov, Virology *54*, 465-475 (1973).

Chemical Linkage of the Tail to the Right-hand End of Bacteriophage Lambda DNA, J. Thomas, J. Mol. Biol. *87*, 1-9 (1974).

Capsid Structure of Bacteriophage Lambda, R. Williams, K. Richards, J. Mol. Biol. *88*, 547-550 (1974).

Comments on the Arrangement of the Morphogenetic Genes of Bacteriophage Lambda, S. Casjens, R. Hendrix, J. Mol. Biol. *90*, 20-23 (1974).

Petite Lambda, A Family of Particles from Coliphage Lambda Infected Cells, T. Hohn, F. Flick, B. Hohn, J. Mol. Biol. *98*, 107-120 (1975).

Symmetry Mismatch and DNA Packaging in Large Bacteriophages, R. Hendrix, Proc. Nat. Acad. Sci. USA *75*, 4779-4783 (1978).

Structure of the Scaffold in Bacteriophage Lambda Preheads, Removal of the Scaffold Leads to a Change of the Prehead Shell, P. Kunzler, H. Berger, J. Mol. Biol. *153*, 961-978 (1981).

Testing Models of the Arrangement of DNA Inside Bacteriophage Lambda by Crosslinking the Packaged DNA, R. Haas, R. Murphy, C. Cantor, J. Mol. Biol. *159*, 71-92 (1982).

Tests of Spool Models for DNA Packaging in Phage Lambda, J. Widom, R. Baldwin, J. Mol. Biol. *171*, 419-437 (1983).

Packaging of DNA into Bacteriophage Heads, a Model, S. Harrison, J. Mol. Biol. *171*, 577-580 (1983).

Length Determination in Bacteriophage Lambda Tails, I. Katsura, R. Hendrix, Cell *39*, 691-698 (1984).

Three-dimensional Structure of Poliovirus at 2.9 Å Resolution, J. Hogle, M. Chow, D. Filman, Science *229*, 1258-1365 (1985).

The Structure of the Adenovirus Capsid II. The Packing Symmetry of Hexon and its Implications for Viral Architecture, R. Burnett, J. Mol. Biol *185*, 125-143 (1985).

Ion Etching of Bacteriophage T4: Support for a Spiral-fold Model of Packaged DNA, L. Black, W. Newcomb, J. Boring, J. Brown, Proc. Nat. Acad. Sci. USA *82*, 7960-7964 (1985).

Interaction of Antibiotics with Functional Sites in 16S Ribosomal RNA, D. Moazed, H. Noller, Nature *327*, 389-394 (1987).Structure and Inherent Properties of the Bacteriophage Lambda Head Shell, I. Katsura, J. Mol. Biol. *205*, 397-405 (1989).

Virus Assembly and Maturation

Ter, A Function which Generates the Ends of the Mature Lambda Chromosome, S. Mousset, R. Thomas, Nature *221*, 242-244 (1969).

Intracellular Pools of Bacteriophage Lambda DNA, B. Carter, M. Smith, J. Mol. Biol. *50*, 713-718 (1970).

Head Assembly Steps Controlled by Genes F and W in Bacteriophage Lambda, S. Casjens, T. Hohn, A. Kaiser, J. Mol. Biol. *64*, 551-563 (1972).

Host Participation in Bacteriophage Lambda Head Assembly, C. Georgopoulos, R. Hendrix, S. Casjens, D. Kaiser, J. Mol. Biol. *76*, 45-60 (1973).

In vitro Assembly of Bacteriophage Lambda Heads, A. Kaiser, T. Masuda, Proc. Nat. Acad. Sci. USA *70*, 260-264 (1973).

Evidence that the Cohesive Ends of Mature Lambda DNA are Generated by the Gene A Product, J. Wang, A. Kaiser, Nature New Biology *241*, 16-17 (1973).

Bacteriophage Lambda F$_{II}$ Gene Protein: Role in Head Assembly, S. Casjens, J. Mol. Biol. *90*, 1-23 (1974).

Protein Fusion: A Novel Reaction in Bacteriophage Lambda Head Assembly, R. Hendrix, S. Casjens, Proc. Nat. Acad. Sci. USA *71*, 1451-1455 (1974).

Protein Cleavage in Bacteriophage Lambda Tail Assembly, R. Hendrix, S. Casjens, Virology *61*, 156-159 (1974).

Morphogenesis of Bacteriophage Lambda Tail, Polymorphism in the Assembly of the Major Tail Proteins, I. Katsura, J. Mol. Biol. *107*, 307-326 (1976).

Early Events in the *in vitro* Packaging of Bacteriophage Lambda DNA, A. Becker, M. Marko, M. Gold, Virology *78*, 291-305 (1977).

Packaging of the Bacteriophage Lambda Chromosome: A Role for Base Sequences Outside *cos*, M. Feiss, R. Fisher, D. Siegele, B. Nicols, J. Donelson, Virology *92*, 56-67 (1979).

DNA Packaging by the Double-stranded DNA Bacteriophages, W. Earnshaw, S. Casjens, Cell *21*, 319-331 (1980).

Bacteriophage Lambda DNA Packaging: Scanning for the Terminal Cohesive End Site During Packaging, M. Feiss, W. Widner, Proc. Nat. Acad. Sci. USA *79*, 3498-3562 (1982).

Separate Sites for Binding and Nicking of Bacteriophage Lambda DNA by Terminase, M. Feiss, I. Kobayashi, W. Widner, Proc. Nat. Acad. Sci. *80*, 955-959 (1983).

A Small Viral RNA is Required for *in vitro* Packaging of Bacteriophage φ29 DNA, P. Guo, S. Erickson, D. Anderson, Science *236*, 690-694 (1987).

Transient Association of Newly Synthesized Unfolded Proteins with the Heat-shock GroEL Proteins, E. Bochkareva, N. Lissin, A. Girshovich, Nature *336*, 254-257 (1988).

Homologous Plant and Bacterial Proteins Chaperone Oligomeric Protein Assembly, S. Hemmingsen, C. Woolford, S. Vander-Vies, K. Tilly, D. Dennis, C. Georgopoulos, R. Hendrix, R. Ellis, Nature *333*, 330-334 (1988).

Structure of the Bacteriophage λ Cohesive End Site, S. Xu, M. Feiss, J. Mol. Biol. *220*, 281-292 (1991).

Chemotaxis

22

Higher organisms have the ability to sense their environment and to process and store information. Additionally, higher organisms frequently show an excitatory response to a stimulus, but they adapt to continued application of the stimulus and ultimately produce a diminished response. We shall see in this chapter that bacteria perform these functions using a relatively small number of proteins. Some of these are similar in structure to proteins that appear to perform similar function in higher cells. Therefore the principles learned about these much simpler systems should be most helpful in understanding similar systems found in higher organisms.

Genetics, physiology, and biochemistry, the tools of molecular biology that have been so useful in the study of relatively simple questions, can also be applied to the study of biological phenomena as complicated as behavior. Bacteria have been used in these studies because they respond to their environment and are amenable to genetic and biochemical studies.

Bacteria respond to their environment by swimming toward some chemicals like sugars, amino acids, and oxygen and away from others like phenol. This response is called chemotaxis. It involves a number of steps. The bacteria must be able to sense the chemical in question, choose the appropriate direction to swim, and be able to swim. That is, they must contain elementary sensors, computers, and motors. How such elements can be assembled from simple components is a major reason for interest in such studies. A second interest is, as mentioned above, that analogs of some of the individual components of a bacterial system are used in other more complicated systems. It will be much easier to learn how the bacterial systems work since genetic, physiological, and biochemical experiments are all much quicker and less expensive for these systems than for systems from higher cells.

In the last century, microbiologists observed chemotactic behavior of bacteria in the microscope. Motile bacteria swim rapidly, in contrast to nonmotile bacteria strains which vibrate slightly due to Brownian motion. Chemotactic bacteria are motile and direct their motility to move themselves toward attractants and away from repellents. The swimming is accomplished by moving long, slender flagella that are attached to the cell wall.

To swim in the correct direction, bacteria must be capable of determining in which direction the concentration of an attractant or repellent increases or decreases. This means being able to measure concentration differences and relating them to direction in space. It might appear that one way a bacterium could do this would be to compare the concentration of an attractant at its two ends. This simple method has a major flaw, however. If the swimming bacterium is catabolizing the attractant, then the cell surface at the front end encounters a higher average concentration than the back end. Thus the bacterium loses the ability to sense the external concentration of the attractant. Whatever direction it is currently swimming seems to be the direction of increasing concentration. To obtain a better determination of attractant or repellent gradient, the bacterium can measure the average concentration of attractant over its entire surface and then move to a different point and measure the concentration again. If the bacterium can retain memory of the direction it moved between the measurements, then it can tell whether attractant or repellent has a concentration gradient in this direction. Complicating the situation is random rotation of the cell. Brownian motion will significantly jostle the cell and reorient it on a time scale of 3-10 seconds. Thus, in order for the "measure-swim-measure" process to work, both the "current" as well as the "less current" concentration measurements must be completed in several seconds.

Assaying Chemotaxis

Sensitive and simple characterization of a cell's chemotactic ability is necessary for efficient study of the phenomenon. One straightforward assay is the motility plates or swarm plates developed by Adler. In one application, the medium in these plates contains the usual salts necessary for cell growth, a low concentration of galactose, and a concentration of agar such that the medium has the consistency of vichyssoise. Five hours after a drop of chemotactic cells is placed in the center, the plate will contain a ring about two centimeters in diameter surrounding many of the cells that remain where they were spotted. The ring expands outward with time and ultimately reaches the edge of the plate. Usually, a second ring also forms and moves behind the first.

The origin of the rings is straightforward. Sensitive chemical tests show that the cells spotted in the center of the plate gradually consume the galactose where they are placed (Fig. 22.1), but just beyond the edge of the spot, galactose remains at its original concentration. That is, the consumption of galactose creates a concentration gradient in galactose at the edge of the spot. The bacteria on the edge of the spot detect this

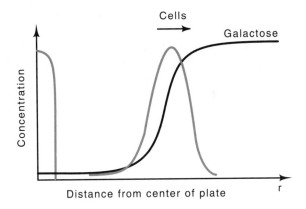

Figure 22.1 The concentrations of galactose and cells as a function of distance r from the center of the plate several hours after spotting cells.

gradient, swim toward higher galactose concentrations, and, at the same time, consume the galactose in the new position. Consequently, the cells move outward pursuing and consuming the galactose. The second ring that forms on the galactose plates derives from bacteria utilizing their endogenous energy sources to swim toward oxygen. If tryptone medium is substituted for galactose salts medium, then multiple rings form, the first ring from bacteria swimming toward the most active attractant, the second toward a second attractant, and so on. In this case also, one of the rings is formed by bacteria pursuing oxygen.

The capillary tube assay is more sensitive and quantitative than the swarm plate assay. Instead of requiring the bacteria to produce their own gradient of attractant concentration, the gradient is produced by diffusion of the substrate. A short capillary tube is filled with medium containing an attractant and is sealed at one end. The open end is pushed into a drop of medium lacking attractant but containing cells, and the tube is allowed to rest there an hour. As attractant diffuses out of the mouth of the capillary, a gradient of attractant is produced in the region

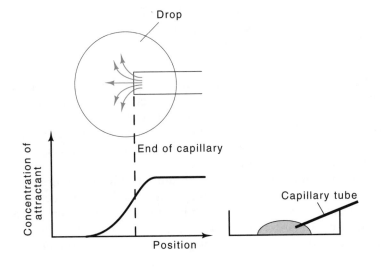

Figure 22.2 Diagram of the capillary tube assay in which the capillary is inserted in a drop of medium in a petri plate, a top view showing diffusion out of a capillary to generate a gradient, and the effective concentration as a function of position of the attractant after several hours of diffusion.

(Fig. 22.2). The chemotactic bacteria in the drop near the capillary end detect and swim up the gradient a short distance into the capillary. At the end of the assay the capillary tube is removed, rinsed from the outside, and the bacteria inside are blown out into a tube. These are diluted and plated on petri plates. Since the number of bacteria having entered the capillary can be accurately measured by counting the number of colonies on the plates after incubation, the chemotactic ability of the cells can be quantitated. Typically less than 100 nonmotile bacteria enter the capillary. Several thousand motile but nonchemotactic bacteria normally enter a capillary with or without attractant, and as many as 500,000 chemotactic bacteria will enter a capillary tube that contains attractant.

Fundamental Properties of Chemotaxis

The capillary tube assay permits several simple measurements delimiting methods by which cells accomplish chemotaxis. The first is that metabolism of attractant is not required for chemotaxis. An analog of galactose, fucose, is not metabolized by *Escherichia coli*, and yet this serves as an attractant in the capillary tube assay. Another line of evidence leading to this same conclusion is that some mutants that are unable to metabolize galactose are still able to swim toward galactose.

Since cells can swim toward or away from a wide variety of chemicals, it is likely that they possess a number of receptors, each with different specificity. Therefore the question arises as to how many different types of receptors a cell possesses. This question can be answered by a cross-inhibition test. Consider the case of fucose and galactose. Since their structures are similar, it seems likely that cells detect both chemicals with the same receptor protein. This conjecture

D-galactose   D-fucose

can be tested by using a high concentration of fucose in both the tube and the drop so as to saturate the galactose receptor and blind cells to the galactose that is placed only in the tube. Fucose does blind cells to a galactose gradient, but it does not blind them to a serine gradient. These findings prove that galactose and fucose share the same receptor, but that serine uses a different receptor.

By blinding experiments and the use of mutants, nine receptors have been found for sugars and three for amino acids (Table 22.1). The three amino acid receptors are not highly selective, and they allow chemotaxis of *E. coli* toward 10 different amino acids. Except for the glucose receptor, synthesis of the sugar receptors is inducible, as is the proline

Table 22.1 Compounds to which Cells Respond

Sensor	Response	Sensor	Response
N-acetyl glucosamine	Attractant	Aspartate	Attractant
Fructose	Attractant	Glutamate	
Galactose	Attractant	Methionine	
Glucose, Fucose		Proline	Attractant
Glucose	Attractant	Serine	Attractant
Mannose	Attractant	Cysteine	
Glucose		Alanine	
Mannitol	Attractant	Glycine	
Ribose	Attractant	Asparagine	
Sorbitol	Attractant	Threonine	
Trehalose	Attractant	Alcohols	Repellent
Indole	Repellent	Fatty acids	Repellent
Skatole		Hydrophobic amino acids	Repellent

receptor. The glucose, serine, and aspartate receptors are synthesized constitutively.

The capillary assay also allows a convenient determination of the sensitivity of the receptors. By varying the concentration of attractant within the capillary, the ranges over which the receptor will respond can be easily determined. Typically, the lowest concentration is about 10^{-7} M, and the highest concentration is about 10^{-1} or 10^{-2} M (Fig. 22.3). It is not at all surprising that a detection system is no more sensitive than 10^{-7} M. First, at about 10^{-6} M the rate of diffusion of a sugar to a bacterium is just adequate to support a 30-minute doubling time if every sugar molecule reaching the cell is utilized. The ability to chemotact would not change this lower limit by much.

The second limitation on detection sensitivity is the lowest concentration at which a sufficiently accurate measurement may be made in the one second measurement window set by the rate of random rotations. At a concentration of 10^{-7} M, the number of attractant molecules

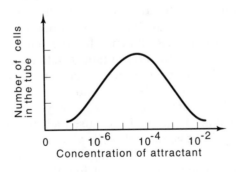

Figure 22.3 A typical response curve showing the number of cells that have entered the capillary tube as a function of the concentration of attractant placed in the tube.

reaching the cell surface is so low that statistical fluctuations in their number necessitate averaging for a full second to obtain sufficient accuracy. Thus 10^{-7} M is about the lowest practical concentration to which bacteria can be expected to chemotact.

The actual receptor proteins for the different sugars have been sought, and several have been identified biochemically. The receptors for galactose, maltose, and ribose are found in the periplasm outside the inner membrane but inside the peptidoglycan layer. These and other periplasmic proteins can be removed from cells by osmotic shock.

No periplasmic binding proteins have been found for the glucose, mannitol, and trehalose systems. Instead, these systems use receptors that are tightly bound to or located in the inner membrane. These receptors serve a double purpose as they also function in the group translocation of their substrates into the cell via the phosphotransferase transport system.

Genetics of Motility and Chemotaxis

Mutants permit dissection of the chemotaxis system, both by allowing observation of the behavior of cells containing damaged components of the system and by facilitating biochemical study of the components. One method for isolation of chemotaxis mutants utilizes swarm plates. About 100 candidate mutant cells are diluted into a plate containing low agar concentration, minimal salts, any necessary growth factors, and a metabolizable attractant such as galactose. As each cell grows into a colony, it generates an outward-moving ring if the cells are able to consume the attractant and swim toward higher concentrations. A motile but nonchemotactic cell makes a moderately large colony with diffuse edges, whereas a nonmotile cell makes a compact colony with sharp edges.

Galactose-specific chemotactic mutants can be detected with the swarm plates. One type possessed an altered galactose-binding protein. The protein had a greater dissociation constant for galactose, as did the galactose transport system, proving that the same protein is involved with both galactose chemotaxis and galactose transport.

A different type of genetic selection proved useful for the isolation of deletions and point mutations in flagellin genes. It is based on the fact that antibodies against flagella stop their motion and block motility. Bacterial mutants with different antigenic determinants on their flagella are resistant to the antibody. Imagine the situation in a merodiploid cell where an episome possesses genes for the mutant flagellin and the chromosome codes for the nonmutant flagellin. Such cells will not be chemotactic in the presence of the antibody because their flagella will be a mosaic of both types of flagellin and their action will be blocked by the antibody. Any mutants synthesizing only the flagellin encoded by the episome will be chemotactic because their flagella will be resistant to the antibody. That is, the only cells able to swim out of a spot on a swarm plate containing the antiflagellin antibody are those whose chromosomal flagellin genes are not functional, but whose episomal

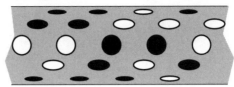

Flagellum of a and b type flagellin

flagellin genes are functional. Apparently deletions are the most common method for accomplishing this.

How Cells Swim

Escherichia coli propels itself through liquid by rotating its flagella. Normally the flagella are left-handed helices and their rotation generates a thrust that moves the cell. In this section we consider the structure of flagella, how it's known they rotate, how the rotation is created, and how the several flagella present on a single cell can function together.

Flagella are too thin to be easily seen by ordinary light microscopes, but they can be visualized with light microscopes operating in the dark-field mode, interference microscopes, video processing, or electron microscopes. Careful isolation of flagella shows that they are attached to a hook-shaped structure connected to a set of rings that is embedded in the cell's membranes. The rings have the appearance of a motor that rotates the flagella (Fig. 22.4). The hook is a flexible connector between the basal structure and the flagella. Such a universal joint

Figure 22.4 The appearance of a flagellum at low magnification in the electron microscope and the structure of the basal body, the motor, at high magnification showing the rings and the membranes of the cell wall.

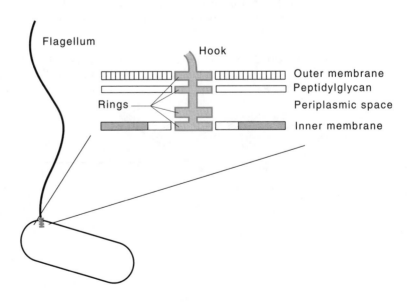

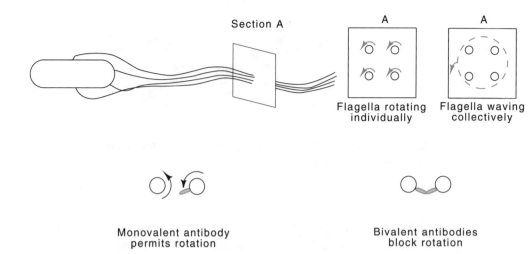

Figure 22.5 A bacterium with flagella sprouting from various locations bending via the hook portion and coming together in a bundle. The section taken at point A shows flagella rotating individually and their movement if the bundle waves as a whole.

is necessary because in *E. coli* the flagella sprout from random points on the cell's surface and point in several different directions. A cell typically contains about six flagella. These must join together in a bundle in a way that permits each to rotate in response to its motor. The hook acts as a universal joint that permits the torque to be transmitted around a bend.

Because of the size of the flagella, indirect means must be used to demonstrate that they rotate. One simple experiment uses the observation that antibody against flagellin can block motility. More precisely, bivalent antibody blocks motility, but monovalent antibody does not (Fig. 22.5). This result can be understood if the flagella form a bundle

Figure 22.6 Demonstration of the hook's rotation. After attachment of the hook or a short flagellum to a cover slip, the cell rotates, and this can easily be observed.

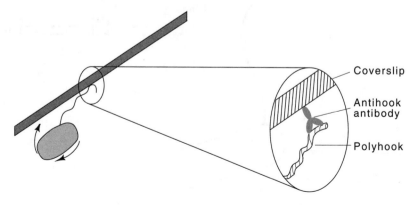

and each flagellum rotates within this bundle. A bivalent antibody molecule can link different flagella and prevent their rotation, but if flagella waved or rotated as a group, bivalent antibodies would have an effect no different from monovalent antibodies.

The most graphic demonstration that flagella rotate is also the basis of many other important experiments on chemotaxis. Simon used one mutation to block flagellin synthesis and another to permit greater than usual growth of the hook. As the synthesis of flagella is sensitive to catabolite repression, growth of cells in glucose reduced the number of the resulting polyhooks from about six per cell to about one. These cells could be bound to a microscope slide by means of antihook antibody that had bound to the hook and nonspecifically bound to the glass as well (Fig. 22.6). Chemotactic cells immobilized in this way rotated at two to nine revolutions per second. This leads to the conclusion that the hook normally rotates, but when it was fastened to the microscope slide instead, the cell rotated. Nonmotile cells did not rotate. Of course, motile but nonchemotactic mutants did rotate because they can swim but do so in a nondirected fashion and are incapable of swimming up a gradient.

The immobilization experiment shows that a single flagellum rotates and that the rotation is associated with chemotaxis. Dark-field microscopy has shown that the bundle of flagella on a cell is stable as long as the flagella rotate counterclockwise. If the flagella reverse their rotation, their left-helical structure compels the bundle to fly apart temporarily. Furthermore, if the reversal is sufficiently vigorous, the flagella snap into a right-helical conformation. This further ensures that the bundle of flagella disperses.

The Mechanism of Chemotaxis

As viewed in the microscope, chemotactic bacteria appear to swim smoothly for about 20 bacterial lengths in 1 second, to tumble for about 0.1 second, and then to swim smoothly in another direction. How this run-tumble-run behavior is converted to overall swimming toward increasing concentrations of an attractant was one of the major problems in chemotaxis research.

Berg constructed an elaborate tracking microscope that quantitated the movement of a chemotactic bacterium in three dimensions. This showed that, indeed, the impression obtained from simple visual observation was correct. Following a tumble, a cell's subsequent run was in a random direction. More important, however, was the behavior of bacteria under conditions simulating the presence of a gradient in attractant. It is difficult to control a gradient in the concentration of an attractant around the cell, but it is likely that a cell detects not a spatial gradient but the change in the concentration of attractant from one time to the next. Therefore, Berg devised a system to vary the concentration of attractant with time.

Alanine aminotransferase catalyzes the interconversion of alanine and α-ketoglutarate to pyruvate and L-glutamate. Both alanine and glutamate are attractants, and therefore the reaction cannot be used to create or destroy total attractant. This problem can be overcome by using a mutant that is blind to alanine but not blind to glutamate. Then the enzyme-catalyzed reaction can be used to vary the concentration with time of attractant surrounding a cell. If alanine, α-ketoglutarate, and alanine aminotransferase are put in the medium bathing the cell

$$\text{Alanine} + \alpha\text{-ketoglutarate} \underset{\xleftarrow{\hspace{1.2cm}}}{\xrightarrow[\text{aminotransferase}]{\text{Alanine}}} \text{Pyruvate} + \text{L-glutamate}$$

being observed, then the concentration of attractant, glutamate, increases with time. In this case, the runs of the bacteria are longer than average. When the glutamate concentration is decreasing as a result of adding glutamate, pyruvate, and alanine aminotransferase to the bathing medium, the distribution of run lengths is the same as when attractant concentration remains constant.

The technique of immobilizing a cell via its hook or flagellum has also been highly useful in characterizing chemotaxis responses. After such an immobilization, a microprobe containing attractant or repellent can be positioned near a cell, and application of a brief electrical pulse will drive a known amount of the chemical into the medium in the immediate area (Fig. 22.7). If the concentration changes are within the linear response range of the cell, the response to a brief pulse of

Figure 22.7 A method for generating small impulses of attractant from the micropipette and observing the cell's responses. The microscope contains additional electronics that record the rotation of the bacterium.

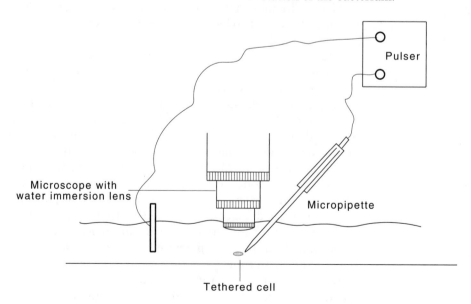

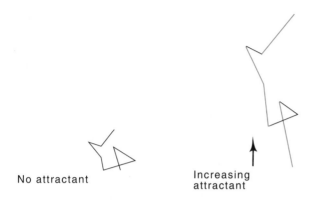

No attractant

Increasing
attractant

Figure 22.8 A random walk and the same walk with the run duration increased when in the direction of increasing attractant concentration.

attractant contains all the information necessary to predict the response of a cell to any function of attractant concentration as a function of time. Any other function can be approximated as a series of impulses of appropriate magnitude, and the responses to each of these can be summed. Such a technique of circuit analysis is well known to electrical engineers, but it has not yet been widely applied to biochemical systems.

All the experiments mentioned above show that cells modulate the duration of their intervals of smooth swimming. Such a modulation is sufficient for the cells to achieve a net drift up a gradient of attractant. When a cell is swimming up a gradient in attractant concentration, that is, experiencing an increase in attractant concentration, it decreases the chances of its tumbling. When the attractant is decreasing, the average run duration is the same as when the cell is in constant attractant concentration. These properties mean that cells drift up gradients in attractant concentration, that is, they undergo a biased random walk, taking lengthened steps when they are moving up the gradient (Fig. 22.8).

The Energy for Chemotaxis

Application of genetic and biochemical methods has allowed determination of the energy source required for chemotaxis. On one hand it seems logical that ATP would be the direct source of the mechanical energy required for swimming since most energy transductions in higher cells appear to use ATP. On the other hand, the flagella originate in the cell membrane, and a substantial proton gradient exists across the inner membrane under most growth conditions. Therefore the direct source of energy could also derive from the membrane potential.

ATP and the proton motive force across the inner membrane are normally interconvertible by means of the membrane-bound ATPase (Fig. 22.9). The membrane potential generates ATP and, conversely, ATP can be used to generate a membrane potential. Therefore, blocking formation of either ATP or the membrane potential could have an effect on the other. Arsenate permits decoupling to determine what actually

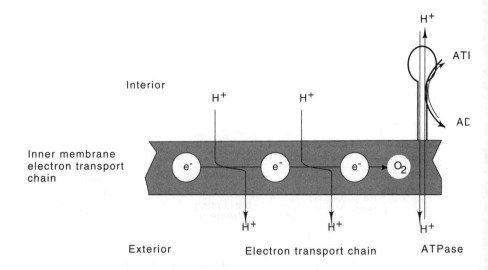

Figure 22.9 A schematic of the inner membrane of a cell showing that electron flow down the electron transport chain leads to export of protons and ultimate transfer of electrons to oxygen. Reentry of protons to the cell through the ATPase generates ATP. Conversely, ATP can be hydrolyzed by ATPase to pump protons out of the cell.

drives the motor of the flagella. It blocks ATP formation both by glycolysis and by ATPase, but does not directly affect the membrane potential. Cells treated with arsenate are found to be motile, but they do not swim up gradients of attractants. This shows that ATP is not required for motility, but probably is involved with the process of modulating run duration.

Next, to investigate the role of the membrane potential, cells can be grown anaerobically. This blocks the usual means of generation of the membrane potential because the electron transfer chain becomes inactive owing to the lack of a terminal electron acceptor. Then, to prevent energy from ATP from being used to create a membrane potential, an ATPase-negative mutant can be used. Anaerobically grown ATPase mutants are found not to be motile, leading to the conclusion that the motors that drive the flagella are run by the cell's membrane potential. Other experiments show that cells swim when transmembrane pH gradient or membrane potentials are generated artificially.

Adaptation

Many sensory systems in animals adapt to a stimulus. That is, excitation with a particular stimulus will evoke a response, but the response weakens with continued application of the stimulus. An increase in the exciting stimulus will then produce a new response, but eventually a new adaptation occurs (Fig. 22.10). Such an adaptation phenomenon is also observed in bacterial chemotaxis and is inherent in the machinery

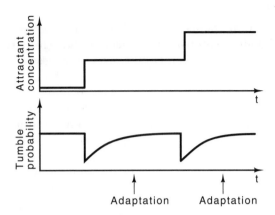

Figure 22.10 Adaptation to increases in attractant concentration.

a cell uses for comparing attractant concentrations at two different times.

Excitation and adaptation can be simply generated by comparing the sensor's present signal to some average of the signal over the recent past (Fig. 22.11). Such averaging is equivalent to a slow response, and this provides the cell's "memory." Hence the instantaneous conditions surrounding a cell must be compared to an average of the conditions taken over the previous few seconds. If the concentration of attractant has increased during that interval, the instantaneous signal will exceed the average signal, and the comparator should produce a signal that suppresses clockwise rotations generated by the flagella motors. If, however, the concentration of attractant has not increased over an interval of a few seconds, the instantaneous and averaged signals will be equal or negative and the comparator signal will fall. Consequently, the motor will be permitted to reverse direction at its natural frequency. In such a direction reversal, the bundles of flagella come apart, and the cells tumble and randomize their subsequent swimming direction.

Figure 22.11 One circuit that will generate a response usable for chemotaxis. The comparator response regulates the direction of flagellum rotation.

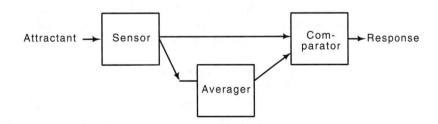

Methylation and Adaptation

The biochemical components in the cell's rapid response, adaptation response, and signaling pathways have been found. They are phosphorylation and methylation of several of the proteins necessary for chemotaxis. First we will discuss methylation. Adler had observed that methionine is necessary for chemotaxis. That is, a methionine auxotroph grown in the presence of methionine would be chemotactic. Upon removal of methionine, however, the cells remained motile, but could no longer swim up gradients of attractant. Auxotrophs for other amino acids remained chemotactic upon amino acid starvation.

The results discussed above indicate that methionine plays some special role in chemotaxis. Later, when the mechanism of chemotaxis was understood to be modulation in the frequency of runs and tumbles, it was sensible to ask what methionine did. Experiments showed that methionine-starved cells are unable to tumble. The next question was whether methionine starvation destroyed the physical ability to tumble or whether it eliminated the signal to tumble. At first this question seems unapproachable. Genetics, however, came to the rescue. Among the many types of nonchemotactic mutants known are some that incessantly tumble. The question then was whether these tumbling mutants continued to tumble during methionine starvation. All that was necessary to answer this question was to make these tumbling mutants methionine auxotrophs and to starve them of methionine. After this treatment they continued to tumble. This shows methionine is not necessary for tumbling, and therefore that the amino acid must be part of a pathway that signals tumbling to occur.

How might methionine be required? Methionine via S-adenosyl methionine is a common source of methyl groups in metabolism. Most likely, then, the methionine requirement hints that methylation is involved in the tumbling signal. Consistent with this notion is the fact discussed above that the addition of arsenate to cells, which blocks ATP formation and hence S-adenosyl methionine synthesis, also blocks chemotaxis but not motility.

With such clues for the involvement of methylation in chemotaxis, it is natural to look for methylated proteins. The level of methylation of several membrane proteins has been found to be correlated with chemotaxis behavior for some attractants. The addition of attractant to cells leads to preferential methylation of one of a set of at least four membrane proteins, products of genes named *tsr*, *tar*, *tap*, and *trg* (Fig. 22.12). These proteins are the receptors and signal transducers for some of the attractants and repellents to which *E. coli* responds. For example the *tar* product is the receptor for the attractants aspartate, maltose, and many repellents.

Signaling between the receptor proteins and the intracellular chemotaxis machinery requires sending a signal through the membrane. The *tar* protein spans the membrane. It has a domain of about one hundred of the N-terminal amino acids outside the membrane, and about two hundred more amino acids inside the membrane. The external portion binds the aspartate, but it is the internal portion that is methylated by

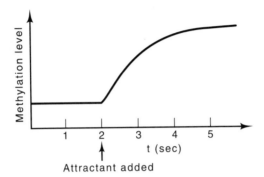

Figure 22.12 Methylation of one of the proteins in the inner membrane as a function of time after attractant addition.

methyl transfer from S-adenosyl methionine. As many as five glutamic acid residues per receptor can be modified as the cells adapt to the continued presence of attractant. The intracellular portion of the protein is highly homologous to other methyl-accepting chemotaxis proteins, whereas the N-terminal portion is only weakly homologous to the other chemotaxis receptors.

Increase in the concentration of an attractant generates a conformation change in one of these membrane-bound proteins. This change is ultimately communicated to the flagellar motor. Increasing methylation of the protein decreases its ability to send a signal for smooth swimming or increases its ability to send a signal for tumbling. It is the change in signal strength resulting from change in methylation that produces adaptation.

For methylation to be part of the cell's "memory," at least two enzymes are required: one is an enzyme to methylate and the other is an enzyme to demethylate. More precisely, a methyltransferase and a methylesterase must exist. Both of these have been found. They act on the proper membrane proteins and are encoded by genes in the set of eight or nine genes that are required for chemotaxis but are not required for motility.

Phosphorylation and the Rapid Response

Chemotaxis requires sensing an extracellular condition and then signaling to some intracellular component. A number of other systems, for example the phosphotransfer system, the nitrogen regulatory system, phosphate uptake system, and porin system, use similar signaling pathways. These pathways contain sensor proteins which often span the membrane and serve to couple extracellular conditions to internal activity. The sensors are similar in structure to one another, phosphorylate themselves on histidine residues, and phosphorylate the regulator proteins on aspartate residues. The regulator proteins also are similar in structure to one another. In order that the stimulation be reversible,

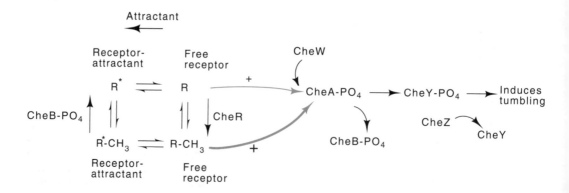

Figure 22.13 Pathway of methylation and phosphorylation. The methylated state of free receptor is most active in stimulating phosphorylation of CheA. CheB phosphate demethylates the receptor and CheR methylates it.

another component to these systems is a phosphatase of the regulator proteins.

In the chemotaxis system the Tar, Tsr, Trg, or Tap proteins transmit a signal across the inner membrane. These signals are received by the CheW and CheA proteins and are translated into varying levels of phosphorylation of CheA. This was found by analogy to the phosphotransfer system and a finding by Magasanik that phosphorylation was involved with regulation of nitrogen metabolism. When the purified chemotaxis proteins are appropriately incubated together *in vitro*, a phosphate is transferred from ATP to CheA protein. To demonstrate that this pathway is involved with chemotaxis, mutants lacking the histidine to which the phosphate is transferred have been found not to be chemotactic.

Two further phosphorylation pathways diverge from CheA. The first is phosphorylation of CheY from CheA. This is a rapid step that leads to signaling the flagellar motor, most likely by direct interaction between phosphorylated CheY protein and the motor. This reaction induces tumbling. Phosphorylated CheA also leads to phosphorylation of CheB, which demethylates the receptors.

The response of the chemotaxis pathway involves the complicated interaction of many components (Fig. 22.13). An increase in the concentration of attractant rapidly decreases the phosphorylating signal from receptor to CheA-CheW, and as a result decreases the concentration of phosphorylated CheY. Tumbling is suppressed and smooth swimming is prolonged. Additionally, the decrease in phosphorylation of CheA decreases the phosphorylation of CheB which decreases its demethylation activity. As a result, but one taking place more slowly than the first response, methylation of the receptor increases, the cells begin to adapt, and the phosphorylation levels begin to increase. Overall, then, the phosphorylation levels respond rapidly and transmit a signal to the motor, and the methylation levels respond more slowly and serve to bring the phosphorylation levels back to the levels characteristic of the absence of attractant.

Chemotaxis is remarkable. A single cell manages to sense attractants with receptors located outside the cytoplasm, transmit the information to the interior, remember the information it has received so as to adapt if the attractant or repellent concentration is no longer changing, build and energize a motor, and synthesize a bundle of flagella that smoothly rotate together or fly apart when necessary. The biochemical bases for parts of the system are known. A few of the receptors have been purified, and the center of the adaptation phenomenon, the methylation and phosphorylation of the proteins, is partially understood. Whether we will be able to understand the underlying physical chemistry of the protein conformational changes and the drive mechanism of the flagellar motor in the years ahead is an interesting question.

Problems

22.1. Suppose a strain of chemotactic cells with no endogenous energy source is starved of all possible sources of metabolic energy. Why should they swim for a minute or two after dilution from a medium containing a high potassium concentration to a medium with a low potassium concentration plus valinomycin? Valinomycin is a potassium ionophore.

22.2. What information do we learn about the chemotaxis system by the finding that the steeper the gradient in attractant concentration, the faster the bacteria swim up the gradient?

22.3. How would you isolate chemotactic mutants (a) that are blind to an attractant, (b) that are temperature-sensitive blind, (c) that have an altered K_m for an attractant?

22.4. What would be the behavior of a singly-flagellated cell when the flagellum changes from counterclockwise to clockwise rotation?

22.5. How fast do bacteria move when they swim?

22.6. In the case of repellent versus attractant, how could you determine whether the response to the simultaneous presence of repellent and attractant is a momentary response appropriate to attractant and repellent or an algebraic summation of the signals so as to give no response at all?

22.7. Through an ordinary monocular light microscope, does a right-handed screw thread look right-handed or left-handed, that is, do bacteria rotating clockwise actually appear to be rotating clockwise when observed through the microscope?

22.8. Is the model proposed on the basis of physiological information in "Sensory Transduction in *Escherichia coli*: Role of a protein Methylation Reaction in Sensory Adaptation," M. Goy, M. Springer, J. Adler, Proc. Natl. Acad. Sci. USA *74*, 4964-4968 (1977), essentially the model proposed in "Transient Response to Chemotactic Stimuli in *E. coli*," H.

Berg, P. Tedesco, Proc. Natl. Acad. Sci. USA *72*, 3235-3239 (1975)? Explain how they are fundamentally the same or fundamentally irreconcilable.

22.9. How would you determine from which end the flagellum, exclusive of the basal parts and hook, grows? Why would anyone want to know?

22.10. In terms of the amount of carbohydrate consumed by a bacterial cell in one doubling time, how much carbohydrate metabolism is required to provide the energy necessary for a cell to swim 5 cm?

22.11. Polymerized flagellin exists in at least two conformations. Must any bonds be made or broken during a transition between conformations? If so, what experiments could confirm their existence?

22.12. In light of the Macnab paper, Proc. Natl. Acad. Sci. USA *74*, 221-225 (1977), is it still reasonable to consider phage chi to be brought to the cell surface by rotation of the flagella as a nut is drawn to the head of a bolt by rotation?

22.13. The chapter has discussed run-tumble, counterclockwise-versus-clockwise flagellar rotation for variations in attractant concentrations. What must the responses be for increases and decreases in repellent concentrations?

22.14. Show that the diffusion of sugar to the surface of a bacterium at 10^{-6} M is just adequate to support a 30-minute cell doubling time.

22.15. What simple experiment proves that it is the unliganded receptor rather than the receptor-attractant complex that sends signals on to the tumble generator?

22.16. Why is the separation of both the Tsr and the Tar proteins according to their methylation on SDS polyacrylamide gels unexpected, and what is a likely explanation for the separation?

Figure 22.14 Probabilities of counterclockwise rotation following a brief exposure to attractant.

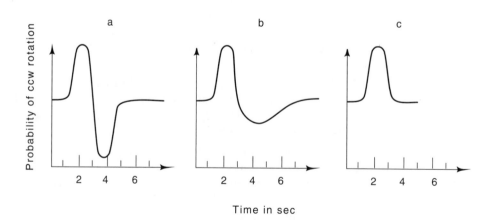

22.17. Both the methyltransferase minus and methylesterase minus mutants are not observed to methylate the Tsr and Tar proteins in the standard protocols of treating cells with chloramphenicol and then adding radioactive methionine. Why is this surprising on first consideration, and what is the likely explanation?

22.18. What is the likely chemotaxis behavior of methyltransferase minus and methylesterase minus mutants?

22.19. The response of the wild-type chemotaxis system to a short pulse of attractant at $t = 0$ produces the response shown in Fig. 22.14a. Describe the chemotactic behavior if the response is as shown in parts b or c.

22.20. Look up Weber's law concerning the magnitudes of stimulation required to evoke responses. Consider a system in which the ratio of the increase in the magnitude of the stimulus to the magnitude itself must be greater than or equal to some constant value for a response to be evoked. Why does or does not such a system follow Weber's law?

References

Recommended Readings

Change in Direction of Flagellar Rotation is the Basis of the Chemotactic Response in *Escherichia coli*, S. Larsen, R. Reader, E. Kort, W. Tso, J. Adler, Nature *249*, 74-77 (1974).

Temporal Comparisons in Bacterial Chemotaxis, J. Segall, S. Block, H. Berg, Proc. Nat. Acad. Sci. USA *83*, 8987-8991 (1986).

Protein Phosphorylation is Involved in Bacterial Chemotaxis, J. Hess, K. Oosawa, P. Matsumura, M. Simon, Proc. Nat. Acad. Sci. USA *84*, 7609-7613 (1987).

Physiology of Chemotaxis

Effect of Amino Acids and Oxygen on Chemotaxis in *Escherichia coli*, J. Adler, J. Bact. *92*, 121-129 (1966).

Chemotaxis in Bacteria, J. Adler, Science *153*, 708-716 (1966).

Chemoreceptors in Bacteria, J. Adler, Science *166*, 1588-1597 (1969).

Role of Galactose Binding Protein in Chemotaxis of *E. coli* toward Galactose, G. Hazelbauer, J. Adler, Nature New Biol. *230*, 101-104 (1971).

Chemotaxis in *E. coli* Analysed by Three-dimensional Tracking, H. Berg, D. Brown, Nature *239*, 500-504 (1972).

Quantitative Analysis of Bacterial Migration in Chemotaxis, F. Dahlquist, P. Lovely, D. Koshland, Nature New Biology *236*, 120-123 (1972).

Chemotaxis Toward Amino Acids in *E. coli*, R. Mesibov, J. Adler, J. Bact. *112*, 315-326 (1972).

Chemotaxis Toward Sugars in *E. coli*, J. Adler, G. Hazelbauer, M. Dahl, J. Bact. *115*, 824-847 (1973).

Temporal Stimulation of Chemotaxis in *Escherichia coli*, D. Brown, H. Berg, Proc. Nat. Acad. Sci. USA 71, 1388-1392 (1974).

"Decision"-making in Bacteria: Chemotactic Response of *Escherichia coli* to Conflicting Stimuli, J. Adler, W. Tso, Science *184*, 1292-1294 (1974).

Quantitation of the Sensory Response in Bacterial Chemotaxis, J. Spudich, D. Koshland, Proc. Nat. Acad. Sci. USA 72, 710-713 (1975).

Transient Response to Chemotactic Stimuli in *E. coli*, H. Berg, P. Tedesco, Proc. Nat. Acad. Sci. USA 72, 3235-3239 (1975).

Bacterial Behavior, H. Berg, Nature 254, 389-392 (1975).

A Protonmotive Force Drives Bacterial Flagella, M. Manson, P. Tedesco, H. Berg, F. Harold, C. van der Drift, Proc. Nat. Acad. Sci. USA 74, 3060-3064 (1977).

Energetics of Flagellar Rotation in Bacteria, M. Manson, P. Tedesco, H. Berg, J. Mol. Biol. 138, 541-561 (1980).

Signal Processing Times in Bacterial Chemotaxis, J. Segall, M. Manson, H. Berg, Nature 296, 855-857 (1982).

Impulse Responses in Bacterial Chemotaxis, S. Block, J. Segall, H. Berg, Cell 31, 215-226 (1982).

Asynchronous Switching of Flagellar Motors on a Single Bacterial Cell, R. Macnab, D. Han, Cell 32, 109-117 (1983).

Isotope and Thermal Effects in Chemiosmotic Coupling to the Flagellar Motor of *Streptococcus*, S. Khan, H. Berg, Cell 32, 913-919 (1983).

Adaptation Kinetics in Bacterial Chemotaxis, S. Block, J. Segall, H. Berg, J. Bact. 154, 312-323 (1983).

Chemotactic Signaling in Filamentous Cells of *Escherichia coli*, J. Segall, A. Ishihara, H. Berg, J. Bact. 161, 51-59 (1985).

Additive and Independent Responses in a Single Receptor: Aspartate and Maltose Stimuli, S. Mowbray, D. Koshland, Cell 50, 171-180 (1987).

Multiple Kinetic States for the Flagellar Motor Switch, S. Kuo, D. Koshland, Jr., J. Bact. 171, 6279-6287 (1989).

Genetics of Chemotaxis

Nonchemotactic Mutants of *Escherichia coli*, J. Armstrong, J. Adler, M. Dahl, J. Bact. 93, 390-398 (1967).

Genetic Analysis of Flagellar Mutants in *E. coli*, M. Silverman, M. Simon, J. Bact. 113, 105-113 (1973).

Genetic Analysis of Bacteriophage Mu-induced Flagellar Mutants in *Escherichia coli*, M. Silverman, M. Simon, J. Bact. 116, 114-122 (1973).

Chemomechanical Coupling without ATP: The Source of Energy for Motility and Chemotaxis in Bacteria, S. Larsen, J. Adler, J. Gargus, R. Hogg, Proc. Nat. Acad. Sci. USA 71, 1239-1243 (1974).

Positioning Flagellar Genes in *E. coli* by Deletion Analysis, M. Silverman, M. Simon, J. Bact. 117, 73-79 (1974).

Isolation and Complementation of Mutants in Galactose Taxis and Transport, G. Ordal, J. Adler, J. Bact. 117, 509-516 (1974).

Properties of Mutants in Galactose Taxis and Transport, G. Ordal, J. Adler, J. Bact. 117, 517-526 (1974).

Isolation, Characterization and Complementation of *Salmonella typhimurium* Chemotaxis Mutants, D. Aswad, D. Koshland, Jr., J. Mol. Biol. 97, 225-235 (1975).

cheA, *cheB* and *cheC* Genes of *E. coli* and their Role in Chemotaxis, J. Parkinson, J. Bact. 126, 758-770 (1976).

Identification of Polypeptides Necessary for Chemotaxis in *Escherichia coli*, M. Silverman, M. Simon, J. Bact. 130, 1317-1325 (1977).

Complementation Analysis and Deletion Mapping of *Escherichia coli* Mutants Defective in Chemotaxis, J. Parkinson, J. Bact. *135*, 45-53 (1978).

Chemotaxis Receptors

Phosphotransferase-system Enzymes as Chemoreceptors for Certain Sugars in *E. coli* Chemotaxis, J. Adler, W. Epstein, Proc. Nat. Acad. Sci. USA *71*, 2895-2899 (1974).

Genetic and Biochemical Properties of *Escherichia coli* Mutants with Defects in Serine Chemotaxis, M. Hedblom, J. Adler, J. Bact. *144*, 1048-1060 (1980).

Sensory Transducers of *E. coli* Are Composed of Discrete Structural and Functional Domains, A. Krikos, N. Mutoh, A. Boyd, M. Simon, Cell *33*, 615-622 (1983).

Structure of the Trg Protein: Homologies with and Differences from Other Sensory Transducers of *Escherichia coli*, J. Ballinger, C. Park, S. Harayama, G. Hazelbauer, Proc. Nat. Acad. Sci. USA *81*, 3287-3291 (1984).

Analysis of Mutations in the Transmembrane Region of the Aspartate Chemoreceptor in *Escherichia coli*, K. Oosawa, M. Simon, Proc. Nat. Acad. Sci. USA *83*, 6930-6934 (1986).

Polar Location of the Chemoreceptor Complex in the *Escherichia coli* Cell, J. Maddock, L. Shapiro, Science *259*, 1717-1723 (1993).

Flagella, their Motors and Rotation

Fine Structure and Isolation of the Hook-basal Body Complex of Flagella from *Escherichia coli* and *Bacillus subtilis*, M. DePamphilis, J. Adler, J. Bact. *105*, 384-395 (1971).

Attachment of Flagellar Basal Bodies to the Cell Envelope: Specific Attachment to the Outer, Lipopolysaccharide Membrane and the Cytoplasmic Membrane, M. DePamphilis, J. Adler, J. Bact. *105*, 396-407 (1971).

Bacteria Swim by Rotating their Flagellar Filaments, H. Berg, R. Anderson, Nature *245*, 380-382 (1973).

Flagellar Rotation and the Mechanism of Bacterial Motility, M. Silverman, M. Simon, Nature *249*, 73-74 (1974).

Bacterial Flagella Rotating in Bundles: A Study in Helical Geometry, R. Macnab, Proc. Nat. Acad. Sci. USA *74*, 221-225 (1977).

Normal to Curly Flagellar Transitions and their Role in Bacterial Tumbling, Stabilization of an Alternative Quaternary Structure by Mechanical Force, R. Macnab, M. Ornston, J. Mol. Biol. *112*, 1-30 (1977).

Flagellar Hook Structures of *Caulobacter* and *Salmonella* and their Relationship to Filament Structure, T. Wagenknecht, D. DeRosier, J. Mol. Biol. *162*, 69-87 (1982).

Correlation between Bacteriophage Chi Adsorption and Mode of Flagellar Rotation of *Escherichia coli* Chemotaxis Mutants, S. Ravid, M. Eisenbach, J. Bact. *154*, 604-611 (1983).

Successive Incorporation of Force-generating Units in the Bacterial Rotary Motor, S. Block, H. Berg, Nature *309*, 470-472 (1984).

Cell Envelopes of Chemotaxis Mutants of *Escherichia coli* Rotate their Flagella Counterclockwise, C. Szupica, J. Adler, J. Bact. *162*, 451-453 (1985).

Identification of Proteins of the Outer (L and P) Rings of the Flagellar Basal Body of *Escherichia coli*, C. Jones, M. Homma, R. Macnab, J. Bact. *169*, 1489-1492 (1987).

Localization of the *Salmonella typhimurium* Flagellar Switch Protein FliG to the Cytoplasmic M-ring Face of the Basal Body, N. Francis, V. Irikura, S. Yamaguchi, D. DeRosier, R. Mcnab, Proc. Natl. Acad. Sci. USA *89*, 6304-6308 (1992).

Methylation in Chemotaxis

Role of Methionine in Bacterial Chemotaxis: Requirement for Tumbling and Involvement in Information Processing, M. Springer, E. Kort, S. Larsen, G. Ordal, R. Reader, J. Adler, Proc. Nat. Acad. Sci. USA *72*, 4640-4644 (1975).

Evidence for an S-adenosylmethionine Requirement in the Chemotactic Behavior of *Salmonella typhimurium*, D. Aswad, D. Koshland, Jr., J. Mol. Biol. *97*, 207-233 (1975).

Sensory Transduction in *Escherichia coli*: Role of a Protein Methylation Reaction in Sensory Adaptation, M. Goy, M. Springer, J. Adler, Proc. Nat. Acad. Sci. USA *74*, 4964-4968 (1977).

Identification of a Protein Methyltransferase as the *cheR* Gene Product in the Bacterial Sensing System, W. Springer, D. Koshland, Jr., Proc. Nat. Acad. Sci. USA *74*, 533-537 (1977).

Sensory Transduction in *Escherichia coli*: Two Complementary Pathways of Information Processing that Involve Methylated Proteins, M. Springer, M. Goy, J. Adler, Proc. Nat. Acad. Sci. USA *74*, 3312-3316 (1977).

A Protein Methylesterase Involved in Bacterial Sensing, J. Stock, D. E. Koshland, Jr., Proc. Nat. Acad. Sci. USA *75*, 3659-3663 (1978).

Multiple Methylation in Processing of Sensory Signals During Bacterial Chemotaxis, A. DeFranco, D. Koshland Jr. Proc. Nat. Acad. Sci. USA *77*, 2429-2433 (1980).

Adaptation in Bacterial Chemotaxis: CheB-dependent Modification Permits Additional Methylations of Sensory Transducer Proteins, M. Kehry, F. Dahlquist, Cell *29*, 761-772 (1982).

Sensory Adaptation in Bacterial Chemotaxis: Regulation of Demethylation, M. Kehry, T. Doak, F. Dahlquist, J. Bact. *163*, 983-990 (1985).

Compensatory Mutations in Receptor Function: A Reevaluation of the Role of Methylation in Bacterial Chemotaxis, J. Stock, A. Borczuk, F. Chiou, J. Burchenal, Proc. Nat. Acad. Sci. USA *82*, 8364-8368 (1985).

Adaptational "Crosstalk" and the Crucial Role of Methylation in Chemotactic Migration by *Escherichia coli*, G. Hazelbauer, C. Park, D. Nowlin, Proc. Natl. Acad. Sci. USA *86*, 1448-1452 (1989).

Phosphorylation in Chemotaxis

Restoration of Flagellar Clockwise Rotation in Bacterial Envelopes by Insertion of the Chemotaxis Protein CheY, S. Ravid, P. Matsumura, M. Eisenbach, Proc. Nat. Acad. Sci. USA *83*, 7157-7161 (1986).

Roles of *cheY* and *cheZ* Gene Products in Controlling Flagellar Rotation in Bacterial Chemotaxis of *Escherichia coli*, S. Kuo, D. Koshland, J. Bact. *169*, 1307-1314 (1986).

Histidine Phosphorylation and Phosphoryl Group Transfer in Bacterial Chemotaxis, J. Hess, R. Bourret, M. Simon, Nature *336*, 139-143 (1988).

Receptor Interactions through Phosphorylation and Methylation Pathways in Bacterial Chemotaxis, D. Sanders, D. Koshland, Jr., Proc. Nat. Acad. Sci. USA *85*, 8425-8429 (1988).

Three-dimensional Structure of CheY, the Response Regulator of Bacterial Chemotaxis, A. Stock, J. Mottonen, J. Stock, C. Schutt, Nature *337*, 745-749 (1989).

Conserved Aspartate Residues and Phosphorylation in Signal Transduction by the Chemotaxis Protein CheY, R. Bourret, J. Hess, M. Simon, Proc. Natl. Acad. Sci. USA *87*, 41-45 (1990).

Oncogenesis, Molecular Aspects

23

We have already seen that the elementary regulatory mechanisms found in bacteria can be damaged by mutations. We would expect, then, that in higher cells growth regulation pathways could also be damaged and the cell would either not grow and divide or would grow without stopping. Such uncontrolled growth is seen, and is called cancer. Such a state is expressed in cell culture as the failure of fibroblast cells to stop growing when they have grown to form a confluent monolayer. These cells are then said to have been transformed to a noninhibited state and have lost their density-dependent regulation of growth. Additional properties like anchorage independent growth, growth factor-independent growth, cell morphology, and immortality as opposed to the normal pattern of growth for only about 40 generations, can also be altered in transformed cells. Cancers, then, give us a window into the normal cellular regulatory processes. The important experimental questions are determining the biochemical changes that cause different cancers and then determining how these changes alter the cell growth patterns.

In an earlier chapter we saw that diffusible growth factors permit cells in a multicelled organism to communicate in order to coordinate growth. Differentiating cells can also regulate growth by utilizing their lineages. For example, the 32 descendants resulting from 5 cycles of division could all be identical. On the next division cycle, one of the daughters of one of the cells could differentiate from the others and start development of a different tissue type. This type of scheme appears to operate in the nematode, *Caenorhabditis elegans*, and portions of developmental pathways in higher organisms make use of this principle as well. It is possible to imagine mutational damage that alters the behavior of both the signaling and lineage growth regulation schemes.

This chapter focuses on the application of principles and techniques discussed in earlier chapters to determine the molecular alterations that lead to the cancerous state in plants and animals. These studies are providing valuable information on the workings of normal cells and they indicate that at least 50 genes play highly important roles in regulating the growth of animal cells. Yet to be determined, however, is whether the information gained from these studies will be directly applicable to the treatment of cancer patients.

Bacterially Induced Tumors in Plants

Plants appear to be simpler than mammals and yet they are susceptible to tumors just like mammals. Therefore, some of the lessons that we learn about plant tumors may be helpful in the analysis of animal tumors. One feature that makes the study of plant cells and plant tumors valuable in research on oncogenesis is that cell cultures from some plants can be maintained indefinitely in cell culture, but when desired, these undifferentiated cells can be induced to differentiate back into normal plants that reproduce sexually. Such techniques permit detailed analysis of the causes of oncogenesis. In addition, these approaches have the potential for yielding valuable mutants by using techniques similar to those used with bacteria for the isolation of single mutant cells from large cultures and then regenerating complete mutant plants.

The bacterium *Agrobacterium tumefaciens* can induce the growth of masses of undifferentiated cells, called crown galls, in susceptible plants. This transformation to the undifferentiated state requires a 200-Kb plasmid carried by the bacterium. In the transformation process at least 8 to 10 Kb of DNA from the plasmid are transferred from the bacterium into the plant cells. There the DNA is integrated into the chromosome of the plant, where it is replicated along with the cellular DNA. As a result, all cells of a crown gall contain fragments of DNA originating from the plasmid. Part of the integrated plasmid DNA directs the plant to synthesize and secrete the compounds octopine or nopaline (Fig. 23.1). In turn, these compounds can be catabolized by the *Agrobacter* bacteria in the crown gall. Few other bacteria or parasites can utilize these compounds for growth. Thus, one *Agrobacterium*

Figure 23.1 Structures of octopine and nopaline.

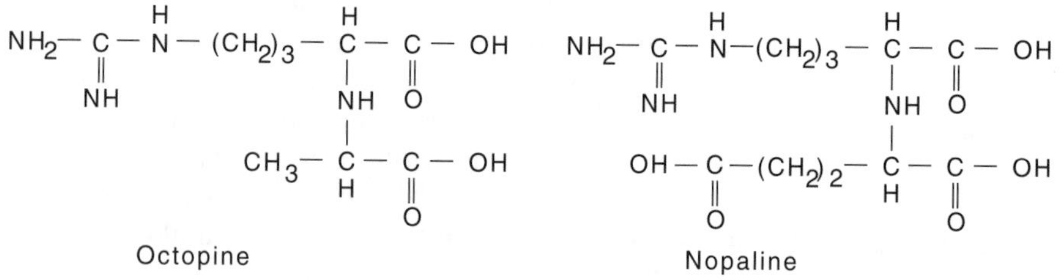

Octopine

Nopaline

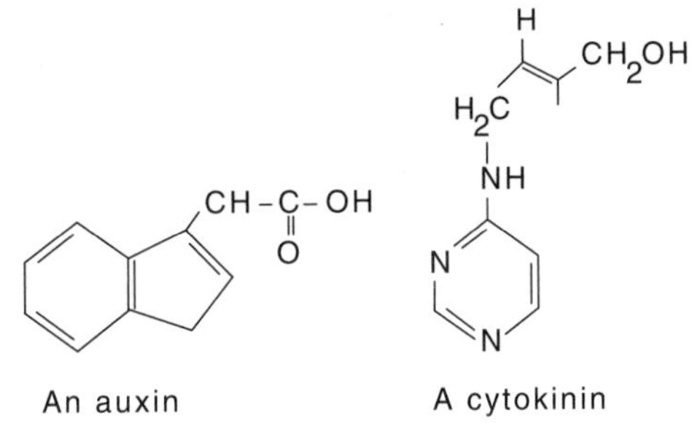

Figure 23.2 Indole 3-acetic acid, an auxin and 6-(4-hydroxy-3-methyl-trans-2-butenylamino)purine, a cytokinin.

subverts part of the plant to produce nutrients for a large bacterial colony.

In contrast to the nontransformed plant cells, the cells from crown galls do not require the growth factors auxin and cytokinin for their continued growth in culture (Fig. 23.2). Ordinarily, medium for plant cells requires the presence of these two small-molecule growth factors in addition to a variety of other metabolites. With the ratio of auxin and cytokinin at one value, plant cells in culture remain undifferentiated, but if the ratio of auxin to cytokinin is increased, stems and leaves tend to develop. Conversely, if the auxin to cytokinin ratio is decreased, the cells become root-like. Without either auxin or cytokinin present, the cells do not grow. In the whole plant, auxin is synthesized in the stem tips and cytokinin in the root tips. A concentration gradient in these molecules from the top to bottom of the plant helps cells identify their positions and develop appropriately. Plant cells transformed with *Agrobacter* do not require either auxin or cytokinin for growth. It is likely that the DNA that was acquired from the plasmid by the transformed cells directs or induces synthesis of auxin and cytokinin-like substances that substitute for these chemicals both in the crown gall and in cell culture. By analogy to the crown galls in plants, it was predicted and found that many types of animal cancers would involve alterations in the cell's synthesis of, or their response to, growth factors.

The DNA transfer mechanism utilized by *Agrobacter* can be utilized for genetic engineering. The DNA to be introduced to the plant cell can be included between the DNA recombination sequences of the transformation plasmid and transformation then occurs much like lambda phage integration.

Transformation and Oncogenesis by Damaging the Chromosome

Epidemiologists have long recognized that certain chemicals induce cancer, and Ames has shown that many bacterial mutagens are human carcinogens. The mutagenicity of chemicals, as detected by the reversion of a set of bacterial histidine mutants, correlates well with their carcinogenicity. Therefore chromosome damage in the form of mutations or small insertions and deletions is one cause of cancer. The next section explains how the exact base change responsible for induction of one particular cancer was determined.

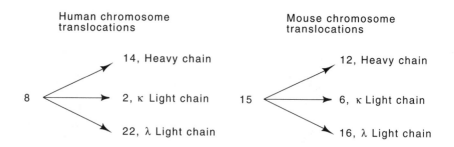

Figure 23.3 Human and mouse translocations associated with cancers of the immune system.

In addition to small lesions, chromosomal rearrangements can alter either the expression or the structure of RNAs or proteins. Such chromosomal rearrangements are frequently observed in a class of cancers called Burkitt's lymphomas in humans and plasmacytomas in mice. These are cancers of immunoglobulin-secreting cells, and therefore it is likely that the chromosome rearrangements in Burkitt's lymphomas actually involve the immunoglobulin genes. Indeed, the chromosomes involved in the rearrangements usually are those that contain the heavy and light immunoglobulin genes (Fig. 23.3). The actual site of the chromosomal translocation in a Burkitt's lymphoma can be identified by cloning the heavy chain constant region from lymphoma and non-lymphoma cells. Many lymphomas contain a translocation that fuses part of chromosome 8 to a heavy chain switch region on chromosome 14 (Fig. 23.4). The junction of the translocation sometimes contains a sequence resembling the DNA it replaced, implying that a recombination event, or an immunoglobulin switch-related event, was responsible for the translocation. The DNA that has been translocated to the immu-

Figure 23.4 Schematic of the heavy chain region of human chromosome 14 and the approximate location of the material from chromosome 8 that is translocated to chromosome 14. Transcripts originating in the region are also shown.

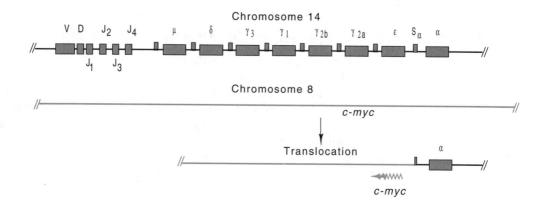

noglobulin region codes for one of the cell's growth regulators known as *c-myc*. As a result of the translocation, its expression is deregulated and its gene is often slightly rearranged. Similar gene translocations of *c-myc* are also involved with mouse plasmacytomas.

Identifying a Nucleotide Change Causing Cancer

As mentioned above, since mutagens are carcinogenic, it is likely that some cancers are caused by mutation. How could we determine the nucleotide change or changes that cause a human cell to be transformed? If such a change could be detected, it would definitively identify the altered gene. This would assist the study of the gene product's normal cellular function. In principle, comparing the nucleotide sequences of all the chromosomes from a normal cell and a transformed cell would reveal the change. Even if such an ill-advised attempt to sequence about 6×10^9 base pairs of DNA could be accomplished, however, the important change would likely be masked by a multitude of extraneous random nucleotide changes.

To reduce the problem of locating the nucleotide change that causes a cancer to manageable proportions, a powerful selection technique for the critical DNA fragment is required. The best selection is to use the ability of the DNA to induce uncontrolled growth. Cells growing in culture can be used for such a screening. For example, one line of mouse cells, NIH 3T3 cells, has been adapted for continuous growth in culture, in contrast to primary cell lines that grow for only a limited number of generations. Nonetheless, the NIH 3T3 cells still display density-dependent growth regulation and cease further growth when a monolayer of cells has grown on the container surface. Thus, the cells can be considered to be part way along the pathway leading to transformation. If one of these cells is transformed to a cancerous state, it loses its growth regulation, and its descendants continue growing when the other cells in the monolayer have stopped. The resulting pile of cells is detectable through a microscope, and is called a focus. Weinberg and others have

Figure 23.5 The use of a tag to permit detection of an oncogene when cloning crude DNA from a transformed cell into *E. coli*. The desired plaque contains material complementary to the tag known to be associated with the oncogene.

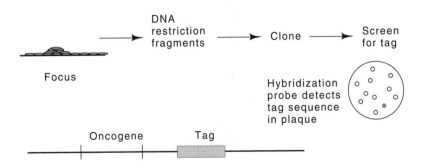

found that DNA extracted from 10% to 20% of different human tumors can transform NIH 3T3 cells to form foci.

Although the focus assay for transforming DNA fragments could be used in conjunction with a straightforward biochemical purification of the desired DNA fragments, genetic engineering methods permitted more rapid isolation of the DNA. The basic trick was to tag restriction fragments of the DNA extracted from a tumor before it was used to transform cells to grow foci. DNA extracted from cells of a focus was then cloned in *E. coli* and the colonies or plaques were screened using the tag sequence as a probe. Of course, the tag had to be a sequence not normally found in the NIH 3T3 cells or in *Escherichia coli* (Fig. 23.5). Therefore, the clone containing the human-derived sequence capable of transforming the cells to density-independent growth could be detected.

Two methods can be used for tagging the oncogene isolated from cancerous cells. One makes use of naturally occurring sequences. As mentioned in Chapter 19, in humans, a particular sequence called the *Alu* sequence is widely scattered throughout the genome and is repeated hundreds of thousands of times. Because this sequence contains two *Alu* restriction enzyme cleavage sites, cleavage of human DNA with *Alu* restriction enzyme generates hundreds of thousands of copies of a specific size of DNA fragment in addition to all the other random sized fragments. The mouse genome lacks this sequence. Simply because of its high repetition number and wide dispersal through the human genome, one of the *Alu* sequences was likely to lie near the mutated gene responsible for the uncontrolled cell growth and transformation of cells to form a focus. Indeed, one did, and Weinberg was able to use this as a tag of the human gene that was mutated to cause the human cancer.

A second method for tagging the oncogene is direct and general. DNA isolated from a tumor can be digested with a restriction enzyme, and a synthetic DNA containing the desired tagging sequence can be ligated to it. Then the mixture of DNAs can be transformed into cultured mouse

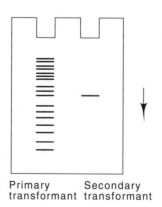

Primary
transformant

Secondary
transformant

Figure 23.6 Appearance of a Southern transfer of DNA extracted from primary and secondary mouse transformants in which the radioactive probe would be the human *Alu* sequence.

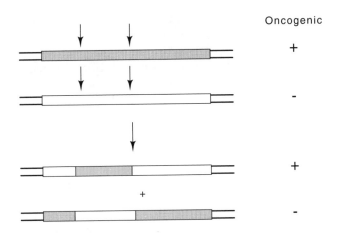

Figure 23.7 Restriction fragment interchange between the cellular homolog of the oncogene and the oncogene itself, which can identify the fragment containing the critical base change.

cells for selection of the fragment containing the oncogene. This is followed by cloning in bacteria and identification of the oncogene clone with the use of the tag sequence as a probe.

A minor technical point must be mentioned. The cultured cells that take up the human DNA usually take up many molecules. Therefore, although a focus of cells will contain the desired transforming gene, it is likely to contain a number of other irrelevant human sequences as well. These also will contain the tagging sequence. This problem is eliminated by using DNA extracted from the primary round of transformation in a second round of transformation. Indeed, after the first cycle, a Southern transfer prepared from DNA of a transformed cell reveals many sizes of restriction fragments containing the human-derived *Alu* tagging sequence (Fig. 23.6), but after a second cycle of transformation, all the transformants contain only a single copy of the tagging sequence. Presumably this is directly linked to the stretch of DNA that is responsible for the transformation.

The techniques described above permitted the cloning and isolation of DNA fragments 5 to 10 Kb long that contained the cellular oncogene from bladder carcinoma cells that could transform NIH 3T3 cells. Once these fragments were obtained, they could be used as probes in screening genomic libraries made from DNA isolated from noncancerous human cells. Thus the normal gene predecessor of the oncogene could also be obtained. Then the question was merely to determine the difference between the two DNAs. Again, they could have been totally sequenced, but ingenuity reduced the work. Restriction fragments between the two were exchanged to find a small DNA segment that was necessary for transformation (Fig. 23.7).

A single nucleotide difference was found to separate the normal cellular prototype and its oncogenic variant. The most common change in the bladder cancers was found to occur in the twelfth amino acid of a protein. Since both Northern transfers of total cellular RNA and immune precipitations of the actual gene product involved show that neither transcription of the gene nor its translation efficiency is appreciably altered by the mutation, the structural change in the protein must be the cause of its oncogenicity, and this has been observed. The gene encodes a protein called Ras that hydrolyzes GTP to GDP. The mutant protein possesses reduced hydrolytic activity.

Retroviruses and Cancer

Just as bacteria can induce a type of cancer in plants, bacteria or viruses could cause cancer in animals. One method to determine whether viruses can do so in animals is to try to induce cancer in animals using a cell-free or bacteria-free extract prepared from tumor cells. One of the first such isolations of a tumor-inducing material was made by Peyton Rous from a cancer of connective tissue—a sarcoma—isolated from a chicken. Hence the virus is called Rous sarcoma virus, or RSV. Repeated propagation of the virus in the laboratory extended its range from chickens to many different animals and increased its potency. RSV is a member of a class of RNA viruses that cause sarcomas in birds, mice, cats, monkeys, and other animals. These viruses are called avian, murine, feline, and simian sarcoma viruses. Similar techniques have been used to detect cancers of various types of blood cells from nonhumans, and the causative viruses are called leukemia viruses. Detection of similar viruses that cause tumors or leukemias in humans has been much more difficult and only a few are known.

The sarcoma and leukemia viruses are both retroviruses. That is, the nucleic acid within the virus particle is single-stranded RNA, and yet this RNA can be duplicated to double-stranded DNA and integrated into the genome of some infected cells (Fig. 23.8). The enzyme reverse transcriptase that performs the replication is packaged within the virus particle along with a virally encoded protease and the integrase.

Initiating the viral replication, much like replication from DNA templates, requires a free hydroxyl group in the proper location for polymerase to use as an initiator. A tRNA molecule of cellular origin

Figure 23.8 Life cycle of a retrovirus.

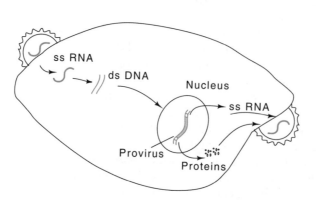

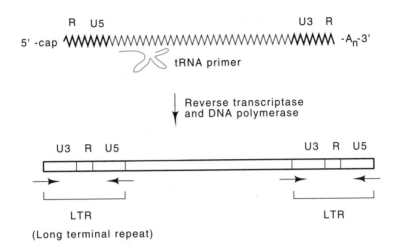

R   U5   U3   R

5'-cap　ΛΛΛΛΛΛΛ/\/\/\/\/\/\/\/\/\/\/\/\/\/\/\/\/\ΛΛΛΛ -A$_n$-3'

tRNA primer

Reverse transcriptase
and DNA polymerase

U3   R   U5   U3   R   U5

LTR   LTR

(Long terminal repeat)

Figure 23.9 The reverse transcription process initiates from tRNA, copies the U3, R, and U5 sequences to generate the LTRs, and ultimately yields double-stranded DNA of the provirus.

hybridized to the virus RNA and packaged within the virus provides the necessary priming hydroxyl. The conversion to double-stranded DNA slightly rearranges and duplicates the ends of the genome to generate the long terminal repeats, LTRs, that are found at each end of the RNA (Fig. 23.9). These are synthesized from the unique U5 and U3 regions that are originally found at each end of the RNA, and the R region that is found at both ends. This replication and synthesis process most likely makes use of the two genetically identical RNA molecules that are encapsidated within the virion and can involve jumping of the reverse transcriptase from one template to the other or from one point on the template to another point. After its duplication, the linear double-stranded DNA migrates to the nucleus where the phage-encoded IN protein, which comes from part of the *pol* gene, integrates at least one copy into the chromosome. If such an integration occurs in a germ line cell, the retroviruses can pass vertically from one generation to the next. If non-germ line cells are infected, the virus is said to spread horizontally.

Three essential viral genes are found in nondefective retroviruses: *gag*, *pol*, and *env*. In Rous sarcoma virus, these three genes encode: 1. four structural proteins found within the virus and a protease; 2. the reverse transcriptase, its associated RNAse H, an enzyme that degrades RNA from the RNA-DNA duplex intermediates, and the integrase; 3. two virus envelope glycoproteins. The virus uses two methods to synthesize this large group of different proteins from just three "genes" and one major promoter.

Recall that eukaryotes rarely use polycistronic messengers. Generally, multiple translation initiations are not used to synthesize different proteins from a single messenger. Retroviruses follow this general rule, but in doing so, they use several different mechanisms to generate at least ten different proteins from three open reading frames on one RNA molecule. From the *gag* gene one polypeptide is synthesized. This is cut

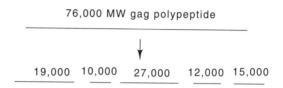

into five different polypeptides. The *pol* gene following the *gag* gene usually is out of frame with *gag*, and sometimes slightly overlapping. Ribosomes translating *gag* occasionally shift their reading frame into that of the *pol* gene and synthesize a combined gag-pol polypeptide. This frameshifting requires an RNA hairpin or pseudoknot shortly downstream from the frameshift point. As a result of the RNA structure and suitable codons at the frameshift site, the frequency of frameshifting approaches 5%. From the long Gag-pol polypeptide, proteases excise segments that constitute the internal proteins and the protease. The protease is encoded by RNA occupying the position between *gag* and pol, or sometimes by RNA constituting the first part of the *pol* gene.

A more elaborate scheme is used for translation of the *env* genes. The retrovirus mRNA is spliced so that the *env* gene occupies the 5′ end of messenger, sometimes using the the initial portion of the *gag* coding region, so that it can be translated. The gag product also is cleaved and yields two polypeptides.

It is the integrated or proviral form of the retroviral DNA that is transcribed to yield capped and polyadenylated mRNA that is ultimately translated into viral proteins. The R region of the virus is defined as the transcription start point of this RNA. The promoter for this RNA must therefore lie in the U3 regions that exist at both ends of the integrated virus. Although the right-hand promoter usually is inactive, in the cases where it is active it is in position to transcribe adjacent cellular DNA. Therefore a retrovirus can act like a portable promoter. Usually this causes no problems, but sometimes the adventitious transcription is harmful.

Acutely transforming retroviruses are capable of inducing tumor formation within several days of injection into animals. These carry a gene in addition to the three discussed above. This additional gene is called an oncogene. Such acutely transforming viruses usually are defective, however, and lack functional *pol* or *env* genes. Therefore they must be grown in association with their nondefective relatives. This is analogous to complementation by lambda of a defective transducing phage. RSV is an exception to the general case, as this virus is nonde-

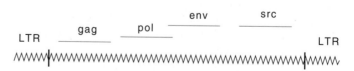

fective and carries an oncogene, called *src*. Translation of the *src* transcript is accomplished after its mRNA is spliced in a reaction that places the AUG codon for this protein near the 5' end of the RNA molecule.

Retroviruses lacking oncogenes are also capable of inducing tumors in susceptible animals, but more slowly than the acutely transforming retroviruses. These slow retroviruses often take several months to generate tumors. Apparently, out of a great number of virus-infected cells, a small number, often only one, contain a retrovirus inserted adjacent to a cellular gene involved with growth regulation. The nearby location of the retrovirus increases expression of the gene, either from readthrough of a viral transcript or from increased expression from the cellular promoter as a result of an enhancer sequence within the LTR of the virus. The virus may also promote rearrangement of the gene so that variant protein is synthesized.

Cellular Counterparts of Retroviral Oncogenes

Are the viral oncogenes of acutely transforming viruses related to any cellular genes? This is a good question for a number of reasons. Since chromosomal rearrangements point to specific genes as being involved in transformation to the cancerous state, genes related to these may be carried on the retroviruses. Also, since cancer can be induced by mutagens, it would seem likely that the mutations responsible could alter cellular genes to resemble their analogs carried on the acutely transforming retroviruses. Finally, a slow retrovirus occasionally becomes acutely transforming by the substitution of part of its genetic material by host material. Most likely the new gene originates from the chromosome of the cells in which the virus has been growing.

The question as to whether the viral oncogenes are related to cellular genes can easily be answered by Southern transfers. Undue fears of dangers from cloning such DNAs have long passed, and viral oncogenes may now be handled, at least in the United States, without unnecessarily elaborate precautions. The viral oncogenes from a number of retroviruses have been cloned from the virus into plasmid or phage vectors. Then the appropriate restriction fragments have been used to probe Southern transfers of DNA extracted from noncancerous animals. Most surprisingly, chickens, mice, and humans all possess a sequence of DNA with high homology to the *src* gene. Clearly this gene has not evolved rapidly in the time since these animals diverged from one another. This implies that the cellular function of this protein is fundamental and closely tied to other cellular functions so that its further evolution is frozen.

The viral form of the *src* gene is called *v-src* and the cellular form is *c-src*. Homology measurements similar to those carried out for Rous sarcoma virus have been done for other acutely transforming retroviruses (Table 23.1). Most of the viral oncogenes in these viruses possess cellular counterparts as well. About 50 different viral and cellular forms of the oncogenes are known. This means that a reasonably small number

Table 23.1 Oncogenes Found on Some Retroviruses

Gene	Virus	Function
src	Rous sarcoma virus	Tyrosine kinase
myc	Avian myelocytomatosis virus	DNA binding
erbA	Avian erythroblastosis virus	Thyroxin receptor
erbB	Avian erythroblastosis virus	EGF receptor
myb	Avian myeloblastosis virus	DNA binding
ras	Kirsten murine sarcoma virus	G protein
jun	Avian sarcoma virus	Transcription factor
fos	Murine sarcoma virus	Transcription factor
sis	Simian sarcoma virus	Plate-derived growth factor

of genes are involved in regulating cell growth and that there is some hope of being able to deduce how the products of these genes function in normal and transformed cells.

The mutated gene that Weinberg found in the human bladder carcinoma cells is carried in a slightly modified form on a rat-derived Harvey murine sarcoma virus. It is called *H-ras*. The *myc* gene which translocates into the heavy chain locus of the immunoglobulin genes in Burkitt's lymphoma, has also been identified on a retrovirus. Thus the same proto-oncogene can be activated through various cellular or viral mechanisms.

Identification of the *src* and *sis* Gene Products

Eukaryotic cells contain thousands of proteins. Most likely, only a few of these are involved in regulation of growth, and these probably are synthesized in small quantities. Therefore it would seem like a difficult task to identify a protein as the product of either *c-src* or *v-src* and virtually impossible to determine the enzymatic activity of such a protein. Nonetheless, Erikson accomplished both for the v-src protein.

The starting point of this work was the fact that animals with sarcomas or lymphomas frequently synthesize antibodies against the retroviral proteins. The v-src protein might also induce antibody synthesis in animals if it were sufficiently different in structure from all other cellular proteins. Therefore rabbits were infected with an avian sarcoma virus that could induce sarcomas even in some other types of animals. The same virus was also used to transform chick embryo fibroblasts *in vitro*. These chicken cells ought to synthesize the same viral proteins as the rabbit. As a result, these proteins would be recognized by the rabbit antibodies. In addition, the chicken embryo fibroblasts ought to contain very few of the other proteins that would be recognized by the rabbit antibodies. Therefore, the *v-src* product ought to be one of the very few proteins recognized by the antibodies.

Extracts from radioactively labeled RSV-transformed chick cells were incubated with the serum from the RSV-infected rabbits. Since the quantities of proteins bound by the antibodies were low, an antigen-antibody lattice was unlikely to form. Another method had to be used for selectively isolating the antigen bound to antibody. Erikson used whole cells of *Staphylococcus aureus* because as mentioned earlier, one of its surface proteins, protein A, specifically binds to IgG. After incubation of rabbit serum with the radioactive chick proteins, the *Staphylococcus* cells were added and the mixture centrifuged to separate antibody-bound radioactive proteins from those not bound to antibody.

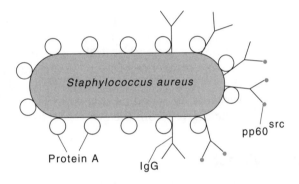

About 20 radioactive proteins from the chick cells were precipitated by the rabbit serum. Among them was one that was present only if the labeled proteins originated from transformed chick cells. Furthermore, this protein was not present if the cells had been infected with a nonacutely transforming retrovirus. In addition, the protein was the same molecular weight as a protein synthesized *in vitro* using the oncogene portion of RSV as a messenger. In this way, the *v-src* gene product was identified as a 60,000 molecular weight protein, which is called p6O^{src}.

An enzymatic activity associated with src was discovered on the basis of an educated guess. Since the protein probably is involved with regulating growth and since phosphorylation of proteins is involved in regulating many activities in eukaryotic cells, the src protein could be a protein kinase. Therefore radioactive ATP and potential kinase substrate proteins were included with the *Staphylococcus*-precipitated IgG-bound src protein. Indeed, a kinase activity was detected. It was shown to be the product of the *src* gene because it was absent from cells grown at high temperature that had been transformed with mutants of RSV that do not transform at high temperature. While most amino acid-linked phosphate in cells is phosphothreonine or phosphoserine, less than 0.03% is phosphotyrosine. Following RSV infection, the levels of phosphotyrosine dramatically increase. These elevated levels fall if the infecting virus possess a temperature-sensitive kinase and the temperature of the cell culture is raised. Together, these pieces of evidence provide solid proof that the kinase is the product of *v-src*.

Another related question is whether the antibodies against the *v-src* gene product could detect the *c-src* gene product. They could, even though the protein is synthesized at a small fraction of the level at which the viral homolog is synthesized. The cellular protein also is a kinase. Similar studies have permitted partial purification of a number of other viral and cellular oncogene products. In many cases the cellular locations of these proteins can also be determined by fractionation of cellular components or with labeled antibodies.

The actual cellular function of the oncogene carried on a simian retrovirus, *v-sis*, has also been identified. This was not done via antibodies and enzyme assays. Instead, the nucleotide sequence of its oncogene identified its product as the platelet-derived growth factor. This had been purified for entirely different reasons from blood plasma. It is released by platelets at the sites of wounds and stimulates growth of fibroblasts, smooth muscle cells, and glial cells. When the amino acid sequence for the growth factor was entered into a data base and homologies were sought, the simian virus oncogene immediately was revealed as encoding a close relative. Almost certainly the retrovirus gene derives from the growth factor gene.

DNA Tumor Viruses

DNA viruses are less closely associated with cancer than are the retroviruses because their infection does not usually yield tumors. Except for papillomaviruses, which cause warts, DNA viruses cannot readily be found in natural tumors. Papovaviruses, including SV40, and the very closely related polyoma virus, as well as adenoviruses and herpes viruses, do have the capability of transforming cells in culture so that they lose their density-dependent growth regulation. These transformed cells can then be injected into animals, where they will cause tumors.

Transformation of mouse cells by SV40 and of hamster cells by polyoma virus involves T proteins that are called T antigens. Normally three are found. They are called small, middle, and large T, for their respective sizes. The messenger for these three different proteins originates from one gene as a result of alternative splicing of the mRNA. The three proteins were identified immunologically in the same way the retroviral oncogene proteins were identified. Antibodies against these proteins are induced in response to the presence of the virus.

The antibodies obtained from animals carrying SV40- and polyoma-induced tumors have been used to show that the T antigen genes are expressed early after infection and that the T antigens bind to proteins in the nucleus, to DNA, and to other cellular structures. The presence of one of the T antigens alone is usually insufficient to transform cells, suggesting that in general the process of transformation requires a number of steps. This conclusion is in accord with the fact that the NIH 3T3 cell foci-inducing DNAs isolated from tumors are not capable of transforming primary cell lines.

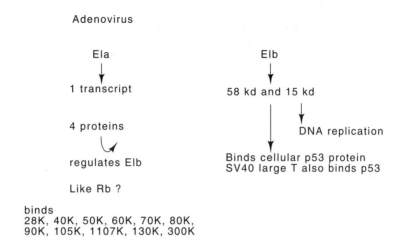

Figure 23.10 Products of adenovirus E1a and E1b bind multiple cellular proteins.

The activity of the T antigens has been a subject of great interest, but the low levels at which they are synthesized *in vivo* has greatly hampered their study. They were therefore perfect targets for genetic engineering. Indeed, by properly fusing the DNA encoding the different proteins to suitable ribosome-binding sequences, these proteins have been hypersynthesized in bacteria, and the proteins can be used for biochemical studies. They also serve as a marker for expression of a promoter in an organism. A promoter and its associated regulatory sequences can be fused to the T antigen gene and inserted into the chromosome of an animal. Expression of the T antigen gene frequently gives rise to tumors that are confined to the tissues in which the gene is expressed.

Adenoviruses and papillomaviruses also possess genes responsible for transformation of cells. Like the SV40 virus, their oncogenes possess no cellular counterparts. Two transcripts are synthesized from the E1a and E1b genes of adenovirus (Fig. 23.10). Through splicing and protein processing, each gives rise to four or five protein products. Some regulate other genes, and two bind to cellular proteins important in transformation. The adenovirus E1a binds to a protein discussed below called Rb for retinoblastoma. The adenovirus E1b protein binds to a cellular protein of 53,000 molecular weight called p53. One of the SV40 T antigens also binds to this same protein, as does a protein encoded by papillomaviruses. Papillomaviruses also encode a protein which binds to Rb. The fact that different tumor viruses synthesize proteins that inactivate Rb and p53 indicates that these two cellular proteins are important for normal cell growth. Indeed, the next section describes experiments leading to the same conclusion.

Recessive Oncogenic Mutations, Tumor Suppressors

Transformation studies like those used to find the DNA change associated with a bladder cancer have shown that only about 10% of tumors can be associated with dominant mutations. Similarly, fusion of transformed tumor cells and normal cells shows that in most cases, the genotype of the normal cells can suppress the oncogene being expressed in the transformed cells. If the fusion cells possess an unstable karyotype and lose chromosomes, then the descendants of these fused cells can again express the transformed phenotype. Thus, oncogenesis can derive from two different pathways. The first is dominant mutations that generate transformed cells independent of the other cellular genes, and the second is a product of two events. One of these is a mutation that would produce uncontrolled cell growth were it not for the presence of another cellular regulator, and the second is the loss of the regulator.

The simplest example of the two-step pathway is a mutation in a growth factor such that it no longer prevents uncontrolled growth. Ordinarily the mutation to inability to repress growth is not revealed because the mutation is recessive and the nonmutated copy of the gene product from the other copy of the chromosome suppresses growth. If, however, the nonmutated copy is lost by recombination or mutated, then the growth inhibition ceases. Retinoblastoma is a form of cancer that develops in young children. Most cases of retinoblastoma are familial, and the pattern of inheritance indicates that the gene involved is not on a sex chromosome. The patterns of retinoblastoma development suggest that in addition to inheriting a defective gene, a second event must occur in the stricken individuals. We now know that this second event frequently is recombination of the defect from one chromosome onto the homologous chromosome in a way that leaves both chromosomes defective for the critical gene. Gene conversion as discussed earlier could do this.

Through good luck and hard work the gene whose defect gives rise to retinoblastoma was cloned. The first step in the process came from the finding that an appreciable fraction of retinoblastoma cell lines possess a gross defect or deletion in part of chromosome 13. This suggests that the gene in question lies on chromosome 13. With this as a starting point it was possible to clone the gene. The original idea was that perhaps a clone could be found containing some of the DNA that is deleted from chromosome 13 in some of the retinoblastoma lines. By chromosome walking, it might then be possible to clone the desired retinoblastoma gene.

A library of about 1,000 clones of segments of chromosome 13 was available. These were then screened against a large panel of retinoblastoma lines to see if any of the clones contained DNA absent from the transformed cell lines. One clone indeed contained DNA absent from two of the retinoblastoma lines. Best of all, it was complementary to an RNA that is synthesized in normal cells, but not in retinoblastoma cells. This suggests that the protein could, in fact, be the gene for the suppressor of retinoblastoma.

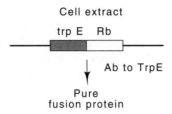

Figure 23.11 The fusion protein TrpE-Rb can be made in *E. coli* and purified for immunization. Some of the resulting antibodies should have Rb specificity.

Further study of the retinoblastoma protein required sensitive assays for its presence. Two approaches were used to make antibodies against the supposed retinoblastoma suppressor protein. Part of the cloned and sequenced gene was fused to the *trp*E gene and a fusion protein was synthesized in *E. coli*, purified by making use of the *trp* tag, and used to immunize animals for antibody synthesis (Fig. 23.11). The second approach was to synthesize chemically peptides corresponding to several regions of the presumed protein and use these to immunize animals.

With antibodies against the protein, the synthesis and properties of the retinoblastoma protein could be studied. It is indeed synthesized in normal cells, but frequently is missing in retinoblastoma cells. The protein is phosphorylated, and has a molecular weight of about 105 kD. This is the molecular weight of one of the cellular proteins to which the adenovirus E1a protein, the papillomavirus E7 protein, and the SV40 T antigen bind. Since the E1a-associated protein is phosphorylated like the Rb protein, and since phosphorylation of proteins is relatively rare, it seemed possible that the retinoblastoma suppressor protein and the E1a-associated protein were one and the same. Indeed, they were. Both antibody specificity and proteolysis patterns confirm the identity. Thus it appears that the normal role of the retinoblastoma suppressor gene is to counteract the positive activity of some other cell growth inducing pathway. Hence one of the steps in transformation by adenovirus, papillomavirus, and SV40 is to inactivate suppressor proteins rather than to synthesize a dominant growth-inducing gene in the cells.

In as many as 50% of human tumors, the p53 protein is found to have been altered. Thus, this protein appears to play a key role in regulating cell growth. What is the nature of the regulation expressed by the p53 protein? Is it required at all times for proper cell growth or regulation of differentiation? A conceptually simple way to answer these questions is to attempt to make a homozygous mouse deficient in p53 protein. Such mice turn out to develop normally, but they are highly likely to develop cancers before they are as old as six months. The tumors are found in many different tissues. Further study has shown that p53 prevents the onset of DNA replication in a cell possessing damaged DNA or a broken chromosome. By blocking DNA synthesis, and allowing time for DNA repair, the spontaneous mutation frequency in cells is greatly reduced.

The *ras-fos-jun* Pathway

The *ras* protein is one well-studied oncogene, partially because it is encoded by the oncogene Weinberg found in the transformation assays of the NIH 3T3 cells. More extensive studies have shown that an activated *ras* product is found in 10 to 20% of all human cancers. Humans possess three different *ras* products, *N-ras*, which is encoded on chromosome 1, *H-ras* from chromosome 11, and *K-ras* from chromosome 12. Even *Saccharomyces cerevisiae* possesses two *ras* products.

In order that these ras proteins be able to respond to extracellular and transmembrane receptors, they are held on the inner surface of the plasma membrane by farnesyl. This is a fatty acid, membrane-loving, intermediate of the cholesterol biosynthetic pathway. The farnesyl is attached to the carboxy terminal cysteine of ras. The same principle of attaching a membrane-soluble molecule to a protein is used to attach the outer membrane of *Escherichia coli* to the peptidoglycan layer. In yeast, the same enzymes that transfer the fatty acid to the cysteine of ras for membrane attachment also modify *a* mating-type factor. Although ras proteins bind nucleotides and have a GTPase activity and are therefore G proteins, they function as monomers, whereas many other G proteins function as heterotrimers.

The G proteins are activated by the binding of a GTP nucleotide. When they hydrolyze this to GDP, the activation ceases. Not surprisingly, then, oncogenic *ras* mutants do not hydrolyze the GTP normally. Some have mutations in the GTP-binding site, and others have lost the ability to bind to an accessory protein, GAP, for GTPase activating protein (Fig. 23.12). It is the GAP protein that interacts with the growth factor receptor. When the receptor is occupied, the GAP protein is phosphorylated. It therefore ceases to stimulate hydrolysis of GTP bound to ras. This leaves ras in an activated state. As a result, it activates a phosphorylation pathway that ultimately phosphorylates and activates the c-jun protein.

Figure 23.12 Pathway for activation of ras protein.

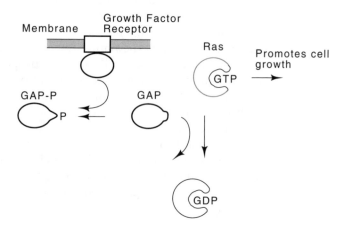

Phosphorylated c-jun then forms a heterodimer with c-fos protein. This heterodimer is also known as AP-1 and it activates transcription from a wide collection of promoters.

The *fos* and *jun* oncogenes were identified on retroviruses and the cellular analogs were then found using their homology to the viral forms. These proteins are typical leucine zipper transcriptional activators and they bind to the same DNA sequence as the yeast general amino acid regulator protein GCN4. The structure of the fos and jun leucine zippers favors a heterodimer rather than homodimer formation due to the presence of opposite charges on the two helices. The theme of heterodimer formation regulating cell growth is common and both leucine zipper proteins and helix-loop-helix proteins participate in this form of regulation.

Directions for Future Research in Molecular Biology

It is clear that knowledge of the eukaryotic cell is far from our understanding of lambda phage or *E. coli*, but our understanding of even these simpler objects is far from complete. Therefore we can perceive three general areas for future research in genetics and molecular biology. In the first, simpler systems in bacteria, yeast, and the fruit fly will be studied in greater depth so that we may understand as much as possible about the physics, chemistry, and biochemistry common to all systems. An ultimate goal of this direction of work is the ability to predict the properties of a protein given its structure and to modify known proteins so that they will fold fold and bind to any desired sequence on a nucleic acid to regulate a cell activity in any way we desire, or possess any reasonable structural or enzymatic property. In the more immediate future, we should learn how to modify known proteins, RNA molecules, or DNA molecules to carry out functions different from, but related to their natural ones.

A second general area of work will be understanding cellular processes. This includes study of the following: the mechanisms used in nature for regulating gene and enzyme activity; the complicated interactions between metabolic pathways; and the mechanisms of many functions such as movement, active transport, RNA splicing, cell division, and receptor function. A third area of future work will be that of understanding organisms as a whole. This will include how cells and tissues develop and differentiate and how they signal to one another and then respond.

Problems

23.1. Recalling that either of the X chromosomes in each cell of a woman is inactivated, how would you determine whether a carcinoma derived from transformation of a single cell in contrast to transformation of multiple cells?

23.2. Rous sarcoma virus contains two identical single-stranded RNA molecules, each of which contains all the genetic information of the virus. Devise an experiment to test the conjecture that recombination between the two genomes, not necessarily in their RNA form, frequently generates an intact genome from two damaged ones.

23.3. What control experiments are necessary to ensure reliability in the series of experiments in which serum from an animal with a sarcoma was used to precipitate proteins from cultured transformed cells and thereby identify the *src* gene product?

23.4. In the experiments examining the kinase activity of *v-src*, what experiments would you do next if you observed that protein phosphorylation was nearly instantaneous and that the rate was independent of the volume of the reaction?

23.5. In experiments to determine which portion of the mutated gene conferred tumorgenicity, restriction fragments were exchanged between the oncogenic and non-oncogenic forms. This sort of experiment is fraught with danger, however, for false positives could arise as a result of incomplete restriction enzyme digestion, minor DNA contamination, and so on. What controls can be put into this experiment to warn of such pitfalls?

23.6. Why is injection of a retrovirus or DNA virus into a newborn or young animal more likely to generate a tumor than injection into a mature animal?

23.7. Suppose you had a collection of tumors whose DNA was capable of transforming cells in culture. What is the most efficient way to determine, not with absolute certainty but with a high degree of confidence, whether the oncogenes involved are identical?

23.8. Remarkably, monoclonal antibody against the oncogene product responsible for the EJ bladder carcinoma was available for comparing the translation efficiency of the cellular prototype and the oncogenic version. By searching through the literature, find how this antibody was obtained.

23.9. In transforming susceptible cells in a search for the DNA sequences involved, one might worry that several disconnected DNA segments may be required. How could this possibility be excluded by varying the transformation conditions?

23.10. Exposure to UV light can induce cancers that can ultimately be traced to retroviruses. Propose a hypothesis to explain this finding and an experiment to test your idea.

23.11. Describe the steps that would be required to generate a homozygous p53 deficient mouse.

23.12. Why is it more sensible for *ras* proteins to use the hydrolysis of GTP to GDP rather than dissociation of GTP from the protein as the signal to end their activation?

References 663

References

Recommended Readings

Carcinogens are Mutagens: A Simple Test System Combining Liver Homogenates for Activation and Bacteria for Detection, B. Ames, W. Durston, E. Yamasaki, F. Lee, Proc. Nat. Acad. Sci. USA *70*, 2281-2285 (1973).

Dietary Carcinogens and Anticarcinogens, B. Ames, Science *221*, 1256-1264 (1983).

Association Between an Oncogene and an Anti-Oncogene: The Adenovirus E1A Proteins Bind to the Retinoblastoma Gene Product, P. Whyte, K. Buchkovich, J. Horowitz, S. Friend, M. Raybuck, R. Weinberg, E. Harlow, Nature *334*, 124-129 (1988).

Mice Deficient for p53 are Developmentally Normal but Susceptible to Spontaneous Tumors, L. Donehower, M. Harvey, B. Slagle, M. McArthur, C. Montgomery, J. Butel, A. Bradley, Nature *356*, 215-221 (1992).

Development

Post-embryonic Cell Lineages of the Nematode *Caenorhabditis elegans*, J. Sulston, H. Horvitz, Dev. Biol. *56*, 110-156 (1977).

The Embryonic Cell Lineage of the Nematode *Caenorhabditis elegans*, J. Sulston, E. Schierenberg, J. White, Dev. Biol. *100*, 64-119 (1983).

Size, Location and Polarity of T-DNA-encoded Transcripts in Nopaline Crown Gall Tumors; Common Transcripts in Octopine and Nopaline Tumors, L. Willmitzer, P. Dhaese, P. Schreier, W. Schmalenbach, M. Van Montagu, J. Schell, Cell *32*, 1045-1056 (1983).

A Tissue-specific Transcription Enhancer Element is Located in the Major Intron of a Rearranged Immunoglobulin Heavy Chain Gene, S. Gillies, S. Morrison, V. Oi, S. Tonegawa, Cell *33*, 717-728 (1983).

Heart and Bone Tumors in Transgenic Mice, R. Behringer, J. Peschon, A. Messing, C. Gartside, S. Hauschka, R. Palmiter, R. Brinster, Proc. Nat. Acad. Sci. USA *85*, 2648-2652 (1988).

DNA, Chromosomes, and DNA Changes Associated with Oncogenesis

Banding in Human Chromosomes Treated with Trypsin, H. Wang, S. Fedoroff, Nature New Biology *235*, 52-54 (1971).

Nonrandom Chromosome Changes Involving the Ig Gene-carrying Chromosomes 12 and 6 in the Pristane-induced Mouse Plasmacytomas, S. Ohno, M. Babonits, F. Wiener, J. Spira, G. Klein, M. Potter, Cell *18*, 1001-1007 (1979).

Translocation of the c-myc Gene into the Immunoglobulin Heavy Chain Locus in Human Burkitt Lymphoma and Murine Plasmacytoma Cells, R. Taub, I. Kirsch, C. Morton, G. Lenoir, D. Swan, S. Tronick, S. Aaronson, P. Leder, Proc. Natl. Acad. Sci. USA *79*, 7838-7841 (1982).

A Point Mutation is Responsible for the Acquisition of Transforming Properties by the T24 Human Bladder Carcinoma Oncogene, E. Reddy, R. Reynolds, E. Santos, M. Barbacid, Nature *300*, 149-152 (1983).

Mechanism of Activation of a Human Oncogene, C. Tabin, S. Bradley, C. Bargmann, R. Weinberg, A. Papageorge, E. Scolnick, R. Dhar, D. Lowy, E. Chang, Nature *300*, 143-149 (1983).

Activation of a Translocated Human c-myc Gene by an Enhancer in the Immunoglobulin Heavy-chain Locus, A. Hayday, S. Gillies, H. Saito, C. Wood, K. Wiman, W. Hayward, S. Tonegawa, Nature *307*, 334-340 (1983).

Cellular *myc* Oncogene is Altered by Chromosome Translocation to an Immunoglobulin Locus in Murine Plasmacytomas and is Rearranged Similarly in Human Burkitt Lymphomas, J. Adams, S. Gerondakis, E. Webb, L. Corcoran, S. Cory, Proc. Nat. Acad. Sci. USA *80*, 1982-1986 (1983).

A Major Inducer of Anticarcinogenic Protective Enzymes from Broccoli: Isolation and Elucidation of Structure, Y. Zhang, P. Talalay, C. Cho, G. Posner, Proc. Natl. Acad. Sci. USA *89*, 2399-2403 (1992).

Retroviruses

Rous Sarcoma Virus: A Function Required for the Maintenance of the Transformed State, G. Martin, Nature *227*, 1021-1023 (1970).

Detection and Isolation of Type C Retrovirus Particles from Fresh and Cultured Lymphocytes of a Patient with Cutaneous T-cell Lymphoma, B. Poiesz, F. Ruscetti, A. Gazdar, P. Bunn, J. Minna, R. Gallo, Proc. Nat. Acad. Sci. USA 77, 7415-7419 (1980).

Evidence That Phosphorylation of Tyrosine is Essential for Cellular Transformation by Rous Sarcoma Virus, B. Sefton, T. Hunter, K. Beemon, W. Eckart, Cell *20*, 807-816 (1980).

Identification of a Functional Promoter in the Long Terminal Repeat of Rous Sarcoma Virus, Y. Yamamoto, B. de Crombrugghe, I. Pastan, Cell *22*, 787-797 (1980).

Structure and Functions of the Kirsten Murine Sarcoma Virus Genome: Molecular Cloning of Biologically Active Kirsten Murine Sarcoma Virus DNA, N. Tsuchida, S. Uesugi, J. Vir. *38*, 720-727 (1981).

Nucleotide Sequence of Rous Sarcoma Virus, D. Schwartz, R. Tizard, W. Gilbert, Cell *32*, 853-869 (1983).

Proviruses are Adjacent to c-myc in Some Murine Leukemia Virus-induced Lymphomas, D. Steffen, Proc. Nat. Acad. Sci. USA *81*, 2097-2101 (1984).

Trans-acting Transcriptional Activation of the Long Terminal Repeat of Human T Lymphotropic Viruses in Infected Cells, J. Sodroski, C. Rosen, W. Haseltine, Science *225*, 381-385 (1984).

Regulatory Pathways Governing HIV-1 Replication, R. Cullen, W. Greene, Cell *58*, 423-426 (1989).

Human Immunodeficiency Virus Integration Protein Expressed in *Escherichia coli* Possesses Selective DNA Cleaving Activity, P. Sherman, J. Fyfe, Proc. Natl. Acad. Sci. USA *87*, 5119-5123 (1990).

Oncogenes, their Activities, and their Cellular Counterparts

Identification of a Polypeptide Encoded by the Avian Sarcoma Virus *src* Gene, A. Purchio, E. Erikson, J. Brugge, R. Erikson, Proc. Nat. Acad. Sci. USA *75*, 1567-1571 (1975).

Protein Kinase Activity Associated with the Avian Sarcoma Virus *src* Gene Product, M. Collett, R. Erikson, Proc. Nat. Acad. Sci. USA *75*, 2021-2024 (1978).

Nucleotide Sequences Related to the Transforming Gene of Avian Sarcoma Virus are Present in DNA of Uninfected Vertebrates, D. Spector, H. Varmus, J. Bishop, Proc. Nat. Acad. Sci. USA 75, 4102-4106 (1978).

Uninfected Vertebrate Cells Contain a Protein that is Closely Related to the Product of the Avian Sarcoma Virus Transforming Gene (src), H. Opperman, A. Levinson, H. Varmus, L. Levintow, J. Bishop, Proc. Nat. Acad. Sci. USA 76, 1804-1808 (1979).

The Protein Encoded by the Transforming Gene of Avian Sarcoma Virus (pp60^{src}) and a Homologous Protein in Normal Cells are Associated with the Plasma Membrane, S. Courtneidge, A. Levinson, J. Bishop, Proc. Nat. Acad. Sci. USA 77, 3783-3787 (1980).

Activation of a Cellular onc Gene by Promoter Insertion in ALV-induced Lymphoid Leukosis, W. Hayward, B. Neel, S. Astrin, Nature 290, 475-480 (1981).

Unique Transforming Gene in Carcinogen-transformed Mouse Cells, B. Shilo, R. Weinberg, Nature 289, 607-609 (1981).

Viral src Gene Products are Related to the Catalytic Chain of Mammalian c-AMP-dependent Protein Kinase, W. Barker, M. Dayhoff, Proc. Nat. Acad. Sci. USA 79, 2836-2839 (1982).

Isolation and Preliminary Characterization of a Human Transforming Gene from T24 Bladder Carcinoma Cells, M. Goldfarb, K. Shimizu, M. Perucho, M. Wigler, Nature 296, 404-409 (1982).

Monoclonal Antibodies to the p21 Products of the Transforming Gene of Harvey Murine Sarcoma Virus and of the Cellular ras Gene Family, M. Furth, L. Davis, B. Fleurdelys, E. Scolnick. J. Vir. 43, 294-304 (1982).

Platelet-derived Growth Factor is Structurally Related to the Putative Transforming Protein p28^{sis} of Simian Sarcoma Virus, M. Waterfield, G. Scrace, N. Whittle, P. Stroobant, A. Johnson, Å. Wasteson, B. Westermark, C. Heldin, J. Huang, T. Deuel, Nature 304, 35-39 (1983).

Simian Sarcoma Virus onc Gene, v-sis, is Derived from the Gene (or Genes) Encoding a Platelet-derived Growth Factor, R. Doolittle, M. Hunkapiller, L. Hood, S. Devare, K. Robbins, S. Aaronson, H. Antoniades, Science 221, 275-277 (1983).

Structure and Sequence of the Cellular Gene Homologous to the RSV src Gene and the Mechanism for Generating the Transforming Virus, T. Takeya, H. Hanafusa, Cell 32, 881-890 (1983).

The Product of the Avian Erythroblastosis Virus erbB Locus is a Glycoprotein, M. Privalsky, L. Sealy, J. Bishop, J. McGrath, A. Levinson, Cell 32, 1257-1267 (1983).

The Retinoblastoma Susceptibility Gene Encodes a Nuclear Phosphoprotein Associated with DNA Binding Activity, W. Lee, J. Shew, F. Hong, T. Sery, L. Donoso, L. Young, R. Bookstein, E. Lee, Nature 329, 642-645 (1987).

Protein Encoded by v-erbA Functions as a Thyroid-hormone Receptor Antagonist, K. Damm, C. Thompson, R. Evans, Nature 339, 593-597 (1989).

The Role of the Leucine Zipper in the fos-jun Interaction, T. Kouzarides, E. Ziff, Nature 336, 646-651 (1989).

Genetic and Pharmacological Suppression of Oncogenic Mutations in RAS Genes of Yeast and Humans, W. Schafer, R. Kim, R. Sterne, J. Thorner, S. Kim., J. Rine, Science 245, 379-385 (1989).

Direct Activation of the Serine/Thereonine Kinase Activity of Raf-1 through Tyrosine Phosphorylation by the PDF β-Receptor, D. Morrison, D. Kaplan, J. Escobedo, U. Rapp, T. Roberts, L. Williams, Cell *58*, 649-657 (1989).

Retinoblastoma in Transgenic Mice, J. Windle, D. Albert, J. O'Brien, D. Marcus, D. Djsteche, R. Bernards, P. Mellon, Nature *343*, 665-669 (1990).

The E6 Oncoprotein Encoded by Human Papillomavirus Types 16 & 18 Promotes the Degradation of p53, M. Scheffner, B. Werness, J. Huilbregtse, A. Levine, P. Howley, Cell *63*, 1129-1136 (1990).

Nonfunctional Mutants of the Retinoblastoma Protein are Characterized by Defects in Phosphorylation, Viral Oncoprotein Association and Nuclear Tethering, D. Templeton, S. Park, L. Canier, R. Weinberg, Proc. Natl. Acad. Sci. USA *88*, 3033-3037 (1991).

The Retinoblastoma Protein Copurifies with E2F-1, an E1A-Regulated Inhibitor of the Transcription Factor E2F, S. Bagchi, R. Weinmann, P. Raychardura, Cell *65*, 1063-1072 (1991).

Repression of the Interleukin 6 Gene Promoter by p53 and the Retinoblastoma Susceptibility Gene Product, U. Santhanom, A. Ray, P. Sehgal. Proc. Natl. Acad. Sci. USA *88*, 7605-7609 (1991).

Retinoblastoma Gene Product Activates Expression of the Human TGF-β2 Gene Through Transcription Factor ATF-2, S. Kim, S. Wagner, F. Liu, M. O'Reilly, P. Robbins, M. Green, Nature *358*, 331-334 (1992).

Mice Eficiient for Rb are Nonviable and Show Defects in Neurogenesis and Haematopoesis, E. Lee, C. Chang, Y. Wang, C. Lai, K. Herrup, W. Lee, A. Bradley, Nature *359*, 288-294 (1992).

Effects on an Rb Mutation in the Mouse, T. Jacks, A. Fazeli, E. Schmitt, R. Bronson, M. Goodell, R. Weinberg, Nature *359*, 295-300 (1992).

Interactions of Myogenic Factors and Retinoblastoma Protein Mediates Muscle Cell Commitment and Differentiation, W. Gu, J. Schneider, G. Condorelli, S. Kaushal, V. Mahdavi. B. Nadal-Ginard, Cell *72*, 309-324 (1993).

DNA Viruses

Tumor DNA Structure in Plant Cells Transformed by *A. tumefaciens*, P. Zambryski, M. Holsters, K. Kruger, A. Depicker, J. Schell, M. Van Montagu, H. Goodman, Science *209*, 1385-1391 (1980).

Adenovirus E1b-58kd Tumor Antigen and SV40 Large Tumor Antigen Are Physically Associated with the Same 54kd Cellular Protein in Transformed Cells, P. Sarnow, Y. Ho, J. Williams, A. Levine, Cell *28*, 387-394 (1982).

Regeneration of Intact Tobacco Plants Containing Full Length Copies of Genetically Engineered T-DNA, and Transmission of T-DNA to R1 Progeny, K. Barton, A. Binns, A. Matzke, M. Chilton, Cell *32*, 1033-1043 (1983).

The Human Papilloma Virus 16 E7 Oncoprotein Is Able to Bind to the Retinoblastoma Gene Product, N. Dyson, P. Howley, K. Münger, E. Harlow, Science *243*, 934-937 (1989).

Molecular Switch for Signal Transduction: Structural Differences between Active and Inactive Forms of Protooncogenic *ras* Proteins, M. Milburn, L. Tong, A. deVos, A. Brünger, Z. Yamaizumi, S. Nishimura, S. Kim, Science *247*, 939-945 (1990).

Hints and Solutions to Odd-Numbered Problems

Chapter 1

1.1. The periplasmic space must possess an osmotic pressure nearly equal to that found within the cell on the other side of the inner membrane. As this osmotic balancer must not escape through pores in the outer membrane, it must be large. It must also be flexible so that its constituents each contribute substantially to the osmotic pressure.

1.3. Solve $e^n = 10^7 e^{n/0.95}$ for n.

1.5. Cells grow as $e^{\mu t}$, and total protein increases as $P_0 e^{\mu t}$. Therefore, beginning at $t = 0$, the rate of the enzyme's synthesis is $\dfrac{dE}{dt} = 0.01\dfrac{dP}{dt} = 0.01\mu P_0 e^{\mu t}$. Integrating between 0 and t gives $E(t) = 0.01 P_0(e^{\mu t} - 1)$, and dividing this by total protein to obtain the fraction which is the enzyme yields $0.01 P_0(1 - e^{-\mu t})$.

1.7. First, express everything in moles or grams. We will use moles. 10^{-13} gm $\approx 10^{-15}$ moles, $10^{-15}/30$ min $= 3 \times 10^{-17}$ moles/min. 10^{-3} molar in 10^{-12} cm^3 (10^{-15} liter) is 10^{-18} moles. 3×10^{-17} moles/min $\times$ $t = 10^{-18}$ moles, $t = 0.3$ min or about 20 sec.

1.9. If each protein molecule were located on a corner point of a cubic lattice with a distance of x cm between the centers of adjacent molecules, there would be $1/x$ molecules along a 1 cm distance, and therefore $(1/x)^3$ molecules per cm^3. One molecule has a mass of 30,000/Avagadro's number $\approx 5 \times 10^{-20}$ gm. Therefore, to make a concentration of 0.2 gm/cm^3, set $(1/x)^3 \times 5 \times 10^{-20} = 0.2$. Solving for x yields $x^3 = 2.5 \times 10^{-19}$, or $x \approx 6 \times 10^{-7}$, giving $x = 60$ Å.

1.11. Place cells or vesicles in a high concentration of KCl for a long time so that the ion concentration inside equilibrates with the outside. Rapidly dilute into low salt buffer and add valinomycin. The potassium will exit the vesicles, leaving them temporarily negatively charged.

1.13. Because the time scale of interest in this problem is short in comparison to the doubling time of the cells, we can take the flow rates and pool size to be constant over the time of the problem. Let the internal pool size be P moles and the flow rate of amino acids from the pool into protein be F moles/min. When the radioactive amino acids are added, the pool does not change in size. Let $C(t)$ be the total radioactive counts in the pool at time t, and A cts/min be the specific activity of the externally added radioactive amino acid. Clearly $C(0) = 0$ and $C(\infty) = AP$. Flow of cts into the pool in a time interval dt at time t is $AFdt$, and flow out is $\frac{C(t)}{P}Fdt$. The difference, $dC(t) = AFdt - \frac{C(t)}{P}Fdt$. Separating the appropriate terms, integrating, and rearranging ultimately yields $C(t) = AP(1 - e^{-Ft/P})$. The pool specific activity is $C(t)/P$. The counts of radioactivity in protein equals the integral of $\frac{C(t)}{P} \times Fdt = FA(t + \frac{P}{F}e^{-Ft/P} - \frac{P}{F})$. To find the small t behavior, expand the exponential in powers of t. The first nonvanishing term is F^2At^2/P, and thus radioactivity in protein first increases as t^2. For later times, the exponential decays away, and cts in protein $= FA(t - \frac{P}{F})$. This is a linear increase with a delay of $\frac{F}{P}$. Hence, knowing the flow rate and delay, P is obtained.

Chapter 2

2.1. Because the bonds from the phosphate to the four oxygens are tetrahedral, the two ways of drawing the situation are equivalent.

2.3. 3×10^9 base pairs $\times 3.4$ Å per base pair.

2.5. Consider a two-stranded rope. If a stick is inserted between the strands and pulled forward, it tends to overwind the rope in front and underwind the rope behind the original insertion point.

2.7. A nick in the DNA could act as a flexible point, and would have an effect similar to a bend at the same position. If nicked near the middle of the DNA length, the DNA would migrate more slowly, hence the DNA extracted from the trailing part of the band, when run on a denaturing gel, contains nicks in the central region of either strand. DNA extracted from the leading part of the band tends to be nicked near either end of either strand.

2.9. Consider a circular DNA molecule containing -200 superhelical turns. These knot up the DNA, and have some difficulty in forming. With respect to supercoiling, the molecule resists having these turns and tries to have, say -190. To keep the linking number constant, a shift from -200 superhelical turns to -190 means a decrease in the linking number, $L = T + W$, 4800 = 5000 + (-200) and a shift to -190 superhelical turns means 4800 = 4990 + (-190). The greater the failure to assume the desired number of superhelical turns, the greater the unwinding.

2.11. a and d.

2.13. Building the structure without twists means $L = W$. Pulling the ends taut removes all writhe from the structure and since L remains invariant in the operation, $L = T$ afterwards. Thus W before $= T$ after.

2.15. 12.

2.17. Two bands, +1 and -1, run identically. To separate the three, perform the electrophoresis in the presence of a low concentration of ethidium bromide or preferably, some weaker binding intercalating agent. Intercalation reduces twist, and hence adds to writhe. In ethidium bromide the three might go to +10, +11, and +12 superhelical turns, which would separate from one another.

2.19. Electron microscopy if the molecules are large, sedimentation where the catenated molecules will sediment more rapidly than the decatenated molecules, or electrophoresis in which catenated molecules will migrate more slowly

than decatenated molecules. If we are assaying the decatenation of two inter-
locked rings to two independent rings, we would see a change in the electrophoresis rate of the product, and neither of the molecules before or after the
decatenation reaction would be sensitive to exonuclease.

Chapter 3

3.1 The numbers of molecules per cell will surely fluctuate. The relative sizes of
the fluctuations will be smaller the larger the total number of molecules
involved.

3.3.

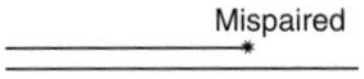

Mispaired

3.5. Have tritiated thymidine present for a minute or two, and then increase the
specific activity four-fold. Bidirectional replication will yield autoradiograph
tracks with heavier track density at both ends.

3.7.

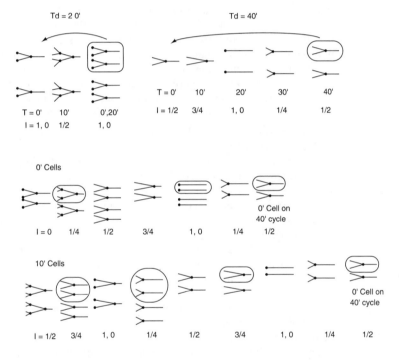

3.9. Average from all the time points is 97.6 nucleotides/sec.

3.11. No.

3.13.

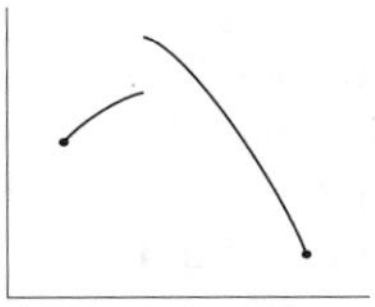

3.15. RNAseH removes RNA from RNA-DNA duplexes. These probably result from
abnormal transcription events. If they are not removed, they could serve as
primers for DNA synthesis, thereby obviating the need for the normal replication origin.

Chapter 4
4.1. If the combined molecular weight of β and β' is 3×10^5, one molecule weighs 5×10^{-19} gm. If total cellular protein is 10^{-13} gm, the amount of β and β' is 5×10^{-16} gm. Thus the cells contain $10^{-16}/10^{-19} = 1000$ molecules.

4.3. The bases could be contacted by polymerase in the closed complex or in some other state between the closed and open complex, or the bases could be required to generate a necessary structure in the DNA.

4.5. Untwisting the DNA by moving a bubble forward without rotation tends to overwind the DNA in front. Overwinding is equivalent to positive supercoils. The positive supercoils thus introduced neutralize the negative supercoils and thus reduce the amount of negative supercoiling.

4.7. A penny is about 1 cm^2. Assume each electron deriving from a decay travels straight through the film and activates at least one silver halide to be reduced to silver upon development. If multiple crystals are formed, assume they will be in a line with the first. This area of blackness will be the cross-sectional area of the crystals in the emulsion. Depending on the film, these are about 10^{-3} cm across, for an area of 10^{-6} cm^2. For a very crude guess ignoring statistics, one might estimate about 10^6 decays would be necessary. In reality, a significant number of the areas of blackness from the decays will overlap. An OD of 1 means 0.1 of the light penetrates or 0.1 of the area has not been hit by an electron. Therefore, consider an area of 10^{-6} cm^2. If the overall OD is to be 1.0, then the probability that our small area has not been hit has to be 0.1. From the Poisson distribution (see a calculus or statistics book) if the average number of hits is m, the probability that a site will not be hit is e^{-m}. Thus $0.1 = e^{-m}$ and $m = 2.3$, giving a total of 2.3×10^6 decays.

4.9. If there were only one functional target on the initiation complex, doubling the number of enhancer proteins could do no more than double the response. The greater than linear response means that there is more than one target for the enchancer proteins so that if one contacts the complex, the other can also bond and stabilize the complex. In turn this requires that the event of the formation of the enhancer-initiator complex stimulate more than one round of transcription, that is, that enhancer complexes can be stable entities and that once formed, they can activate transcription repeatedly.

4.11. Three sites are involved here, two RNA polymerase binding sites, 1 and 2, and the A protein binding site. Polymerase bound to 1 blocks polymerase binding to 2. Bound A protein blocks polymerase binding to site 1, but not 2. If Site 1 binds polymerase rapidly but only very slowly dissociates or initiates, then adding polymerase before A gives a slowly increasing rate of initiations, but adding A before polymerase blocks the inhibitory site 1, and upon polymerase addition, it binds to site 2 and initiations almost immediately begin at the full rate.

4.13. If the repressor binding site is downstream from the promoter, then almost surely the usual situation will have a polymerase stuck behind the repressor. When the repressor dissociates, then at least one polymerase will be released to transcribe the operon. Alternatively, if the repressor binds in the promoter, only after the repressor has dissociated is the promoter accessible for the binding of a polymerase. Many repressor dissociation events will be followed by the rebinding of the repressor or binding of another repressor molecule without the binding of a single polymerase molecule. Hence the repressed level in this case will be lower.

Chapter 5
5.1 Assuming the hybridization and digestion steps were complete, the two bands means two different RNA sizes contain the sequence of the hot probe. They

could be the result of incomplete splicing, or alternative use of two 5' or two 3' splice sites.

5.3

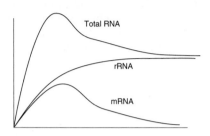

5.5. If the cap site were created by splicing, then the β phosphate could not be radioactive in the final product.

5.7. Such introns will incorporate radioactive guanosine *in vitro* into the RNA. Incubate RNA extracted from the cells in buffer with radioactive guanosine, run a gel and look for radioactive bands.

5.9. The internal acceptor nucleotide forms the bond before cleavage. Consider a nuclease that cleaves the 3'-5' phosphodiester bond. Likely this will be inhibited by the presence of a 2'-5' bond. Thus the nucleotide in question will yield a higher molecular weight and more highly charged product than normal. Further, standard RNA sequencing with digestion and characterization of the products will reveal the anomalous nucleotide. Finally, such a branched site would block reverse transcription past this point.

5.11. There is only one place left where the molecules could store the energy in a phosphodiester bond, another phosphodiester bond. The only such bond possibility subject to the constraints imposed by the problem is a 2'-3' cyclic phosphodiester bond.

Chapter 6

6.1. Donors: R, D, Q, K, S, T, Y, W, H. Acceptors: S, T, Y, D, Q, M, H, D, E.

6.3. If they denature from a structural change and not a chemical change like oxidation, then the active form was not the lowest energy form, but merely one that formed first and later disappears as the protein slowly acquires the more stable but inactive conformation.

6.5. The amino group is tied up in the proline ring and is not free to form the normal H bonds required in an alpha helix.

6.7. Nothing, as the change to D form changes the stereochemistry of the side group which is not involved in the basic hydrogen bonding pattern of an alpha helix.

6.9. 30,000 MW is approximately 5×10^{-20} gm per molecule, or a volume of approximately 4×10^{-20} cm^3. If a cube, its edges are roughly 35 Å. From the structure of an alpha helix, the H bond parallel to the axis is in a group of three bonds, and since most chemical bonds are about 2 Å, the rise per helical turn on an alpha helix is about 6 Å. Thus it takes $^{35}\!/_6 \approx 6$ turns to go from one edge of the protein to the other. This is $6 \times 3.6 = 22$ amino acids, and $300/22 = 14$. Therefore about 14 direction reversals could be expected.

6.11. Consider the helical wheel representation. The leucines and the valines on one helix all line up behind one another. Since both valine and leucine are hydrophobic, both must be involved in the hydrophobic interactions that dimerize the molecule. If the monomers are antiparallel, the leucine-leucine interactions should be equivalent to the valine-valine inter-

Antiparallel

Parallel

actions. This is impossible, but if the monomers are parallel, the valine-leucine interactions must be equivalent to leucine-valine interactions, which is easily satisfied.

6.13. Reflections or inversions convert L-amino acids to D-amino acids, but all amino acids in normal proteins are L-form only.

6.15. Having a smaller side chain, glycine introduces more degrees of freedom that the various binding interactions must overcome.

6.17. (a) Ala, Leu, Met (b) Pro, Gly (c) KAALAMAMMMLLK.

Chapter 7

7.1. As the mass of tRNA per cell is somewhat less than of rRNA, there are about 20 tRNA molecules per ribosome. At a protein elongation rate of 20 amino acids/sec, it would take about one second to deplete all the charged tRNA if most of the tRNA in a cell were charged, which it is.

7.3. The finding, which does not contradict the Shine-Dalgarno hypothesis, means that the fmet-tRNA was blocked from binding not because the trinucleotide homologous to its AUG was covered, but because some other part of the fmet-tRNA encountered a steric clash.

7.5. Gly-Ala, Ser-Thr, Phe-Try.

7.7. The trp codon is UGG, the anticodon is CCA. If the CCA anticodon is mutated to CUG to suppress UAG, it will fail to read trp codons and the cell will die.

7.9. Lys, Gln, Glu, Leu, Ser, Tyr, Trp, Arg, Gly Phe are possible. Lys, Gln, Leu, Ser, Tyr, Trp, and Gly have been found. Most likely the tRNA required is a single species and the former amino acid can no longer be properly decoded.

7.11. If this were the case, we would expect in the electron microscope to observe many of the ribosomes in a polysome to be clustered at the polymerase.

7.13. If you must synthesize twice the ribosomes in half the time, their synthesis rate must be increased by four. Symbolically,

$$R(t) = R(0)e^{\mu t}, \text{ and } \frac{dr}{dt} = \mu R(0)e^{\mu t}. \text{ Since } R(0) \text{ is proportional to growth rate } \mu,$$

$\frac{dr}{dt}$ is proportional to μ^2.

7.15. The synthesis of some of the proteins that help protein folding or the refolding of denatured proteins could be induced by the presence of excess denatured protein. Methanol would denature some proteins and could therefore induce proteins in the heat shock class.

7.17. Ribosomal proteins regulate their own synthesis. Oversynthesis of one protein would likely turn down the synthesis of others in the same operon or which are regulated by the same protein. If these were not also oversynthesized, the cell would be deficient in these proteins and could not synthesize ribosomes to the proper level.

Chapter 8

8.1. Mitotic recombination would shuffle copies of genes from one chromosome to another, but generally would not accomplish anything useful in diploid organisms. If genes could be inactivated by such recombination, then mosaic tissues could be formed, and these probably would not be beneficial.

8.3. After mutagenesis, select Lac constitutives at high temperature with phenyl-galactose, and from these select Lac uninduced by growing on glycerol and orthonitrophenyl-thiogalactoside.

8.5. Suppose the original cross is $aA \times aA$, to yield aa, aA, Aa, and AA. If A is dominant, aA, Aa, and AA cannot be distinguished. If each of these is crossed with aA, then aA and Aa yield some aa progeny whereas AA yields only aA or AA progeny and none with the phenotype of a.

8.7. Suppose the methylation were in the promoter of a gene whose product methylates that particular promoter sequence, and as a result of the methylation, expression of the gene is turned on. Alternatively, suppose that if the base is methylated, a cellular enzyme methylates the complementary strand of this particular sequence, and similarly, if the complementary strand becomes methylated, the enzyme methylates its complementary strand.

8.9. If the markers were on different chromosomes, an offspring receiving one of the markers would have a 50% chance of receiving the other. If the markers were on the same chromosome, the chances could be higher than 50% if the markers were not too widely separated. If they were widely separated, the best approach would be to map each to a chromosome by using markers whose chromosome location is known.

8.11. With contiguous assayable gene material between the two markers, if imprecise recombination occurs, on selecting for $A^+ C^+$ recombinants, B^- should be generated.

8.13. Consider markers A, B, C, D at the corners of a square. The genetic distances A-B, B-C, C-D, D-A =1, and A-C as well as B-D = 1.414.

8.15. 5-E-1-(D, B)-$(3, 4)$-(A, F)-6-C-2. Like most real data, there was an inconsistency. Most likely the cross between deletion E and point mutant 6 should have given recombinants.

8.17. Runs of consecutive A's or T's, since misincorporation of another base that matched the base beyond the homopolymer run might facilitate slipping or small loop formation rather than repair.

8.19. The streptomycin-sensitive ribosomes would synthesize toxic peptides and kill the cells despite the presence of streptomycin-resistant ribosomes.

8.21. When transducing a recessive marker in which the dominant allele is lethal, as are both streptomycin resistance and rifamycin resistance, growth must be permitted so as to allow the dominant and lethal cell component to be replaced by the recessive and nonlethal one. That is, growth must be permitted so that the streptomycin-sensitive ribosomes have been diluted away by the streptomycin-resistant ones in those cells that receive the Sm^r allele.

8.23. A brief treatment with nitrosoguanidine mutagenizes just the region of the chromosome replicated during the treatment. Treat with nitrosoguanidine and select Leu^+ cells, which will be enriched for mutations in the vicinity.

8.25. Temperature mutations were sought. Therefore mutagenized cells were shifted to high temperature, exposed to tritiated leucine for 30 min, washed, and put in the freezer for two months. The survivors were greatly enriched for temperature-sensitive protein synthesis mutations.

8.27. Both will be mutagenic, either from not permitting time for the newly replicated strand to be repaired, and thus either strand would be repaired, or in letting a mispaired base exist until the next replication fork passes.

8.29. UV damaged DNA must be repaired by an error-prone system that is induced by the presence of a significant amount of damaged DNA or by a UV-induced change to some other component in the cell.

8.31. The sex specificity could well reside at a step beyond adsorption, for example at the initiation of protein synthesis or DNA synthesis. Nonetheless, the problem asks for the reasoning if adsorption is the source. In this case the data means that cells synthesize two different surface components, one specific for each state of the cells. When one is synthesized, the other is not synthesized.

8.33. Because XX *Drosophila* exist, no essential genes can be located only on the Y chromosome.

8.35. Repeatedly subject a stock of appropriately mutagenized flies in a long cylinder to light from one end and take the flies from the unilluminated end for the next cycle of enrichment.

8.37. The average number of cells per tube is three, and by the Poisson distribution (see a calculus or statistics book) the probability P that a tube receives n cells when the mean is m is $\dfrac{m^n}{n!}e^{-m}$. For $m = 3$ and $n = 0$, $P = 0.05$. Since cells in more of the tubes than this failed to grow, either not all the cells are capable of growing, or a tube must receive more than one cell to grow. The probability of receiving one cell is 0.15, so the probability of receiving two or more cells is 0.80. Since this is close to the observed value of 0.75, it seems most likely that cells can grow in a tube only if the tube receives two or more cells.

8.39. One expects a temperature-sensitive mutation to result from some component, usually a protein, being nonfunctional only at high temperature. A deletion fails to produce a component at all temperatures. The explanation could be as follows. A protein could be required for the transmission of the impulses, but become much less effective at higher temperatures. If one copy of the gene for the protein were deleted, then the critical temperature at which impulses could no longer be transmitted would be lowered.

Chapter 9

9.1. The restriction and modification systems must be a much more recent evolutionary development than the metabolic pathways.

9.3. When the concentration of transforming DNA is low enough that a small fraction of the competent cells are being transformed, if the number of transformants is proportional to the DNA concentration, a single DNA molecule is responsible, but if it is proportional to the square of the DNA concentration, two molecules are required.

9.5. 0.15^4, 0.25^4, 0.35^4, 0.0005, 0.004, 0.015. Thus, the frequency a site appears can be strongly dependent on the base composition.

9.7. Fill out the sticky end of the *Eco*RI-cut DNA with DNA polymerase so it is flush-ended, and blunt end ligate the two DNAs together so as to regenerate the *Eco*RI site.

9.9. Such a position likely would generate a band in all four lanes because the DNA would cleave at the methylated position. Omitting the chemical modification step and performing just the chemical cleavage step could confirm that the DNA contained a modification at the position in question.

9.11. Infect transformed cells with a phage. Those having acquired a restriction enzyme will survive better. A number of cycles could be required.

9.13. Yes, it is a way to avoid the need for isolating single-stranded DNA and the use of a primer.

9.15. Assume the plasmid contains a selectable drug resistance gene and at least one cleavage site for the restriction enzyme partner of the methylase. Transform, isolate plasmid DNA, cleave it with the restriction enzyme, and transform again. Cycle if necessary. A plasmid that carries the methylase will modify itself when in the cells and this DNA will not be cleaved by the restriction enzyme treatment and hence will transform again.

9.17. Ligate cut *E. coli* DNA to a selectable DNA fragment lacking an origin and transform.

Chapter 10

10.1. If the injected DNA recombines before DNA replication occurs, then all cells of the the progeny are identical. If the recombination occurs after one or more replication cycles, mosaics will be generated. The pattern of mosaicism can reveal the developmental pathways of the mouse. The ability to cross defective genes into the chromosome permits study of their effects.

10.3. Transformation and selection for markers X and Y assure recombination at A' and A" and the disruption of the chromosomal copy of the gene plus the incorporation of the copy carried on the transducing fragment.

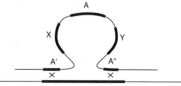

10.5. Don't want extraneous promoters, terminators, splicing sites, binding sites of regulatory proteins, methylation sites, or codons for rare tRNAs. Would like conveniently located restriction sites and unique sequences suitable for primer hybridization for PCR amplification.

10.7. PCR amplify from template A using primers 1 and 2 where the nonhybridizing portion of primer 1 contains restriction sites that will be convenient in cloning, and the nonhybridizing part of primer 2 is the initial sequence of the bottom strand of template B. PCR amplify from template B using the PCR product from the first reaction and primer 3, which also contains useful restriction sites. Note that the PCR amplification from template B uses the top strand of the product from the first reaction as the left-hand primer.

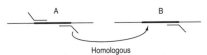

Homologous

10.9. The numbers of individuals, and hence, of recombination events is very low; hence distances cannot be computed directly from the fraction of individuals in a large population in which recombination has or has not occurred.

10.11.

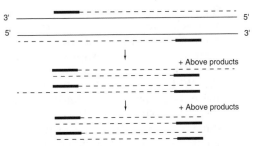

10.13. PCR amplify a short segment of the gene including the region that expands.

10.15. Left-end PCR primer should be complementary to the bottom strand. Right-end PCR primer should have the same sequence as the portion of primer B that was not homologous to the top strand.

Chapter 11

11.1. 1-4-6-4-1, 1-40-600-4000-10,000. The probabilities can be found from the binomial expansion of $(A + B)^4$ and $(A + 10B)^4$.

11.3. For β-galactosidase, cells will be Lac$^+$ at low temperature, Lac$^-$ at high temperature, and β-galactosidase synthesized at low temperature will be functional at high temperature. Perform selections for the first two properties, then grow at low temperature on lactose and shift to high temperature for two to three generation times. For repressor, select for Lac repressed at low temperature, for example by using phenyl-galactose and glycerol in a GalE$^-$ strain in which the constitutive expression of β-galactosidase will generate galactose which will be lethal in the GalE$^-$ background. Then grow cells at low temperature, shift to high temperature for one doubling time, and again select for Lac repressed cells.

11.5. Mate an episome with a LacY$^-$ mutation into cells with another LacY$^-$ mutation. After mating, grow cells with and without IPTG and then score for the fraction of Lac$^+$ cells. Need to show the mutations do not affect transcription beyond the point of the mutation.

11.7. For fully induced cells with the "a" response, in 10^{-6} M IPTG, C_i is 10^{-3} M, and cells remain fully induced. A slight down fluctuation in C_i still leaves cells 100% induced and the transport system concentrating IPTG by 1000-fold. If their response were "b", a slight down fluctuation in C_i reduces the induction, the transport system no longer concentrates by 1000-fold, and the average value of C_i falls, further reducing the induction level. Estimating numbers from the graphs shows that once the system begins to fall down curve "b", it eventually reaches a point of zero induction. Similar reasoning applies to uninduced cells upon being put in 10^{-6} M IPTG.

11.9. 0.1 ml of 10^{-6} = 10^{-10} moles = 10^{-7} mmoles. Activity is 50 mCi/mmole so radioactivity is 10^{-7} mmoles $\times$ 50 mCi/mmoles = 5×10^{-6} mCi. Total dpm = 1.2 $\times 10^9$ dpm/mCi $\times 5 \times 10^{-6}$ mCi = 6000 dpm from the outside sample. For the inside sample we first need the repressor concentration. One molecule per cell volume of 10^{-12} cm^3 is 10^{-9} M, therefore inside the cell repressor concentration was 10^{-7} M, but the cytoplasm was diluted from 100-200 mg/ml to 50 mg/ml of the extract, so $R_t = 5 \times 10^{-8}$. Dpm inside = $6000 \times 1 + \dfrac{R_t}{K_D + IPTG}$ = 6000 $\times$

$$1 + 5 \times \dfrac{10^{-8}}{4 \times 10^{-6} + 1 \times 10^{-6}} = 60.$$ Ratio of inside to outside was 1.01. At the lower IPTG concentration it becomes 1.0125. It is not of use to lower IPTG concentration.

11.11. Cells should remain uninduced at IPTG concentrations that normally induce. Select for nonexpression with tONPG or GalE$^-$ as in problem 11.3, and then select for expression by growing in lactose. Could score for the desired mutants by first growing in lactose to select those that induce normally and then plating on X-gal with marginally inducing concentrations of IPTG and taking the colonies that are white.

11.13. With little increase in the amount of DNA required to code for protein, twice the number of protein-DNA interactions can be generated if a repeat of the binding site is used. If the site is an inverted repeat, a finite oligomer-like dimer or tetramer will possess the necessary symmetry in its subunit arrangement. If the site were a direct repeat, the protein would also have to contain a direct repeat, and could infinitely polymerize.

11.15. If the *lacI* product were RNA, the mutations that made cells LacI$^-$ either reduced the synthesis of repressor or reduced its stability. The introduction of nonsense suppression to a cell could increase RNA synthesis by altering the activity or specificity of a transcription factor, or reducing termination. The latter, though unlikely, is more likely than the former. Just as protein can be helped to fold by specialized proteins, suppose an analogous RNA folding protein existed. Suppose it had been useful to cells, but was no longer essential and was being evolved away. If the cellular gene for an "RNA-foldase" had acquired a nonsense mutation, then the introduction of nonsense suppression could increase the amount of an RNA. Another possibility is that modification of repressor stabilizes it, and the modifying protein is nonessential and happens to possess a nonsense mutation.

11.17. At equilibrium, the rate of formation of repressor-DNA complex equals the rate of dissociation. If k_1 and k_{-1} are the association and dissociation rates and K_D is the equilibrium dissociation constant, setting the rates equal we find $\dfrac{R \times D}{RD} = \dfrac{k_{-1}}{k_1} = K_D.$

11.19. One must be careful in *in vitro* experiments to use components at concentrations near their *in vivo* values. If the mutation reduced the binding constant, but components were at too high a concentration, the results described would have been found.

11.21. The repressor has an increased affinity for DNA such that it binds to nonspecific DNA and rarely makes its way within a doubling time to operator to repress. If the system could reach equilibrium, eventually repressor would partition itself between operator and nonspecific DNA in the ratio of its affinities for these sequences and their relative numbers. If equilibration is slow compared to a doubling time, this partitioning will not occur. As IPTG concentration is increased, repressor's affinity for DNA falls and eventually matches the wild-type affinity and *lac* the basal level becomes normal.

Chapter 12

12.1. Assume the functional form of AraC is an oligomer with a large number of subunits, say eight or more, and that the presence of a single-wild type subunit in an oligomer can force the C^c subunits into a wild-type state.

12.3. Arabinose isomerase is an oligomer, and it is conceivable that in an oligomer the presence of a subunit with one mutation could be compensated by the presence of a subunit carrying the other mutation. The compensation of an AraC mutation by a CRP mutation suggests that the two proteins contact one another and the mutations compensate for a bad contact. This is a reasonably likely inference and certainly not a proof.

12.5. One simple generic mechanism is for AraC-arabinose and CRP to push the balance toward induction and AraC and the weak promoter to oppose it. If the promoter were altered to facilitate induction by tipping the balance toward induction, then perhaps removing CRP would tip the balance back. Alternatively, if the $AraI^c$ mutation facilitates the same step normally aided by CRP, removing CRP would have no effect.

12.7. Repressor bound there likely would interfere with repression and therefore induction would be higher. Addition of IPTG to such a system would have a repressive effect!

12.9. The key here is to devise a growth medium on which C^c mutations do not grow, but C^+ mutations do. If you had such a medium, you could mate an episome carrying one mutation into female cells carrying another mutation, and plate on the selective medium as well as nonselective medium to determine the recombination frequency. Fucose + ribitol + glycerol is a suitable selective medium; fucose further induces C^c mutations, ribitol is phosphorylated by ribulokinase and inhibits cell growth, and glycerol provides a carbon source to the C^+ recombinants.

12.11. In possibility a, CRP would be implicated in breaking a repression loop between *araI* and *araO₁* or *araO₂*. In possibility b CRP would appear to act more directly on AraC and independent of any upstream site.

12.13. Suppose the AraC product were an RNA. There is no reason why temperature-sensitive RNA mutations cannot exist. Indeed, they are known in tRNAs.

12.15. Nitrosoguanidine mutagenizes DNA at the replication fork. When growing in rich medium, multiple replication forks are present on a chromosome, and they are spaced such that one is crossing the *rif* region at the same time the other is crossing the *ara* region. Hence the double mutations are not particularly rare.

12.17. AraC directly contacts RNA polymerase in the induction process, but does not contact CRP.

Chapter 13

13.1. Hybridize an oligo that would bind to region 1 of the newly synthesized RNA so the 2-3 hairpin could form and attenuation would be blocked.

13.3. Secondary structure could cover the ribosome binding site. A region beyond the site could cover the site once this downstream region has been synthesized and the site is accessible.

13.5. Fuse a TrpR-regulated promoter to the β-galactosidase gene. If a multicopy plasmid carrying a *trp* operator enters the cells, repressor will be titrated and *trp* promoters will be derepressed. On x-gal such a cell would form a blue colony.

13.7. We need to change the repression and attenuation mechanisms and assure the availability of chorismate. Eliminate the *trp* operator. If the leucine operon possesses a repressor, place its operator so as to repress transcription. Replace the *trp* leader and attenuation region with the analogous region from the *leu* operon. Introduce a DAHP synthetase that is not feedback-inhibited by phenylalanine or tyrosine or tryptophan.

13.9. Perhaps the multiple alternative hairpins are unnecessary. Suppose just the analog of the *trp* 3-4 termination hairpin might form. Suppose also this region of the leader RNA is translatable and the corresponding RNA is rich in purines. If the cell were low in purines, the polymerase elongation rate would be reduced, and the ribosome translating the leader RNA would keep right up with the polymerase. This would prevent formation of the termination hairpin. If purines were at high concentration, polymerase could outrun the ribosome translating the leader RNA, could form the termination hairpin, and terminate before entering the structural genes.

13.11. The termination hairpin might jam in the polymerase thereby blocking elongation and eventually the RNA dissociates. Base and sequence-specific interactions are made between bases in the region and RNA polymerase. Formation of the G-C hairpin forces the RNA through a channel in polymerase that pulls the RNA off the DNA template.

13.13. Place the hypersynthesizing deletion on a plasmid or episome and assay the behavior of the wild-type chromosomal operon, for example by assaying TrpE, D, or C synthesis, since the deletion lacks genes for these proteins.

13.15. You would see equal amounts of RNA upstream and downstream from the attenuation site and you would conclude that if attenuation occurred, it required some component missing from the *in vitro* reaction mixture.

Chapter 14

14.1. Cross λ and 434 and plate on a λ lysogen. Grow up a plaque, cross against λ and again plate on a λ lysogen. Repeatedly cycle this procedure.

14.3. The footprinting data is incompatible with the cloverleaf model. See the footprinting paper.

14.5. Although the *A* and *R* genes are adjacent in circular or concatenated DNA, the adjacent copies do not end within the same phage particle. The effective distance between the *A* and *R* genes is the entire length of the phage genome. Note this is true only of sites flanking the *ter* or *cos* cleavage site.

14.7. Make a lysogen of $\lambda v_1 v_3 P^-$. Infect with a $\lambda \text{imm}^{434} O^-$. If O protein is present as a result of nonrepression of $v_1 v_3$, complementation will occur. Do analogous experiments using N protein for testing v_2.

14.9. The existence of the membrane potential keeps S protein from allowing R protein to reach the peptidoglycan layer. Blocking protein synthesis also allows S protein to function. Perhaps continuous cellular protein synthesis is necessary to keep S from working.

14.11. After the heat treatment the lysogen is expressing *cro*, and upon infection the presence of Cro protein directs all the incoming phage down the lytic pathway.

14.13. There is a striking tendency for the DNA target site of proteins to be located close to the gene encoding the protein. Evolutionarily this makes sense as recombining one or the other only into a different phage would be of no value.

14.15. Destruction of N protein likely would be dominant whereas alteration of the cellular target of N would be recessive. Therefore make cells diploid for the *nus* allele and test.

14.17. The outer membrane of the cell provides protection. The larger the holes, the more types of molecules have access to the inner membrane of the cell.

14.19. Repeatedly coinfect a λ lysogen with λ and λimm^{434}. Finally plate on a imm^{434} lysogen.

14.21. (a) Lysogens and nonlysogens. (b) Practically no cells. Mutants to which λ cannot adsorb or cells in which λ cannot grow. (c) Lysogens. (d) λ^r cells.

14.23. The basic mechanism of action is likely to be the same, differing only in the structures that read sequence. It is unlikely that ten different antitermination mechanisms could exist.

14.25. Although it would appear that the effective concentration of repressor would be increased, the auto-regulation system would take care of this aspect. Instead, the consequences would be reduced cooperativity in the overall response system. Thus the spontaneous induction rate would be increased, and/or the effectiveness of inducing would be decreased.

Chapter 15

15.1. Try using cell and nuclear extracts from yeast and human cells as a source of most of the transcription apparatus and eggs as a source of specific transcription factors that would stimulate the synthesis of the 5S RNA. Less satisfactory would have been to take eggs that did stimulate transcription and purify or isolate one component after another that would permit synthesis. This would be similar to the approach Kornberg took to isolating proteins involved in DNA synthesis before the use of mutants streamlined the process.

15.3. The time scales of development are relatively long compared to the times required for proteins to find and bind to specific sites on DNA. Therefore equilibrium binding constants generally will be more relevant.

15.5. Thoroughly remove DNA with enzymes and then remove the enzymes. Then make copies of the 5' end of the desired RNA molecules by hybridizing a primer to the RNA and using reverse transcriptase. Further amplify these with PCR.

15.7. The problem is finding the proteins and genes involved in turning off one gene and turning on another. Make β-galactosidase fusions to the promoters and then isolate mutants that no longer shut off oocyte RNA or which still synthesize somatic RNA. Map genetically to see if they were in known or likely regulatory factors or protein binding sites, purify the relevant proteins using *in vitro* complementation between extracts, and then biochemically look at the roles of the involved components.

15.9. After binding polymerase, flood the system with a huge excess of TFIIIA binding site on a short oligonucleotide.

Chapter 16

16.1. They must be shorter than the life cycle of the yeast.

16.3. A diploid homozygous for a *SIR* mutation would not yield fertile spores because the resulting haploids would express both mating types. If all four spores produced by a diploid heterozygous for a *SIR* mutation were fertile immediately upon sporulation, then the wild-type allele would be dominant since it generated repressor that could work in the spore with the mutant gene before it was diluted away.

16.5. Mix extracts used in the *in vitro* cutting assay.

16.7. An *HO* strain is likely to express the enzyme 10^6-fold higher than an *ho* strain. Such cells could constantly be shifting their mating type and therefore there would be much mating between *a* and *α* cells, which would shut off synthesis of the enzyme. To prevent the mating type switch, you would need mutations in *HML* and *MAT* or *HMR* and *MAT*.

16.9. Plate cells on a semi-starvation plate so colonies grow, then form spores if they are diploid. *HO* mutants would mate with their own progeny and become diploid so they could form spores. Haploids would not form spores. The spores from *HO* colonies will mate with both mating type testers while *ho* colonies will mate only with one type of tester.

16.11. The question is whether the mutant *MATa* cells do not respond to α mating factor or whether mutant *MATα* cells do not synthesize α factor. This can be tested by micromanipulation: place the mutant *MATa* cells near normal α mating type cells and observe whether they elongate towards the *MATα* cells.

16.13. Results mean that 94% of the time *a* mating type cells accept information from *HML*, independent of what it contains, and 94% of the time α mating type cells take information from *HMR*, independent of what it contains.

Chapter 17

17.1. Mutagenize males (XY) to generate X'Y and keep each separate. Mate with females (XX), take any female offspring, which could be X'X and mate with its male progenitor. Test the female offspring of this mating for fertility of the eggs.

17.3. The time required should be on the order of the time required for the protein to diffuse the length of the egg. Taking the egg to be 0.1 cm long and using the material from Chap. 1, the time is about 4000 sec, or 1 hr.

17.5. Preservation of protein sequence but not DNA sequence.

17.7. You couldn't map the mutations genetically.

Chapter 18

18.1. The probability of a molecule segregating into a specific daughter is $\frac{1}{2}$. For *n* molecules the probability is $\frac{1}{2}^{n} = 10^{-6}$, or $n \approx 20$.

18.3. Show that the restriction fragment spanning the λ insertion site disappears when λ integrates and that the new fragments are generated that consist of bacterial and phage DNA and that these are of the expected sizes.

18.5. λ*b2* cannot easily integrate, but λ can. A simple way for λ*b2* to integrate would be to recombine with λ to form a double-sized circle. When this integrates, a double lysogen of λ and λ*b2* is formed.

18.7. Use two PCR reactions. In one, use a primer from the beginning of the gene and a primer from one end of the phage and in another PCR reaction, change to a primer from the other end of the phage and the other strand.

18.9. Thermodynamics says that the reverse of any reaction must also proceed. It does not say the rates must be the same. The relative rates will be determined by the equilibrium constant of the reaction. The excision reaction is too slow in the absence of Xis to be biologically useful and thus an entirely new reaction pathway is opened by the presence of Xis.

18.11. A single excision event of the type described integrates the episome into the chromosome. This is rarely observed because when lambda is induced, multiple excision events can occur, so that if the integration event did occur, it would soon be followed by further excision events and lysis of the cell. Of course, the sites are not near one another in the cell due to folding of the chromosome, so the event would be rare. One might wonder about heat pulse curing and integration of the episome, and there currently is no good explanation for the low frequency of this type of event.

18.13. Measure the supercoiling sensitivity of the binding of Int and Xis. If it is not simple binding, the supercoiling might be required to bring distant sites into proximity. This might be shown if the supercoiling requirement were overcome by putting *attL* and *attR* on small circular DNA.

18.15. λimm^{21} must not induce its *int* gene well since phage 21 has a different *CII* protein from lambda. Therefore λimm^{21} is unlikely to make enough Int protein from its own genome or from the lysogen's genome to give facile excision.

18.17. P1 transduce the cells to Gal$^+$ λ lysogenic with P1 grown on a Gal$^+$ λ lysogen. This is possible since P1 will carry DNA fragments twice the size of λ DNA.

18.19. In one orientation the region between the *att* sites is removed to form a smaller circle, and in the other the direction of the segment is inverted. Of course the excision is useful for assaying integration or excision. When it is performed with supercoiled DNA, the products are catenated. This is useful for checking whether a reaction can proceed in *trans*. The inversion reaction is useful for separating two sequences, as in turning a promoter on or off.

18.21. The substrate can act as a suicide substrate inhibitor because the reaction can proceed to the point where it needs to reform the phosphodiester bond. It cannot reform the bond because the initial phosphotransfer reaction could not occur.

Chapter 19

19.1. Show that a particular drug resistance carried on an R factor can be transferred at high frequency to an episome. For example, mate the episome into a population of cells carrying the R factor and then mate again into a third strain and show that the frequency of transfer is much higher than the 10^{-6} or lower that might be expected for illegitimate recombination.

19.3. The old fashioned way was utilizing the fact that the strands do not possess equal amounts of purines and pyrimidines. Denature the DNA, mix with poly(dG), which anneals to many regions on one strand and not so many on the other, and separate DNA by density by equilibrium density centrifugation. The more modern method is to denature the DNA, quick chill and run on an acrylamide gel. For reasons unclear, the strands either do not form identical secondary structures or do, but do not migrate with equal velocities, and the two strands separate. The most modern method to obtain one or the other strand would be to place the desired region of the lambda in a plasmid that contains an origin for a single-stranded phage. Phage infection of the cells containing the plasmid would yield the desired DNA. A short segment of single-stranded lambda could also be made by asymmetrical PCR in which one primer is limiting and the other is present in five to tenfold excess.

19.5. Repeated and highly repeated sequences will have rehybridized. Therefore the rRNA genes and IS sequences will be present in double-stranded DNA. That is, there should be half a dozen bands of size from 1,000 to 5,000 base pairs.

19.7. Soon it may be possible merely to read off the sequence to see where anything is located. Currently the best approach is to screen the clones of a physically mapped set using a fragment of the IS as a probe. Once the clones have been identified, the position of the IS within the clones can be determined with Southern transfer or restriction enzyme cleavage site mapping.

19.9. Cut the Mu DNA with a restriction enzyme that has a cleavage site near the end of the molecule. Run the cut Mu DNA on an acrylamide gel and do a Southern transfer using as a probe this end fragment of DNA.

19.11. Mu replicatively transposes itself into new positions. We would therefore expect a similar replicative transposition into the chromosome upon infection, but instead, a direct transposition occurs by another mechanism.

19.13. As there seems to be no step of restriction that Mu could block, it is more likely that the phage would make its DNA look modified. Therefore it is likely to methylate its DNA extensively in the course of induction and replication. This could be checked directly by looking for a drastic increase in the methylated bases of DNA from such cells, or indirectly by assaying the restriction enzyme susceptibility of DNA extracted from Mu-infected cells.

19.15. The time scales would be those of an intestinal infection, the time of an immune response, and possibly the time for synthesis of an entire flagellum. It is, therefore, somewhere between once per week and once per hour.

19.17. The transposable elements serve as portable sites of homology. The recombination events between such elements are equivalent to integration and excision events between lambda *att* sites as covered in the previous chapter.

Depending on the orientation of the elements, excision or inversion occurs, and if between different chromosomes, fusion or translocation.

19.19. The transposon moves from a position where it will take an entire doubling time until it is replicated again to a position elsewhere on the chromosome. On average the wait for replication from this new position will be about half a doubling time.

Chapter 20

20.1. Precipitation requires formation of aggregates or lattices in which one antibody molecule cross joins two different antigen molecules. Being monovalent, F_{ab} cannot crossjoin.

20.3. $300 \times 10 \times 300 \times 10 \times 10$, or about 10^8. Not all combinations are compatible with one another, and some of the combinations duplicate one another in specificity.

20.5. If these residues were changed by somatic mutation, the resultant antibody was nonfunctional or even toxic to the cell.

20.7. Pick the clone with the highest melting temperature when complexed with segments of the cDNA probe not including spliced sites.

20.9. Somatic mutations are a fine-tuning to an antibody. Once an antibody has been generated that is capable of binding the antigen and clonal expansion has begun, then it is sensible to begin fine-tuning. Therefore the first antibodies, IgM, have not been refined.

20.11. As the chromosomal rearrangements to create an antibody gene take place, is the intervening DNA inverted or deleted? Which it is depends on the orientations. If deletions do occur, then are any genes of interest or importance deleted from a rearranging chromosome?

20.13. This region of the embryonic DNA is likely to be separated from the coding region for the main portion of the C region by an intron. Since only a fraction of the antibody is membrane-bound, we can expect alternative splicing to generate two forms of the coding DNA. We could use any cDNA clone to identify a clone of embryonic DNA, then search this DNA for the additional coding region. Knowing the anchor sequence of the protein would permit building oligonucleotide probes, or we could look for cDNA clones that contained the anchor region and then use this to find the corresponding region in the embryonic DNA.

20.15. Isolate the RNAs used in the PCR amplifications from animals injected with molecules related to those to which you seek the antibodies.

20.17. As long as the DNA doesn't saturate the cell's ability to generate recombinants, the more of the vector DNA that can be isolated from the cells, the easier it is to perform the experiment.

Chapter 21

21.1. (a) Isolate drug-resistant mutants. They might be mapped, and could lie in known ribosomal genes. More sensible and less work would be to prepare *in vitro* protein synthesis reactions using ribosome free extracts mixed with isolated ribosomes. Mixing between components prepared from sensitive and resistant cells will reveal the drug-resistant component. (b) Same as (a), but separate subunits. (c) Reconstitute the ribosome from isolated RNA, and protein from sensitive and resistant cells. (d) The change would have to be a modification or the absence of a modification, for example a methylation. Most likely the effect would be the absence of a methylation.

21.3. On the chromosome have a Sm^r allele and an episome with the Sm^s allele. It would be good to have these cells RecA$^-$. Isolate Sm^r mutants and mate the episomes into Su$^+$ and Su$^-$ Sm^r females. The desired episome will yield Sm^s cells in Su$^+$ background and Sm^r in Su$^-$ background.

21.5. 10 μgm/ml of a 20,000 MW protein is 3×10^{14} molecules/cm^3. The volume of a ribosome is roughly 10^{-17} cm^3, and the volume of a sphere of twice the radius is roughly 10^{-16} cm^3. Thus the probability of a single protein being in the volume is $3 \times 10^{14} \times 10^{-16} = 3 \times 10^{-2}$, and the probability for 21 proteins is $(3 \times 10^{-2})^{21} = 10^{-32}$.

21.7. Crosslinking might work, or see which proteins lose accessibility to antibodies against individual ribosomal proteins when the other subunit is added, or see which ribosomal protein antibodies prevent subunit association.

21.9. Adjacent proteins likely have specific interactions. It makes sense for such proteins to move together upon recombination with other cells.

21.11. Six steps separate two fivefold vertices. T = 36 and the number of triangular faces is S/3 = 720.

21.13. With a radioactive amino acid label for about 15 minutes cells infected with A$^-$ phage. Remove the label and infect these same cells with A$^+$ E$^-$ phage. Isolate intact phage from the cells and examine the radioactivity in the candidate protein. If it is radioactive, it is not a product of the A gene.

21.15. A lambda with *ter* or *cos* duplicated at both ends that has circularized from the sticky ends.

21.17. Infect cells deleted of *att*λ with lambda and plate on *groE* cells. Plaques will be from *groE* transducing phage.

21.19. Infect cells with V$^-$ phage that are F$_{II}^-$ or W$^-$, and isolate the heads. Add these heads to extracts prepared from cells infected with E$^-$ phage that are also F$_{II}^-$ or W$^-$. If the order is F$_{II}$ then W, only the W$^-$ heads mixed with F$_{II}^-$ tails will yield infective phage.

21.21. The number of subunits will be proportional to the surface area which is proportional to the square of the distance between fivefold vertices denoted by A and B in the drawing. The lattice numbers are m and n. Since the side opposite the 30 degree angle in a 30, 60, 90 degree right triangle is one half the hypotenuse, the side opposite B is $m/2$. Applying the Pythagorean theorem, the third side of this triangle, BC, is $\sqrt{\dfrac{3m^2}{4}}$. Applying the theorem again to the sides AC $= n + \dfrac{m}{2}$ and BC, we

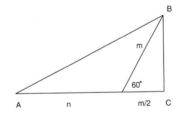

obtain $(AB)^2 = n^2 + nm + m^2$.

21.23. 1500 bases of RNA at about 10 Å per base since the RNA is extended is 15,000 Å. The diameter of the subunit is about 150 Å. Since 20,000 gm of protein is 6×10^{23} molecules, one molecule is 3.33×10^{-20} gm. Since protein has a density of about 1.3 gm/cm^3, the volume of a single protein molecule is about 2.6×10^{-20} cm^3, for a diameter of about 30 Å.

Chapter 22

22.1. The treatment generates a membrane potential which then powers the flagellular motors.

22.3. (a) Use the mini-swarm plates and pick colonies that don't expand. Check for candidates that can chemotact toward other substances. (b) Same as (a), but use two temperatures. (c) Pick candidates from mini-swarm plates that don't chemotact or only weakly chemotact. Test these in the capillary assay with a high concentration of attractant in the capillary.

22.5. Approximately 20 cell lengths per second, or 2×10^{-3} cm/sec.

22.7. There are three lenses in the optical path of standard microscopes. Since each inverts the image, the result is an inverted image and the screw threads are reversed.

22.9. Find an antibody that binds to the end of the flagellum and see if it prevents growth. Another approach might be to label cells for 30 sec with very radioactive amino acids, and then do autoradiography to see which end is labeled. There exist mutant curly flagellins. Shear off the cells' flagella, let synthesis proceed for half the time required to regenerate the flagella, turn off the wild-type flagellin gene and turn on the curly gene and observe which end of the flagella is curly.

22.11. No covalent bonds are made or broken. Hydrogen bonds likely are, and their existence could be shown with tritium exchange from water.

22.13. Decrease in repellent equals increase in attractant and lengthens average run length.

22.15. If the liganded receptor sent signals on to the tumble generator, then flooding the system with high concentrations of fucose would tend to blind the cells to all attractants, not just galactose.

22.17. This is the expected result for the methyl transferase mutants. In methylesterase mutants, the proteins would be methylated soon after synthesis, and since they are not ever removed, the Tar and Tsr proteins would be saturated with methyl groups and could not be further methylated by the standard protocol.

22.19. The first response compares the most previous second of attractant concentration to the concentration a second before that. While this will generate a chemotactic response, it would be better for the cell to average over a period of time closer to the time interval over which it swims in one direction before random rotation throws it off course. The second response shows this better behavior. Note that in the first two, the area of the positive lobe equals that of the negative lobe. This is necessary in order that no signal be generated by constant attractant concentrations. The third response shows such a nonphysiological behavior.

Chapter 23   23.1. The same X chromosome will be active in all cells of the carcinoma.

23.3. Check that the protein in question is not precipitated from nontransformed cells.

23.5. First, perform the same steps without the addition of the homologous fragments from the other piece of DNA. No transformants should be obtained. Next, once a fragment that appears responsible for transformation has been identified, clone it in a bacterial plasmid, cut it out and ligate it into the remainder of the gene to make the gene transform cells.

23.7. The spectrum of sensitivity to restriction enzymes ought to be similar.

23.9. Vary the DNA concentration in the transformation. If more than one gene is involved, the variation in the number of transformants as a function of the DNA concentration should be quadratic or higher.

23.11. Infect fertilized eggs with DNA fragments containing a defective *p53* gene. Screen the resulting mice with Southern transfers to find those that carry the defective gene. Screen their progeny to find those that have the defect in the germ line. Mate a pair of such mice to find homozygous *p53* defective progeny.

Index